Introductory Algebra:
An Applied Approach

Seventh Edition

Introductory Algebra:
An Applied Approach

Richard N. Aufmann
Palomar College, California

Vernon C. Barker
Palomar College, California

Joanne S. Lockwood
Plymouth State University, New Hampshire

Houghton Mifflin Company
Boston   New York

Editor-in-Chief: Jack Shira
Senior Sponsoring Editor: Lynn Cox
Associate Editor: Melissa Parkin
Editorial Assistant: Noel Kamm
Editorial Assistant: Julia Keller
Manufacturing Manager: Karen Banks
Senior Marketing Manager: Ben Rivera
Marketing Associate: Lisa Lawler

Cover photo: © Dave Robertson/Masterfile

Printed in the U.S.A.

Library of Congress Control Number: 2004113920

ISBN Numbers

Student Edition: 0-618-50307-2

Instructor's Annotated Edition: 0-618-50682-9

2 3 4 5 6 7 8 9-WC-09 08 07 06 05

Contents

Preface   xiii
AIM for Success   xxv

1   **Prealgebra Review**   **1**

Prep Test   2

Section 1.1   Introduction to Integers   3
Objective A   *To use inequality symbols with integers* 3
Objective B   *To use opposites and absolute value* 5

Section 1.2   Addition and Subtraction of Integers   9
Objective A   *To add integers* 9
Objective B   *To subtract integers* 10
Objective C   *To solve application problems* 12

Section 1.3   Multiplication and Division of Integers   17
Objective A   *To multiply integers* 17
Objective B   *To divide integers* 19
Objective C   *To solve application problems* 21

Section 1.4   Exponents and the Order of Operations Agreement   24
Objective A   *To evaluate exponential expressions* 24
Objective B   *To use the Order of Operations Agreement to simplify expressions* 25

Section 1.5   Factoring Numbers and Prime Factorization   30
Objective A   *To factor numbers* 30
Objective B   *To find the prime factorization of a number* 31
Objective C   *To find the least common multiple and greatest common factor* 32

Section 1.6   Addition and Subtraction of Rational Numbers   35
Objective A   *To write a rational number in simplest form and as a decimal* 35
Objective B   *To add rational numbers* 37
Objective C   *To subtract rational numbers* 39
Objective D   *To solve application problems* 40

Section 1.7   Multiplication and Division of Rational Numbers   46
Objective A   *To multiply rational numbers* 46
Objective B   *To divide rational numbers* 48
Objective C   *To convert among percents, fractions, and decimals* 50
Objective D   *To solve application problems* 51

Section 1.8   Concepts from Geometry   57
Objective A   *To find the measures of angles* 57
Objective B   *To solve perimeter problems* 59
Objective C   *To solve area problems* 60

Focus on Problem Solving: Inductive Reasoning 67 • **Projects and Group Activities:** The +/− Key on a Calculator 68 • **Chapter 1 Summary 69** • **Chapter 1 Review Exercises 73** • **Chapter 1 Test 75**

v

2   Variable Expressions   77

Prep Test   78

Section 2.1   Evaluating Variable Expressions   79
Objective A   *To evaluate a variable expression* 79

Section 2.2   Simplifying Variable Expressions   83
Objective A   *To simplify a variable expression using the Properties of Addition* 83
Objective B   *To simplify a variable expression using the Properties of Multiplication* 85
Objective C   *To simplify a variable expression using the Distributive Property* 86
Objective D   *To simplify general variable expressions* 88

Section 2.3   Translating Verbal Expressions into Variable Expressions   93
Objective A   *To translate a verbal expression into a variable expression, given the variable* 93
Objective B   *To translate a verbal expression into a variable expression and then simplify* 94
Objective C   *To translate application problems* 96

Focus on Problem Solving: From Concrete to Abstract 101 • **Projects and Group Activities:** Prime and Composite Numbers 102 • **Chapter 2 Summary 104** • **Chapter 2 Review Exercises 106** • **Chapter 2 Test 109** • **Cumulative Review Exercises 111**

3   Solving Equations   113

Prep Test   114

Section 3.1   Introduction to Equations   115
Objective A   *To determine whether a given number is a solution of an equation* 115
Objective B   *To solve an equation of the form $x + a = b$* 116
Objective C   *To solve an equation of the form $ax = b$* 117
Objective D   *To solve application problems using the basic percent equation* 119
Objective E   *To solve uniform motion problems* 123

Section 3.2   General Equations—Part I   134
Objective A   *To solve an equation of the form $ax + b = c$* 134
Objective B   *To solve application problems using formulas* 137

Section 3.3   General Equations—Part II   146
Objective A   *To solve an equation of the form $ax + b = cx + d$* 146
Objective B   *To solve an equation containing parentheses* 147
Objective C   *To solve application problems using formulas* 149

Section 3.4   Translating Sentences into Equations   154
Objective A   *To solve integer problems* 154
Objective B   *To translate a sentence into an equation and solve* 156

Section 3.5   Geometry Problems   161
Objective A   *To solve problems involving angles* 161
Objective B   *To solve problems involving the angles of a triangle* 163

Section 3.6   Mixture and Uniform Motion Problems   168
Objective A   *To solve value mixture problems* 168
Objective B   *To solve percent mixture problems* 170
Objective C   *To solve uniform motion problems* 172

Focus on Problem Solving: Trial-and-Error Approach to Problem Solving 180 • **Projects and Group Activities:** Nielsen Ratings 181 • **Chapter 3 Summary 182 • Chapter 3 Review Exercises 185 • Chapter 3 Test 187 • Cumulative Review Exercises 189**

4   **Polynomials   191**

Prep Test   192

Section 4.1   Addition and Subtraction of Polynomials   193
Objective A   *To add polynomials* 193
Objective B   *To subtract polynomials* 194

Section 4.2   Multiplication of Monomials   197
Objective A   *To multiply monomials* 197
Objective B   *To simplify powers of monomials* 198

Section 4.3   Multiplication of Polynomials   201
Objective A   *To multiply a polynomial by a monomial* 201
Objective B   *To multiply two polynomials* 201
Objective C   *To multiply two binomials* 202
Objective D   *To multiply binomials that have special products* 203
Objective E   *To solve application problems* 204

Section 4.4   Integer Exponents and Scientific Notation   209
Objective A   *To divide monomials* 209
Objective B   *To write a number in scientific notation* 214

Section 4.5   Division of Polynomials   219
Objective A   *To divide a polynomial by a monomial* 219
Objective B   *To divide polynomials* 219

Focus on Problem Solving: Dimensional Analysis 223 • **Projects and Group Activities:** Diagramming the Square of a Binomial 225 • **Chapter 4 Summary 226 • Chapter 4 Review Exercises 229 • Chapter 4 Test 231 • Cumulative Review Exercises 233**

5   **Factoring   235**

Prep Test   236

Section 5.1   Common Factors   237
Objective A   *To factor a monomial from a polynomial* 237
Objective B   *To factor by grouping* 239

Section 5.2   Factoring Polynomials of the Form $x^2 + bx + c$ 243
Objective A   *To factor a trinomial of the form $x^2 + bx + c$* 243
Objective B   *To factor completely* 245

Section 5.3   Factoring Polynomials of the Form $ax^2 + bx + c$ 251
Objective A   *To factor a trinomial of the form $ax^2 + bx + c$ by using trial factors* 251
Objective B   *To factor a trinomial of the form $ax^2 + bx + c$ by grouping* 253

Section 5.4   Special Factoring   259
Objective A   *To factor the difference of two squares and perfect-square trinomials* 259
Objective B   *To factor completely* 261

Section 5.5   Solving Equations   267

Objective A   *To solve equations by factoring* 267
Objective B   *To solve application problems* 269

Focus on Problem Solving: Making a Table 275 • **Projects and Group Activities:**
Evaluating Polynomials Using a Graphing Calculator 276 • **Chapter 5 Summary 278** •
Chapter 5 Review Exercises 281 • **Chapter 5 Test 283** • **Cumulative Review
Exercises 285**

6   **Rational Expressions   287**

Prep Test   288

Section 6.1   Multiplication and Division of Rational Expressions   289

Objective A   *To simplify a rational expression* 289
Objective B   *To multiply rational expressions* 290
Objective C   *To divide rational expressions* 292

Section 6.2   Expressing Fractions in Terms of the Least Common Multiple (LCM)   297

Objective A   *To find the least common multiple (LCM) of two or more
              polynomials* 297
Objective B   *To express two fractions in terms of the LCM of their
              denominators* 298

Section 6.3   Addition and Subtraction of Rational Expressions   301

Objective A   *To add or subtract rational expressions with the same
              denominators* 301
Objective B   *To add or subtract rational expressions with different
              denominators* 302

Section 6.4   Complex Fractions   310

Objective A   *To simplify a complex fraction* 310

Section 6.5   Solving Equations Containing Fractions   315

Objective A   *To solve an equation containing fractions* 315

Section 6.6   Ratio and Proportion   319

Objective A   *To solve a proportion* 319
Objective B   *To solve application problems* 320
Objective C   *To solve problems involving similar triangles* 320

Section 6.7   Literal Equations   327

Objective A   *To solve a literal equation for one of the variables* 327

Section 6.8   Application Problems   331

Objective A   *To solve work problems* 331
Objective B   *To use rational expressions to solve uniform motion problems* 333

Focus on Problem Solving: Negations and If . . . then Sentences 339 • **Projects and
Group Activities:** Intensity of Illumination 340 • **Chapter 6 Summary 342** • **Chapter 6
Review Exercises 345** • **Chapter 6 Test 347** • **Cumulative Review Exercises 349**

7   **Linear Equations in Two Variables   351**

Prep Test   352

Section 7.1   The Rectangular Coordinate System   353

Objective A   *To graph points in a rectangular coordinate system* 353

Objective B *To determine ordered-pair solutions of an equation in two variables* 355
Objective C *To determine whether a set of ordered pairs is a function* 357
Objective D *To evaluate a function written in functional notation* 360

Section 7.2 Linear Equations in Two Variables 365

Objective A *To graph an equation of the form $y = mx + b$* 365
Objective B *To graph an equation of the form $Ax + By + C$* 367
Objective C *To solve application problems* 370

Section 7.3 Intercepts and Slopes of Straight Lines 375

Objective A *To find the x- and y-intercepts of a straight line* 375
Objective B *To find the slope of a straight line* 376
Objective C *To graph a line using the slope and the y-intercept* 379

Section 7.4 Equations of Straight Lines 385

Objective A *To find the equation of a line given a point and the slope* 385
Objective B *To find the equation of a line given two points* 386
Objective C *To solve application problems* 388

Focus on Problem Solving: Counterexamples 392 • **Projects and Group Activities:** Graphing Linear Equations with a Graphing Utility 392 • **Chapter 7 Summary 394 • Chapter 7 Review Exercises 397 • Chapter 7 Test 399 • Cumulative Review Exercises 401**

8 Systems of Linear Equations 403

Prep Test 404

Section 8.1 Solving Systems of Linear Equations by Graphing 405

Objective A *To solve a system of linear equations by graphing* 405

Section 8.2 Solving Systems of Linear Equations by the Substitution Method 413

Objective A *To solve a system of linear equations by the substitution method* 413
Objective B *To solve investment problems* 417

Section 8.3 Solving Systems of Equations by the Addition Method 423

Objective A *To solve a system of linear equations by the addition method* 423

Section 8.4 Application Problems in Two Variables 429

Objective A *To solve rate-of-wind or rate-of-current problems* 429
Objective B *To solve application problems using two variables* 430

Focus on Problem Solving: Using a Table and Searching for a Pattern 435 • **Projects and Group Activities:** Finding a Pattern 436 • **Chapter 8 Summary 437 • Chapter 8 Review Exercises 439 • Chapter 8 Test 441 • Cumulative Review Exercises 443**

9 Inequalities 445

Prep Test 446

Section 9.1 Sets 447

Objective A *To write a set using the roster method* 447
Objective B *To write a set using set-builder notation* 448
Objective C *To graph an inequality on the number line* 449

Section 9.2 The Addition and Multiplication Properties of Inequalities 453

Objective A *To solve an inequality using the Addition Property of Inequalities* 453

Objective B   *To solve an inequality using the Multiplication Property of Inequalities* 454
Objective C   *To solve application problems* 456

Section 9.3   General Inequalities   461
Objective A   *To solve general inequalities* 461
Objective B   *To solve application problems* 462

Section 9.4   Graphing Linear Inequalities   465
Objective A   *To graph an inequality in two variables* 465

Focus on Problem Solving: Graphing Data 469 • **Projects and Group Activities:** Mean and Standard Deviation 469 • **Chapter 9 Summary 472** • **Chapter 9 Review Exercises 473** • **Chapter 9 Test 475** • **Cumulative Review Exercises 477**

10   **Radical Expressions   479**

Prep Test   480

Section 10.1   Introduction to Radical Expressions   481
Objective A   *To simplify numerical radical expressions* 481
Objective B   *To simplify variable radical expressions* 483

Section 10.2   Addition and Subtraction of Radical Expressions   487
Objective A   *To add and subtract radical expressions* 487

Section 10.3   Multiplication and Division of Radical Expressions   491
Objective A   *To multiply radical expressions* 491
Objective B   *To divide radical expressions* 492

Section 10.4   Solving Equations Containing Radical Expressions   497
Objective A   *To solve an equation containing a radical expression* 497
Objective B   *To solve application problems* 499

Focus on Problem Solving: Deductive Reasoning 503 • **Projects and Group Activities:** Distance to the Horizon 504 • **Chapter 10 Summary 505** • **Chapter 10 Review Exercises 507** • **Chapter 10 Test 509** • **Cumulative Review Exercises 511**

11   **Quadratic Equations   513**

Prep Test   514

Section 11.1   Solving Quadratic Equations by Factoring or by Taking Square Roots   515
Objective A   *To solve a quadratic equation by factoring* 515
Objective B   *To solve a quadratic equation by taking square roots* 517

Section 11.2   Solving Quadratic Equations by Completing the Square   521
Objective A   *To solve a quadratic equation by completing the square* 521

Section 11.3   Solving Quadratic Equations by Using the Quadratic Formula   527
Objective A   *To solve a quadratic equation by using the quadratic formula* 527

Section 11.4   Graphing Quadratic Equations in Two Variables   531
Objective A   *To graph a quadratic equation of the form $y = ax^2 + bx + c$* 531

Section 11.5   Application Problems   536
Objective A   *To solve application problems* 536

Focus on Problem Solving: Algebraic Manipulation and Graphing Techniques 540 •
Projects and Group Activities: Graphical Solutions of Quadratic Equations 541 •
Chapter 11 Summary 543 • **Chapter 11 Review Exercises 545** • **Chapter 11 Test 547** •
Cumulative Review Exercises 549

Final Exam   551
Appendix   555
    Keystroke Guide for the TI-83 and TI-83 Plus 555
    Tables 563
Solutions to You Try Its   S1
Answers to Selected Exercises   A1
Glossary   G1
Index   I1

Preface

The seventh edition of *Introductory Algebra: An Applied Approach* provides mathematically sound and comprehensive coverage of the topics considered essential in an introductory algebra course. The text has been designed not only to meet the needs of the traditional college student, but also to serve the needs of returning students whose mathematical proficiency may have declined during years away from formal education.

In this new edition of *Introductory Algebra: An Applied Approach*, we have continued to integrate some of the approaches suggested by AMATYC. Each chapter opens with an illustration and a reference to a mathematical application within the chapter. At the end of each section there are Applying the Concepts exercises, which include writing, synthesis, critical thinking, and challenge problems. At the end of each chapter there is a "Focus on Problem Solving," which introduces students to various problem-solving strategies. This is followed by "Projects and Group Activities," which can be used for cooperative-learning activities.

NEW! Changes to This Edition

Chapter 1, now titled "Prealgebra Review," has been rewritten to present a more complete review of topics from prealgebra. Coverage of operations with integers and with rational numbers has been expanded. Included in this chapter is a section on geometry, in which students solve problems involving angle measurement, perimeter, and area. There are now a greater number of applications throughout the sections of this chapter.

The in-text examples are now highlighted by a prominent HOW TO bar. Students looking for a worked-out example can easily locate one of these problems.

As another aid for students, more annotations have been added to the Examples provided in the paired Example/You Try It boxes. This will assist students in understanding what is happening in key steps of the solution to an exercise.

In response to user requests, Section 4 of Chapter 6 now presents two methods of simplifying a complex fraction: (1) multiplying the numerator and denominator of the complex fraction by the least common multiple of the denominators and (2) multiplying the numerator by the reciprocal of the denominator of the complex fraction.

Section 3 in Chapter 7 now introduces the concept of slopes of perpendicular lines.

Throughout the text, data problems have been updated to reflect current data and trends. Also, titles have been added to the application exercises in the exercise sets. These changes emphasize the relevance of mathematics and the variety of problems in real life that require mathematical analysis.

The Chapter Summaries have been remodeled and expanded. Students are provided with definitions, rules, and procedures, along with examples of each. An objective reference and a page reference accompany each entry. We are confident that these will be valuable aids as students review material and study for exams.

In many chapters, the number of exercises in the Chapter Review Exercises has been increased. This will provide students with more practice on the concepts presented in the chapter.

The calculator appendix has been expanded to include instruction on more functions of the graphing calculator. Notes entitled Integrating Technology appear throughout the book and many refer the student to this appendix. Annotated illustrations of both a scientific calculator and a graphing calculator appear on the inside back cover of this text.

Chapter Opening Features

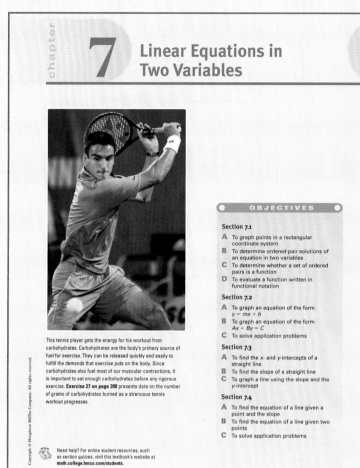

chapter

7 Linear Equations in Two Variables

OBJECTIVES

Section 7.1
A To graph points in a rectangular coordinate system
B To determine ordered-pair solutions of an equation in two variables
C To determine whether a set of ordered pairs is a function
D To evaluate a function written in functional notation

Section 7.2
A To graph an equation of the form $y = mx + b$
B To graph an equation of the form $Ax + By = C$
C To solve application problems

Section 7.3
A To find the x- and y-intercepts of a straight line
B To find the slope of a straight line
C To graph a line using the slope and the y-intercept

Section 7.4
A To find the equation of a line given a point and the slope
B To find the equation of a line given two points
C To solve application problems

This tennis player gets the energy for his workout from carbohydrates. Carbohydrates are the body's primary source of fuel for exercise. They can be released quickly and easily to fulfill the demands that exercise puts on the body. Since carbohydrates also fuel most of our muscular contractions, it is important to eat enough carbohydrates before any rigorous exercise. **Exercise 27 on page 390** presents data on the number of grams of carbohydrates burned as a strenuous tennis workout progresses.

Need help? For online student resources, such as section quizzes, visit this textbook's website at **math.college.hmco.com/students.**

Page 351

NEW! Chapter Opener

New, motivating chapter opener photos and captions have been added, illustrating and referencing a specific application from the chapter.

The 🌐 at the bottom of the page lets students know of additional on-line resources at math.college.hmco.com/students.

Objective-Specific Approach

Each chapter begins with a list of learning objectives that form the framework for a complete learning system. The objectives are woven throughout the text (i.e., Exercises, Prep Tests, Chapter Review Exercises, Chapter Tests, Cumulative Review Exercises) as well as through the print and multimedia ancillaries. This results in a seamless learning system delivered in one consistent voice.

Page 78

Prep Test and Go Figure

Prep Tests occur at the beginning of each chapter and test students on previously covered concepts that are required in the coming chapter. Answers are provided in the Answer Section. Objective references are also provided if a student needs to review specific concepts.

The **Go Figure** problem that follows the *Prep Test* is a playful puzzle problem designed to engage students in problem solving.

PREP TEST • • •

Do these exercises to prepare for Chapter 2.

1. Subtract: $-12 - (-15)$

2. Divide: $-36 \div (-9)$

3. Add: $-\frac{3}{4} + \frac{5}{6}$

4. What is the reciprocal of $-\frac{9}{4}$?

5. Divide: $-\frac{3}{4} \div \left(-\frac{5}{2}\right)$

6. Evaluate: -2^4

7. Evaluate: $\left(\frac{2}{3}\right)^3$

8. Evaluate: $3 \cdot 4^2$

9. Evaluate: $7 - 2 \cdot 3$

10. Evaluate: $5 - 7(3 - 2^2)$

GO FIGURE • • •

Two fractions are inserted between $\frac{1}{4}$ and $\frac{1}{2}$ so that the difference between any two successive fractions is the same. Find the sum of the four fractions.

Aufmann Interactive Method (AIM)

Page 121

Page S5

An Interactive Approach

Introductory Algebra: An Applied Approach uses an interactive style that provides a student with an opportunity to try a skill as it is presented. Each section is divided into objectives, and every objective contains one or more sets of matched-pair examples. The first example in each set is worked out; the second example, called "You Try It," is for the student to work. By solving this problem, the student actively practices concepts as they are presented in the text.

There are complete worked-out solutions to these examples in an appendix. By comparing their solution to the solution in the appendix, students obtain immediate feedback on, and reinforcement of, the concept.

Page xxv

AIM for Success Student Preface

This student 'how to use this book' preface explains what is required of a student to be successful and how this text has been designed to foster student success, including the Aufmann Interactive Method (AIM). *AIM for Success* can be used as a lesson on the first day of class or as a project for students to complete to strengthen their study skills. There are suggestions for teaching this lesson in the *Instructor's Resource Manual*.

AIM for Success

Welcome to *Introductory Algebra: An Applied Approach*. As you begin this course, we know two important facts: (1) We want you to succeed. (2) You want to succeed. To do that requires an effort from each of us. For the next few pages, we are going to show you what is required of you to achieve that success and how you can use the features of this text to be successful.

Motivation   One of the most important keys to success is motivation. We can try to motivate you by offering interesting or important ways mathematics can benefit you. But, in the end, the motivation must come from you. On the first day of class, it is easy to be motivated. Eight weeks into the term, it is harder to keep that motivation.

Problem Solving

Focus on Problem Solving

At the end of each chapter is a Focus on Problem Solving feature that introduces the student to various successful problem-solving strategies. Strategies such as drawing a diagram, applying solutions to other problems, working backwards, inductive reasoning, and trial and error are some of the techniques that are demonstrated.

Page 171

Focus on Problem Solving

Negations and If ... then Sentences

The sentence "George Washington was the first president of the United States" is a true sentence. The **negation** of that sentence is "George Washington was **not** the first president of the United States." That sentence is false. In general, the negation of a true sentence is a false statement.

The negation of a false sentence is a true sentence. For instance, the sentence "The moon is made of green cheese" is a false statement. The negation of that sentence, "The moon is **not** made of green cheese," is true.

The words *all*, *no* (or **none**), and *some* are called **quantifiers.** Writing the negation of a sentence that contains these words requires special attention. Consider the sentence "All pets are dogs." This sentence is not true because there are pets that are not dogs; cats, for example, are pets. Because the sentence is false, its negation must be true. You might be tempted to write "All pets are not dogs," but that sentence is not true because some pets are dogs. The correct negation of "All pets are dogs" is "Some pets are not dogs." Note the use of the word *some* in the negation.

Now consider the sentence "Some computers are portable." Because that sentence is true, its negation must be false. Writing "Some computers are not portable" as the negation is not correct, because that sentence is true. The negation of "Some computers are portable" is "No computers are portable."

The sentence "No flowers have red blooms" is false, because there is at least one flower (some roses, for example) that has red blooms. Because the sentence is false, its negation must be true. The negation is "Some flowers have red blooms."

Statement	Negation
All *A* are *B*.	Some *A* are not *B*.
No *A* are *B*.	Some *A* are *B*.
Some *A* are *B*.	No *A* are *B*.
Some *A* are not *B*.	All *A* are *B*.

Write the negation of the sentence.

1. All cats like milk.

2. All computers need people.

3. Some trees are tall.

4. No politicians are honest.

5. No houses have kitchens.

6. All police officers are tall.

7. All lakes are not polluted.

8. Some drivers are unsafe.

9. Some speeches are interesting.

10. All laws are good.

11. All businesses are not profitable.

12. All motorcycles are not large.

13. Some vegetables are good for you to eat.

14. Some banks are not open on Sunday.

Page 339

Problem-Solving Strategies

The text features a carefully developed approach to problem solving that emphasizes the importance of *strategy* when solving problems. Students are encouraged to develop their own strategies—to draw diagrams, to write out the solution steps in words—as part of their solution to a problem. In each case, model strategies are presented as guides for students to follow as they attempt the "You Try It" problem. Having students provide strategies is a natural way to incorporate writing into the math curriculum.

Example 2

A chemist wishes to make 2 L of an 8% acid solution by mixing a 10% acid solution and a 5% acid solution. How many liters of each solution should the chemist use?

Strategy

* Liters of 10% solution: x
  Liters of 5% solution: $2 - x$

	Amount	Percent	Quantity
10% solution	x	0.10	$0.10x$
5% solution	$2 - x$	0.05	$0.05(2 - x)$
8% solution	2	0.08	$0.08(2)$

* The sum of the quantities before mixing is equal to the quantity after mixing.

Solution

$$0.10x + 0.05(2 - x) = 0.08(2)$$
$$0.10x + 0.10 - 0.05x = 0.16$$
$$0.05x + 0.10 = 0.16$$
$$0.05x = 0.06$$
$$x = 1.2$$

$$2 - x = 2 - 1.2 = 0.8$$

The chemist needs 1.2 L of the 10% solution and 0.8 L of the 5% solution.

You Try It 2

A pharmacist dilutes 5 L of a 12% solution with a 6% solution. How many liters of the 6% solution are added to make an 8% solution?

Your strategy

Your solution

Solution on p. S9

Real Data and Applications

Applications

One way to motivate an interest in mathematics is through applications. Wherever appropriate, the last objective of a section presents applications that require the student to use problem-solving strategies, along with the skills covered in that section, to solve practical problems. This carefully integrated applied approach generates student awareness of the value of algebra as a real-life tool.

Applications are taken from many disciplines, including agriculture, business, carpentry, chemistry, construction, education, finance, nutrition, real estate, sports, and telecommunications.

Page 16

Real Data

Real data examples and exercises, identified by , ask students to analyze and solve problems taken from actual situations. Students are often required to work with tables, graphs, and charts drawn from a variety of disciplines.

Page 12

Objective C   To solve application problems

Positive and negative numbers are used to express the profitability of a company. A profit is recorded as a positive number; a loss is recorded as a negative number.

HOW TO   The bar graph below shows the net earnings of Apple Computer Corporation for the years 1996 through 2002. Calculate the total net earnings for Apple Computer for the years 1996 to 2002.

Source: Apple Computer Corporation

Strategy
To determine the total net earnings, add the profits and losses for the years shown.

Solution
$-816 + (-1045) + 309 + 601 + 522 + (-344) + 17 = -756$

The total net earnings for 1996 to 2002 were $-$756 million.

Example 7

The average temperature on Mercury's sunlit side is 950°F. The average temperature on Mercury's dark side is -346°F. Find the difference between these two average temperatures.

Strategy
To find the difference, subtract the average temperature on the dark side (-346) from the average temperature on the sunlit side (950).

Solution
$$950 - (-346) = 950 + 346$$
$$= 1296$$
The difference between these average temperatures is 1296°F.

You Try It 7

The average daytime temperature on Mars is -17°F. The average nighttime temperature on Mars is -130°F. Find the difference between these two average temperatures.

Your strategy

Your solution

Solution on p. S1

Page 16

Geography   The graph at the right shows Earth's three deepest ocean trenches and its three tallest mountains. Use this graph for Exercises 87 to 89.

87.   Could Mt. Everest fit in the Tonga Trench?

88.   What is the difference between the depth of the Philippine Trench and the Mariana Trench?

89.   What is the difference between the height of Mt. Everest and the depth of the Mariana Trench?

Meteorology   A meteorologist may report a wind-chill temperature. This is the equivalent temperature, including the effects of wind and temperature, that a person would feel in calm air conditions. The table below gives the wind-chill temperature for various wind speeds and temperatures. For instance, when the temperature is 5°F and the wind is blowing at 15 mph, the wind-chill temperature is -13°F. Use this table for Exercises 90 and 91.

Wind Chill Factors

Wind Speed (in mph)	\multicolumn{18}{c|}{Thermometer Reading (in degrees Fahrenheit)}																	
	25	20	15	10	5	0	−5	−10	−15	−20	−25	−30	−35	−40	−45
5	19	13	7	1	−5	−11	−16	−22	−28	−34	−40	−46	−52	−57	−63
10	15	9	3	−4	−10	−16	−22	−28	−35	−41	−47	−53	−59	−66	−72
15	13	6	0	−7	−13	−19	−26	−32	−39	−45	−51	−58	−64	−71	−77
20	11	4	−2	−9	−15	−22	−29	−35	−42	−48	−55	−61	−68	−74	−81
25	9	3	−4	−11	−17	−24	−31	−37	−44	−51	−58	−64	−71	−78	−84
30	8	1	−5	−12	−19	−26	−33	−39	−46	−53	−60	−67	−73	−80	−87
35	7	0	−7	−14	−21	−27	−34	−41	−48	−55	−62	−69	−76	−82	−89
40	6	−1	−8	−15	−22	−29	−36	−43	−50	−57	−64	−71	−78	−84	−91
45	5	−2	−9	−16	−23	−30	−37	−44	−51	−58	−65	−72	−79	−86	−93

90.   When the thermometer reading is -5°F, what is the difference between the wind-chill factor when the wind is blowing 10 mph and when the wind is blowing 25 mph?

91.   When the thermometer reading is -20°F, what is the difference between the wind-chill factor when the wind is blowing 15 mph and when the wind is blowing 25 mph?

APPLYING THE CONCEPTS

92.   If a and b are integers, is the expression $|a + b| = |a| + |b|$ always true, sometimes true, or never true?

93.   Is the difference between two integers always smaller than either one of the numbers in the difference? If not, give an example for which the difference between two integers is greater than either integer.

Student Pedagogy

Icons

The at each objective head remind students that both a video and a tutorial lesson are available for that objective.

Key Terms and Concepts

Key terms, in bold, emphasize important terms. The key terms are also provided in a **Glossary** at the back of the text.

Key Concepts are presented in orange boxes in order to highlight these important concepts and to provide for easy reference.

Study Tips

These margin notes remind students of study skills presented in the *AIM for Success*; some notes provide page references to the original descriptions. They also provide students with reminders of how to practice good study habits.

HOW TO Examples

HOW TO examples use annotations to explain what is happening in key steps of the complete, worked-out solutions.

Page 531

Page 9

Take Note

These margin notes alert students to a point requiring special attention or are used to amplify the concept under discussion.

Point of Interest

These margin notes contain interesting sidelights about mathematics, its history, or its application.

Integrating Technology

These margin notes provide suggestions for using a calculator or refer the student to an appendix for more complete instructions on using a calculator.

Exercises and Projects

Exercises

The exercise sets of *Introductory Algebra: An Applied Approach* emphasize skill building, skill maintenance, and applications. Concept-based writing or developmental exercises have been integrated with the exercise sets. Icons identify appropriate writing , data analysis , and calculator exercises.

Included in each exercise set are **Applying the Concepts** that present extensions of topics, require analysis, or offer challenge problems. The writing exercises ask students to explain answers, write about a topic in the section, or research and report on a related topic.

Page 374

374   Chapter 7 / Linear Equations in Two Variables

Objective C   To solve application problems

37.   **Emergency Response**   A rescue helicopter is rushing at a constant speed of 150 mph to reach several people stranded in the ocean 11 mi away after their boat sank. The rescuers can determine how far they are from the victims using the equation $D = 11 - 2.5t$, where D is the distance in miles and t is the time elapsed in minutes. Graph this equation for $0 \le t \le 4$. The point $(3, 3.5)$ is on the graph. Write a sentence that describes the meaning of this ordered pair.

38.   **Business**   A custom illustrated sign or banner can be commissioned for a cost of $25 for the material and $10.50 per square foot for the artwork. The equation that represents this cost is given by $y = 10.50x + 25$, where y is the cost and x is the number of square feet in the sign. Graph this equation for $0 \le x \le 20$. The point $(15, 182.5)$ is on the graph. Write a sentence that describes the meaning of this ordered pair.

39.   **Veterinary Science**   According to some veterinarians, the age, x, of a dog can be translated to "human years" by using the equation $H = 4x + 16$, where H is the human equivalent age for the dog. Graph this equation for $2 \le x \le 21$. The point whose coordinates are $(6, 40)$ is on this graph. Write a sentence that explains the meaning of this ordered pair.

40.   **Business**   Judging on the basis of data from the Consumer Electronics Association, the projected number, N (in millions), of sales of high definition televisions (HDTVs) can be approximated by $N = 3t + 4$, where $0 \le t \le 4$ and $t = 0$ corresponds to the year 2003. Graph this equation. The point whose coordinates are $(3, 13)$ is on this graph. Write a sentence that explains the meaning of this ordered pair in the context of the problem.

APPLYING THE CONCEPTS

41.   Graph $y = 2x - 2$, $y = 2x$, and $y = 2x + 3$. What observation can you make about the graphs?

42.   Graph $y = x + 3$, $y = 2x + 3$, and $y = -\frac{1}{2}x + 3$. What observation can you make about the graphs?

43.   For the equation $y = 3x + 2$, when the value of x changes from 1 to 2, does the value of y increase or decrease? What is the change in y? Suppose that the value of x chan...

44.   For the equation $y = -2x +$ does the value of y increase pose the value of x changes f

45.   **Telecommunications**   A lon rate of $.99 for the first 15 each additional minute. The of the graphs of two linear $C = 0.15(t - 15) + 0.99$ whe
   a. What is the cost of a telep
   b. What is the cost of a tele

Page 127

59. $d + 1.3619 = 2.0148$ 60. $w + 2.932 = 4.801$

61. $-0.813 + x = -1.096$ 62. $-1.926 + t = -1.042$

63. $6.149 = -3.108 + z$ 64. $5.237 = -2.014 + x$

Objective C   To solve an equation of the form $ax = b$

65.   Without solving $-\frac{15}{41}x = -\frac{23}{25}$, determine whether x is less than or greater than 0. Explain your answer.

66.   Explain why multiplying each side of an equation by the reciprocal of the coefficient of the variable is the same as dividing each side of the equation by the coefficient.

For Exercises 67 to 110, solve and check.

67. $5x = -15$ 68. $4y = -28$ 69. $3b = 0$ 70. $2a = 0$

71. $-3x = 6$ 72. $-5m = 20$ 73. $-3x = -27$ 74. $-\frac{1}{6}n = -30$

Page 225

Projects and Group Activities

The Projects and Group Activities featured at the end of each chapter can be used as extra credit or for cooperative learning activities. The projects cover various aspects of mathematics, including the use of calculators, collecting data from the Internet, data analysis, and extended applications.

Projects and Group Activities   225

Projects and Group Activities

Diagramming the Square of a Binomial

1. Explain why the diagram at the right represents $(a + b)^2 = a^2 + 2ab + b^2$

2. Draw similar diagrams representing each of the following.

$$(x + 2)^2$$
$$(x + 4)^2$$

Pascal's Triangle   Simplifying the power of a binomial is called *expanding the binomial*. The expansions of the first three powers of a binomial are shown below.

$$(a + b)^1 = a + b$$
$$(a + b)^2 = (a + b)(a + b) = a^2 + 2ab + b^2$$
$$(a + b)^3 = (a + b)^2(a + b) = (a^2 + 2ab + b^2)(a + b) = a^3 + 3a^2b + 3ab^2 + b^3$$

Point of Interest

Pascal did not invent the triangle of numbers known as Pascal's Triangle. It was known to mathematicians in China probably as early as 1050 A.D. But Pascal's *Traite du triangle arithmetique* (*Treatise Concerning the Arithmetical Triangle*) brought together all the different aspects of the numbers for the first time.

Find $(a + b)^4$. [*Hint:* $(a + b)^4 = (a + b)^3(a + b)$]

Find $(a + b)^5$. [*Hint:* $(a + b)^5 = (a + b)^4(a + b)$]

If we continue in this way, the results for $(a + b)^6$ are

$$(a + b)^6 = a^6 + 6a^5b + 15a^4b^2 + 20a^3b^3 + 15a^2b^4 + 6ab^5 + b^6$$

Now expand $(a + b)^8$. Before you begin, see whether you can find a pattern that will help you write the expansion of $(a + b)^8$ without having to multiply it out. Here are some hints.

End of Chapter

Chapter Summary

At the end of each chapter there is a Chapter Summary that includes Key Words, Essential Rules and Procedures, and an example of each. Each entry includes an objective reference and a page reference indicating where the concept is introduced. These chapter summaries provide a single point of reference as the student prepares for a test.

Chapter 2 Summary

Key Words	Examples
A *variable* is a letter that is used for a quantity that is unknown or that can change. A *variable expression* is an expression that contains one or more variables. [2.1A, p. 79]	$4x + 2y - 6z$ is a variable expression. It contains the variables x, y, and z.
The *terms* of a variable expression are the addends of the expression. Each term is a *variable term* or a *constant term*. [2.1A, p. 79]	The expression $2a^2 - 3b^3 + 7$ has three terms. $2a^2$, $-3b^3$, and 7. $2a^2$ and $-3b^3$ are variable terms. 7 is a constant term.

Page 104

Essential Rules and Procedures	Examples
The Distributive Property [2.2A, p. 83] If a, b, and c are real numbers, then $a(b + c) = ab + ac$.	$5(4 + 7) = 5 \cdot 4 + 5 \cdot 7$ $= 20 + 35 = 55$
The Associative Property of Addition [2.2A, p. 83] If a, b, and c are real numbers, then $(a + b) + c = a + (b + c)$.	$-4 + (2 + 7) = -4 + 9 = 5$ $(-4 + 2) + 7 = -2 + 7 = 5$

Page 104

Chapter Review Exercises

Chapter Review Exercises are found at the end of each chapter. These exercises are selected to help the student integrate all of the topics presented in the chapter.

Page 106

Chapter 2 Review Exercises

1. Simplify: $3(x^2 - 8x - 7)$ 2. Simplify: $7x + 4x$

3. Simplify: $6a - 4b + 2a$ 4. Simplify: $(-50n)\left(\dfrac{1}{10}\right)$

Chapter Test

Each Chapter Test is designed to simulate a possible test of the material in the chapter.

Page 109

Chapter 2 Test

1. Simplify: $3x - 5x + 7x$ 2. Simplify: $-3(2x^2 - 7y^2)$

3. Simplify: $2x - 3(x - 2)$ 4. Simplify: $2x + 3[4 - (3x - 7)]$

Cumulative Review Exercises

Cumulative Review Exercises, which appear at the end of each chapter (beginning with Chapter 2), help students maintain skills learned in previous chapters.

The answers to all Chapter Review Exercises, all Chapter Test exercises, and all Cumulative Review Exercises are given in the Answer Section. Along with the answer, there is a reference to the objective that pertains to the exercise.

Page 111

Cumulative Review Exercises

1. Add: $-4 + 7 + (-10)$ 2. Subtract: $-16 - (-25) - 4$

3. Multiply: $(-2)(3)(-4)$ 4. Divide: $(-60) \div 12$

Page A4

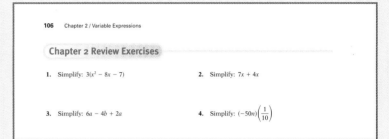

CUMULATIVE REVIEW EXERCISES

1. -7 [1.2A] 2. 5 [1.2B] 3. 24 [1.3A] 4. -5 [1.3B] 5. $53°$ [1.8A] 6. $\dfrac{11}{48}$ [1.6C] 7. $-\dfrac{1}{6}$ [1.7B]

8. $\dfrac{1}{4}$ [1.7A] 9. 75% [1.7C] 10. -5 [1.4B] 11. $-\dfrac{27}{26}$ [1.7B] 12. 16 [2.1A] 13. $5x^2$ [2.2A]

14. $-7a - 10b$ [2.2A] 15. 153.86 cm² [1.8C] 16. 96 ft [1.8B] 17. $24 - 6x$ [2.2C] 18. $6y - 18$ [2.2C]

Instructor Resources

Introductory Algebra: An Applied Approach has a complete set of support materials for the instructor.

Instructor's Annotated Edition This edition contains a replica of the student text and additional resources just for the instructor. These include: *Instructor Notes, New Vocabulary/Symbols, etc., Vocabulary/Symbols, etc. to Review, In-Class Examples, Discuss the Concepts, Concept Checks, Optional Student Activities, Suggested Assignments, Quick Quizzes, Answers to Writing Exercises/Focus on Problem Solving/Projects and Group Activities,* and *PowerPoints*. Answers to all exercises are also provided.

Instructor's Solutions Manual The *Instructor's Solutions Manual* contains worked-out solutions for all exercises in the text.

Instructor's Resource Manual with Testing This resource includes eight ready-to-use printed *Chapter Tests* per chapter, suggested *Course Sequences*, and a printout of the *AIM for Success* PowerPoint slide show. All resources are also available on the Instructor website and *Class Prep* CD.

HM ClassPrep with HM Testing CD-ROM *HM ClassPrep* contains a multitude of text-specific resources for instructors to use to enhance the classroom experience. These resources can be easily accessed by chapter or resource type and can also link you to the text's web site. *HM Testing* is our computerized test generator and contains a database of algorithmic test items as well as providing **on-line testing** and **gradebook** functions.

Instructor Text-Specific Website The resources available on the *Class Prep* CD are also available on the instructor website at math.college.hmco.com/instructors. Appropriate items are password protected. Instructors also have access to the student part of the text's website.

WebCT ePacks *WebCT ePacks* provide instructors with a flexible, Internet-based education platform providing multiple ways to present learning materials. The *WebCT ePacks* come with a full array of features to enrich the on-line learning experience.

Blackboard Cartridges The *Houghton Mifflin Blackboard cartridge* allows flexible, efficient, and creative ways to present learning materials and opportunities. In addition to course management benefits, instructors may make use of an electronic grade book, receive papers from students enrolled in the course via the Internet, and track student use of the communication and collaboration functions.

NEW! HM Eduspace® is a powerful course management system powered by Blackboard that makes preparing, presenting, and managing courses easier. You can use this distance-learning platform to customize, create, and deliver course materials and tests online, and easily maintain student portfolios using the grade book, where grades for all assignments are automatically scored, averaged, and saved.

Student Resources

Student Solutions Manual The *Student Solutions Manual* contains complete solutions to all odd-numbered exercises in the text.

Math Study Skills Workbook by Paul D. Nolting This workbook is designed to reinforce skills and minimize frustration for students in any math class, lab, or study skills course. It offers a wealth of study tips and sound advice on note taking, time management, and reducing math anxiety. In addition, numerous opportunities for self-assessment enable students to track their own progress.

HM Eduspace® Online Learning Environment *Eduspace* is a text-specific, web-based learning environment which combines an algorithmic tutorial program with homework capabilities. Specific content is available 24 hours a day to help you further understand your textbook.

HM mathSpace® Tutorial CD-ROM This tutorial CD-ROM allows students to practice skills and review concepts as many times as necessary by providing algorithmically-generated exercises and step-by-step solutions for practice.

SMARTHINKING™ Live, Online Tutoring Houghton Mifflin has partnered with SMARTHINKING to provide an easy-to-use and effective on-line tutorial service. **Whiteboard Simulations** and **Practice Area** promote real-time visual interaction.

Three levels of service are offered.

- **Text-specific Tutoring** provides real-time, one-on-one instruction with a specially qualified 'e-structor.'

- **Questions Any Time** allows students to submit questions to the tutor outside the scheduled hours and receive a reply within 24 hours.

- **Independent Study Resources** connect students with around-the-clock access to additional educational services, including interactive websites, diagnostic tests, and Frequently Asked Questions posed to SMARTHINKING e-structors.

Houghton Mifflin Instructional Videos and DVDs Text-specific videos and DVDs, hosted by Dana Mosely, cover all sections of the text and provide a valuable resource for further instruction and review.

Student Text-Specific Website On-line student resources can be found at this text's website at math.college.hmco.com/students.

Acknowledgments

The authors would like to thank the people who have reviewed this manuscript and provided many valuable suggestions.

Dr. Mark L. Campbell, *Slippery Rock University of Pennsylvania, PA*
Lynn M. Irons, *College of Southern Idaho, ID*
Jim Matovina, *Community College of Southern Nevada, NV*
Helen Medley
Christopher P. Reisch, *Jamestown Community College, NY*
Daniel Russow, *Arizona Western College, AZ*
Lauri Semarne
Kenneth Takvorian, *Mount Wachusett Community College, MA*
Joseph Verret, *DeVry University, CA*
Mary L. Wolyniak, *Broome Community College, NY*

With special thanks to Dawn Nuttall for her contributions in the developmental editing of three editions of this textbook series.

AIM for Success

Welcome to *Introductory Algebra: An Applied Approach*. As you begin this course, we know two important facts: (1) We want you to succeed. (2) You want to succeed. To do that requires an effort from each of us. For the next few pages, we are going to show you what is required of you to achieve that success and how you can use the features of this text to be successful.

Motivation

One of the most important keys to success is motivation. We can try to motivate you by offering interesting or important ways mathematics can benefit you. But, in the end, the motivation must come from you. On the first day of class, it is easy to be motivated. Eight weeks into the term, it is harder to keep that motivation.

To stay motivated, there must be outcomes from this course that are worth your time, money, and energy.

List some reasons you are taking this course.

_____

_____

_____

Although we hope that one of the reasons you listed was an interest in mathematics, we know that many of you are taking this course because it is required to graduate, it is a prerequisite for a course you must take, or because it is required for your major. Although you may not agree that this course is necessary, it is! If you are motivated to graduate or complete the requirements for your major, then use that motivation to succeed in this course. Do not become distracted from your goal to complete your education!

Commitment

To be successful, you must make a commitment to succeed. This means devoting time to math so that you achieve a better understanding of the subject.

List some activities (sports, hobbies, talents such as dance, art, or music) that you enjoy and at which you would like to become better.

ACTIVITY	TIME SPENT	TIME WISHED SPENT

Thinking about these activities, put the number of hours that you spend each week practicing these activities next to the activity. Next to that number, indicate the number of hours per week you would like to spend on these activities.

Whether you listed surfing or sailing, aerobics or restoring cars, or any other activity you enjoy, note how many hours a week you spend doing it. To succeed in math, you must be willing to commit the same amount of time. Success requires some sacrifice.

The "I Can't Do Math" Syndrome

There may be things you cannot do, such as lift a two-ton boulder. You can, however, do math. It is much easier than lifting the two-ton boulder. When you first learned the activities you listed above, you probably could not do them well. With practice, you got better. With practice, you will be better at math. Stay focused, motivated, and committed to success.

It is difficult for us to emphasize how important it is to overcome the "I Can't Do Math" Syndrome. If you listen to interviews of very successful athletes after a particularly bad performance, you will note that they focus on the positive aspect of what they did, not the negative. Sports psychologists encourage athletes to always be positive—to have a "Can Do" attitude. Develop this attitude toward math.

Strategies for Success

Textbook Review   Right now, do a 15-minute "textbook review" of this book. Here's how:

First, read the table of contents. Do it in three minutes or less. Next, look through the entire book, page by page. Move quickly. Scan titles, look at pictures, notice diagrams.

A textbook reconnaissance shows you where a course is going. It gives you the big picture. That's useful because brains work best when going from the general to the specific. Getting the big picture before you start makes details easier to recall and understand later on.

Your textbook reconnaissance will work even better if, as you scan, you look for ideas or topics that are interesting to you. List three facts, topics, or problems that you found interesting during your textbook reconnaissance.

_____

_____

_____

The idea behind this technique is simple: It's easier to work at learning material if you know it's going to be useful to you.

Not all the topics in this book will be "interesting" to you. But that is true of any subject. Surfers find that on some days the waves are better than others, musicians find some music more appealing than other music, computer gamers find some computer games more interesting than others, car enthusiasts find some cars more exciting than others. Some car enthusiasts would rather have a completely restored 1957 Chevrolet than a new Ferrari.

Know the Course Requirements   To do your best in this course, you must know exactly what your instructor requires. Course requirements may be stated in a *syllabus*, which is a printed outline of the main topics of the course, or they may be presented orally. When they are listed in a syllabus or on other printed pages, keep them in a safe place. When they are presented orally, make sure to take complete notes. In either case, it is important that you understand them completely and follow them exactly. Be sure you know the answer to each of the following questions.

1. What is your instructor's name?
2. Where is your instructor's office?
3. At what times does your instructor hold office hours?
4. Besides the textbook, what other materials does your instructor require?
5. What is your instructor's attendance policy?
6. If you must be absent from a class meeting, what should you do before returning to class? What should you do when you return to class?

7. What is the instructor's policy regarding collection or grading of homework assignments?

8. What options are available if you are having difficulty with an assignment? Is there a math tutoring center?

9. If there is a math lab at your school, Where is it located? What hours is it open?

10. What is the instructor's policy if you miss a quiz?

11. What is the instructor's policy if you miss an exam?

12. Where can you get help when studying for an exam?

Remember: Your instructor wants to see you succeed. If you need help, ask! Do not fall behind. If you are running a race and fall behind by 100 yards, you may be able to catch up but it will require more effort than had you not fallen behind.

Time Management   We know that there are demands on your time. Family, work, friends, and entertainment all compete for your time. We do not want to see you receive poor job evaluations because you are studying math. However, it is also true that we do not want to see you receive poor math test scores because you devoted too much time to work. When several competing and important tasks require your time and energy, the only way to manage the stress of being successful at both is to manage your time efficiently.

Instructors often advise students to spend twice the amount of time outside of class studying as they spend in the classroom. Time management is important if you are to accomplish this goal and succeed in school. The following activity is intended to help you structure your time more efficiently.

List the name of each course you are taking this term, the number of class hours each course meets, and the number of hours you should spend studying each subject outside of class. Then fill in a weekly schedule like the one printed below. Begin by writing in the hours spent in your classes, the hours spent at work (if you have a job), and any other commitments that are not flexible with respect to the time that you do them. Then begin to write down commitments that are more flexible, including hours spent studying. Remember to reserve time for activities such as meals and exercise. You should also schedule free time.

> **TAKE NOTE**
>
> Besides time management, there must be realistic ideas of how much time is available. There are very few people who can *successfully* work full-time and go to school full-time. If you work 40 hours a week, take 15 units, spend the recommended study time given at the right, and sleep 8 hours a day, you will use over 80% of the available hours in a week. That leaves less than 20% of the hours in a week for family, friends, eating, recreation, and other activities.

	Monday	Tuesday	Wednesday	Thursday	Friday	Saturday	Sunday
7–8 a.m.							
8–9 a.m.							
9–10 a.m.							
10–11 a.m.							
11–12 p.m.							
12–1 p.m.							
1–2 p.m.							
2–3 p.m.							
3–4 p.m.							
4–5 p.m.							
5–6 p.m.							
6–7 p.m.							
7–8 p.m.							
8–9 p.m.							
9–10 p.m.							
10–11 p.m.							
11–12 a.m.							

We know that many of you must work. If that is the case, realize that working 10 hours a week at a part-time job is equivalent to taking a three-unit class. If you must work, consider letting your education progress at a slower rate to allow you to be successful at both work and school. There is no rule that says you must finish school in a certain time frame.

Schedule Study Time   As we encouraged you to do by filling out the time management form above, schedule a certain time to study. You should think of this time the way you would the time for work or class—that is, reasons for missing study time should be as compelling as reasons for missing work or class. "I just didn't feel like it" is not a good reason to miss your scheduled study time.

Although this may seem like an obvious exercise, list a few reasons you might want to study.

_____

_____

_____

Of course we have no way of knowing the reasons you listed, but from our experience one reason given quite frequently is "To pass the course." There is nothing wrong with that reason. If that is the most important reason for you to study, then use it to stay focused.

One method of keeping to a study schedule is to form a ***study group***. Look for people who are committed to learning, who pay attention in class, and who are punctual. Ask them to join your group. Choose people with similar educational goals but different methods of learning. You can gain insight from seeing the material from a new perspective. Limit groups to four or five people; larger groups are unwieldy.

There are many ways to conduct a study group. Begin with the following suggestions and see what works best for your group.

1. Test each other by asking questions. Each group member might bring two or three sample test questions to each meeting.
2. Practice teaching each other. Many of us who are teachers learned a lot about our subject when we had to explain it to someone else.
3. Compare class notes. You might ask other students about material in your notes that is difficult for you to understand.
4. Brainstorm test questions.
5. Set an agenda for each meeting. Set approximate time limits for each agenda item and determine a quitting time.

And finally, probably the most important aspect of studying is that it should be done in relatively small chunks. If you can only study three hours a week for this course (probably not enough for most people), do it in blocks of one hour on three separate days, preferably after class. Three hours of studying on a Sunday is not as productive as three hours of paced study.

Text Features That Promote Success

There are 11 chapters in this text. Each chapter is divided into sections, and each section is subdivided into learning objectives. Each learning objective is labeled with a letter from A to E.

Preparing for a Chapter   Before you begin a new chapter, you should take some time to review previously learned skills. There are two ways to do this. The first is to complete the *Cumulative Review Exercises*, which occurs after every chapter (except Chapter 1). For instance, turn to page 285. The questions in this review are taken from the previous chapters. The answers for all these exercises can be found on page A11. Turn to that page now and locate the answers for the Chapter 5 Cumulative Review Exercises. After the answer to the first exercise, which is 7, you will see the objective reference [1.2B]. This means that this question was taken from Chapter 1, Section 2, Objective B. If you missed this question, you should return to that objective and restudy the material.

A second way of preparing for a new chapter is to complete the *Prep Test*. This test focuses on the particular skills that will be required for the new chapter. Turn to page 236 to see a Prep Test. The answers for the Prep Test are the first set of answers in the answer section for a chapter. Turn to page A9 to see the answers for the Chapter 5 Prep Test. Note that an objective reference is given for each question. If you answer a question incorrectly, restudy the objective from which the question was taken.

Before the class meeting in which your professor begins a new section, you should read each objective statement for that section. Next, browse through the objective material, being sure to note each word in bold type. These words indicate important concepts that you must know in order to learn the material. Do not worry about trying to understand all the material. Your professor is there to assist you with that endeavor. The purpose of browsing through the material is so that your brain will be prepared to accept and organize the new information when it is presented to you.

Turn to page 3. Write down the title of the first objective in Section 1.1. Under the title of the objective, write down the words in the objective that are in bold print. It is not necessary for you to understand the meaning of these words. You are in this class to learn their meaning.

Math Is Not a Spectator Sport   To learn mathematics you must be an active participant. Listening and watching your professor do mathematics is not enough. Mathematics requires that you interact with the lesson you are studying. If you filled in the blanks above, you were being interactive. There are other ways this textbook has been designed to help you be an active learner.

Annotated Examples   The HOW TO feature indicates an example with explanatory remarks to the right of the work. Using paper and pencil, you should work along as you go through the example.

$$\frac{3}{4}x - 2 = -11$$

$$\frac{3}{4}x - 2 + 2 = -11 + 2$$

$$\frac{3}{4}x = -9$$

$$\frac{4}{3} \cdot \frac{3}{4}x = \frac{4}{3}(-9)$$

$$x = -12$$

HOW TO Solve: $\frac{3}{4}x - 2 = -11$

The goal is to write the equation in the form *variable = constant*.

$$\frac{3}{4}x - 2 = -11$$

$$\frac{3}{4}x - 2 + 2 = -11 + 2 \qquad \bullet \text{ Add 2 to each side of the equation.}$$

$$\frac{3}{4}x = -9 \qquad \bullet \text{ Simplify.}$$

$$\frac{4}{3} \cdot \frac{3}{4}x = \frac{4}{3}(-9) \qquad \bullet \text{ Multiply each side of the equation by } \frac{4}{3}$$

$$x = -12 \qquad \bullet \text{ The equation is in the form } \textit{variable = constant.}$$

The solution is -12.

TAKE NOTE

Check:
$$\frac{\frac{3}{4}x - 2 = -11}{\frac{3}{4}(-12) - 2 \;\big|\; -11}$$
$$\frac{-9 - 2 \;\big|\; -11}{-11 = -11}$$
A true equation

Page 134

When you complete the example, get a clean sheet of paper. Write down the problem and then try to complete the solution without referring to your notes or the book. When you can do that, move on to the next part of the objective.

Leaf through the book now and write down the page numbers of two other occurrences of a HOW TO example.

You Try Its One of the key instructional features of this text is the paired examples. Notice that in each example box, the example on the left is completely worked out and the "You Try It" example on the right is not. Study the worked-out example carefully by working through each step. Then work the You Try It. If you get stuck, refer to the page number at the end of the example, which directs you to the place where the You Try It is solved—a complete worked-out solution is provided. Try to use the given solution to get a hint for the step you are stuck on. Then try to complete your solution.

Example 5

Solve: $2x + 4 - 5x = 10$

Solution

$$2x + 4 - 5x = 10$$
$$-3x + 4 = 10 \qquad \bullet \text{ Combine like terms.}$$
$$-3x + 4 - 4 = 10 - 4$$
$$-3x = 6$$
$$\frac{-3x}{-3} = \frac{6}{-3}$$
$$x = -2$$

The solution is -2.

You Try It 5

Solve: $x - 5 + 4x = 25$

Your solution

$$x - 5 + 4x = 25$$
$$5x - 5 = 25$$
$$5x - 5 + 5 = 25 + 5$$
$$5x = 30$$
$$\frac{5x}{5} = \frac{30}{5}$$
$$x = 6$$

The solution is 6.

Solution on p. S7

Page 137

When you have completed your solution, check your work against the solution we provided. (Turn to page S7 to see the solution of You Try It 5.) Be aware that frequently there is more than one way to solve a problem. Your answer, however, should be the same as the given answer. If you have any question as to whether your method will "always work," check with your instructor or with someone in the math center.

Browse through the textbook and write down the page numbers where two other paired example features occur.

Remember: Be an active participant in your learning process. When you are sitting in class watching and listening to an explanation, you may think that you understand. However, until you actually try to do it, you will have no confirmation of the new knowledge or skill. Most of us have had the experience of sitting in class thinking we knew how to do something only to get home and realize that we didn't.

TAKE NOTE

There is a strong connection between reading and being a successful student in math or any other subject. If you have difficulty reading, consider taking a reading course. Reading is much like other skills. There are certain things you can learn that will make you a better reader.

Word Problems Word problems are difficult because we must read the problem, determine the quantity we must find, think of a method to do that, and then actually solve the problem. In short, we must formulate a *strategy* to solve the problem and then devise a *solution*.

Note in the paired example below that part of every word problem is a strategy and part is a solution. The strategy is a written description of how we will solve the problem. In the corresponding You Try It, you are asked to formulate a strategy. Do not skip this step, and be sure to write it out.

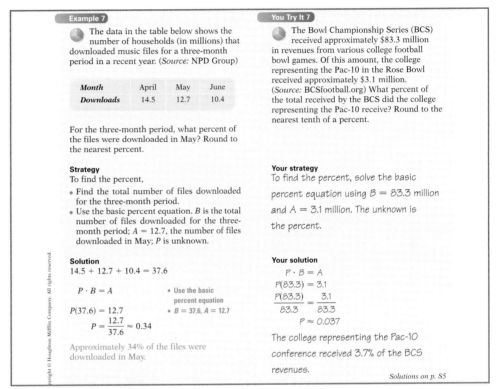

Example 7

The data in the table below shows the number of households (in millions) that downloaded music files for a three-month period in a recent year. (*Source:* NPD Group)

Month	April	May	June
Downloads	14.5	12.7	10.4

For the three-month period, what percent of the files were downloaded in May? Round to the nearest percent.

Strategy
To find the percent,
• Find the total number of files downloaded for the three-month period.
• Use the basic percent equation. *B* is the total number of files downloaded for the three-month period; *A* = 12.7, the number of files downloaded in May; *P* is unknown.

Solution
14.5 + 12.7 + 10.4 = 37.6

$P \cdot B = A$ • Use the basic percent equation

$P(37.6) = 12.7$ • *B* = 37.6, *A* = 12.7

$P = \dfrac{12.7}{37.6} \approx 0.34$

Approximately 34% of the files were downloaded in May.

You Try It 7

The Bowl Championship Series (BCS) received approximately $83.3 million in revenues from various college football bowl games. Of this amount, the college representing the Pac-10 in the Rose Bowl received approximately $3.1 million. (*Source:* BCSfootball.org) What percent of the total received by the BCS did the college representing the Pac-10 receive? Round to the nearest tenth of a percent.

Your strategy
To find the percent, solve the basic percent equation using B = 83.3 million and A = 3.1 million. The unknown is the percent.

Your solution
$P \cdot B = A$
$P(83.3) = 3.1$
$\dfrac{P(83.3)}{83.3} = \dfrac{3.1}{83.3}$
$P \approx 0.037$

The college representing the Pac-10 conference received 3.7% of the BCS revenues.

Solutions on p. S5

Page 121

Rule Boxes Pay special attention to rules placed in boxes. These rules give you the reasons certain types of problems are solved the way they are. When you see a rule, try to rewrite the rule in your own words.

The equations $2x = 6$, $10x = 30$, and $-8x = -24$ are equivalent equations; each equation has 3 as its solution. These examples suggest that multiplying each side of an equation by the same nonzero number produces an equivalent equation.

Multiplication Property of Equations

Each side of an equation can be multiplied by the same *nonzero* number without changing the solution of the equation. In symbols, if $c \neq 0$, then the equation $a = b$ has the same solutions as the equation $ac = bc$.

Page 117

Chapter Exercises When you have completed studying an objective, do the exercises in the exercise set that correspond with that objective. The exercises are labeled with the same letter as the objective. Math is a subject that needs to be learned in small sections and practiced continually in order to be mastered. Doing all of the exercises in each exercise set will help you master the problem-solving techniques necessary for success. As you work through the exercises for an objective, check your answers to the odd-numbered exercises with those in the back of the book.

Preparing for a Test   There are important features of this text that can be used to prepare for a test.

- Chapter Summary
- Chapter Review Exercises
- Chapter Test

After completing a chapter, read the Chapter Summary. (See page 226 for the Chapter 4 Summary.) This summary highlights the important topics covered in the chapter. The page number following each topic refers you to the page in the text on which you can find more information about the concept.

Following the Chapter Summary are Chapter Review Exercises (see page 229) and a Chapter Test (see page 231). Doing the review exercises is an important way of testing your understanding of the chapter. The answer to each review exercise is given at the back of the book, along with its objective reference. After checking your answers, restudy any objective from which a question you missed was taken. It may be helpful to retry some of the exercises for that objective to reinforce your problem-solving techniques.

The Chapter Test should be used to prepare for an exam. We suggest that you try the Chapter Test a few days before your actual exam. Take the test in a quiet place and try to complete the test in the same amount of time you will be allowed for your exam. When taking the Chapter Test, practice the strategies of successful test takers: (1) scan the entire test to get a feel for the questions; (2) read the directions carefully; (3) work the problems that are easiest for you first; and perhaps most importantly, (4) try to stay calm.

When you have completed the Chapter Test, check your answers. If you missed a question, review the material in that objective and rework some of the exercises from that objective. This will strengthen your ability to perform the skills in that objective.

Is it difficult to be successful? YES! Successful music groups, artists, professional athletes, chefs, and Write your major here have to work very hard to achieve their goals. They focus on their goals and ignore distractions. The things we ask you to do to achieve success take time and commitment. We are confident that if you follow our suggestions, you will succeed.

1 Prealgebra Review

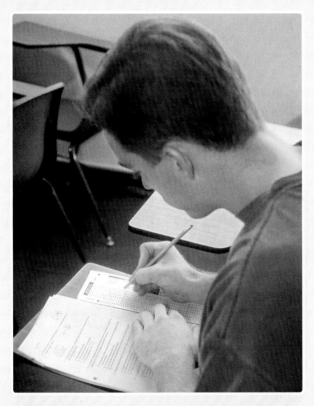

When you take a multiple-choice test, such as a class exam, the ACT, or the SAT, there is usually a point system for scoring your answers. Correct answers receive a positive number of points, and incorrect answers receive a negative number of points. For the ACT, the SAT, and most multiple-choice tests, it is better for your score to leave a question blank if you are unsure of the answer. An unanswered question will cause fewer points to be deducted from your score; sometimes it will not cost you any points at all. **Exercises 67 and 68 on page 23** show how professors can adjust the grading of multiple-choice exams to discourage students from guessing randomly.

Need help? For online student resources, such as section quizzes, visit this textbook's website at **math.college.hmco.com/students**.

OBJECTIVES

Section 1.1

A To use inequality symbols with integers
B To use opposites and absolute value

Section 1.2

A To add integers
B To subtract integers
C To solve application problems

Section 1.3

A To multiply integers
B To divide integers
C To solve application problems

Section 1.4

A To evaluate exponential expressions
B To use the Order of Operations Agreement to simplify expressions

Section 1.5

A To factor numbers
B To find the prime factorization of a number
C To find the least common multiple and greatest common factor

Section 1.6

A To write a rational number in simplest form and as a decimal
B To add rational numbers
C To subtract rational numbers
D To solve application problems

Section 1.7

A To multiply rational numbers
B To divide rational numbers
C To convert among percents, fractions, and decimals
D To solve application problems

Section 1.8

A To find the measures of angles
B To solve perimeter problems
C To solve area problems

Do these exercises to prepare for Chapter 1.

1. What is 127.1649 rounded to the nearest hundredth?

2. Add: 3416 + 42,561 + 537

3. Subtract: 5004 − 487

4. Multiply: 407 × 28

5. Divide: 11,684 ÷ 23

6. What is the smallest number that both 8 and 12 divide evenly?

7. What is the greatest number that divides both 16 and 20 evenly?

8. Without using 1, write 21 as a product of two whole numbers.

9. Represent the shaded portion of the figure as a fraction.

10. Which of the following, if any, is not possible?
   a. 6 + 0 **b.** 6 − 0 **c.** 6 × 0 **d.** 6 ÷ 0

GO FIGURE • • •

In a group of 9 coins, one is counterfeit and weighs less than the 8 genuine coins. Using a balance scale, determine the counterfeit coin in two weighings.

1.1 Introduction to Integers

Objective A   **To use inequality symbols with integers**

It seems to be a human characteristic to group similar items. For instance, a biologist places similar animals in groups called *species*. Nutritionists classify foods according to *food groups*; for example, pasta, crackers, and rice are among the foods in the bread group.

Mathematicians place objects with similar properties in groups called *sets*. A **set** is a collection of objects. The objects in a set are called the **elements of the set.**

The **roster method** of writing sets encloses a list of the elements in braces. Thus the set of sections within an orchestra is written {brass, percussion, string, woodwind}. When the elements of a set are listed, each element is listed only once. For instance, if the list of numbers 1, 2, 3, 2, 3 were placed in a set, the set would be $\{1, 2, 3\}$.

The numbers that we use to count objects, such as the students in a classroom or the horses on a ranch, are the *natural numbers*.

$$\textbf{Natural numbers} = \{1, 2, 3, 4, 5, 6, 7, 8, 9, 10, \ldots\}$$

The three dots mean that the list of natural numbers continues on and on and that there is no largest natural number.

The natural numbers alone do not provide all the numbers that are useful in applications. For instance, a meteorologist also needs the number zero and numbers below zero.

$$\textbf{Integers} = \{\ldots, -5, -4, -3, -2, -1, 0, 1, 2, 3, 4, 5, \ldots\}$$

Each integer can be shown on a number line. The integers to the left of zero on the number line are called **negative integers.** The integers to the right of zero are called **positive integers,** or natural numbers. Zero is neither a positive nor a negative integer.

Point of Interest

The Alexandrian astronomer Ptolemy began using *omicron*, *0*, the first letter of the Greek word that means "nothing," as the symbol for zero in 150 A.D. It was not until the 13th century, however, that Fibonacci introduced 0 to the Western world as a placeholder so that we could distinguish, for example, 45 from 405.

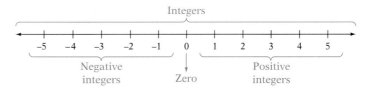

The **graph** of an integer is shown by placing a heavy dot on the number line directly above the number. The graphs of -3 and 4 are shown on the number line below.

Consider the following sentences.

> The quarterback threw the football and the receiver caught *it*.
> A student purchased a computer and used *it* to write history papers.

In the first sentence, *it* is used to mean the football; in the second sentence, *it* means the computer. In language, the word *it* can stand for many different objects. Similarly, in mathematics, a letter of the alphabet can be used to stand for a number. Such a letter is called a **variable.** Variables are used in the following definition of inequality symbols.

Point of Interest

The symbols for "is less than" and "is greater than" were introduced by Thomas Harriot around 1630. Before that, ⊏ and ⊐ were used for > and <, respectively.

Inequality Symbols

If *a* and *b* are two numbers and *a* is to the left of *b* on the number line, then *a* **is less than** *b*. This is written $a < b$.

If *a* and *b* are two numbers and *a* is to the right of *b* on the number line, then *a* **is greater than** *b*. This is written $a > b$.

Negative 4 is less than negative 1.

$-4 < -1$

5 is greater than 0.

$5 > 0$

There are also inequality symbols for **is less than or equal to (≤)** and **is greater than or equal to (≥)**.

$7 \le 15$ 7 is less than or equal to 15.
               This is true because $7 < 15$.

$6 \le 6$ 6 is less than or equal to 6.
               This is true because $6 = 6$.

Point of Interest

The Greek influence is again felt with the symbol ∈ for "is an element of." Giuseppe Peano used this symbol in 1889 as an abbreviation for the Greek word for "is."

The symbol ∈ means "is an element of." $2 \in B$ is read "2 is an element of set *B*."

Given $C = \{3, 5, 9\}$, then $3 \in C$, $5 \in C$, and $9 \in C$. $7 \notin C$ is read "7 is not an element of set *C*."

Example 1

Use the roster method to write the set of negative integers greater than or equal to -4.

Solution

$A = \{-4, -3, -2, -1\}$ • A set is designated by a capital letter.

You Try It 1

Use the roster method to write the set of positive integers less than 7.

Your solution

Solution on p. S1

Example 2

Given $A = \{-6, -2, 0\}$, which elements of set A are less than or equal to -2?

Solution

Find the order relation between each element of set A and -2.

$-6 < -2$
$-2 = -2$
$0 > -2$

The elements -6 and -2 are less than or equal to -2.

You Try It 2

Given $B = \{-5, -1, 5\}$, which elements of set B are greater than -1?

Your solution

Solution on p. S1

Objective B   **To use opposites and absolute value**

Two numbers that are the same distance from zero on the number line but are on opposite sides of zero are **opposite numbers,** or **opposites.** The opposite of a number is also called its **additive inverse.**

The opposite of 5 is -5.

The opposite of -5 is 5.

The negative sign can be read "the opposite of."

$-(2) = -2$ The opposite of 2 is -2.
$-(-2) = 2$ The opposite of -2 is 2.

Study Tip

Some students think that they can "coast" at the beginning of this course because the topic of Chapter 1 is a review of pre-algebra. However, this chapter lays the foundation for the entire course. Be sure you know and understand all the concepts presented. For example, study the properties of absolute value presented in this lesson.

The **absolute value of a number** is its distance from zero on the number line. Therefore, the absolute value of a number is a positive number or zero. The symbol for absolute value is two vertical bars, $|\ |$.

The distance from 0 to 3 is 3. Therefore, the absolute value of 3 is 3.

$|3| = 3$

The distance from 0 to -3 is 3. Therefore, the absolute value of -3 is 3.

$|-3| = 3$

Point of Interest

The definition of *absolute value* given in the box is written in what is called rhetorical style. That is, it is written without the use of variables. This is how *all* mathematics was written prior to the Renaissance. During that period from the 14th to the 16th century, the idea of expressing a variable symbolically was developed. In terms of that symbolism, the definition of absolute value is

$$|x| = \begin{cases} x, & x > 0 \\ 0, & x = 0 \\ -x, & x < 0 \end{cases}$$

Absolute Value

The absolute value of a positive number is the number itself. For example, $|9| = 9$.

The absolute value of zero is zero. $|0| = 0$

The absolute value of a negative number is the opposite of the negative number. For example, $|-7| = 7$.

HOW TO Evaluate: $-|-12|$

$-|-12| = -12$ • The absolute value sign does not affect the negative sign in front of the absolute value sign.

Example 3

Evaluate $|-4|$ and $-|-10|$.

Solution
$|-4| = 4$
$-|-10| = -10$

You Try It 3

Evaluate $|-5|$ and $-|-23|$.

Your solution

Example 4

Given $A = \{-12, 0, 4\}$, find the additive inverse of each element of set A.

Solution
$-(-12) = 12$
$-0 = 0$ • Zero is neither positive nor negative.

$-(4) = -4$

You Try It 4

Given $B = \{-11, 0, 8\}$, find the additive inverse of each element of set B.

Your solution

Example 5

Given $C = \{-17, 0, 14\}$, find the absolute value of each element of set C.

Solution
$|-17| = 17$
$|0| = 0$
$|14| = 14$

You Try It 5

Given $D = \{-37, 0, 29\}$, find the absolute value of each element of set D.

Your solution

Solutions on p. S1

1.1 Exercises

Objective A   **To use inequality symbols with integers**

1. Explain the difference between the natural numbers and the integers.

2. Name the smallest integer that is larger than any negative integer.

For Exercises 3 to 12, place the correct symbol, < or >, between the two numbers.

3. 8   −6   **4.** −14   16   **5.** −12   1   **6.** 35   28   **7.** 42   19

8. −42   27   **9.** 0   −31   **10.** −17   0   **11.** 53   −46   **12.** −27   −38

For Exercises 13 to 22, answer True or False.

13. −13 > 0   **14.** −20 > 3   **15.** 12 > −31   **16.** 9 > 7   **17.** −5 > −2

18. −44 > −21   **19.** −4 > −120   **20.** 0 > −8   **21.** −1 ≥ −1   **22.** −10 ≤ −10

For Exercises 23 to 28, use the roster method to write the set.

23. The natural numbers less than 9

24. The natural numbers less than or equal to 6

25. The positive integers less than or equal to 8

26. The positive integers less than 4

27. The negative integers greater than −7

28. The negative integers greater than or equal to −5

29. Given $A = \{-7, 0, 2, 5\}$, which elements of set A are greater than 2?

30. Given $B = \{-8, 0, 7, 15\}$, which elements of set B are greater than 7?

31. Given $D = \{-23, -18, -8, 0\}$, which elements of set D are less than −8?

32. Given $C = \{-33, -24, -10, 0\}$, which elements of set C are less than −10?

33. Given $E = \{-35, -13, 21, 37\}$, which elements of set E are greater than −10?

34. Given $F = \{-27, -14, 14, 27\}$, which elements of set F are greater than −15?

35. Given $B = \{-52, -46, 0, 39, 58\}$, which elements of set B are less than or equal to 0?

36. Given $A = \{-12, -9, 0, 12, 34\}$, which elements of set A are greater than or equal to 0?

37. Given $C = \{-23, -17, 0, 4, 29\}$, which elements of set C are greater than or equal to −17?

38. Given $D = \{-31, -12, 0, 11, 45\}$, which elements of set D are less than or equal to −12?

39. Given that set A is the positive integers less than 10, which elements of set A are greater than or equal to 5?

40. Given that set B is the positive integers less than or equal to 12, which elements of set B are greater than 6?

41. Given that set D is the negative integers greater than or equal to -10, which elements of set D are less than -4?

42. Given that set C is the negative integers greater than -8, which elements of set C are less than or equal to -3?

Objective B   **To use opposites and absolute value**

For Exercises 43 to 47, find the additive inverse.

43. 4

44. 8

45. -9

46. -28

47. -36

For Exercises 48 to 57, evaluate.

48. $-(-14)$

49. $-(-40)$

50. $-(77)$

51. $-(39)$

52. $-(-13)$

53. $|-74|$

54. $|-96|$

55. $-|-82|$

56. $-|-53|$

57. $-|81|$

For Exercises 58 to 65, place the correct symbol, $<$ or $>$, between the values of the two numbers.

58. $|-83|$ $|58|$

59. $|22|$ $|-19|$

60. $|43|$ $|-52|$

61. $|-71|$ $|-92|$

62. $|-68|$ $|-42|$

63. $|12|$ $|-31|$

64. $|-45|$ $|-61|$

65. $|-28|$ $|43|$

66. Use the set $A = \{-8, -5, -2, 1, 3\}$.
   a. Find the opposite of each element of set A.
   b. Find the absolute value of each element of set A.

67. Use the set $B = \{-11, -7, -3, 1, 5\}$.
   a. Find the opposite of each element of set B.
   b. Find the absolute value of each element of set B.

APPLYING THE CONCEPTS

68. If x represents a negative integer, then $-x$ represents a _____ integer.

69. If x is an integer, is the inequality $|x| < -3$ always true, sometimes true, or never true?

70. In your own words, explain the meaning of the absolute value of a number and the additive inverse of a number.

1.2 Addition and Subtraction of Integers

Objective A To add integers

A number can be represented anywhere along the number line by an arrow. A positive number is represented by an arrow pointing to the right, and a negative number is represented by an arrow pointing to the left. The size of the number is represented by the length of the arrow.

Addition is the process of finding the total of two numbers. The numbers being added are called **addends.** The total is called the **sum.** Addition of integers can be shown on the number line. To add integers, start at zero and draw, above the number line, an arrow representing the first number. At the tip of the first arrow, draw a second arrow representing the second number. The sum is below the tip of the second arrow.

$4 + 2 = 6$ $-4 + (-2) = -6$

$-4 + 2 = -2$ $4 + (-2) = 2$

The pattern for addition shown on the number lines above is summarized in the following rules for adding integers.

> **Addition of Integers**
>
> **To add two numbers with the same sign,** add the absolute values of the numbers. Then attach the sign of the addends.
>
> **To add two numbers with different signs,** find the absolute value of each number. Subtract the smaller of the two numbers from the larger. Then attach the sign of the number with the larger absolute value.

Study Tip

The HOW TO feature indicates an example with explanatory remarks. Using paper and pencil, you should work through the example. See *AIM for Success,* page xxv.

HOW TO Add: $-12 + (-26)$

$-12 + (-26) = -38$ • The signs are the same. Add the absolute values of the numbers (12 + 26). Attach the sign of the addends.

HOW TO Add: $-19 + 8$

$|-19| = 19$ $|8| = 8$ • The signs are different. Find the absolute value of each number.

$19 - 8 = 11$ • Subtract the smaller number from the larger.

$-19 + 8 = -11$ • Attach the sign of the number with the larger absolute value.

HOW TO Find the sum of -23, 47, -18, and -10.

Recall that a *sum* is the answer to an addition problem.

$-23 + 47 + (-18) + (-10)$

$= 24 + (-18) + (-10)$

$= 6 + (-10)$

$= -4$

• To add more than two numbers, add the first two numbers. Then add the sum to the third number. Continue until all the numbers are added.

The phrase *the sum of* in the example above indicates the operation of addition. All of the phrases below indicate addition.

added to	-6 added to 9	$9 + (-6) = 3$
more than	3 more than -8	$-8 + 3 = -5$
the sum of	the sum of -2 and -8	$-2 + (-8) = -10$
increased by	-7 increased by 5	$-7 + 5 = -2$
the total of	the total of 4 and -9	$4 + (-9) = -5$
plus	6 plus -10	$6 + (-10) = -4$

Example 1 Add: $-52 + (-39)$

Solution
$-52 + (-39) = -91$

You Try It 1 Add: $100 + (-43)$

Your solution

Example 2 Add: $37 + (-52) + (-14)$

Solution
$37 + (-52) + (-14) = -15 + (-14)$
$\qquad\qquad\qquad\quad = -29$

You Try It 2 Add: $-51 + 42 + 17 + (-102)$

Your solution

Example 3 Find 11 more than -23.

Solution $-23 + 11 = -12$

You Try It 3 Find -8 increased by 7.

Your solution

Solutions on p. S1

Objective B To subtract integers

Look at the expressions below. Note that each expression equals the same number.

$$8 - 3 = 5 \quad \text{8 minus 3 is 5.}$$
$$8 + (-3) = 5 \quad \text{8 plus the opposite of 3 is 5.}$$

This example suggests the following.

Subtraction of Integers

To subtract one number from another, add the opposite of the second number to the first number.

HOW TO   Subtract: $-21 - (-40)$

Change this sign to plus.

$-21 - (-40) = -21 + 40 = 19$ • Rewrite each subtraction as addition of the opposite. Then add.

Change -40 to the opposite of -40.

HOW TO   Subtract: $15 - 51$

Change this sign to plus.

$15 - 51 = 15 + (-51) = -36$ • Rewrite each subtraction as addition of the opposite. Then add.

Change 51 to the opposite of 51.

HOW TO   Subtract: $-12 - (-21) - 15$

$-12 - (-21) - 15 = -12 + 21 + (-15)$ • Rewrite each subtraction as addition of the opposite. Then add.

$= 9 + (-15) = -6$

HOW TO   Find the difference between -8 and 7.

A *difference* is the answer to a subtraction problem.

$-8 - 7 = -8 + (-7)$ • Rewrite the subtraction as addition of the opposite.

$= -15$

The phrase *the difference of* in the example above indicates the operation of subtraction. All of the phrases below indicate subtraction.

TAKE NOTE

Note the order in which numbers are subtracted when the phrase *less than* is used. If you have $10 and a friend has $6 less than you do, then your friend has $6 less than $10, or $10 − $6 = $4.

minus	−5 minus 11	$-5 - 11 = -16$
less	−3 less 5	$-3 - 5 = -8$
less than	−8 less than −2	$-2 - (-8) = 6$
the difference between	the difference between −5 and 4	$-5 - 4 = -9$
decreased by	−4 decreased by 9	$-4 - 9 = -13$
subtract . . . from	subtract 8 from −3	$-3 - 8 = -11$

Example 4   Subtract: $-11 - 15$

Solution $-11 - 15 = -11 + (-15) = -26$

You Try It 4   Subtract: $19 - (-32)$

Your solution

Example 5   Subtract: $-14 - 18 - (-21) - 4$

Solution
$-14 - 18 - (-21) - 4$
$= -14 + (-18) + 21 + (-4)$
$= -32 + 21 + (-4)$
$= -11 + (-4) = -15$

You Try It 5   Subtract: $-9 - (-12) - 17 - 4$

Your solution

Example 6   Find 9 less than -4.

Solution $-4 - 9 = -4 + (-9) = -13$

You Try It 6   Subtract -12 from -11.

Your solution

Solutions on p. S1

Objective C   **To solve application problems**

Positive and negative numbers are used to express the profitability of a company. A profit is recorded as a positive number; a loss is recorded as a negative number.

HOW TO   The bar graph below shows the net earnings of Apple Computer Corporation for the years 1996 through 2002. Calculate the total net earnings for Apple Computer for the years 1996 to 2002.

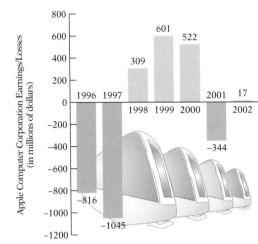

Source: Apple Computer Corporation

Strategy
To determine the total net earnings, add the profits and losses for the years shown.

Solution
$$-816 + (-1045) + 309 + 601 + 522 + (-344) + 17 = -756$$

The total net earnings for 1996 to 2002 were −$756 million.

Example 7

The average temperature on Mercury's sunlit side is 950°F. The average temperature on Mercury's dark side is −346°F. Find the difference between these two average temperatures.

Strategy
To find the difference, subtract the average temperature on the dark side (−346) from the average temperature on the sunlit side (950).

Solution
$$950 - (-346) = 950 + 346$$
$$= 1296$$
The difference between these average temperatures is 1296°F.

You Try It 7

The average daytime temperature on Mars is −17°F. The average nighttime temperature on Mars is −130°F. Find the difference between these two average temperatures.

Your strategy

Your solution

Solution on p. S1

1.2 Exercises

Objective A   **To add integers**

1. ✎ Explain how to add two integers with the same sign.

2. ✎ Explain how to add two integers with different signs.

For Exercises 3 to 34, add.

3. $-3 + (-8)$

4. $-6 + (-9)$

5. $-8 + 3$

6. $-9 + 2$

7. $-3 + (-80)$

8. $-12 + (-1)$

9. $-23 + (-23)$

10. $-12 + (-12)$

11. $16 + (-16)$

12. $-17 + 17$

13. $48 + (-53)$

14. $19 + (-41)$

15. $-17 + (-3) + 29$

16. $13 + 62 + (-38)$

17. $-3 + (-8) + 12$

18. $-27 + (-42) + (-18)$

19. $13 + (-22) + 4 + (-5)$

20. $-14 + (-3) + 7 + (-21)$

21. $-22 + 20 + 2 + (-18)$

22. $-6 + (-8) + 14 + (-4)$

23. $-16 + (-17) + (-18) + 10$

24. $-25 + (-31) + 24 + 19$

25. $26 + (-15) + (-11) + (-12)$

26. $-32 + 40 + (-8) + (-19)$

27. $-17 + (-18) + 45 + (-10)$

28. $23 + (-15) + 9 + (-15)$

29. $46 + (-17) + (-13) + (-50)$

30. $-37 + (-17) + (-12) + (-15)$

31. $-14 + (-15) + (-11) + 40$

32. $28 + (-19) + (-8) + (-1)$

33. $-23 + (-22) + (-21) + 5$

34. $-31 + 9 + (-16) + (-15)$

35. Find the sum of -42 and -23.

36. What is 4 more than -8?

37. What is 16 more than -31?

38. Find -17 increased by 12.

39. Find the total of -17, -23, 43, and 19.

40. What is -8 added to -21?

Objective B   **To subtract integers**

41. ✏ What is the difference between the terms *minus* and *negative*?

42. ✏ Explain how to subtract two integers.

For Exercises 43 to 72, subtract.

43. $16 - 8$ **44.** $12 - 3$ **45.** $7 - 14$ **46.** $6 - 9$

47. $-7 - 2$ **48.** $-9 - 4$ **49.** $7 - (-2)$ **50.** $3 - (-4)$

51. $-6 - (-3)$ **52.** $-4 - (-2)$ **53.** $6 - (-12)$ **54.** $-12 - 16$

55. $-4 - 3 - 2$ **56.** $4 - 5 - 12$ **57.** $12 - (-7) - 8$

58. $-12 - (-3) - (-15)$ **59.** $-19 - (-19) - 18$ **60.** $-8 - (-8) - 14$

61. $-17 - (-8) - (-9)$ **62.** $7 - 8 - (-1)$ **63.** $-30 - (-65) - 29 - 4$

64. $42 - (-82) - 65 - 7$ **65.** $-16 - 47 - 63 - 12$ **66.** $42 - (-30) - 65 - (-11)$

67. $-47 - (-67) - 13 - 15$ **68.** $-18 - 49 - (-84) - 27$ **69.** $-19 - 17 - (-36) - 12$

70. $48 - 19 - 29 - 51$ **71.** $21 - (-14) - 43 - 12$ **72.** $17 - (-17) - 14 - 21$

73. Find the difference between -21 and -36.   **74.** What is 9 less than -12?

75. What is 12 less than -27?   **76.** Find -21 decreased by 19.

77. What is -21 minus -37?   **78.** Subtract 41 from -22.

Objective C **To solve application problems**

79. Chemistry   The temperature at which mercury boils is 360°C. Mercury freezes at −39°C. Find the difference between the temperature at which mercury boils and the temperature at which it freezes.

80. Chemistry   The temperature at which radon boils is −62°C. Radon freezes at −71°C. Find the difference between the temperature at which radon boils and the temperature at which it freezes.

Geography   The elcvation, or height, of places on Earth is measured in relation to sea level, or the average level of the ocean's surface. The table below shows height above sea level as a positive number and depth below sea level as a negative number. Use the table for Exercises 81 to 84.

Continent	Highest Elevation (in meters)		Lowest Elevation (in meters)	
Africa	Mt. Kilimanjaro	5895	Qattara Depression	−133
Asia	Mt. Everest	8850	Dead Sea	−400
Europe	Mt. Elbrus	5634	Caspian Sea	−28
America	Mt. Aconcagua	6960	Death Valley	−86

Mt. Everest

81. Find the difference in elevation between Mt. Aconcagua and Death Valley.

82. What is the difference in elevation between Mt. Kilimanjaro and the Qattara Depression?

83. For which continent shown is the difference between the highest and lowest elevations greatest?

84. For which continent shown is the difference between the highest and lowest elevations smallest?

Chemistry   The table at the right shows the boiling point and the melting point in degrees Celsius of three chemical elements. Use this table for Exercises 85 and 86.

Chemical Element	Boiling Point	Melting Point
Mercury	357	−39
Radon	−62	−71
Xenon	−107	−112

85. Find the difference between the boiling point and the melting point of mercury.

86. Find the difference between the boiling point and the melting point of xenon.

Geography The graph at the right shows Earth's three deepest ocean trenches and its three tallest mountains. Use this graph for Exercises 87 to 89.

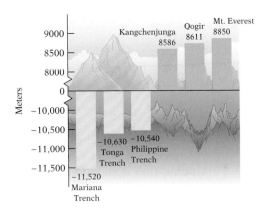

87. Could Mt. Everest fit in the Tonga Trench?

88. What is the difference between the depth of the Philippine Trench and the Mariana Trench?

89. What is the difference between the height of Mt. Everest and the depth of the Mariana Trench?

Meteorology A meteorologist may report a wind-chill temperature. This is the equivalent temperature, including the effects of wind and temperature, that a person would feel in calm air conditions. The table below gives the wind-chill temperature for various wind speeds and temperatures. For instance, when the temperature is 5°F and the wind is blowing at 15 mph, the wind-chill temperature is −13°F. Use this table for Exercises 90 and 91.

Wind Chill Factors															
Wind Speed (in mph)	Thermometer Reading (in degrees Fahrenheit)														
	25	20	15	10	5	0	−5	−10	−15	−20	−25	−30	−35	−40	−45
5	19	13	7	1	−5	−11	−16	−22	−28	−34	−40	−46	−52	−57	−63
10	15	9	3	−4	−10	−16	−22	−28	−35	−41	−47	−53	−59	−66	−72
15	13	6	0	−7	−13	−19	−26	−32	−39	−45	−51	−58	−64	−71	−77
20	11	4	−2	−9	−15	−22	−29	−35	−42	−48	−55	−61	−68	−74	−81
25	9	3	−4	−11	−17	−24	−31	−37	−44	−51	−58	−64	−71	−78	−84
30	8	1	−5	−12	−19	−26	−33	−39	−46	−53	−60	−67	−73	−80	−87
35	7	0	−7	−14	−21	−27	−34	−41	−48	−55	−62	−69	−76	−82	−89
40	6	−1	−8	−15	−22	−29	−36	−43	−50	−57	−64	−71	−78	−84	−91
45	5	−2	−9	−16	−23	−30	−37	−44	−51	−58	−65	−72	−79	−86	−93

90. When the thermometer reading is −5°F, what is the difference between the wind-chill factor when the wind is blowing 10 mph and when the wind is blowing 25 mph?

91. When the thermometer reading is −20°F, what is the difference between the wind-chill factor when the wind is blowing 15 mph and when the wind is blowing 25 mph?

APPLYING THE CONCEPTS

92. If a and b are integers, is the expression $|a + b| = |a| + |b|$ always true, sometimes true, or never true?

93. Is the difference between two integers always smaller than either one of the numbers in the difference? If not, give an example for which the difference between two integers is greater than either integer.

1.3 Multiplication and Division of Integers

Objective A   ## To multiply integers

Several different symbols are used to indicate multiplication. The numbers being multiplied are called **factors;** for instance, 3 and 2 are factors in each of the examples at the right. The result is called the **product.** Note that when parentheses are used and there is no arithmetic symbol, the operation is multiplication.

$$3 \times 2 = 6$$
$$3 \cdot 2 = 6$$
$$(3)(2) = 6$$
$$3(2) = 6$$
$$(3)2 = 6$$

Multiplication is repeated addition of the same number. The product 3×5 is shown on the number line below.

5 is added 3 times.

$$3 \times 5 = 5 + 5 + 5 = 15$$

Now consider the product of a positive and a negative number.

−5 is added 3 times.

$$3(-5) = (-5) + (-5) + (-5) = -15$$

This suggests that the product of a positive number and a negative number is negative. Here are a few more examples.

$$4(-7) = -28 \qquad -6 \cdot 7 = -42 \qquad (-8)7 = -56$$

To find the product of two negative numbers, look at the pattern at the right. As −5 multiplies a sequence of decreasing integers, the products increase by 5.

The pattern can be continued by requiring that the product of two negative numbers be positive.

These numbers decrease by 1. ↓   ↓ These numbers increase by 5.

$$-5 \times 3 = -15$$
$$-5 \times 2 = -10$$
$$-5 \times 1 = -5$$
$$-5 \times 0 = 0$$
$$-5 \times (-1) = 5$$
$$-5 \times (-2) = 10$$
$$-5 \times (-3) = 15$$

> **Multiplication of Integers**
>
> **To multiply two numbers with the same sign,** multiply the absolute values of the numbers. The product is positive.
>
> **To multiply two numbers with different signs,** multiply the absolute values of the numbers. The product is negative.

HOW TO   Multiply: $5(-12)$.

$$5(-12) = -60$$ • The signs are different. The product is negative.

HOW TO Find the product of −8 and −16.

A *product* is the answer to a multiplication problem.

$-8(-16) = 128$ • The signs are the same. The product is positive.

The phrase *the product of* in the example above indicates the operation of multiplication. All of the phrases below indicate multiplication.

times	−7 times −9	$-7(-9) = 63$
the product of	the product of 12 and −8	$12(-8) = -96$
multiplied by	−15 multiplied by 11	$-15(11) = -165$
twice	twice −14	$2(-14) = -28$

HOW TO Multiply: $-2(5)(-7)(-4)$

$-2(5)(-7)(-4) = -10(-7)(-4)$

$= 70(-4) = -280$

• To multiply more than two numbers, multiply the first two. Then multiply the product by the third number. Continue until all the numbers are multiplied.

Consider the products shown at the right. Note that when there is an even number of negative factors, the product is positive. When there is an odd number of negative factors, the product is negative.

$(-3)(-5) = 15$
$(-2)(-5)(-6) = -60$
$(-4)(-3)(-5)(-7) = 420$
$(-3)(-3)(-5)(-4)(-5) = -900$
$(-6)(-3)(-4)(-2)(-10)(-5) = 7200$

This idea can be summarized by the following useful rule: **The product of an even number of negative factors is positive; the product of an odd number of negative factors is negative.**

Example 1 Multiply: $-24(-18)$

Solution $-24(-18) = 432$

You Try It 1 Multiply: $-7(32)$

Your solution

Example 2 Multiply: $(-3)4(-5)$

Solution $(-3)4(-5) = (-12)(-5) = 60$

You Try It 2 Multiply: $8(-9)10$

Your solution

Example 3 Multiply: $12(-4)(-3)(-5)$

Solution $12(-4)(-3)(-5) = (-48)(-3)(-5)$
$= 144(-5)$
$= -720$

You Try It 3 Multiply: $(-2)3(-8)7$

Your solution

Example 4 Find the product of −13 and −9.

Solution $-13(-9) = 117$

You Try It 4 What is −9 times 34?

Your solution

Solutions on p. S1

Objective B   **To divide integers**

For every division problem there is a related multiplication problem.

$$\underset{\text{Division}}{\frac{8}{2} = 4} \qquad \text{because} \qquad \underset{\text{Related multiplication}}{4 \cdot 2 = 8.}$$

This fact and the rules for multiplying integers can be used to illustrate the rules for dividing integers.

Note in the following examples that the quotient of two numbers with the same sign is positive.

$$\frac{12}{3} = 4 \text{ because } 4 \cdot 3 = 12. \qquad \frac{-12}{-3} = 4 \text{ because } 4(-3) = -12.$$

The next two examples illustrate that the quotient of two numbers with different signs is negative.

$$\frac{12}{-3} = -4 \text{ because } (-4)(-3) = 12. \qquad \frac{-12}{3} = -4 \text{ because } (-4)3 = -12.$$

Point of Interest

There was quite a controversy over the date the new millennium started because of the number zero. When our current calendar was created, numbering began with the year 1 because 0 had not yet been invented. Thus at the beginning of year 2, 1 year had elapsed; at the beginning of year 3, 2 years had elapsed; and so on. This means that at the beginning of year 2000, 1999 years had elapsed and it was not until the beginning of year 2001 that 2000 years had elapsed and a new millennium began.

Division of Integers

To divide two numbers with the same sign, divide the absolute values of the numbers. The quotient is positive.

To divide two numbers with different signs, divide the absolute values of the numbers. The quotient is negative.

HOW TO   Divide: $-36 \div 9$.

$-36 \div 9 = -4$ • The signs are different. The quotient is negative.

HOW TO   Find the quotient of -63 and -7.

A *quotient* is the answer to a division problem.

$\frac{-63}{-7} = 9$ • The signs are the same. The quotient is positive.

The phrase *the quotient of* in the example above indicates the operation of division. All of the phrases below indicate division.

divided by	15 divided by -3	$15 \div (-3) = -5$
the quotient of	the quotient of -56 and -8	$(-56) \div (-8) = 7$
the ratio of	the ratio of 45 and -5	$45 \div (-5) = -9$
divide . . . by . . .	divide -100 by -20	$-100 \div (-20) = 5$

HOW TO Simplify: $-\dfrac{-56}{7}$

$$-\dfrac{-56}{7} = -\left(\dfrac{-56}{7}\right) = -(-8) = 8$$

The properties of division are stated below. In these statements, the symbol $\neq$ is read "is not equal to."

Properties of Zero and One in Division

If $a \neq 0$, $\dfrac{0}{a} = 0$.   Zero divided by any number other than zero is zero.

If $a \neq 0$, $\dfrac{a}{a} = 1$.   Any number other than zero divided by itself is one.

$\dfrac{a}{1} = a$.   A number divided by one is the number.

$\dfrac{a}{0}$ is undefined.   Division by zero is not defined.

The fact that $\dfrac{-12}{3} = -4$, $\dfrac{12}{-3} = -4$, and $-\dfrac{12}{3} = -4$ suggests the following rule.

If a and b are integers, and $b \neq 0$, then $\dfrac{-a}{b} = \dfrac{a}{-b} = -\dfrac{a}{b}$.

Example 5 Divide: $(-120) \div (-8)$

Solution $(-120) \div (-8) = 15$

You Try It 5 Divide: $(-135) \div (-9)$

Your solution

Example 6 Divide: $\dfrac{95}{-5}$

Solution $\dfrac{95}{-5} = -19$

You Try It 6 Divide: $\dfrac{-72}{4}$

Your solution

Example 7 Simplify: $-\dfrac{-81}{3}$

Solution $-\dfrac{-81}{3} = -(-27) = 27$

You Try It 7 Simplify: $-\dfrac{36}{-12}$

Your solution

Example 8 Find the quotient of 98 and -14.

Solution $98 \div (-14) = -7$

You Try It 8 What is the ratio of -72 and -8?

Your solution

Solutions on p. S1

Objective C   **To solve application problems**

In many courses, your course grade depends on the *average* of all your test scores. You compute the average by calculating the sum of all your test scores and then dividing that result by the number of tests. Statisticians call this average an **arithmetic mean.** Besides its application to finding the average of your test scores, the arithmetic mean is used in many other situations.

Stock market analysts calculate the **moving average** of a stock. This is the arithmetic mean of the changes in the value of a stock for a given number of days. To illustrate the procedure, we will calculate the 5-day moving average of a stock. In actual practice, a stock market analyst may use 15 days, 30 days, or some other number.

The table below shows the amount of increase or decrease, in cents, in the closing price of a stock for a 10-day period.

Day 1	Day 2	Day 3	Day 4	Day 5	Day 6	Day 7	Day 8	Day 9	Day 10
+50	−175	+225	0	−275	−75	−50	+50	−475	−50

To calculate the 5-day moving average of this stock, determine the average of the stock for days 1 through 5, days 2 through 6, days 3 through 7, and so on.

Days 1–5	Days 2–6	Days 3–7	Days 4–8	Days 5–9	Days 6–10
+50	−175	+225	0	−275	−75
−175	+225	0	−275	−75	−50
+225	0	−275	−75	−50	+50
0	−275	−75	−50	+50	−475
−275	−75	−50	+50	−475	−50
Sum = −175	Sum = −300	Sum = −175	Sum = −350	Sum = −825	Sum = −600
$Av = \dfrac{-175}{5} = -35$	$Av = \dfrac{-300}{5} = -60$	$Av = \dfrac{-175}{5} = -35$	$Av = \dfrac{-350}{5} = -70$	$Av = \dfrac{-825}{5} = -165$	$Av = \dfrac{-600}{5} = -120$

The 5-day moving average is the list of means: -35, -60, -35, -70, -165, and -120. If the list tends to increase, the price of the stock is showing an upward trend; if it decreases, the price of the stock is showing a downward trend. These trends help an analyst recommend stocks.

Example 9

The daily high temperatures (in degrees Celsius) for six days in Anchorage, Alaska, were $-14°$, $3°$, $0°$, $-8°$, $2°$, and $-1°$. Find the average daily high temperature.

Strategy

To find the average daily high temperature:

- Add the six temperature readings.
- Divide the sum by 6.

Solution

$-14 + 3 + 0 + (-8) + 2 + (-1) = -18$

$-18 \div 6 = -3$

The average daily high temperature was $-3°C$.

You Try It 9

The daily low temperatures (in degrees Celsius) during one week were recorded as $-6°$, $-7°$, $0°$, $-5°$, $-8°$, $-1°$, and $-1°$. Find the average daily low temperature.

Your strategy

Your solution

Solution on p. S1

1.3 Exercises

To multiply integers

1. ✎ Explain how to multiply two integers with the same sign.

2. ✎ Explain how to multiply two integers with different signs.

For Exercises 3 to 22, multiply.

3. (14)3

4. 17(6)

5. (−12)(−5)

6. (−13)(−9)

7. −11(23)

8. −8(21)

9. 6(−19)

10. 17(−13)

11. 7(5)(−3)

12. (−3)(−2)8

13. −3(−8)(−9)

14. −7(−6)(−5)

15. (−9)7(5)

16. (−8)7(10)

17. (−3)7(−2)8

18. −9(−4)(−8)(−10)

19. 7(9)(−11)4

20. −12(−4)7(−2)

21. (−14)9(−11)0

22. (−13)(15)(−19)0

23. Find the product of −14 and −25.

24. What is 4 times −8?

25. What is 16 multiplied by −21?

26. Find −22 multiplied by 15.

27. Find the product of 4, −8, and 11.

28. Find the product of −2, −3, −4, and −5.

To divide integers

For Exercises 29 to 56, divide.

29. 12 ÷ (−6)

30. 18 ÷ (−3)

31. (−72) ÷ (−9)

32. (−64) ÷ (−8)

33. −42 ÷ 6

34. (−56) ÷ 8

35. (−144) ÷ 12

36. (−93) ÷ (−3)

37. 48 ÷ (−8)

38. 57 ÷ (−3)

39. $\dfrac{-44}{-4}$

40. $\dfrac{-36}{-9}$

41. $\dfrac{98}{-7}$

42. $\dfrac{85}{-5}$

43. $-\dfrac{-120}{8}$

44. $-\dfrac{-72}{4}$

45. $-\dfrac{-80}{-5}$

46. $-\dfrac{-114}{-6}$

47. 0 ÷ (−9)

48. 0 ÷ (−14)

49. $\dfrac{-261}{9}$

50. $\dfrac{-128}{4}$

51. 9 ÷ 0

52. (−21) ÷ 0

53. $\dfrac{132}{-12}$

54. $\dfrac{250}{-25}$

55. $\dfrac{0}{0}$

56. $\dfrac{-58}{0}$

57. Find the quotient of −132 and 11.

58. What is 15 divided by −15?

59. What is −96 divided by −4?

60. What is the ratio of −175 and 25?

61. Divide −196 by −7.

62. Find the quotient of 342 and −9.

Objective C   **To solve application problems**

63.   **Meteorology**   The high temperatures for a 6-day period in Barrow, Alaska, were $-23°F$, $-29°F$, $-21°F$, $-28°F$, $-28°F$, and $-27°F$. Calculate the average daily high temperature.

64.   **Meteorology**   The low temperatures for a 10-day period in a midwestern city were $-4°F$, $-9°F$, $-5°F$, $-2°F$, $4°F$, $-1°F$, $-1°F$, $-2°F$, $-2°F$, and $2°F$. Calculate the average daily low temperature for this city.

65.   **Investments**   The value of a share of Yahoo's stock on August 21, 2003, was \$32.82. The table below shows the approximate increase or decrease, in cents, from the August 21 closing price of the stock for a 10-day period. Calculate the 5-day moving average for this stock.

Day 1	Day 2	Day 3	Day 4	Day 5	Day 6	Day 7	Day 8	Day 9	Day 10
-100	20	0	30	-10	100	80	-60	130	10

66.   **Investments**   The value of a share of Ford Motor Company's stock on August 22, 2003, was \$11.13. The table below shows the approximate increase or decrease, in cents, from the August 22 closing price of the stock for a 10-day period. Calculate the 5-day moving average for this stock.

Day 1	Day 2	Day 3	Day 4	Day 5	Day 6	Day 7	Day 8	Day 9	Day 10
-20	30	20	-20	10	90	-40	-10	-10	10

67.   **Testing**   To discourage random guessing on a multiple-choice exam, a professor assigns 5 points for a correct answer, -2 points for an incorrect answer, and 0 points for leaving the question blank. What is the score for a student who had 20 correct answers, had 13 incorrect answers, and left 7 questions blank?

68.   **Testing**   To discourage random guessing on a multiple-choice exam, a professor assigns 7 points for a correct answer, -3 points for an incorrect answer, and -1 point for leaving the question blank. What is the score for a student who had 17 correct answers, had 8 incorrect answers, and left 2 questions blank?

APPLYING THE CONCEPTS

69.   If $x \in \{-6, -2, 7\}$, for which value of x does the expression $-3x$ have the greatest value?

70.   Explain why $0 \div 0$ is not defined.

71.   If $-4x$ equals a positive integer, is x a positive or a negative integer? Explain your answer.

1.4 Exponents and the Order of Operations Agreement

Objective A To evaluate exponential expressions

Repeated multiplication of the same factor can be written using an exponent.

$$2 \cdot 2 \cdot 2 \cdot 2 \cdot 2 = 2^5 \leftarrow \text{Exponent}$$
$$\underset{\text{Base}}{}$$

$$a \cdot a \cdot a \cdot a = a^4 \leftarrow \text{Exponent}$$
$$\underset{\text{Base}}{}$$

Point of Interest

Rene Descartes (1596–1650) was the first mathematician to use exponential notation extensively as it is used today. However, for some unknown reason, he always used *xx* for x^2.

The **exponent** indicates how many times the factor, which is called the **base**, occurs in the multiplication. The multiplication $2 \cdot 2 \cdot 2 \cdot 2 \cdot 2$ is in **factored form.** The exponential expression 2^5 is in **exponential form.**

2^1 is read "2 to the first power" or just "2." Usually the exponent 1 is not written.

2^2 is read "2 to the second power" or "2 squared."

2^3 is read "2 to the third power" or "2 cubed."

2^4 is read "2 to the fourth power."

a^4 is read "*a* to the fourth power."

There is a geometric interpretation of the first three natural-number powers.

$4^1 = 4$
Length: 4 ft

$4^2 = 16$
Area: 16 ft²

$4^3 = 64$
Volume: 64 ft³

To evaluate an exponential expression, write each factor as many times as indicated by the exponent. Then multiply.

HOW TO Evaluate $(-2)^4$.

$(-2)^4 = (-2)(-2)(-2)(-2)$ • Write −2 as a factor 4 times.

$\quad = 16$ • Multiply.

TAKE NOTE

Note the difference between $(-2)^4$ and -2^4.
$(-2)^4 = 16$
$-2^4 = -16$

HOW TO Evaluate -2^4.

$-2^4 = -(2 \cdot 2 \cdot 2 \cdot 2)$ • Write 2 as a factor 4 times.

$\quad = -16$ • Multiply.

Example 1   Evaluate -5^3.

Solution $-5^3 = -(5 \cdot 5 \cdot 5) = -125$

You Try It 1   Evaluate -6^3.

Your solution

Example 2   Evaluate $(-4)^4$.

Solution $(-4)^4 = (-4)(-4)(-4)(-4)$
            $= 256$

You Try It 2   Evaluate $(-3)^4$.

Your solution

Example 3   Evaluate $(-3)^2 \cdot 2^3$.

Solution $(-3)^2 \cdot 2^3 = (-3)(-3) \cdot (2)(2)(2)$
                  $= 9 \cdot 8 = 72$

You Try It 3   Evaluate $(3^3)(-2)^3$.

Your solution

Example 4   Evaluate $(-1)^6$.

Solution   The product of an even number of negative factors is positive. Therefore, $(-1)^6 = 1$.

You Try It 4   Evaluate $(-1)^7$.

Your solution

Example 5   Evaluate $-2 \cdot (-3)^2 \cdot (-1)^9$.

Solution $-2 \cdot (-3)^2 \cdot (-1)^9$
       $= -2[(-3) \cdot (-3)](-1)$ • $(-1)^9 = -1$
       $= -2 \cdot 9 \cdot (-1) = 18$

You Try It 5   Evaluate $-2^2 \cdot (-1)^{12} \cdot (-3)^2$.

Your solution

Solutions on pp. S1–S2

Objective B   **To use the Order of Operations Agreement to simplify expressions**

Let's evaluate $2 + 3 \cdot 5$.

There are two arithmetic operations, addition and multiplication, in this expression. The operations could be performed in different orders.

Multiply first. $2 + \underline{3 \cdot 5}$ Add first. $2 + 3 \cdot 5$

Then add. $\underline{2 + 15}$ Then multiply. $\underline{5 \cdot 5}$

                 17 25

To prevent there being more than one answer for a numerical expression, an Order of Operations Agreement has been established.

> **The Order of Operations Agreement**
>
> **Step 1**   Perform operations inside grouping symbols. Grouping symbols include parentheses ( ), brackets [ ], braces { }, and the fraction bar.
> **Step 2**   Simplify exponential expressions.
> **Step 3**   Do multiplication and division as they occur from left to right.
> **Step 4**   Do addition and subtraction as they occur from left to right.

Integrating Technology

See the Keystroke Guide: *Basic Operations* for instruction on using a calculator to evaluate a numerical expression.

HOW TO Evaluate $12 - 24(8 - 5) \div 2^2$.

$$12 - 24(8 - 5) \div 2^2 = 12 - 24(3) \div 2^2$$

- Perform operations inside grouping symbols.

$$= 12 - 24(3) \div 4$$

- Simplify exponential expressions.

$$= 12 - 72 \div 4$$

- Do multiplication and division as they occur from left to right.

$$= 12 - 18$$

$$= -6$$

- Do addition and subtraction as they occur from left to right.

One or more of the steps listed above may not be needed to evaluate an expression. In that case, proceed to the next step in the Order of Operations Agreement.

HOW TO Evaluate $\dfrac{4 + 8}{2 + 1} - (3 - 1) + 2$.

$$\frac{4 + 8}{2 + 1} - (3 - 1) + 2 = \frac{12}{3} - 2 + 2$$

- Perform operations above and below the fraction bar and inside parentheses.

$$= 4 - 2 + 2$$

- Do multiplication and division as they occur from left to right.

$$= 2 + 2$$

- Do addition and subtraction as they occur from left to right.

$$= 4$$

Example 6

Evaluate $6 \div [4 - (6 - 8)] - 2^3$.

Solution

$6 \div [4 - (6 - 8)] - 2^3$

$= 6 \div [4 - (-2)] - 2^3$ • Perform operations inside grouping symbols.

$= 6 \div 6 - 2^3$

$= 6 \div 6 - 8$ • Simplify exponential expressions.

$= 1 - 8$ • Do multiplication and division from left to right.

$= -7$ • Do addition and subtraction from left to right.

You Try It 6

Evaluate $7 - 2[2 \cdot 3 - 7 \cdot 2]^2$.

Your solution

Solution on p. S2

Example 7

Evaluate $4 - 3[4 - 2(6 - 3)] \div 2$.

Solution

$4 - 3[4 - 2(6 - 3)] \div 2$

$= 4 - 3[4 - 2 \cdot 3] \div 2$ • Perform operations inside grouping symbols.

$= 4 - 3[4 - 6] \div 2$

$= 4 - 3[-2] \div 2$

$= 4 + 6 \div 2$ • Do multiplication and division from left to right.

$= 4 + 3$

$= 7$ • Do addition and subtraction from left to right.

You Try It 7

Evaluate $18 - 5[8 - 2(2 - 5)] \div 10$.

Your solution

Example 8

Evaluate $27 \div (5 - 2)^2 + (-3)^2 \cdot 4$.

Solution

$27 \div (5 - 2)^2 + (-3)^2 \cdot 4$

$= 27 \div 3^2 + (-3)^2 \cdot 4$ • Perform operations inside grouping symbols.

$= 27 \div 9 + 9 \cdot 4$ • Simplify exponential expressions.

$= 3 + 9 \cdot 4$ • Do multiplication and division from left to right.

$= 3 + 36$

$= 39$ • Do addition and subtraction from left to right.

You Try It 8

Evaluate $36 \div (8 - 5)^2 - (-3)^2 \cdot 2$.

Your solution

Solutions on p. S2

1.4 Exercises

Objective A   To evaluate exponential expressions

For Exercises 1 to 27, evaluate.

1. 6^2

2. 7^4

3. -7^2

4. -4^3

5. $(-3)^2$

6. $(-2)^3$

7. $(-3)^4$

8. $(-5)^3$

9. -4^4

10. $(-4)^4$

11. $2 \cdot (-3)^2$

12. $-2 \cdot (-4)^2$

13. $(-1)^9 \cdot 3^3$

14. $(-1)^8(-8)^2$

15. $(3)^3 \cdot 2^3$

16. $(5)^2 \cdot 3^3$

17. $(-3) \cdot 2^2$

18. $(-5) \cdot 3^4$

19. $(-2) \cdot (-2)^3$

20. $(-2) \cdot (-2)^2$

21. $2^3 \cdot 3^3 \cdot (-4)$

22. $(-3)^3 \cdot 5^2 \cdot 10$

23. $(-7) \cdot 4^2 \cdot 3^2$

24. $(-2) \cdot 2^3 \cdot (-3)^2$

25. $-3^2 \cdot (-3)^2$

26. $(-2)^3(-3)^2(-1)^7$

27. $8^2 \cdot (-3)^5 \cdot 5$

Objective B   To use the Order of Operations Agreement to simplify expressions

28. Explain why an Order of Operations Agreement is necessary.

29. Write out the steps of the Order of Operations Agreement.

For Exercises 30 to 62, evaluate by using the Order of Operations Agreement.

30. $4 - 8 \div 2$

31. $2^2 \cdot 3 - 3$

32. $2(3 - 4) - (-3)^2$

33. $16 - 32 \div 2^3$

34. $24 - 18 \div 3 + 2$

35. $8 - (-3)^2 - (-2)$

36. $8 - 2(3)^2$

37. $16 - 16 \cdot 2 \div 4$

38. $12 + 16 \div 4 \cdot 2$

39. $16 - 2 \cdot 4^2$

40. $27 - 18 \div (-3^2)$

41. $4 + 12 \div 3 \cdot 2$

42. $16 + 15 \div (-5) - 2$

43. $14 - 2^2 - (4 - 7)$

44. $3 - 2[8 - (3 - 2)]$

45. $-2^2 + 4[16 \div (3 - 5)]$

46. $6 + \dfrac{16 - 4}{2^2 + 2} - 2$

47. $24 \div \dfrac{3^2}{8 - 5} - (-5)$

48. $96 \div 2[12 + (6 - 2)] - 3^2$

49. $4[16 - (7 - 1)] \div 10$

50. $18 \div 2 - 4^2 - (-3)^2$

51. $18 \div (9 - 2^3) + (-3)$

52. $16 - 3(8 - 3)^2 \div 5$

53. $4(-8) \div [2(7 - 3)^2]$

54. $\dfrac{(-19) + (-2)}{6^2 - 29} \div (2 - 5)$

55. $16 - 4 \cdot \dfrac{3^3 - 7}{2^3 + 2} - (-2)^2$

56. $7 - 3[1 - (2 - (-3))^2]$

57. $-18 \div (-6) \div (1 - (-2))$

58. $-4 \cdot 2^3 - \dfrac{1 - 13}{2^2 \cdot 3}$

59. $(8 - 3^2)^{10} + (2 \cdot 3 - 7)^{11}$

60. $(2^3 - (-3))(2^3 + (-3))$

61. $-6^2 \cdot 3 - 2^2(1 - 5)^2$

62. $14 - 2^2(3 - 4^2)$

APPLYING THE CONCEPTS

63. ✏️ The following was offered as the simplification of $6 + 2(4 - 9)$.

$$
\begin{aligned}
6 + 2(4 - 9) &= 6 + 2(-5) \\
&= 8(-5) \\
&= -40
\end{aligned}
$$

If this is a correct simplification, write yes for the answer. If it is incorrect, write no and explain the incorrect step.

64. ✏️ The following was offered as the simplification of $2 \cdot 3^3$.

$$2 \cdot 3^3 = 6^3 = 216$$

If this is a correct simplification, write yes for the answer. If it is incorrect, write no and explain the incorrect step.

1.5 Factoring Numbers and Prime Factorization

Objective A   To factor numbers

A **factor of a number** is a natural number that divides the number with a remainder of 0.

The factors of 12 are 1, 2, 3, 4, 6, and 12 because each of those numbers divides 12 with a remainder of 0. Note that both the divisor and the quotient are factors of the dividend.

$12 \div 1 = 12 \qquad 12 \div 4 = 3$
$12 \div 2 = 6 \qquad 12 \div 6 = 2$
$12 \div 3 = 4 \qquad 12 \div 12 = 1$

To find the factors of a number, try dividing the number by 1, 2, 3, 4, 5, ... Those numbers that divide the number are its factors. Continue this process until the factors start to repeat.

HOW TO   Find all the factors of 40.

$40 \div 1 = 40$ 1 and 40 are factors.
$40 \div 2 = 20$ 2 and 20 are factors.
$40 \div 3$ Remainder is not 0.
$40 \div 4 = 10$ 4 and 10 are factors.
$40 \div 5 = 8$ 5 and 8 are factors. ←
$40 \div 6$ Remainder is not 0.   } Factors are repeating. All the
$40 \div 7$ Remainder is not 0.   factors of 40 have been found.
$40 \div 8 = 5$ 8 and 5 are factors. ←

1, 2, 4, 5, 8, 10, 20, and 40 are factors of 40.

The following rules are helpful in finding the factors of a number.

2 is a factor of a number if the last digit of the number is 0, 2, 4, 6, or 8.	528 ends in 8; therefore, 2 is a factor of 528. ($528 \div 2 = 264$)
3 is a factor of a number if the sum of digits of the number is divisible by 3.	The sum of the digits of 378 is $3 + 7 + 8 = 18$. 18 is divisible by 3. Therefore, 3 is a factor of 378. ($378 \div 3 = 126$)
5 is a factor of a number if the last digit of the number is a 0 or a 5.	495 ends in 5; therefore, 5 is a factor of 495. ($495 \div 5 = 99$)

Example 1   Find all the factors of 18.

Solution
$18 \div 1 = 18$
$18 \div 2 = 9$
$18 \div 3 = 6$
$18 \div 4$ Remainder is not 0.
$18 \div 5$ Remainder is not 0.
$18 \div 6 = 3$ The factors are repeating.

1, 2, 3, 6, 9, and 18 are the factors of 18.

You Try It 1   Find the factors of 24.

Your solution

Solution on p. S2

Objective B   **To find the prime factorization of a number**

A natural number greater than 1 is a **prime number** if its only factors are 1 and the number. For instance, 11 is a prime number because the only factors of 11 are 1 and 11. A natural number greater than 1 that is not a prime number is a **composite number.** An example of a composite number is 6. It has factors of 1, 2, 3, and 6. The number 1 is neither a prime nor a composite number.

Prime numbers less than 50 = 2, 3, 5, 7, 11, 13, 17, 19, 23, 29, 31, 37, 41, 43, 47

The **prime factorization** of a number is the expression of the number as a product of its prime factors. We use a "T-diagram" to find the prime factors of a number. Begin with the smallest prime number as a trial divisor, and continue to use prime numbers as trial divisors until the final quotient is 1.

TAKE NOTE

A prime number that is a factor of a number is called a **prime factor** of the number. For instance, 3 is a prime factor of 18. However, 6 is a factor of 18 but is not a prime factor of 18.

HOW TO   Find the prime factorization of 84.

$$
\begin{array}{r|r}
\multicolumn{2}{c}{84} \\
\hline
2 & 42 \\
2 & 21 \\
3 & 7 \\
7 & 1 \\
\end{array}
\qquad
\begin{array}{l}
84 \div 2 = 42 \\
42 \div 2 = 21 \\
21 \div 3 = 7 \\
7 \div 7 = 1 \\
\end{array}
$$

The prime factorization of 84 is $2^2 \cdot 3 \cdot 7$.

Finding the prime factorization of larger numbers can be difficult. Try each prime number as a trial divisor until the square of the trial divisor exceeds the number.

Point of Interest

Prime numbers are an important part of cryptology, the study of secret codes. Codes based on prime numbers with hundreds of digits are used to send sensitive information over the Internet.

HOW TO   Find the prime factorization of 177.

$$
\begin{array}{r|r}
\multicolumn{2}{c}{177} \\
\hline
3 & 59 \\
59 & 1 \\
\end{array}
$$

• For 59, only try prime numbers up to 11 because $11^2 = 121 > 59$.

The prime factorization of 177 is $3 \cdot 59$.

Example 2   Find the prime factorization of 132.

Solution

$$
\begin{array}{r|r}
\multicolumn{2}{c}{132} \\
\hline
2 & 66 \\
2 & 33 \\
3 & 11 \\
11 & 1 \\
\end{array}
$$

$132 = 2^2 \cdot 3 \cdot 11$

You Try It 2   Find the prime factorization of 315.

Your solution

Example 3   Find the prime factorization of 141.

Solution

$$
\begin{array}{r|r}
\multicolumn{2}{c}{141} \\
\hline
3 & 47 \\
47 & 1 \\
\end{array}
$$

• For 47, try prime numbers up to 7 because $7^2 > 47$.

$141 = 3 \cdot 47$

You Try It 3   Find the prime factorization of 326.

Your solution

Solutions on p. S2

Objective C   **To find the least common multiple and greatest common factor**

The **least common multiple (LCM)** of two or more numbers is the smallest number that is a multiple of all the numbers. For instance, 24 is the LCM of 6 and 8 because it is the smallest number that is divisible by both 6 and 8.

The LCM can be found by first writing each number as a product of prime factors. The LCM must contain all the prime factors of each number.

HOW TO   Find the LCM of 10 and 12.

Determine the prime factorization of each number.

$$10 = 2 \cdot 5$$

Factors of 10

$$12 = 2 \cdot 2 \cdot 3 \qquad \text{LCM} = \underbrace{2 \cdot 2 \cdot 3}_{\text{Factors of 12}} \cdot 5 = 60$$

The LCM of 10 and 12 is 60.

HOW TO   Find the LCM of 8, 14, and 18.

Determine the prime factorization of each number.

$$8 = 2 \cdot 2 \cdot 2 \qquad 14 = 2 \cdot 7 \qquad 18 = 2 \cdot 3 \cdot 3$$

The LCM must contain the prime factors of 8, 14, and 18.

$$\text{LCM} = 2 \cdot 2 \cdot 2 \cdot 3 \cdot 3 \cdot 7 = 504$$

The **greatest common factor (GCF)** of two or more numbers is the greatest number that divides evenly into all the numbers. For instance, the GCF of 12 and 18 is 6, the largest number that divides evenly into 12 and 18.

The GCF can be found by first writing each number as a product of prime factors. The GCF contains the prime factors common to each number.

HOW TO   Find the GCF of 36 and 90.

Determine the prime factorization of each number.

$$36 = 2 \cdot 2 \cdot 3 \cdot 3$$ • The common factors
$$90 = 2 \cdot 3 \cdot 3 \cdot 5$$ are shown in red.

The GCF is the product of the prime factors common to each number.

The GCF of 36 and 90 is $2 \cdot 3 \cdot 3 = 18$.

Example 4   Find the LCM of 15, 20, and 30.

Solution
$$15 = 3 \cdot 5 \quad 20 = 2 \cdot 2 \cdot 5 \quad 30 = 2 \cdot 3 \cdot 5$$
$$\text{LCM} = 2 \cdot 2 \cdot 3 \cdot 5 = 60$$

You Try It 4   Find the LCM of 20 and 21.

Your solution

Example 5   Find the GCF of 30, 45, and 60.

Solution
$$30 = 2 \cdot 3 \cdot 5 \quad 45 = 3 \cdot 3 \cdot 5 \quad 60 = 2 \cdot 2 \cdot 3 \cdot 5$$
$$\text{GCF} = 3 \cdot 5 = 15$$

You Try It 5   Find the GCF of 42 and 63.

Your solution

Solutions on p. S2

1.5 Exercises

Objective A   **To factor numbers**

For Exercises 1 to 30, find the factors of the number.

1. 4	**2.** 20	**3.** 12	**4.** 7	**5.** 8	**6.** 9
7. 13	**8.** 30	**9.** 56	**10.** 28	**11.** 45	**12.** 33
13. 29	**14.** 22	**15.** 52	**16.** 37	**17.** 82	**18.** 69
19. 57	**20.** 64	**21.** 48	**22.** 46	**23.** 50	**24.** 54
25. 77	**26.** 66	**27.** 100	**28.** 80	**29.** 85	**30.** 96

Objective B   **To find the prime factorization of a number**

For Exercises 31 to 60, find the prime factorization of the number.

31. 14	**32.** 6	**33.** 72	**34.** 17	**35.** 24	**36.** 27
37. 36	**38.** 115	**39.** 26	**40.** 18	**41.** 49	**42.** 42
43. 31	**44.** 81	**45.** 62	**46.** 39	**47.** 89	**48.** 101
49. 86	**50.** 66	**51.** 95	**52.** 74	**53.** 78	**54.** 67
55. 144	**56.** 120	**57.** 175	**58.** 160	**59.** 400	**60.** 625

Objective C   **To find the least common multiple and greatest common factor**

For Exercises 61 to 90, find the LCM.

61. 3, 8	**62.** 5, 11	**63.** 4, 6	**64.** 6, 8	**65.** 9, 12
66. 8, 14	**67.** 14, 20	**68.** 7, 21	**69.** 12, 36	**70.** 6, 10

71. 48, 60   **72.** 16, 24   **73.** 80, 90   **74.** 35, 42   **75.** 72, 108

76. 5, 12   **77.** 24, 45   **78.** 8, 20   **79.** 32, 80   **80.** 20, 28

81. 3, 8, 12   **82.** 6, 12, 18   **83.** 3, 5, 10   **84.** 6, 12, 24   **85.** 3, 8, 9

86. 4, 10, 14   **87.** 10, 15, 25   **88.** 8, 12, 18   **89.** 18, 27, 36   **90.** 14, 28, 35

For Exercises 91 to 120, find the GCF.

91. 4, 10   **92.** 9, 15   **93.** 5, 11   **94.** 11, 19   **95.** 6, 8

96. 7, 28   **97.** 6, 12   **98.** 14, 42   **99.** 8, 28   **100.** 24, 36

101. 60, 70   **102.** 72, 108   **103.** 40, 56   **104.** 48, 60   **105.** 35, 42

106. 45, 63   **107.** 60, 90   **108.** 45, 55   **109.** 20, 63   **110.** 28, 45

111. 6, 12, 20   **112.** 12, 18, 24   **113.** 6, 12, 18   **114.** 30, 45, 75   **115.** 24, 36, 60

116. 10, 30, 45   **117.** 26, 52, 78   **118.** 100, 150, 200   **119.** 36, 54, 360   **120.** 18, 27, 36

APPLYING THE CONCEPTS

121. Twin primes are two prime numbers that differ by 2. For instance, 17 and 19 are twin primes. Find three sets of twin primes, not including 17 and 19.

122. In 1742, Christian Goldbach conjectured that every even number greater than 2 could be expressed as the sum of two prime numbers. Show this conjecture is true for 12, 24, and 72. (*Note:* Mathematicians have not yet been able to determine whether Goldbach's conjecture is true or false.)

123. Explain why 2 is the only even prime number.

124. Choose some prime numbers and find the square of each number. Now determine the number of factors in the square of the prime number. Make a conjecture as to the number of factors in the square of any prime number.

125. What is the LCM of two consecutive natural numbers?

126. What is the GCF of two consecutive natural numbers?

127. Explain how the LCM of two numbers can be found by using the GCF of the two numbers.

1.6 Addition and Subtraction of Rational Numbers

Objective A

To write a rational number in simplest form and as a decimal

> **TAKE NOTE**
>
> The numbers $-\frac{4}{9}, \frac{-4}{9}$, and $\frac{4}{-9}$ all represent the same rational number.

A **rational number** is the quotient of two integers. A rational number written in this way is commonly called a fraction. Some examples of rational numbers are shown at the right.

$$\frac{3}{4}, \quad \frac{-4}{9}, \quad \frac{15}{-4}, \quad \frac{8}{1}, \quad -\frac{5}{6}$$

> **Rational Numbers**
>
> A **rational number** is a number that can be written in the form $\frac{a}{b}$, where a and b are integers and $b \neq 0$.

Point of Interest

As early as 630 A.D., the Hindu mathematician Brahmagupta wrote a fraction as one number over another separated by a space. The Arab mathematician al Hassar (around 1050 A.D.) was the first to show a fraction with the horizontal bar separating the numerator and denominator.

Because an integer can be written as the quotient of the integer and 1, every integer is a rational number.

$$6 = \frac{6}{1} \qquad -8 = \frac{-8}{1}$$

A fraction is in **simplest form** when there are no common factors in the numerator and the denominator. The fractions $\frac{4}{6}$ and $\frac{2}{3}$ are equivalent fractions because they represent the same part of a whole. However, the fraction $\frac{2}{3}$ is in simplest form because there are no common factors (other than 1) in the numerator and denominator.

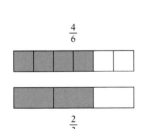

$$\frac{4}{6}$$

$$\frac{2}{3}$$

To write a fraction in simplest form, eliminate the common factors from the numerator and denominator by using the fact that $1 \cdot \frac{a}{b} = \frac{a}{b}$.

$$\frac{4}{6} = \frac{\cancel{2} \cdot 2}{\cancel{2} \cdot 3} = \frac{\cancel{2}}{\cancel{2}} \cdot \frac{2}{3} = 1 \cdot \frac{2}{3} = \frac{2}{3}$$

The process of eliminating common factors is usually written as shown at the right.

$$\frac{4}{6} = \frac{\overset{1}{\cancel{2}} \cdot 2}{\underset{1}{\cancel{2}} \cdot 3} = \frac{2}{3}$$

> **HOW TO** Write $\frac{18}{30}$ in simplest form.
>
> $$\frac{18}{30} = \frac{\overset{1}{\cancel{2}} \cdot \overset{1}{\cancel{3}} \cdot 3}{\underset{1}{\cancel{2}} \cdot \underset{1}{\cancel{3}} \cdot 5} = \frac{3}{5}$$
>
> • To find the common factors, write the numerator and denominator in terms of prime factors.

A rational number can also be written in **decimal notation.**

three tenths $0.3 = \frac{3}{10}$ forty-three thousandths $0.043 = \frac{43}{1000}$

A rational number written as a fraction can be written in decimal notation by dividing the numerator of the fraction by the denominator. Think of the fraction bar as meaning "divided by."

HOW TO   Write $\frac{5}{8}$ as a decimal.

$$
\begin{array}{r}
0.625 \\
8\overline{)5.000} \\
-4\ 8 \\
\hline
20 \\
-16 \\
\hline
40 \\
-40 \\
\hline
0
\end{array}
$$

• Divide the numerator, 5, by the denominator, 8.

When a resulting remainder is zero, the decimal is called a **terminating decimal.** The decimal 0.625 is a terminating decimal.

$$\frac{5}{8} = 0.625$$

HOW TO   Write $\frac{4}{11}$ as a decimal.

$$
\begin{array}{r}
0.3636 \\
11\overline{)4.0000} \\
-3\ 3 \\
\hline
70 \\
-66 \\
\hline
40 \\
-33 \\
\hline
70 \\
-66 \\
\hline
4
\end{array}
$$

• Divide the numerator, 4, by the denominator, 11.

No matter how long we continue to divide, the remainder is never zero. The decimal $0.\overline{36}$ is a **repeating decimal.** The bar over the 36 indicates that these digits repeat.

$$\frac{4}{11} = 0.\overline{36}$$

Every rational number can be written as a terminating or a repeating decimal. Some numbers—for example, $\sqrt{7}$ and π—have decimal representations that never terminate or repeat. These numbers are called **irrational numbers.**

$$\sqrt{7} \approx 2.6457513\ldots \qquad \pi \approx 3.1415926\ldots$$

The rational numbers and the irrational numbers taken together are called the **real numbers.**

The diagram below shows the relationship between some of the sets of numbers we have discussed. The arrows indicate that one set is contained completely within the other set.

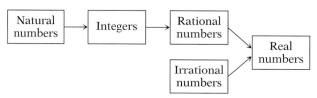

Note that there is no arrow between the rational numbers and the irrational numbers. Any given real number is either a rational number or an irrational number. It cannot be both. However, a natural number such as 7 can also be called an integer, a rational number, and a real number.

TAKE NOTE

Rational numbers are fractions, such as $-\frac{6}{7}$ or $\frac{10}{3}$, in which the numerator and denominator are integers. Rational numbers are also represented by repeating decimals such as 0.25767676... or terminating decimals such as 1.73. An irrational number is neither a terminating decimal nor a repeating decimal. For instance, 2.45445444544445... is an irrational number.

Example 1   Write $\frac{90}{168}$ in simplest form.

Solution $\dfrac{90}{168} = \dfrac{\overset{1}{\cancel{2}} \cdot \overset{1}{\cancel{3}} \cdot 3 \cdot 5}{\underset{1}{\cancel{2}} \cdot 2 \cdot 2 \cdot \underset{1}{\cancel{3}} \cdot 7} = \dfrac{15}{28}$

You Try It 1   Write $\frac{60}{140}$ in simplest form.

Your solution

Solution on p. S2

Example 2 Write $\frac{3}{20}$ as a decimal.

Solution $\frac{3}{20} = 3 \div 20 = 0.15$

You Try It 2 Write $\frac{4}{9}$ as a decimal. Place a bar over the repeating digits.

Your solution

Solution on p. S2

Objective B   **To add rational numbers**

Point of Interest

One of the earliest written mathematical documents is the Rhind Papyrus. It was discovered in Egypt in 1858 but it is estimated to date from 1650 B.C. The Papyrus shows that the earlier Egyptian method of calculating with fractions was much different from the methods used today. The early Egyptians used *unit fractions*, which are fractions with a numerator of one. With the exception of $\frac{2}{3}$, fractions with numerators other than 1 were written as the sum of two unit fractions. For instance, $\frac{2}{11}$ was written as $\frac{1}{6} + \frac{1}{66}$.

Two of the 7 squares in the rectangle have dark shading. This is $\frac{2}{7}$ of the entire rectangle. Three of the 7 squares in the rectangle have light shading. This is $\frac{3}{7}$ of the entire rectangle. A total of 5 squares are shaded. This is $\frac{5}{7}$ of the entire rectangle.

$$\frac{2}{7} + \frac{3}{7} = \frac{2+3}{7} = \frac{5}{7}$$

Addition of Fractions

To add two fractions with the same denominator, add the numerators and place the sum over the common denominator.

HOW TO   Find the sum of $\frac{3}{8}$ and $\frac{1}{8}$.

$$\frac{3}{8} + \frac{1}{8} = \frac{3+1}{8}$$

• The denominators are the same.
  Add the numerators.

$$= \frac{4}{8} = \frac{1}{2}$$

• Write the answer in simplest form.

To add fractions with different denominators, first rewrite the fractions as equivalent fractions with a common denominator. Then add the fractions. The common denominator is the least common multiple (LCM) of the denominators. The least common multiple of the denominators is frequently called the **least common denominator.**

TAKE NOTE

In this text, we will normally leave improper fractions as the answer and not change them to mixed numbers.

HOW TO   Add: $\frac{7}{10} + \frac{11}{12}$

The LCM of 10 and 12 is 60.

$$\frac{7}{10} + \frac{11}{12} = \frac{42}{60} + \frac{55}{60}$$

• Rewrite each fraction as an equivalent fraction with a denominator of 60.

$$= \frac{42 + 55}{60} = \frac{97}{60}$$

• Add the fractions.

If one of the addends is a negative rational number, use the same rules as for addition of integers.

HOW TO Add: $-\dfrac{5}{6} + \dfrac{3}{10}$

The LCM of 6 and 10 is 30.

$$-\dfrac{5}{6} + \dfrac{3}{10} = -\dfrac{25}{30} + \dfrac{9}{30}$$

• Rewrite each fraction as an equivalent fraction with a denominator of 30.

$$= \dfrac{-25 + 9}{30}$$

• Add the fractions.

$$= \dfrac{-16}{30} = -\dfrac{8}{15}$$

To add decimals, write the numbers so that the decimal points are in a vertical line. Then proceed as in the addition of integers. Write the decimal point in the answer directly below the decimal points in the problem.

HOW TO Add: $-114.03 + 89.254$

$$\begin{array}{r} 114.030 \\ -\ \ 89.254 \\ \hline 24.776 \end{array}$$

$-114.03 + 89.254$
$= -24.776$

• The signs are different. Find the difference between the absolute values of the numbers $|-114.03| = 114.03;\ |89.254| = 89.254$

• Attach the sign of the number with the larger absolute value. Because $|-114.03| > |89.254|$, use the sign of -114.03.

Example 3 Add: $\dfrac{5}{16} + \left(-\dfrac{7}{40}\right)$

Solution The LCM of 16 and 40 is 80.

$$\dfrac{5}{16} + \left(-\dfrac{7}{40}\right) = \dfrac{25}{80} + \left(-\dfrac{14}{80}\right) = \dfrac{25 + (-14)}{80} = \dfrac{11}{80}$$

You Try It 3 Add: $\dfrac{5}{9} + \left(-\dfrac{11}{12}\right)$

Your solution

Example 4 Find the total of $\dfrac{3}{4}$, $\dfrac{1}{6}$, and $\dfrac{5}{8}$.

Solution The LCM of 4, 6, and 8 is 24.

$$\dfrac{3}{4} + \dfrac{1}{6} + \dfrac{5}{8} = \dfrac{18}{24} + \dfrac{4}{24} + \dfrac{15}{24}$$

$$= \dfrac{18 + 4 + 15}{24} = \dfrac{37}{24}$$

You Try It 4 Find $\dfrac{7}{8}$ more than $-\dfrac{5}{6}$.

Your solution

Example 5 Add: $-4 + 2.37$

Solution $-4 + 2.37 = -1.63$

You Try It 5 Add: $-6.12 + (-12.881)$

Your solution

Solutions on p. S2

Objective C   **To subtract rational numbers**

Subtracting fractions is similar to adding fractions in that the denominators must be the same.

> **Subtraction of Fractions**
>
> **To subtract two fractions with the same denominator,** subtract the numerators and place the difference over the common denominator.

HOW TO   What is $\frac{3}{10}$ less than $\frac{4}{15}$?

Translate this as the subtraction problem $\frac{4}{15} - \frac{3}{10}$.

$$\frac{4}{15} - \frac{3}{10} = \frac{8}{30} - \frac{9}{30}$$

- The LCM of 15 and 10 is **30**. Rewrite each fraction as an equivalent fraction with a denominator of 30.

$$= \frac{8-9}{30} = \frac{8 + (-9)}{30}$$

- Subtract the fractions.

$$= \frac{-1}{30} = -\frac{1}{30}$$

HOW TO   Subtract: $-\frac{7}{8} - \left(-\frac{5}{12}\right)$

$$-\frac{7}{8} - \left(-\frac{5}{12}\right) = -\frac{21}{24} - \left(-\frac{10}{24}\right)$$

- The LCM of 8 and 12 is **24**.

$$= \frac{-21 - (-10)}{24} = \frac{-21 + 10}{24}$$

- Subtract the fractions.

$$= \frac{-11}{24} = -\frac{11}{24}$$

HOW TO   Subtract: $-2.984 - (-1.45)$

$$-2.984 - (-1.45) = -2.984 + 1.45$$
$$= -1.534$$

Example 6   Subtract: $-\frac{1}{2} - \frac{5}{6} - \left(-\frac{3}{4}\right)$

Solution   The LCM of 2, 6, and 4 is 12.

$$-\frac{1}{2} - \frac{5}{6} - \left(-\frac{3}{4}\right) = \frac{6}{12} - \frac{10}{12} - \left(-\frac{9}{12}\right)$$
$$= \frac{-6 - 10 - (-9)}{12} = \frac{-6 - 10 + 9}{12}$$
$$= -\frac{7}{12}$$

You Try It 6   Subtract: $\frac{7}{8} - \left(-\frac{5}{12}\right) - \frac{1}{9}$

Your solution

Example 7   Subtract: $45.2 - 56.89$

Solution $45.2 - 56.89 = -11.69$

You Try It 7   Subtract: $-12.03 - 19.117$

Your solution

Solutions on p. S2

Objective D **To solve application problems**

 The graph at the right shows the number of millions of barrels of oil and natural gas that are produced and exported per day for five selected countries. Use this graph for Example 8 and You Try It 8.

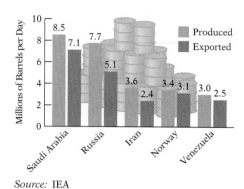

Source: IEA

Example 8

Using the graph above, find the total number of barrels of oil and natural gas that are exported by the five countries.

Strategy

To find the total number of barrels:

• Read the numbers from the graph that correspond to exporting (7.1, 5.1, 2.4, 3.1, 2.5).
• Add the numbers.

Solution

Total = 7.1 + 5.1 + 2.4 + 3.1 + 2.5 = 20.2

The five countries export 20.2 million barrels of oil and gas per day.

You Try It 8

For the five countries in the graph above, find the difference between the total amount of oil and gas produced and the total amount of oil and gas exported.

Your strategy

Your solution

Example 9

A cabinet maker is joining two pieces of wood. What is the measure of the cut from the left side of the board so that the pieces fit as shown.

Strategy

To find the measure of the cut, subtract $\frac{5}{16}$ in. from $\frac{7}{8}$ in.

Solution

$$\frac{7}{8} - \frac{5}{16} = \frac{14}{16} - \frac{5}{16} = \frac{9}{16}$$

The cut must be made $\frac{9}{16}$ in. from the left side of the board.

You Try It 9

Barbara Walsh spent $\frac{1}{6}$ of her day studying, $\frac{1}{8}$ of her day in class, and $\frac{1}{4}$ of her day working. What fraction of her day did she spend on these three activities?

Your strategy

Your solution

Solutions on pp. S2–S3

1.6 Exercises

Objective A   **To write a rational number in simplest form and as a decimal**

For Exercises 1 to 18, write each fraction in simplest form.

1. $\dfrac{7}{21}$ **2.** $\dfrac{10}{15}$ **3.** $-\dfrac{8}{22}$ **4.** $-\dfrac{8}{60}$ **5.** $-\dfrac{50}{75}$ **6.** $-\dfrac{20}{44}$

7. $\dfrac{12}{8}$ **8.** $\dfrac{36}{4}$ **9.** $\dfrac{0}{36}$ **10.** $\dfrac{12}{18}$ **11.** $-\dfrac{60}{100}$ **12.** $-\dfrac{14}{45}$

13. $-\dfrac{28}{20}$ **14.** $-\dfrac{20}{5}$ **15.** $-\dfrac{45}{3}$ **16.** $\dfrac{44}{60}$ **17.** $\dfrac{23}{46}$ **18.** $\dfrac{31}{93}$

For Exercises 19 to 36, write as a decimal. Place a bar over repeating digits.

19. $\dfrac{4}{5}$ **20.** $\dfrac{1}{8}$ **21.** $\dfrac{1}{6}$ **22.** $\dfrac{5}{6}$ **23.** $-\dfrac{1}{3}$ **24.** $-\dfrac{1}{20}$

25. $-\dfrac{2}{9}$ **26.** $-\dfrac{5}{11}$ **27.** $-\dfrac{7}{12}$ **28.** $\dfrac{7}{8}$ **29.** $\dfrac{11}{12}$ **30.** $\dfrac{4}{11}$

31. $-\dfrac{7}{18}$ **32.** $-\dfrac{17}{18}$ **33.** $\dfrac{9}{16}$ **34.** $\dfrac{15}{16}$ **35.** $-\dfrac{6}{7}$ **36.** $\dfrac{5}{13}$

Objective B   **To add rational numbers**

For Exercises 37 to 66, add.

37. $\dfrac{2}{3} + \dfrac{5}{12}$ **38.** $\dfrac{1}{2} + \dfrac{3}{8}$ **39.** $\dfrac{5}{8} + \dfrac{5}{6}$ **40.** $\dfrac{1}{9} + \dfrac{5}{27}$

41. $\dfrac{5}{12} + \left(-\dfrac{3}{8}\right)$ **42.** $-\dfrac{5}{6} + \dfrac{5}{9}$ **43.** $-\dfrac{6}{13} + \left(-\dfrac{17}{26}\right)$ **44.** $\dfrac{3}{5} + \left(-\dfrac{11}{12}\right)$

45. $-\dfrac{3}{4} + \left(-\dfrac{5}{6}\right)$ **46.** $-\dfrac{5}{8} + \dfrac{11}{12}$ **47.** $\dfrac{1}{3} + \dfrac{5}{6} + \dfrac{2}{9}$ **48.** $\dfrac{1}{2} + \dfrac{2}{3} + \dfrac{1}{6}$

49. $-\dfrac{3}{8} + \dfrac{5}{12} + \left(-\dfrac{3}{16}\right)$ **50.** $\dfrac{5}{16} + \left(-\dfrac{3}{4}\right) + \left(-\dfrac{7}{8}\right)$ **51.** $-\dfrac{1}{8} + \left(-\dfrac{11}{12}\right) + \dfrac{1}{3}$

52. $\dfrac{3}{8} + \left(-\dfrac{7}{12}\right) + \left(-\dfrac{5}{9}\right)$ **53.** $7.56 + 0.462$ **54.** $1.09 + 6.2$

55. $-32.1 + 6.7$ **56.** $5.138 + (-8.41)$ **57.** $-16.92 + 6.956$

58. $48 + (-34.12)$ **59.** $-19.84 + 17$ **60.** $-3.739 + (-2.03)$

61. $2.34 + (-3.7) + (-5.601)$ **62.** $-5.507 + (-4.91) + 15.2$

63. $-7.89 + 12.041 + (-4.151)$ **64.** $-3.04 + (-2.191) + (-0.06)$

65. $-91.2 + 24.56 + (-42.037)$ **66.** $81.02 + (-75.603) + (-17.8)$

67. What is $\dfrac{3}{4}$ more than $-\dfrac{5}{6}$?   **68.** Find the total of $\dfrac{5}{8}$ and $-\dfrac{5}{16}$.

69. Find $-\dfrac{5}{9}$ increased by $\dfrac{1}{6}$.   **70.** What is $-\dfrac{3}{8}$ added to $-\dfrac{5}{12}$?

71. Find 1.45 more than -7.   **72.** What is the sum of -4.23 and 3.06?

Objective C   **To subtract rational numbers**

For Exercises 73 to 102, subtract.

73. $\dfrac{1}{9} - \dfrac{5}{27}$ **74.** $\dfrac{5}{8} - \dfrac{5}{6}$ **75.** $\dfrac{1}{2} - \dfrac{5}{8}$

76. $\dfrac{2}{3} - \dfrac{1}{12}$ **77.** $-\dfrac{11}{12} - \dfrac{5}{8}$ **78.** $-\dfrac{7}{13} - \left(-\dfrac{11}{26}\right)$

79. $-\dfrac{5}{6} - \dfrac{4}{9}$ **80.** $-\dfrac{3}{4} - \left(-\dfrac{5}{6}\right)$ **81.** $\dfrac{4}{5} - \left(-\dfrac{5}{12}\right)$

82. $\dfrac{3}{8} - \left(-\dfrac{3}{4}\right)$ **83.** $\dfrac{7}{16} - \left(-\dfrac{3}{4}\right) - \left(-\dfrac{5}{8}\right)$ **84.** $-\dfrac{1}{8} - \dfrac{5}{12} - \left(-\dfrac{5}{16}\right)$

85. $\dfrac{1}{2} - \dfrac{5}{6} - \dfrac{2}{3}$ **86.** $-\dfrac{19}{18} - \left(-\dfrac{5}{6}\right) - \left(-\dfrac{2}{9}\right)$ **87.** $\dfrac{5}{8} - \left(-\dfrac{7}{12}\right) + \dfrac{7}{9}$

88. $-\dfrac{1}{8} - \left(-\dfrac{11}{12}\right) - \dfrac{1}{3}$

89. $6.322 - 9.123$

90. $-43.1 - 19.37$

91. $-3.04 - (-5.128)$

92. $-25 - (-34.12)$

93. $-20.04 - (-41.2)$

94. $0.354 - 16$

95. $-1.023 - (-1.023)$

96. $-5.0614 - 2.31$

97. $4.32 - (-6.1) - (-4.032)$

98. $-1.204 - (-5.027) - 12.3$

99. $-9.2 - 15.02 - (-6.614)$

100. $-6.97 - (-3.258) - (-3.712)$

101. $16.48 - 19.283 - (-6.804)$

102. $12.401 - 17.23 - (-2.1)$

103. What number is $\dfrac{5}{6}$ less than $-\dfrac{3}{8}$?

104. Find the difference between $\dfrac{1}{2}$ and $-\dfrac{5}{16}$.

105. What is $\dfrac{2}{3}$ less $\dfrac{3}{4}$?

106. What number is $\dfrac{4}{5}$ less than $-\dfrac{2}{15}$?

107. Find the difference between $-\dfrac{5}{9}$ and $-\dfrac{1}{6}$.

108. Find $\dfrac{5}{16}$ less $\dfrac{7}{12}$.

Objective D   **To solve application problems**

 Weights of Coins   For Exercises 109 to 114, use the following table showing the weight, diameter, and thickness of various U.S. coins.

	Penny	Nickel	Dime	Quarter
Weight	2.5 g	5 g	2.268 g	5.67 g
Diameter	19.05 mm	21.21 mm	17.91 mm	24.26 mm
Thickness	1.55 mm	1.95 mm	1.35 mm	1.75 mm

Source: The U.S. Mint

109. List the coins from thinnest to thickest.

110. List the weight of the coins from heaviest to lightest.

111. What is the sum of the weights of the four coins?

112. What is the largest difference in weight between two coins?

113. What is the largest difference in diameter between two coins?

114. What is the smallest difference in thickness between two coins?

115. Food Science   A recipe calls for $\frac{3}{4}$ c vegetable broth. If a chef has $\frac{2}{3}$ c vegetable broth, how much additional broth is needed for the recipe?

116. Carpentry   A piece of lath $\frac{1}{16}$ in. thick is glued to the edges of a wood strip that is $\frac{3}{4}$ in. width. What is the width of the wood and lath?

Lath

Finance   The graph at the right shows the earnings per share for Apple Computer for the years 1996 through 2002. Use this graph for Exercises 117 to 120.

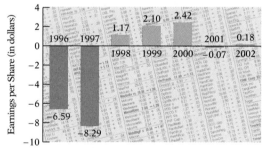

Source: Apple Computer Corporation

117. In which years did Apple Computer have negative earnings per share?

118. In which year did Apple Computer have its lowest earnings per share?

119. What was the decrease in earnings per share between 1996 and 1997?

120. What was the difference in earnings per share between 2000 and 1997?

Oil Consumption The graph at the right shows the number of millions of barrels of oil per day that are consumed by various countries and the number of millions of barrels of oil per day those countries import. Use this graph for Exercises 121 to 124.

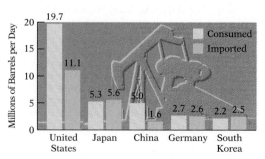

Source: IEA

121. How many millions of barrels of oil per day are consumed by these five countries?

122. How many millions of barrels of oil per day are imported by these five countries?

123. For these five countries, what is the difference between the number of millions of barrels of oil per day consumed and the number imported?

124. What is the largest difference in the numbers of millions of barrels of oil consumed by these five countries?

APPLYING THE CONCEPTS

For Exercises 125 to 130, simplify.

125. $\dfrac{1}{2} - \dfrac{5}{6} + \dfrac{7}{8}$

126. $-\dfrac{2}{3} + \dfrac{1}{4} - \dfrac{5}{6}$

127. $-\dfrac{3}{4} - \left(-\dfrac{2}{3}\right) + \left(-\dfrac{1}{8}\right)$

128. $-\dfrac{5}{6} + \left(-\dfrac{3}{8}\right) - \dfrac{5}{16}$

129. $\dfrac{4}{5} - \left(\dfrac{1}{4} - \left(-\dfrac{2}{3}\right)\right)$

130. $-\dfrac{5}{6} - \left(\dfrac{2}{3} - \dfrac{5}{12}\right)$

For Exercises 131 to 134, replace the question mark with <, >, or = to make a true statement.

131. $\dfrac{1}{4} - \dfrac{1}{3}$? $0.25 - 0.33$

132. $\dfrac{7}{8} - \dfrac{5}{6}$? $0.875 - 0.83$

133. $0.15 - \dfrac{3}{4}$? $0.75 - \dfrac{27}{20}$

134. $\dfrac{2}{3} - \dfrac{5}{6}$? $0.67 - 0.83$

For Exercises 135 to 138, use a calculator. By experimenting with different fractions, find a fraction as the quotient of integers that is equivalent to each of the following.

135. $0.\overline{1}$

136. $0.08\overline{3}$

137. $0.41\overline{6}$

138. $0.\overline{45}$

139. There are only two possible prime factors of the denominator of a terminating decimal. What are those prime factors?

140. The numerator of a fraction is 1. If the denominator is replaced by -2, -3, -4, -5, . . . , are the resulting fractions getting smaller or larger?

141. The numerator of a fraction is -1. If the denominator is replaced by 2, 3, 4, 5, . . . , are the resulting fractions getting smaller or larger?

1.7 Multiplication and Division of Rational Numbers

Objective A   **To multiply rational numbers**

The product $\frac{2}{3} \times \frac{4}{5}$ can be read "$\frac{2}{3}$ times $\frac{4}{5}$" or "$\frac{2}{3}$ of $\frac{4}{5}$."

Reading the times sign as "of" can help with understanding the procedure for multiplying fractions.

$\frac{4}{5}$ of the bar is shaded.

Shade $\frac{2}{3}$ of the $\frac{4}{5}$ already shaded.

$\frac{8}{15}$ of the bar is then shaded dark yellow.

$$\frac{2}{3} \text{ of } \frac{4}{5} = \frac{2}{3} \times \frac{4}{5} = \frac{2 \times 4}{3 \times 5} = \frac{8}{15}$$

> **Multiplication of Fractions**
>
> **The product of two fractions** is the product of the numerators over the product of the denominators.

After multiplying two fractions, write the product in simplest form. Use the rules for multiplying integers to determine the sign of the product.

HOW TO   Multiply: $-\dfrac{3}{4} \cdot \dfrac{10}{21}$

The signs are different. The product is negative.

$$-\frac{3}{4} \cdot \frac{10}{21} = \frac{3 \cdot 10}{4 \cdot 21}$$

• Multiply the numerators. Multiply the denominators.

$$= -\frac{3 \cdot 2 \cdot 5}{2 \cdot 2 \cdot 3 \cdot 7}$$

• Write the prime factorization of each number.

$$= -\frac{\overset{1}{\cancel{3}} \cdot \overset{1}{\cancel{2}} \cdot 5}{\underset{1}{\cancel{2}} \cdot 2 \cdot \underset{1}{\cancel{3}} \cdot 7} = -\frac{5}{14}$$

• Divide by the common factors. Then multiply the remaining factors in the numerator and in the denominator.

This problem can also be worked by using the greatest common factor (GCF) of the numerator and the denominator.

HOW TO   Multiply: $-\dfrac{3}{4} \cdot \dfrac{10}{21}$

$$-\frac{3}{4} \cdot \frac{10}{21} = \frac{30}{84}$$

• Multiply the numerators.
• Multiply the denominators.

$$= \frac{\overset{1}{\cancel{6}} \cdot 5}{\underset{1}{\cancel{6}} \cdot 14} = -\frac{5}{14}$$

• Divide the numerator and denominator by the GCF.

To multiply decimals, multiply as with integers. Write the decimal point in the product in such a way that the number of decimal places in the product equals the sum of the number of decimal places in the factors.

HOW TO   Find the product of 7.43 and -0.00025.

$$\begin{array}{r} 7.43 \\ \times\ 0.00025 \\ \hline 3715 \\ 1486 \\ \hline 0.0018575 \end{array}$$

2 decimal places
5 decimal places

7 decimal places

- Multiply the absolute values.

$7.43(-0.00025) = -0.0018575$

- The signs are different. The product is negative.

Example 1   Multiply: $-\dfrac{3}{8}\left(-\dfrac{12}{17}\right)$

Solution

$-\dfrac{3}{8}\left(-\dfrac{12}{17}\right) = \dfrac{3 \cdot 12}{8 \cdot 17}$ • The signs are the same. The product is positive.

$= \dfrac{3 \cdot \overset{1}{\cancel{2}} \cdot \overset{1}{\cancel{2}} \cdot 3}{\underset{1}{\cancel{2}} \cdot \underset{1}{\cancel{2}} \cdot 2 \cdot 17}$

$= \dfrac{9}{34}$ • Write the answer in simplest form.

You Try It 1   Multiply: $\dfrac{5}{8}\left(-\dfrac{4}{25}\right)$

Your solution

Example 2

Find the product of $\dfrac{4}{9}$, $\dfrac{3}{10}$ and $-\dfrac{5}{18}$.

Solution

$\dfrac{4}{9} \cdot \dfrac{3}{10} \cdot \left(-\dfrac{5}{18}\right) = -\dfrac{4 \cdot 3 \cdot 5}{9 \cdot 10 \cdot 18}$ • The product is negative.

$= -\dfrac{\overset{1}{\cancel{2}} \cdot \overset{1}{\cancel{2}} \cdot \overset{1}{\cancel{3}} \cdot \overset{1}{\cancel{5}}}{\underset{1}{\cancel{3}} \cdot 3 \cdot 2 \cdot \underset{1}{\cancel{5}} \cdot \underset{1}{\cancel{2}} \cdot 3 \cdot 3}$

$= -\dfrac{1}{27}$ • Write the answer in simplest form.

You Try It 2

Find the product of $-\dfrac{4}{5}$, $-\dfrac{3}{8}$, and $-\dfrac{10}{27}$.

Your solution

Example 3   Multiply: $-4.06(-0.065)$

Solution
The product is positive.

$-4.06(-0.065) = 0.2639$

You Try It 3   Multiply: $0.034(-2.14)$

Your solution

Solutions on p. S3

Objective B **To divide rational numbers**

The **reciprocal of a fraction** is the fraction with the numerator and denominator interchanged. For instance, the reciprocal of $\frac{3}{4}$ is $\frac{4}{3}$, and the reciprocal of $-\frac{5}{2}$ is $-\frac{2}{5}$.

The product of a number and its reciprocal is 1. This fact is used in the procedure for dividing fractions.

$$\frac{3}{4} \cdot \frac{4}{3} = \frac{12}{12} = 1 \qquad -\frac{5}{2} \cdot \left(-\frac{2}{5}\right) = \frac{10}{10} = 1$$

Study the example below to see how reciprocals are used when dividing fractions.

Divide: $\frac{3}{5} \div \frac{5}{6}$

$$\frac{3}{5} \div \frac{5}{6} = \frac{\dfrac{3}{5}}{\dfrac{5}{6}} = \frac{\dfrac{3}{5} \cdot \dfrac{6}{5}}{\dfrac{5}{6} \cdot \dfrac{6}{5}}$$

- Multiply the numerator and denominator by the **reciprocal of the divisor.**

$$= \frac{\dfrac{3}{5} \cdot \dfrac{6}{5}}{1}$$

- The product of a number and its reciprocal is **1.**

$$= \frac{3}{5} \cdot \frac{6}{5} = \frac{18}{25}$$

- A number divided by 1 is the number.

These steps are summarized by $\frac{3}{5} \div \frac{5}{6} = \frac{3}{5} \cdot \frac{6}{5} = \frac{18}{25}$.

> **Division of Fractions**
>
> **To divide two fractions,** multiply the dividend by the reciprocal of the divisor.

TAKE NOTE

The method of dividing fractions is sometimes stated, "To divide fractions, invert the divisor and then multiply." Inverting the divisor means writing its reciprocal.

HOW TO Divide: $\dfrac{3}{10} \div \left(-\dfrac{18}{25}\right)$

The signs are different. The quotient is negative.

$$\frac{3}{10} \div \left(-\frac{18}{25}\right) = -\left(\frac{3}{10} \div \frac{18}{25}\right) = -\left(\frac{3}{10} \cdot \frac{25}{18}\right)$$

$$= -\frac{3 \cdot 25}{10 \cdot 18} = -\frac{\overset{1}{3} \cdot \overset{1}{5} \cdot 5}{2 \cdot 5 \cdot 2 \cdot 3 \cdot 3} = -\frac{5}{12}$$

To divide decimals, move the decimal point in the divisor to the right so that the divisor becomes a whole number. Move the decimal point in the dividend the same number of places to the right. Place the decimal point in the quotient directly over the decimal point in the dividend. Then divide as with whole numbers.

TAKE NOTE

The procedure for dividing decimals as we do at the right can be justified as follows.

$$(-1.4) \div (-0.36)$$
$$= \frac{1.4}{0.36}$$
$$= \frac{1.4}{0.36} \cdot \frac{100}{100}$$
$$= \frac{140}{36}$$
$$\approx 3.9$$

HOW TO   Divide: $(-1.4) \div (-0.36)$. Round to the nearest tenth.

The signs are the same. The quotient is positive.

$$
\begin{array}{r}
3.88 \approx 3.9 \\
0.36\overline{)1.40.00} \\
\underline{1\ 08} \\
32\ 0 \\
\underline{28\ 8} \\
3\ 20 \\
\underline{2\ 88} \\
32
\end{array}
$$

- **Move the decimal point** 2 places to the right in the divisor and then in the dividend. Place the decimal point in the quotient directly over the decimal point in the dividend.

- Note that the symbol $\approx$ is used to indicate that the quotient is an approximate value that has been rounded off.

$$(-1.4) \div (-0.36) \approx 3.9$$

Example 4 Divide: $\left(-\dfrac{3}{10}\right) \div \dfrac{9}{16}$

Solution   The quotient is negative.

$$\left(-\frac{3}{10}\right) \div \frac{9}{14} = -\left(\frac{3}{10} \cdot \frac{14}{9}\right)$$

- **Multiply by the reciprocal of the divisor.**

$$= -\frac{3 \cdot 2 \cdot 7}{2 \cdot 5 \cdot 3 \cdot 3}$$

$$= -\frac{7}{15}$$

- **Write the answer in simplest form.**

You Try It 4 Divide: $\dfrac{5}{8} \div \left(-\dfrac{10}{11}\right)$

Your solution

Example 5 Find the quotient of $-\dfrac{5}{8}$ and $-\dfrac{7}{16}$.

Solution   The quotient is positive.

$$-\frac{5}{8} \div \left(-\frac{7}{16}\right) = \frac{5}{8} \cdot \frac{16}{7}$$

- **Multiply by the reciprocal of the divisor.**

$$= \frac{5 \cdot 16}{8 \cdot 7}$$

$$= \frac{10}{7}$$

- **Write the answer in simplest form.**

You Try It 5 What is $-\dfrac{1}{3}$ divided by $\dfrac{7}{15}$?

Your solution

Example 6 Divide: $4.152 \div (-25.2)$. Round to the nearest thousandth.

Solution
Divide the absolute values. The quotient is negative.

$$4.152 \div (-25.2) \approx -0.165$$

You Try It 6 Divide: $(-34) \div (-9.02)$. Round to the nearest hundredth.

Your solution

Solutions on p. S3

Objective C   **To convert among percents, fractions, and decimals**

"A population growth rate of 3%," "a manufacturer's discount of 25%," and "an 8% increase in pay" are typical examples of the many ways in which *percent* is used in applied problems. **Percent** means "parts of 100." Thus 27% means 27 parts of 100.

In applied problems involving a percent, it may be necessary to rewrite a percent as a fraction or as a decimal, or to rewrite a fraction or a decimal as a percent.

To write a percent as a fraction, remove the percent sign and multiply by $\frac{1}{100}$.

$$27\% = 27\left(\frac{1}{100}\right) = \frac{27}{100}$$

To write a percent as a decimal, remove the percent sign and multiply by 0.01.

$$33\% \qquad = \qquad 33(0.01) \qquad = \qquad 0.33$$

> Move the decimal point two places to the left. Then remove the percent sign.

TAKE NOTE

The decimal equivalent of 100% is 1. Therefore, multiplying by 100% is the same as multiplying by 1 and does not change the value of the fraction.

$$\frac{5}{8} = \frac{5}{8}(1) = \frac{5}{8}(100\%)$$

To write a fraction as a percent, multiply by 100%. For example, $\frac{5}{8}$ is changed to a percent as follows:

$$\frac{5}{8} = \frac{5}{8}(100\%) = \frac{500}{8}\% = 62.5\%, \qquad \text{or} \qquad 62\frac{1}{2}\%$$

To write a decimal as a percent, multiply by 100%.

$$0.82 \qquad = \qquad 0.82(100\%) \qquad = \qquad 82\%$$

> Move the decimal point two places to the right. Then write the percent sign.

Example 7

Write 130% as a fraction and as a decimal.

Solution

$$130\% = 130\left(\frac{1}{100}\right) = \frac{130}{100} = \frac{13}{10}$$
$$130\% = 130(0.01) = 1.30$$

You Try It 7

Write 125% as a fraction and as a decimal.

Your solution

Example 8   Write $33\frac{1}{3}\%$ as a fraction.

Solution

$$33\frac{1}{3}\% = 33\frac{1}{3}\left(\frac{1}{100}\right) = \frac{100}{3}\left(\frac{1}{100}\right) = \frac{1}{3}$$

You Try It 8   Write $16\frac{2}{3}\%$ as a fraction.

Your solution

Solutions on p. S3

Example 9 Write $\frac{5}{6}$ as a percent.

Solution

$$\frac{5}{6} = \frac{5}{6}(100\%) = \frac{500}{6}\% = 83\frac{1}{3}\%$$

You Try It 9 Write $\frac{9}{16}$ as a percent.

Your solution

Example 10 Write 0.027 as a percent.

Solution

$$0.027 = 0.027(100\%) = 2.7\%$$

You Try It 10 Write 0.043 as a percent.

Your solution

Solutions on p. S3

Objective D **To solve application problems**

Example 11

A picture frame is supported by two hooks that are $\frac{1}{3}$ and $\frac{2}{3}$ of the distance from the left-hand side of the frame. If the frame is $31\frac{1}{2}$ in. wide, how far from the left side of the frame are the hooks placed?

Strategy

To find the location of the hooks, multiply the width of the frame, $31\frac{1}{2}$ in., by $\frac{1}{3}$ and $\frac{2}{3}$. Recall that to multiply a mixed number by a fraction, first write the mixed number as an improper fraction: $31\frac{1}{2} = \frac{2 \times 31 + 1}{2} = \frac{63}{2}$

Solution

$$31\frac{1}{2} \cdot \frac{1}{3} = \frac{63}{2} \cdot \frac{1}{3} = \frac{21}{2} = 10\frac{1}{2}$$

$$31\frac{1}{2} \cdot \frac{2}{3} = \frac{63}{2} \cdot \frac{2}{3} = 21$$

The hooks are placed $10\frac{1}{2}$ in. and 21 in. from the left of the frame.

You Try It 11

A piece of fabric 20 ft long is being used to make cushions for outdoor furniture. If each cushion requires $1\frac{1}{2}$ ft of fabric, how many cushions can be cut from the fabric?

Your strategy

Your solution

Solution on p. S3

1.7 Exercises

Objective A   **To multiply rational numbers**

For Exercises 1 to 32, multiply.

1. $\dfrac{2}{3} \cdot \dfrac{5}{7}$

2. $\dfrac{1}{2}\left(\dfrac{3}{8}\right)$

3. $\dfrac{5}{8} \cdot \left(-\dfrac{3}{10}\right)$

4. $-\dfrac{5}{16} \cdot \dfrac{7}{15}$

5. $\dfrac{5}{12}\left(-\dfrac{3}{10}\right)$

6. $\left(-\dfrac{11}{12}\right)\left(-\dfrac{6}{7}\right)$

7. $\dfrac{6}{13}\left(-\dfrac{26}{27}\right)$

8. $\dfrac{1}{6}\left(-\dfrac{6}{11}\right)$

9. $\left(-\dfrac{3}{5}\right)\left(-\dfrac{3}{10}\right)$

10. $\dfrac{3}{5}\left(-\dfrac{11}{12}\right)$

11. $\left(-\dfrac{3}{4}\right)^2$

12. $\left(-\dfrac{5}{8}\right)^2$

13. $\left(-\dfrac{3}{4}\right)\dfrac{5}{6}\left(-\dfrac{2}{9}\right)$

14. $\left(-\dfrac{1}{2}\right)\left(-\dfrac{2}{3}\right)\left(-\dfrac{6}{7}\right)$

15. $\left(-\dfrac{3}{8}\right)\left(-\dfrac{5}{12}\right)\left(\dfrac{3}{10}\right)$

16. $\dfrac{5}{16}\left(-\dfrac{4}{5}\right)\left(-\dfrac{7}{8}\right)$

17. $\left(-\dfrac{15}{2}\right)\left(-\dfrac{4}{3}\right)\left(-\dfrac{7}{10}\right)$

18. $\left(-\dfrac{5}{8}\right)\left(\dfrac{5}{12}\right)\left(-\dfrac{16}{25}\right)$

19. $\left(\dfrac{8}{9}\right)\left(-\dfrac{11}{12}\right)\left(\dfrac{3}{4}\right)$

20. $\dfrac{3}{8}\left(-\dfrac{7}{10}\right)\left(-\dfrac{5}{9}\right)$

21. $0.46(-3.9)$

22. $-0.78(6.8)$

23. $(-8.23)(-0.09)$

24. $(-0.003)(-0.189)$

25. $-0.48(0.85)$

26. $0.056(-3.425)$

27. $-6.5(0.0341)$

28. $4.237(-0.54)$

29. $(-8.004)(-3.4)$

30. $(-3.739)(-2.03)$

31. $0.089(-1.098)$

32. $(-3.004)(-2.31)$

33. Find the product of $-\dfrac{3}{4}$ and $\dfrac{4}{5}$.

34. Find the product of $-\dfrac{7}{10}$ and $-\dfrac{5}{14}$.

35. Multiply -0.23 by -4.5.

36. Multiply -7.06 by 0.034.

Objective B   **To divide rational numbers**

For Exercises 37 to 48, divide.

37. $\dfrac{3}{8} \div \left(-\dfrac{9}{10}\right)$

38. $\left(-\dfrac{2}{15}\right) \div \dfrac{3}{5}$

39. $\left(-\dfrac{8}{9}\right) \div \left(-\dfrac{4}{5}\right)$

40. $\left(-\dfrac{11}{15}\right) \div \left(-\dfrac{22}{5}\right)$

41. $\left(-\dfrac{11}{12}\right) \div \left(-\dfrac{7}{6}\right)$

42. $\left(-\dfrac{3}{10}\right) \div \dfrac{5}{12}$

43. $\left(-\dfrac{6}{11}\right) \div 6$

44. $\left(-\dfrac{26}{27}\right) \div \dfrac{13}{6}$

45. $\left(-\dfrac{11}{12}\right) \div \dfrac{5}{3}$

46. $\left(-\dfrac{3}{10}\right) \div \left(-\dfrac{5}{3}\right)$

47. $\left(-\dfrac{5}{8}\right) \div \dfrac{15}{16}$

48. $\dfrac{8}{9} \div \left(-\dfrac{4}{3}\right)$

49. Find the quotient of $-\dfrac{7}{9}$ and $-\dfrac{5}{18}$.

50. Find the quotient of $\dfrac{5}{8}$ and $-\dfrac{7}{12}$.

51. What is $\dfrac{1}{2}$ divided by $-\dfrac{1}{4}$?

52. What is $-\dfrac{5}{18}$ divided by $\dfrac{15}{16}$?

For Exercises 53 to 56, divide.

53. $25.61 \div (-5.2)$

54. $(-0.1035) \div (-0.023)$

55. $(-0.2205) \div (-0.21)$

56. $(-0.357) \div 1.02$

For Exercises 57 to 60, divide. Round to the nearest hundredth.

57. $-0.0647 \div 0.75$

58. $-27.981 \div 59.2$

59. $-2.45 \div (-21.44)$

60. $3.2 \div (-45.12)$

For Exercises 61 to 64, divide. Round to the nearest thousandth.

61. $(-75.469) \div (-77.8)$

62. $6.14 \div (-27.1)$

63. $-0.1142 \div (-17.2)$

64. $-0.2246 \div 12.34$

65. Find the quotient of -0.3045 and -0.203.

66. Find the quotient of 3.672 and -3.6.

67. What is -0.00552 divided by 1.2?

68. What is -0.01925 divided by 0.077?

For Exercises 69 to 86, use the Order of Operations Agreement to simplify the expression.

69. $\dfrac{2}{3} - \dfrac{1}{4}\left(-\dfrac{2}{5}\right)$

70. $-\dfrac{7}{8} - \dfrac{3}{4} \div \dfrac{9}{8}$

71. $\dfrac{3}{4}\left(\dfrac{1}{2}\right)^2 - \dfrac{5}{16}$

72. $\left(-\dfrac{5}{6}\right) \div \dfrac{5}{12} - \dfrac{9}{14}$

73. $\dfrac{7}{12} - \left(\dfrac{2}{3}\right)^2 + \left(-\dfrac{3}{4}\right)$

74. $\left(\dfrac{1}{2} - \dfrac{3}{4}\right)^2 - \left(\dfrac{5}{18} - \dfrac{15}{24}\right)$

75. $\left(\dfrac{2}{3}\right)\left(\dfrac{3}{4}\right) - \left(\dfrac{4}{5}\right)\left(\dfrac{5}{8}\right)$

76. $\dfrac{1}{8} - \left(\dfrac{9}{4}\right)\left(-\dfrac{2}{3}\right)^2$

77. $-\dfrac{3}{4} \div \dfrac{5}{8} - \dfrac{4}{5}$

78. $\dfrac{5}{9}\left(\dfrac{3}{4}\right) - \dfrac{1}{2} \div (-2)$

79. $\left(\dfrac{2}{3}\right)^3 - \left(\dfrac{2}{3}\right)^2$

80. $\left(\dfrac{5}{6} - \dfrac{2}{3}\right)^2 \div \left(\dfrac{7}{18} - \dfrac{7}{12}\right)^2$

81. $1.2 - 2.3^2$

82. $4.01 - 0.2(8.1 - 6.4)$

83. $0.03 \cdot 0.2^2 - 0.5^3$

84. $8.1 - 5.2(3.4 - 5.9)^2$

85. $\dfrac{3.8 - 5.2}{-0.35} - \left(\dfrac{1.2}{0.6}\right)^2$

86. $-0.3^2 + 3.4(-2.01) - (-1.75)$

Objective C   **To convert among percents, fractions, and decimals**

87. ✎ Explain how to write a percent as a fraction.

88. ✎ Explain how to write a fraction as a percent.

For Exercises 89 to 98, write as a fraction and a decimal.

89. 75%

90. 40%

91. 64%

92. 88%

93. 175%

94. 160%

95. 19%

96. 87%

97. 5%

98. 8%

For Exercises 99 to 108, write as a fraction.

99. $11\dfrac{1}{9}\%$

100. $4\dfrac{2}{7}\%$

101. $12\dfrac{1}{2}\%$

102. $37\dfrac{1}{2}\%$

103. $66\dfrac{2}{3}\%$

104. $\dfrac{1}{4}\%$

105. $\dfrac{1}{2}\%$

106. $6\dfrac{1}{4}\%$

107. $83\dfrac{1}{3}\%$

108. $5\dfrac{3}{4}\%$

For Exercises 109 to 118, write as a decimal.

109. 7.3%   **110.** 9.1%   **111.** 15.8%   **112.** 16.7%   **113.** 0.3%

114. 0.9%   **115.** 9.9%   **116.** 9.15%   **117.** 121.2%   **118.** 18.23%

For Exercises 119 to 138, write as a percent.

119. 0.15   **120.** 0.37   **121.** 0.05   **122.** 0.02   **123.** 0.175

124. 0.125   **125.** 1.15   **126.** 1.36   **127.** 0.008   **128.** 0.004

129. $\dfrac{27}{50}$ **130.** $\dfrac{83}{100}$ **131.** $\dfrac{1}{3}$ **132.** $\dfrac{3}{8}$ **133.** $\dfrac{5}{11}$

134. $\dfrac{4}{9}$ **135.** $\dfrac{7}{8}$ **136.** $\dfrac{9}{20}$ **137.** $1\dfrac{2}{3}$ **138.** $2\dfrac{1}{2}$

Objective D   **To solve application problems**

139. **Art**   A picture frame measures $18\dfrac{1}{4}$ in. wide by $24\dfrac{1}{2}$ in. high. How far from the left side and how far from the bottom of the frame is its center?

140. **Carpentry**   A carpenter has a board that is 14 ft long. How many pieces $\dfrac{3}{4}$ ft long can the carpenter cut from the board?

141. **Carpentry**   A board $36\dfrac{5}{8}$ in. long is cut into two pieces of equal length. If the saw blade makes a cut $\dfrac{1}{8}$ in. wide, how far from the left side of the board should the cut be made?

142. **Food Science**   A recipe calls for $\frac{3}{4}$ c of butter. If a chef wants to increase the recipe by one-half, how much butter should the chef use?

143. **Construction**   A stair is made from an 8-inch riser and a $\frac{3}{4}$-inch foot plate. How many inches high is a staircase made from 10 of these stairs?

Foot plate
Riser

144. **Interior Design**   An interior designer needs $12\frac{1}{2}$ yd of fabric that costs $5.43 per yard and $5\frac{3}{4}$ yd of fabric that costs $6.94 per yard to recover a large sofa. Find the total cost of the two fabrics.

APPLYING THE CONCEPTS

145. Find a rational number that is one-half the difference between $\frac{5}{11}$ and $\frac{4}{11}$.

146. Find $\left(1 + \frac{1}{1}\right)\left(1 + \frac{1}{2}\right)$, $\left(1 + \frac{1}{1}\right)\left(1 + \frac{1}{2}\right)\left(1 + \frac{1}{3}\right)$, and $\left(1 + \frac{1}{1}\right)\left(1 + \frac{1}{2}\right)\left(1 + \frac{1}{3}\right)\left(1 + \frac{1}{4}\right)$. Based on the pattern of the answers, make a conjecture as to $\left(1 + \frac{1}{1}\right)\left(1 + \frac{1}{2}\right)\left(1 + \frac{1}{3}\right)\left(1 + \frac{1}{4}\right)\left(1 + \frac{1}{5}\right)$.

147. A frog is 2 ft from a wall and jumps halfway to the wall. At that point, the frog jumps one-half of the remaining distance to the wall. If the frog continues to jump one-half the remaining distance to the wall, how far from the wall will the frog be after the fifth jump?

148. $-\frac{2}{3} > -\frac{3}{4}$. Is $-\frac{2 + 3}{3 + 4}$ less than $-\frac{3}{4}$, greater than $-\frac{2}{3}$, or between $-\frac{2}{3}$ and $-\frac{3}{4}$?

149. Consider the fraction $-\frac{2}{5}$. Suppose that the same positive number is added to the numerator and denominator. What is the relationship between the new fraction and the original fraction?

150. Given any two different rational numbers, is it always possible to find a rational number between the two given numbers? If so, explain how to find such a number. If not, give two rational numbers for which there is no rational number between them.

1.8 Concepts from Geometry

To find the measures of angles

The word geometry comes from the Greek words for "earth" (geo) and "measure." The original purpose of geometry was to measure land. Today, geometry is used in many disciplines such as physics, biology, geology, architecture, art, and astronomy.

Here are some basic geometric concepts.

A **plane** is a flat surface such as a table top that extends indefinitely. Figures that lie entirely in a plane are called **plane figures.**

Space extends in all directions. Objects in space, such as a baseball, house, or a tree, are called **solids.**

A **line** extends indefinitely in two directions in a plane. A line has no width.

A **ray** starts at a point and extends indefinitely in one direction. By placing a point on the ray at the right, we can name the ray *AB*.

A **line segment** is part of a line and has two endpoints. The line segment *AB* is designated by its two endpoints.

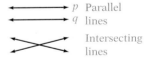

Lines in a plane can be parallel or intersect. **Parallel lines** never meet. The distance between parallel lines in a plane is always the same. We write $p \parallel q$ to indicate line *p* is parallel to line *q*. **Intersecting lines** cross at a point in the plane.

An **angle** is formed when two rays start from the same point. Rays *AB* and *AC* start from the same point *A*. The point at which the rays meet is called the **vertex** of the angle. The symbol ∠ is read "angle" and is used to name an angle. We can refer to the angle at the right as ∠*A*, ∠*BAC*, or ∠*x*.

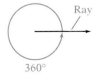

Point of Interest

The Babylonians chose 360° for the measure of one full rotation, probably because they knew there were 365 days in a year and the closest number to 365 with many divisors was 360.

An angle can be measured in **degrees.** The symbol for degree is °. A ray rotated 1 revolution about its beginning point creates an angle of 360°.

The measure of an angle is symbolized by $m\angle$. For instance, $m\angle C = 40°$. Read this as "the measure of angle *C* is 40°."

$\frac{1}{4}$ of a revolution is $\frac{1}{4}$ of 360°, or 90°. A 90° angle is called a **right angle.** The symbol $\llcorner$ is used to represent a right angle. **Perpendicular lines** are intersecting lines that form right angles. We write $p \perp q$ to indicate line p is perpendicular to line q.

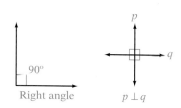

Right angle $p \perp q$

Complementary angles are two angles whose sum is 90°.

$$m\angle A + m\angle B = 35° + 55° = 90°$$

$\angle A$ and $\angle B$ are complementary angles.

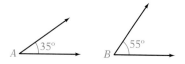

$\frac{1}{2}$ of a revolution is $\frac{1}{2}$ of 360°, or 180°. A 180° angle is called a **straight angle.**

Supplementary angles are two angles whose sum is 180°.

$$m\angle A + m\angle B = 123° + 57° = 180°$$

$\angle A$ and $\angle B$ are supplementary angles.

Straight angle

Example 1 Find the complement of 39°.

Solution To find the complement of 39°, subtract 39° from 90°.

$90° - 39° = 51°$

51° is the complement of 39°.

You Try It 1 Find the complement of 87°.

Your solution

Example 2 Find the supplement of 122°.

Solution To find the supplement of 122°, subtract 122° from 180°.

$180° - 122° = 58°$

58° is the complement of 122°.

You Try It 2 Find the supplement of 87°.

Your solution

Example 3

For the figure at the right, find $m\angle AOB$.

Solution
$m\angle AOB$ is the difference between $m\angle AOC$ and $m\angle BOC$.

$m\angle AOB = 95° - 62° = 33°$

$m\angle AOB = 33°$

You Try It 3

For the figure at the right, find $m\angle x$.

Your solution

Solutions on p. S3

Objective B **To solve perimeter problems**

Perimeter is the distance around a plane figure. Perimeter is used in buying fencing for a yard, wood for the frame of a painting, and rain gutters around a house. The perimeter of a plane figure is the sum of the lengths of the sides of the figure. Formulas for four common geometric figures are given below.

A **triangle** is a three-sided plane figure.

     Perimeter = side 1 + side 2 + side 3

An **isosceles triangle** has two sides of the same length. An **equilateral triangle** has all three sides the same length.

A **parallelogram** is a four-sided plane figure with opposite sides parallel.

A **rectangle** is a parallelogram that has four right angles.

     Perimeter = 2 · length + 2 · width

A **square** is a rectangle with four equal sides.

     Perimeter = 4 · side

A **circle** is a plane figure in which all points are the same distance from point *O*, the **center of the circle.** The **diameter** of a circle is a line segment across the circle passing through the center. *AB* is a diameter of the circle at the right. The **radius** of a circle is a line segment from the center of the circle to a point on the circle. *OC* is a radius of the circle at the right. The perimeter of a circle is called its **circumference.**

     Diameter = 2 · radius   or   Radius = $\frac{1}{2}$ · diameter

     Circumference = 2 · π · radius   or   Circumference = π · diameter,

where $\pi \approx 3.14$ or $\frac{22}{7}$.

HOW TO   The diameter of the circle is 25 cm. Find the radius of the circle.

Radius = $\frac{1}{2}$ · diameter

     = $\frac{1}{2}$ · 25 = 12.5

The radius is 12.5 cm.

Example 4 Find the perimeter of a rectangle with a width of 6 ft and a length of 18 feet.

Solution
Perimeter = 2 · length + 2 · width
= 2 · 18 ft + 2 · 6 ft
= 36 ft + 12 ft = 48 ft

You Try It 4 Find the perimeter of a square that has a side of length 4.2 m.

Your solution

Example 5 Find the circumference of a circle with a radius of 23 cm. Use 3.14 for π.

Solution
Circumference = 2 · π · radius
≈ 2 · 3.14 · 23 cm
= 144.44 cm

You Try It 5 Find the circumference of a circle with a diameter of 5 in. Use 3.14 for π.

Your solution

Example 6

A chain-link fence costs $4.37 per foot. How much will it cost to fence a rectangular playground that is 108 ft wide and 195 ft long?

Strategy
To find the cost of the fence:
• Find the perimeter of the playground.
• Multiply the perimeter by the per-foot cost of the fencing.

Solution
Perimeter = 2 · length + 2 · width
= 2 · 195 ft + 2 · 108 ft
= 390 ft + 216 ft = 606 ft
Cost = 606 × $4.37 = $2648.22
The cost is $2648.22.

You Try It 6

A metal strip is being installed around a circular table that has a diameter of 36 in. If the per-foot cost of the metal strip is $3.21, find the cost for the metal strip. Use 3.14 for π. Round to the nearest cent.

Your strategy

Your solution

Solutions on pp. S3–S4

Objective C **To solve area problems**

Area is a measure of the amount of surface in a region. Area is used to describe the size of a rug, a farm, a house, or a national park.

Area is measured in square units.

A square that is 1 in. on each side has an area of 1 square inch, which is written 1 in².

1 in²   1 in.

1 in.

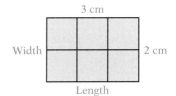

A square that is 1 cm on each side has an area of 1 square centimeter, which is written 1 cm².

Areas of common geometric figures are given by the following formulas.

Rectangle

$$\text{Area} = \textbf{length} \cdot \textbf{width}$$
$$= 3 \text{ cm} \cdot 2 \text{ cm}$$
$$= 6 \text{ cm}^2$$

Square

$$\text{Area} = \textbf{side} \cdot \textbf{side}$$
$$= 2 \text{ cm} \cdot 2 \text{ cm}$$
$$= 4 \text{ cm}^2$$

Parallelogram

The **base of a parallelogram** is one of the parallel sides. The **height of a parallelogram** is the distance between the base and the opposite parallel side. It is perpendicular to the base.

$$\text{Area} = \textbf{base} \cdot \textbf{height}$$
$$= 5 \text{ ft} \cdot 4 \text{ ft}$$
$$= 20 \text{ ft}^2$$

Circle

$$\text{Area} = \pi(\textbf{radius})^2$$
$$\approx 3.14(4 \text{ in.})^2 = 50.24 \text{ in}^2$$

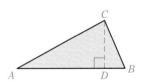

> **TAKE NOTE**
>
> The height of a triangle is always perpendicular to the base. Sometimes it is necessary to extend the base so that a perpendicular line segment can be drawn. The extension is *not* part of the base.

Triangle

For the triangle at the right, the **base of the triangle** is *AB*; the **height of the triangle** is *CD*. Note that the height is perpendicular to the base.

$$\text{Area} = \frac{1}{2} \cdot \textbf{base} \cdot \textbf{height}$$
$$= \frac{1}{2} \cdot 5 \text{ in.} \cdot 4 \text{ in.} = 10 \text{ in}^2$$

Example 7 Find the area of a rectangle whose length is 8 in. and whose width is 6 in.

You Try It 7 Find the area of a triangle whose base is 5 ft and whose height is 3 ft.

Solution Area = length × width
= 8 in. × 6 in. = 48 in²

Your solution

Solution on p. S4

Example 8

Find the area of a circle whose diameter is 5 cm. Use 3.14 for π.

Solution

$$\text{Radius} = \frac{1}{2} \cdot \text{diameter}$$

$$= \frac{1}{2} \cdot 5 \text{ cm} = 2.5 \text{ cm}$$

$$\text{Area} = \pi \cdot (\text{radius})^2$$
$$\approx 3.14(2.5 \text{ cm})^2 = 19.625 \text{ cm}^2$$

You Try It 8

Find the area of a circle whose radius is 6 in. Use 3.14 for π.

Your solution

Example 9

Find the area of the parallelogram shown below.

Solution
Area = base · height
= 12 ft × 7 ft = 84 ft²

You Try It 9

Find the area of the parallelogram shown below.

Your solution

Example 10

To conserve water during a drought, a city's water department is offering homeowners a rebate on their water bill of $1.27 per square foot of lawn that is removed from a yard and replaced with drought-resistant plants. What rebate would a homeowner receive who replaced a rectangular lawn area that is 15 ft wide and 25 ft long?

Strategy   To find the amount of the rebate:
• Find the area of the lawn.
• Multiply the area by the per-square-foot rebate.

Solution
Area = length × width
= 25 ft × 15 ft = 375 ft²

Rebate = 375 × $1.27 = $476.25

The rebate is $476.25.

You Try It 10

An interior designer is choosing from two hallway rugs. A nylon rug costs $1.25 per square foot and a wool rug costs $1.93 per square foot. If the dimensions of the carpet are 4 ft by 15 ft, how much more expensive is the wool rug than the nylon rug?

Your strategy

Your solution

Solutions on p. S4

1.8 Exercises

Objective A   **To find the measures of angles**

1. How many degrees are in a right angle?

2. How many degrees are in a straight angle?

3. Find the complement of a 62° angle.

4. Find the complement of a 13° angle.

5. Find the supplement of a 48° angle.

6. Find the supplement of a 106° angle.

7. Find the complement of a 7° angle.

8. Find the complement of a 76° angle.

9. Find the supplement of an 89° angle.

10. Find the supplement of a 21° angle.

11. Angle *AOB* is a straight angle. Find $m\angle AOC$.

12. Angle *AOB* is a straight angle. Find $m\angle COB$.

13. Find $m\angle x$.

14. Find $m\angle x$.

15. Find $m\angle AOB$.

16. Find $m\angle AOB$.

17. Find $m\angle AOC$.

18. Find $m\angle AOC$.

19. Find $m\angle A$.

20. Find $m\angle A$.

Objective B   **To solve perimeter problems**

21. Find the perimeter of a triangle with sides 2.51 cm, 4.08 cm, and 3.12 cm.

22. Find the perimeter of a triangle with sides 4 ft 5 in., 5 ft 3 in., and 6 ft. 11 in.

23. Find the perimeter of a rectangle whose length is 4 ft 8 in. and whose width is 2 ft 5 in.

24. Find the perimeter of a rectangle whose dimensions are 5 m by 8 m.

25. Find the perimeter of a square whose side is 13 in.

26. Find the perimeter of a square whose side is 34 cm.

27. Find the circumference of a circle whose radius is 21 cm. Use 3.14 for π.

28. Find the circumference of a circle whose radius is 3.4 m. Use 3.14 for π.

29. Find the circumference of a circle whose diameter is 1.2 m. Use 3.14 for π.

30. Find the circumference of a circle whose diameter is 15 in. Use 3.14 for π.

31. Art The wood framing for an art canvas costs $4.81 per foot. How much would the wood framing cost for a rectangular picture that measures 3 ft by 5 ft?

32. Ceramics A decorative mosaic tile is being installed on the border of a square wall behind a stove. If one side of the square measures 5 ft and the cost of installing the mosaic tile is $4.86 per foot, find the cost to install the decorative border.

33. Sewing To prevent fraying, a binding is attached to the outside of a circular rug whose radius is 3 ft. If the binding costs $1.05 per foot, find the cost of the binding. Use 3.14 for π.

34. Landscaping A drip irrigation system is installed around a circular flower garden that is 4 ft in diameter. If the irrigation system costs $2.46 per foot, find the cost to place the irrigation system around the flower garden. Use 3.14 for π.

Objective C **To solve area problems**

35. Find the area of a rectangle that measures 4 ft by 8 ft.

36. Find the area of a rectangle that measures 3.4 cm by 5.6 cm.

37. Find the area of a parallelogram whose height is 14 cm and whose base is 27 cm.

38. Find the area of a parallelogram whose height is 7 ft and whose base is 18 ft.

39. Find the area of a circle whose radius is 4 in. Use 3.14 for π.

40. Find the area of a circle whose radius is 8.2 m. Use 3.14 for π.

41. Find the area of a square whose side measures 4.1 m.

42. Find the area of a square whose side measures 5 yd.

43. Find the area of a triangle whose height is 7 cm and whose base is 15 cm.

44. Find the area of a triangle whose height is 8 in. and whose base is 13 in.

45. Find the area of a circle whose diameter is 17 in. Use 3.14 for π.

46. Find the area of a circle whose diameter is 3.6 m. Use 3.14 for π.

47. **Landscaping**   A landscape architect recommends 0.1 gallon of water per day for each square foot of lawn. How many gallons of water should be used per day on a rectangular lawn area that is 33 ft by 42 ft?

48. **Interior Design**   One side of a square room measures 18 ft. How many square yards of carpet are necessary to carpet the room? *Hint*: $1 \text{ yd}^2 = 9 \text{ ft}^2$.

49. **Carpentry**   A circular, inlaid-wood design for a dining table cost $35 per square foot to build. If the radius of the design is 15 in., find the cost to build the design. Use 3.14 for π. Round to the nearest dollar. *Hint*: $144 \text{ in}^2 = 1 \text{ ft}^2$.

50. **Interior Design**   A circular stained glass window cost $48 per square foot to build. If the diameter of the window is 4 ft, find the cost to build the window. Round to the nearest dollar.

51. **Construction**   The cost of plastering the walls of a rectangular room that is 18 ft long, 14 ft wide, and 8 ft high is $2.56 per square foot. If 125 ft^2 are not plastered because of doors and windows, find the cost to plaster the room.

52. **Interior Design**   A room is 12 ft long, 9 ft wide, and 9 ft high. Two adjacent walls of the rooms are going to be wallpapered using wallpaper that costs $15.25 per square yard. What is the cost to wallpaper the two walls? *Hint*: 1 yd² = 9 ft².

APPLYING THE CONCEPTS

53. Find the perimeter and area of the figure. Use 3.14 for π.

54. Find the perimeter and area of the figure.

55. Find the outside perimeter and area of the shaded portion of the figure.

56. Find the area of the shaded portion of the figure.

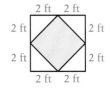

57. A trapezoid is a four-sided plane figure with two parallel sides. The area of a trapezoid is given by Area $= \frac{1}{2}\cdot$ height(base 1 + base 2). See the figure at the right.

    a. Find the area of a trapezoid for which base 1 is 5 in., base 2 is 8 in., and the height is 6 in.

    b. Find the area of the trapezoid shown at the right.

58. Draw parallelogram *ABCD* or one similar to it and then cut it out. Cut along the dotted line to form the shaded triangle. Slide the triangle so that the slanted side corresponds to the slanted side of the parallelogram as shown. Explain how this demonstrates that the area of a parallelogram is the product of the base and the height.

    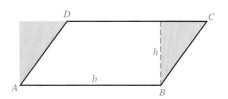

59. Explain how to draw the height of a triangle.

Focus on Problem Solving

Inductive Reasoning   Suppose you take 9 credit hours each semester. The total number of credit hours you have taken at the end of each semester can be described by a list of numbers.

$$9, 18, 27, 36, 45, 54, 63,\ldots$$

The list of numbers that indicates the total credit hours is an ordered list of numbers called a **sequence.** Each number in a sequence is called a **term** of the sequence. The list is ordered because the position of a number in the list indicates the semester in which that number of credit hours has been taken. For example, the 7th term of the sequence is 63, and a total of 63 credit hours have been taken after the 7th semester.

Assuming the pattern is continued, find the next three numbers in the pattern

$$-6, -10, -14, -18,\ldots$$

This list of numbers is a sequence. The first step in solving this problem is to observe the pattern in the list of numbers. In this case, each number in the list is 4 less than the previous number. The next three numbers are $-22, -26, -30$.

This process of discovering the pattern in a list of numbers is inductive reasoning. **Inductive reasoning** involves making generalizations from specific examples; in other words, we reach a conclusion by making observations about particular facts or cases.

Try the following exercises. Each exercise requires inductive reasoning.

For Exercises 1 to 4, name the next two terms in the sequence.

1. 1, 3, 5, 7, 1, 3, 5, 7, 1,…

2. 1, 4, 2, 5, 3, 6, 4,…

3. 1, 2, 4, 7, 11, 16,…

4. A, B, C, G, H, I, M,…

For Exercises 5 and 6, draw the next shape in the sequence.

5.

6. |• ||• ||••• |||•• |||•••

For Exercises 7 and 8, solve.

7. Convert $\frac{1}{11}, \frac{2}{11}, \frac{3}{11}, \frac{4}{11}$, and $\frac{5}{11}$ to decimals. Then use the pattern you observe to convert $\frac{6}{11}, \frac{7}{11}$, and $\frac{9}{11}$ to decimals.

8. Convert $\frac{1}{33}, \frac{2}{33}, \frac{4}{33}, \frac{5}{33}$, and $\frac{7}{33}$ to decimals. Then use the pattern you observe to convert $\frac{8}{33}, \frac{13}{33}$, and $\frac{19}{33}$ to decimals.

Projects and Group Activities

The +/− Key on a Calculator

Using your calculator to simplify numerical expressions sometimes requires use of the +/− key or, on some calculators, the negative key, which is frequently shown as (−) . To enter −4:

- For those calculators with +/− , press 4 and then +/− .
- For those calculators with (−) , press (−) and then 4.

Here are the keystrokes for evaluating the expression $3(-4) - (-5)$.

Calculators with +/− key: 3 × 4 +/− − 5 +/− =

Calculators with (−) key: 3 × (−) 4 − (−) 5 =

This example illustrates that calculators make a distinction between negative and minus. To perform the operation $3 - (-3)$, you cannot enter 3 − − 3. This would result in 0, which is not the correct answer. You must enter

3 − 3 +/− =   or   3 − (−) 3 =

For Exercises 1 to 6, use a calculator to evaluate.

1. $-16 \div 2$ **2.** $3(-8)$ **3.** $47 - (-9)$

4. $-50 - (-14)$ **5.** $4 - (-3)^2$ **6.** $-8 + (-6)^2 - 7$

Moving Averages

Objective 1.3C on page 21 describes how to find the moving average of a stock. Use this method to calculate the 5-day moving average for at least three different stocks. Discuss and compare the results for the different stocks.

For this project, you will need to use stock tables, which are printed in the business section of major newspapers. Your college library should have copies of these publications. In a stock table, the column headed "Chg." provides the change in the price of a share of the stock; that is, it gives the difference between the closing price for the day shown and the closing price for the previous day. The symbol + indicates that the change was an increase in price; the symbol − indicates that the change was a decrease in price.

www.fedstats.gov

Information on the history of the federal budget can be found on the web site **www.fedstats.gov.** Find tables entitled "Federal Budget." In the table, find the column that lists each year's surplus or deficit. You will see that a negative sign (−) is used to show a deficit. Note that it states near the top of the screen that the figures in the table are in millions of dollars.

1. During which years shown in the table was there a surplus?

2. During which year was the deficit the greatest?

3. Find the difference between the surplus or deficit this year and the surplus or deficit 5 years ago.

4. What is the difference between the surplus or deficit this year and the surplus or deficit a decade ago?

5. Determine what two numbers in the table are being subtracted in each row in order to arrive at the number in the surplus or deficit column.

6. Describe the trend of the federal deficit over the last 10 years.

Chapter 1 Summary

Key Words	**Examples**
The set of *natural numbers* is {1, 2, 3, 4, 5, . . .}. The set of *integers* is {. . . , −3, −2, −1, 0, 1, 2, 3, . . .}. [1.1A, p. 3]	
A number *a is less than* a number *b*, written $a < b$, if *a* is to the left of *b* on a number line. A number *a is greater than* a number *b*, written $a > b$, if *a* is to the right of *b* on a number line. The symbol ≤ means *is less than or equal to*. The symbol ≥ means *is greater than or equal to*. [1.1A, p. 4]	$-5 < -3$ $9 > 0$ $3 \le 3$ $4 \le 7$ $5 \ge 5$ $-6 \ge -9$
Two numbers that are the same distance from zero on the number line but on opposite sides of zero are *opposite numbers* or *opposites*. [1.1B, p. 5]	7 and −7 are opposites. $-\frac{3}{4}$ and $\frac{3}{4}$ are opposites.
The *absolute value* of a number is its distance from 0 on the number line. [1.1B, p. 5]	$|5| = 5$ $|-2.3| = 2.3$ $|0| = 0$
An expression of the form a^n is in *exponential form*. The *base* is *a* and the *exponent* is *n*. [1.4A, p. 24]	5^4 is an exponential expression. The base is 5 and the exponent is 4.
A natural number greater than 1 is a *prime number* if its only factors are 1 and the number. [1.5B, p. 31]	3, 17, 23, and 97 are prime numbers.
The *prime factorization* of a number is the expression of the number as a product of its prime factors. [1.5B, p. 31]	$2^3 \cdot 3^2 \cdot 7$ is the prime factorization of 504.
The *least common multiple (LCM)* of two or more numbers is the smallest number that is a multiple of all the numbers. [1.5C, p. 32]	The LCM of 4, 8, and 12 is 24.
The *greatest common factor (GCF)* of two or more numbers is the greatest number that divides evenly into all of the numbers. [1.5C, p. 32]	The GCF of 4, 8, and 12 is 4.
A *rational number* (or *fraction*) is a number that can be written in the form $\frac{a}{b}$, where *a* and *b* are integers and $b \neq 0$. A fraction is in *simplest form* when there are no common factors in the numerator and denominator. A rational number can be represented as a *terminating* or *repeating decimal*. [1.6A, pp. 35–36]	$\frac{3}{8}$, $-\frac{9}{2}$ and 4 are rational numbers written in simplest form. $\frac{3}{8}$ is a fraction in simplest form. 1.13 and $0.4\overline{73}$ are also rational numbers.
An *irrational number* is a number that has a decimal representation that never terminates or repeats. [1.6A, p. 36]	π, $\sqrt{2}$, and 1.34334333433334 . . . are irrational numbers.

The rational numbers and the irrational numbers taken together are the *real numbers*. [1.6A, p. 36]

$\frac{3}{8}$, $-\frac{9}{2}$, 4, 1.13, 0.4$\overline{73}$, π, $\sqrt{2}$, and 1.34334333433334 . . . are real numbers.

The *reciprocal of a fraction* is the fraction with the numerator and denominator interchanged. [1.7B, p. 48]

The reciprocal of $\frac{5}{6}$ is $\frac{6}{5}$.

The reciprocal of $-\frac{1}{3}$ is $-\frac{3}{1}$ or -3.

Percent means "parts of 100." [1.7C, p. 50]

72% means 72 of 100 equal parts.

A *plane* is a flat surface that extends indefinitely. A *line* extends indefinitely in two directions in a plane. A *ray* starts at a point and extends indefinitely in one direction. A *line segment* is part of a line and has two endpoints. [1.8A, p. 57]

Lines in a plane can be parallel or intersect. *Parallel lines* never meet. The distance between parallel lines in a plane is always the same. *Intersecting lines* cross at a point in the plane. [1.8A, p. 57]

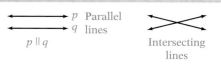

An *angle* is formed when two rays start from the same point. The point at which the rays meet is called the *vertex* of the angle. An angle can be measured in *degrees*. The measure of an angle is symbolized by $m\angle$. [1.8A, p. 57]

This angle can be named $\angle A$, $\angle BAC$, $\angle CAB$, or $\angle x$.

A *right angle* has a measure of 90°. *Perpendicular lines* are intersecting lines that form right angles. *Complementary angles* are two angles whose sum is 90°. A *straight angle* has a measure of 180°. *Supplementary angles* are two angles whose sum is 180°. [1.8A, p. 58]

$\angle A$ and $\angle B$ are complementary angles.

$\angle C$ and $\angle D$ are supplementary angles.

A *circle* is a plane figure in which all points are the same distance from point O, the *center* of the circle. The *diameter* of a circle is a line segment across the circle passing through the center. The *radius* of a circle is a line segment from the center of the circle to a point on the circle. The perimeter of a circle is called its *circumference*. [1.8B, p. 59]

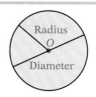

Essential Rules and Procedures	Examples
To add two numbers with the same sign, add the absolute values of the numbers. Then attach the sign of the addends. [1.2A, p. 9]	$7 + 15 = 22$ $-7 + (-15) = -22$
To add two numbers with different signs, find the absolute value of each number. Subtract the smaller of the two numbers from the larger. Then attach the sign of the number with the larger absolute value. [1.2A, p. 9]	$7 + (-15) = -8$ $-7 + 15 = 8$
To subtract one number from another, add the opposite of the second number to the first number. [1.2B, p. 10]	$7 - 19 = 7 + (-19) = -12$ $-6 - (-13) = -6 + 13 = 7$
To multiply two numbers with the same sign, multiply the absolute values of the numbers. The product is positive. [1.3A, p. 17]	$7 \cdot 8 = 56$ $-7(-8) = 56$
To multiply two numbers with different signs, multiply the absolute values of the numbers. The product is negative. [1.3A, p. 17]	$-7 \cdot 8 = -56$ $7(-8) = -56$
To divide two numbers with the same sign, divide the absolute values of the numbers. The quotient is positive. [1.3B, p. 19]	$54 \div 9 = 6$ $(-54) \div (-9) = 6$
To divide two numbers with different signs, divide the absolute values of the numbers. The quotient is negative. [1.3B, p. 19]	$(-54) \div 9 = -6$ $54 \div (-9) = -6$

Properties of Zero and One in Division [1.3B, p. 20]

If $a \neq 0$, $\dfrac{0}{a} = 0$.

$\dfrac{0}{-5} = 0$

If $a \neq 0$, $\dfrac{a}{a} = 1$.

$\dfrac{-12}{-12} = 1$

$\dfrac{a}{1} = a$

$\dfrac{7}{1} = 7$

$\dfrac{a}{0}$ is undefined.

$\dfrac{8}{0}$ is undefined.

Order of Operations Agreement [1.4B, p. 25]

Step 1 Perform operations inside grouping symbols. Grouping symbols include parentheses ( ), brackets [ ], braces { }, and the fraction bar.

Step 2 Simplify exponential expressions.

Step 3 Do multiplication and division as they occur from left to right.

Step 4 Do addition and subtraction as they occur from left to right.

$50 \div (-5)^2 + 2(7 - 16)$
$= 50 \div (-5)^2 + 2(-9)$
$= 50 \div 25 + 2(-9)$
$= 2 + (-18)$
$= -16$

To add two fractions with the same denominator, add the numerators and place the sum over the common denominator. [1.6B, p. 37]	$\dfrac{7}{10} + \dfrac{1}{10} = \dfrac{7+1}{10} = \dfrac{8}{10} = \dfrac{4}{5}$
To subtract two fractions with the same denominator, subtract the numerators and place the difference over the common denominator. [1.6C, p. 39]	$\dfrac{7}{10} - \dfrac{1}{10} = \dfrac{7-1}{10} = \dfrac{6}{10} = \dfrac{3}{5}$
To multiply two fractions, place the product of the numerators over the product of the denominators. [1.7A, p. 46]	$-\dfrac{2}{3} \cdot \dfrac{5}{6} = -\dfrac{2 \cdot 5}{3 \cdot 6} = -\dfrac{10}{18} = -\dfrac{5}{9}$
To divide two fractions, multiply the dividend by the reciprocal of the divisor. [1.7B, p. 48]	$-\dfrac{4}{5} \div \dfrac{2}{3} = -\dfrac{4}{5} \cdot \dfrac{3}{2} = -\dfrac{2 \cdot 2 \cdot 3}{5 \cdot 2} = -\dfrac{6}{5}$
To write a percent as a fraction, remove the percent sign and multiply by $\dfrac{1}{100}$. [1.7C, p. 50]	$60\% = 60\left(\dfrac{1}{100}\right) = \dfrac{60}{100} = \dfrac{3}{5}$
To write a percent as a decimal, remove the percent sign and multiply by 0.01. [1.7C, p. 50]	$73\% = 73(0.01) = 0.73$ $1.3\% = 1.3(0.01) = 0.013$
To write a decimal or a fraction as a percent, multiply by 100%. [1.7C, p. 50]	$0.3 = 0.3(100\%) = 30\%$ $\dfrac{5}{8} = \dfrac{5}{8}(100\%) = \dfrac{500}{8}\% = 62.5\%$
Diameter $= 2 \cdot$ radius **Radius** $= \dfrac{1}{2} \cdot$ diameter [1.8B, p. 59]	Find the diameter of a circle whose radius is 10 in. Diameter $= 2 \cdot$ radius $ = 2(10 \text{ in.}) = 20 \text{ in.}$

Perimeter is the distance around a plane figure. [1.8B, p. 59]

Triangle:	Perimeter $=$ side 1 $+$ side 2 $+$ side 3
Rectangle:	Perimeter $= 2 \cdot$ length $+ 2 \cdot$ width
Square:	Perimeter $= 4 \cdot$ side
Circle:	Circumference $= 2 \cdot \pi \cdot$ radius

Find the perimeter of a rectangle whose width is 12 m and whose length is 15 m.

perimeter $= 2 \cdot 15 \text{ m} + 2 \cdot 12 \text{ m} = 54 \text{ m}$

Find the circumference of a circle whose radius is 3 in. Use 3.14 for π.

circumference $= 2 \cdot \pi \cdot 3 \text{ in.} \approx 18.84 \text{ in.}$

Area is a measure of the amount of surface in a region. [1.8C, pp. 60–61]

Triangle:	Area $= \dfrac{1}{2} \cdot$ base $\cdot$ height
Rectangle:	Area $=$ length $\cdot$ width
Square:	Area $=$ side $\cdot$ side
Parallelogram:	Area $=$ base $\cdot$ height
Circle:	Area $= \pi(\text{radius})^2$

Find the area of a triangle whose base is 13 m and whose height is 11 m.

area $= \dfrac{1}{2} \cdot 13 \text{ m} \cdot 11 \text{ m} = 71.5 \text{ m}^2$

Find the area of a circle whose radius is 9 cm.

area $= \pi \cdot (9 \text{ cm})^2 \approx 254.34 \text{ cm}^2$

Chapter 1 Review Exercises

1. Add: $-13 + 7$

2. Write $\frac{7}{25}$ as a decimal.

3. Evaluate -5^2.

4. Evaluate $5 - 2^2 + 9$.

5. Find $m\angle AOB$ for the figure at the right.

6. Write 6.2% as a decimal.

7. Multiply: $(-6)(7)$

8. Simplify: $\frac{1}{3} - \frac{1}{6} + \frac{5}{12}$

9. Find the complement of a 56° angle.

10. Given $A = \{-4, 0, 11\}$, which elements of set A are less than -1?

11. Find all of the factors of 56.

12. Subtract: $5.17 - 6.238$

13. Write $\frac{5}{8}$ as a percent.

14. Write $\frac{2}{15}$ as a decimal. Place a bar over the repeating digits of the decimal.

15. Subtract: $9 - 13$

16. What is $\frac{2}{5}$ less than $\frac{4}{15}$?

17. Find the additive inverse of -4.

18. Find the area of a triangle whose base is 4 cm and whose height is 9 cm.

19. Divide: $-100 \div 5$

20. Write $79\frac{1}{2}\%$ as a fraction.

21. Find the prime factorization of 280.

22. Evaluate $-3^2 + 4[18 + (12 - 20)]$.

23. Add: $-3 + (-12) + 6 + (-4)$

24. Find the sum of $\frac{4}{5}$ and $-\frac{3}{8}$.

25. Write $\frac{19}{35}$ as a percent. Write the remainder in fractional form.

26. Find the area of a circle whose diameter is 6 m. Use 3.14 for π.

27. Multiply: $4.32(-1.07)$

28. Evaluate $-|-5|$.

29. Subtract: $16 - (-3) - 18$

30. Divide: $-\frac{18}{35} \div \frac{27}{28}$

31. Find the supplement of a 28° angle.

32. Find the perimeter of a rectangle whose length is 12 in. and whose width is 10 in.

33. Place the correct symbol, $>$ or $<$, between the two numbers.
$-|6| \quad |-10|$

34. Evaluate $\frac{5^2 + 11}{2^2 + 5} \div (2^3 - 2^2)$.

35. **Education** To discourage random guessing on a multiple-choice exam, a professor assigns 6 points for a correct answer, -4 points for an incorrect answer, and -2 points for leaving a question blank. What is the score for a student who had 21 correct answers, 5 incorrect answers, and left 4 questions blank?

36. 🥧 **Business** The graph at the right shows the number (in thousands) of personal digital assistants (PDAs) shipped worldwide in the first quarter of 2003. What percent of the total number of PDAs shipped did Palm ship? Round to the nearest tenth of a percent.

Worldwide First-Quarter PDA Shipments in 2003 (in thousands)

Source: Gartner, Inc.

37. 🥧 **Chemistry** The temperature at which mercury boils is 357°C. The temperature at which mercury freezes is -39°C. Find the difference between the boiling point and the freezing point of mercury.

38. **Landscaping** A landscape company is proposing to replace a rectangular flower bed that measures 8 ft by 12 ft with sod that costs $2.51 per square foot. Find the cost to replace the flower bed with the sod.

Chapter 1 Test

1. Divide: $-561 \div (-33)$

2. Write $\frac{5}{6}$ as a percent. Write the remainder in fractional form.

3. Find the complement of a 28° angle.

4. Multiply: $6.02(-0.89)$

5. Subtract: $16 - 30$

6. Write $37\frac{1}{2}\%$ as a fraction.

7. Subtract: $-\frac{5}{6} - \left(-\frac{7}{8}\right)$

8. Evaluate $\frac{-10 + 2}{2 + (-4)} \div 2 + 6$.

9. Multiply: $-5(-6)(3)$

10. Find the circumference of a circle whose diameter is 27 in. Use 3.14 for π.

11. Evaluate $(-3^3) \cdot 2^2$.

12. Find the area of a parallelogram whose base is 10 cm and whose height is 9 cm.

13. Place the correct symbol, $<$ or $>$, between the two numbers.
    $-2 \quad -40$

14. What is $\frac{2}{5}$ more than $-\frac{3}{4}$?

15. Evaluate $-|-4|$.

16. Write 45% as a fraction and as a decimal.

17. Add: $-22 + 14 + (-8)$

18. Multiply: $-4 \cdot 12$

19. Find the prime factorization of 990.

20. Evaluate $16 \div 2[8 - 3(4 - 2)] + 1$.

21. Subtract: $16 - (-30) - 42$

22. Divide: $\dfrac{5}{12} \div \left(-\dfrac{5}{6}\right)$

23. Find $m\angle x$ for the figure at the right.

24. Evaluate $3^2 - 4 + 20 \div 5$.

25. Write $\dfrac{7}{9}$ as a decimal. Place a bar over the repeating digit of the decimal.

26. **Finance** The table below shows the first-quarter profits and losses for 2003 for five companies in the computer chip industry. Profits are shown as positive numbers; losses are shown as negative numbers. One-quarter of a year is 3 months.

a. If earnings were to continue throughout the year at the same level, what would be the annual profit or loss for LSI Logic Corp.?

b. For the quarter shown, what was the average monthly profit or loss for National Semiconductor Corp.? Round to the nearest thousand dollars.

Computer Chip Company	First Quarter 2003 Profits (in millions of dollars)
Advanced Micro Devices, Inc.	-125.5
LSI Logic Corp.	-114.4
Micron Technology, Inc.	-183.7
Motorola, Inc.	171.0
National Semiconductor Corp.	-0.2

Source: Yahoo! Finance

27. **Recreation** The recreation department for a city is enclosing a rectangular playground that measures 150 ft by 200 ft with new fencing that costs $6.52 per foot. Find the cost of the new fencing.

2 Variable Expressions

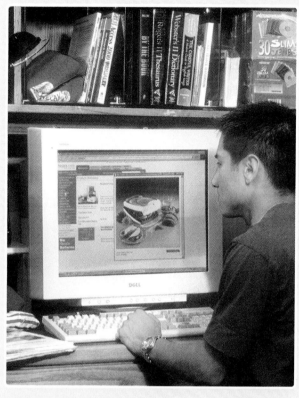

Have you ever purchased a product online? How satisfied were you with the experience? A recent survey conducted by Consumer Internet Barometer found that approximately 25% of people who had purchased a product online were extremely satisfied with their experience. See **Exercise 64 on page 100.** In this exercise, you are asked to express the number of people who were extremely satisfied with their online purchase in terms of the number of people who purchased a product on line. To do this, you must use a variable expression. Variable expressions are the topic of this chapter.

OBJECTIVES

Section 2.1

A To evaluate a variable expression

Section 2.2

A To simplify a variable expression using the Properties of Addition

B To simplify a variable expression using the Properties of Multiplication

C To simplify a variable expression using the Distributive Property

D To simplify general variable expressions

Section 2.3

A To translate a verbal expression into a variable expression, given the variable

B To translate a verbal expression into a variable expression and then simplify

C To translate application problems

Need help? For online student resources, such as section quizzes, visit this textbook's website at **math.college.hmco.com/students.**

Do these exercises to prepare for Chapter 2.

1. Subtract: $-12 - (-15)$

2. Divide: $-36 \div (-9)$

3. Add: $-\dfrac{3}{4} + \dfrac{5}{6}$

4. What is the reciprocal of $-\dfrac{9}{4}$?

5. Divide: $-\dfrac{3}{4} \div \left(-\dfrac{5}{2}\right)$

6. Evaluate: -2^4

7. Evaluate: $\left(\dfrac{2}{3}\right)^3$

8. Evaluate: $3 \cdot 4^2$

9. Evaluate: $7 - 2 \cdot 3$

10. Evaluate: $5 - 7(3 - 2^2)$

GO FIGURE • • •

Two fractions are inserted between $\dfrac{1}{4}$ and $\dfrac{1}{2}$ so that the difference between any two successive fractions is the same. Find the sum of the four fractions.

2.1 Evaluating Variable Expressions

Objective A   **To evaluate a variable expression**

Often we discuss a quantity without knowing its exact value—for example, the price of gold next month, the cost of a new automobile next year, or the tuition cost for next semester. Recall that a letter of the alphabet, called a **variable,** is used to stand for a quantity that is unknown or that can change, or *vary*. An expression that contains one or more variables is called a **variable expression.**

A variable expression is shown at the right. The expression can be rewritten by writing subtraction as the addition of the opposite.

$$3x^2 - 5y + 2xy - x - 7$$

$$3x^2 + (-5y) + 2xy + (-x) + (-7)$$

Note that the expression has 5 addends. The **terms** of a variable expression are the addends of the expression. The expression has 5 terms.

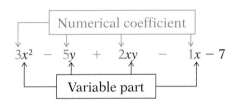

The terms $3x^2$, $-5y$, $2xy$, and $-x$ are **variable terms.**

The term -7 is a **constant term,** or simply a **constant.**

Each variable term is composed of a **numerical coefficient** and a **variable part** (the variable or variables and their exponents).

When the numerical coefficient is 1 or -1, the 1 is usually not written ($x = 1x$ and $-x = -1x$).

Variable expressions occur naturally in science. In a physics lab, a student may discover that a weight of 1 pound will stretch a spring $\frac{1}{2}$ inch. Two pounds will stretch the spring 1 inch. By experimenting, the student can discover that the distance the spring will stretch is found by multiplying the weight by $\frac{1}{2}$. By letting W represent the weight attached to the spring, the student can represent the distance the spring stretches by the variable expression $\frac{1}{2}W$.

With a weight of W pounds, the spring will stretch $\frac{1}{2} \cdot W = \frac{1}{2}W$ inches.

With a weight of 10 pounds, the spring will stretch $\frac{1}{2} \cdot 10 = 5$ inches. The number 10 is called the **value of the variable** W.

With a weight of 3 pounds, the spring will stretch $\frac{1}{2} \cdot 3 = 1\frac{1}{2}$ inches.

Replacing each variable by its value and then simplifying the resulting numerical expression is called **evaluating a variable expression.**

Integrating Technology

See the Keystroke Guide: *Evaluating Variable Expressions* for instructions on using a graphing calculator to evaluate variable expressions.

HOW TO Evaluate $ab - b^2$ when $a = 2$ and $b = -3$.

Replace each variable in the expression by its value. Then use the Order of Operations Agreement to simplify the resulting numerical expression.

$ab - b^2$

$2(-3) - (-3)^2 = -6 - 9$
$$= -15$$

When $a = 2$ and $b = -3$, the value of $ab - b^2$ is -15.

Example 1 Name the variable terms of the expression $2a^2 - 5a + 7$.

Solution $2a^2$ and $-5a$

You Try It 1 Name the constant term of the expression $6n^2 + 3n - 4$.

Your solution

Example 2 Evaluate $x^2 - 3xy$ when $x = 3$ and $y = -4$.

Solution
$x^2 - 3xy$
$3^2 - 3(3)(-4) = 9 - 3(3)(-4)$
$$= 9 - 9(-4)$$
$$= 9 - (-36)$$
$$= 9 + 36 = 45$$

You Try It 2 Evaluate $2xy + y^2$ when $x = -4$ and $y = 2$.

Your solution

Example 3 Evaluate $\frac{a^2 - b^2}{a - b}$ when $a = 3$ and $b = -4$.

Solution $\frac{a^2 - b^2}{a - b}$

$\frac{3^2 - (-4)^2}{3 - (-4)} = \frac{9 - 16}{3 - (-4)}$

$$= \frac{-7}{7} = -1$$

You Try It 3 Evaluate $\frac{a^2 + b^2}{a + b}$ when $a = 5$ and $b = -3$.

Your solution

Example 4 Evaluate $x^2 - 3(x - y) - z^2$ when $x = 2$, $y = -1$, and $z = 3$.

Solution
$x^2 - 3(x - y) - z^2$
$2^2 - 3[2 - (-1)] - 3^2$
$$= 2^2 - 3(3) - 3^2$$
$$= 4 - 3(3) - 9$$
$$= 4 - 9 - 9$$
$$= -5 - 9$$
$$= -14$$

You Try It 4 Evaluate $x^3 - 2(x + y) + z^2$ when $x = 2$, $y = -4$, and $z = -3$.

Your solution

Solutions on p. S4

2.1 Exercises

Objective A   **To evaluate a variable expression**

For Exercises 1 to 3, name the terms of the variable expression. Then underline the constant term.

1. $2x^2 + 5x - 8$ **2.** $-3n^2 - 4n + 7$ **3.** $6 - a^4$

For Exercises 4 to 6, name the variable terms of the expression. Then underline the variable part of each term.

4. $9b^2 - 4ab + a^2$ **5.** $7x^2y + 6xy^2 + 10$ **6.** $5 - 8n - 3n^2$

For Exercises 7 to 9, name the coefficients of the variable terms.

7. $x^2 - 9x + 2$ **8.** $12a^2 - 8ab - b^2$ **9.** $n^3 - 4n^2 - n + 9$

10. What is the numerical coefficient of a variable term?   **11.** Explain the meaning of the phrase "evaluate a variable expression."

For Exercises 12 to 32, evaluate the variable expression when $a = 2$, $b = 3$, and $c = -4$.

12. $3a + 2b$ **13.** $a - 2c$ **14.** $-a^2$

15. $2c^2$ **16.** $-3a + 4b$ **17.** $3b - 3c$

18. $b^2 - 3$ **19.** $-3c + 4$ **20.** $16 \div (2c)$

21. $6b \div (-a)$ **22.** $bc \div (2a)$ **23.** $b^2 - 4ac$

24. $a^2 - b^2$ **25.** $b^2 - c^2$ **26.** $(a + b)^2$

27. $a^2 + b^2$ **28.** $2a - (c + a)^2$ **29.** $(b - a)^2 + 4c$

30. $b^2 - \dfrac{ac}{8}$ **31.** $\dfrac{5ab}{6} - 3cb$ **32.** $(b - 2a)^2 + bc$

For Exercises 33 to 50, evaluate the variable expression when $a = -2$, $b = 4$, $c = -1$, and $d = 3$.

33. $\dfrac{b + c}{d}$

34. $\dfrac{d - b}{c}$

35. $\dfrac{2d + b}{-a}$

36. $\dfrac{b + 2d}{b}$

37. $\dfrac{b - d}{c - a}$

38. $\dfrac{2c - d}{-ad}$

39. $(b + d)^2 - 4a$

40. $(d - a)^2 - 3c$

41. $(d - a)^2 \div 5$

42. $3(b - a) - bc$

43. $\dfrac{b - 2a}{bc^2 - d}$

44. $\dfrac{b^2 - a}{ad + 3c}$

45. $\dfrac{1}{3}d^2 - \dfrac{3}{8}b^2$

46. $\dfrac{5}{8}a^4 - c^2$

47. $\dfrac{-4bc}{2a - b}$

48. $-\dfrac{3}{4}b + \dfrac{1}{2}(ac + bd)$

49. $-\dfrac{2}{3}d - \dfrac{1}{5}(bd - ac)$

50. $(b - a)^2 - (d - c)^2$

51. The value of z is the value of $a^2 - 2a$ when $a = -3$. Find the value of z^2.

52. The value of a is the value of $3x^2 - 4x - 5$ when $x = -2$. Find the value of $3a - 4$.

53. The value of c is the value of $a^2 + b^2$ when $a = 2$ and $b = -2$. Find the value of $c^2 - 4$.

APPLYING THE CONCEPTS

For Exercises 54 to 57, evaluate the following expressions for $x = 2$, $y = 3$, and $z = -2$.

54. $3^x - x^3$

55. z^x

56. $x^x - y^y$

57. $y^{(x^2)}$

58. For each of the following, determine the first natural number x, greater than 1, for which the second expression is larger than the first.
   a. $x^3, 3^x$
   b. $x^4, 4^x$
   c. $x^5, 5^x$
   d. $x^6, 6^x$

59. On the basis of your answer to Exercise 58, make a conjecture that appears to be true about the two expressions x^n and n^x, where $n = 3, 4, 5, 6, 7, \ldots$ and x is a natural number greater than 1.

2.2 Simplifying Variable Expressions

Objective A   **To simplify a variable expression using the Properties of Addition**

Like terms of a variable expression are terms with the same variable part. (Because $x^2 = x \cdot x$, x^2 and x are not like terms.)

Constant terms are like terms. 4 and 9 are like terms.

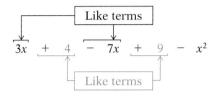

To simplify a variable expression, use the Distributive Property to combine like terms by adding the numerical coefficients. The variable part remains unchanged.

> **Distributive Property**
>
> If *a*, *b*, and *c* are real numbers, then $a(b + c) = ab + ac$.

The Distributive Property can also be written $ba + ca = (b + c)a$. This form is used to simplify a variable expression.

To simplify $2x + 3x$, use the Distributive Property to add the numerical coefficients of the like variable terms. This is called **combining like terms.**

$$2x + 3x = (2 + 3)x$$
$$= 5x$$

HOW TO   Simplify: $5y - 11y$

$5y - 11y = (5 - 11)y$ • Use the **Distributive Property.**
$= -6y$

HOW TO   Simplify: $5 + 7p$

The terms 5 and $7p$ are not like terms.

The expression $5 + 7p$ is in simplest form.

> **The Associative Property of Addition**
>
> If *a*, *b*, and *c* are real numbers, then $(a + b) + c = a + (b + c)$.

When three or more terms are added, the terms can be grouped (with parentheses, for example) in any order. The sum is the same. For example,

$$(5 + 7) + 15 = 5 + (7 + 15)$$
$$12 + 15 = 5 + 22$$
$$27 = 27$$

$$(3x + 5x) + 9x = 3x + (5x + 9x)$$
$$8x + 9x = 3x + 14x$$
$$17x = 17x$$

> **The Commutative Property of Addition**
>
> If a and b are real numbers, then $a + b = b + a$.

When two like terms are added, the terms can be added in either order. The sum is the same. For example,

$$15 + (-28) = (-28) + 15 \qquad\qquad 2x + (-4x) = -4x + 2x$$
$$-13 = -13 \qquad\qquad\qquad -2x = -2x$$

> **The Addition Property of Zero**
>
> If a is a real number, then $a + 0 = 0 + a = a$.

The sum of a term and zero is the term. For example,

$$-9 + 0 = 0 + (-9) = -9 \qquad\qquad 5x + 0 = 0 + 5x = 5x$$

> **The Inverse Property of Addition**
>
> If a is a real number, then $a + (-a) = (-a) + a = 0$.

The sum of a term and its opposite is zero. Recall that the opposite of a number is called its **additive inverse.**

$$12 + (-12) = (-12) + 12 = 0 \qquad\qquad 7x + (-7x) = -7x + 7x = 0$$

HOW TO   Simplify: $8x + 4y - 8x + y$

$8x + 4y - 8x + y$
$= (8x - 8x) + (4y + y)$

$= 0 + 5y = 5y$

- Use the Commutative and Associative Properties of Addition to rearrange and group like terms.
- Combine like terms.

HOW TO   Simplify: $4x^2 + 5x - 6x^2 - 2x + 1$

$4x^2 + 5x - 6x^2 - 2x + 1$
$= (4x^2 - 6x^2) + (5x - 2x) + 1$

$= -2x^2 + 3x + 1$

- Use the Commutative and Associative Properties of Addition to rearrange and group like terms.
- Combine like terms.

Example 1   Simplify: $3x + 4y - 10x + 7y$

Solution $3x + 4y - 10x + 7y = -7x + 11y$

You Try It 1   Simplify: $3a - 2b - 5a + 6b$

Your solution

Example 2   Simplify: $x^2 - 7 + 4x^2 - 16$

Solution $x^2 - 7 + 4x^2 - 16 = 5x^2 - 23$

You Try It 2   Simplify: $-3y^2 + 7 + 8y^2 - 14$

Your solution

Solutions on p. S4

Objective B   **To simplify a variable expression using the Properties of Multiplication**

In simplifying variable expressions, the following Properties of Multiplication are used.

> **The Associative Property of Multiplication**
>
> If a, b, and c are real numbers, then $(ab)c = a(bc)$.

When three or more factors are multiplied, the factors can be grouped in any order. The product is the same.

$$3(5 \cdot 6) = (3 \cdot 5)6 \qquad\qquad 2(3x) = (2 \cdot 3)x$$
$$3(30) = (15)6 \qquad\qquad\qquad = 6x$$
$$90 = 90$$

> **The Commutative Property of Multiplication**
>
> If a and b are real numbers, then $ab = ba$.

Two factors can be multiplied in either order. The product is the same.

$$5(-7) = -7(5) \qquad (5x) \cdot 3 = 3 \cdot (5x) \qquad$$ • Commutative Property of Multiplication
$$-35 = -35 \qquad\qquad = (3 \cdot 5)x \qquad$$ • Associative Property of Multiplication
$$= 15x$$

> **The Multiplication Property of One**
>
> If a is a real number, then $a \cdot 1 = 1 \cdot a = a$.

The product of a term and one is the term.

$$9 \cdot 1 = 1 \cdot 9 = 9 \qquad\qquad (8x) \cdot 1 = 1 \cdot (8x) = 8x$$

> **The Inverse Property of Multiplication**
>
> If a is a real number, and a is not equal to zero, then
> $$a \cdot \frac{1}{a} = \frac{1}{a} \cdot a = 1$$

$\frac{1}{a}$ is called the **reciprocal** of a. $\frac{1}{a}$ is also called the **multiplicative inverse** of a. The product of a number and its reciprocal is one.

$$7 \cdot \frac{1}{7} = \frac{1}{7} \cdot 7 = 1 \qquad\qquad x \cdot \frac{1}{x} = \frac{1}{x} \cdot x = 1, \quad x \neq 0$$

The multiplication properties just discussed are used to simplify variable expressions.

HOW TO   Simplify: $2(-x)$

$$2(-x) = 2(-1 \cdot x) \qquad$$ • Use the Associative Property of
$$= [2(-1)]x \qquad\qquad$$ Multiplication to group factors.
$$= -2x$$

HOW TO Simplify: $\frac{3}{2}\left(\frac{2x}{3}\right)$

$\frac{3}{2}\left(\frac{2x}{3}\right) = \frac{3}{2}\left(\frac{2}{3}x\right)$ • Note that $\frac{2x}{3} = \frac{2}{3}x$.

$= \left(\frac{3}{2} \cdot \frac{2}{3}\right)x$ • Use the Associative Property of Multiplication to group factors.

$= 1 \cdot x$

$= x$

HOW TO Simplify: $(16x)2$

$(16x)2 = 2(16x)$ • Use the Commutative and Associative Properties of Multiplication to rearrange and group factors.

$= (2 \cdot 16)x$

$= 32x$

Example 3 Simplify: $-2(3x^2)$

Solution $-2(3x^2) = -6x^2$

You Try It 3 Simplify: $-5(4y^2)$

Your solution

Example 4 Simplify: $-5(-10x)$

Solution $-5(-10x) = 50x$

You Try It 4 Simplify: $-7(-2a)$

Your solution

Example 5 Simplify: $-\frac{3}{4}\left(\frac{2}{3}x\right)$

Solution $-\frac{3}{4}\left(\frac{2}{3}x\right) = -\frac{1}{2}x$

You Try It 5 Simplify: $-\frac{3}{5}\left(-\frac{7}{9}a\right)$

Your solution

Solutions on p. S4

Objective C   **To simplify a variable expression using the Distributive Property**

Recall that the Distributive Property states that if a, b, and c are real numbers, then

$$a(b + c) = ab + ac$$

The Distributive Property is used to remove parentheses from a variable expression.

HOW TO Simplify: $3(2x + 7)$

$3(2x + 7) = 3(2x) + 3(7)$ • Use the **Distributive Property**. Multiply each term inside the parentheses by **3**.

$= 6x + 21$

HOW TO Simplify: $-5(4x + 6)$

$-5(4x + 6) = -5(4x) + (-5)(6)$ • Use the **Distributive Property**.
$\qquad\qquad = -20x - 30$

HOW TO Simplify: $-(2x - 4)$

$-(2x - 4) = -1(2x - 4)$ • Use the **Distributive Property**.
$\qquad\quad\; = -1(2x) - (-1)(4)$
$\qquad\quad\; = -2x + 4$

Note: When a negative sign immediately precedes the parentheses, the sign of each term inside the parentheses is changed.

HOW TO Simplify: $-\dfrac{1}{2}(8x - 12y)$

$-\dfrac{1}{2}(8x - 12y) = -\dfrac{1}{2}(8x) - \left(-\dfrac{1}{2}\right)(12y)$ • Use the **Distributive Property**.
$\qquad\qquad\qquad = -4x + 6y$

An extension of the Distributive Property is used when an expression contains more than two terms.

HOW TO Simplify: $3(4x - 2y - z)$

$3(4x - 2y - z) = 3(4x) - 3(2y) - 3(z)$ • Use the **Distributive Property**.
$\qquad\qquad\qquad = 12x - 6y - 3z$

Example 6

Simplify: $7(4 + 2x)$

Solution
$7(4 + 2x) = 28 + 14x$

You Try It 6

Simplify: $5(3 + 7b)$

Your solution

Example 7

Simplify: $(2x - 6)2$

Solution
$(2x - 6)2 = 4x - 12$

You Try It 7

Simplify: $(3a - 1)5$

Your solution

Example 8

Simplify: $-3(-5a + 7b)$

Solution
$-3(-5a + 7b) = 15a - 21b$

You Try It 8

Simplify: $-8(-2a + 7b)$

Your solution

Solutions on p. S4

Example 9   Simplify: $3(x^2 - x - 5)$

Solution $3(x^2 - x - 5) = 3x^2 - 3x - 15$

You Try It 9   Simplify: $3(12x^2 - x + 8)$

Your solution

Example 10   Simplify: $-2(x^2 + 5x - 4)$

Solution
$$-2(x^2 + 5x - 4)$$
$$= -2x^2 - 10x + 8$$

You Try It 10   Simplify: $3(-a^2 - 6a + 7)$

Your solution

Solutions on p. S4

Objective D   **To simplify general variable expressions**

When simplifying variable expressions, use the Distributive Property to remove parentheses and brackets used as grouping symbols.

> **HOW TO**   Simplify: $4(x - y) - 2(-3x + 6y)$
>
> $4(x - y) - 2(-3x + 6y)$
>
> $\quad = 4x - 4y + 6x - 12y$ • Use the Distributive Property.
>
> $\quad = 10x - 16y$ • Combine like terms.

Example 11   Simplify: $2x - 3(2x - 7y)$

Solution
$$2x - 3(2x - 7y) = 2x - 6x + 21y$$
$$= -4x + 21y$$

You Try It 11   Simplify: $3y - 2(y - 7x)$

Your solution

Example 12   Simplify:
$$7(x - 2y) - (-x - 2y)$$

Solution
$$7(x - 2y) - (-x - 2y)$$
$$= 7x - 14y + x + 2y$$
$$= 8x - 12y$$

You Try It 12   Simplify:
$$-2(x - 2y) - (-x + 3y)$$

Your solution

Example 13   Simplify: $2x - 3[2x - 3(x + 7)]$

Solution
$$2x - 3[2x - 3(x + 7)]$$
$$= 2x - 3[2x - 3x - 21]$$
$$= 2x - 3[-x - 21]$$
$$= 2x + 3x + 63$$
$$= 5x + 63$$

You Try It 13   Simplify: $3y - 2[x - 4(2 - 3y)]$

Your solution

Solutions on pp. S4–S5

2.2 Exercises

Objective A   **To simplify a variable expression using the Properties of Addition**

1. ✎ What are *like terms*? Give an example of two like terms. Give an example of two terms that are not like terms.

2. ✎ Explain the meaning of the phrase "simplify the variable expression."

For Exercises 3 to 38, simplify.

3. $6x + 8x$

4. $12x + 13x$

5. $9a - 4a$

6. $12a - 3a$

7. $4y - 10y$

8. $8y - 6y$

9. $7 - 3b$

10. $5 + 2a$

11. $-12a + 17a$

12. $-3a + 12a$

13. $5ab - 7ab$

14. $9ab - 3ab$

15. $-12xy + 17xy$

16. $-15xy + 3xy$

17. $-3ab + 3ab$

18. $-7ab + 7ab$

19. $-\dfrac{1}{2}x - \dfrac{1}{3}x$

20. $-\dfrac{2}{5}y + \dfrac{3}{10}y$

21. $2.3x + 4.2x$

22. $6.1y - 9.2y$

23. $x - 0.55x$

24. $0.65A - A$

25. $5a - 3a + 5a$

26. $10a - 17a + 3a$

27. $-5x^2 - 12x^2 + 3x^2$

28. $-y^2 - 8y^2 + 7y^2$

29. $\dfrac{3}{4}x - \dfrac{1}{3}x - \dfrac{7}{8}x$

30. $-\dfrac{2}{5}a - \left(-\dfrac{3}{10}a\right) - \dfrac{11}{15}a$

31. $7x - 3y + 10x$

32. $8y + 8x - 8y$

33. $3a + (-7b) - 5a + b$

34. $-5b + 7a - 7b + 12a$

35. $3x + (-8y) - 10x + 4x$

36. $3y + (-12x) - 7y + 2y$

37. $x^2 - 7x + (-5x^2) + 5x$

38. $3x^2 + 5x - 10x^2 - 10x$

Objective B **To simplify a variable expression using the Properties of Multiplication**

For Exercises 39 to 78, simplify.

39. $4(3x)$

40. $12(5x)$

41. $-3(7a)$

42. $-2(5a)$

43. $-2(-3y)$

44. $-5(-6y)$

45. $(4x)2$

46. $(6x)12$

47. $(3a)(-2)$

48. $(7a)(-4)$

49. $(-3b)(-4)$

50. $(-12b)(-9)$

51. $-5(3x^2)$

52. $-8(7x^2)$

53. $\frac{1}{3}(3x^2)$

54. $\frac{1}{6}(6x^2)$

55. $\frac{1}{5}(5a)$

56. $\frac{1}{8}(8x)$

57. $-\frac{1}{2}(-2x)$

58. $-\frac{1}{4}(-4a)$

59. $-\frac{1}{7}(-7n)$

60. $-\frac{1}{9}(-9b)$

61. $(3x)\left(\frac{1}{3}\right)$

62. $(12x)\left(\frac{1}{12}\right)$

63. $(-6y)\left(-\frac{1}{6}\right)$

64. $(-10n)\left(-\frac{1}{10}\right)$

65. $\frac{1}{3}(9x)$

66. $\frac{1}{7}(14x)$

67. $-\frac{1}{5}(10x)$

68. $-\frac{1}{8}(16x)$

69. $-\frac{2}{3}(12a^2)$

70. $-\frac{5}{8}(24a^2)$

71. $-\frac{1}{2}(-16y)$

72. $-\frac{3}{4}(-8y)$

73. $(16y)\left(\frac{1}{4}\right)$

74. $(33y)\left(\frac{1}{11}\right)$

75. $(-6x)\left(\frac{1}{3}\right)$

76. $(-10x)\left(\frac{1}{5}\right)$

77. $(-8a)\left(-\frac{3}{4}\right)$

78. $(21y)\left(-\frac{3}{7}\right)$

Objective C **To simplify a variable expression using the Distributive Property**

For Exercises 79 to 117, simplify.

79. $-(x + 2)$

80. $-(x + 7)$

81. $2(4x - 3)$

82. $5(2x - 7)$

83. $-2(a + 7)$

84. $-5(a + 16)$

85. $-3(2y - 8)$

86. $-5(3y - 7)$

87. $(5 - 3b)7$ **88.** $(10 - 7b)2$ **89.** $\dfrac{1}{3}(6 - 15y)$ **90.** $\dfrac{1}{2}(-8x + 4y)$

91. $3(5x^2 + 2x)$ **92.** $6(3x^2 + 2x)$ **93.** $-2(-y + 9)$ **94.** $-5(-2x + 7)$

95. $(-3x - 6)5$ **96.** $(-2x + 7)7$ **97.** $2(-3x^2 - 14)$ **98.** $5(-6x^2 - 3)$

99. $-3(2y^2 - 7)$ **100.** $-8(3y^2 - 12)$ **101.** $3(x^2 - y^2)$ **102.** $5(x^2 + y^2)$

103. $-\dfrac{2}{3}(6x - 18y)$ **104.** $-\dfrac{1}{2}(x - 4y)$ **105.** $-(6a^2 - 7b^2)$

106. $3(x^2 + 2x - 6)$ **107.** $4(x^2 - 3x + 5)$ **108.** $-2(y^2 - 2y + 4)$

109. $\dfrac{3}{4}(2x - 6y + 8)$ **110.** $-\dfrac{2}{3}(6x - 9y + 1)$ **111.** $4(-3a^2 - 5a + 7)$

112. $-5(-2x^2 - 3x + 7)$ **113.** $-3(-4x^2 + 3x - 4)$ **114.** $3(2x^2 + xy - 3y^2)$

115. $5(2x^2 - 4xy - y^2)$ **116.** $-(3a^2 + 5a - 4)$ **117.** $-(8b^2 - 6b + 9)$

Objective D   **To simplify general variable expressions**

For Exercises 118 to 138, simplify.

118. $4x - 2(3x + 8)$ **119.** $6a - (5a + 7)$ **120.** $9 - 3(4y + 6)$

121. $10 - (11x - 3)$ **122.** $5n - (7 - 2n)$ **123.** $8 - (12 + 4y)$

124. $3(x + 2) - 5(x - 7)$ **125.** $2(x - 4) - 4(x + 2)$ **126.** $12(y - 2) + 3(7 - 3y)$

127. $6(2y - 7) - (3 - 2y)$ **128.** $3(a - b) - (a + b)$ **129.** $2(a + 2b) - (a - 3b)$

130. $4[x - 2(x - 3)]$ **131.** $2[x + 2(x + 7)]$ **132.** $-2[3x + 2(4 - x)]$

133. $-5[2x + 3(5 - x)]$ **134.** $-3[2x - (x + 7)]$ **135.** $-2[3x - (5x - 2)]$

136. $2x - 3[x - (4 - x)]$ **137.** $-7x + 3[x - (3 - 2x)]$ **138.** $-5x - 2[2x - 4(x + 7)] - 6$

APPLYING THE CONCEPTS

139. Determine whether the statement is true or false. If the statement is false, give an example that illustrates that it is false.
  a. Division is a commutative operation.
  b. Division is an associative operation.
  c. Subtraction is an associative operation.
  d. Subtraction is a commutative operation.

140. Is the statement "any number divided by itself is 1" a true statement? If not, for what number or numbers is the statement not true?

141. Does every number have a multiplicative inverse? If not, which real number or numbers do not have a multiplicative inverse?

142. In your own words, explain the Distributive Property.

143. Give examples of two operations that occur in everyday experience that are not commutative (for example, putting on socks and then shoes).

144. Define an operation $\otimes$ as $a \otimes b = (a \cdot b) - (a + b)$. For example, $7 \otimes 5 = (7 \cdot 5) - (7 + 5) = 35 - 12 = 23$.
  a. Is $\otimes$ a commutative operation? Support your answer.

  b. Is $\otimes$ an associative operation? Support your answer.

2.3 Translating Verbal Expressions into Variable Expressions

Objective A   **To translate a verbal expression into a variable expression, given the variable**

One of the major skills required in applied mathematics is to translate a verbal expression into a variable expression. This requires recognizing the verbal phrases that translate into mathematical operations. A partial list of the verbal phrases used to indicate the different mathematical operations follows.

Point of Interest

The way in which expressions are symbolized has changed over time. Here are how some of the expressions shown at the right may have appeared in the early 16th century.

R p. 9 for $x + 9$. The symbol R was used for a variable to the first power. The symbol p. was used for plus.

R m. 3 for $x - 3$. The symbol R is still used for the variable. The symbol m. was used for minus.

The square of a variable was designated by Q and the cube was designated by C. The expression $x^2 + x^3$ was written **Q p. C.**

Addition	added to	6 added to y	$y + 6$
	more than	8 more than x	$x + 8$
	the sum of	the sum of x and z	$x + z$
	increased by	t increased by 9	$t + 9$
	the total of	the total of 5 and y	$5 + y$
Subtraction	minus	x minus 2	$x - 2$
	less than	7 less than t	$t - 7$
	decreased by	m decreased by 3	$m - 3$
	the difference between	the difference between y and 4	$y - 4$
	subtract...from...	subtract 9 from z	$z - 9$
Multiplication	times	10 times t	$10t$
	twice	twice w	$2w$
	of	one-half of x	$\dfrac{1}{2}x$
	the product of	the product of y and z	yz
	multiplied by	y multiplied by 11	$11y$
Division	divided by	x divided by 12	$\dfrac{x}{12}$
	the quotient of	the quotient of y and z	$\dfrac{y}{z}$
	the ratio of	the ratio of t to 9	$\dfrac{t}{9}$
Power	the square of	the square of x	x^2
	the cube of	the cube of a	a^3

HOW TO   Translate "14 less than the cube of x" into a variable expression.

14 <u>less than</u> the <u>cube</u> of x • Identify the words that indicate the mathematical operations.

$x^3 - 14$ • Use the identified operations to write the variable expression.

Translating a phrase that contains the word *sum, difference, product,* or *quotient* can sometimes cause a problem. In the examples at the right, note where the operation symbol is placed.

the *sum* of x and y $\qquad$ $x + y$

the *difference* between x and y $\qquad$ $x - y$

the *product* of x and y $\qquad$ $x \cdot y$

the *quotient* of x and y $\qquad$ $\dfrac{x}{y}$

HOW TO Translate "the difference between the square of x and the sum of y and z" into a variable expression.

the difference between the square of x and the sum of y and z

$x^2 - (y + z)$

- Identify the words that indicate the mathematical operations.

- Use the identified operations to write the variable expression.

Example 1

Translate "the total of 3 times n and n" into a variable expression.

Solution

the total of 3 times n and n

$3n + n$

You Try It 1

Translate "the difference between twice n and one-third of n" into a variable expression.

Your solution

Example 2

Translate "m decreased by the sum of n and 12" into a variable expression.

Solution

m decreased by the sum of n and 12

$m - (n + 12)$

You Try It 2

Translate "the quotient of 7 less than b and 15" into a variable expression.

Your solution

Solutions on p. S5

Objective B   **To translate a verbal expression into a variable expression and then simplify**

In most applications that involve translating phrases into variable expressions, the variable to be used is not given. To translate these phrases, a variable must be assigned to an unknown quantity before the variable expression can be written.

HOW TO　Translate "a number multiplied by the total of six and the cube of the number" into a variable expression.

the unknown number: n

• Assign a variable to one of the unknown quantities.

the cube of the number: n^3
the total of six and the cube of the number: $6 + n^3$

• Use the assigned variable to write an expression for any other unknown quantity.

$n(6 + n^3)$

• Use the assigned variable to write the variable expression.

Example 3

Translate "a number added to the product of four and the square of the number" into a variable expression.

Solution
the unknown number: n
the square of the number: n^2
the product of four and the square of the number: $4n^2$
$4n^2 + n$

You Try It 3

Translate "negative four multiplied by the total of ten and the cube of a number" into a variable expression.

Your solution

Example 4

Translate "four times the sum of one-half of a number and fourteen" into a variable expression. Then simplify.

Solution
the unknown number: n

one-half of the number: $\frac{1}{2}n$

the sum of one-half of the number and

fourteen: $\frac{1}{2}n + 14$

$4\left(\frac{1}{2}n + 14\right)$
$2n + 56$

You Try It 4

Translate "five times the difference between a number and sixty" into a variable expression. Then simplify.

Your solution

Solutions on p. S5

Objective C **To translate application problems**

Many of the applications of mathematics require that you identify the unknown quantity, assign a variable to that quantity, and then attempt to express other unknown quantities in terms of the variable.

HOW TO The height of a triangle is 10 ft longer than the base of the triangle. Express the height of the triangle in terms of the base of the triangle.

the base of the triangle: b

• Assign a variable to the base of the triangle.

the height is 10 more than the base: $b + 10$

• Express the height of the triangle in terms of b.

Example 5

The length of a swimming pool is 4 ft less than two times the width. Express the length of the pool in terms of the width.

Solution
the width of the pool: w
the length is 4 ft less than two times the width: $2w - 4$

You Try It 5

The speed of a new jet plane is twice the speed of an older model. Express the speed of the new model in terms of the speed of the older model.

Your solution

Example 6

A banker divided $5000 between two accounts, one paying 10% annual interest and the second paying 8% annual interest. Express the amount invested in the 10% account in terms of the amount invested in the 8% account.

Solution
the amount invested at 8%: x
the amount invested at 10%: $5000 - x$

You Try It 6

A guitar string 6 ft long was cut into two pieces. Express the length of the shorter piece in terms of the length of the longer piece.

Your solution

Solutions on p. S5

2.3 Exercises

Objective A   **To translate a verbal expression into a variable expression, given the variable**

For Exercises 1 to 26, translate into a variable expression.

1. the sum of 8 and y

2. a less than 16

3. t increased by 10

4. p decreased by 7

5. z added to 14

6. q multiplied by 13

7. 20 less than the square of x

8. 6 times the difference between m and 7

9. the sum of three-fourths of n and 12

10. b decreased by the product of 2 and b

11. 8 increased by the quotient of n and 4

12. the product of -8 and y

13. the product of 3 and the total of y and 7

14. 8 divided by the difference between x and 6

15. the product of t and the sum of t and 16

16. the quotient of 6 less than n and twice n

17. 15 more than one-half of the square of x

18. 19 less than the product of n and -2

19. the total of 5 times the cube of n and the square of n

20. the ratio of 9 more than m to m

21. r decreased by the quotient of r and 3

22. four-fifths of the sum of w and 10

23. the difference between the square of x and the total of x and 17

24. s increased by the quotient of 4 and s

25. the product of 9 and the total of z and 4

26. n increased by the difference between 10 times n and 9

Objective B **To translate a verbal expression into a
variable expression and then simplify**

For Exercises 27 to 58, translate into a variable expression. Then simplify.

27. twelve minus a number

28. a number divided by eighteen

29. two-thirds of a number

30. twenty more than a number

31. the quotient of twice a number and nine

32. ten times the difference between a number and fifty

33. eight less than the product of eleven and a number

34. the sum of five-eighths of a number and six

35. nine less than the total of a number and two

36. the difference between a number and three more than the number

37. the quotient of seven and the total of five and a number

38. four times the sum of a number and nineteen

39. five increased by one-half of the sum of a number and three

40. the quotient of fifteen and the sum of a number and twelve

41. a number added to the difference between twice the number and four

42. the product of two-thirds and the sum of a number and seven

43. the product of five less than a number and seven

44. the difference between forty and the quotient of a number and twenty

45. the quotient of five more than twice a number and the number

46. the sum of the square of a number and twice the number

47. a number decreased by the difference between three times the number and eight

48. the sum of eight more than a number and one-third of the number

49. a number added to the product of three and the number

50. a number increased by the total of the number and nine

51. five more than the sum of a number and six

52. a number decreased by the difference between eight and the number

53. a number minus the sum of the number and ten

54. the difference between one-third of a number and five-eighths of the number

55. the sum of one-sixth of a number and four-ninths of the number

56. two more than the total of a number and five

57. the sum of a number divided by three and the number

58. twice the sum of six times a number and seven

Objective C   **To translate application problems**

59. **Internet**   According to Brightmail, Inc., one-half of all email filtered by their email program in August 2003 was unsolicited email (spam). Express the amount of spam in terms of the number of emails filtered.

60. **Telecommunications**   In 1951, phone companies began using area codes. According to information found at **www.area-code.com,** at the beginning of 2004 there were 207 more area codes than there were in 1951. Express the number of area codes in 2004 in terms of the number of area codes in 1951.

61. **Sports**   A halyard 12 ft long was cut into two pieces of different lengths. Use one variable to express the lengths of the two pieces.

62. **Natural Resources**   Twenty gallons of crude oil were poured into two containers of different sizes. Use one variable to express the amount of oil poured into each container.

63. Rates of Cars   Two cars start at the same place and travel at different rates in opposite directions. Two hours later the cars are 200 mi apart. Express the distance traveled by the slower car in terms of the distance traveled by the faster car.

64. **Online Sales**   A recent survey conducted by Consumer Internet Barometer found that approximately 25% of those people who purchased a product online were extremely satisfied with their experience. Express the number of people that were extremely satisfied with their online purchases in terms of the number of people surveyed.

65. **Medicine**   According to the American Podiatric Medical Association, the bones in your foot account for one-fourth of all the bones in your body. Express the number of bones in your foot in terms of the total number of bones in your body.

66. **Sports**   The diameter of a basketball is approximately 4 times the diameter of a baseball. Express the diameter of a basketball in terms of the diameter of a baseball.

67. **Cost of Living**   A cost-of-living calculator provided by **Realtor.com** shows that a person living in San Francisco, California, would need approximately twice the salary of a person living in Daytona Beach, Florida, to maintain the same standard of living. Express the salary needed in Daytona Beach in terms of the salary needed in San Francisco.

APPLYING THE CONCEPTS

68. Metalwork   A wire whose length is given as x inches is bent into a square. Express the length of a side of the square in terms of x.

69. **Chemistry**   The chemical formula for glucose (sugar) is $C_6H_{12}O_6$. This formula means that there are 12 hydrogen atoms for every 6 carbon atoms and 6 oxygen atoms in each molecule of glucose (see the figure at the right). If x represents the number of atoms of oxygen in a pound of sugar, express the number of hydrogen atoms in the pound of sugar.

70. Translate the expressions $5x + 8$ and $5(x + 8)$ into phrases.

71. In your own words, explain how variables are used.

72. Explain the similarities and the differences between the expressions "the difference between x and 5" and "5 less than x."

Focus on Problem Solving

From Concrete to Abstract

In your study of algebra, you will find that the problems are less concrete than those you studied in arithmetic. Problems that are concrete provide information pertaining to a specific instance. Algebra is more abstract. Abstract problems are theoretical; they are stated without reference to a specific instance. Let's look at an example of an abstract problem.

How many minutes are in h hours?

A strategy that can be used to solve this problem is to solve the same problem after substituting a number for the variable.

How many minutes are in 5 hours?

You know that there are 60 minutes in 1 hour. To find the number of minutes in 5 hours, multiply 5 by 60.

$$60 \cdot 5 = 300 \qquad \text{There are 300 minutes in 5 hours.}$$

Use the same procedure to find the number of minutes in h hours: multiply h by 60.

$$60 \cdot h = 60h \qquad \text{There are } 60h \text{ minutes in } h \text{ hours.}$$

This problem might be taken a step further:

If you walk 1 mile in x minutes, how far can you walk in h hours?

Consider the same problem using numbers in place of the variables.

If you walk 1 mile in 20 minutes, how far can you walk in 3 hours?

To solve this problem, you need to calculate the number of minutes in 3 hours (multiply 3 by 60), and divide the result by the number of minutes it takes to walk 1 mile (20 minutes).

$$\frac{60 \cdot 3}{20} = \frac{180}{20} = 9 \qquad \text{If you walk 1 mile in 20 minutes, you can walk 9 miles in 3 hours.}$$

Use the same procedure to solve the related abstract problem. Calculate the number of minutes in h hours (multiply h by 60), and divide the result by the number of minutes it takes to walk 1 mile (x minutes).

$$\frac{60 \cdot h}{x} = \frac{60h}{x} \qquad \text{If you walk 1 mile in } x \text{ minutes, you can walk } \frac{60h}{x} \text{ miles in } h \text{ hours.}$$

At the heart of the study of algebra is the use of variables. It is the variables in the problems above that make them abstract. But it is variables that allow us to generalize situations and state rules about mathematics.

Try each of the following problems.

1. How many hours are in d days?

2. You earn *d* dollars an hour. What are your wages for working *h* hours?

3. If *p* is the price of one share of stock, how many shares can you purchase with *d* dollars?

4. A company pays a television station *d* dollars to air a commercial lasting *s* seconds. What is the cost per second?

5. After every *v* videotape rentals, you are entitled to one free rental. You have rented *t* tapes, where *t* < *v*. How many more do you need to rent before you are entitled to a free rental?

6. Your car gets *g* miles per gallon. How many gallons of gasoline does your car consume traveling *t* miles?

7. If you drink *j* ounces of juice each day, how many days will *q* quarts of the juice last?

8. A TV station has *m* minutes of commercials each hour. How many ads lasting *s* seconds each can be sold for each hour of programming?

9. A factory worker can assemble *p* products in *m* minutes. How many products can the factory worker assemble in *h* hours?

10. If one candy bar costs *n* nickels, how many candy bars can be purchased with *q* quarters?

Projects and Group Activities

Prime and Composite Numbers

Recall that a prime number is a natural number greater than 1 whose only natural-number factors are itself and 1. The number 11 is a prime number because the only natural-number factors of 11 are 11 and 1.

Eratosthenes, a Greek philosopher and astronomer who lived from 270 to 190 B.C., devised a method of identifying prime numbers. It is called the **Sieve of Eratosthenes.** The procedure is illustrated below.

1̶	②	③	4̶	⑤	6̶	⑦	8̶	9̶	1̶0̶
⑪	1̶2̶	⑬	1̶4̶	1̶5̶	1̶6̶	⑰	1̶8̶	⑲	2̶0̶
2̶1̶	2̶2̶	㉓	2̶4̶	2̶5̶	2̶6̶	2̶7̶	2̶8̶	㉙	3̶0̶
㉛	3̶2̶	3̶3̶	3̶4̶	3̶5̶	3̶6̶	㊲	3̶8̶	3̶9̶	4̶0̶
㊶	4̶2̶	㊸	4̶4̶	4̶5̶	4̶6̶	㊼	4̶8̶	4̶9̶	5̶0̶
5̶1̶	5̶2̶	㊻	5̶4̶	5̶5̶	5̶6̶	5̶7̶	5̶8̶	㊾	6̶0̶
㉽	6̶2̶	6̶3̶	6̶4̶	6̶5̶	6̶6̶	㉿	6̶8̶	6̶9̶	7̶0̶
㋛	7̶2̶	㋓	7̶4̶	7̶5̶	7̶6̶	7̶7̶	7̶8̶	㋘	8̶0̶
8̶1̶	8̶2̶	㋛	8̶4̶	8̶5̶	8̶6̶	8̶7̶	8̶8̶	㋈	9̶0̶
9̶1̶	9̶2̶	9̶3̶	9̶4̶	9̶5̶	9̶6̶	㋙	9̶8̶	9̶9̶	1̶0̶0̶

List all the natural numbers from 1 to 100. Cross out the number 1, because it is not a prime number. The number 2 is prime; circle it. Cross out all the other multiples of 2 (4, 6, 8,…), because they are not prime. The number 3 is prime; circle it. Cross out all the other multiples of 3 (6, 9, 12,…) that are not already crossed out. The number 4, the next consecutive number in the list, has already been crossed out. The number 5 is prime; circle it. Cross out all the other multiples of 5 that are not already crossed out. Continue in this manner until all the prime numbers less than 100 are circled.

A composite number is a natural number greater than 1 that has a natural-number factor other than itself and 1. The number 21 is a composite number because it has factors of 3 and 7. All the numbers crossed out in the preceding table, except the number 1, are composite numbers.

1. Use the Sieve of Eratosthenes to find the prime numbers between 100 and 200.

2. How many prime numbers are even numbers?

3. Find the "twin primes" between 100 and 200. Twin primes are two prime numbers whose difference is 2. For instance, 3 and 5 are twin primes; 5 and 7 are also twin primes.

4. **a.** List two prime numbers that are consecutive natural numbers.
   b. Can there be any other pairs of prime numbers that are consecutive natural numbers?

5. Some primes are the sum of a square and 1. For example, $5 = 2^2 + 1$. Find another prime p such that $p = n^2 + 1$, where n is a natural number.

6. Find a prime number p such that $p = n^2 - 1$, where n is a natural number.

7. **a.** 4! (which is read "4 factorial") is equal to $4 \cdot 3 \cdot 2 \cdot 1$. Show that 4! + 2, 4! + 3, and 4! + 4 are all composite numbers.
   b. 5! (which is read "5 factorial") is equal to $5 \cdot 4 \cdot 3 \cdot 2 \cdot 1$. Will 5! + 2, 5! + 3, 5! + 4, and 5! + 5 generate four consecutive composite numbers?
   c. Use the notation 6! to represent a list of five consecutive composite numbers.

Investigation into Operations with Even and Odd Integers

For Exercises 1 to 10, complete each statement with the word *even* or *odd*.

1. If k is an odd integer, then $k + 1$ is an _____ integer.

2. If k is an odd integer, then $k - 2$ is an _____ integer.

3. If n is an integer, then $2n$ is an _____ integer.

4. If m and n are even integers, then $m - n$ is an _____ integer.

5. If m and n are even integers, then mn is an _____ integer.

6. If m and n are odd integers, then $m + n$ is an _____ integer.

7. If m and n are odd integers, then $m - n$ is an _____ integer.

8. If m and n are odd integers, then mn is an _____ integer.

9. If m is an even integer and n is an odd integer, then $m - n$ is an _____ integer.

10. If m is an even integer and n is an odd integer, then $m + n$ is an _____ integer.

Study Tip

Three important features of this text that can be used to prepare for a test are the:

- Chapter Summary
- Chapter Review Exercises
- Chapter Test

See *AIM for Success,* page xxv.

Chapter 2 Summary

Key Words	**Examples**
A *variable* is a letter that is used for a quantity that is unknown or that can change. A *variable expression* is an expression that contains one or more variables. [2.1A, p. 79]	$4x + 2y - 6z$ is a variable expression. It contains the variables x, y, and z.
The *terms* of a variable expression are the addends of the expression. Each term is a *variable term* or a *constant term*. [2.1A, p. 79]	The expression $2a^2 - 3b^3 + 7$ has three terms. $2a^2$, $-3b^3$, and 7. $2a^2$ and $-3b^3$ are variable terms. 7 is a constant term.
A variable term is composed of a *numerical coefficient* and a *variable part*. [2.1A, p. 79]	For the expression $-7x^3y^2$, -7 is the coefficient and x^3y^2 is the variable part.
In a variable expression, replacing each variable by its value and then simplifying the resulting numerical expression is called *evaluating the variable expression*. [2.1A, p. 80]	To evaluate $2ab - b^2$ when $a = 3$ and $b = -2$, replace a by 3 and b by -2 then simplify the numerical expression. $2(3)(-2) - (-2)^2 = -16$
Like terms of a variable expression are terms with the same variable part. Constant terms are like terms. [2.2A, p. 83]	For the expressions $3a^2 + 2b - 3$ and $2a^2 - 3a + 4$, $3a^2$ and $2a^2$ are like terms; -3 and 4 are like terms.
To simplify the sum of like variable terms, use the Distributive Property to add the numerical coefficients. This is called *combining like terms*. [2.2A, p. 83]	$5y + 3y = (5 + 3)y$ $= 8y$
The *additive inverse* of a number is the opposite of the number. [2.2A, p. 84]	-4 is the additive inverse of 4. $\frac{2}{3}$ is the additive inverse of $-\frac{2}{3}$. 0 is the additive inverse of 0.
The *multiplicative inverse* of a number is the *reciprocal of the number*. [2.2B, p. 85]	$\frac{3}{4}$ is the multiplicative inverse of $\frac{4}{3}$. $-\frac{1}{4}$ is the multiplicative inverse of -4.

Essential Rules and Procedures	**Examples**
The Distributive Property [2.2A, p. 83] If a, b, and c are real numbers, then $a(b + c) = ab + ac$.	$5(4 + 7) = 5 \cdot 4 + 5 \cdot 7$ $= 20 + 35 = 55$
The Associative Property of Addition [2.2A, p. 83] If a, b, and c are real numbers, then $(a + b) + c = a + (b + c)$.	$-4 + (2 + 7) = -4 + 9 = 5$ $(-4 + 2) + 7 = -2 + 7 = 5$

The Commutative Property of Addition [2.2A, p. 84]
If a and b are real numbers, then $a + b = b + a$.

$2 + 5 = 7$ and $5 + 2 = 7$

The Addition Property of Zero [2.2A, p. 84]
If a is a real number, then $a + 0 = 0 + a = a$.

$-8 + 0 = -8$ and $0 + (-8) = -8$

The Inverse Property of Addition [2.2A, p. 84]
If a is a real number, then $a + (-a) = (-a) + a = 0$.

$5 + (-5) = 0$ and $(-5) + 5 = 0$

The Associative Property of Multiplication [2.2B, p. 85]
If a, b, and c are real numbers, then $(ab)c = a(bc)$.

$-3 \cdot (5 \cdot 4) = -3(20) = -60$
$(-3 \cdot 5) \cdot 4 = -15 \cdot 4 = -60$

The Commutative Property of Multiplication [2.2B, p. 85]
If a and b are real numbers, then $ab = ba$.

$-3(7) = -21$ and $7(-3) = -21$

The Multiplication Property of One [2.2B, p. 85]
If a is a real number, then $a \cdot 1 = 1 \cdot a = a$.

$-3(1) = -3$ and $1(-3) = -3$

The Inverse Property of Multiplication [2.2B, p. 85]
If a is a real number, and a is not equal to zero, then
$a \cdot \dfrac{1}{a} = \dfrac{1}{a} \cdot a = 1$.

$-3 \cdot -\dfrac{1}{3} = 1$ and $-\dfrac{1}{3} \cdot -3 = 1$

Chapter 2 Review Exercises

1. Simplify: $3(x^2 - 8x - 7)$

2. Simplify: $7x + 4x$

3. Simplify: $6a - 4b + 2a$

4. Simplify: $(-50n)\left(\dfrac{1}{10}\right)$

5. Evaluate $(5c - 4a)^2 - b$ when $a = -1$, $b = 2$, and $c = 1$.

6. Simplify: $5(2x - 7)$

7. Simplify: $2(6y^2 + 4y - 5)$

8. Simplify: $\dfrac{1}{4}(-24a)$

9. Simplify: $-6(7x^2)$

10. Simplify: $-9(7 + 4x)$

11. Simplify: $12y - 17y$

12. Evaluate $2bc \div (a + 7)$ when $a = 3$, $b = -5$, and $c = 4$.

13. Simplify: $7 - 2(3x + 4)$

14. Simplify: $6 + 2[2 - 5(4a - 3)]$

15. Simplify: $6(8y - 3) - 8(3y - 6)$

16. Simplify: $5c + (-2d) - 3d - (-4c)$

17. Simplify: $5(4x)$

18. Simplify: $-4(2x - 9) + 5(3x + 2)$

19. Evaluate $(b - a)^2 + c$ when $a = -2$, $b = 3$, and $c = 4$.

20. Simplify: $-9r + 2s - 6s + 12s$

21. Evaluate $(2x - y)^2 + (2x + y)^2$ when $x = -2$ and $y = -3$.

22. Evaluate $b^2 - 4ac$ when $b = -4$, $a = 1$, and $c = -3$.

23. Simplify: $4x - 3x^2 + 2x - x^2$

24. Simplify: $5[2 - 3(6x - 1)]$

25. Simplify: $0.4x + 0.6(250 - x)$

26. Simplify: $\dfrac{2}{3}x - \dfrac{3}{4}x$

27. Simplify: $(7a^2 - 2a + 3)4$

28. Simplify: $18 - (4x - 2)$

29. Evaluate $a^2 - b^2$ when $a = 3$ and $b = 4$.

30. Simplify: $-3(-12y)$

31. Translate "two-thirds of the total of x and 10" into a variable expression.

32. Translate "the product of 4 and x" into a variable expression.

33. Translate "6 less than x" into a variable expression.

34. Translate "a number plus twice the number" into a variable expression. Then simplify.

35. Translate "the difference between twice a number and one-half of the number" into a variable expression. Then simplify.

36. Translate "three times a number plus the product of five and one less than the number" into a variable expression. Then simplify.

37. Sports   A baseball card collection contains five times as many National League players' cards as American League players' cards. Express the number of National League players' cards in the collection in terms of the number of American League players' cards.

38. Finance   A club treasurer has some five-dollar bills and some ten-dollar bills. The treasurer has a total of 35 bills. Express the number of five-dollar bills in terms of the number of ten-dollar bills.

39. Nutrition   A candy bar contains eight more calories than twice the number of calories in an apple. Express the number of calories in a candy bar in terms of the number of calories in an apple.

40. Architecture   The length of the Parthenon is approximately 1.6 times the width. Express the length of the Parthenon in terms of the width.

41. Anatomy   Leonardo DaVinci studied various proportions of human anatomy. One of his findings was that the standing height of a person is approximately 1.3 times the kneeling height of the same person. Represent the standing height of a person in terms of kneeling height.

Chapter 2 Test

1. Simplify: $3x - 5x + 7x$

2. Simplify: $-3(2x^2 - 7y^2)$

3. Simplify: $2x - 3(x - 2)$

4. Simplify: $2x + 3[4 - (3x - 7)]$

5. Simplify: $3x - 7y - 12x$

6. Evaluate $b^2 - 3ab$ when $a = 3$ and $b = -2$.

7. Simplify: $\frac{1}{5}(10x)$

8. Simplify: $5(2x + 4) - 3(x - 6)$

9. Simplify: $-5(2x^2 - 3x + 6)$

10. Simplify: $3x + (-12y) - 5x - (-7y)$

11. Evaluate $\frac{-2ab}{2b - a}$ when $a = -4$ and $b = 6$.

12. Simplify: $(12x)\left(\frac{1}{4}\right)$

13. Simplify: $-7y^2 + 6y^2 - (-2y^2)$

14. Simplify: $-2(2x - 4)$

15. Simplify: $\frac{2}{3}(-15a)$

16. Simplify: $-2[x - 2(x - y)] + 5y$

17.   Simplify: $(-3)(-12y)$

18.   Simplify: $5(3 - 7b)$

19.   Translate "the difference between the squares of a and b" into a variable expression.

20.   Translate "ten times the difference between a number and 3" into a variable expression. Then simplify.

21.   Translate "the sum of a number and twice the square of the number" into a variable expression.

22.   Translate "three less than the quotient of six and a number" into a variable expression.

23.   Translate "b decreased by the product of b and seven" into a variable expression.

24.   **Sports**   The speed of a pitcher's fastball is twice the speed of the catcher's return throw. Express the speed of the fastball in terms of the speed of the return throw.

25.   **Metalwork**   A wire is cut into two lengths. The length of the longer piece is 3 in. less than four times the length of the shorter piece. Express the length of the longer piece in terms of the length of the shorter piece.

Cumulative Review Exercises

1. Add: $-4 + 7 + (-10)$

2. Subtract: $-16 - (-25) - 4$

3. Multiply: $(-2)(3)(-4)$

4. Divide: $(-60) \div 12$

5. Find the complement of a $37°$ angle.

6. Simplify: $\dfrac{7}{12} - \dfrac{11}{16} - \left(-\dfrac{1}{3}\right)$

7. Simplify: $-\dfrac{5}{12} \div \dfrac{5}{2}$

8. Simplify: $\left(-\dfrac{9}{16}\right) \cdot \left(\dfrac{8}{27}\right) \cdot \left(-\dfrac{3}{2}\right)$

9. Write $\dfrac{3}{4}$ as a percent.

10. Simplify: $-2^5 \div (3 - 5)^2 - (-3)$

11. Simplify: $\left(-\dfrac{3}{4}\right)^2 \div \left(\dfrac{3}{8} - \dfrac{11}{12}\right)$

12. Evaluate $a^2 - 3b$ when $a = 2$ and $b = -4$.

13. Simplify: $-2x^2 - (-3x^2) + 4x^2$

14. Simplify: $5a - 10b - 12a$

15. Find the area of a circle whose radius is 7 cm. Use 3.14 for π.

16. Find the perimeter of a square whose side measures 24 ft.

17. Simplify: $3(8 - 2x)$

18. Simplify: $-2(-3y + 9)$

19. Write $37\frac{1}{2}\%$ as a fraction.

20. Write 1.05% as a decimal.

21. Simplify: $-4(2x^2 - 3y^2)$

22. Simplify: $-3(3y^2 - 3y - 7)$

23. Simplify: $-3x - 2(2x - 7)$

24. Simplify: $4(3x - 2) - 7(x + 5)$

25. Simplify: $2x + 3[x - 2(4 - 2x)]$

26. Simplify: $3[2x - 3(x - 2y)] + 3y$

27. Translate "the sum of one-half of b and b" into a variable expression.

28. Translate "10 divided by the difference between y and 2" into a variable expression.

29. Translate "the difference between eight and the quotient of a number and twelve" into a variable expression.

30. Translate "the sum of a number and two more than the number" into a variable expression. Then simplify.

31. **Sports**   A softball diamond is a square with each side measuring 60 ft. Find the area enclosed by the sides of the softball diamond.

32. **Telecommunications**   The speed of a DSL (Digital Subscriber Line) Internet connection is ten times faster than that of a dial-up connection. Express the speed of the DSL connection in terms of the speed of the dial-up connection.

chapter

3 Solving Equations

Suppose that this bus is traveling the posted speed limit of 45 mph and is making a 180-mile drive between two cities. How long will the bus ride last? To figure out the answer, you will need to use the formula $d = rt$, where d is the distance traveled, r is the rate of speed, and t is the time spent traveling. In this case, d is 180, and r is 45. When you solve for t, you get the correct answer of 4 hours. The formula $d = rt$ is also used to solve **Exercises 52 and 53 on page 179,** which are more advanced motion problems involving bus travel.

Need help? For online student resources, such as section quizzes, visit this textbook's website at **math.college.hmco.com/students.**

OBJECTIVES

Section 3.1

A To determine whether a given number is a solution of an equation

B To solve an equation of the form $x + a = b$

C To solve an equation of the form $ax = b$

D To solve application problems using the basic percent equation

E To solve uniform motion problems

Section 3.2

A To solve an equation of the form $ax + b = c$

B To solve application problems using formulas

Section 3.3

A To solve an equation of the form $ax + b = cx + d$

B To solve an equation containing parentheses

C To solve application problems using formulas

Section 3.4

A To solve integer problems

B To translate a sentence into an equation and solve

Section 3.5

A To solve problems involving angles

B To solve problems involving the angles of a triangle

Section 3.6

A To solve value mixture problems

B To solve percent mixture problems

C To solve uniform motion problems

PREP TEST • • •

Do these exercises to prepare for Chapter 3.

1. Write $\frac{9}{100}$ as a decimal.

2. Write $\frac{3}{4}$ as a percent.

3. Evaluate $3x^2 - 4x - 1$ when $x = -4$.

4. Simplify: $R - 0.35R$

5. Simplify: $\frac{1}{2}x + \frac{2}{3}x$

6. Simplify: $6x - 3(6 - x)$

7. Simplify: $0.22(3x + 6) + x$

8. Translate into a variable expression: "The difference between 5 and twice a number."

9. A new graphics card for computer games is five times faster than a graphics card made two years ago. Express the speed of the new card in terms of the speed of the old card.

10. A board 5 ft long is cut into two pieces. If x represents the length of the longer piece, write an expression for the shorter piece in terms of x.

GO FIGURE • • •

How can a donut be cut into 8 equal pieces with three cuts of a knife?

3.1 Introduction to Equations

Objective A   **To determine whether a given number is a solution of an equation**

An **equation** expresses the equality of two mathematical expressions. The expressions can be either numerical or variable expressions.

$$\left. \begin{array}{l} 9 + 3 = 12 \\ 3x - 2 = 10 \\ y^2 + 4 = 2y - 1 \\ z = 2 \end{array} \right\} \text{Equations}$$

The equation at the right is true if the variable is replaced by 5.

$x + 8 = 13$
$5 + 8 = 13$ A true equation

The equation is false if the variable is replaced by 7.

$7 + 8 = 13$ A false equation

A **solution of an equation** is a number that, when substituted for the variable, results in a true equation. 5 is a solution of the equation $x + 8 = 13$. 7 is not a solution of the equation $x + 8 = 13$.

HOW TO   Is -2 a solution of $2x + 5 = x^2 - 3$?

$$\begin{array}{c|c} \multicolumn{2}{c}{2x + 5 = x^2 - 3} \\ \hline 2(-2) + 5 & (-2)^2 - 3 \\ -4 + 5 & 4 - 3 \\ 1 & = 1 \end{array}$$

- Replace x by **−2**.
- Evaluate the numerical expressions.
- If the results are equal, **−2** is a solution of the equation. If the results are not equal, **−2** is not a solution of the equation.

Yes, -2 is a solution of the equation.

TAKE NOTE

The Order of Operations Agreement applies to evaluating $2(-2) + 5$ and $(-2)^2 - 3$.

Example 1   Is -4 a solution of
$5x - 2 = 6x + 2$?

Solution

$$\begin{array}{c|c} \multicolumn{2}{c}{5x - 2 = 6x + 2} \\ \hline 5(-4) - 2 & 6(-4) + 2 \\ -20 - 2 & -24 + 2 \\ -22 & = -22 \end{array}$$

Yes, -4 is a solution.

You Try It 1   Is $\frac{1}{4}$ a solution of
$5 - 4x = 8x + 2$?

Your solution

Example 2   Is -4 a solution of
$4 + 5x = x^2 - 2x$?

Solution

$$\begin{array}{c|c} \multicolumn{2}{c}{4 + 5x = x^2 - 2x} \\ \hline 4 + 5(-4) & (-4)^2 - 2(-4) \\ 4 + (-20) & 16 - (-8) \\ -16 & \neq 24 \end{array}$$

($\neq$ means "is not equal to")

No, -4 is not a solution.

You Try It 2   Is 5 a solution of
$10x - x^2 = 3x - 10$?

Your solution

Solutions on p. S5

Objective B   **To solve an equation of the form** $x + a = b$

Study Tip

To learn mathematics, you must be an active participant. Listening and watching your professor do mathematics is not enough. Take notes in class, mentally think through every question your instructor asks, and try to answer it even if you are not called on to answer it verbally. Ask questions when you have them. See *AIM for Success*, page xxv, for other ways to be an active learner.

To **solve an equation** means to find a solution of the equation. The simplest equation to solve is an equation of the form *variable = constant*, because the constant is the solution.

The solution of the equation $x = 5$ is 5 because $5 = 5$ is a true equation.

The solution of the equation at the right is 7 because $7 + 2 = 9$ is a true equation.

$$x + 2 = 9 \qquad\qquad 7 + 2 = 9$$

Note that if 4 is added to each side of the equation $x + 2 = 9$, the solution is still 7.

$$\begin{aligned} x + 2 &= 9 \\ x + 2 + 4 &= 9 + 4 \\ x + 6 &= 13 \qquad\qquad 7 + 6 = 13 \end{aligned}$$

If -5 is added to each side of the equation $x + 2 = 9$, the solution is still 7.

$$\begin{aligned} x + 2 &= 9 \\ x + 2 + (-5) &= 9 + (-5) \\ x - 3 &= 4 \qquad\qquad 7 - 3 = 4 \end{aligned}$$

Equations that have the same solution are **equivalent equations.** The equations $x + 2 = 9$, $x + 6 = 13$, and $x - 3 = 4$ are equivalent equations; each equation has 7 as its solution. These examples suggest that adding the same number to each side of an equation produces an equivalent equation. This is called the *Addition Property of Equations.*

Addition Property of Equations

The same number can be added to each side of an equation without changing its solution. In symbols, the equation $a = b$ has the same solution as the equation $a + c = b + c$.

In solving an equation, the goal is to rewrite the given equation in the form *variable = constant*. The Addition Property of Equations is used to remove a *term* from one side of the equation by adding the opposite of that term to each side of the equation.

HOW TO   Solve: $x - 4 = 2$

$$x - 4 = 2$$ • The goal is to rewrite the equation as *variable = constant*.

$$x - 4 + 4 = 2 + 4$$ • **Add 4** to each side of the equation.

$$x + 0 = 6$$ • Simplify.

$$x = 6$$ • The equation is in the form *variable = constant*.

Check: $\dfrac{x - 4 = 2}{6 - 4 \ | \ 2}$
$$\qquad\qquad 2 = 2 \qquad \text{A true equation}$$

The solution is 6.

Because subtraction is defined in terms of addition, the Addition Property of Equations also makes it possible to subtract the same number from each side of an equation without changing the solution of the equation.

HOW TO Solve: $y + \dfrac{3}{4} = \dfrac{1}{2}$

$$y + \frac{3}{4} = \frac{1}{2}$$

- The goal is to rewrite the equation in the form *variable = constant*.

$$y + \frac{3}{4} - \frac{3}{4} = \frac{1}{2} - \frac{3}{4}$$

- Subtract $\dfrac{3}{4}$ from each side of the equation.

$$y + 0 = \frac{2}{4} - \frac{3}{4}$$

- Simplify.

$$y = -\frac{1}{4}$$

- The equation is in the form *variable = constant*.

The solution is $-\dfrac{1}{4}$. You should check this solution.

Example 3   Solve: $x + \dfrac{2}{5} = \dfrac{1}{3}$

Solution

$$x + \frac{2}{5} = \frac{1}{3}$$

$$x + \frac{2}{5} - \frac{2}{5} = \frac{1}{3} - \frac{2}{5}$$ • Subtract $\dfrac{2}{5}$ from each side.

$$x + 0 = \frac{5}{15} - \frac{6}{15}$$

$$x = -\frac{1}{15}$$

The solution is $-\dfrac{1}{15}$.

You Try It 3   Solve: $\dfrac{5}{6} = y - \dfrac{3}{8}$

Your solution

Solution on p. S5

Objective C   **To solve an equation of the form *ax = b***

The solution of the equation at the right is 3 because $2 \cdot 3 = 6$ is a true equation.

$$2x = 6 \qquad\qquad 2 \cdot 3 = 6$$

Note that if each side of $2x = 6$ is multiplied by 5, the solution is still 3.

$$2x = 6$$
$$5(2x) = 5 \cdot 6$$
$$10x = 30 \qquad\qquad 10 \cdot 3 = 30$$

If each side of $2x = 6$ is multiplied by -4, the solution is still 3.

$$2x = 6$$
$$(-4)(2x) = (-4)6$$
$$-8x = -24 \qquad\qquad -8 \cdot 3 = -24$$

The equations $2x = 6$, $10x = 30$, and $-8x = -24$ are equivalent equations; each equation has 3 as its solution. These examples suggest that multiplying each side of an equation by the same nonzero number produces an equivalent equation.

> **Multiplication Property of Equations**
>
> Each side of an equation can be multiplied by the same *nonzero* number without changing the solution of the equation. In symbols, if $c \neq 0$, then the equation $a = b$ has the same solutions as the equation $ac = bc$.

The Multiplication Property of Equations is used to remove a coefficient by multiplying each side of the equation by the reciprocal of the coefficient.

HOW TO Solve: $\dfrac{3}{4}z = 9$

$$\dfrac{3}{4}z = 9$$ • The goal is to rewrite the equation in the form *variable = constant*.

$$\dfrac{4}{3} \cdot \dfrac{3}{4}z = \dfrac{4}{3} \cdot 9$$ • Multiply each side of the equation by $\dfrac{4}{3}$.

$$1 \cdot z = 12$$ • Simplify.

$$z = 12$$ • The equation is in the form *variable = constant*.

The solution is 12. You should check this solution.

Because division is defined in terms of multiplication, each side of an equation can be divided by the same nonzero number without changing the solution of the equation.

HOW TO Solve: $6x = 14$

$$6x = 14$$ • The goal is to rewrite the equation in the form *variable = constant*.

$$\dfrac{6x}{6} = \dfrac{14}{6}$$ • Divide each side of the equation by **6**.

$$x = \dfrac{7}{3}$$ • Simplify. The equation is in the form *variable = constant*.

The solution is $\dfrac{7}{3}$.

When using the Multiplication Property of Equations, multiply each side of the equation by the reciprocal of the coefficient when the coefficient is a fraction. Divide each side of the equation by the coefficient when the coefficient is an integer or a decimal.

Example 4 Solve: $\dfrac{3x}{4} = -9$

Solution
$$\dfrac{3x}{4} = -9$$
$$\dfrac{4}{3} \cdot \dfrac{3}{4}x = \dfrac{4}{3}(-9) \qquad \bullet \ \dfrac{3x}{4} = \dfrac{3}{4}x$$
$$x = -12$$

The solution is -12.

You Try It 4 Solve: $-\dfrac{2x}{5} = 6$

Your solution

Example 5 Solve: $5x - 9x = 12$

Solution
$$5x - 9x = 12$$
$$-4x = 12 \qquad \bullet \text{ Combine like terms.}$$
$$\dfrac{-4x}{-4} = \dfrac{12}{-4}$$
$$x = -3$$

The solution is -3.

You Try It 5 Solve: $4x - 8x = 16$

Your solution

Solutions on p. S5

Objective D   **To solve application problems**
using the basic percent equation

An equation that is used frequently in mathematics applications is the basic percent equation.

> **Basic Percent Equation**
>
> Percent · Base = Amount
>
> P · B = A

In many application problems involving percent, the base follows the word *of*.

HOW TO   20% of what number is 30?

$P \cdot B = A$ • Use the basic percent equation.

$0.20B = 30$ • $P = 20\% = 0.20$, $A = 30$, and B is unknown.

$\dfrac{0.20B}{0.20} = \dfrac{30}{0.20}$ • Solve for B.

$B = 150$

The number is 150.

TAKE NOTE

We have written
$P(80) = 70$ because that
is the form of the basic
percent equation. We
could have written
$80P = 70$. The important
point is that each side of
the equation is divided
by 80, the coefficient
of P.

HOW TO   70 is what percent of 80?

$P \cdot B = A$ • Use the basic percent equation.

$P(80) = 70$ • $B = 80$, $A = 70$, and P is unknown.

$\dfrac{P(80)}{80} = \dfrac{70}{80}$ • Solve for P.

$P = 0.875$ • The question asked for a percent.

$P = 87.5\%$ Convert the decimal to a percent.

70 is 87.5% of 80.

HOW TO   The world's production of cocoa for a recent year was 2928 metric tons. Of this, 1969 metric tons came from Africa. (*Source: World Cocoa Foundation*) What percent of the world's cocoa production came from Africa? Round to the nearest tenth of a percent.

Strategy   To find the percent, use the basic percent equation. $B = 2928$, $A = 1969$, P is unknown.

Solution $P \cdot B = A$

$P(2928) = 1969$

$P = \dfrac{1969}{2928} \approx 0.672$

Approximately 67.2% of the world's cocoa production comes from Africa.

The simple interest that an investment earns is given by the **simple interest equation** $I = Prt$, where I is the simple interest, P is the principal, or amount invested, r is the simple interest rate, and t is the time.

> **HOW TO** A $1500 investment has an annual simple interest rate of 7%. Find the simple interest earned on the investment after 18 months.
>
> The time is given in months but the interest rate is an annual rate. Therefore, we must convert 18 months to years.
>
> $$18 \text{ months} = \frac{18}{12} \text{ years} = 1.5 \text{ years}$$
>
> To find the interest, solve $I = Prt$ for I.
>
> $I = Prt$
> $I = 1500(0.07)(1.5)$ • $P = 1500, r = 0.07, t = 1.5$
> $I = 157.5$
>
> The investment earned $157.50.

Point of Interest

In the jewelry industry, the amount of gold in a piece of jewelry is measured by the *karat*. Pure gold is 24 karats. A necklace that is 18 karats is $\frac{18}{24} = 0.75 = 75\%$ gold.

The amount of a substance in a solution can be given as a percent of the total solution. For instance, if a certain fruit juice drink is advertised as containing 27% cranberry juice, then 27% of the contents of the bottle must be cranberry juice.

Solving problems involving mixtures is based on the **percent mixture equation** $Q = Ar$, where Q is the quantity of a substance in the solution, A is the amount of the solution, and r is the percent concentration of the substance.

> **HOW TO** Part of the formula for a perfume requires that the concentration of jasmine be 1.2% of the total amount of perfume. How many ounces of jasmine are in a 2-ounce bottle of this perfume?
>
> The amount of perfume is 2 oz. Therefore, $A = 2$. The percent concentration is 1.2% so $r = 0.012$. To find the number of ounces of jasmine, solve $Q = Ar$ for Q.
>
> $Q = Ar$
> $Q = 2(0.012)$ • $A = 2, r = 0.012$
> $Q = 0.024$
>
> There is 0.024 ounce of jasmine in the perfume.

In most cases, you should write the percent as a decimal before solving the basic percent equation. However, some percents are more easily written as a fraction. For example,

$$33\frac{1}{3}\% = \frac{1}{3} \qquad 66\frac{2}{3}\% = \frac{2}{3} \qquad 16\frac{2}{3}\% = \frac{1}{6} \qquad 83\frac{1}{3}\% = \frac{5}{6}$$

Example 6

12 is $33\frac{1}{3}$% of what number?

Solution

$P \cdot B = A$ • Use the basic percent equation.

$\frac{1}{3}B = 12$ • $33\frac{1}{3}\% = \frac{1}{3}$

$3 \cdot \frac{1}{3}B = 3 \cdot 12$

$B = 36$

12 is $33\frac{1}{3}$% of 36.

You Try It 6

18 is $16\frac{2}{3}$% of what number?

Your solution

Example 7

The data in the table below shows the number of households (in millions) that downloaded music files for a three-month period in a recent year. (*Source:* NPD Group)

Month	April	May	June
Downloads	14.5	12.7	10.4

For the three-month period, what percent of the files were downloaded in May? Round to the nearest percent.

Strategy

To find the percent,

• Find the total number of files downloaded for the three-month period.
• Use the basic percent equation. B is the total number of files downloaded for the three-month period; $A = 12.7$, the number of files downloaded in May; P is unknown.

Solution

$14.5 + 12.7 + 10.4 = 37.6$

$P \cdot B = A$ • Use the basic percent equation

$P(37.6) = 12.7$ • $B = 37.6$, $A = 12.7$

$P = \frac{12.7}{37.6} \approx 0.34$

Approximately 34% of the files were downloaded in May.

You Try It 7

The Bowl Championship Series (BCS) received approximately $83.3 million in revenues from various college football bowl games. Of this amount, the college representing the Pac-10 in the Rose Bowl received approximately $3.1 million. (*Source:* BCSfootball.org) What percent of the total received by the BCS did the college representing the Pac-10 receive? Round to the nearest tenth of a percent.

Your strategy

Your solution

Solutions on p. S5

Example 8

In April, Marshall Wardell was charged an interest fee of $8.72 on an unpaid credit card balance of $545. Find the annual interest rate on this credit card.

Strategy

The interest is $8.72. Therefore, $I = 8.72$. The unpaid balance is $545. This is the principal on which interest is calculated. Therefore, $P = 545$. The time is 1 month. Because the *annual* interest rate must be found and the time is given as 1 month, we write 1 month as $\frac{1}{12}$ year so $t = \frac{1}{12}$. To find the interest rate, solve $I = Prt$ for r.

Solution

$$I = Prt$$

· Use the simple interest equation.

$$8.72 = 545r\left(\frac{1}{12}\right)$$

· $I = 8.72, P = 545, t = \frac{1}{12}$.

$$8.72 = \frac{545}{12}r$$

$$\frac{12}{545}(8.72) = \frac{12}{545}\left(\frac{545}{12}r\right)$$

$$0.192 = r$$

The annual interest rate is 19.2%.

You Try It 8

Clarissa Adams purchased a municipal bond for $1000 that earns an annual simple interest rate of 6.4%. How much must she deposit into a bank account that earns 8% annual simple interest so that the interest earned from each account after one year is the same?

Your strategy

Your solution

Example 9

To make a certain color of blue, 4 oz of cyan must be contained in 1 gal of paint. What is the percent concentration of cyan in the paint?

Strategy

The cyan is given in ounces and the amount of paint is given in gallons. We must convert ounces to gallons or gallons to ounces. For this problem, we will convert gallons to ounces: 1 gal = 128 oz. Solve $Q = Ar$ for r with $Q = 4$ and $A = 128$.

Solution

$$Q = Ar$$

· Use the percent mixture equation.

$$4 = 128r$$

· $Q = 4, A = 128$.

$$\frac{4}{128} = \frac{128r}{128}$$

$$0.03125 = r$$

The percent concentration of cyan is 3.125%.

You Try It 9

The concentration of sugar in a certain breakfast cereal is 25%. If there are 2 oz of sugar contained in the cereal in a bowl, how many ounces of cereal are in the bowl?

Your strategy

Your solution

Solutions on p. S6

Objective E **To solve uniform motion problems**

Any object that travels at a constant speed in a straight line is said to be in *uniform motion*. **Uniform motion** means that the speed and direction of an object do not change. For instance, a car traveling at a constant speed of 45 mph on a straight road is in uniform motion.

The solution of a uniform motion problem is based on the **uniform motion equation** $d = rt$, where d is the distance traveled, r is the rate of travel, and t is the time spent traveling. For instance, suppose a car travels at 50 mph for 3 h. Because the rate (50 mph) and time (3 h) are known, we can find the distance traveled by solving the equation $d = rt$ for d.

$$d = rt$$
$$d = 50(3) \qquad \bullet \ r = 50, t = 3.$$
$$d = 150$$

The car travels a distance of 150 mi.

HOW TO A jogger runs 3 mi in 45 min. What is the rate of the jogger in miles per hour?

Strategy
- Because the answer must be in miles per *hour* and the given time is in *minutes*, convert 45 min to hours.

- To find the rate of the jogger, solve the equation $d = rt$ for r.

Solution

$$45 \text{ min} = \frac{45}{60} \text{ h} = \frac{3}{4} \text{ h}$$

$$d = rt$$

$$3 = r\left(\frac{3}{4}\right) \qquad \bullet \ d = 3, t = \frac{3}{4}.$$

$$3 = \frac{3}{4}r$$

$$\left(\frac{4}{3}\right)3 = \left(\frac{4}{3}\right)\frac{3}{4}r \qquad \bullet \ \text{Multiply each side of the equation by the \textbf{reciprocal of} } \frac{3}{4}.$$

$$4 = r$$

The rate of the jogger is 4 mph.

If two objects are moving in opposite directions, then the rate at which the distance between them is increasing is the sum of the speeds of the two objects. For instance, in the diagram below, two cars start from the same point and travel in opposite directions. The distance between them is changing at 70 mph.

30 mph   40 mph

30 + 40 = 70 mph

Similarly, if two objects are moving toward each other, the distance between them is decreasing at a rate that is equal to the sum of the speeds. The rate at which the two planes at the right are approaching one another is 800 mph.

HOW TO   Two cars start from the same point and move in opposite directions. The car moving west is traveling 45 mph, and the car moving east is traveling 60 mph. In how many hours will the cars be 210 mi apart?

Strategy   The distance is 210 mi. Therefore, $d = 210$. The cars are moving in opposite directions so the rate at which the distance between them is changing is the sum of the rates of each of the cars. The rate is 45 mph + 60 mph = 105 mph. Therefore, $r = 105$. To find the time, solve the equation $d = rt$ for t.

Solution

$$d = rt$$
$$210 = 105t \qquad \bullet \; d = 210, r = 105.$$
$$\frac{210}{105} = \frac{105t}{105} \qquad \bullet \; \text{Solve for } t.$$
$$2 = t$$

In 2 h, the cars will be 210 mi apart.

If a motorboat is on a river that is flowing at a rate of 4 mph, then the boat will float down the river at a speed of 4 mph when the motor is not on. Now suppose the motor is turned on and the power adjusted so that the boat would travel 10 mph without the aid of the current. Then, if the boat is moving with the current, its effective speed is the speed of the boat using power plus of the speed of the current: 10 mph + 4 mph = 14 mph. (See the figure below.)

However, if the boat is moving against the current, the current slows the boat down, and the effective speed of the boat is the speed of the boat using power minus the speed of the current: 10 mph − 4 mph = 6 mph. (See the figure below.)

There are other situations in which the preceding concepts may be applied.

HOW TO   An airline passenger is walking between two airline terminals and decides to get on a moving sidewalk that is 150 ft long. If the passenger walks at a rate of 7 ft/s and the moving sidewalk moves at a rate of 9 ft/s, how long, in seconds, will it take the passenger to walk from one end of the moving sidewalk to the other? Round to the nearest thousandth.

Strategy   The distance is 150 ft. Therefore, $d = 150$. The passenger is traveling at 7 ft/s and the moving sidewalk is traveling at 9 ft/s. The rate of the passenger is the sum of the two rates, or 16 ft/s. Therefore, $r = 16$. To find the time, solve the equation $d = rt$ for t.

Solution

$$d = rt$$
$$150 = 16t \qquad \bullet \; d = 150, r = 16.$$
$$\frac{150}{16} = \frac{16t}{16} \qquad \bullet \; \text{Solve for } t.$$
$$9.375 = t$$

It will take 9.375 s for the passenger to travel the length of the moving sidewalk.

Example 10

Two cyclists start at the same time at opposite ends of an 80-mile course. One cyclist is traveling 18 mph, and the second cyclist is traveling 14 mph. How long after they begin will they meet?

Strategy
The distance is 80 mi. Therefore, $d = 80$. The cyclists are moving toward each other so the rate at which the distance between them is changing is the sum of the rates of each of the cyclists. The rate is 18 mph + 14 mph = 32 mph. Therefore, $r = 32$. To find the time, solve the equation $d = rt$ for t.

Solution

$$d = rt$$
$$80 = 32t \qquad \bullet \; d = 80, r = 32.$$
$$\frac{80}{32} = \frac{32t}{32} \qquad \bullet \; \text{Solve for } t.$$
$$2.5 = t$$

The cyclists will meet in 2.5 h.

You Try It 10

A plane that can normally travel at 250 mph in calm air is flying into a headwind of 25 mph. How far can the plane fly in 3 h?

Your strategy

Your solution

Solution on p. S6

3.1 Exercises

To determine whether a given number is a solution of an equation

1. What is the difference between an equation and an expression?

2. Explain how to determine whether a given number is a solution of an equation.

3. Is 4 a solution of $2x = 8$?

4. Is 3 a solution of $y + 4 = 7$?

5. Is -1 a solution of $2b - 1 = 3$?

6. Is -2 a solution of $3a - 4 = 10$?

7. Is 1 a solution of $4 - 2m = 3$?

8. Is 2 a solution of $7 - 3n = 2$?

9. Is 5 a solution of $2x + 5 = 3x$?

10. Is 4 a solution of $3y - 4 = 2y$?

11. Is -2 a solution of $3a + 2 = 2 - a$?

12. Is 3 a solution of $z^2 + 1 = 4 + 3z$?

13. Is 2 a solution of $2x^2 - 1 = 4x - 1$?

14. Is -1 a solution of $y^2 - 1 = 4y + 3$?

15. Is 4 a solution of $x(x + 1) = x^2 + 5$?

16. Is 3 a solution of $2a(a - 1) = 3a + 3$?

17. Is $-\frac{1}{4}$ a solution of $8t + 1 = -1$?

18. Is $\frac{1}{2}$ a solution of $4y + 1 = 3$?

19. Is $\frac{2}{5}$ a solution of $5m + 1 = 10m - 3$?

20. Is $\frac{3}{4}$ a solution of $8x - 1 = 12x + 3$?

To solve an equation of the form $x + a = b$

21. Can 0 ever be the solution of an equation? If so, give an example of an equation for which 0 is a solution.

22. Without solving $x + \frac{13}{15} = -\frac{21}{43}$, determine whether x is less than or greater than $-\frac{21}{43}$. Explain your answer.

For Exercises 23 to 64, solve and check.

23. $x + 5 = 7$

24. $y + 3 = 9$

25. $b - 4 = 11$

26. $z - 6 = 10$

27. $2 + a = 8$

28. $5 + x = 12$

29. $n - 5 = -2$

30. $x - 6 = -5$

31. $b + 7 = 7$

32. $y - 5 = -5$

33. $z + 9 = 2$

34. $n + 11 = 1$

35. $10 + m = 3$

36. $8 + x = 5$

37. $9 + x = -3$

38. $10 + y = -4$

39. $2 = x + 7$

40. $-8 = n + 1$

41. $4 = m - 11$

42. $-6 = y - 5$

43. $12 = 3 + w$

44. $-9 = 5 + x$

45. $4 = -10 + b$

46. $-7 = -2 + x$

47. $m + \dfrac{2}{3} = -\dfrac{1}{3}$

48. $c + \dfrac{3}{4} = -\dfrac{1}{4}$

49. $x - \dfrac{1}{2} = \dfrac{1}{2}$

50. $x - \dfrac{2}{5} = \dfrac{3}{5}$

51. $\dfrac{5}{8} + y = \dfrac{1}{8}$

52. $\dfrac{4}{9} + a = -\dfrac{2}{9}$

53. $m + \dfrac{1}{2} = -\dfrac{1}{4}$

54. $b + \dfrac{1}{6} = -\dfrac{1}{3}$

55. $x + \dfrac{2}{3} = \dfrac{3}{4}$

56. $n + \dfrac{2}{5} = \dfrac{2}{3}$

57. $-\dfrac{5}{6} = x - \dfrac{1}{4}$

58. $-\dfrac{1}{4} = c - \dfrac{2}{3}$

59. $d + 1.3619 = 2.0148$

60. $w + 2.932 = 4.801$

61. $-0.813 + x = -1.096$

62. $-1.926 + t = -1.042$

63. $6.149 = -3.108 + z$

64. $5.237 = -2.014 + x$

Objective C   **To solve an equation of the form $ax = b$**

65. Without solving $-\dfrac{15}{41}x = -\dfrac{23}{25}$, determine whether x is less than or greater than 0. Explain your answer.

66. Explain why multiplying each side of an equation by the reciprocal of the coefficient of the variable is the same as dividing each side of the equation by the coefficient.

For Exercises 67 to 110, solve and check.

67. $5x = -15$

68. $4y = -28$

69. $3b = 0$

70. $2a = 0$

71. $-3x = 6$

72. $-5m = 20$

73. $-3x = -27$

74. $-\dfrac{1}{6}n = -30$

75. $20 = \dfrac{1}{4}c$ **76.** $18 = 2t$ **77.** $0 = -5x$ **78.** $0 = -8a$

79. $49 = -7t$ **80.** $\dfrac{x}{3} = 2$ **81.** $\dfrac{x}{4} = 3$ **82.** $-\dfrac{y}{2} = 5$

83. $-\dfrac{b}{3} = 6$ **84.** $\dfrac{3}{4}y = 9$ **85.** $\dfrac{2}{5}x = 6$ **86.** $-\dfrac{2}{3}d = 8$

87. $-\dfrac{3}{5}m = 12$ **88.** $\dfrac{2n}{3} = 0$ **89.** $\dfrac{5x}{6} = 0$ **90.** $\dfrac{-3z}{8} = 9$

91. $\dfrac{3x}{4} = 2$ **92.** $\dfrac{3}{4}c = \dfrac{3}{5}$ **93.** $\dfrac{2}{9} = \dfrac{2}{3}y$ **94.** $-\dfrac{6}{7} = -\dfrac{3}{4}b$

95. $\dfrac{1}{5}x = -\dfrac{1}{10}$ **96.** $-\dfrac{2}{3}y = -\dfrac{8}{9}$ **97.** $-1 = \dfrac{2n}{3}$ **98.** $-\dfrac{3}{4} = \dfrac{a}{8}$

99. $-\dfrac{2}{5}m = -\dfrac{6}{7}$ **100.** $5x + 2x = 14$ **101.** $3n + 2n = 20$

102. $7d - 4d = 9$ **103.** $10y - 3y = 21$ **104.** $2x - 5x = 9$

105. $\dfrac{x}{1.46} = 3.25$ **106.** $\dfrac{z}{2.95} = -7.88$ **107.** $3.47a = 7.1482$

108. $2.31m = 2.4255$ **109.** $-3.7x = 7.881$ **110.** $\dfrac{n}{2.65} = 9.08$

Objective D   **To solve application problems using the basic percent equation**

111. Without solving an equation, indicate whether 40% of 80 is less than, equal to, or greater than 80% of 40.

112. Without solving an equation, indicate whether $\frac{1}{4}$% of 80 is less than, equal to, or greater than 25% of 80.

113. What is 35% of 80?

114. What percent of 8 is 0.5?

115. Find 1.2% of 60.

116. 8 is what percent of 5?

117. 125% of what is 80?

118. What percent of 20 is 30?

119. 12 is what percent of 50?

120. What percent of 125 is 50?

121. Find 18% of 40.

122. What is 25% of 60?

123. 12% of what is 48?

124. 45% of what is 9?

125. What is $33\frac{1}{3}$% of 27?

126. Find $16\frac{2}{3}$% of 30.

127. What percent of 12 is 3?

128. 10 is what percent of 15?

129. 12 is what percent of 6?

130. 20 is what percent of 16?

131. $5\frac{1}{4}$% of what is 21?

132. $37\frac{1}{2}$% of what is 15?

133. Find 15.4% of 50.

134. What is 18.5% of 46?

135. 1 is 0.5% of what?

136. 3 is 1.5% of what?

137. $\frac{3}{4}$% of what is 3?

138. $\frac{1}{2}$% of what is 3?

139. What is 250% of 12?

140. 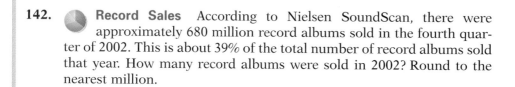 **Education** The stacked-bar graph at the right shows the number of people age 25 or older in the U.S. that have attained some type of degree beyond high school.

    a. In 2002, there were approximately 182.7 million people age 25 or older. What percent of the people age 25 or older had received an associate degree or a bachelor's degree in 2002? Round to the nearest tenth of a percent.

    b. In 2000, there were approximately 177.5 million people age 25 or older. Was the percent of people in 2000 with a graduate degree less than or greater than the percent of people in 2002 with a graduate degree?

141. Chemistry Approximately 21% of air is oxygen. Using this estimate, determine how many liters of oxygen there are in a room containing 21,600 L of air.

142. **Record Sales** According to Nielsen SoundScan, there were approximately 680 million record albums sold in the fourth quarter of 2002. This is about 39% of the total number of record albums sold that year. How many record albums were sold in 2002? Round to the nearest million.

143. **Income** According to the U.S. Census Bureau, the median income fell 1.1% between two successive years. If the median income before the decline was $42,900, what was the median income the next year? Round to the nearest dollar.

144. Government To override a presidential veto, at least $66\frac{2}{3}\%$ of the Senate must vote to override the veto. There are 100 senators in the Senate. What is the minimum number of votes needed to override a veto?

145. **Sports** According to **www.superbowl.com,** approximately 138.9 million people watched Super Bowl XXXVIII. What percent of the U.S. population watched Super Bowl XXXVIII? Use a figure of 290 million for the U.S. population. Round to the nearest tenth of a percent.

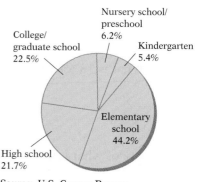

146. **Advertising** Suppose 9.4 million people watch a 30-second commercial for a new cellular phone during a broadcast of the TV show *CSI*. The cost of that commercial was approximately $470,000. If the cellular phone manufacturer makes a profit of $10 on every phone sold, what percent of the people watching the commercial would have to buy one phone for the company to recover the cost of the commercial? *Source:* Nielsen Media Research/ San Diego Union

147. **School Enrollment** The circle graph at the right shows the percent of the U.S. population over three years old that are enrolled in school. To answer the question "How many people are enrolled in college or graduate school?," what additional piece of information is necessary?

Source: U.S. Census Bureau

148. **Investment** If Kachina Caron invested $1200 in a simple interest account and earned $72 in 8 months, what is the annual interest rate?

149. **Investment** How much money must Andrea invest for two years in an account that earns an annual interest rate of 8% if she wants to earn $300 from the investment?

150. **Investment** Sal Boxer decided to divide a gift of $3000 into two different accounts. He placed $1000 in one account that earns an annual simple interest rate of 7.5%. The remaining money was placed in an account that earns an annual simple interest rate of 8.25%. How much interest will Sal earn from the two accounts after one year?

151. **Investment** If Americo invests $2500 at an 8% annual simple interest rate and Octavia invests $3000 at a 7% annual simple interest rate, which of the two will earn the greater amount of interest after one year?

152. **Investment** Makana invested $900 in a simple interest account that had an interest rate that was 1% more than that of her friend Marlys. If Marlys earned $51 after one year from an investment of $850, how much would Makana earn in one year?

153. **Investment** A $2000 investment at an annual simple interest rate of 6% earned as much interest after one year as another investment in an account that earns 8% simple interest. How much was invested at 8%?

154. **Investment** An investor placed $1000 in an account that earns 9% annual simple interest and $1000 in an account that earns 6% annual simple interest. If each investment is left in the account for the same period of time, is the interest rate on the combined investment less than 6%, between 6% and 9%, or greater than 9%?

155. **Metallurgy** The concentration of platinum in a necklace is 15%. If the necklace weighs 12 g, find the amount of platinum in the necklace.

156. **Dye Mixtures** A 250-milliliter solution of a fabric dye contains 5 ml of hydrogen peroxide. What is the percent concentration of the hydrogen peroxide?

157. **Fabric Mixtures** A carpet is made with a blend of wool and other fibers. If the concentration of wool in the carpet is 75% and the carpet weighs 175 lb, how much wool is in the carpet?

158. **Juice Mixtures** Apple Dan's 32-ounce apple-flavored fruit drink contains 8 oz of apple juice. A 40-ounce generic brand of an apple-flavored fruit drink contains 9 oz of apple juice. Which of the two brands has the greater concentration of apple juice?

159. Food Mixtures Bakers use simple syrup in many of their recipes. Simple syrup is made by combining 500 g of sugar with 500 g of water and mixing it well until the sugar dissolves. What is the percent concentration of sugar in the simple syrup?

160. Pharmacology A pharmacist has 50 g of a topical cream that contains 75% glycerine. How many grams of the cream is not glycerine?

161. Chemistry A chemist has 100 ml of a solution that is 9% acetic acid. If the chemist adds 50 ml of pure water to this solution, what is the percent concentration of the resulting mixture?

162. Chemistry A 500-gram salt and water solution contains 50 g of salt. This mixture is left in the open air and 100 g of water evaporates from the solution. What is the percent concentration of salt in the remaining solution?

Objective E **To solve uniform motion problems**

163. As part of the training program for the Boston Marathon, a runner wants to build endurance by running at a rate of 9 mph for 20 min. How far will the runner travel in that time period?

164. It takes a hospital dietician 40 min to drive from home to the hospital, a distance of 20 mi. What is the dietician's average rate of speed?

165. Marcella leaves home at 9:00 A.M. and drives to school, arriving at 9:45 A.M. If the distance between home and school is 27 mi, what is Marcella's average rate of speed?

166. The Ride for Health Bicycle Club has chosen a 36-mile course for this Saturday's ride. If the riders plan on averaging 12 mph while they are riding, and they have a 1-hour lunch break planned, how long will it take them to complete the trip?

167. Palmer's average running speed is 3 kilometers per hour faster than his walking speed. If Palmer can run around a 30-kilometer course in 2 h, how many hours would it take for Palmer to walk the same course?

168. A shopping mall has a moving sidewalk that takes shoppers from the shopping area to the parking garage, a distance of 250 ft. If your normal walking rate is 5 ft/s and the moving sidewalk is traveling at 3 ft/s, how many seconds would it take for you to walk from one end of the moving sidewalk to the other end?

169. Two joggers start at the same time from opposite ends of an 8-mile jogging trail and begin running toward each other. One jogger is running at the rate of 5 mph, and the other jogger is running at a rate of 7 mph. How long, in minutes, after they start will the two joggers meet?

170. Two cyclists start from the same point at the same time and move in opposite directions. One cyclist is traveling at 8 mph, and the other cyclist is traveling at 9 mph. After 30 min, how far apart are the two cyclists?

171. Petra and Celine can paddle their canoe at a rate of 10 mph in calm water. How long will it take them to travel 4 mi against the 2 mph current of the river?

172. At 8:00 A.M., a train leaves a station and travels at a rate of 45 mph. At 9:00 A.M., a second train leaves the same station on the same track and travels in the direction of the first train at a speed of 60 mph. At 10:00 A.M., how far apart are the two trains?

APPLYING THE CONCEPTS

173. Geometry Solve for x.

174. Geometry Solve for x.

175. Geometry Solve for x.

176. Geometry Solve for x.

177. a. Make up an equation of the form $x + a = b$ that has 2 as a solution.
 b. Make up an equation of the form $ax = b$ that has -1 as a solution.

178. Write out the steps for solving the equation $\frac{1}{2}x = -3$. Identify each Property of Real Numbers or Property of Equations as you use it.

179. In your own words, state the Addition Property of Equations and the Multiplication Property of Equations.

180. If a quantity increases by 100%, how many times its original value is the new value?

3.2 General Equations — Part I

Objective A   **To solve an equation of the form $ax + b = c$**

In solving an equation of the form $ax + b = c$, the goal is to rewrite the equation in the form *variable = constant*. This requires the application of both the Addition and the Multiplication Properties of Equations.

HOW TO   Solve: $\dfrac{3}{4}x - 2 = -11$

The goal is to write the equation in the form *variable = constant*.

$$\frac{3}{4}x - 2 = -11$$

$$\frac{3}{4}x - 2 + 2 = -11 + 2 \qquad \bullet \text{ Add 2 to each side of the equation.}$$

$$\frac{3}{4}x = -9 \qquad \bullet \text{ Simplify.}$$

$$\frac{4}{3} \cdot \frac{3}{4}x = \frac{4}{3}(-9) \qquad \bullet \text{ Multiply each side of the equation by } \frac{4}{3}.$$

$$x = -12 \qquad \bullet \text{ The equation is in the form } variable = constant.$$

The solution is -12.

TAKE NOTE

Check: $\dfrac{3}{4}x - 2 = -11$

$\dfrac{3}{4}(-12) - 2 \,\big|\, -11$

$-9 - 2 \,\big|\, -11$

$-11 = -11$

A true equation

Here is an example of solving an equation that contains more than one fraction.

HOW TO   Solve: $\dfrac{2}{3}x + \dfrac{1}{2} = \dfrac{3}{4}$

$$\frac{2}{3}x + \frac{1}{2} = \frac{3}{4}$$

$$\frac{2}{3}x + \frac{1}{2} - \frac{1}{2} = \frac{3}{4} - \frac{1}{2} \qquad \bullet \text{ Subtract } \frac{1}{2} \text{ from each side of the equation.}$$

$$\frac{2}{3}x = \frac{1}{4} \qquad \bullet \text{ Simplify.}$$

$$\frac{3}{2}\left(\frac{2}{3}x\right) = \frac{3}{2}\left(\frac{1}{4}\right) \qquad \bullet \text{ Multiply each side of the equation by } \frac{3}{2},$$

$$\text{the reciprocal of } \frac{2}{3}.$$

$$x = \frac{3}{8}$$

The solution is $\dfrac{3}{8}$.

It may be easier to solve an equation containing two or more fractions by multiplying each side of the equation by the least common multiple (LCM) of the denominators. For the equation above, the LCM of 3, 2, and 4 is 12. The LCM has the property that 3, 2, and 4 will divide evenly into it. Therefore, if both sides of the equation are multiplied by 12, the denominators will divide evenly into 12. The result is an equation that does not contain any fractions. Multiplying each side of an equation that contains fractions by the LCM of the denominators is called **clearing denominators.** It is an alternative method, as we show in the next example, of solving an equation that contains fractions.

HOW TO   Solve: $\dfrac{2}{3}x + \dfrac{1}{2} = \dfrac{3}{4}$

$$\dfrac{2}{3}x + \dfrac{1}{2} = \dfrac{3}{4}$$

$$12\left(\dfrac{2}{3}x + \dfrac{1}{2}\right) = 12\left(\dfrac{3}{4}\right)$$ • Multiply each side of the equation by **12**, the LCM of 3, 2, and 4.

$$12\left(\dfrac{2}{3}x\right) + 12\left(\dfrac{1}{2}\right) = 12\left(\dfrac{3}{4}\right)$$ • Use the Distributive Property.

$$8x + 6 = 9$$ • Simplify.

$$8x + 6 - 6 = 9 - 6$$ • **Subtract 6** from each side of the equation.

$$8x = 3$$

$$\dfrac{8x}{8} = \dfrac{3}{8}$$ • **Divide** each side of the equation **by 8**.

$$x = \dfrac{3}{8}$$

The solution is $\dfrac{3}{8}$.

Note that both methods give exactly the same solution. You may use either method to solve an equation containing fractions.

Example 1   Solve: $3x - 7 = -5$

Solution
$$\begin{aligned}
3x - 7 &= -5 \\
3x - 7 + 7 &= -5 + 7 \qquad \text{• Add 7 to each side.} \\
3x &= 2 \\
\dfrac{3x}{3} &= \dfrac{2}{3} \qquad \text{• Divide each side by 3.} \\
x &= \dfrac{2}{3}
\end{aligned}$$

The solution is $\dfrac{2}{3}$.

You Try It 1   Solve: $5x + 7 = 10$

Your solution

Example 2   Solve: $5 = 9 - 2x$

Solution
$$\begin{aligned}
5 &= 9 - 2x \\
5 - 9 &= 9 - 9 - 2x \qquad \text{• Subtract 9 from each side.} \\
-4 &= -2x \\
\dfrac{-4}{-2} &= \dfrac{-2x}{-2} \qquad \text{• Divide each side by −2.} \\
2 &= x
\end{aligned}$$

The solution is 2.

You Try It 2   Solve: $2 = 11 + 3x$

Your solution

Solutions on p. S6

Example 3 Solve: $\dfrac{2}{3} - \dfrac{x}{2} = \dfrac{3}{4}$

You Try It 3 Solve: $\dfrac{5}{8} - \dfrac{2x}{3} = \dfrac{5}{4}$

Solution

$$\dfrac{2}{3} - \dfrac{x}{2} = \dfrac{3}{4}$$

$$\dfrac{2}{3} - \dfrac{2}{3} - \dfrac{x}{2} = \dfrac{3}{4} - \dfrac{2}{3}$$ • Subtract $\dfrac{2}{3}$ from each side.

$$-\dfrac{x}{2} = \dfrac{1}{12}$$

$$-2\left(-\dfrac{x}{2}\right) = -2\left(\dfrac{1}{12}\right)$$ • Multiply each side by −2.

$$x = -\dfrac{1}{6}$$

The solution is $-\dfrac{1}{6}$.

Your solution

Example 4 Solve $\dfrac{4}{5}x - \dfrac{1}{2} = \dfrac{3}{4}$ by first clearing denominators.

You Try It 4 Solve $\dfrac{2}{3}x + 3 = \dfrac{7}{2}$ by first clearing denominators.

Solution
The LCM of 5, 2, and 4 is 20.

$$\dfrac{4}{5}x - \dfrac{1}{2} = \dfrac{3}{4}$$

$$20\left(\dfrac{4}{5}x - \dfrac{1}{2}\right) = 20\left(\dfrac{3}{4}\right)$$ • Multiply each side by 20.

$$20\left(\dfrac{4}{5}x\right) - 20\left(\dfrac{1}{2}\right) = 20\left(\dfrac{3}{4}\right)$$ • Use the Distributive Property.

$$16x - 10 = 15$$

$$16x - 10 + 10 = 15 + 10$$ • Add 10 to each side.

$$16x = 25$$

$$\dfrac{16x}{16} = \dfrac{25}{16}$$ • Divide each side by 16.

$$x = \dfrac{25}{16}$$

The solution is $\dfrac{25}{16}$.

Your solution

Solutions on p. S6

Example 5

Solve: $2x + 4 - 5x = 10$

Solution

$$2x + 4 - 5x = 10$$
$$-3x + 4 = 10 \qquad \bullet \text{ Combine like terms.}$$
$$-3x + 4 - 4 = 10 - 4$$
$$-3x = 6$$
$$\frac{-3x}{-3} = \frac{6}{-3}$$
$$x = -2$$

The solution is -2.

You Try It 5

Solve: $x - 5 + 4x = 25$

Your solution

Solution on p. S7

Objective B   **To solve application problems using formulas**

In this objective we will be solving application problems using formulas. Two of the formulas we will use are related to markup and discount.

Markup

Cost

Selling price

Cost is the price a business pays for a product. **Selling price** is the price for which a business sells a product to a customer. The difference between selling price and cost is called **markup.** Markup is added to the cost to cover the expenses of operating a business. The diagram at the left illustrates these terms. The total length is the selling price. One part of the diagram is the cost, and the other part is the markup.

When the markup is expressed as a percent of the retailer's cost, it is called the **markup rate.**

The **basic markup equations** used by a business are

Selling price = cost + markup   Markup = markup rate · cost

$$S \quad = C + M \qquad\qquad M = r \cdot C$$

Substituting $r \cdot C$ for M in the first equation results in $S = C + (r \cdot C)$, or $S = C + rC$.

HOW TO   The manager of a clothing store buys a jacket for $80 and sells the jacket for $116. Find the markup rate.

$$S = C + rC \qquad\qquad \bullet \text{ Use the equation } S = C + rC.$$

$$116 = 80 + 80r \qquad\quad \bullet \text{ Given: } C = \$80 \text{ and } S = \$116$$

$$36 = 80r \qquad\qquad\quad \bullet \text{ Subtract 80 from each side of the equation.}$$

$$\frac{36}{80} = \frac{80r}{80} \qquad\qquad \bullet \text{ Divide both sides of the equation by 80.}$$

$$0.45 = r$$

The markup rate is 45%.

Discount or markdown

Sale price

Regular price

A retailer may reduce the regular price of a product because the goods are damaged, odd sizes, or discontinued items. The **discount,** or **markdown,** is the amount by which a retailer reduces the regular price of a product. The percent discount is called the **discount rate** and is usually expressed as a percent of the original selling price (the regular price).

The **basic discount equations** used by a business are

$$\begin{array}{ccc} \text{Sale} \\ \text{price} \end{array} = \begin{array}{c} \text{regular} \\ \text{price} \end{array} - \text{discount} \qquad\qquad \text{Discount} = \begin{array}{c} \text{discount} \\ \text{rate} \end{array} \cdot \begin{array}{c} \text{regular} \\ \text{price} \end{array}$$

$$S \;=\; R \;-\; D \qquad\qquad\qquad D \;=\; r \;\cdot\; R$$

Substituting $r \cdot R$ for D in the first equation results in $S = R - (r \cdot R)$, or $S = R - rR$.

HOW TO A portable computer that regularly sells for \$1850 is on sale for \$1480. Find the discount rate.

$$S = R - rR$$ • Use the equation $S = R - rR$.

$$1480 = 1850 - 1850r$$ • Given: $S = \$1480$ and $R = \$1850$

$$-370 = -1850r$$ • Subtract 1850 from each side of the equation.

$$\frac{-370}{-1850} = \frac{-1850r}{-1850}$$ • Divide each side of the equation by -1850.

$$0.2 = r$$

The discount rate on the portable computer is 20%.

Example 6

A markup rate of 40% was used on a refrigerator that has a selling price of \$749. Find the cost of the refrigerator. Use the formula $S = C + rC$.

Strategy

Given: $S = \$749$
$\qquad\quad r = 40\% = 0.40$
Unknown: C

Solution

$$S = C + rC$$
$$749 = C + 0.40C \qquad • \; C + 0.40C = 1C + 0.40C$$
$$749 = 1.40C \qquad\quad • \; \text{Combine like terms.}$$
$$\frac{749}{1.40} = \frac{1.40C}{1.40}$$
$$535 = C$$

The cost of the refrigerator is \$535.

You Try It 6

A markup rate of 45% was used on an outboard motor that has a selling price of \$986. Find the cost of the outboard motor. Use the formula $S = C + rC$.

Your strategy

Your solution

Solution on p. S7

Example 7

A necklace that is marked down 35% has a sale price of $292.50. Find the regular price of the necklace. Use the formula $S = R - rR$.

Strategy

Given: $S = 292.50$
        $r = 35\% = 0.35$
Unknown: R

Solution

$$S = R - rR$$
$$292.50 = R - 0.35R \quad \bullet \ R - 0.35R = 1R - 0.35R$$
$$292.50 = 0.65R \quad \bullet \ \text{Combine like terms.}$$
$$\frac{292.50}{0.65} = \frac{0.65R}{0.65}$$
$$450 = R$$

The regular price of the necklace is $450.

You Try It 7

A garage door opener, marked down 25%, is on sale for $159. Find the regular price of the garage door opener. Use the formula $S = R - rR$.

Your strategy

Your solution

Example 8

To determine the total cost of production, an economist uses the equation $T = U \cdot N + F$, where T is the total cost, U is the unit cost, N is the number of units made, and F is the fixed cost. Use this equation to find the number of units made during a month when the total cost was $9000, the unit cost was $25, and the fixed cost was $3000.

Strategy

Given: $T = \$9000$
        $U = \$25$
        $F = \$3000$
Unknown: N

Solution

$$T = U \cdot N + F$$
$$9000 = 25N + 3000$$
$$6000 = 25N$$
$$\frac{6000}{25} = \frac{25N}{25}$$
$$240 = N$$

There were 240 units made.

You Try It 8

The pressure at a certain depth in the ocean can be approximated by the equation

$P = 15 + \frac{1}{2}D$, where P is the pressure in

pounds per square inch and D is the depth in feet. Use this equation to find the depth when the pressure is 45 pounds per square inch.

Your strategy

Your solution

Solutions on p. S7

3.2 Exercises

To solve an equation of the form $ax + b = c$

For Exercises 1 to 80, solve and check.

1. $3x + 1 = 10$ **2.** $4y + 3 = 11$ **3.** $2a - 5 = 7$ **4.** $5m - 6 = 9$

5. $5 = 4x + 9$ **6.** $2 = 5b + 12$ **7.** $2x - 5 = -11$ **8.** $3n - 7 = -19$

9. $4 - 3w = -2$ **10.** $5 - 6x = -13$ **11.** $8 - 3t = 2$ **12.** $12 - 5x = 7$

13. $4a - 20 = 0$ **14.** $3y - 9 = 0$ **15.** $6 + 2b = 0$ **16.** $10 + 5m = 0$

17. $-2x + 5 = -7$ **18.** $-5d + 3 = -12$ **19.** $-12x + 30 = -6$ **20.** $-13 = -11y + 9$

21. $2 = 7 - 5a$ **22.** $3 = 11 - 4n$ **23.** $-35 = -6b + 1$ **24.** $-8x + 3 = -29$

25. $-3m - 21 = 0$ **26.** $-5x - 30 = 0$ **27.** $-4y + 15 = 15$ **28.** $-3x + 19 = 19$

29. $9 - 4x = 6$ **30.** $3t - 2 = 0$ **31.** $9x - 4 = 0$ **32.** $7 - 8z = 0$

33. $1 - 3x = 0$ **34.** $9d + 10 = 7$ **35.** $12w + 11 = 5$ **36.** $6y - 5 = -7$

37. $8b - 3 = -9$ **38.** $5 - 6m = 2$ **39.** $7 - 9a = 4$ **40.** $9 = -12c + 5$

41. $10 = -18x + 7$ **42.** $2y + \dfrac{1}{3} = \dfrac{7}{3}$ **43.** $4a + \dfrac{3}{4} = \dfrac{19}{4}$ **44.** $2n - \dfrac{3}{4} = \dfrac{13}{4}$

45. $3x - \dfrac{5}{6} = \dfrac{13}{6}$ **46.** $5y + \dfrac{3}{7} = \dfrac{3}{7}$ **47.** $9x + \dfrac{4}{5} = \dfrac{4}{5}$ **48.** $8 = 7d - 1$

49. $8 = 10x - 5$ **50.** $4 = 7 - 2w$ **51.** $7 = 9 - 5a$ **52.** $8t + 13 = 3$

53. $12x + 19 = 3$ **54.** $-6y + 5 = 13$ **55.** $-4x + 3 = 9$ **56.** $\dfrac{1}{2}a - 3 = 1$

57. $\dfrac{1}{3}m - 1 = 5$ **58.** $\dfrac{2}{5}y + 4 = 6$ **59.** $\dfrac{3}{4}n + 7 = 13$ **60.** $-\dfrac{2}{3}x + 1 = 7$

61. $-\dfrac{3}{8}b + 4 = 10$ **62.** $\dfrac{x}{4} - 6 = 1$ **63.** $\dfrac{y}{5} - 2 = 3$ **64.** $\dfrac{2x}{3} - 1 = 5$

65. $\dfrac{2}{3}x - \dfrac{5}{6} = -\dfrac{1}{3}$ **66.** $\dfrac{5}{4}x + \dfrac{2}{3} = \dfrac{1}{4}$ **67.** $\dfrac{1}{2} - \dfrac{2}{3}x = \dfrac{1}{4}$ **68.** $\dfrac{3}{4} - \dfrac{3}{5}x = \dfrac{19}{20}$

69. $\dfrac{3}{2} = \dfrac{5}{6} + \dfrac{3x}{8}$ **70.** $-\dfrac{1}{4} = \dfrac{5}{12} + \dfrac{5x}{6}$ **71.** $\dfrac{11}{27} = \dfrac{4}{9} - \dfrac{2x}{3}$ **72.** $\dfrac{37}{24} = \dfrac{7}{8} - \dfrac{5x}{6}$

73. $7 = \dfrac{2x}{5} + 4$ **74.** $5 - \dfrac{4c}{7} = 8$ **75.** $7 - \dfrac{5}{9}y = 9$ **76.** $6a + 3 + 2a = 11$

77. $5y + 9 + 2y = 23$ **78.** $7x - 4 - 2x = 6$ **79.** $11z - 3 - 7z = 9$ **80.** $2x - 6x + 1 = 9$

81. Solve $3x + 4y = 13$ when $y = -2$.   **82.** Solve $2x - 3y = 8$ when $y = 0$.

83. Solve $-4x + 3y = 9$ when $x = 0$.   **84.** Solve $5x - 2y = -3$ when $x = -3$.

85. If $2x - 3 = 7$, evaluate $3x + 4$.

86. If $3x + 5 = -4$, evaluate $2x - 5$.

87. If $4 - 5x = -1$, evaluate $x^2 - 3x + 1$.

88. If $2 - 3x = 11$, evaluate $x^2 + 2x - 3$.

89. If $5x + 3 - 2x = 12$, evaluate $4 - 5x$.

90. If $2x - 4 - 7x = 16$, evaluate $x^2 + 1$.

Objective B **To solve application problems using formulas**

Business   For Exercises 91 to 98, solve. Use the markup equation $S = C + rC$, where S is selling price, C is cost, and r is the markup rate.

91. A watch costing $98 is sold for $156.80. Find the markup rate on the watch.

92. A set of golf clubs costing $360 is sold for $630. Find the markup rate on the set of golf clubs.

93. A markup rate of 40% was used on a basketball with a selling price of $82.60. Find the cost of the basketball.

94. A portable tape player with a selling price of $57 has a markup rate of 50%. Find the cost of the tape player.

95. A freezer costing $360 is sold for $520. Find the markup rate. Round to the nearest tenth of a percent.

96. A sofa costing $320 is sold for $479. Find the markup rate. Round to the nearest tenth of a percent.

97. A digitally recorded compact disc has a selling price of $11.90. The markup rate is 40%. Find the cost of the CD.

98. A markup rate of 25% is used on a computer that has a selling price of $2187.50. Find the cost of the computer.

Business   For Exercises 99 to 106, solve. Use the discount equation $S = R - rR$, where S is the sale price, R is the regular price, and r is the discount rate.

99.   An oak bedroom set with a regular price of $1295 is on sale for $995. Find the discount rate. Round to the nearest tenth of a percent.

100.   A stereo system with a regular price of $495 is on sale for $395. Find the markdown rate. Round to the nearest tenth of a percent.

101.   A mechanic's tool set is on sale for $180 after a markdown of 40% off the regular price. Find the regular price.

102.   A battery with a discount price of $65 is on sale for 22% off the regular price. Find the regular price. Round to the nearest cent.

103.   A compact disc player with a regular price of $325 is on sale for $201.50. Find the markdown rate.

104.   A luggage set with a regular price of $178 is on sale for $103.24. Find the discount rate.

105.   A telescope is on sale for $165 after a markdown of 40% off the regular price. Find the regular price.

106.   An exercise bike is on sale for $390, having been marked down 25% off the regular price. Find the regular price.

Physics   The distance s, in feet, that an object will fall in t seconds is given by $s = 16t^2 + vt$, where v is the initial velocity of the object in feet per second. Use this equation for Exercises 107 and 108.

107.   Find the initial velocity of an object that falls 80 ft in 2 s.

108.   Find the initial velocity of an object that falls 144 ft in 3 s.

Depreciation   A company uses the equation $V = C - 6000t$ to determine the depreciated value V, after t years, of a milling machine that originally cost C dollars. Equations like this are used in accounting for straight-line depreciation. Use this equation for Exercises 109 and 110.

109.   A milling machine originally cost $50,000. In how many years will the depreciated value of the machine be $38,000?

110. A milling machine originally cost $78,000. In how many years will the depreciated value of the machine be $48,000?

Anthropology   Anthropologists approximate the height of a primate by the size of its humerus (the bone from the elbow to the shoulder) using the equation $H = 1.2L + 27.8$, where L is the length of the humerus and H is the height, in inches, of the primate. Use this equation for Exercises 111 and 112.

111. An anthropologist estimates the height of a primate to be 66 in. What is the approximate length of the humerus of this primate? Round to the nearest tenth of an inch.

112. An anthropologist estimates the height of a primate to be 62 in. What is the approximate length of the humerus of this primate?

Car Safety   Black ice is an ice covering on roads that is especially difficult to see and therefore extremely dangerous for motorists. The distance that a car traveling 30 mph will slide after its brakes are applied is related to the outside temperature by the formula $C = \frac{1}{4}D - 45$, where C is the Celsius temperature and D is the distance in feet that the car will slide. Use this equation for Exercises 113 and 114.

113. Determine the distance a car will slide on black ice when the outside temperature is $-3°C$.

114. Determine the distance a car will slide on black ice when the outside temperature is $-11°C$.

Telecommunications   A telephone company estimates that the number N of phone calls per day between two cities of populations P_1 and P_2 that are d miles apart is given by the equation $N = \frac{2.51 P_1 P_2}{d^2}$. Use this equation for Exercises 115 and 116.

115. Estimate the population (P_1) of a city, given that the population of a second city (P_2) is 48,000, the number of phone calls per day between the two cities is 1,100,000, and the distance between the cities is 75 mi. Round to the nearest thousand.

116. Estimate the population (P_1) of a city given that the population of a second city (P_2) is 125,000, the number of phone calls per day between the two cities is 2,500,000, and the distance between the cities is 50 mi. Round to the nearest thousand.

APPLYING THE CONCEPTS

117. If $2x + 1 = a$ and $3x - 2 = a$, find a.   **118.** If $1 - 4x = y$ and $2x - 5 = y$, find y.

119. **Business**   The sale price of a television was $180. Find the regular price if the sale price was computed by taking one-third off the regular price, followed by an additional 15% discount on the reduced price.

120. **Business**   A customer buys four tires, three at the regular price and one for 20% off the regular price. The four tires cost $323. What was the regular price of a tire?

121. **Geometry**   The area of the triangle at the right is 40 m². Find x.

122. **Geometry**   The area of the parallelogram at the right is 364 m². Find h.

123. Solve: $x \div 15 = 25$ remainder 10

124. Does the sentence "Solve $3x - 4(x - 1)$" make sense? Why or why not?

125. Explain what is wrong with the following demonstration, which suggests that $2 = 3$.

$$2x + 5 = 3x + 5$$
$$2x + 5 - 5 = 3x + 5 - 5 \qquad \bullet \text{ Subtract 5 from each side of the equation.}$$
$$2x = 3x$$
$$2 = 3 \qquad \bullet \text{ Divide each side of the equation by } x.$$

126. The following problem does not contain enough information for us to find only one solution. Supply some additional information so that the problem has exactly one solution. Then write and solve an equation.

The sum of two numbers is 15. Find the numbers.

127. The following problem does not contain enough information. What additional information is needed to answer the question?

How many hours does it take to fly from Los Angeles to New York?

3.3 General Equations—Part II

Objective A   **To solve an equation of the form *ax* + *b* = *cx* + *d***

In solving an equation of the form $ax + b = cx + d$, the goal is to rewrite the equation in the form *variable = constant*. Begin by rewriting the equation so that there is only one variable term in the equation. Then rewrite the equation so that there is only one constant term.

HOW TO   Solve: $2x + 3 = 5x - 9$

$$2x + 3 = 5x - 9$$

$$2x - 5x + 3 = 5x - 5x - 9$$ • **Subtract 5x** from each side of the equation.

$$-3x + 3 = -9$$ • **Simplify.** There is only one variable term.

$$-3x + 3 - 3 = -9 - 3$$ • **Subtract 3** from each side of the equation.

$$-3x = -12$$ • **Simplify.** There is only one constant term.

$$\frac{-3x}{-3} = \frac{-12}{-3}$$ • **Divide** each side of the equation by **−3**.

$$x = 4$$ • The equation is in the form *variable = constant*.

The solution is 4. You should verify this by checking this solution.

Example 1   Solve: $4x - 5 = 8x - 7$

Solution

$$4x - 5 = 8x - 7$$

$$4x - 8x - 5 = 8x - 8x - 7$$ • **Subtract 8x from** each side.

$$-4x - 5 = -7$$

$$-4x - 5 + 5 = -7 + 5$$ • **Add 5 to each side.**

$$-4x = -2$$

$$\frac{-4x}{-4} = \frac{-2}{-4}$$ • **Divide each side by −4.**

$$x = \frac{1}{2}$$

The solution is $\frac{1}{2}$.

You Try It 1   Solve: $5x + 4 = 6 + 10x$

Your solution

Solution on p. S7

Example 2   Solve: $3x + 4 - 5x = 2 - 4x$

You Try It 2   Solve: $5x - 10 - 3x = 6 - 4x$

Solution

$$3x + 4 - 5x = 2 - 4x$$

$$-2x + 4 = 2 - 4x$$ • Combine like terms.

$$-2x + 4x + 4 = 2 - 4x + 4x$$ • Add **4x** to each side.

$$2x + 4 = 2$$

$$2x + 4 - 4 = 2 - 4$$ • Subtract **4** from each side.

$$2x = -2$$

$$\frac{2x}{2} = \frac{-2}{2}$$ • Divide each side by **2**.

$$x = -1$$

The solution is -1.

Your solution

Solution on p. S7

Objective B   **To solve an equation containing parentheses**

When an equation contains parentheses, one of the steps in solving the equation requires the use of the Distributive Property. The Distributive Property is used to remove parentheses from a variable expression.

HOW TO   Solve: $4 + 5(2x - 3) = 3(4x - 1)$

$$4 + 5(2x - 3) = 3(4x - 1)$$

$$4 + 10x - 15 = 12x - 3$$ • Use the Distributive Property. Then simplify.

$$10x - 11 = 12x - 3$$

$$10x - 12x - 11 = 12x - 12x - 3$$ • Subtract **12x** from each side of the equation.

$$-2x - 11 = -3$$ • Simplify.

$$-2x - 11 + 11 = -3 + 11$$ • Add **11** to each side of the equation.

$$-2x = 8$$ • Simplify.

$$\frac{-2x}{-2} = \frac{8}{-2}$$ • Divide each side of the equation by **−2**.

$$x = -4$$ • The equation is in the form *variable = constant*.

The solution is -4. You should verify this by checking this solution.

In the next example, we solve an equation with parentheses and decimals.

HOW TO Solve: $16 + 0.55x = 0.75(x + 20)$

$$16 + 0.55x = 0.75(x + 20)$$
$$16 + 0.55x = 0.75x + 15$$ • Use the Distributive Property.
$$16 + 0.55x - 0.75x = 0.75x - 0.75x + 15$$ • Subtract **0.75x** from each side of the equation.

$$16 - 0.20x = 15$$ • Simplify.
$$16 - 16 - 0.20x = 15 - 16$$ • Subtract **16** from each side of the equation.

$$-0.20x = -1$$ • Simplify.
$$\frac{-0.20x}{-0.20} = \frac{-1}{-0.20}$$ • Divide each side of the equation by **−0.20**.

$$x = 5$$ • The equation is in the form *variable = constant*.

The solution is 5.

Example 3

Solve: $3x - 4(2 - x) = 3(x - 2) - 4$

Solution

$$3x - 4(2 - x) = 3(x - 2) - 4$$
$$3x - 8 + 4x = 3x - 6 - 4 \quad \bullet \text{ Distributive Property}$$
$$7x - 8 = 3x - 10$$
$$7x - 3x - 8 = 3x - 3x - 10 \quad \bullet \text{ Subtract 3x.}$$
$$4x - 8 = -10$$
$$4x - 8 + 8 = -10 + 8 \quad \bullet \text{ Add 8.}$$
$$4x = -2$$
$$\frac{4x}{4} = \frac{-2}{4} \quad \bullet \text{ Divide by 4.}$$
$$x = -\frac{1}{2}$$

The solution is $-\frac{1}{2}$.

You Try It 3

Solve: $5x - 4(3 - 2x) = 2(3x - 2) + 6$

Your solution

Example 4

Solve: $3[2 - 4(2x - 1)] = 4x - 10$

Solution

$$3[2 - 4(2x - 1)] = 4x - 10$$
$$3[2 - 8x + 4] = 4x - 10 \quad \bullet \text{ Distributive Property}$$
$$3[6 - 8x] = 4x - 10$$
$$18 - 24x = 4x - 10 \quad \bullet \text{ Distributive Property}$$
$$18 - 24x - 4x = 4x - 4x - 10 \quad \bullet \text{ Subtract 4x.}$$
$$18 - 28x = -10$$
$$18 - 18 - 28x = -10 - 18 \quad \bullet \text{ Subtract 18.}$$
$$-28x = -28$$
$$\frac{-28x}{-28} = \frac{-28}{-28} \quad \bullet \text{ Divide by −28.}$$
$$x = 1$$

The solution is 1.

You Try It 4

Solve: $-2[3x - 5(2x - 3)] = 3x - 8$

Your solution

Solutions on p. S7

Objective C   **To solve application problems using formulas**

TAKE NOTE

This system balances because

$$F_1x = F_2(d - x)$$
$$60(6) = 90(10 - 6)$$
$$60(6) = 90(4)$$
$$360 = 360$$

A lever system is shown at the right. It consists of a lever, or bar; a fulcrum; and two forces, F_1 and F_2. The distance d represents the length of the lever, x represents the distance from F_1 to the fulcrum, and $d - x$ represents the distance from F_2 to the fulcrum.

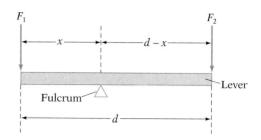

A principle of physics states that when the lever system balances, $F_1x = F_2(d - x)$.

Example 5

A lever is 15 ft long. A force of 50 lb is applied to one end of the lever, and a force of 100 lb is applied to the other end. Where is the fulcrum located when the system balances?

Strategy
Make a drawing.

Given: $F_1 = 50$
$F_2 = 100$
$d = 15$
Unknown: x

Solution
$$F_1x = F_2(d - x)$$
$$50x = 100(15 - x)$$
$$50x = 1500 - 100x$$
$$50x + 100x = 1500 - 100x + 100x$$
$$150x = 1500$$
$$\frac{150x}{150} = \frac{1500}{150}$$
$$x = 10$$

The fulcrum is 10 ft from the 50-pound force.

You Try It 5

A lever is 25 ft long. A force of 45 lb is applied to one end of the lever, and a force of 80 lb is applied to the other end. Where is the location of the fulcrum when the system balances?

Your strategy

Your solution

Solution on p. S8

3.3 Exercises

Objective A To solve an equation of the form $ax + b = cx + d$

For Exercises 1 to 27, solve and check.

1. $8x + 5 = 4x + 13$

2. $6y + 2 = y + 17$

3. $5x - 4 = 2x + 5$

4. $13b - 1 = 4b - 19$

5. $15x - 2 = 4x - 13$

6. $7a - 5 = 2a - 20$

7. $3x + 1 = 11 - 2x$

8. $n - 2 = 6 - 3n$

9. $2x - 3 = -11 - 2x$

10. $4y - 2 = -16 - 3y$

11. $2b + 3 = 5b + 12$

12. $m + 4 = 3m + 8$

13. $4y - 8 = y - 8$

14. $5a + 7 = 2a + 7$

15. $6 - 5x = 8 - 3x$

16. $10 - 4n = 16 - n$

17. $5 + 7x = 11 + 9x$

18. $3 - 2y = 15 + 4y$

19. $2x - 4 = 6x$

20. $2b - 10 = 7b$

21. $8m = 3m + 20$

22. $9y = 5y + 16$

23. $8b + 5 = 5b + 7$

24. $6y - 1 = 2y + 2$

25. $7x - 8 = x - 3$

26. $2y - 7 = -1 - 2y$

27. $2m - 1 = -6m + 5$

28. If $5x = 3x - 8$, evaluate $4x + 2$.

29. If $7x + 3 = 5x - 7$, evaluate $3x - 2$.

30. If $2 - 6a = 5 - 3a$, evaluate $4a^2 - 2a + 1$.

31. If $1 - 5c = 4 - 4c$, evaluate $3c^2 - 4c + 2$.

32. If $2y + 3 = 5 - 4y$, evaluate $6y - 7$.

33. If $3z + 1 = 1 - 5z$, evaluate $3z^2 - 7z + 8$.

Objective B   **To solve an equation containing parentheses**

For Exercises 34 to 54, solve and check.

34. $5x + 2(x + 1) = 23$

35. $6y + 2(2y + 3) = 16$

36. $9n - 3(2n - 1) = 15$

37. $12x - 2(4x - 6) = 28$

38. $7a - (3a - 4) = 12$

39. $9m - 4(2m - 3) = 11$

40. $5(3 - 2y) + 4y = 3$

41. $4(1 - 3x) + 7x = 9$

42. $5y - 3 = 7 + 4(y - 2)$

43. $0.22(x + 6) = 0.2x + 1.8$

44. $0.05(4 - x) + 0.1x = 0.32$

45. $0.3x + 0.3(x + 10) = 300$

46. $2a - 5 = 4(3a + 1) - 2$

47. $5 - (9 - 6x) = 2x - 2$

48. $7 - (5 - 8x) = 4x + 3$

49. $3[2 - 4(y - 1)] = 3(2y + 8)$

50. $5[2 - (2x - 4)] = 2(5 - 3x)$

51. $3a + 2[2 + 3(a - 1)] = 2(3a + 4)$

52. $5 + 3[1 + 2(2x - 3)] = 6(x + 5)$

53. $-2[4 - (3b + 2)] = 5 - 2(3b + 6)$

54. $-4[x - 2(2x - 3)] + 1 = 2x - 3$

55. If $4 - 3a = 7 - 2(2a + 5)$, evaluate $a^2 + 7a$.

56. If $9 - 5x = 12 - (6x + 7)$, evaluate $x^2 - 3x - 2$.

57. If $2z - 5 = 3(4z + 5)$, evaluate $\dfrac{z^2}{z - 2}$.

58. If $3n - 7 = 5(2n + 7)$, evaluate $\dfrac{n^2}{2n - 6}$.

Objective C   **To solve application problems using formulas**

Physics   For Exercises 59 to 65, solve. Use the lever system equation $F_1x = F_2(d - x)$.

59. A lever 10 ft long is used to move a 100-pound rock. The fulcrum is placed 2 ft from the rock. What force must be applied to the other end of the lever to move the rock?

60. An adult and a child are on a seesaw 14 ft long. The adult weighs 175 lb and the child weighs 70 lb. How many feet from the child must the fulcrum be placed so that the seesaw balances?

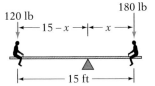

61. Two people are sitting 15 ft apart on a seesaw. One person weighs 180 lb. The second person weighs 120 lb. How far from the 180-pound person should the fulcrum be placed so that the seesaw balances?

62. Two children are sitting on a seesaw that is 12 ft long. One child weighs 60 lb. The other child weighs 90 lb. How far from the 90-pound child should the fulcrum be placed so that the seesaw balances?

63. In preparation for a stunt, two acrobats are standing on a plank 18 ft long. One acrobat weighs 128 lb and the second acrobat weighs 160 lb. How far from the 128-pound acrobat must the fulcrum be placed so that the acrobats are balanced on the plank?

64. A screwdriver 9 in. long is used as a lever to open a can of paint. The tip of the screwdriver is placed under the lip of the can with the fulcrum 0.15 in. from the lip. A force of 30 lb is applied to the other end of the screwdriver. Find the force on the lip of the can.

65. A metal bar 8 ft long is used to move a 150-pound rock. The fulcrum is placed 1.5 ft from the rock. What minimum force must be applied to the other end of the bar to move the rock? Round to the nearest tenth.

Business   To determine the break-even point, or the number of units that must be sold so that no profit or loss occurs, an economist uses the formula $Px = Cx + F$, where P is the selling price per unit, x is the number of units that must be sold to break even, C is the cost to make each unit, and F is the fixed cost. Use this equation for Exercises 66 to 71.

66. A business analyst has determined that the selling price per unit for a laser printer is $1600. The cost to make one laser printer is $950, and the fixed cost is $211,250. Find the break-even point.

67. A business analyst has determined that the selling price per unit for a gas barbecue is $325. The cost to make one gas barbecue is $175, and the fixed cost is $39,000. Find the break-even point.

68. A manufacturer of thermostats determines that the cost per unit for a programmable thermostat is $38 and that the fixed cost is $24,400. The selling price for the thermostat is $99. Find the break-even point.

69. A manufacturing engineer determines that the cost per unit for a desk lamp is $12 and that the fixed cost is $19,240. The selling price for the desk lamp is $49. Find the break-even point.

70. A manufacturing engineer determines the cost to make one compact disc to be $3.35 and the fixed cost to be $6180. The selling price for each compact disc is $8.50. Find the number of compact discs that must be sold to break even.

71. To manufacture a softball bat requires two steps. The first step is to cut a rough shape. The second step is to sand the bat to its final form. The cost to rough-shape a bat is $.45, and the cost to sand a bat to final form is $1.05. The total fixed cost for the two steps is $16,500. How many softball bats must be sold at a price of $7.00 to break even?

APPLYING THE CONCEPTS

72. Write an equation of the form $ax + b = cx + d$ that has 4 as the solution.

For Exercises 73 to 76, solve. If the equation has no solution, write "no solution."

73. $3(2x - 1) - (6x - 4) = -9$

74. $7(3x + 6) - 4(3 + 5x) = 13 + x$

75. $\dfrac{1}{5}(25 - 10a) + 4 = \dfrac{1}{3}(12a - 15) + 14$

76. $5[m + 2(3 - m)] = 3[2(4 - m) - 5]$

77. ✎ The equation $x = x + 1$ has no solution, whereas the solution of the equation $2x + 3 = 3$ is zero. Is there a difference between no solution and a solution of zero? Explain your answer.

3.4 Translating Sentences into Equations

Objective A To solve integer problems

An equation states that two mathematical expressions are equal. Therefore, to **translate** a sentence into an equation requires recognition of the words or phrases that mean "equals." Some of these phrases are listed below.

$$\left.\begin{array}{l} \text{equals} \\ \text{is} \\ \text{is equal to} \\ \text{amounts to} \\ \text{represents} \end{array}\right\} \text{translate to } =$$

Once the sentence is translated into an equation, the equation can be solved by rewriting the equation in the form *variable = constant*.

HOW TO Translate "five less than a number is thirteen" into an equation and solve.

The unknown number: n
• Assign a variable to the unknown number.

| Five less than a number | is | thirteen |

• Find two verbal expressions for the same value.

$$n - 5 = 13$$
• Write a mathematical expression for each verbal expression. Write the equals sign.

$$n - 5 + 5 = 13 + 5$$
• Solve the equation.

$$n = 18$$

The number is 18.

TAKE NOTE
You can check the solution to a translation problem.

Check:

5 less than 18 is 13

$$\begin{array}{c|c} 18 - 5 & 13 \\ 13 & = 13 \end{array}$$

Recall that the integers are the numbers $\{\ldots, -4, -3, -2, -1, 0, 1, 2, 3, 4, \ldots\}$. An **even integer** is an integer that is divisible by 2. Examples of even integers are -8, 0, and 22. An **odd integer** is an integer that is not divisible by 2. Examples of odd integers are -17, 1, and 39.

Consecutive integers are integers that follow one another in order. Examples of consecutive integers are shown at the right. (Assume that the variable n represents an integer.)

11, 12, 13
$-8, -7, -6$
$n, n + 1, n + 2$

Examples of **consecutive even integers** are shown at the right. (Assume that the variable n represents an even integer.)

24, 26, 28
$-10, -8, -6$
$n, n + 2, n + 4$

TAKE NOTE
Both consecutive even and consecutive odd integers are represented using $n, n + 2, n + 4, \ldots$.

Examples of **consecutive odd integers** are shown at the right. (Assume that the variable n represents an odd integer.)

19, 21, 23
$-1, 1, 3$
$n, n + 2, n + 4$

HOW TO The sum of three consecutive odd integers is forty-five. Find the integers.

Strategy

- First odd integer: n
  Second odd integer: $n + 2$
  Third odd integer: $n + 4$
- The sum of the three odd integers is 45.

- Represent three consecutive odd integers.

Solution

$$n + (n + 2) + (n + 4) = 45$$

$$3n + 6 = 45$$

$$3n = 39$$

$$n = 13$$

$$n + 2 = 13 + 2 = 15$$

$$n + 4 = 13 + 4 = 17$$

- Write an equation.
- Solve the equation.

- The first odd integer is 13.
- Find the second odd integer.
- Find the third odd integer.

The three consecutive odd integers are 13, 15, and 17.

Example 1

The sum of two numbers is sixteen. The difference between four times the smaller number and two is two more than twice the larger number. Find the two numbers.

Solution

The smaller number: n
The larger number: $16 - n$

The difference between four times the smaller and two	is	two more than twice the larger

$$4n - 2 = 2(16 - n) + 2$$
$$4n - 2 = 32 - 2n + 2$$
$$4n - 2 = 34 - 2n$$
$$4n + 2n - 2 = 34 - 2n + 2n$$
$$6n - 2 = 34$$
$$6n - 2 + 2 = 34 + 2$$
$$6n = 36$$
$$\frac{6n}{6} = \frac{36}{6}$$
$$n = 6$$

$$16 - n = 16 - 6 = 10$$

The smaller number is 6.
The larger number is 10.

You Try It 1

The sum of two numbers is twelve. The total of three times the smaller number and six amounts to seven less than the product of four and the larger number. Find the two numbers.

Your solution

Solution on p. S8

Example 2

Find three consecutive even integers such that three times the second equals four more than the sum of the first and third.

Strategy

- First even integer: n
  Second even integer: $n + 2$
  Third even integer: $n + 4$
- Three times the second equals four more than the sum of the first and third.

Solution

$$3(n + 2) = n + (n + 4) + 4$$
$$3n + 6 = 2n + 8$$
$$3n - 2n + 6 = 2n - 2n + 8$$
$$n + 6 = 8$$
$$n = 2$$
$$n + 2 = 2 + 2 = 4$$
$$n + 4 = 2 + 4 = 6$$

The three integers are 2, 4, and 6.

You Try It 2

Find three consecutive integers whose sum is negative six.

Your strategy

Your solution

Solution on p. S8

Objective B **To translate a sentence into an equation and solve**

Example 3

A wallpaper hanger charges a fee of $25 plus $12 for each roll of wallpaper used in a room. If the total charge for hanging wallpaper is $97, how many rolls of wallpaper were used?

Strategy

To find the number of rolls of wallpaper used, write and solve an equation using n to represent the number of rolls of wallpaper used.

Solution

| $25 plus $12 for each roll of wallpaper | is | $97 |

$$25 + 12n = 97$$
$$12n = 72$$
$$\frac{12n}{12} = \frac{72}{12}$$
$$n = 6$$

6 rolls of wallpaper were used.

You Try It 3

The fee charged by a ticketing agency for a concert is $3.50 plus $17.50 for each ticket purchased. If your total charge for tickets is $161, how many tickets are you purchasing?

Your strategy

Your solution

Solution on p. S8

Example 4

A board 20 ft long is cut into two pieces. Five times the length of the shorter piece is 2 ft more than twice the length of the longer piece. Find the length of each piece.

Strategy

Let x represent the length of the shorter piece. Then $20 - x$ represents the length of the longer piece.

Make a drawing.

To find the lengths, write and solve an equation using x to represent the length of the shorter piece and $20 - x$ to represent the length of the longer piece.

Solution

Five times the length of the shorter piece	is	2 ft more than twice the length of the longer

$$5x = 2(20 - x) + 2$$
$$5x = 40 - 2x + 2$$
$$5x = 42 - 2x$$
$$5x + 2x = 42 - 2x + 2x$$
$$7x = 42$$
$$\frac{7x}{7} = \frac{42}{7}$$
$$x = 6$$

$20 - x = 20 - 6 = 14$

The length of the shorter piece is 6 ft.
The length of the longer piece is 14 ft.

You Try It 4

A wire 22 in. long is cut into two pieces. The length of the longer piece is 4 in. more than twice the length of the shorter piece. Find the length of each piece.

Your strategy

Your solution

Solution on p. S8

3.4 Exercises

Objective A **To solve integer problems**

For Exercises 1 to 18, translate into an equation and solve.

1. The difference between a number and fifteen is seven. Find the number.

2. The sum of five and a number is three. Find the number.

3. The product of seven and a number is negative twenty-one. Find the number.

4. The quotient of a number and four is two. Find the number.

5. The difference between nine and a number is seven. Find the number.

6. Three-fifths of a number is negative thirty. Find the number.

7. The difference between five and twice a number is one. Find the number.

8. Four more than three times a number is thirteen. Find the number.

9. The sum of twice a number and five is fifteen. Find the number.

10. The difference between nine times a number and six is twelve. Find the number.

11. Six less than four times a number is twenty-two. Find the number.

12. Four times the sum of twice a number and three is twelve. Find the number.

13. Three times the difference between four times a number and seven is fifteen. Find the number.

14. Twice the difference between a number and twenty-five is three times the number. Find the number.

15. The sum of two numbers is twenty. Three times the smaller is equal to two times the larger. Find the two numbers.

16. The sum of two numbers is fifteen. One less than three times the smaller is equal to the larger. Find the two numbers.

17. The sum of two numbers is fourteen. The difference between two times the smaller and the larger is one. Find the two numbers.

18. The sum of two numbers is eighteen. The total of three times the smaller and twice the larger is forty-four. Find the two numbers.

19. The sum of three consecutive odd integers is fifty-one. Find the integers.

20. Find three consecutive even integers whose sum is negative eighteen.

21. Find three consecutive odd integers such that three times the middle integer is one more than the sum of the first and third.

22. Twice the smallest of three consecutive odd integers is seven more than the largest. Find the integers.

23. Find two consecutive even integers such that three times the first equals twice the second.

24. Find two consecutive even integers such that four times the first is three times the second.

25. Seven times the first of two consecutive odd integers is five times the second. Find the integers.

26. Find three consecutive even integers such that three times the middle integer is four more than the sum of the first and third.

Objective B   **To translate a sentence into an equation and solve**

27. **Computer Science**   The processor speed of a personal computer is 3.2 gigahertz (GHz). This is three-fourths of the processor speed of a newer model personal computer. Find the processor speed of the newer personal computer.

28. **Computer Science**   The storage capacity of a hard-disk drive is 60 gigabytes. This is one-fourth of the storage capacity of a second hard-disk drive. Find the storage capacity of the second hard-disk drive.

29. **Geometry**   An isosceles triangle has two sides of equal length. The length of the third side is 1 ft less than twice the length of an equal side. Find the length of each side when the perimeter is 23 ft.

30. **Geometry**   An isosceles triangle has two sides of equal length. The length of one of the equal sides is two more than 3 times the length of the third side. If the perimeter is 46 m, find the length of each side.

31. **Union Dues**   A union charges monthly dues of $4.00 plus $.15 for each hour worked during the month. A union member's dues for March were $29.20. How many hours did the union member work during the month of March?

32. **Technical Support**   A technical information hotline charges a customer $15.00 plus $2.00 per minute to answer questions about software. How many minutes did a customer who received a bill for $37 use this service?

33. **Construction**   The total cost to paint the inside of a house was $1346. This cost included $125 for materials and $33 per hour for labor. How many hours of labor were required to paint the inside of the house?

34. **Telecommunications**   The cellular phone service for a business executive is $35 per month plus $.40 per minute of phone use. In a month when the executive's cellular phone bill was $99.80, how many minutes did the executive use the phone?

35. **Computer Science**   A computer screen consists of tiny dots of light called pixels. In a certain graphics mode, there are 1280 horizontal pixels. This is 768 less than twice the number of vertical pixels. Find the number of vertical pixels.

36. **Energy**   The cost of electricity in a certain city is $.08 for each of the first 300 kWh (kilowatt-hours) and $.13 for each kilowatt-hour over 300 kWh. Find the number of kilowatt-hours used by a family with a $51.95 electric bill.

37. **Geometry**   The perimeter of a rectangle is 42 m. The length of the rectangle is 3 m less than twice the width. Find the length and width of the rectangle.

38. **Geometry**   A rectangular vegetable garden has a perimeter of 64 ft. The length of the garden is 20 ft. Find the width of the garden.

39. **Carpentry**   A 12-foot board is cut into two pieces. Twice the length of the shorter piece is 3 ft less than the length of the longer piece. Find the length of each piece.

40. **Sports**   A 14-yard fishing line is cut into two pieces. Three times the length of the longer piece is four times the length of the shorter piece. Find the length of each piece.

41. **Education**   Seven thousand dollars is divided into two scholarships. Twice the amount of the smaller scholarship is $1000 less than the larger scholarship. What is the amount of the larger scholarship?

42. **Investing**   An investment of $10,000 is divided into two accounts, one for stocks and one for mutual funds. The value of the stock account is $2000 less than twice the value of the mutual fund account. Find the amount in each account.

APPLYING THE CONCEPTS

43.   Make up two word problems; one that requires solving the equation $6x = 123$ and one that requires solving the equation $8x + 100 = 300$ to find the answer to the problem.

44.   A formula is an equation that relates variables in a known way. Find two examples of formulas that are used in your college major. Explain what each of the variables represents.

45.   It is always important to check the answer to an application problem to be sure the answer makes sense. Consider the following problem. A 4-quart mixture of fruit juices is made from apple juice and cranberry juice. There are 6 more quarts of apple juice than of cranberry juice. Write and solve an equation for the number of quarts of each juice used. Does the answer to this question make sense? Explain.

3.5 Geometry Problems

Objective A To solve problems involving angles

In Section 1.8, we discussed some basic properties of angles. Recall that a ray that is rotated one complete revolution about its starting point creates an angle of 360°. Recall also that a 90° angle is called a right angle and a 180° angle is called a straight angle.

Point of Interest

The word *degree* first appeared in Chaucer's *Canterbury Tales*, which was written in 1386.

An **acute angle** is an angle whose measure is between 0° and 90°. ∠A at the right is an acute angle. An **obtuse angle** is an angle whose measure is between 90° and 180°. ∠B at the right is an obtuse angle.

HOW TO Given the diagram at the left, find x.

$$3x + 4x + 5x = 360°$$

• The sum of the measures of three angles is 360°.

$$12x = 360°$$
$$x = 30°$$

The measure of x is 30°.

Four angles are formed by the intersection of two lines. If the two lines are perpendicular, then each of the four angles is a right angle. If the two lines are not perpendicular, then two of the angles formed are acute angles and two of the angles are obtuse angles. The two acute angles are always opposite each other, and the two obtuse angles are always opposite each other.

In the figure at the right, ∠w and ∠y are acute angles, and ∠x and ∠z are obtuse angles.

Vertical angles have the same measure.

$$m∠w = m∠y$$
$$m∠x = m∠z$$

Two angles that are on opposite sides of the intersection of two lines are called **vertical angles.** In the figure above ∠w and ∠y are vertical angles. ∠x and ∠z are vertical angles.

Two angles that share a common side are called **adjacent angles.** In the figure above, ∠x and ∠y are adjacent angles, as are ∠y and ∠z, ∠z and ∠w, and ∠w and ∠x.

Adjacent angles of intersecting lines are supplementary.

$$m∠x + m∠y = 180°   \quad m∠z + m∠w = 180°$$
$$m∠y + m∠z = 180°   \quad m∠w + m∠x = 180°$$

HOW TO In the diagram at the left, $m∠b = 115°$. Find $m∠a, m∠c,$ and $m∠d$.

$$m∠a + m∠b = 180°$$
$$m∠a + 115° = 180°$$
$$m∠a = 65°$$

• ∠a is supplementary to ∠b because ∠a and ∠b are adjacent angles of intersecting lines.

$$m∠c = 65°$$
$$m∠d = 115°$$

• $m∠c = m∠a$ because ∠c and ∠a are vertical angles.
• $m∠d = m∠b$ because ∠d and ∠b are vertical angles.

A line that intersects two other lines at different points is called a **transversal.** If the lines cut by a transversal t are parallel lines and the transversal is not perpendicular to the parallel lines, then all four acute angles have the same measure and all four obtuse angles have the same measure.

$$m\angle b = m\angle d = m\angle x = m\angle z$$
$$m\angle a = m\angle c = m\angle w = m\angle y$$

Alternate interior angles are two non-adjacent angles that are on opposite sides of the transversal and between the parallel lines. In the figure above, $\angle c$ and $\angle w$ are alternate interior angles, and $\angle d$ and $\angle x$ are alternate interior angles.

Alternate interior angles have the same measure.

$$m\angle c = m\angle w$$
$$m\angle d = m\angle x$$

Alternate exterior angles are two non-adjacent angles that are on opposite sides of the transversal and outside the parallel lines. In the figure above, $\angle a$ and $\angle y$ are alternate exterior angles, and $\angle b$ and $\angle z$ are alternate exterior angles.

Alternate exterior angles have the same measure.

$$m\angle a = m\angle y$$
$$m\angle b = m\angle z$$

Corresponding angles are two angles that are on the same side of the transversal and are both acute angles or are both obtuse angles. In the figure above, there are four pairs of corresponding angles: $\angle a$ and $\angle w$, $\angle d$ and $\angle z$, $\angle b$ and $\angle x$, and $\angle c$ and $\angle y$.

Corresponding angles have the same measure.

$$m\angle a = m\angle w$$
$$m\angle d = m\angle z$$
$$m\angle b = m\angle x$$
$$m\angle c = m\angle y$$

HOW TO In the diagram at the left, $\ell_1 \parallel \ell_2$ and $m\angle f = 58°$. Find $m\angle a$, $m\angle c$, and $m\angle d$.

$m\angle a = m\angle f = 58°$ • $\angle a$ and $\angle f$ are corresponding angles.

$m\angle c = m\angle f = 58°$ • $\angle c$ and $\angle f$ are alternate interior angles.

$m\angle d + m\angle a = 180°$ • $\angle d$ is supplementary to $\angle a$.
$m\angle d + 58° = 180°$
$m\angle d = 122°$

Example 1

Find x.

Strategy

The angles labeled are adjacent angles of intersecting lines and are therefore supplementary angles. To find x, write an equation and solve for x.

Solution
$$x + (x + 70°) = 180°$$
$$2x + 70° = 180°$$
$$2x = 110°$$
$$x = 55°$$

You Try It 1

Find x.

Your strategy

Your solution

Solution on p. S9

Example 2

Given $\ell_1 \parallel \ell_2$, find x.

Strategy

$3x = y$ because corresponding angles have the same measure. $y + (x + 40°) = 180°$ because adjacent angles of intersecting lines are supplementary angles. Substitute $3x$ for y and solve for x.

Solution

$$y + (x + 40°) = 180°$$
$$3x + (x + 40°) = 180°$$
$$4x + 40° = 180°$$
$$4x = 140°$$
$$x = 35°$$

You Try It 2

Given $\ell_1 \parallel \ell_2$, find x.

Your strategy

Your solution

Solution on p. S9

Objective B   **To solve problems involving the angles of a triangle**

If the lines cut by a transversal are not parallel lines, then the three lines will intersect at three points, forming a triangle. The angles within the region enclosed by the triangle are called **interior angles**. In the figure at the right, angles a, b, and c are interior angles. **The sum of the measures of the interior angles of a triangle is 180°.**

$$m\angle a + m\angle b + m\angle c = 180°$$

An angle adjacent to an interior angle is an **exterior angle**. In the figure at the right, angles m and n are exterior angles for angle a. The sum of the measures of an interior angle of a triangle and an adjacent exterior angle is 180°.

$$m\angle a + m\angle m = 180°$$
$$m\angle a + m\angle n = 180°$$

HOW TO   Given that $m\angle c = 40°$ and $m\angle e = 60°$, find $m\angle d$.

$$m\angle a = m\angle e = 60°$$ • $\angle a$ and $\angle e$ are vertical angles.

$$m\angle c + m\angle a + m\angle b = 180°$$ • The sum of the interior angles is 180°.
$$40° + 60° + m\angle b = 180°$$
$$100° + m\angle b = 180°$$
$$m\angle b = 80°$$

$$m\angle b + m\angle d = 180°$$ • $\angle b$ and $\angle d$ are supplementary angles.
$$80° + m\angle d = 180°$$
$$m\angle d = 100°$$

Example 3

Given that $m\angle a = 45°$ and $m\angle x = 100°$, find the measures of angles b, c, and y.

Strategy

• To find the measure of $\angle b$, use the fact that $\angle b$ and $\angle x$ are supplementary angles.
• To find the measure of $\angle c$, use the fact that the sum of the measures of the interior angles of a triangle is 180°.
• To find the measure of $\angle y$, use the fact that $\angle c$ and $\angle y$ are vertical angles.

Solution

$$m\angle b + m\angle x = 180°$$
$$m\angle b + 100° = 180°$$
$$m\angle b = 80°$$

$$m\angle a + m\angle b + m\angle c = 180°$$
$$45° + 80° + m\angle c = 180°$$
$$125° + m\angle c = 180°$$
$$m\angle c = 55°$$

$$m\angle y = m\angle c = 55°$$

You Try It 3

Given that $m\angle y = 55°$, find the measures of angles a, b, and d.

Your strategy

Your solution

Example 4

Two angles of a triangle measure 43° and 86°. Find the measure of the third angle.

Strategy

To find the measure of the third angle, use the fact that the sum of the measures of the interior angles of a triangle is 180°. Write an equation using x to represent the measure of the third angle. Solve the equation for x.

Solution

$$x + 43° + 86° = 180°$$
$$x + 129° = 180°$$
$$x = 51°$$

The measure of the third angle is 51°.

You Try It 4

One angle in a triangle is a right angle, and one angle measures 27°. Find the measure of the third angle.

Your strategy

Your solution

Solutions on p. S9

3.5 Exercises

Objective A   **To solve problems involving angles**

For Exercises 1 and 2, find the measure of ∠a.

1.

2.

For Exercises 3 to 8, find x.

3.

4.

5.

6.

7.

8.

For Exercises 9 to 12, find x.

9.

10.

11.

12.

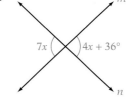

For Exercises 13 to 16, given that $\ell_1 \parallel \ell_2$, find the measures of angles a and b.

13.

14.

15.

16.

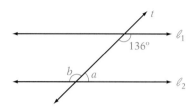

For Exercises 17 to 20, given that $\ell_1 \parallel \ell_2$, find x.

17.

18.

19.

20.

21. Given that $m\angle a = 51°$, find $m\angle b$.

22. Given that $m\angle a = 38°$, find $m\angle b$.

Objective B   **To solve problems involving the angles of a triangle**

23. Given that $m\angle a = 95°$ and $m\angle b = 70°$, find $m\angle x$ and $m\angle y$.

24. Given that $m\angle a = 35°$ and $m\angle b = 55°$, find $m\angle x$ and $m\angle y$.

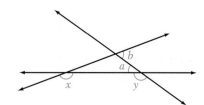

25. Given that $m\angle y = 45°$, find $m\angle a$ and $m\angle b$.

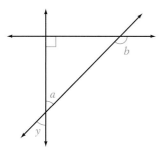

26. Given that $m\angle y = 130°$, find $m\angle a$ and $m\angle b$.

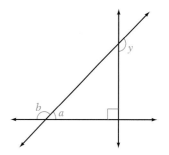

27. Given that $AO \perp OB$, express $m\angle BOC$ in terms of x.

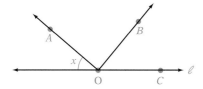

28. Given that $AO \perp OB$, express $m\angle AOC$ in terms of x.

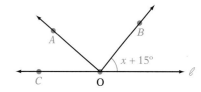

29. A triangle has a 30° angle and a right angle. What is the measure of the third angle?

30. A triangle has a 45° angle and a right angle. Find the measure of the third angle.

31. Two angles of a triangle measure 42° and 103°. Find the measure of the third angle.

32. Two angles of a triangle measure 62° and 45°. Find the measure of the third angle.

33. A triangle has a 13° angle and a 65° angle. What is the measure of the third angle?

34. A triangle has a 105° angle and a 32° angle. What is the measure of the third angle?

APPLYING THE CONCEPTS

35. **Geometry** A carpenter is carrying a ladder around the corner of a hallway as shown in the figure below. Find $m\angle x$.

36. **Geometry** For the figure below, which of the following is true?
(a) $m\angle a < m\angle b$
(b) $m\angle a = m\angle b$
(c) $m\angle a > m\angle b$

37. **Geometry** For the figure at the right, find the sum of the measures of angles x, y, and z.

38. **Geometry** For the figure at the right, explain why $m\angle a + m\angle b = m\angle x$. Write a rule that describes the relationship between an exterior angle of a triangle and the opposite interior angles. Use the rule to write an equation involving the measures of angles a, c, and z.

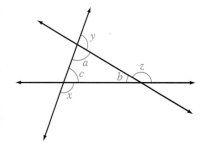

3.6 Mixture and Uniform Motion Problems

Objective A   To solve value mixture problems

A value mixture problem involves combining two ingredients that have different prices into a single blend. For example, a coffee merchant may blend two types of coffee into a single blend, or a candy manufacturer may combine two types of candy to sell as a variety pack.

The solution of a value mixture problem is based on the **value mixture equation** $AC = V$, where A is the amount of an ingredient, C is the cost per unit of the ingredient, and V is the value of the ingredient.

> **TAKE NOTE**
> The equation $AC = V$ is used to find the value of an ingredient. For example, the value of 4 lb of cashews costing $6 per pound is
> $$AC = V$$
> $$4 \cdot \$6 = V$$
> $$\$24 = V$$

HOW TO   A coffee merchant wants to make 6 lb of a blend of coffee costing $5 per pound. The blend is made using a $6-per-pound grade and a $3-per-pound grade of coffee. How many pounds of each of these grades should be used?

> **Strategy for Solving a Value Mixture Problem**
>
> 1. For each ingredient in the mixture, write a numerical or variable expression for the amount of the ingredient used, the unit cost of the ingredient, and the value of the amount used. For the blend, write a numerical or variable expression for the amount, the unit cost of the blend, and the value of the amount. The results can be recorded in a table.

The sum of the amounts is 6 lb.

Amount of $3 coffee: $6 - x$
Amount of $6 coffee: x

> **TAKE NOTE**
> Use the information given in the problem to fill in the amount and unit cost columns of the table. Fill in the value column by multiplying the two expressions you wrote in each row. Use the expressions in the last column to write the equation.

	Amount, A	·	Unit Cost, C	=	Value, V
$6 grade	x	·	6	=	$6x$
$3 grade	$6 - x$	·	3	=	$3(6 - x)$
$5 blend	6	·	5	=	$5(6)$

> 2. Determine how the values of the ingredients are related. Use the fact that the sum of the values of all the ingredients is equal to the value of the blend.

The sum of the values of the $6 grade and the $3 grade is equal to the value of the $5 blend.

$$6x + 3(6 - x) = 5(6)$$
$$6x + 18 - 3x = 30$$
$$3x + 18 = 30$$
$$3x = 12$$
$$x = 4$$

$6 - x = 6 - 4 = 2$ • Find the amount of the $3 grade coffee.

The merchant must use 4 lb of the $6 coffee and 2 lb of the $3 coffee.

Example 1

How many ounces of a silver alloy that costs $4 an ounce must be mixed with 10 oz of an alloy that costs $6 an ounce to make a mixture that costs $4.32 an ounce?

Strategy

- Ounces of $4 alloy: x

	Amount	*Cost*	*Value*
$4 alloy	x	4	$4x$
$6 alloy	10	6	6(10)
$4.32 mixture	$10 + x$	4.32	4.32(10 + x)

- The sum of the values before mixing equals the value after mixing.

Solution

$$4x + 6(10) = 4.32(10 + x)$$

$$4x + 60 = 43.2 + 4.32x$$

$$-0.32x + 60 = 43.2$$

$$-0.32x = -16.8$$

$$x = 52.5$$

52.5 oz of the $4 silver alloy must be used.

You Try It 1

A gardener has 20 lb of a lawn fertilizer that costs $.80 per pound. How many pounds of a fertilizer that costs $.55 per pound should be mixed with this 20 lb of lawn fertilizer to produce a mixture that costs $.75 per pound?

Your strategy

Your solution

Solution on p. S9

> **Objective B**   To solve percent mixture problems

Recall from Section 3.1 that a percent mixture problem can be solved using the equation $Ar = Q$, where A is the amount of a solution, r is the percent concentration of a substance in the solution, and Q is the quantity of the substance in the solution.

For example, a 500-milliliter bottle is filled with a 4% solution of hydrogen peroxide.

$$Ar = Q$$
$$500(0.04) = Q$$
$$20 = Q$$

The bottle contains 20 ml of hydrogen peroxide.

HOW TO   How many gallons of a 20% salt solution must be mixed with 6 gal of a 30% salt solution to make a 22% salt solution?

> **Strategy for Solving a Percent Mixture Problem**
>
> **1.** For each solution, write a numerical or variable expression for the amount of solution, the percent concentration, and the quantity of the substance in the solution. The results can be recorded in a table.

The unknown quantity of 20% solution: x

TAKE NOTE

Use the information given in the problem to fill in the amount and percent columns of the table. Fill in the quantity column by multiplying the two expressions you wrote in each row. Use the expressions in the last column to write the equation.

	Amount of Solution, A	·	Percent Concentration, r	=	Quantity of Substance, Q
20% solution	x	·	0.20	=	$0.20x$
30% solution	6	·	0.30	=	$0.30(6)$
22% solution	$x + 6$	·	0.22	=	$0.22(x + 6)$

> **2.** Determine how the quantities of the substances in the solutions are related. Use the fact that the sum of the quantities of the substances being mixed is equal to the quantity of the substance after mixing.

The sum of the quantities of the substances in the 20% solution and the 30% solution is equal to the quantity of the substance in the 22% solution.

$$0.20x + 0.30(6) = 0.22(x + 6)$$
$$0.20x + 1.80 = 0.22x + 1.32$$
$$-0.02x + 1.80 = 1.32$$
$$-0.02x = -0.48$$
$$x = 24$$

24 gal of the 20% solution are required.

Example 2

A chemist wishes to make 2 L of an 8% acid solution by mixing a 10% acid solution and a 5% acid solution. How many liters of each solution should the chemist use?

Strategy

x L of   (2 − x) L of   2 L of
10% acid   +   5% acid   =   8% acid

- Liters of 10% solution: x
  Liters of 5% solution: $2 - x$

	Amount	*Percent*	*Quantity*
10% solution	x	0.10	$0.10x$
5% solution	$2 - x$	0.05	$0.05(2 - x)$
8% solution	2	0.08	$0.08(2)$

- The sum of the quantities before mixing is equal to the quantity after mixing.

Solution

$$0.10x + 0.05(2 - x) = 0.08(2)$$

$$0.10x + 0.10 - 0.05x = 0.16$$

$$0.05x + 0.10 = 0.16$$

$$0.05x = 0.06$$

$$x = 1.2$$

$$2 - x = 2 - 1.2 = 0.8$$

The chemist needs 1.2 L of the 10% solution and 0.8 L of the 5% solution.

You Try It 2

A pharmacist dilutes 5 L of a 12% solution with a 6% solution. How many liters of the 6% solution are added to make an 8% solution?

Your strategy

Your solution

Solution on p. S9

Objective C   **To solve uniform motion problems**

Recall from Section 3.1 that an object traveling at a constant speed in a straight line is in *uniform motion*. The solution of a uniform motion problem is based on the equation $rt = d$, where r is the rate of travel, t is the time spent traveling, and d is the distance traveled.

HOW TO   A car leaves a town traveling at 40 mph. Two hours later, a second car leaves the same town, on the same road, traveling at 60 mph. In how many hours will the second car pass the first car?

> **Strategy for Solving a Uniform Motion Problem**
> **1.** For each object, write a numerical or variable expression for the rate, time, and distance. The results can be recorded in a table.

The first car traveled 2 h longer than the second car.

Unknown time for the second car: t
Time for the first car: $t + 2$

	Rate, r	·	Time, t	=	Distance, d
First car	40	·	$t + 2$	=	$40(t + 2)$
Second car	60	·	t	=	$60t$

TAKE NOTE
Use the information given in the problem to fill in the rate and time columns of the table. Find the expression in the distance column by multiplying the two expressions you wrote in each row.

> **2.** Determine how the distances traveled by the two objects are related. For example, the total distance traveled by both objects may be known, or it may be known that the two objects traveled the same distance.

The two cars travel the same distance.

$$40(t + 2) = 60t$$
$$40t + 80 = 60t$$
$$80 = 20t$$
$$4 = t$$

The second car will pass the first car in 4 h.

Example 3

Two cars, one traveling 10 mph faster than the other, start at the same time from the same point and travel in opposite directions. In 3 h they are 300 mi apart. Find the rate of each car.

Strategy

• Rate of 1st car: r
  Rate of 2nd car: $r + 10$

	Rate	Time	Distance
1st car	r	3	$3r$
2nd car	$r + 10$	3	$3(r + 10)$

• The total distance traveled by the two cars is 300 mi.

Solution

$$3r + 3(r + 10) = 300$$
$$3r + 3r + 30 = 300$$
$$6r + 30 = 300$$
$$6r = 270$$
$$r = 45$$

$$r + 10 = 45 + 10 = 55$$

The first car is traveling 45 mph.
The second car is traveling 55 mph.

You Try It 3

Two trains, one traveling at twice the speed of the other, start at the same time on parallel tracks from stations that are 288 mi apart and travel toward each other. In 3 h, the trains pass each other. Find the rate of each train.

Your strategy

Your solution

Example 4

How far can the members of a bicycling club ride out into the country at a speed of 12 mph and return over the same road at 8 mph if they travel a total of 10 h?

Strategy

• Time spent riding out: t
  Time spent riding back: $10 - t$

	Rate	Time	Distance
Out	12	t	$12t$
Back	8	$10 - t$	$8(10 - t)$

• The distance out equals the distance back.

Solution

$$12t = 8(10 - t)$$
$$12t = 80 - 8t$$
$$20t = 80$$
$$t = 4 \quad \text{(The time is 4 h.)}$$

The distance out = $12t = 12(4) = 48$ mi.

The club can ride 48 mi into the country.

You Try It 4

A pilot flew out to a parcel of land and back in 5 h. The rate out was 150 mph, and the rate returning was 100 mph. How far away was the parcel of land?

Your strategy

Your solution

Solutions on p. S10

3.6 Exercises

Objective A **To solve value mixture problems**

1. An herbalist has 30 oz of herbs costing $2 per ounce. How many ounces of herbs costing $1 per ounce should be mixed with the 30 oz to produce a mixture costing $1.60 per ounce?

2. The manager of a farmer's market has 500 lb of grain that costs $1.20 per pound. How many pounds of meal costing $.80 per pound should be mixed with the 500 lb of grain to produce a mixture that costs $1.05 per pound?

3. Find the cost per pound of a meatloaf mixture made from 3 lb of ground beef costing $1.99 per pound and 1 lb of ground turkey costing $1.39 per pound.

4. Find the cost per ounce of a sunscreen made from 100 oz of a lotion that costs $2.50 per ounce and 50 oz of a lotion that costs $4.00 per ounce.

5. A snack food is made by mixing 5 lb of popcorn that costs $.80 per pound with caramel that costs $2.40 per pound. How much caramel is needed to make a mixture that costs $1.40 per pound?

6. A wild birdseed mix is made by combining 100 lb of millet seed costing $.60 per pound with sunflower seeds costing $1.10 per pound. How many pounds of sunflower seeds are needed to make a mixture that costs $.70 per pound?

7. Ten cups of a restaurant's house Italian dressing is made by blending olive oil costing $1.50 per cup with vinegar that costs $.25 per cup. How many cups of each are used if the cost of the blend is $.50 per cup?

8. A high-protein diet supplement that costs $6.75 per pound is mixed with a vitamin supplement that costs $3.25 per pound. How many pounds of each should be used to make 5 lb of a mixture that costs $4.65 per pound?

9. Find the cost per ounce of a mixture of 200 oz of a cologne that costs $5.50 per ounce and 500 oz of a cologne that costs $2.00 per ounce.

10. Find the cost per pound of a trail mix made from 40 lb of raisins that cost $4.40 per pound and 100 lb of granola that costs $2.30 per pound.

11. A 20-ounce alloy of platinum that costs $220 per ounce is mixed with an alloy that costs $400 per ounce. How many ounces of the $400 alloy should be used to make an alloy that costs $300 per ounce?

12. How many liters of a blue dye that costs $1.60 per liter must be mixed with 18 L of anil that costs $2.50 per liter to make a mixture that costs $1.90 per liter?

13. The manager of a specialty food store combined almonds that cost $4.50 per pound with walnuts that cost $2.50 per pound. How many pounds of each were used to make a 100-pound mixture that costs $3.24 per pound?

14. A goldsmith combined an alloy that cost $4.30 per ounce with an alloy that cost $1.80 per ounce. How many ounces of each were used to make a mixture of 200 oz costing $2.50 per ounce?

15. Adult tickets for a play cost $6.00 and children's tickets cost $2.50. For one performance, 370 tickets were sold. Receipts for the performance were $1723. Find the number of adult tickets sold.

16. Tickets for a piano concert sold for $4.50 for each adult. Student tickets sold for $2.00 each. The total receipts for 1720 tickets were $5980. Find the number of adult tickets sold.

17. Find the cost per pound of sugar-coated breakfast cereal made from 40 lb of sugar that costs $1.00 per pound and 120 lb of corn flakes that cost $.60 per pound.

18. Find the cost per pound of a coffee mixture made from 8 lb of coffee that costs $9.20 per pound and 12 lb of coffee that costs $5.50 per pound.

Objective B   **To solve percent mixture problems**

19. Forty ounces of a 30% gold alloy are mixed with 60 oz of a 20% gold alloy. Find the percent concentration of the resulting gold alloy.

20. One hundred ounces of juice that is 50% tomato juice is added to 200 oz of a vegetable juice that is 25% tomato juice. What is the percent concentration of tomato juice in the resulting mixture?

21. How many gallons of a 15% acid solution must be mixed with 5 gal of a 20% acid solution to make a 16% acid solution?

22. How many pounds of a chicken feed that is 50% corn must be mixed with 400 lb of a feed that is 80% corn to make a chicken feed that is 75% corn?

23. A rug is made by weaving 20 lb of yarn that is 50% wool with a yarn that is 25% wool. How many pounds of the yarn that is 25% wool are used if the finished rug is 35% wool?

24. Five gallons of a light green latex paint that is 20% yellow paint is combined with a darker green latex paint that is 40% yellow paint. How many gallons of the darker green paint must be used to create a green paint that is 25% yellow paint?

25. How many gallons of a plant food that is 9% nitrogen must be combined with another plant food that is 25% nitrogen to make 10 gal of a solution that is 15% nitrogen?

26. A chemist wants to make 50 ml of a 16% acid solution by mixing a 13% acid solution and an 18% acid solution. How many milliliters of each solution should the chemist use?

x ml of 13% acid $+$ $(50 - x)$ ml of 18% acid $=$ 50 ml of 16% acid

27. Five grams of sugar are added to a 45-gram serving of a breakfast cereal that is 10% sugar. What is the percent concentration of sugar in the resulting mixture?

28. A goldsmith mixes 8 oz of a 30% gold alloy with 12 oz of a 25% gold alloy. What is the percent concentration of the resulting alloy?

29. How many pounds of coffee that is 40% java beans must be mixed with 80 lb of coffee that is 30% java beans to make a coffee blend that is 32% java beans?

30. The manager of a garden shop mixes grass seed that is 60% rye grass with 70 lb of grass seed that is 80% rye grass to make a mixture that is 74% rye grass. How much of the 60% rye grass is used?

31. A hair dye is made by blending a 7% hydrogen peroxide solution and a 4% hydrogen peroxide solution. How many milliliters of each are used to make a 300-milliliter solution that is 5% hydrogen peroxide?

32. A tea that is 20% jasmine is blended with a tea that is 15% jasmine. How many pounds of each tea are used to make 5 lb of tea that is 18% jasmine?

33. How many ounces of pure chocolate must be added to 150 oz of chocolate topping that is 50% chocolate to make a topping that is 75% chocolate?

34. How many ounces of pure bran flakes must be added to 50 oz of cereal that is 40% bran flakes to produce a mixture that is 50% bran flakes?

35. Thirty ounces of pure silver are added to 50 oz of a silver alloy that is 20% silver. What is the percent concentration of the resulting alloy?

36. A clothing manufacturer has some pure silk thread and some thread that is 85% silk. How many kilograms of each must be woven together to make 75 kg of cloth that is 96% silk?

Objective C   **To solve uniform motion problems**

37. Two small planes start from the same point and fly in opposite directions. The first plane is flying 25 mph slower than the second plane. In 2 h, the planes are 470 mi apart. Find the rate of each plane.

470 mi

38. Two cyclists start from the same point and ride in opposite directions. One cyclist rides twice as fast as the other. In 3 h, they are 81 mi apart. Find the rate of each cyclist.

39. Two planes leave an airport at 8 A.M., one flying north at 480 km/h and the other flying south at 520 km/h. At what time will they be 3000 km apart?

40. A long-distance runner started on a course running at an average speed of 6 mph. One-half hour later, a second runner began the same course at an average speed of 7 mph. How long after the second runner started will the second runner overtake the first runner?

41. A motorboat leaves a harbor and travels at an average speed of 9 mph toward a small island. Two hours later a cabin cruiser leaves the same harbor and travels at an average speed of 18 mph toward the same island. In how many hours after the cabin cruiser leaves the harbor will it be alongside the motorboat?

42. A 555-mile, 5-hour plane trip was flown at two speeds. For the first part of the trip, the average speed was 105 mph. For the remainder of the trip, the average speed was 115 mph. How long did the plane fly at each speed?

43. An executive drove from home at an average speed of 30 mph to an airport where a helicopter was waiting. The executive boarded the helicopter and flew to the corporate offices at an average speed of 60 mph. The entire distance was 150 mi. The entire trip took 3 h. Find the distance from the airport to the corporate offices.

44. After a sailboat had been on the water for 3 h, a change in the wind direction reduced the average speed of the boat by 5 mph. The entire distance sailed was 57 mi. The total time spent sailing was 6 h. How far did the sailboat travel in the first 3 h?

45. A car and a bus set out at 3 P.M. from the same point headed in the same direction. The average speed of the car is twice the average speed of the bus. In 2 h the car is 68 mi ahead of the bus. Find the rate of the car.

46. A passenger train leaves a train depot 2 h after a freight train leaves the same depot. The freight train is traveling 20 mph slower than the passenger train. Find the rate of each train if the passenger train overtakes the freight train in 3 h.

47. As part of flight training, a student pilot was required to fly to an airport and then return. The average speed on the way to the airport was 100 mph, and the average speed returning was 150 mph. Find the distance between the two airports if the total flying time was 5 h.

48. A ship traveling east at 25 mph is 10 mi from a harbor when another ship leaves the harbor traveling east at 35 mph. How long does it take the second ship to catch up to the first ship?

49. At 10 A.M. a plane leaves Boston, Massachusetts, for Seattle, Washington, a distance of 3000 mi. One hour later a plane leaves Seattle for Boston. Both planes are traveling at a speed of 500 mph. How many hours after the plane leaves Seattle will the planes pass each other?

50. At noon a train leaves Washington, D.C., headed for Charleston, South Carolina, a distance of 500 mi. The train travels at a speed of 60 mph. At 1 P.M. a second train leaves Charleston headed for Washington, D.C., traveling at 50 mph. How long after the train leaves Charleston will the two trains pass each other?

51. Two cyclists start at the same time from opposite ends of a course that is 51 mi long. One cyclist is riding at a rate of 16 mph, and the second cyclist is riding at a rate of 18 mph. How long after they begin will they meet?

52. A bus traveled on a level road for 2 h at an average speed that was 20 mph faster than its average speed on a winding road. The time spent on the winding road was 3 h. Find the average speed on the winding road if the total trip was 210 mi.

53. A bus traveling at a rate of 60 mph overtakes a car traveling at a rate of 45 mph. If the car had a 1-hour head start, how far from the starting point does the bus overtake the car?

54. A car traveling at 48 mph overtakes a cyclist who, riding at 12 mph, had a 3-hour head start. How far from the starting point does the car overtake the cyclist?

APPLYING THE CONCEPTS

55. **Chemistry**   How many grams of pure water must be added to 50 g of pure acid to make a solution that is 40% acid?

56. **Chemistry**   How many ounces of water must be evaporated from 50 oz of a 12% salt solution to produce a 15% salt solution?

57. **Automotive Technology**   A radiator contains 15 gal of a 20% antifreeze solution. How many gallons must be drained from the radiator and replaced by pure antifreeze so that the radiator will contain 15 gal of a 40% antifreeze solution?

58. **Travel**   At 10 A.M., two campers left their campsite by canoe and paddled downstream at an average speed of 12 mph. They then turned around and paddled back upstream at an average rate of 4 mph. The total trip took 1 h. At what time did the campers turn around downstream?

59. **Transportation**   A bicyclist rides for 2 h at a speed of 10 mph and then returns at a speed of 20 mph. Find the cyclist's average speed for the trip.

60. **Travel**   A car travels a 1-mile track at an average speed of 30 mph. At what average speed must the car travel the next mile so that the average speed for the 2 mi is 60 mph?

Focus on Problem Solving

Trial-and-Error Approach to Problem Solving

The questions below require an answer of always true, sometimes true, or never true. These problems are best solved by the trial-and-error method. The trial-and-error method of arriving at a solution to a problem involves repeated tests or experiments.

For example, consider the statement

> Both sides of an equation can be divided by the same number without changing the solution of the equation.

The solution of the equation $6x = 18$ is 3. If we divide both sides of the equation by 2, the result is $3x = 9$ and the solution is still 3. So the answer "never true" has been eliminated. We still need to determine whether there is a case for which the statement is not true. Is there a number that we could divide both sides of the equation by and the result would be an equation for which the solution is not 3?

If we divide both sides of the equation by 0, the result is $\frac{6x}{0} = \frac{18}{0}$; the solution of this equation is not 3 because the expressions on either side of the equals sign are undefined. Thus the statement is true for some numbers and not true for 0. The statement is sometimes true.

For Exercises 1 to 18, determine whether the statement is always true, sometimes true, or never true.

1. Both sides of an equation can be multiplied by the same number without changing the solution of the equation.

2. For an equation of the form $ax = b$, $a \neq 0$, multiplying both sides of the equation by the reciprocal of a will result in an equation of the form $x = constant$.

3. The Multiplication Property of Equations is used to remove a term from one side of an equation.

4. Adding -3 to each side of an equation yields the same result as subtracting 3 from each side of the equation.

5. An equation contains an equals sign.

6. The same variable term can be added to both sides of an equation without changing the solution of the equation.

7. An equation of the form $ax + b = c$ cannot be solved if a is a negative number.

8. The solution of the equation $\frac{x}{0} = 0$ is 0.

9. Given that $\ell_1 \parallel \ell_2$ in the diagram at the left, $\angle b = \angle d = \angle e = \angle g$.

10. In solving an equation of the form $ax + b = cx + d$, subtracting cx from each side of the equation results in an equation with only one variable term in it.

11. If a rope 8 m long is cut into two pieces and one of the pieces has length x meters, then the length of the other piece can be represented as $(x - 8)$ meters.

12. An even integer is a multiple of 2.

13. If the first of three consecutive odd integers is n, then the second and third consecutive odd integers are represented by $n + 1$ and $n + 3$.

14. Suppose we are mixing two salt solutions. Then the variable Q in the percent mixture equation $Q = Ar$ represents the amount of salt in a solution.

15. If 100 oz of a silver alloy is 25% silver, then the alloy contains 25 oz of silver.

16. If we combine an alloy that costs $8 an ounce with an alloy that costs $5 an ounce, the cost of the resulting mixture will be greater than $8 an ounce.

17. If we combine a 9% acid solution with a solution that is 4% acid, the resulting solution will be less than 4% acid.

18. If the speed of one train is 20 mph slower than that of a second train, then the speeds of the two trains can be represented as r and $20 - r$.

Projects and Group Activities

Nielsen Ratings

Point of Interest

The 10 top-ranked programs in prime time for the week of September 15 to September 21, 2003, ranked by Nielsen Media Research, were

NFL Monday Night Football
Survivor—Pearl Islands
Emmy Awards
NFL Monday Showcase
Everybody Loves Raymond
ABC News Special
CSI
Friends
Law and Order
CSI: Miami

Nielsen Media Research surveys television viewers to determine the numbers of people watching particular shows. There are an estimated 107.9 million U.S. households with televisions. Each **rating point** represents 1% of that number, or 1,079,000 households. Therefore, for instance, if *CSI* received a rating of 9.2, then 9.2%, or $(0.092)(107,900,000) = 9,926,800$ households, watched that program.

A rating point does not mean that 1,079,000 people are watching a program. A rating point refers to the number of TV sets tuned to that program; there may be more than one person watching a television set in the household.

Nielsen Media Research also describes a program's *share* of the market. **Share** is the percent of television sets in use that are tuned to a program. Suppose the same week that *CSI* received 9.2 rating points, the show received a share of 25. This would mean that 25% of all households with a television *turned on* were tuned to *CSI*, whereas 9.2% of all households with a television were tuned to the program.

1. If *Law and Order* received a Nielsen rating of 8.8 and a share of 15, how many TV households watched the program that week? How many TV households were watching television during that hour? Round to the nearest hundred thousand.

2. Suppose *Everybody Loves Raymond* received a rating of 9.7 and a share of 15. How many TV households watched the program that week? How many TV households were watching television during that hour? Round to the nearest hundred-thousand.

3. Suppose *NFL Monday Night Football* received a rating of 12.9 during a week in which 34,750,000 people were watching the show. Find the average number of people per TV household who watched the program. Round to the nearest tenth.

The cost to advertise during a program is related to its Nielsen rating. The sponsor (the company paying for the advertisement) pays a certain number of dollars for each rating point a show receives.

4. Suppose a television network charges $35,000 per rating point for a 30-second commercial on a daytime talk show. Determine the cost for three 30-second commercials if the Nielsen rating of the show is 11.5.

Nielsen Media Research also tracks the exposure of advertisements. For example, it might be reported that commercials for McDonald's had 500,000,000 household exposures during a week when its advertisement was aired 90 times.

5. Information regarding household exposure of advertisements can be found in *USA Today* each Monday. For a recent week, find the information for the top four advertised brands. For each brand, calculate the average household exposure for each time the ad was aired.

Nielsen Media Research has a web site on the Internet. You can locate the site by using a search engine. The site does not list rating points and market share, but these statistics can be found on other web sites by using a search engine.

6. Find the top two prime-time television shows for last week. Calculate the number of TV households that watched each program. Compare these figures with the top two sports programs for last week.

Chapter 3 Summary

Key Words	Examples
An *equation* expresses the equality of two mathematical expressions. [3.1A, p. 115]	$3 + 2(4x - 5) = x + 4$ is an equation.
A *solution of an equation* is a number that, when substituted for the variable, results in a true equation. [3.1A, p. 115]	-2 is a solution of $2 - 3x = 8$ because $2 - 3(-2) = 8$ is a true equation.
To *solve an equation* means to find a solution of the equation. The goal is to rewrite the equation in the form *variable = constant*, because the constant is the solution. [3.1B, p. 116]	The equation $x = -3$ is in the form *variable = constant*. The constant, -3, is the solution of the equation.
Cost is the price that a business pays for a product. *Selling price* is the price for which a business sells a product to a customer. *Markup* is the difference between selling price and cost. *Markup rate* is the markup expressed as a percent of the retailer's cost. [3.2B, p. 137]	If a business pays $50 for a product and sells that product for $70, then the cost of the product is $50, the selling price is $70, the markup is $70 − $50 = $20, and the markup rate is $\frac{20}{50} = 40\%$.
Discount is the amount by which a retailer reduces the regular price of a product. *Discount rate* is the discount expressed as a percent of the regular price. [3.2B, p. 138]	The regular price of a product is $25. The product is now on sale for $20. The discount is $25 − $20 = $5. The discount rate is $\frac{5}{25} = 20\%$.

Consecutive integers follow one another in order. [3.4A, p. 154]

5, 6, 7 are consecutive integers.
$-9, -8, -7$ are consecutive integers.

An *acute angle* is an angle whose measure is between 0° and 90°. An *obtuse angle* is an angle whose measure is between 90° and 180°. [3.5A, p. 161]

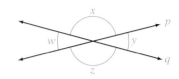

Acute angle   Obtuse angle

Two angles that are on the opposite sides of the intersection of two lines are *vertical angles*. Vertical angles have the same measure. Two angles that share a common side are *adjacent angles*. [3.5A, p. 161]

$$m\angle w = m\angle y$$
$$m\angle x = m\angle z$$

A line that intersects two other lines at two different points is a *transversal*. If the lines cut by a transversal are parallel lines, pairs of equal angles are formed: *alternate exterior angles*, *alternate interior angles*, and *corresponding angles*. [3.5A, p. 162]

$$m\angle b = m\angle d = m\angle x = m\angle z$$
$$m\angle a = m\angle c = m\angle w = m\angle y$$

Essential Rules and Procedures

Examples

Addition Property of Equations [3.1B, p. 116]
The same number can be added to each side of an equation without changing the solution of the equation.

If $a = b$, then $a + c = b + c$.

Multiplication Property of Equations [3.1C, p. 117]
Each side of an equation can be multiplied by the same *nonzero* number without changing the solution of the equation.

If $a = b$ and $c \neq 0$, then $ac = bc$.

Basic Percent Equation [3.1D, p. 119]
Percent · Base = Amount
$$P \cdot B = A$$

30% of what number is 24?

$$PB = A$$
$$0.30B = 24$$
$$\frac{0.30B}{0.30} = \frac{24}{0.30}$$
$$B = 80$$

Simple Interest Equation [3.1D, p. 120]
Interest = Principle · Rate · Time
$$I = Prt$$

A credit card company charges an annual interest rate of 21% on the monthly unpaid balance on a card. Find the amount of interest charged on an unpaid balance of $232 for April.

$$I = Prt$$
$$I = 232(0.21)\left(\frac{1}{12}\right)$$
$$= 4.06$$

Basic Markup Equation   [3.2B, p. 137]

$S = C + rC$

The manager of a electronics store buys an MP3 player for $200 and sells the player for $250. Find the markup rate.

$250 = 200 + 200r$

Basic Discount Equation   [3.2B, p. 138]

$S = R - rR$

The sale price for a digital phone is $49. This price is 25% off the regular price. Find the regular price.

$49 = R - 0.25R$

Consecutive Integers   [3.4A, p. 154]

$n, n + 1, n + 2, \ldots$

The sum of three consecutive integers is 33.

$n + (n + 1) + (n + 2) = 33$

Consecutive Even or Consecutive Odd Integers   [3.4A, p. 154]

$n, n + 2, n + 4$

The sum of three consecutive odd integers is 33.

$n + (n + 2) + (n + 4) = 33$

Sum of the Angles of a Triangle   [3.5B, p. 163]

The sum of the measures of the angles of a triangle is 180°.

$m\angle a + m\angle b + m\angle c = 180°$

In a right triangle, the measure of one acute angle is twice the measure of the other acute angle.

$90° + x + 2x = 180°$

Value Mixture Equation   [3.6A, p. 168]

Amount · Unit Cost = Value

$\qquad AC = V$

An herbalist has 30 oz of herbs costing $4 per ounce. How many ounces of herbs costing $2 per ounce should be mixed with the 30 oz to produce a mixture costing $3.20 per ounce?

$30(4) + 2x = 3.20(30 + x)$

Percent Mixture Equation   [3.1D, 3.6B, p. 170]

Quantity = Amount · Percent Concentration

$\qquad Q = Ar$

Forty ounces of a 30% gold alloy are mixed with 60 oz of a 20% gold alloy. Find the percent concentration of the resulting gold alloy.

$0.30(40) + 0.20(60) = x(100)$

Uniform Motion Equation   [3.1E, 3.6C, p. 172]

Distance = Rate · Time

$\qquad d = rt$

A boat traveled from a harbor to an island at an average speed of 20 mph. The average speed on the return trip was 15 mph. The total trip took 3.5 h. How long did it take to travel to the island?

$20t = 15(3.5 - t)$

Chapter 3 Review Exercises

1. Solve: $x + 3 = 24$

2. Solve: $x + 5(3x - 20) = 10(x - 4)$

3. Solve: $5x - 6 = 29$

4. Is 3 a solution of $5x - 2 = 4x + 5$?

5. Solve: $\dfrac{3}{5}a = 12$

6. Solve: $6x + 3(2x - 1) = -27$

7. 30 is what percent of 12?

8. Solve: $5x + 3 = 10x - 17$

9. Solve: $7 - [4 + 2(x - 3)] = 11(x + 2)$

10. Solve: $-6x + 16 = -2x$

11. **Business** A furniture store uses a markup rate of 60%. The store sells a solid oak curio cabinet for $1074. Find the cost of the curio cabinet. Use the formula $S = C + rC$, where S is the selling price, C is the cost, and r is the markup rate.

12. **Geometry** Find the measure of x.

13. **Geometry** Find the measure of x.

14. **Physics** A lever is 12 ft long. At a distance of 2 ft from the fulcrum, a force of 120 lb is applied. How large a force must be applied to the other end so that the system will balance? Use the lever system equation $F_1x = F_2(d - x)$.

15. **Travel** A bus traveled on a level road for 2 h at an average speed of 20 mph faster than it traveled on a winding road. The time spent on the winding road was 3 h. Find the average speed on the winding road if the total trip was 200 mi.

16. **Business**   A ceiling fan, which regularly sells for $60, is on sale for $40. Find the discount rate. Use the formula $S = R - rR$, where S is the sale price, R is the regular price, and r is the discount rate.

17. **Geometry**   Given that $m\angle a = 74°$ and $m\angle b = 52°$, find the measures of angles x and y.

18. **Juice Mixtures**   A health food store combined cranberry juice that cost $1.79 per quart with apple juice that cost $1.19 per quart. How many quarts of each were used to make 10 qt of cranapple juice costing $1.61 per quart?

19. **Number Sense**   Four times the second of three consecutive integers equals the sum of the first and third integers. Find the integers.

20. **Geometry**   One angle of a triangle is 15° more than the measure of the second angle. The third angle is 15° less than the measure of the second angle. Find the measure of each angle.

21. **Number Sense**   Translate "four less than the product of five and a number is sixteen" into an equation and solve.

22.   **Building Height**   The Empire State Building is 1472 ft tall. This is 654 ft less than twice the height of the Eiffel Tower. Find the height of the Eiffel Tower.

23. **Geometry**   Given $m\angle y = 115°$, find $m\angle x$.

24. **Geometry**   Given $OA \perp OB$ and $m\angle x = 30°$, find $m\angle y$.

25. **Travel**   A jet plane traveling at 600 mph overtakes a propeller-driven plane that had a 2-hour head start. The propeller-driven plane is traveling at 200 mph. How far from the starting point does the jet overtake the propeller-driven plane?

26. **Number Sense**   The sum of two numbers is twenty-one. Three times the smaller number is two less than twice the larger number. Find the two numbers.

27. **Food Mixtures**   A dairy owner mixed 5 gal of cream containing 30% butterfat with 8 gal of milk containing 4% butterfat. What is the percent of butterfat in the resulting mixture?

Chapter 3 Test

1. Solve: $3x - 2 = 5x + 8$

2. Solve: $x - 3 = -8$

3. Solve: $3x - 5 = -14$

4. Solve: $4 - 2(3 - 2x) = 2(5 - x)$

5. Is -2 a solution of $x^2 - 3x = 2x - 6$?

6. Solve: $7 - 4x = -13$

7. What is 0.5% of 8?

8. Solve: $5x - 2(4x - 3) = 6x + 9$

9. Solve: $5x + 3 - 7x = 2x - 5$

10. Solve: $\dfrac{3}{4}x = -9$

11. **Flour Mixtures** A baker wants to make a 15-pound blend of flour that costs $.60 per pound. The blend is made using a rye flour that costs $.70 per pound and a wheat flour that costs $.40 per pound. How many pounds of each flour should be used?

12. **Geometry** Find x.

13. **Business** A television that regularly sells for $450 is on sale for $360. Find the discount rate. Use the formula $S = R - rR$, where S is the sale price, R is the regular price, and r is the discount rate.

14. **Finance** A financial manager has determined that the cost per unit for a calculator is $15 and that the fixed cost per month is $2000. Find the number of calculators produced during a month in which the total cost was $5000. Use the equation $T = U \cdot N + F$, where T is the total cost, U is the cost per unit, N is the number of units produced, and F is the fixed cost.

15. **Geometry** In an isosceles triangle, two angles are equal. The third angle of the triangle is 30° less than one of the equal angles. Find the measure of one of the equal angles.

16. **Consecutive Integers** Find three consecutive even integers whose sum is 36.

17. **Chemistry** How many gallons of water must be mixed with 5 gal of a 20% salt solution to make a 16% salt solution?

18. **Geometry** Given that $\ell_1 \parallel \ell_2$, find the measures of angles a and b.

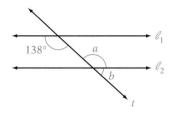

19. **Number Sense** Translate "The difference between three times a number and fifteen is twenty-seven" into an equation and solve.

20. **Sports** A cross-country skier leaves a camp to explore a wilderness area. Two hours later a friend leaves the camp in a snowmobile, traveling 4 mph faster than the skier. This friend meets the skier 1 h later. Find the rate of the snowmobile.

21. **Business** A company makes 140 televisions per day. Three times the number of 15-inch TVs made equals 20 less than the number of 25-inch TVs made. Find the number of 25-inch TVs made each day.

22. **Number Sense** The sum of two numbers is eighteen. The difference between four times the smaller number and seven is equal to the sum of two times the larger number and five. Find the two numbers.

23. **Aviation** As part of flight training, a student pilot was required to fly to an airport and then return. The average speed to the airport was 90 mph, and the average speed returning was 120 mph. Find the distance between the two airports if the total flying time was 7 h.

24. **Geometry** Given that $m\angle a = 50°$ and $m\angle b = 92°$, find the measures of angles x and y.

25. **Chemistry** A chemist mixes 100 g of water at 80°C with 50 g of water at 20°C. To find the final temperature of the water after mixing, use the equation $m_1(T_1 - T) = m_2(T - T_2)$, where m_1 is the quantity of water at the hotter temperature, T_1 is the temperature of the hotter water, m_2 is the quantity of water at the cooler temperature, T_2 is the temperature of the cooler water, and T is the final temperature of the water after mixing.

Cumulative Review Exercises

1. Subtract: $-6 - (-20) - 8$

2. Multiply: $(-2)(-6)(-4)$

3. Subtract: $-\dfrac{5}{6} - \left(-\dfrac{7}{16}\right)$

4. Divide: $-2\dfrac{1}{3} \div 1\dfrac{1}{6}$

5. Simplify: $-4^2 \cdot \left(-\dfrac{3}{2}\right)^3$

6. Simplify: $25 - 3\dfrac{(5-2)^2}{2^3 + 1} - (-2)$

7. Evaluate $3(a - c) - 2ab$ when $a = 2$, $b = 3$, and $c = -4$.

8. Simplify: $3x - 8x + (-12x)$

9. Simplify: $2a - (-3b) - 7a - 5b$

10. Simplify: $(16x)\left(\dfrac{1}{8}\right)$

11. Simplify: $-4(-9y)$

12. Simplify: $-2(-x^2 - 3x + 2)$

13. Simplify: $-2(x - 3) + 2(4 - x)$

14. Simplify: $-3[2x - 4(x - 3)] + 2$

15. Is -3 a solution of $x^2 + 6x + 9 = x + 3$?

16. Is $\dfrac{1}{2}$ a solution of $3 - 8x = 12x - 2$?

17. Find 32% of 60.

18. Solve: $\dfrac{3}{5}x = -15$

19. Solve: $7x - 8 = -29$

20. Solve: $13 - 9x = -14$

21. Solve: $8x - 3(4x - 5) = -2x - 11$

22. Solve: $6 - 2(5x - 8) = 3x - 4$

23. Solve: $5x - 8 = 12x + 13$

24. Solve: $11 - 4x = 2x + 8$

25. Chemistry A chemist mixes 300 g of water at 75°C with 100 g of water at 15°C. To find the final temperature of the water after mixing, use the equation $m_1(T_1 - T) = m_2(T - T_2)$, where m_1 is the quantity of water at the hotter temperature, T_1 is the temperature of the hotter water, m_2 is the quantity of water at the cooler temperature, T_2 is the temperature of the cooler water, and T is the final temperature of the water after mixing.

26. Number Sense Translate "The difference between twelve and the product of five and a number is negative eighteen" into an equation and solve.

27. Construction The area of a cement foundation of a house is 2000 ft². This is 200 ft² more than three times the area of the garage. Find the area of the garage.

28. Flour Mixtures How many pounds of an oat flour that costs $.80 per pound must be mixed with 40 lb of a wheat flour that costs $.50 per pound to make a blend that costs $.60 per pound?

29. Metallurgy How many grams of pure gold must be added to 100 g of a 20% gold alloy to make an alloy that is 36% gold?

30. Geometry The perimeter of a rectangular office is 44 ft. The length of the office is 2 ft more than the width. Find the dimensions of the office.

31. Geometry Find the measure of $\angle x$.

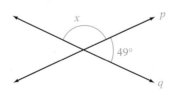

32. Geometry In an equilateral triangle, all three angles are equal. Find the measure of one of the angles of an equilateral triangle.

33. Sports A sprinter ran to the end of a track at an average rate of 8 m/s and then jogged back to the starting point at an average rate of 3 m/s. The sprinter took 55 s to run to the end of the track and jog back. Find the length of the track.

4 Polynomials

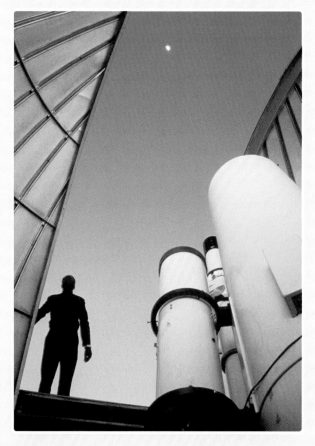

High-powered telescopes, such as the one shown here, allow scientists to look at objects that are extremely large but very far away. As an astronomer, this man uses scientific notation to describe distances in space. Scientific notation replaces very large or very small numbers with more concise expressions, making these numbers easier to read and write. **Exercises 115 to 122 on pages 217 and 218** describe some situations in which scientific notation would be used.

OBJECTIVES

Section 4.1

A To add polynomials
B To subtract polynomials

Section 4.2

A To multiply monomials
B To simplify powers of monomials

Section 4.3

A To multiply a polynomial by a monomial
B To multiply two polynomials
C To multiply two binomials
D To multiply binomials that have special products
E To solve application problems

Section 4.4

A To divide monomials
B To write a number in scientific notation

Section 4.5

A To divide a polynomial by a monomial
B To divide polynomials

Need help? For online student resources, such as section quizzes, visit this textbook's website at **math.college.hmco.com/students.**

PREP TEST • • •

Do these exercises to prepare for Chapter 4.

1. Subtract: $-2 - (-3)$

2. Multiply: $-3(6)$

3. Simplify: $-\dfrac{24}{-36}$

4. Evaluate $3n^4$ when $n = -2$.

5. If $\dfrac{a}{b}$ is a fraction in simplest form, what number is not a possible value of b?

6. Are $2x^2$ and $2x$ like terms?

7. Simplify: $3x^2 - 4x + 1 + 2x^2 - 5x - 7$

8. Simplify: $-4y + 4y$

9. Simplify: $-3(2x - 8)$

10. Simplify: $3xy - 4y - 2(5xy - 7y)$

GO FIGURE • • •

If $x + y$, xy, and $\dfrac{x}{y}$ all equal the same number, find the values of x and y.

4.1 Addition and Subtraction of Polynomials

Objective A ▶ **To add polynomials**

A **monomial** is a number, a variable, or a product of numbers and variables. For instance,

7	b	$\dfrac{2}{3}a$	$12xy^2$
A number	A variable	A product of a number and a variable	A product of a number and variables

A **polynomial** is a variable expression in which the terms are monomials.

A polynomial of *one* term is a **monomial.** $-7x^2$ is a monomial.
A polynomial of *two* terms is a **binomial.** $4x + 2$ is a binomial.
A polynomial of *three* terms is a **trinomial.** $7x^2 + 5x - 7$ is a trinomial.

The **degree of a polynomial in one variable** is the greatest exponent on a variable. The degree of $4x^3 - 5x^2 + 7x - 8$ is 3; the degree of $2y^4 + y^2 - 1$ is 4. The degree of a nonzero constant is zero. For instance, the degree of 7 is zero. The number zero has no degree.

The terms of a polynomial in one variable are usually arranged so that the exponents of the variable decrease from left to right. This is called **descending order.**

$$5x^3 - 4x^2 + 6x - 1$$
$$7z^4 + 4z^3 + z - 6$$
$$2y^4 + y^3 - 2y^2 + 4y - 1$$

Polynomials can be added, using either a horizontal or a vertical format, by combining like terms.

HOW TO   Add $(3x^3 - 7x + 2) + (7x^2 + 2x - 7)$. Use a horizontal format.

$(3x^3 - 7x + 2) + (7x^2 + 2x - 7)$

$= 3x^3 + 7x^2 + (-7x + 2x) + (2 - 7)$ • Use the Commutative and Associative Properties of Addition to rearrange and group like terms.

$= 3x^3 + 7x^2 - 5x - 5$ • Then combine like terms.

HOW TO   Add $(-4x^2 + 6x - 9) + (12 - 8x + 2x^3)$. Use a vertical format.

$$\begin{array}{r} -4x^2 + 6x - 9 \\ 2x^3 - 8x + 12 \\ \hline 2x^3 - 4x^2 - 2x + 3 \end{array}$$

• Arrange the terms of each polynomial in descending order with like terms in the same column.

• Combine the terms in each column.

Example 1

Use a horizontal format to add
$(8x^2 - 4x - 9) + (2x^2 + 9x - 9)$.

Solution

$(8x^2 - 4x - 9) + (2x^2 + 9x - 9)$
$= (8x^2 + 2x^2) + (-4x + 9x) + (-9 - 9)$
$= 10x^2 + 5x - 18$

You Try It 1

Use a horizontal format to add
$(-4x^3 + 2x^2 - 8) + (4x^3 + 6x^2 - 7x + 5)$.

Your solution

Solution on p. S10

Example 2

Use a vertical format to add
$(-5x^3 + 4x^2 - 7x + 9) + (2x^3 + 5x - 11)$.

Solution

$$
\begin{array}{r}
-5x^3 + 4x^2 - 7x + 19 \\
2x^3 + 4x^2 + 5x - 11 \\
\hline
-3x^3 + 4x^2 - 2x - 2
\end{array}
$$

You Try It 2

Use a vertical format to add
$(6x^3 + 2x + 8) + (-9x^3 + 2x^2 - 12x - 8)$.

Your solution

Solution on p. S10

Objective B   **To subtract polynomials**

The **opposite of the polynomial** $(3x^2 - 7x + 8)$ is $-(3x^2 - 7x + 8)$.

To simplify the opposite of a polynomial, $\quad -(3x^2 - 7x + 8) = -3x^2 + 7x - 8$
change the sign of each term to its opposite.

Polynomials can be subtracted using either a horizontal or a vertical format. To subtract, add the opposite of the second polynomial to the first.

> **TAKE NOTE**
>
> This is the same definition used for subtraction of integers: Subtraction is addition of the opposite.

HOW TO Subtract $(4y^2 - 6y + 7) - (2y^3 - 5y - 4)$. Use a horizontal format.

$(4y^2 - 6y + 7) - (2y^3 - 5y - 4)$
$= (4y^2 - 6y + 7) + (-2y^3 + 5y + 4)$ • Add the opposite of the second
$= -2y^3 + 4y^2 + (-6y + 5y) + (7 + 4)$ polynomial to the first.
$= -2y^3 + 4y^2 - y + 11$ • Combine like terms.

HOW TO Subtract $(9 + 4y + 3y^3) - (2y^2 + 4y - 21)$. Use a vertical format.

The opposite of $2y^2 + 4y - 21$ is $-2y^2 - 4y + 21$.

$$
\begin{array}{r}
3y^3 + 4y + 9 \\
- 2y^2 - 4y + 21 \\
\hline
3y^3 - 2y^2 + 30
\end{array}
$$

• Arrange the terms of each polynomial in descending order with like terms in the same column.

• Note that $4y - 4y = 0$, but 0 is not written.

Example 3

Use a horizontal format to subtract
$(7c^2 - 9c - 12) - (9c^2 + 5c - 8)$.

Solution

$(7c^2 - 9c - 12) - (9c^2 + 5c - 8)$
$= (7c^2 - 9c - 12) + (-9c^2 - 5c + 8)$
$= -2c^2 - 14c - 4$

You Try It 3

Use a horizontal format to subtract
$(-4w^3 + 8w - 8) - (3w^3 - 4w^2 - 2w - 1)$.

Your solution

Example 4

Use a vertical format to subtract
$(3k^2 - 4k + 1) - (k^3 + 3k^2 - 6k - 8)$.

Solution

$$
\begin{array}{r}
3k^2 - 4k + 1 \\
-k^3 - 3k^2 + 6k + 8 \\
\hline
-k^3 + 2k + 9
\end{array}
$$

• Add the opposite of $(k^3 + 3k^2 - 6k - 8)$ to the first polynomial.

You Try It 4

Use a vertical format to subtract
$(13y^3 - 6y - 7) - (4y^2 - 6y - 9)$.

Your solution

Solutions on p. S10

4.1 Exercises

Objective A   **To add polynomials**

For Exercises 1 to 8, state whether the expression is a monomial.

1. 17 **2.** $3x^4$ **3.** $\dfrac{17}{\sqrt{x}}$ **4.** xyz

5. $\dfrac{2}{3}y$ **6.** $\dfrac{xy}{z}$ **7.** $\sqrt{5}\,x$ **8.** πx

For Exercises 9 to 16, state whether the expression is a monomial, a binomial, a trinomial, or none of these.

9. $3x + 5$ **10.** $2y - 3\sqrt{y}$ **11.** $9x^2 - x - 1$ **12.** $x^2 + y^2$

13. $\dfrac{2}{x} - 3$ **14.** $\dfrac{ab}{4}$ **15.** $6x^2 + 7x$ **16.** $12a^4 - 3a + 2$

For Exercises 17 to 26, add. Use a vertical format.

17. $(x^2 + 7x) + (-3x^2 - 4x)$ **18.** $(3y^2 - 2y) + (5y^2 + 6y)$

19. $(y^2 + 4y) + (-4y - 8)$ **20.** $(3x^2 + 9x) + (6x - 24)$

21. $(2x^2 + 6x + 12) + (3x^2 + x + 8)$ **22.** $(x^2 + x + 5) + (3x^2 - 10x + 4)$

23. $(-7x + x^3 + 4) + (2x^2 + x - 10)$ **24.** $(y^2 + 3y^3 + 1) + (-4y^3 - 6y - 3)$

25. $(2a^3 - 7a + 1) + (1 - 4a - 3a^2)$ **26.** $(5r^3 - 6r^2 + 3r) + (-3 - 2r + r^2)$

For Exercises 27 to 36, add. Use a horizontal format.

27. $(4x^2 + 2x) + (x^2 + 6x)$ **28.** $(-3y^2 + y) + (4y^2 + 6y)$

29. $(4x^2 - 5xy) + (3x^2 + 6xy - 4y^2)$ **30.** $(2x^2 - 4y^2) + (6x^2 - 2xy + 4y^2)$

31. $(2a^2 - 7a + 10) + (a^2 + 4a + 7)$ **32.** $(-6x^2 + 7x + 3) + (3x^2 + x + 3)$

33. $(7x + 5x^3 - 7) + (10x^2 - 8x + 3)$ **34.** $(4y + 3y^3 + 9) + (2y^2 + 4y - 21)$

35. $(7 - 5r + 2r^2) + (3r^3 - 6r)$ **36.** $(14 + 4y + 3y^3) + (-4y^2 + 21)$

> Objective B   **To subtract polynomials**

For Exercises 37 to 46, subtract. Use a vertical format.

37. $(x^2 - 6x) - (x^2 - 10x)$

38. $(y^2 + 4y) - (y^2 + 10y)$

39. $(2y^2 - 4y) - (-y^2 + 2)$

40. $(-3a^2 - 2a) - (4a^2 - 4)$

41. $(x^2 - 2x + 1) - (x^2 + 5x + 8)$

42. $(3x^2 + 2x - 2) - (5x^2 - 5x + 6)$

43. $(4x^3 + 5x + 2) - (1 + 2x - 3x^2)$

44. $(5y^2 - y + 2) - (-3 + 3y - 2y^3)$

45. $(-2y + 6y^2 + 2y^3) - (4 + y^2 + y^3)$

46. $(4 - x - 2x^2) - (-2 + 3x - x^3)$

For Exercises 47 to 56, subtract. Use a horizontal format.

47. $(y^2 - 10xy) - (2y^2 + 3xy)$

48. $(x^2 - 3xy) - (-2x^2 + xy)$

49. $(3x^2 + x - 3) - (4x + x^2 - 2)$

50. $(5y^2 - 2y + 1) - (-y - 2 - 3y^2)$

51. $(-2x^3 + x - 1) - (-x^2 + x - 3)$

52. $(2x^2 + 5x - 3) - (3x^3 + 2x - 5)$

53. $(1 - 2a + 4a^3) - (a^3 - 2a + 3)$

54. $(7 - 8b + b^2) - (4b^3 - 7b - 8)$

55. $(-1 - y + 4y^3) - (3 - 3y - 2y^2)$

56. $(-3 - 2x + 3x^2) - (4 - 2x^2 + 2x^3)$

APPLYING THE CONCEPTS

57. What polynomial must be added to $3x^2 - 6x + 9$ so that the sum is $4x^2 + 3x - 2$?

58. What polynomial must be subtracted from $2x^2 - x - 2$ so that the difference is $5x^2 + 3x + 1$?

59. ✎ In your own words, explain the terms *monomial*, *binomial*, *trinomial*, and *polynomial*. Give an example of each.

60. ✎ Is it possible to subtract two polynomials, each of degree 3, and have the difference be a polynomial of degree 2? If so, give an example. If not, explain why not.

61. ✎ Is it possible to add two polynomials, each of degree 3, and have the sum be a polynomial of degree 2? If so, give an example. If not, explain why not.

4.2 Multiplication of Monomials

Objective A   ## To multiply monomials

Recall that in an exponential expression such as x^6, x is the base and 6 is the exponent. The exponent indicates the number of times the base occurs as a factor.

The product of exponential expressions with the *same* base can be simplified by writing each expression in factored form and writing the result with an exponent.

$$x^3 \cdot x^2 = \overbrace{(x \cdot x \cdot x)}^{\text{3 factors}} \cdot \overbrace{(x \cdot x)}^{\text{2 factors}}$$
$$\underbrace{\qquad\qquad\qquad}_{\text{5 factors}}$$
$$= x^5$$

Note that adding the exponents results in the same product.

$$x^3 \cdot x^2 = x^{3+2} = x^5$$

Rule for Multiplying Exponential Expressions

If m and n are positive integers, then $x^m \cdot x^n = x^{m+n}$.

HOW TO   Simplify: $y^4 \cdot y \cdot y^3$

$y^4 \cdot y \cdot y^3 = y^{4+1+3}$ • The bases are the same. **Add the**
$\qquad\qquad\;\; = y^8$ **exponents.** Recall that $y = y^1$.

HOW TO   Simplify: $(-3a^4b^3)(2ab^4)$

$(-3a^4b^3)(2ab^4) = (-3 \cdot 2)(a^4 \cdot a)(b^3 \cdot b^4)$ • Use the Commutative and Associative Properties of Multiplication to rearrange and group factors.

$\qquad\qquad\qquad = -6(a^{4+1})(b^{3+4})$ • To multiply expressions with the same base, **add the exponents.**

$\qquad\qquad\qquad = -6a^5b^7$ • Simplify.

TAKE NOTE

The Rule for Multiplying Exponential Expressions requires the bases to be the same. The expression a^5b^7 cannot be simplified.

Example 1   Simplify: $(-5ab^3)(4a^5)$

Solution
$(-5ab^3)(4a^5)$
$= (-5 \cdot 4)(a \cdot a^5)b^3$ • Multiply coefficients. Add exponents with same base.
$= -20a^6b^3$

You Try It 1   Simplify: $(8m^3n)(-3n^5)$

Your solution

Example 2   Simplify: $(6x^3y^2)(4x^4y^5)$

Solution
$(6x^3y^2)(4x^4y^5)$
$= (6 \cdot 4)(x^3 \cdot x^4)(y^2 \cdot y^5)$ • Multiply coefficients. Add exponents with same base.
$= 24x^7y^7$

You Try It 2   Simplify: $(12p^4q^3)(-3p^5q^2)$

Your solution

Solutions on p. S10

Objective B   **To simplify powers of monomials**

The power of a monomial can be simplified by writing the power in factored form and then using the Rule for Multiplying Exponential Expressions.

$(x^4)^3 = x^4 \cdot x^4 \cdot x^4$ $(a^2b^3)^2 = (a^2b^3)(a^2b^3)$ • Write in factored form.

$\qquad = x^{4+4+4}$ $\qquad = a^{2+2}b^{3+3}$ • Use the Rule for Multiplying Exponential Expressions.

$\qquad = x^{12}$ $\qquad = a^4b^6$

Note that multiplying each exponent inside the parentheses by the exponent outside the parentheses results in the same product.

$(x^4)^3 = x^{4 \cdot 3} = x^{12}$ $(a^2b^3)^2 = a^{2 \cdot 2}b^{3 \cdot 2} = a^4b^6$ • Multiply each exponent inside the parentheses by the exponent outside the parentheses.

> **Rule for Simplifying the Power of an Exponential Expression**
>
> If m and n are positive integers, then $(x^m)^n = x^{mn}$.

> **Rule for Simplifying the Power of a Product**
>
> If m, n, and p are positive integers, then $(x^m y^n)^p = x^{mp} y^{np}$.

HOW TO   Simplify: $(5x^2y^3)^3$

$(5x^2y^3)^3 = 5^{1 \cdot 3}x^{2 \cdot 3}y^{3 \cdot 3}$ • Use the Rule for Simplifying the Power of a Product. Note that $5 = 5^1$.

$\qquad = 5^3x^6y^9$

$\qquad = 125x^6y^9$ • Evaluate 5^3.

Example 3   Simplify: $(-2p^3r)^4$

Solution
$(-2p^3r)^4 = (-2)^{1 \cdot 4}p^{3 \cdot 4}r^{1 \cdot 4}$ • Use the Rule for Simplifying the Power of a Product.
$\qquad = (-2)^4p^{12}r^4 = 16p^{12}r^4$

You Try It 3   Simplify: $(-3a^4bc^2)^3$

Your solution

Example 4   Simplify: $(2a^2b)(2a^3b^2)^3$

Solution
$(2a^2b)(2a^3b^2)^3$
$\quad = (2a^2b)(2^{1 \cdot 3}a^{3 \cdot 3}b^{2 \cdot 3})$ • Use the Rule for Simplifying the Power of a Product.
$\quad = (2a^2b)(2^3a^9b^6)$
$\quad = (2a^2b)(8a^9b^6) = 16a^{11}b^7$

You Try It 4   Simplify: $(-xy^4)(-2x^3y^2)^2$

Your solution

Solutions on p. S10

4.2 Exercises

Objective A **To multiply monomials**

1. ✏ Explain how to multiply two monomials. Provide an example.

2. ✏ Explain how to simplify the power of a monomial. Provide an example.

For Exercises 3 to 35, simplify.

3. $(6x^2)(5x)$

4. $(-4y^3)(2y)$

5. $(7c^2)(-6c^4)$

6. $(-8z^5)(5z^8)$

7. $(-3a^3)(-3a^4)$

8. $(-5a^6)(-2a^5)$

9. $(x^2)(xy^4)$

10. $(x^2y^4)(xy^7)$

11. $(-2x^4)(5x^5y)$

12. $(-3a^3)(2a^2b^4)$

13. $(-4x^2y^4)(-3x^5y^4)$

14. $(-6a^2b^4)(-4ab^3)$

15. $(2xy)(-3x^2y^4)$

16. $(-3a^2b)(-2ab^3)$

17. $(x^2yz)(x^2y^4)$

18. $(-ab^2c)(a^2b^5)$

19. $(-a^2b^3)(-ab^2c^4)$

20. $(-x^2y^3z)(-x^3y^4)$

21. $(-5a^2b^2)(6a^3b^6)$

22. $(7xy^4)(-2xy^3)$

23. $(-6a^3)(-a^2b)$

24. $(-2a^2b^3)(-4ab^2)$

25. $(-5y^4z)(-8y^6z^5)$

26. $(3x^2y)(-4xy^2)$

27. $(x^2y)(yz)(xyz)$

28. $(xy^2z)(x^2y)(z^2y^2)$

29. $(3ab^2)(-2abc)(4ac^2)$

30. $(-2x^3y^2)(-3x^2z^2)(-5y^3z^3)$

31. $(4x^4z)(-yz^3)(-2x^3z^2)$

32. $(-a^3b^4)(-3a^4c^2)(4b^3c^4)$

33. $(-2x^2y^3)(3xy)(-5x^3y^4)$

34. $(4a^2b)(-3a^3b^4)(a^5b^2)$

35. $(3a^2b)(-6bc)(2ac^2)$

Objective B **To simplify powers of monomials**

For Exercises 36 to 66, simplify.

36. $(z^4)^3$

37. $(x^3)^5$

38. $(y^4)^2$

39. $(x^7)^2$

40. $(-y^5)^3$

41. $(-x^2)^4$

42. $(-x^2)^3$

43. $(-y^3)^4$

44. $(-3y)^3$

45. $(-2x^2)^3$

46. $(a^3b^4)^3$ **47.** $(x^2y^3)^2$ **48.** $(2x^3y^4)^5$ **49.** $(3x^2y)^2$ **50.** $(-2ab^3)^4$

51. $(-3x^3y^2)^5$ **52.** $(3b^2)(2a^3)^4$ **53.** $(-2x)(2x^3)^2$ **54.** $(2y)(-3y^4)^3$

55. $(3x^2y)(2x^2y^2)^3$ **56.** $(a^3b)^2(ab)^3$ **57.** $(ab^2)^2(ab)^2$ **58.** $(-x^2y^3)^2(-2x^3y)^3$

59. $(-2x)^3(-2x^3y)^3$ **60.** $(-3y)(-4x^2y^3)^3$ **61.** $(-2x)(-3xy^2)^2$ **62.** $(-3y)(-2x^2y)^3$

63. $(ab^2)(-2a^2b)^3$ **64.** $(a^2b^2)(-3ab^4)^2$ **65.** $(-2a^3)(3a^2b)^3$ **66.** $(-3b^2)(2ab^2)^3$

APPLYING THE CONCEPTS

For Exercises 67 to 74, simplify.

67. $3x^2 + (3x)^2$ **68.** $4x^2 - (4x)^2$ **69.** $2x^6y^2 + (3x^2y)^2$ **70.** $(x^2y^2)^3 + (x^3y^3)^2$

71. $(2a^3b^2)^3 - 8a^9b^6$ **72.** $4y^2z^4 - (2yz^2)^2$ **73.** $(x^2y^4)^2 + (2xy^2)^4$ **74.** $(3a^3)^2 - 4a^6 + (2a^2)^3$

For Exercises 75 to 78, answer true or false. If the answer is false, correct the right-hand side of the equation.

75. $(-a)^5 = -a^5$ **76.** $(-b)^8 = b^8$ **77.** $(x^2)^5 = x^{2+5} = x^7$ **78.** $x^3 + x^3 = 2x^{3+3} = 2x^6$

79. Evaluate $(2^3)^2$ and $2^{(3^2)}$. Are the results the same? If not, which expression has the larger value?

80. ✎   If n is a positive integer and $x^n = y^n$, does $x = y$? Explain your answer.

81. ✎   The distance a rock will fall in t seconds is $16t^2$ (neglecting air resistance). Find other examples of quantities that can be expressed in terms of an exponential expression, and explain where the expression is used.

Multiplication of Polynomials

Objective A   **To multiply a polynomial by a monomial**

To multiply a polynomial by a monomial, use the Distributive Property and the Rule for Multiplying Exponential Expressions.

> **HOW TO**   Multiply: $-3a(4a^2 - 5a + 6)$
>
> $-3a(4a^2 - 5a + 6) = -3a(4a^2) - (-3a)(5a) + (-3a)(6)$ • Use the **Distributive Property**.
> $= -12a^3 + 15a^2 - 18a$

Example 1

Multiply: $(5x + 4)(-2x)$

Solution
$(5x + 4)(-2x) = -10x^2 - 8x$

You Try It 1

Multiply: $(-2y + 3)(-4y)$

Your solution

Example 2

Multiply: $2a^2b(4a^2 - 2ab + b^2)$

Solution
$2a^2b(4a^2 - 2ab + b^2)$
$= 8a^4b - 4a^3b^2 + 2a^2b^3$

You Try It 2

Multiply: $-a^2(3a^2 + 2a - 7)$

Your solution

Solutions on p. S10

Objective B   **To multiply two polynomials**

Multiplication of two polynomials requires the repeated application of the Distributive Property.

$$(y - 2)(y^2 + 3y + 1) = (y - 2)(y^2) + (y - 2)(3y) + (y - 2)(1)$$
$$= y^3 - 2y^2 + 3y^2 - 6y + y - 2$$
$$= y^3 + y^2 - 5y - 2$$

A convenient method of multiplying two polynomials is to use a vertical format similar to that used for multiplication of whole numbers.

$$
\begin{array}{r}
y^2 + 3y + 1 \\
y - 2 \\
\hline
-2y^2 - 6y - 2 \\
y^3 + 3y^2 + y \\
\hline
y^3 + y^2 - 5y - 2
\end{array}
$$

$-2y^2 - 6y - 2 = -2(y^2 + 3y + 1)$ • Multiply by **−2**.
$y^3 + 3y^2 + y = y(y + 3y + 1)$ • Multiply by **y**.
                                        • Add the terms in each column.

HOW TO Multiply: $(2a^3 + a - 3)(a + 5)$

$$
\begin{array}{r}
2a^3 \qquad + \; a \; - \; 3 \\
a \; + \; 5 \\
\hline
10a^3 \qquad + 5a - 15 \\
2a^4 \qquad + a^2 - 3a \\
\hline
2a^4 + 10a^3 + a^2 + 2a - 15
\end{array}
$$

- Note that spaces are provided in each product so that like terms are in the same column.

- Add the terms in each column.

Example 3

Multiply: $(2b^3 - b + 1)(2b + 3)$

Solution

$$
\begin{array}{r}
2b^3 \qquad - \; b + 1 \\
2b + 3 \\
\hline
6b^3 \qquad - 3b + 3 \; = 3(2b^3 - b + 1) \\
4b^4 + \qquad - 2b^2 + 2b \quad = 2b(2b^3 - b + 1) \\
\hline
4b^4 + 6b^3 - 2b^2 - \; b + 3
\end{array}
$$

You Try It 3

Multiply: $(2y^3 + 2y^2 - 3)(3y - 1)$

Your solution

Solution on p. S10

Objective C   **To multiply two binomials**

It is frequently necessary to find the product of two binomials. The product can be found using a method called **FOIL**, which is based on the Distributive Property. The letters of FOIL stand for **F**irst, **O**uter, **I**nner, and **L**ast. To find the product of two binomials, add the products of the **F**irst terms, the **O**uter terms, the **I**nner terms, and the **L**ast terms.

TAKE NOTE

FOIL is not really a different way of multiplying. It is based on the Distributive Property.

$(2x + 3)(x + 5)$
$= 2x(x + 5) + 3(x + 5)$
$\quad\;\; \text{F} \quad\;\; \text{O} \quad\; \text{I} \quad\;\; \text{L}$
$= 2x^2 + 10x + 3x + 15$
$= 2x^2 + 13x + 15$

HOW TO Multiply: $(2x + 3)(x + 5)$

Multiply the First terms.	$(2x + 3)(x + 5)$	$2x \cdot x = 2x^2$
Multiply the Outer terms.	$(2x + 3)(x + 5)$	$2x \cdot 5 = 10x$
Multiply the Inner terms.	$(2x + 3)(x + 5)$	$3 \cdot x = 3x$
Multiply the Last terms.	$(2x + 3)(x + 5)$	$3 \cdot 5 = 15$

$$\qquad\qquad\qquad\qquad\qquad\qquad \text{F} \quad\;\; \text{O} \quad\;\; \text{I} \quad\;\; \text{L}$$

Add the products. $(2x + 3)(x + 5)$ $= 2x^2 + 10x + 3x + 15$

Combine like terms. $= 2x^2 + 13x + 15$

HOW TO Multiply: $(4x - 3)(3x - 2)$

$$(4x - 3)(3x - 2) = 4x(3x) + 4x(-2) + (-3)(3x) + (-3)(-2)$$
$$= 12x^2 - 8x - 9x + 6$$
$$= 12x^2 - 17x + 6$$

HOW TO Multiply: $(3x - 2y)(x + 4y)$

$$(3x - 2y)(x + 4y) = 3x(x) + 3x(4y) + (-2y)(x) + (-2y)(4y)$$
$$= 3x^2 + 12xy - 2xy - 8y^2$$
$$= 3x^2 + 10xy - 8y^2$$

Example 4

Multiply: $(2a - 1)(3a - 2)$

Solution

$(2a - 1)(3a - 2) = 6a^2 - 4a - 3a + 2$
$= 6a^2 - 7a + 2$

You Try It 4

Multiply: $(4y - 5)(2y - 3)$

Your solution

Example 5

Multiply: $(3x - 2)(4x + 3)$

Solution

$(3x - 2)(4x + 3) = 12x^2 + 9x - 8x - 6$
$= 12x^2 + x - 6$

You Try It 5

Multiply: $(3b + 2)(3b - 5)$

Your solution

Solutions on p. S11

Objective D   **To multiply binomials that have special products**

Using FOIL, it is possible to find a pattern for the product of the sum and difference of two terms and for the square of a binomial.

Product of the Sum and Difference of the Same Terms

$$(a + b)(a - b) = a^2 - ab + ab - b^2$$
$$= a^2 - b^2$$

Square of the first term ————┘
Square of the second term ————┘

Square of a Binomial

$$(a + b)^2 = (a + b)(a + b) = a^2 + ab + ab + b^2$$
$$= a^2 + 2ab + b^2$$

Square of the first term ——┘
Twice the product of the two terms ——┘
Square of the last term ——┘

HOW TO   Multiply: $(2x + 3)(2x - 3)$

$(2x + 3)(2x - 3) = (2x)^2 - 3^2$ • This is the product of the sum and
$= 4x^2 - 9$ difference of the same terms.

TAKE NOTE

The word *expand* is used frequently to mean "multiply out a power."

HOW TO   Expand: $(3x - 2)^2$

$(3x - 2)^2 = (3x)^2 + 2(3x)(-2) + (-2)^2$ • This is the square of a
$= 9x^2 - 12x + 4$ binomial.

Example 6

Multiply: $(4z - 2w)(4z + 2w)$

Solution
$(4z - 2w)(4z + 2w) = 16z^2 - 4w^2$

You Try It 6

Multiply: $(2a + 5c)(2a - 5c)$

Your solution

Example 7

Expand: $(2r - 3s)^2$

Solution
$(2r - 3s)^2 = 4r^2 - 12rs + 9s^2$

You Try It 7

Expand: $(3x + 2y)^2$

Your solution

Solutions on p. S11

Objective E   **To solve application problems**

Example 8

The length of a rectangle is $(x + 7)$ m. The width is $(x - 4)$ m. Find the area of the rectangle in terms of the variable x.

$x + 7$

$x - 4$

Strategy
To find the area, replace the variables L and W in the equation $A = L \cdot W$ by the given values and solve for A.

Solution
$A = L \cdot W$
$A = (x + 7)(x - 4)$
$A = x^2 - 4x + 7x - 28$
$A = x^2 + 3x - 28$

The area is $(x^2 + 3x - 28)$ m².

You Try It 8

The radius of a circle is $(x - 4)$ ft. Use the equation $A = \pi r^2$, where r is the radius, to find the area of the circle in terms of x. Leave the answer in terms of π.

$x - 4$

Your strategy

Your solution

Solution on p. S11

4.3 Exercises

To multiply a polynomial by a monomial

For Exercises 1 and 2, replace the question marks to make a true statement.

1. $3(4x - 5) = (?)(4x) - (?)(5) = ?$

2. $2x^2(3x + 7) = (?)(3x) + (?)(7) = ?$

For Exercises 3 to 34, multiply.

3. $x(x - 2)$

4. $y(3 - y)$

5. $-x(x + 7)$

6. $-y(7 - y)$

7. $3a^2(a - 2)$

8. $4b^2(b + 8)$

9. $-5x^2(x^2 - x)$

10. $-6y^2(y + 2y^2)$

11. $-x^3(3x^2 - 7)$

12. $-y^4(2y^2 - y^6)$

13. $2x(6x^2 - 3x)$

14. $3y(4y - y^2)$

15. $(2x - 4)3x$

16. $(3y - 2)y$

17. $(3x + 4)x$

18. $(2x + 1)2x$

19. $-xy(x^2 - y^2)$

20. $-x^2y(2xy - y^2)$

21. $x(2x^3 - 3x + 2)$

22. $y(-3y^2 - 2y + 6)$

23. $-a(-2a^2 - 3a - 2)$

24. $-b(5b^2 + 7b - 35)$

25. $x^2(3x^4 - 3x^2 - 2)$

26. $y^3(-4y^3 - 6y + 7)$

27. $2y^2(-3y^2 - 6y + 7)$

28. $4x^2(3x^2 - 2x + 6)$

29. $(a^2 + 3a - 4)(-2a)$

30. $(b^3 - 2b + 2)(-5b)$

31. $-3y^2(-2y^2 + y - 2)$

32. $-5x^2(3x^2 - 3x - 7)$

33. $xy(x^2 - 3xy + y^2)$

34. $ab(2a^2 - 4ab - 6b^2)$

To multiply two polynomials

For Exercises 35 to 52, multiply.

35. $(x^2 + 3x + 2)(x + 1)$

36. $(x^2 - 2x + 7)(x - 2)$

37. $(a^2 - 3a + 4)(a - 3)$

38. $(x^2 - 3x + 5)(2x - 3)$ **39.** $(-2b^2 - 3b + 4)(b - 5)$ **40.** $(-a^2 + 3a - 2)(2a - 1)$

41. $(-2x^2 + 7x - 2)(3x - 5)$ **42.** $(-a^2 - 2a + 3)(2a - 1)$ **43.** $(x^2 + 5)(x - 3)$

44. $(y^2 - 2y)(2y + 5)$ **45.** $(x^3 - 3x + 2)(x - 4)$ **46.** $(y^3 + 4y^2 - 8)(2y - 1)$

47. $(5y^2 + 8y - 2)(3y - 8)$ **48.** $(3y^2 + 3y - 5)(4y - 3)$ **49.** $(5a^3 - 5a + 2)(a - 4)$

50. $(3b^3 - 5b^2 + 7)(6b - 1)$ **51.** $(y^3 + 2y^2 - 3y + 1)(y + 2)$ **52.** $(2a^3 - 3a^2 + 2a - 1)(2a - 3)$

Objective C   **To multiply two binomials**

For Exercises 53 to 84, multiply.

53. $(x + 1)(x + 3)$ **54.** $(y + 2)(y + 5)$ **55.** $(a - 3)(a + 4)$ **56.** $(b - 6)(b + 3)$

57. $(y + 3)(y - 8)$ **58.** $(x + 10)(x - 5)$ **59.** $(y - 7)(y - 3)$ **60.** $(a - 8)(a - 9)$

61. $(2x + 1)(x + 7)$ **62.** $(y + 2)(5y + 1)$ **63.** $(3x - 1)(x + 4)$ **64.** $(7x - 2)(x + 4)$

65. $(4x - 3)(x - 7)$ **66.** $(2x - 3)(4x - 7)$ **67.** $(3y - 8)(y + 2)$ **68.** $(5y - 9)(y + 5)$

69. $(3x + 7)(3x + 11)$ **70.** $(5a + 6)(6a + 5)$ **71.** $(7a - 16)(3a - 5)$ **72.** $(5a - 12)(3a - 7)$

73. $(3a - 2b)(2a - 7b)$ **74.** $(5a - b)(7a - b)$ **75.** $(a - 9b)(2a + 7b)$

76. $(2a + 5b)(7a - 2b)$

77. $(10a - 3b)(10a - 7b)$

78. $(12a - 5b)(3a - 4b)$

79. $(5x + 12y)(3x + 4y)$

80. $(11x + 2y)(3x + 7y)$

81. $(2x - 15y)(7x + 4y)$

82. $(5x + 2y)(2x - 5y)$

83. $(8x - 3y)(7x - 5y)$

84. $(2x - 9y)(8x - 3y)$

Objective D   **To multiply binomials that have special products**

For Exercises 85 to 92, multiply.

85. $(y - 5)(y + 5)$

86. $(y + 6)(y - 6)$

87. $(2x + 3)(2x - 3)$

88. $(4x - 7)(4x + 7)$

89. $(3x - 7)(3x + 7)$

90. $(9x - 2)(9x + 2)$

91. $(4 - 3y)(4 + 3y)$

92. $(4x - 9y)(4x + 9y)$

For Exercises 93 to 102, expand.

93. $(x + 1)^2$

94. $(y - 3)^2$

95. $(3a - 5)^2$

96. $(6x - 5)^2$

97. $(2a + b)^2$

98. $(x + 3y)^2$

99. $(x - 2y)^2$

100. $(2x - 3y)^2$

101. $(5x + 2y)^2$

102. $(2a - 9b)^2$

Objective E   **To solve application problems**

103. **Geometry**   The length of a rectangle is $(5x)$ ft. The width is $(2x - 7)$ ft. Find the area of the rectangle in terms of the variable x.

104. **Geometry**   The width of a rectangle is $(3x + 1)$ in. The length of the rectangle is twice the width. Find the area of the rectangle in terms of the variable x.

105. Geometry The length of a side of a square is $(2x + 1)$ km. Find the area of the square in terms of the variable x.

106. Geometry The radius of a circle is $(x + 4)$ cm. Find the area of the circle in terms of the variable x. Leave the answer in terms of π.

107. Geometry The base of a triangle is $(4x)$ m and the height is $(2x + 5)$ m. Find the area of the triangle in terms of the variable x.

108. Sports A softball diamond has dimensions 45 ft by 45 ft. A base-path border x feet wide lies on both the first-base side and third-base side of the diamond. Express the total area of the softball diamond and the base paths in terms of the variable x.

109. Sports An athletic field has dimensions 30 yd by 100 yd. An end zone that is w yards wide borders each end of the field. Express the total area of the field and the end zones in terms of the variable w.

APPLYING THE CONCEPTS

110. Simplify: $(a + b)^2 - (a - b)^2$ **111.** Expand: $(x^2 + x - 3)^2$ **112.** Expand: $(a + 3)^3$

113. What polynomial has quotient $3x - 4$ when divided by $4x + 5$?

114. Add $x^2 + 2x - 3$ to the product of $2x - 5$ and $3x + 1$.

115. Subtract $4x^2 - x - 5$ from the product of $x^2 + x + 3$ and $x - 4$.

116. If a polynomial of degree 3 is multiplied by a polynomial of degree 2, what is the degree of the resulting polynomial?

117. Is it possible to multiply a polynomial of degree 2 by a polynomial of degree 2 and have the product be a polynomial of degree 3? If so, give an example. If not, explain why not.

4.4 Integer Exponents and Scientific Notation

Objective A   ## To divide monomials

The quotient of two exponential expressions with the same base can be simplified by writing each expression in factored form, dividing by the common factors, and then writing the result with an exponent.

$$\frac{x^5}{x^2} = \frac{\overset{1}{\cancel{x}} \cdot \overset{1}{\cancel{x}} \cdot x \cdot x \cdot x}{\underset{1}{\cancel{x}} \cdot \underset{1}{\cancel{x}}} = x^3$$

Note that subtracting the exponents gives the same result.

$$\frac{x^5}{x^2} = x^{5-2} = x^3$$

To divide two monomials with the same base, subtract the exponents of the like bases.

HOW TO   Simplify: $\dfrac{a^7}{a^3}$

$\dfrac{a^7}{a^3} = a^{7-3}$ • The bases are the same. **Subtract the exponents.**

$\quad = a^4$

HOW TO   Simplify: $\dfrac{r^8 t^6}{r^7 t}$

$\dfrac{r^8 t^6}{r^7 t} = r^{8-7} t^{6-1}$ • **Subtract the exponents** of the like bases.

$\quad = r t^5$

HOW TO   Simplify: $\dfrac{p^7}{z^4}$

Because the bases are not the same, $\dfrac{p^7}{z^4}$ is already in simplest form.

Consider the expression $\dfrac{x^4}{x^4}$, $x \neq 0$. This expression can be simplified, as shown below, by subtracting exponents or by dividing by common factors.

$$\frac{x^4}{x^4} = x^{4-4} = x^0 \qquad\qquad \frac{x^4}{x^4} = \frac{\overset{1}{\cancel{x}} \cdot \overset{1}{\cancel{x}} \cdot \overset{1}{\cancel{x}} \cdot \overset{1}{\cancel{x}}}{\underset{1}{\cancel{x}} \cdot \underset{1}{\cancel{x}} \cdot \underset{1}{\cancel{x}} \cdot \underset{1}{\cancel{x}}} = 1$$

The equations $\dfrac{x^4}{x^4} = x^0$ and $\dfrac{x^4}{x^4} = 1$ suggest the following definition of x^0.

Definition of Zero as an Exponent

If $x \neq 0$, then $x^0 = 1$. The expression 0^0 is not defined.

TAKE NOTE

In the example at the right, we indicated that $a \neq 0$. If we try to evaluate $(12a^3)^0$ when $a = 0$, we have

$$[12(0)^3]^0 = [12(0)]^0 = 0^0$$

However, 0^0 is not defined. Therefore, we must assume that $a \neq 0$. To avoid stating this for every example or exercise, we will assume that variables do not have values that result in the expression 0^0.

HOW TO   Simplify: $(12a^3)^0$, $a \neq 0$

$(12a^3)^0 = 1$ • **Any nonzero expression to the zero power is 1.**

HOW TO   Simplify: $-(4x^3y^7)^0$

$-(4x^3y^7)^0 = -(1) = -1$

Consider the expression $\dfrac{x^4}{x^6}$, $x \neq 0$. This expression can be simplified, as shown below, by subtracting exponents or by dividing by common factors.

$$\frac{x^4}{x^6} = x^{4-6} = x^{-2} \qquad\qquad \frac{x^4}{x^6} = \frac{\overset{1}{\cancel{x}} \cdot \overset{1}{\cancel{x}} \cdot \overset{1}{\cancel{x}} \cdot \overset{1}{\cancel{x}}}{\underset{1}{\cancel{x}} \cdot \underset{1}{\cancel{x}} \cdot \underset{1}{\cancel{x}} \cdot \underset{1}{\cancel{x}} \cdot x \cdot x} = \frac{1}{x^2}$$

Point of Interest

In the 15th century, the expression $12^{\overline{2m}}$ was used to mean $12x^{-2}$. The use of $\overline{m}$ reflected an Italian influence. In Italy, m was used for minus and p was used for plus. It was understood that $2\overline{m}$ referred to an unnamed variable. Issac Newton, in the 17th century, advocated the negative exponent we currently use.

The equations $\dfrac{x^4}{x^6} = x^{-2}$ and $\dfrac{x^4}{x^6} = \dfrac{1}{x^2}$ suggest that $x^{-2} = \dfrac{1}{x^2}$.

Definition of a Negative Exponent

If $x \neq 0$ and n is a positive integer, then

$$x^{-n} = \frac{1}{x^n} \qquad \text{and} \qquad \frac{1}{x^{-n}} = x^n$$

An exponential expression is in simplest form when it is written with only positive exponents.

TAKE NOTE

Note from the example at the right that 2^{-4} is a *positive* number. A negative exponent does not change the sign of a number.

HOW TO   Evaluate 2^{-4}.

$2^{-4} = \dfrac{1}{2^4}$ • **Use the Definition of a Negative Exponent.**

$= \dfrac{1}{16}$ • **Evaluate the expression.**

TAKE NOTE

For the expression $3n^{-5}$ the exponent on n is -5 (*negative* 5). The n^{-5} is written in the denominator as n^5. The exponent on 3 is 1 (*positive* 1). The 3 remains in the numerator. Also, we indicated that $n \neq 0$. This is done because division by zero is not defined. In this textbook, we will assume that values of the variables are chosen so that division by zero does not occur.

HOW TO   Simplify: $3n^{-5}$, $n \neq 0$

$3n^{-5} = 3 \cdot \dfrac{1}{n^5} = \dfrac{3}{n^5}$ • **Use the Definition of a Negative Exponent to rewrite the expression with a positive exponent.**

HOW TO   Simplify: $\dfrac{2}{5a^{-4}}$

$\dfrac{2}{5a^{-4}} = \dfrac{2}{5} \cdot \dfrac{1}{a^{-4}} = \dfrac{2}{5} \cdot a^4 = \dfrac{2a^4}{5}$ • **Use the Definition of a Negative Exponent to rewrite the expression with a positive exponent.**

The expression $\left(\dfrac{x^4}{y^3}\right)^2$, $y \neq 0$, can be simplified by squaring $\dfrac{x^4}{y^3}$ or by multiplying each exponent in the quotient by the exponent outside the parentheses.

$$\left(\frac{x^4}{y^3}\right)^2 = \left(\frac{x^4}{y^3}\right)\left(\frac{x^4}{y^3}\right) = \frac{x^4 \cdot x^4}{y^3 \cdot y^3} = \frac{x^{4+4}}{y^{3+3}} = \frac{x^8}{y^6} \qquad \left(\frac{x^4}{y^3}\right)^2 = \frac{x^{4 \cdot 2}}{y^{3 \cdot 2}} = \frac{x^8}{y^6}$$

> **Rule for Simplifying the Power of a Quotient**
>
> If m, n, and q are integers and $y \neq 0$, then $\left(\dfrac{x^m}{y^n}\right)^p = \dfrac{x^{mp}}{y^{np}}$.

HOW TO Simplify: $\left(\dfrac{a^3}{b^2}\right)^{-2}$

$$\left(\frac{a^3}{b^2}\right)^{-2} = \frac{a^{3(-2)}}{b^{2(-2)}}$$
- Use the **Rule for Simplifying the Power of a Quotient**.

$$= \frac{a^{-6}}{b^{-4}} = \frac{b^4}{a^6}$$
- Use the **Definition of a Negative Exponent** to write the expression with positive exponents.

The example above suggests the following rule.

> **Rule for Negative Exponents on Fractional Expressions**
>
> If $a \neq 0$, $b \neq 0$, and n is a positive integer, then
>
> $$\left(\frac{a}{b}\right)^{-n} = \left(\frac{b}{a}\right)^n$$

Now that zero as an exponent and negative exponents have been defined, a rule for dividing exponential expressions can be stated.

> **Rule for Dividing Exponential Expressions**
>
> If m and n are integers and $x \neq 0$, then $\dfrac{x^m}{x^n} = x^{m-n}$.

HOW TO Evaluate $\dfrac{5^{-2}}{5}$.

$$\frac{5^{-2}}{5} = 5^{-2-1} = 5^{-3}$$
- Use the **Rule for Dividing Exponential Expressions**.

$$= \frac{1}{5^3} = \frac{1}{125}$$
- Use the **Definition of a Negative Exponent** to rewrite the expression with a positive exponent. Then evaluate.

HOW TO Simplify: $\dfrac{x^4}{x^9}$

$\dfrac{x^4}{x^9} = x^{4-9}$ • Use the **Rule for Dividing Exponential Expressions**.

$\quad\quad = x^{-5}$ • Subtract the exponents.

$\quad\quad = \dfrac{1}{x^5}$ • Use the Definition of a Negative Exponent to rewrite the expression with a positive exponent.

The rules for simplifying exponential expressions and powers of exponential expressions are true for all integers. These rules are restated here, along with the rules for dividing exponential expressions.

Rules of Exponents

If m, n, and p are integers, then

$$x^m \cdot x^n = x^{m+n} \quad\quad\quad (x^m)^n = x^{mn} \quad\quad\quad (x^m y^n)^p = x^{mp} y^{np}$$

$$\frac{x^m}{x^n} = x^{m-n},\, x \neq 0 \quad\quad \left(\frac{x^m}{y^n}\right)^p = \frac{x^{mp}}{y^{np}},\, y \neq 0 \quad\quad x^{-n} = \frac{1}{x^n},\, x \neq 0$$

$$x^0 = 1,\, x \neq 0$$

HOW TO Simplify: $(3ab^{-4})(-2a^{-3}b^7)$

$(3ab^{-4})(-2a^{-3}b^7) = [3 \cdot (-2)](a^{1+(-3)}b^{-4+7})$ • When multiplying expressions, add the exponents on like bases.

$\quad\quad = -6a^{-2}b^3$

$\quad\quad = -\dfrac{6b^3}{a^2}$

HOW TO Simplify: $\left[\dfrac{6m^2n^3}{8m^7n^2}\right]^{-3}$

$\left[\dfrac{6m^2n^3}{8m^7n^2}\right]^{-3} = \left[\dfrac{3m^{2-7}n^{3-2}}{4}\right]^{-3}$ • Simplify inside the brackets.

$\quad\quad = \left[\dfrac{3m^{-5}n}{4}\right]^{-3}$ • Subtract the exponents.

$\quad\quad = \dfrac{3^{-3}m^{15}n^{-3}}{4^{-3}}$ • Use the Rule for Simplifying the Power of a Quotient.

$\quad\quad = \dfrac{4^3m^{15}}{3^3n^3} = \dfrac{64m^{15}}{27n^3}$ • Use the Definition of a Negative Exponent to rewrite the expression with positive exponents. Then simplify.

HOW TO   Simplify: $\dfrac{4a^{-2}b^5}{6a^5b^2}$

$$\dfrac{4a^{-2}b^5}{6a^5b^2} = \dfrac{2a^{-2}b^5}{3a^5b^2}$$

- Divide the coefficients by their common factor.

$$= \dfrac{2a^{-2-5}b^{5-2}}{3}$$

- Use the Rule for Dividing Exponential Expressions.

$$= \dfrac{2a^{-7}b^3}{3} = \dfrac{2b^3}{3a^7}$$

- Use the Definition of a Negative Exponent to rewrite the expression with positive exponents.

Example 1   Simplify: $(-2x)(3x^{-2})^{-3}$

Solution
$(-2x)(3x^{-2})^{-3} = (-2x)(3^{-3}x^6)$ • Rule for Simplifying the Power of a Product

$$= \dfrac{-2x^{1+6}}{3^3}$$

$$= -\dfrac{2x^7}{27}$$

You Try It 1   Simplify: $(-2x^2)(x^{-3}y^{-4})^{-2}$

Your solution

Example 2   Simplify: $\dfrac{(2r^2t^{-1})^{-3}}{(r^{-3}t^4)^2}$

Solution
$$\dfrac{(2r^2t^{-1})^{-3}}{(r^{-3}t^4)^2} = \dfrac{2^{-3}r^{-6}t^3}{r^{-6}t^8}$$

- Rule for Simplifying the Power of a Product

$$= 2^{-3}r^{-6-(-6)}t^{3-8}$$

- Rule for Dividing Exponential Expressions

$$= 2^{-3}r^0t^{-5}$$

$$= \dfrac{1}{2^3t^5}$$

$$= \dfrac{1}{8t^5}$$

- Write answer in simplest form.

You Try It 2   Simplify: $\dfrac{(6a^{-2}b^3)^{-1}}{(4a^3b^{-2})^{-2}}$

Your solution

Example 3   Simplify: $\left[\dfrac{4a^{-2}b^3}{6a^4b^{-2}}\right]^{-3}$

Solution
$$\left[\dfrac{4a^{-2}b^3}{6a^4b^{-2}}\right]^{-3} = \left[\dfrac{2a^{-6}b^5}{3}\right]^{-3}$$

- Simplify inside brackets.

$$= \dfrac{2^{-3}a^{18}b^{-15}}{3^{-3}}$$

- Rule for Simplifying the Power of a Quotient

$$= \dfrac{27a^{18}}{8b^{15}}$$

- Write answer in simplest form.

You Try It 3   Simplify: $\left[\dfrac{6r^3s^{-3}}{9r^3s^{-1}}\right]^{-2}$

Your solution

Solutions on p. S11

Objective B   **To write a number in scientific notation**

Integrating Technology

See the Keystroke Guide: *Scientific Notation* for instruction on entering scientific notation on a calculator.

Point of Interest

An electron microscope uses wavelengths that are approximately 4×10^{-12} meter to make images of viruses.

The human eye can detect wavelengths between 4.3×10^{-7} meter and 6.9×10^{-7} meter. Although these are very short, they are approximately 10^5 times longer than the waves used in an electron microscope.

Very large and very small numbers abound in the natural sciences. For example, the mass of an electron is 0.00000000000000000000000000000911 kg. Numbers such as this are difficult to read, so a more convenient system called **scientific notation** is used. In scientific notation, a number is expressed as the product of two factors, one a number between 1 and 10, and the other a power of 10.

To express a number in scientific notation, write it in the form $a \times 10^n$, where a is a number between 1 and 10, and n is an integer.

For numbers greater than or equal to 10, move the decimal point to the right of the first digit. The exponent n is positive and equal to the number of places the decimal point has been moved.

$$240{,}000 = 2.4 \times 10^5$$

$$93{,}000{,}000 = 9.3 \times 10^7$$

For numbers less than 1, move the decimal point to the right of the first nonzero digit. The exponent n is negative. The absolute value of the exponent is equal to the number of places the decimal point has been moved.

$$0.0003 = 3 \times 10^{-4}$$

$$0.0000832 = 8.32 \times 10^{-5}$$

Changing a number written in scientific notation to decimal notation also requires moving the decimal point.

When the exponent is positive, move the decimal point to the right the same number of places as the exponent.

$$3.45 \times 10^6 = 3{,}450{,}000$$

$$2.3 \times 10^8 = 230{,}000{,}000$$

When the exponent is negative, move the decimal point to the left the same number of places as the absolute value of the exponent.

$$8.1 \times 10^{-3} = 0.0081$$

$$6.34 \times 10^{-7} = 0.000000634$$

Example 4   Write the number 824,300,000 in scientific notation.

Solution $824{,}300{,}000 = 8.243 \times 10^8$

You Try It 4   Write the number 0.000000961 in scientific notation.

Your solution

Example 5   Write the number 6.8×10^{-10} in decimal notation.

Solution $6.8 \times 10^{-10} = 0.00000000068$

You Try It 5   Write the number 7.329×10^6 in decimal notation.

Your solution

Solutions on p. S11

4.4 Exercises

Objective A **To divide monomials**

For Exercises 1 and 2, replace the question marks to make a true statement.

1. $\dfrac{x^7}{x^5} = x^{?-?} = x^?$

2. $\dfrac{a^9}{a^{12}} = a^{?-?} = \dfrac{1}{a^?}$

For Exercises 3 to 10, evaluate.

3. 5^{-2}

4. 3^{-3}

5. $\dfrac{1}{8^{-2}}$

6. $\dfrac{1}{12^{-1}}$

7. $\dfrac{3^{-2}}{3}$

8. $\dfrac{5^{-3}}{5}$

9. $\dfrac{2^{-2}}{2^{-3}}$

10. $\dfrac{3^2}{3^2}$

For Exercises 11 to 94, simplify.

11. x^{-2}

12. y^{-10}

13. $\dfrac{1}{a^{-6}}$

14. $\dfrac{1}{b^{-4}}$

15. $4x^{-7}$

16. $-6y^{-1}$

17. $\dfrac{2}{3}z^{-2}$

18. $\dfrac{4}{5}a^{-4}$

19. $\dfrac{5}{b^{-8}}$

20. $\dfrac{-3}{v^{-3}}$

21. $\dfrac{1}{3x^{-2}}$

22. $\dfrac{2}{5c^{-6}}$

23. $(ab^5)^0$

24. $(32x^3y^4)^0$

25. $-(3p^2q^5)^0$

26. $-\left(\dfrac{2}{3}xy\right)^0$

27. $\dfrac{y^7}{y^3}$

28. $\dfrac{z^9}{z^2}$

29. $\dfrac{a^8}{a^5}$

30. $\dfrac{c^{12}}{c^5}$

31. $\dfrac{p^5}{p}$

32. $\dfrac{w^9}{w}$

33. $\dfrac{4x^8}{2x^5}$

34. $\dfrac{12z^7}{4z^3}$

35. $\dfrac{22k^5}{11k^4}$

36. $\dfrac{14m^{11}}{7m^{10}}$

37. $\dfrac{m^9n^7}{m^4n^5}$

38. $\dfrac{y^5z^6}{yz^3}$

39. $\dfrac{6r^4}{4r^2}$

40. $\dfrac{8x^9}{12x^6}$

41. $\dfrac{-16a^7}{24a^6}$

42. $\dfrac{-18b^5}{27b^4}$

43. $\dfrac{y^3}{y^8}$

44. $\dfrac{z^4}{z^6}$

45. $\dfrac{a^5}{a^{11}}$

46. $\dfrac{m}{m^7}$

47. $\dfrac{4x^2}{12x^5}$ **48.** $\dfrac{6y^8}{8y^9}$ **49.** $\dfrac{-12x}{-18x^6}$ **50.** $\dfrac{-24c^2}{-36c^{11}}$

51. $\dfrac{x^6y^5}{x^8y}$ **52.** $\dfrac{a^3b^2}{a^2b^3}$ **53.** $\dfrac{2m^6n^2}{5m^9n^{10}}$ **54.** $\dfrac{5r^3t^7}{6r^5t^7}$

55. $\dfrac{pq^3}{p^4q^4}$ **56.** $\dfrac{a^4b^5}{a^5b^6}$ **57.** $\dfrac{3x^4y^5}{6x^4y^8}$ **58.** $\dfrac{14a^3b^6}{21a^5b^6}$

59. $\dfrac{14x^4y^6z^2}{16x^3y^9z}$ **60.** $\dfrac{24a^2b^7c^9}{36a^7b^5c}$ **61.** $\dfrac{15mn^9p^3}{30m^4n^9p}$ **62.** $\dfrac{25x^4y^7z^2}{20x^5y^9z^{11}}$

63. $(-2xy^{-2})^3$ **64.** $(-3x^{-1}y^2)^2$ **65.** $(3x^{-1}y^{-2})^2$

66. $(5xy^{-3})^{-2}$ **67.** $(2x^{-1})(x^{-3})$ **68.** $(-2x^{-5})x^7$

69. $(-5a^2)(a^{-5})^2$ **70.** $(2a^{-3})(a^7b^{-1})^3$ **71.** $(-2ab^{-2})(4a^{-2}b)^{-2}$

72. $(3ab^{-2})(2a^{-1}b)^{-3}$ **73.** $(-5x^{-2}y)(-2x^{-2}y^2)$ **74.** $\dfrac{a^{-3}b^{-4}}{a^2b^2}$

75. $\dfrac{3x^{-2}y^2}{6xy^2}$ **76.** $\dfrac{2x^{-2}y}{8xy}$ **77.** $\dfrac{3x^{-2}y}{xy}$ **78.** $\dfrac{2x^{-1}y^4}{x^2y^3}$

79. $\dfrac{2x^{-1}y^{-4}}{4xy^2}$ **80.** $\dfrac{(x^{-1}y)^2}{xy^2}$ **81.** $\dfrac{(x^{-2}y)^2}{x^2y^3}$ **82.** $\dfrac{(x^{-3}y^{-2})^2}{x^6y^8}$

83. $\dfrac{(a^{-2}y^3)^{-3}}{a^2y}$

84. $\dfrac{12a^2b^3}{-27a^2b^2}$

85. $\dfrac{-16xy^4}{96x^4y^4}$

86. $\dfrac{-8x^2y^4}{44y^2z^5}$

87. $\dfrac{22a^2b^4}{-132b^3c^2}$

88. $\dfrac{-(8a^2b^4)^3}{64a^3b^8}$

89. $\dfrac{-(14ab^4)^2}{28a^4b^2}$

90. $\dfrac{(2a^{-2}b^3)^{-2}}{(4a^2b^{-4})^{-1}}$

91. $\dfrac{(3^{-1}r^4s^{-3})^{-2}}{(6r^2s^{-1}t^{-2})^2}$

92. $\left(\dfrac{6x^{-4}yz^{-1}}{14xy^{-4}z^2}\right)^{-3}$

93. $\left(\dfrac{15m^3n^{-2}p^{-1}}{25m^{-2}n^{-4}}\right)^{-3}$

94. $\left(\dfrac{18a^4b^{-2}c^4}{12ab^{-3}d^2}\right)^{-2}$

Objective B   **To write a number in scientific notation**

95. Why might a number be written in scientific notation instead of decimal notation?

96. **a.** Explain how to write 0.00000076 in scientific notation.
    b. Explain how to write 4.3×10^8 in decimal notation.

For Exercises 97 to 105, write in scientific notation.

97. 0.00000000324

98. 0.00000012

99. 0.000000000000000003

100. 1,800,000,000

101. 32,000,000,000,000,000

102. 76,700,000,000,000

103. 0.000000000000000000122

104. 0.00137

105. 547,000,000

For Exercises 106 to 114, write in decimal notation.

106. 2.3×10^{-12}

107. 1.67×10^{-4}

108. 2×10^{15}

109. 6.8×10^7

110. 9×10^{-21}

111. 3.05×10^{-5}

112. 9.05×10^{11}

113. 1.02×10^{-9}

114. 7.2×10^{-3}

115. **Chemistry**   Avogadro's number is used in chemistry, and its value is approximately 602,300,000,000,000,000,000,000. Express this number in scientific notation.

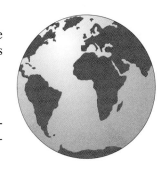

116. **Geology**   5,980,000,000,000,000,000,000,000 kg is the approximate mass of the planet Earth. Write the mass of Earth in scientific notation.

117.   **Physics**   The length of an infrared light wave is approximately 0.0000037 m. Write this number in scientific notation.

118.   **Physics**   Light travels approximately 16,000,000,000 mi in one day. Write this number in scientific notation.

119.   **Computer Science**   One unit used to measure the speed of a computer is the picosecond. One picosecond is 0.000000001 s. Write this number in scientific notation.

120.   **Astronomy**   One light-year is the distance traveled by light in 1 year. One light-year is 5,880,000,000,000 mi. Write this number in scientific notation.

121.   **Electricity**   The electric charge on an electron is 0.00000000000000000016 coulomb. Write this number in scientific notation.

122.   **Chemistry**   Approximately 35 teragrams (3.5×10^{13} g) of sulfur in the atmosphere is converted to sulfate each year. Write this number in decimal notation.

APPLYING THE CONCEPTS

123.   Evaluate 2^x when $x = -2, -1, 0, 1,$ and 2.

124.   Evaluate 3^x when $x = -2, -1, 0, 1,$ and 2.

125.   Evaluate 2^{-x} when $x = -2, -1, 0, 1,$ and 2.

126.   Evaluate 3^{-x} when $x = -2, -1, 0, 1,$ and 2.

For Exercises 127 to 129, determine whether each equation is true or false. If the equation is false, change the right-hand side of the equation to make a true equation.

127. $(2a)^{-3} = \dfrac{2}{a^3}$

128. $\dfrac{x^{-3}}{y^{-3}} = \left(\dfrac{x}{y}\right)^{-3}$

129. $(2 + 3)^{-1} = 2^{-1} + 3^{-1}$

130.   Simplify: $\left(\dfrac{6x^4yz^3}{2x^2y^3}\right)\left(\dfrac{2x^2z^3}{4y^2z}\right) \div \left(\dfrac{6x^2y^3}{x^4y^2z}\right)$

131.   If x is a nonzero real number, is x^{-2} always positive, always negative, or positive or negative depending on whether x is positive or negative? Explain your answer.

4.5 Division of Polynomials

Objective A **To divide a polynomial by a monomial**

To divide a polynomial by a monomial, divide each term in the numerator by the denominator and write the sum of the quotients.

HOW TO Divide: $\dfrac{6x^3 - 3x^2 + 9x}{3x}$

$$\frac{6x^3 - 3x^2 + 9x}{3x} = \frac{6x^3}{3x} - \frac{3x^2}{3x} + \frac{9x}{3x}$$

• Divide each term of the polynomial by the monomial.

$$= 2x^2 - x + 3$$

• Simplify each expression.

Example 1 Divide: $\dfrac{12x^2y - 6xy + 4x^2}{2xy}$

Solution

$$\frac{12x^2y - 6xy + 4x^2}{2xy} = \frac{12x^2y}{2xy} - \frac{6xy}{2xy} + \frac{4x^2}{2xy} = 6x - 3 + \frac{2x}{y}$$

You Try It 1 Divide: $\dfrac{24x^2y^2 - 18xy + 6y}{6xy}$

Your solution

Solution on p. S11

Objective B **To divide polynomials**

Study Tip

An important element of success is practice. We cannot do anything well if we do not practice it repeatedly. Practice is crucial to success in mathematics. In this objective you are learning a new skill, how to divide polynomials. You will need to practice this skill over and over again in order to be successful at it.

The procedure for dividing two polynomials is similar to the one for dividing whole numbers. The same equation used to check division of whole numbers is used to check polynomial division.

(Quotient × divisor) + remainder = dividend

HOW TO Divide: $(x^2 - 5x + 8) \div (x - 3)$

Step 1

$$\begin{array}{r} x \\ x - 3 \overline{)x^2 - 5x + 8} \\ \underline{x^2 - 3x} \\ -2x + 8 \end{array}$$

• Think: $x\overline{)x^2} = \dfrac{x^2}{x} = x$

• Multiply: $x(x - 3) = x^2 - 3x$

• Subtract: $(x^2 - 5x) - (x^2 - 3x) = -2x$
Bring down the 8.

Step 2

$$\begin{array}{r} x - 2 \\ x - 3 \overline{)x^2 - 5x + 8} \\ \underline{x^2 - 3x} \\ -2x + 8 \\ \underline{-2x + 6} \\ 2 \end{array}$$

• Think: $x\overline{)-2x} = \dfrac{-2x}{x} = -2$

• Multiply: $-2(x - 3) = -2x + 6$

• Subtract: $(-2x + 8) - (-2x + 6) = 2$
• The remainder is 2.

Check: $(x - 2)(x - 3) + 2 = x^2 - 5x + 6 + 2 = x^2 - 5x + 8$

$$(x^2 - 5x + 8) \div (x - 3) = x - 2 + \frac{2}{x - 3}$$

If a term is missing from the dividend, a zero can be inserted for that term. This helps keep like terms in the same column.

TAKE NOTE

Recall that a fraction bar means "divided by." Therefore, $6 \div 2$ can be written $\frac{6}{2}$, and $a \div b$ can be written $\frac{a}{b}$.

HOW TO Divide: $\dfrac{6x + 26 + 2x^3}{2 + x}$

$$\frac{2x^3 + 6x + 26}{x + 2}$$

- Arrange the terms of each polynomial in descending order.

$$
\begin{array}{r}
2x^2 - 4x + 14 \\
x + 2 \overline{\smash{)}2x^3 + 0\ \ + 6x + 26} \\
\underline{2x^3 + 4x^2} \\
-4x^2 + 6x \\
\underline{-4x^2 - 8x} \\
14x + 26 \\
\underline{14x + 28} \\
-2
\end{array}
$$

- There is no x^2 term in $2x^3 + 6x + 26$. Insert a **zero** for the missing term.

Check:
$(2x^2 - 4x + 14)(x + 2) + (-2) = (2x^3 + 6x + 28) + (-2) = 2x^3 + 6x + 26$

$(2x^3 + 6x + 26) \div (x + 2) = 2x^2 - 4x + 14 - \dfrac{2}{x + 2}$

Example 2

Divide: $(8x^2 + 4x^3 + x - 4) \div (2x + 3)$

Solution

$$
\begin{array}{r}
2x^2 + \ x - 1 \\
2x + 3 \overline{\smash{)}4x^3 + 8x^2 + \ x - 4} \\
\underline{4x^3 + 6x^2} \\
2x^2 + \ x \\
\underline{2x^2 + 3x} \\
-2x - 4 \\
\underline{-2x - 3} \\
-1
\end{array}
$$

- Write the dividend in descending powers of x.

$(4x^3 + 8x^2 + x - 4) \div (2x + 3)$

$\quad = 2x^2 + x - 1 - \dfrac{1}{2x + 3}$

You Try It 2

Divide: $(2x^3 + x^2 - 8x - 3) \div (2x - 3)$

Your solution

Example 3 Divide: $\dfrac{x^2 - 1}{x + 1}$

Solution

$$
\begin{array}{r}
x - 1 \\
x + 1 \overline{\smash{)}x^2 + 0 - 1} \\
\underline{x^2 + x} \\
-x - 1 \\
\underline{-x - 1} \\
0
\end{array}
$$

- Insert a zero for the missing term.

$(x^2 - 1) \div (x + 1) = x - 1$

You Try It 3 Divide: $\dfrac{x^3 - 2x + 1}{x - 1}$

Your solution

Solutions on p. S11

4.5 Exercises

Objective A To divide a polynomial by a monomial

For Exercises 1 to 24, divide.

1. $\dfrac{10a - 25}{5}$

2. $\dfrac{16b - 40}{8}$

3. $\dfrac{6y^2 + 4y}{y}$

4. $\dfrac{4b^3 - 3b}{b}$

5. $\dfrac{3x^2 - 6x}{3x}$

6. $\dfrac{10y^2 - 6y}{2y}$

7. $\dfrac{5x^2 - 10x}{-5x}$

8. $\dfrac{3y^2 - 27y}{-3y}$

9. $\dfrac{x^3 + 3x^2 - 5x}{x}$

10. $\dfrac{a^3 - 5a^2 + 7a}{a}$

11. $\dfrac{x^6 - 3x^4 - x^2}{x^2}$

12. $\dfrac{a^8 - 5a^5 - 3a^3}{a^2}$

13. $\dfrac{5x^2y^2 + 10xy}{5xy}$

14. $\dfrac{8x^2y^2 - 24xy}{8xy}$

15. $\dfrac{9y^6 - 15y^3}{-3y^3}$

16. $\dfrac{4x^4 - 6x^2}{-2x^2}$

17. $\dfrac{3x^2 - 2x + 1}{x}$

18. $\dfrac{8y^2 + 2y - 3}{y}$

19. $\dfrac{-3x^2 + 7x - 6}{x}$

20. $\dfrac{2y^2 - 6y + 9}{y}$

21. $\dfrac{16a^2b - 20ab + 24ab^2}{4ab}$

22. $\dfrac{22a^2b - 11ab - 33ab^2}{11ab}$

23. $\dfrac{9x^2y + 6xy - 3xy^2}{xy}$

24. $\dfrac{5a^2b - 15ab + 30ab^2}{5ab}$

Objective B To divide polynomials

For Exercises 25 and 26, replace the question marks to make a true statement.

25. If $\dfrac{x^2 - x - 6}{x - 3} = x + 2$, then
$x^2 - x - 6 = (?)(?)$.

26. If $\dfrac{x^2 + 2x - 3}{x - 2} = x + 4 + \dfrac{5}{x - 2}$, then
$x^2 + 2x - 3 = (?)(?) + ?$.

For Exercises 27 to 53, divide.

27. $(b^2 - 14b + 49) \div (b - 7)$

28. $(x^2 - x - 6) \div (x - 3)$

29. $(y^2 + 2y - 35) \div (y + 7)$

30. $(2x^2 + 5x + 2) \div (x + 2)$

31. $(2y^2 - 13y + 21) \div (y - 3)$

32. $(4x^2 - 16) \div (2x + 4)$

33. $\dfrac{2y^2 + 7}{y - 3}$

34. $\dfrac{x^2 + 1}{x - 1}$

35. $\dfrac{x^2 + 4}{x + 2}$

36. $\dfrac{6x^2 - 7x}{3x - 2}$

37. $\dfrac{6y^2 + 2y}{2y + 4}$

38. $\dfrac{5x^2 + 7x}{x - 1}$

39. $(6x^2 - 5) \div (x + 2)$

40. $(a^2 + 5a + 10) \div (a + 2)$

41. $(b^2 - 8b - 9) \div (b - 3)$

42. $\dfrac{2y^2 - 9y + 8}{2y + 3}$

43. $\dfrac{3x^2 + 5x - 4}{x - 4}$

44. $(8x + 3 + 4x^2) \div (2x - 1)$

45. $(10 + 21y + 10y^2) \div (2y + 3)$

46. $\dfrac{15a^2 - 8a - 8}{3a + 2}$

47. $\dfrac{12a^2 - 25a - 7}{3a - 7}$

48. $(5 - 23x + 12x^2) \div (4x - 1)$

49. $(24 + 6a^2 + 25a) \div (3a - 1)$

50. $\dfrac{5x + 3x^2 + x^3 + 3}{x + 1}$

51. $\dfrac{7x + x^3 - 6x^2 - 2}{x - 1}$

52. $(x^4 - x^2 - 6) \div (x^2 + 2)$

53. $(x^4 + 3x^2 - 10) \div (x^2 - 2)$

APPLYING THE CONCEPTS

54. ✎ In your own words, explain how to divide exponential expressions.

55. The product of a monomial and $4b$ is $12ab^2$. Find the monomial.

Focus on Problem Solving

Dimensional Analysis

In solving application problems, it may be useful to include the units in order to organize the problem so that the answer is in the proper units. Using units to organize and check the correctness of an application is called **dimensional analysis.** We use the operations of multiplying units and dividing units in applying dimensional analysis to application problems.

The Rule for Multiplying Exponential Expressions states that we multiply two expressions with the same base by adding the exponents.

$$x^4 \cdot x^6 = x^{4+6} = x^{10}$$

In calculations that involve quantities, the units are operated on algebraically.

HOW TO A rectangle measures 3 m by 5 m. Find the area of the rectangle.

$$A = LW = (3 \text{ m})(5 \text{ m}) = (3 \cdot 5)(\text{m} \cdot \text{m}) = 15 \text{ m}^2$$

The area of the rectangle is 15 m² (square meters).

HOW TO A box measures 10 cm by 5 cm by 3 cm. Find the volume of the box.

$$V = LWH = (10 \text{ cm})(5 \text{ cm})(3 \text{ cm}) = (10 \cdot 5 \cdot 3)(\text{cm} \cdot \text{cm} \cdot \text{cm}) = 150 \text{ cm}^3$$

The volume of the box is 150 cm³ (cubic centimeters).

HOW TO Find the area of a square whose side measures $(3x + 5)$ in.

$$A = s^2 = [(3x + 5) \text{ in.}]^2 = (3x + 5)^2 \text{ in}^2 = (9x^2 + 30x + 25) \text{ in}^2$$

The area of the square is $(9x^2 + 30x + 25)$ in² (square inches).

Dimensional analysis is used in the conversion of units.

The following example converts the unit miles to feet. The equivalent measures 1 mi = 5280 ft are used to form the following rates, which are called conversion

factors: $\dfrac{1 \text{ mi}}{5280 \text{ ft}}$ and $\dfrac{5280 \text{ ft}}{1 \text{ mi}}$. Because 1 mi = 5280 ft, both of the conversion factors

$\dfrac{1 \text{ mi}}{5280 \text{ ft}}$ and $\dfrac{5280 \text{ ft}}{1 \text{ mi}}$ are equal to 1.

To convert 3 mi to feet, multiply 3 mi by the conversion factor $\dfrac{5280 \text{ ft}}{1 \text{ mi}}$.

$$3 \text{ mi} = 3 \text{ mi} \cdot 1 = \frac{3 \text{ mi}}{1} \cdot \frac{5280 \text{ ft}}{1 \text{ mi}} = \frac{3 \text{ mi} \cdot 5280 \text{ ft}}{1 \text{ mi}} = 3 \cdot 5280 \text{ ft} = 15{,}840 \text{ ft}$$

There are two important points in the above illustration. **First,** you can think of dividing the numerator and denominator by the common unit "mile" just as you would divide the numerator and denominator of a fraction by a common factor.

Second, the conversion factor $\dfrac{5280 \text{ ft}}{1 \text{ mi}}$ is equal to 1, and multiplying an expression by 1 does not change the value of the expression.

In the application problem that follows, the units are kept in the problem while the problem is worked.

In 2003, a horse named Funny Cide ran a 1.25-mile race in 2.01 min. Find Funny Cide's average speed for that race in miles per hour. Round to the nearest tenth.

Strategy   To find the average speed, use the formula $r = \dfrac{d}{t}$, where r is the speed, d is the distance, and t is the time. Use the conversion factor $\dfrac{60 \text{ min}}{1 \text{ h}}$.

Solution $r = \dfrac{d}{t} = \dfrac{1.25 \text{ mi}}{2.01 \text{ min}} = \dfrac{1.25 \text{ mi}}{2.01 \text{ min}} \cdot \dfrac{60 \text{ min}}{1 \text{ h}}$

$= \dfrac{75 \text{ mi}}{2.01 \text{ h}} \approx 37.3 \text{ mph}$

Funny Cide's average speed was 37.3 mph.

Try each of the following problems. Round to the nearest tenth or nearest cent.

1. Convert 88 ft/s to miles per hour.

2. Convert 8 m/s to kilometers per hour (1 km = 1000 m).

3. A carpet is to be placed in a meeting hall that is 36 ft wide and 80 ft long. At $21.50 per square yard, how much will it cost to carpet the meeting hall?

4. A carpet is to be placed in a room that is 20 ft wide and 30 ft long. At $22.25 per square yard, how much will it cost to carpet the area?

5. Find the number of gallons of water in a fish tank that is 36 in. long and 24 in. wide and is filled to a depth of 16 in. (1 gal = 231 in³).

6. Find the number of gallons of water in a fish tank that is 24 in. long and 18 in. wide and is filled to a depth of 12 in. (1 gal = 231 in³).

7. A $\frac{1}{4}$-acre commercial lot is on sale for $2.15 per square foot. Find the sale price of the commercial lot (1 acre = 43,560 ft²).

8. A 0.75-acre industrial parcel was sold for $98,010. Find the parcel's price per square foot (1 acre = 43,560 ft²).

9. A new driveway will require 800 ft³ of concrete. Concrete is ordered by the cubic yard. How much concrete should be ordered?

10. A piston-engined dragster traveled 440 yd in 4.936 s at Ennis, Texas, on October 9, 1988. Find the average speed of the dragster in miles per hour.

11. The Marianas Trench in the Pacific Ocean is the deepest part of the ocean. Its depth is 6.85 mi. The speed of sound under water is 4700 ft/s. Find the time it takes sound to travel from the surface to the bottom of the Marianas Trench and back.

Projects and Group Activities

Diagramming the Square of a Binomial

1. Explain why the diagram at the right represents $(a + b)^2 = a^2 + 2ab + b^2$

2. Draw similar diagrams representing each of the following.

$$(x + 2)^2$$

$$(x + 4)^2$$

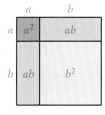

Pascal's Triangle

Simplifying the power of a binomial is called *expanding the binomial.* The expansions of the first three powers of a binomial are shown below.

$$(a + b)^1 = a + b$$

$$(a + b)^2 = (a + b)(a + b) = a^2 + 2ab + b^2$$

$$(a + b)^3 = (a + b)^2(a + b) = (a^2 + 2ab + b^2)(a + b) = a^3 + 3a^2b + 3ab^2 + b^3$$

Point of Interest

Pascal did not invent the triangle of numbers known as Pascal's Triangle. It was known to mathematicians in China probably as early as 1050 A.D. But Pascal's *Traite du triangle arithmetique* (*Treatise Concerning the Arithmetical Triangle*) brought together all the different aspects of the numbers for the first time.

Find $(a + b)^4$. [*Hint*: $(a + b)^4 = (a + b)^3(a + b)$]

Find $(a + b)^5$. [*Hint*: $(a + b)^5 = (a + b)^4(a + b)$]

If we continue in this way, the results for $(a + b)^6$ are

$$(a + b)^6 = a^6 + 6a^5b + 15a^4b^2 + 20a^3b^3 + 15a^2b^4 + 6ab^5 + b^6$$

Now expand $(a + b)^8$. Before you begin, see whether you can find a pattern that will help you write the expansion of $(a + b)^8$ without having to multiply it out. Here are some hints.

1. Write out the variable terms of each binomial expansion from $(a + b)^1$ through $(a + b)^6$. Observe how the exponents on the variables change.

2. Write out the coefficients of all the terms without the variable parts. It will be helpful to make a triangular arrangement as shown at the left. Note that each row begins and ends with a 1. Also note (in the two shaded regions, for example) that any number in a row is the sum of the two closest numbers above it. For instance, $1 + 5 = 6$ and $6 + 4 = 10$.

The triangle of numbers shown at the left is called Pascal's Triangle. To find the expansion of $(a + b)^8$, you need to find the eighth row of Pascal's Triangle. First find row seven. Then find row eight and use the patterns you have observed to write the expansion $(a + b)^8$.

Pascal's Triangle has been the subject of extensive analysis, and many patterns have been found. See whether you can find some of them.

Chapter 4 Summary

Key Words	Examples
A *monomial* is a number, a variable, or a product of numbers and variables. [4.1A, p. 193]	5 is a number, y is a variable. $2a^3b^2$ is a product of numbers and variables. 5, y, and $2a^3b^2$ are monomials.
A *polynomial* is a variable expression in which the terms are monomials. [4.1A, p. 193]	$5x^2y - 3xy^2 + 2$ is a polynomial. Each term of this expression is a monomial.
A polynomial of two terms is a *binomial*. [4.1A, p. 193]	$x + 2$, $y^2 - 3$, and $6a + 5b$ are binomials.
A polynomial of three terms is a *trinomial*. [4.1A, p. 193]	$x^2 - 6x + 7$ is a trinomial.
The *degree of a polynomial in one variable* is the greatest exponent on a variable. [4.1A, p. 193]	The degree of $3x - 4x^3 + 17x^2 + 25$ is 3.
A polynomial in one variable is usually written in *descending order* where the exponents of the variable terms decrease from left to right. [4.1A, p. 193]	The polynomial $2x^4 + 3x^2 - 4x - 7$ is written in descending order.
The *opposite of a polynomial* is the polynomial with the sign of every term changed to its opposite. [4.1B, p. 194]	The opposite of the polynomial $x^2 - 3x + 4$ is $-x^2 + 3x - 4$.

Essential Rules and Procedures	Examples
Addition of Polynomials [4.1A, p. 193] To add polynomials, add the coefficients of the like terms.	$(2x^2 + 3x - 4) + (3x^3 - 4x^2 + 2x - 5)$ $= 3x^3 + (2x^2 - 4x^2) + (3x + 2x)$ $\quad + (-4 - 5)$ $= 3x^3 - 2x^2 + 5x - 9$
Subtraction of Polynomials [4.1B, p. 194] To subtract polynomials, add the opposite of the second polynomial to the first.	$(3y^2 - 8y - 9) - (5y^2 - 10y + 3)$ $= (3y^2 - 8y - 9) + (-5y^2 + 10y - 3)$ $= (3y^2 - 5y^2) + (-8y + 10y)$ $\quad + (-9 - 3)$ $= -2y^2 + 2y - 12$
Rule for Multiplying Exponential Expressions [4.2A, p. 197] If m and n are integers, then $x^m \cdot x^n = x^{m+n}$.	$a^3 \cdot a^6 = a^{3+6} = a^9$

Rule for Simplifying the Power of an Exponential Expression [4.2B, p. 198]
If m and n are integers, then $(x^m)^n = x^{mn}$.

$(c^3)^4 = c^{3 \cdot 4} = c^{12}$

Rule for Simplifying the Power of a Product [4.2B, p. 198]
If m, n, and p are integers, then $(x^m y^n)^p = x^{mp} y^{np}$.

$(a^3 b^2)^4 = a^{3 \cdot 4} b^{2 \cdot 4} = a^{12} b^8$

To multiply a polynomial by a monomial, use the Distributive Property and the Rule for Multiplying Exponential Expressions. [4.3A, p. 201]

$(-4y)(5y^2 + 3y - 8)$
$= (-4y)(5y^2) + (-4y)(3y) - (-4y)(8)$
$= -20y^3 - 12y^2 + 32y$

To multiply two polynomials, multiply each term of one polynomial by each term of the other polynomial. [4.3B, p. 201]

$$\begin{array}{r} x^2 - 5x + 6 \\ x + 4 \\ \hline 4x^2 - 20x + 24 \\ x^3 - 5x^2 + 6x \\ \hline x^3 - x^2 - 14x + 24 \end{array}$$

FOIL Method [4.3C, p. 202]
To find the product of two binomials, add the products of the **F**irst terms, the **O**uter terms, the **I**nner terms, and the **L**ast terms.

$(2x - 5)(3x + 4)$
$= (2x)(3x) + (2x)(4) + (-5)(3x)$
$\quad + (-5)(4)$
$= 6x^2 + 8x - 15x - 20$
$= 6x^2 - 7x - 20$

Product of the Sum and Difference of the Same Terms [4.3D, p. 203]
$(a + b)(a - b) = a^2 - b^2$

$(3x + 4)(3x - 4) = (3x)^2 - 4^2$
$= 9x^2 - 16$

Square of a Binomial [4.3D, p. 203]
$(a + b)^2 = a^2 + 2ab + b^2$
$(a - b)^2 = a^2 - 2ab - b^2$

$(2x + 5)^2 = (2x^2) + 2(2x)(5) + 5^2$
$= 4x^2 + 20x + 25$
$(3x - 4)^2 = (3x)^2 - 2(3x)(4) + (-4)^2$
$= 9x^2 - 24x + 16$

Definition of Zero as an Exponent [4.4A, p. 209]
If $x \neq 0$ then $x^0 = 1$.

$17^0 = 1$, $(-6c)^0 = 1$, $c \neq 0$

Definition of a Negative Exponent [4.4A, p. 210]
If $x \neq 0$ and n is a positive integer, then $x^{-n} = \dfrac{1}{x^n}$ and $\dfrac{1}{x^{-n}} = x^n$.

$x^{-6} = \dfrac{1}{x^6}$ and $\dfrac{1}{x^{-6}} = x^6$

Rule for Simplifying the Power of a Quotient [4.4A, p. 211]

If m, n, and p are integers and $y \neq 0$, then $\left(\dfrac{x^m}{y^n}\right)^p = \dfrac{x^{mp}}{y^{np}}$.

$\left(\dfrac{c^3}{a^5}\right)^2 = \dfrac{c^{3 \cdot 2}}{a^{5 \cdot 2}} = \dfrac{c^6}{a^{10}}$

Rule for Negative Exponents on Fractional Expressions [4.4A, p. 211]

If $a \neq 0$, $b \neq 0$, and n is a positive integer, then $\left(\dfrac{a}{b}\right)^{-n} = \left(\dfrac{b}{a}\right)^n$.

$\left(\dfrac{x}{y}\right)^{-3} = \left(\dfrac{y}{x}\right)^3$

Rule for Dividing Exponential Expressions [4.4A, p. 211]

If m and n are integers and $x \neq 0$, then $\dfrac{x^m}{x^n} = x^{m-n}$.

$\dfrac{a^7}{a^2} = a^{7-2} = a^5$

To Express a Number in Scientific Notation [4.4B, p. 214]
To express a number in scientific notation, write it in the form $a \times 10^n$ where $1 \le a < 10$, and n is an integer. If the number is greater than 10, then n is a positive integer. If the number is between 0 and 1, then n is a negative integer.

$367{,}000{,}000 = 3.67 \times 10^8$
$0.0000078 = 7.8 \times 10^{-6}$

To Change a Number in Scientific Notation to Decimal Notation [4.4B, p. 214]
To change a number in scientific notation to decimal notation, move the decimal point to the right if n is positive and to the left if n is negative. Move the decimal point the same number of places as the absolute value of the exponent on 10.

$2.418 \times 10^7 = 24{,}180{,}000$
$9.06 \times 10^{-5} = 0.0000906$

To divide a polynomial by a monomial, divide each term in the numerator by the denominator and write the sum of the quotients. [4.5A, p. 219]

$\dfrac{8xy^3 - 4y^2 + 12y}{4y}$

$= \dfrac{8xy^3}{4y} - \dfrac{4y^2}{4y} + \dfrac{12y}{4y}$

$= 2xy^2 - y + 3$

To check polynomial division, use the same equation used to check division of whole numbers:

$$(\text{Quotient} \times \text{divisor}) + \text{remainder} = \text{dividend}$$

[4.5B, p. 219]

$$
\begin{array}{r}
x - 4 \\
x + 3 \overline{\big)\, x^2 - x - 10} \\
\underline{x^2 + 3x } \\
-4x - 10 \\
\underline{-4x - 12} \\
2
\end{array}
$$

Check:

$(x - 4)(x + 3) + 2 = x^2 - x - 12 + 2$
$ = x^2 - x - 10$

$(x^2 - x - 10) \div (x + 3) = x - 4 + \dfrac{2}{x + 3}$

Chapter 4 Review Exercises

1. Multiply: $(2b - 3)(4b + 5)$

2. Add: $(12y^2 + 17y - 4) + (9y^2 - 13y + 3)$

3. Simplify: $(xy^5z^3)(x^3y^3z)$

4. Simplify: $\dfrac{8x^{12}}{12x^9}$

5. Multiply: $-2x(4x^2 + 7x - 9)$

6. Simplify: $\dfrac{3ab^4}{-6a^2b^4}$

7. Simplify: $(-2u^3v^4)^4$

8. Evaluate: $(2^3)^2$

9. Subtract: $(5x^2 - 2x - 1) - (3x^2 - 5x + 7)$

10. Simplify: $\dfrac{a^{-1}b^3}{a^3b^{-3}}$

11. Simplify: $(-2x^3)^2(-3x^4)^3$

12. Expand: $(5y - 7)^2$

13. Simplify: $(5a^7b^6)^2(4ab)$

14. Divide: $\dfrac{12b^7 + 36b^5 - 3b^3}{3b^3}$

15. Evaluate: -4^{-2}

16. Subtract: $(13y^3 - 7y - 2) - (12y^2 - 2y - 1)$

17. Divide: $\dfrac{7 - x - x^2}{x + 3}$

18. Multiply: $(2a - b)(x - 2y)$

19. Multiply: $(3y^2 + 4y - 7)(2y + 3)$

20. Divide: $(b^3 - 2b^2 - 33b - 7) \div (b - 7)$

21. Multiply: $2ab^3(4a^2 - 2ab + 3b^2)$

22. Multiply: $(2a - 5b)(2a + 5b)$

23. Multiply: $(6b^3 - 2b^2 - 5)(2b^2 - 1)$

24. Add: $(2x^3 + 7x^2 + x) + (2x^2 - 4x - 12)$

25. Divide: $\dfrac{16y^2 - 32y}{-4y}$

26. Multiply: $(a + 7)(a - 7)$

27. Write 37,560,000,000 in scientific notation.

28. Write 1.46×10^7 in decimal notation.

29. Simplify: $(2a^{12}b^3)(-9b^2c^6)(3ac)$

30. Divide: $(6y^2 - 35y + 36) \div (3y - 4)$

31. Simplify: $(-3x^{-2}y^{-3})^{-2}$

32. Multiply: $(5a - 7)(2a + 9)$

33. Write 0.000000127 in scientific notation.

34. Write 3.2×10^{-12} in decimal notation.

35. **Geometry**   The length of a table-tennis table is 1 ft less than twice the width of the table. Let w represent the width of the table-tennis table. Express the area of the table in terms of the variable w.

36. **Geometry**   The side of a checkerboard is $(3x - 2)$ in. Express the area of the checkerboard in terms of the variable x.

Chapter 4 Test

1. Multiply: $2x(2x^2 - 3x)$

2. Divide: $\dfrac{12x^3 - 3x^2 + 9}{3x^2}$

3. Simplify: $\dfrac{12x^2}{-3x^8}$

4. Simplify: $(-2xy^2)(3x^2y^4)$

5. Divide: $(x^2 + 1) \div (x + 1)$

6. Multiply: $(x - 3)(x^2 - 4x + 5)$

7. Simplify: $(-2a^2b)^3$

8. Simplify: $\dfrac{(3x^{-2}y^3)^3}{3x^4y^{-1}}$

9. Multiply: $(a - 2b)(a + 5b)$

10. Divide: $\dfrac{16x^5 - 8x^3 + 20x}{4x}$

11. Divide: $(x^2 + 6x - 7) \div (x - 1)$

12. Multiply: $-3y^2(-2y^2 + 3y - 6)$

13. Multiply: $(-2x^3 + x^2 - 7)(2x - 3)$

14. Multiply: $(4y - 3)(4y + 3)$

15. Simplify: $(ab^2)(a^3b^5)$

16. Simplify: $\dfrac{2a^{-1}b}{2^{-2}a^{-2}b^{-3}}$

17. Divide: $\dfrac{20a - 35}{5}$

18. Subtract: $(3a^2 - 2a - 7) - (5a^3 + 2a - 10)$

19. Expand: $(2x - 5)^2$

20. Divide: $(4x^2 - 7) \div (2x - 3)$

21. Simplify: $\dfrac{-(2x^2y)^3}{4x^3y^3}$

22. Multiply: $(2x - 7y)(5x - 4y)$

23. Add: $(3x^3 - 2x^2 - 4) + (8x^2 - 8x + 7)$

24. Write 0.00000000302 in scientific notation.

25. **Geometry**   The radius of a circle is $(x - 5)$ m. Use the equation $A = \pi r^2$, where r is the radius, to find the area of the circle in terms of the variable x. Leave the answer in terms of π.

Cumulative Review Exercises

1. Simplify: $\dfrac{3}{16} - \left(-\dfrac{5}{8}\right) - \dfrac{7}{9}$

2. Evaluate $-3^2 \cdot \left(\dfrac{2}{3}\right)^3 \cdot \left(-\dfrac{5}{8}\right)$.

3. Simplify: $\left(-\dfrac{1}{2}\right)^3 \div \left(\dfrac{3}{8} - \dfrac{5}{6}\right) + 2$

4. Evaluate $\dfrac{b - (a - b)^2}{b^2}$ when $a = -2$ and $b = 3$.

5. Simplify: $-2x - (-xy) + 7x - 4xy$

6. Simplify: $(12x)\left(-\dfrac{3}{4}\right)$

7. Simplify: $-2[3x - 2(4 - 3x) + 2]$

8. Solve: $12 = -\dfrac{3}{4}x$

9. Solve: $2x - 9 = 3x + 7$

10. Solve: $2 - 3(4 - x) = 2x + 5$

11. 35.2 is what percent of 160?

12. Add: $(4b^3 - 7b^2 - 7) + (3b^2 - 8b + 3)$

13. Subtract: $(3y^3 - 5y + 8) - (-2y^2 + 5y + 8)$

14. Simplify: $(a^3b^5)^3$

15. Simplify: $(4xy^3)(-2x^2y^3)$

16. Multiply: $-2y^2(-3y^2 - 4y + 8)$

17. Multiply: $(2a - 7)(5a^2 - 2a + 3)$

18. Multiply: $(3b - 2)(5b - 7)$

19. Simplify: $\dfrac{(-2a^2b^3)^2}{8a^4b^8}$

20. Divide: $(a^2 - 4a - 21) \div (a + 3)$

21. Write 6.09×10^{-5} in decimal notation.

22. Translate "the difference between eight times a number and twice the number is eighteen" into an equation and solve.

23. **Juice Mixtures**   Fifty ounces of orange juice are added to 200 oz of a fruit punch that is 10% orange juice. What is the percent concentration of orange juice in the resulting mixture?

24. **Transportation**   A car traveling at 50 mph overtakes a cyclist who, riding at 10 mph, has had a 2-hour head start. How far from the starting point does the car overtake the cyclist?

25. **Geometry**   The width of a rectangle is 40% of the length. The perimeter of the rectangle is 42 m. Find the length and width of the rectangle.

chapter

5 Factoring

A University of Florida receiver is being tackled during the play in the photo above. The University of Florida is part of the Southeastern conference of the NCAA. Imagine you have just gotten a job as the manager for this conference. One of your first tasks would be to organize the game schedule for the upcoming season. There are 12 teams in the Southeastern conference. How would you go about creating this schedule? **Exercises 71 and 72 on page 273** show you how to use a formula to determine the number of league games that must be scheduled if each team is to play each other team once.

OBJECTIVES

Section 5.1

A To factor a monomial from a polynomial
B To factor by grouping

Section 5.2

A To factor a trinomial of the form $x^2 + bx + c$
B To factor completely

Section 5.3

A To factor a trinomial of the form $ax^2 + bx + c$ by using trial factors
B To factor a trinomial of the form $ax^2 + bx + c$ by grouping

Section 5.4

A To factor the difference of two squares and perfect-square trinomials
B To factor completely

Section 5.5

A To solve equations by factoring
B To solve application problems

 Need help? For online student resources, such as section quizzes, visit this textbook's website at **math.college.hmco.com/students.**

Do these exercises to prepare for Chapter 5.

1. Write 30 as a product of prime numbers.

2. Simplify: $-3(4y - 5)$

3. Simplify: $-(a - b)$

4. Simplify: $2(a - b) - 5(a - b)$

5. Solve: $4x = 0$

6. Solve: $2x + 1 = 0$

7. Multiply: $(x + 4)(x - 6)$

8. Multiply: $(2x - 5)(3x + 2)$

9. Simplify: $\dfrac{x^5}{x^2}$

10. Simplify: $\dfrac{6x^4y^3}{2xy^2}$

GO FIGURE • • •

Without using a calculator (it probably won't work anyway), how many digits are in the product $4^{54} \cdot 5^{100}$?

5.1 Common Factors

Objective A   **To factor a monomial from a polynomial**

In Section 1.5C we discussed how to find the greatest common factor (GCF) of two or more integers. The **greatest common factor (GCF) of two or more monomials** is the product of the GCF of the coefficients and the common variable factors.

$$6x^3y = 2 \cdot 3 \cdot x \cdot x \cdot x \cdot y$$
$$8x^2y^2 = 2 \cdot 2 \cdot 2 \cdot x \cdot x \cdot y \cdot y$$
$$\text{GCF} = 2 \cdot x \cdot x \cdot y = 2x^2y$$

Note that the exponent of each variable in the GCF is the same as the *smallest* exponent of that variable in either of the monomials.

The GCF of $6x^3y$ and $8x^2y^2$ is $2x^2y$.

HOW TO   Find the GCF of $12a^4b$ and $18a^2b^2c$.

The common variable factors are a^2 and b; c is not a common variable factor.

$$12a^4b = 2 \cdot 2 \cdot 3 \cdot a^4 \cdot b$$
$$18a^2b^2c = 2 \cdot 3 \cdot 3 \cdot a^2 \cdot b^2 \cdot c$$
$$\text{GCF} = 2 \cdot 3 \cdot a^2 \cdot b = 6a^2b$$

To **factor a polynomial** means to write the polynomial as a product of other polynomials. In the example at the right, $2x$ is the GCF of the terms $2x^2$ and $10x$.

HOW TO   Factor: $5x^3 - 35x^2 + 10x$

Find the GCF of the terms of the polynomial.

$$5x^3 = 5 \cdot x^3$$
$$35x^2 = 5 \cdot 7 \cdot x^2$$
$$10x = 2 \cdot 5 \cdot x$$

The GCF is $5x$.

Rewrite the polynomial, expressing each term as a product with the GCF as one of the factors.

$$5x^3 - 35x^2 + 10x = 5x(x^2) + 5x(-7x) + 5x(2)$$
$$= 5x(x^2 - 7x + 2)$$

• Use the Distributive Property to write the polynomial as a product of factors.

TAKE NOTE

At the right, the factors in parentheses are determined by dividing each term of the trinomial by the GCF, $5x$.

$$\frac{5x^3}{5x} = x^2,$$
$$\frac{-35x^2}{5x} = -7x, \text{ and}$$
$$\frac{10x}{5x} = 2$$

> **HOW TO**   Factor: $21x^2y^3 - 6xy^5 + 15x^4y^2$
>
> Find the GCF of the terms of the polynomial.
>
> $21x^2y^3 = 3 \cdot 7 \cdot x^2 \cdot y^3$
>
> $6xy^5 = 2 \cdot 3 \cdot x \cdot y^5$
>
> $15x^4y^2 = 3 \cdot 5 \cdot x^4 \cdot y^2$
>
> The GCF is $3xy^2$.
>
> Rewrite the polynomial, expressing each term as a product with the GCF as one of the factors.
>
> $21x^2y^3 - 6xy^5 + 15x^4y^2 = 3xy^2(7xy) + 3xy^2(-2y^3) + 3xy^2(5x^3)$
>
> $= 3xy^2(7xy - 2y^3 + 5x^3)$ • **Use the Distributive Property to write the polynomial as a product of factors.**

Example 1

Factor: $8x^2 + 2xy$

Solution
The GCF is $2x$.

$8x^2 + 2xy = 2x(4x) + 2x(y)$
$= 2x(4x + y)$

You Try It 1

Factor: $14a^2 - 21a^4b$

Your solution

Example 2

Factor: $n^3 - 5n^2 + 2n$

Solution
The GCF is n.

$n^3 - 5n^2 + 2n = n(n^2) + n(-5n) + n(2)$
$= n(n^2 - 5n + 2)$

You Try It 2

Factor: $27b^2 + 18b + 9$

Your solution

Example 3

Factor: $16x^2y + 8x^4y^2 - 12x^4y^5$

Solution
The GCF is $4x^2y$.

$16x^2y + 8x^4y^2 - 12x^4y^5$
$= 4x^2y(4) + 4x^2y(2x^2y) + 4x^2y(-3x^2y^4)$
$= 4x^2y(4 + 2x^2y - 3x^2y^4)$

You Try It 3

Factor: $6x^4y^2 - 9x^3y^2 + 12x^2y^4$

Your solution

Solutions on pp. S11–S12

Objective B   **To factor by grouping**

A factor that has two terms is called a **binomial factor.** In the examples at the right, the binomials in parentheses are binomial factors.

$$2a(a + b)^2$$
$$3xy(x - y)$$

The Distributive Property is used to factor a common binomial factor from an expression.

The common binomial factor of the expression $6x(x - 3) + y(x - 3)$ is $(x - 3)$. To factor that expression, use the Distributive Property to write the expression as a product of factors.

$$6x(x - 3) + y(x - 3) = (x - 3)(6x + y)$$

Consider the following simplification of $-(a - b)$.

$$-(a - b) = -1(a - b) = -a + b = b - a$$

Thus

$$b - a = -(a - b)$$

This equation is sometimes used to factor a common binomial from an expression.

HOW TO   Factor: $2x(x - y) + 5(y - x)$

$$2x(x - y) + 5(y - x) = 2x(x - y) - 5(x - y)$$
$$= (x - y)(2x - 5)$$

- $5(y - x) = 5[(-1)(x - y)]$
  $= -5(x - y)$

A polynomial can be **factored by grouping** if its terms can be grouped and factored in such a way that a common binomial factor is found.

HOW TO   Factor: $ax + bx - ay - by$

$$ax + bx - ay - by = (ax + bx) - (ay + by)$$

- Group the first two terms and the last two terms. Note that $-ay - by = -(ay + by)$.

$$= x(a + b) - y(a + b)$$
$$= (a + b)(x - y)$$

- Factor each group.
- Factor the GCF, $(a + b)$, from each group.

HOW TO   Factor: $6x^2 - 9x - 4xy + 6y$

$$6x^2 - 9x - 4xy + 6y = (6x^2 - 9x) - (4xy - 6y)$$

- Group the first two terms and the last two terms. Note that $-4xy + 6y = -(4xy - 6y)$.

$$= 3x(2x - 3) - 2y(2x - 3)$$
$$= (2x - 3)(3x - 2y)$$

- Factor each group.
- Factor the GCF, $(2x - 3)$, from each group.

Example 4

Factor: $4x(3x - 2) - 7(3x - 2)$

Solution

$4x(3x - 2) - 7(3x - 2)$ • $3x - 2$ is the common binomial factor.

$= (3x - 2)(4x - 7)$

You Try It 4

Factor: $2y(5x - 2) - 3(2 - 5x)$

Your solution

Example 5

Factor: $9x^2 - 15x - 6xy + 10y$

Solution

$9x^2 - 15x - 6xy + 10y$

$= (9x^2 - 15x) - (6xy - 10y)$ • $-6xy + 10y = -(6xy - 10y)$

$= 3x(3x - 5) - 2y(3x - 5)$ • $3x - 5$ is the common factor.

$= (3x - 5)(3x - 2y)$

You Try It 5

Factor: $a^2 - 3a + 2ab - 6b$

Your solution

Example 6

Factor: $3x^2y - 4x - 15xy + 20$

Solution

$3x^2y - 4x - 15xy + 20$

$= (3x^2y - 4x) - (15xy - 20)$ • $-15xy + 20 = -(15xy - 20)$

$= x(3xy - 4) - 5(3xy - 4)$ • $3xy - 4$ is the common factor.

$= (3xy - 4)(x - 5)$

You Try It 6

Factor: $2mn^2 - n + 8mn - 4$

Your solution

Example 7

Factor: $4ab - 6 + 3b - 2ab^2$

Solution

$4ab - 6 + 3b - 2ab^2$

$= (4ab - 6) + (3b - 2ab^2)$

$= 2(2ab - 3) + b(3 - 2ab)$

$= 2(2ab - 3) - b(2ab - 3)$ • $3 - 2ab = -(2ab - 3)$

$= (2ab - 3)(2 - b)$ • $2ab - 3$ is the common factor.

You Try It 7

Factor: $3xy - 9y - 12 + 4x$

Your solution

Solutions on p. S12

5.1 Exercises

Objective A   **To factor a monomial from a polynomial**

1. Explain the meaning of "a common monomial factor of a polynomial."

2. Explain the meaning of "a factor" and the meaning of "to factor."

For Exercises 3 to 41, factor.

3. $5a + 5$

4. $7b - 7$

5. $16 - 8a^2$

6. $12 + 12y^2$

7. $8x + 12$

8. $16a - 24$

9. $30a - 6$

10. $20b + 5$

11. $7x^2 - 3x$

12. $12y^2 - 5y$

13. $3a^2 + 5a^5$

14. $9x - 5x^2$

15. $14y^2 + 11y$

16. $6b^3 - 5b^2$

17. $2x^4 - 4x$

18. $3y^4 - 9y$

19. $10x^4 - 12x^2$

20. $12a^5 - 32a^2$

21. $8a^8 - 4a^5$

22. $16y^4 - 8y^7$

23. $x^2y^2 - xy$

24. $a^2b^2 + ab$

25. $3x^2y^4 - 6xy$

26. $12a^2b^5 - 9ab$

27. $x^2y - xy^3$

28. $3x^3 + 6x^2 + 9x$

29. $5y^3 - 20y^2 + 5y$

30. $2x^4 - 4x^3 + 6x^2$

31. $3y^4 - 9y^3 - 6y^2$

32. $2x^3 + 6x^2 - 14x$

33. $3y^3 - 9y^2 + 24y$

34. $2y^5 - 3y^4 + 7y^3$

35. $6a^5 - 3a^3 - 2a^2$

36. $x^3y - 3x^2y^2 + 7xy^3$

37. $2a^2b - 5a^2b^2 + 7ab^2$

38. $5y^3 + 10y^2 - 25y$

39. $4b^5 + 6b^3 - 12b$

40. $3a^2b^2 - 9ab^2 + 15b^2$

41. $8x^2y^2 - 4x^2y + x^2$

Objective B   **To factor by grouping**

For Exercises 42 to 68, factor.

42. $x(b + 4) + 3(b + 4)$

43. $y(a + z) + 7(a + z)$

44. $a(y - x) - b(y - x)$

45. $3r(a - b) + s(a - b)$

46. $x(x - 2) + y(2 - x)$

47. $t(m - 7) + 7(7 - m)$

48. $2x(7 + b) - y(b + 7)$

49. $2y(4a - b) - (b - 4a)$

50. $8c(2m - 3n) + (3n - 2m)$

51. $x^2 + 2x + 2xy + 4y$

52. $x^2 - 3x + 4ax - 12a$

53. $p^2 - 2p - 3rp + 6r$

54. $t^2 + 4t - st - 4s$

55. $ab + 6b - 4a - 24$

56. $xy - 5y - 2x + 10$

57. $2z^2 - z + 2yz - y$

58. $2y^2 - 10y + 7xy - 35x$

59. $8v^2 - 12vy + 14v - 21y$

60. $21x^2 + 6xy - 49x - 14y$

61. $2x^2 - 5x - 6xy + 15y$

62. $4a^2 + 5ab - 10b - 8a$

63. $3y^2 - 6y - ay + 2a$

64. $2ra + a^2 - 2r - a$

65. $3xy - y^2 - y + 3x$

66. $2ab - 3b^2 - 3b + 2a$

67. $3st + t^2 - 2t - 6s$

68. $4x^2 + 3xy - 12y - 16x$

APPLYING THE CONCEPTS

69. **Number Sense**   A natural number is a *perfect number* if it is the sum of all its factors less than itself. For example, 6 is a perfect number because all the factors of 6 that are less than 6 are 1, 2, and 3, and $1 + 2 + 3 = 6$.
a. Find the one perfect number between 20 and 30.
b. Find the one perfect number between 490 and 500.

70. **Geometry**   In the equation $P = 2L + 2W$, what is the effect on P when the quantity $L + W$ doubles?

71. **Geometry**   Write an expression in factored form for each of the shaded portions in the following diagrams. Use the equation for the area of a rectangle ($A = LW$) and the equation for the area of a circle ($A = \pi r^2$).

a.

b.

c.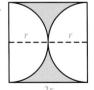

5.2 Factoring Polynomials of the Form $x^2 + bx + c$

Objective A   **To factor a trinomial of the form $x^2 + bx + c$**

Trinomials of the form $x^2 + bx + c$, where b and c are integers, are shown at the right.

$x^2 + 8x + 12; b = 8, c = 12$
$x^2 - 7x + 12; b = -7, c = 12$
$x^2 - 2x - 15; b = -2, c = -15$

To factor a trinomial of this form means to express the trinomial as the product of two binomials.

Trinomials expressed as the product of binomials are shown at the right.

$x^2 + 8x + 12 = (x + 6)(x + 2)$
$x^2 - 7x + 12 = (x - 3)(x - 4)$
$x^2 - 2x - 15 = (x + 3)(x - 5)$

The method by which factors of a trinomial are found is based on FOIL. Consider the following binomial products, noting the relationship between the constant terms of the binomials and the terms of the trinomials.

The signs in the binomials are the same.

$(x + 6)(x + 2) = x^2 + 2x + 6x + (6)(2) = x^2 + 8x + 12$
Sum of 6 and 2
Product of 6 and 2

$(x - 3)(x - 4) = x^2 - 4x - 3x + (-3)(-4) = x^2 - 7x + 12$
Sum of -3 and -4
Product of -3 and -4

The signs in the binomials are opposites.

$(x + 3)(x - 5) = x^2 - 5x + 3x + (3)(-5) = x^2 - 2x - 15$
Sum of 3 and -5
Product of 3 and -5

$(x - 4)(x + 6) = x^2 + 6x - 4x + (-4)(6) = x^2 + 2x - 24$
Sum of -4 and 6
Product of -4 and 6

Factoring $x^2 + bx + c$: IMPORTANT RELATIONSHIPS

1. When the constant term of the trinomial is positive, the constant terms of the binomials have the same sign. They are both positive when the coefficient of the x term in the trinomial is positive. They are both negative when the coefficient of the x term in the trinomial is negative.
2. When the constant term of the trinomial is negative, the constant terms of the binomials have opposite signs.
3. In the trinomial, the coefficient of x is the sum of the constant terms of the binomials.
4. In the trinomial, the constant term is the product of the constant terms of the binomials.

HOW TO Factor: $x^2 - 7x + 10$

Because the constant term is positive and the coefficient of x is negative, the binomial constants will be negative. Find two negative factors of 10 whose sum is -7. The results can be recorded in a table.

Negative Factors of 10	Sum
$-1, -10$	-11
$-2, -5$	-7

• These are the correct factors.

$x^2 - 7x + 10 = (x - 2)(x - 5)$ • Write the trinomial as a product of its factors.

You can check the proposed factorization by multiplying the two binomials.

$Check: (x - 2)(x - 5) = x^2 - 5x - 2x + 10$
$= x^2 - 7x + 10$

HOW TO Factor: $x^2 - 9x - 36$

The constant term is negative. The binomial constants will have opposite signs. Find two factors of -36 whose sum is -9.

Factors of -36	Sum
$+1, -36$	-35
$-1, +36$	35
$+2, -18$	-16
$-2, +18$	16
$+3, -12$	-9

• Once the correct factors are found, it is not necessary to try the remaining factors.

$x^2 - 9x - 36 = (x + 3)(x - 12)$ • Write the trinomial as a product of its factors.

For some trinomials it is not possible to find integer factors of the constant term whose sum is the coefficient of the middle term. A polynomial that does not factor using only integers is **nonfactorable over the integers.**

HOW TO Factor: $x^2 + 7x + 8$

The constant term is positive and the coefficient of x is positive. The binomial constants will be positive. Find two positive factors of 8 whose sum is 7.

Positive Factors of 8	Sum
$1, 8$	9
$2, 4$	6

• There are no positive integer factors of 8 whose sum is 7.

$x^2 + 7x + 8$ is nonfactorable over the integers.

Example 1 Factor: $x^2 - 8x + 15$

Solution
Find two negative factors of 15 whose sum is -8.

Factors	Sum
$-1, -15$	-16
$-3, -5$	-8

$x^2 - 8x + 15 = (x - 3)(x - 5)$

You Try It 1 Factor: $x^2 + 9x + 20$

Your solution

Solution on p. S12

Example 2   Factor: $x^2 + 6x - 27$

Solution
Find two factors of
-27 whose sum is 6.

Factors	Sum
$+1, -27$	-26
$-1, +27$	26
$+3, -9$	-6
$-3, +9$	6

$x^2 + 6x - 27 = (x - 3)(x + 9)$

You Try It 2   Factor: $x^2 + 7x - 18$

Your solution

Solution on p. S12

Objective B   **To factor completely**

A polynomial is **factored completely** when it is written as a product of factors that are nonfactorable over the integers.

TAKE NOTE

The first step in *any* factoring problem is to determine whether the terms of the polynomial have a *common factor*. If they do, factor it out first.

HOW TO   Factor: $4y^3 - 4y^2 - 24y$

$4y^3 - 4y^2 - 24y = 4y(y^2) - 4y(y) - 4y(6)$ • The GCF is **4y**.

$= 4y(y^2 - y - 6)$ • Use the Distributive Property to factor out the GCF.

$= 4y(y + 2)(y - 3)$ • Factor $y^2 - y - 6$. The two factors of -6 whose sum is -1 are 2 and -3.

It is always possible to check the proposed factorization by multiplying the polynomials. Here is the check for the last example.

Check: $4y(y + 2)(y - 3) = 4y(y^2 - 3y + 2y - 6)$
$= 4y(y^2 - y - 6)$
$= 4y^3 - 4y^2 - 24y$ • This is the original polynomial.

HOW TO   Factor: $5x^2 + 60xy + 100y^2$

$5x^2 + 60xy + 100y^2 = 5(x^2) + 5(12xy) + 5(20y^2)$ • The GCF is **5**.

$= 5(x^2 + 12xy + 20y^2)$ • Use the Distributive Property to factor out the GCF.

$= 5(x + 2y)(x + 10y)$ • Factor $x^2 + 12xy + 20y^2$. The two factors of 20 whose sum is 12 are 2 and 10.

TAKE NOTE

$2y$ and $10y$ are placed in the binomials. This is necessary so that the middle term contains xy and the last term contains y^2.

Note that $2y$ and $10y$ were placed in the binomials. The following check shows that this was necessary.

Check: $5(x + 2y)(x + 10y) = 5(x^2 + 10xy + 2xy + 20y^2)$
$= 5(x^2 + 12xy + 20y^2)$
$= 5x^2 + 60xy + 100y^2$ • This is the original polynomial.

HOW TO   Factor: $15 - 2x - x^2$

Because the coefficient of x^2 is -1, factor -1 from the trinomial and then write the resulting trinomial in descending order.

$15 - 2x - x^2 = -(x^2 + 2x - 15)$

$= -(x + 5)(x - 3)$

- $15 - 2x - x^2 = -1(-15 + 2x + x^2)$
  $= -(x^2 + 2x - 15)$
- Factor $x^2 + 2x - 15$. The two factors of -15 whose sum is 2 are 5 and -3.

Check: $-(x + 5)(x - 3) = -(x^2 + 2x - 15)$
$= -x^2 - 2x + 15$
$= 15 - 2x - x^2$

- This is the original polynomial.

Example 3

Factor: $-3x^3 + 9x^2 + 12x$

Solution
The GCF is $-3x$.
$-3x^3 + 9x^2 + 12x = -3x(x^2 - 3x - 4)$
Factor the trinomial $x^2 - 3x - 4$. Find two factors of -4 whose sum is -3.

Factors	Sum
$+1, -4$	-3

$-3x^3 + 9x^2 + 12x = -3x(x + 1)(x - 4)$

You Try It 3

Factor: $-2x^3 + 14x^2 - 12x$

Your solution

Example 4

Factor: $4x^2 - 40xy + 84y^2$

Solution
The GCF is 4.
$4x^2 - 40xy + 84y^2 = 4(x^2 - 10xy + 21y^2)$
Factor the trinomial $x^2 - 10xy + 21y^2$.
Find two negative factors of 21 whose sum is -10.

Factors	Sum
$-1, -21$	-22
$-3, -7$	-10

$4x^2 - 40xy + 84y^2 = 4(x - 3y)(x - 7y)$

You Try It 4

Factor: $3x^2 - 9xy - 12y^2$

Your solution

Solutions on p. S12

5.2 Exercises

Objective A   **To factor a trinomial of the form $x^2 + bx + c$**

1. Fill in the blank. In factoring a trinomial, if the constant term is positive, then the signs in both binomial factors will be _____ .

2. Fill in the blanks. To factor $x^2 + 8x - 48$, we must find two numbers whose product is _____ and whose sum is _____ .

For Exercises 3 to 75, factor.

3. $x^2 + 3x + 2$

4. $x^2 + 5x + 6$

5. $x^2 - x - 2$

6. $x^2 + x - 6$

7. $a^2 + a - 12$

8. $a^2 - 2a - 35$

9. $a^2 - 3a + 2$

10. $a^2 - 5a + 4$

11. $a^2 + a - 2$

12. $a^2 - 2a - 3$

13. $b^2 - 6b + 9$

14. $b^2 + 8b + 16$

15. $b^2 + 7b - 8$

16. $y^2 - y - 6$

17. $y^2 + 6y - 55$

18. $z^2 - 4z - 45$

19. $y^2 - 5y + 6$

20. $y^2 - 8y + 15$

21. $z^2 - 14z + 45$

22. $z^2 - 14z + 49$

23. $z^2 - 12z - 160$

24. $p^2 + 2p - 35$

25. $p^2 + 12p + 27$

26. $p^2 - 6p + 8$

27. $x^2 + 20x + 100$

28. $x^2 + 18x + 81$

29. $b^2 + 9b + 20$

30. $b^2 + 13b + 40$

31. $x^2 - 11x - 42$

32. $x^2 + 9x - 70$

33. $b^2 - b - 20$

34. $b^2 + 3b - 40$

35. $y^2 - 14y - 51$

36. $y^2 - y - 72$

37. $p^2 - 4p - 21$

38. $p^2 + 16p + 39$

39. $y^2 - 8y + 32$

40. $y^2 - 9y + 81$

41. $x^2 - 20x + 75$

42. $x^2 - 12x + 11$

43. $p^2 + 24p + 63$

44. $x^2 - 15x + 56$

45. $x^2 + 21x + 38$

46. $x^2 + x - 56$

47. $x^2 + 5x - 36$

48. $a^2 - 21a - 72$

49. $a^2 - 7a - 44$

50. $a^2 - 15a + 36$

51. $a^2 - 21a + 54$

52. $z^2 - 9z - 136$

53. $z^2 + 14z - 147$

54. $c^2 - c - 90$

55. $c^2 - 3c - 180$

56. $z^2 + 15z + 44$

57. $p^2 + 24p + 135$

58. $c^2 + 19c + 34$

59. $c^2 + 11c + 18$

60. $x^2 - 4x - 96$

61. $x^2 + 10x - 75$

62. $x^2 - 22x + 112$

63. $x^2 + 21x - 100$

64. $b^2 + 8b - 105$

65. $b^2 - 22b + 72$

66. $a^2 - 9a - 36$

67. $a^2 + 42a - 135$

68. $b^2 - 23b + 102$

69. $b^2 - 25b + 126$

70. $a^2 + 27a + 72$

71. $z^2 + 24z + 144$

72. $x^2 + 25x + 156$

73. $x^2 - 29x + 100$

74. $x^2 - 10x - 96$

75. $x^2 + 9x - 112$

Objective B   **To factor completely**

For Exercises 76 to 135, factor.

76. $2x^2 + 6x + 4$

77. $3x^2 + 15x + 18$

78. $18 + 7x - x^2$

79. $12 - 4x - x^2$

80. $ab^2 + 2ab - 15a$

81. $ab^2 + 7ab - 8a$

82. $xy^2 - 5xy + 6x$

83. $xy^2 + 8xy + 15x$

84. $z^3 - 7z^2 + 12z$

85. $-2a^3 - 6a^2 - 4a$

86. $-3y^3 + 15y^2 - 18y$

87. $4y^3 + 12y^2 - 72y$

88. $3x^2 + 3x - 36$

89. $2x^3 - 2x^2 + 4x$

90. $5z^2 - 15z - 140$

91. $6z^2 + 12z - 90$

92. $2a^3 + 8a^2 - 64a$

93. $3a^3 - 9a^2 - 54a$

94. $x^2 - 5xy + 6y^2$

95. $x^2 + 4xy - 21y^2$

96. $a^2 - 9ab + 20b^2$

97. $a^2 - 15ab + 50b^2$

98. $x^2 - 3xy - 28y^2$

99. $s^2 + 2st - 48t^2$

100. $y^2 - 15yz - 41z^2$

101. $x^2 + 85xy + 36y^2$

102. $z^4 - 12z^3 + 35z^2$

103. $z^4 + 2z^3 - 80z^2$

104. $b^4 - 22b^3 + 120b^2$

105. $b^4 - 3b^3 - 10b^2$

106. $2y^4 - 26y^3 - 96y^2$

107. $3y^4 + 54y^3 + 135y^2$

108. $-x^4 - 7x^3 + 8x^2$

109. $-x^4 + 11x^3 + 12x^2$

110. $4x^2y + 20xy - 56y$

111. $3x^2y - 6xy - 45y$

112. $c^3 + 18c^2 - 40c$ **113.** $-3x^3 + 36x^2 - 81x$ **114.** $-4x^3 - 4x^2 + 24x$

115. $x^2 - 8xy + 15y^2$ **116.** $y^2 - 7xy - 8x^2$ **117.** $a^2 - 13ab + 42b^2$

118. $y^2 + 4yz - 21z^2$ **119.** $y^2 + 8yz + 7z^2$ **120.** $y^2 - 16yz + 15z^2$

121. $3x^2y + 60xy - 63y$ **122.** $4x^2y - 68xy - 72y$ **123.** $3x^3 + 3x^2 - 36x$

124. $4x^3 + 12x^2 - 160x$ **125.** $4z^3 + 32z^2 - 132z$ **126.** $5z^3 - 50z^2 - 120z$

127. $4x^3 + 8x^2 - 12x$ **128.** $5x^3 + 30x^2 + 40x$ **129.** $5p^2 + 25p - 420$

130. $4p^2 - 28p - 480$ **131.** $p^4 + 9p^3 - 36p^2$ **132.** $p^4 + p^3 - 56p^2$

133. $t^2 - 12ts + 35s^2$ **134.** $a^2 - 10ab + 25b^2$ **135.** $a^2 - 8ab - 33b^2$

APPLYING THE CONCEPTS

For Exercises 136 to 138, factor.

136. $2 + c^2 + 9c$ **137.** $x^2y - 54y - 3xy$ **138.** $45a^2 + a^2b^2 - 14a^2b$

For Exercises 139 to 141, find all integers k such that the trinomial can be factored over the integers.

139. $x^2 + kx + 35$ **140.** $x^2 + kx + 18$ **141.** $x^2 + kx + 21$

For Exercises 142 to 147, determine the positive integer values of k for which the following polynomials are factorable over the integers.

142. $y^2 + 4y + k$ **143.** $z^2 + 7z + k$ **144.** $a^2 - 6a + k$

145. $c^2 - 7c + k$ **146.** $x^2 - 3x + k$ **147.** $y^2 + 5y + k$

148. In Exercises 142 to 147, there was the stated requirement that $k > 0$. If k is allowed to be any integer, how many different values of k are possible for each polynomial?

5.3 Factoring Polynomials of the Form $ax^2 + bx + c$

Objective A

To factor a trinomial of the form $ax^2 + bx + c$ by using trial factors

Trinomials of the form $ax^2 + bx + c$, where a, b, and c are integers, are shown at the right.

$3x^2 - x + 4$; $a = 3, b = -1, c = 4$
$6x^2 + 2x - 3$; $a = 6, b = 2, c = -3$

These trinomials differ from those in the preceding section in that the coefficient of x^2 is not 1. There are various methods of factoring these trinomials. The method described in this objective is factoring polynomials using trial factors.

To reduce the number of trial factors that must be considered, remember the following:

1. Use the signs of the constant term and the coefficient of x in the trinomial to determine the signs of the binomial factors. If the constant term is positive, the signs of the binomial factors will be the same as the sign of the coefficient of x in the trinomial. If the sign of the constant term is negative, the constant terms in the binomials have opposite signs.

2. If the terms of the trinomial do not have a common factor, then the terms of neither of the binomial factors will have a common factor.

HOW TO Factor: $2x^2 - 7x + 3$

The terms have no common factor. The constant term is positive. The coefficient of x is negative. The binomial constants will be negative.

Positive Factors of 2 (coefficient of x^2)	Negative Factors of 3 (constant term)
1, 2	−1, −3

Write trial factors. Use the **O**uter and **I**nner products of FOIL to determine the middle term, $-7x$, of the trinomial.

Trial Factors	Middle Term
$(x - 1)(2x - 3)$	$-3x - 2x = -5x$
$(x - 3)(2x - 1)$	$-x - 6x = -7x$

Write the factors of the trinomial. $2x^2 - 7x + 3 = (x - 3)(2x - 1)$

HOW TO Factor: $3x^2 + 14x + 15$

The terms have no common factor. The constant term is positive. The coefficient of x is positive. The binomial constants will be positive.

Positive Factors of 3 (coefficient of x^2)	Positive Factors of 15 (constant term)
1, 3	1, 15
	3, 5

Write trial factors. Use the **O**uter and **I**nner products of FOIL to determine the middle term, $14x$, of the trinomial.

Trial Factors	Middle Term
$(x + 1)(3x + 15)$	Common factor
$(x + 15)(3x + 1)$	$x + 45x = 46x$
$(x + 3)(3x + 5)$	$5x + 9x = 14x$
$(x + 5)(3x + 3)$	Common factor

Write the factors of the trinomial. $3x^2 + 14x + 15 = (x + 3)(3x + 5)$

HOW TO Factor: $6x^3 + 14x^2 - 12x$

Factor the GCF, $2x$, from the terms.

$$6x^3 + 14x^2 - 12x = 2x(3x^2 + 7x - 6)$$

Factor the trinomial. The constant term is negative. The binomial constants will have opposite signs.

Positive Factors of 3	Factors of −6
1, 3	1, −6
	−1, 6
	2, −3
	−2, 3

Write trial factors. Use the **O**uter and **I**nner products of FOIL to determine the middle term, $7x$, of the trinomial.

It is not necessary to test trial factors that have a common factor.

Trial Factors	Middle Term
$(x + 1)(3x - 6)$	Common factor
$(x - 6)(3x + 1)$	$x - 18x = -17x$
$(x - 1)(3x + 6)$	Common factor
$(x + 6)(3x - 1)$	$-x + 18x = 17x$
$(x + 2)(3x - 3)$	Common factor
$(x - 3)(3x + 2)$	$2x - 9x = -7x$
$(x - 2)(3x + 3)$	Common factor
$(x + 3)(3x - 2)$	$-2x + 9x = 7x$

Write the factors of the trinomial.

$$6x^3 + 14x^2 - 12x = 2x(x + 3)(3x - 2)$$

For this example, all the trial factors were listed. Once the correct factors have been found, however, the remaining trial factors can be omitted. For the examples and solutions in this text, all trial factors except those that have a common factor will be listed.

Example 1 Factor: $3x^2 + x - 2$

Solution
Positive factors of 3: 1, 3   Factors of −2: 1, −2 ; −1, 2

Trial Factors	Middle Term
$(x + 1)(3x - 2)$	$-2x + 3x = x$
$(x - 2)(3x + 1)$	$x - 6x = -5x$
$(x - 1)(3x + 2)$	$2x - 3x = -x$
$(x + 2)(3x - 1)$	$-x + 6x = 5x$

$3x^2 + x - 2 = (x + 1)(3x - 2)$

You Try It 1 Factor: $2x^2 - x - 3$

Your solution

Example 2 Factor: $-12x^3 - 32x^2 + 12x$

Solution
The GCF is $-4x$.
$-12x^3 - 32x^2 + 12x = -4x(3x^2 + 8x - 3)$
Factor the trinomial.
Positive factors of 3: 1, 3   Factors of −3: 1, −3 ; −1, 3

Trial Factors	Middle Term
$(x - 3)(3x + 1)$	$x - 9x = -8x$
$(x + 3)(3x - 1)$	$-x + 9x = 8x$

$-12x^3 - 32x^2 + 12x = -4x(x + 3)(3x - 1)$

You Try It 2 Factor: $-45y^3 + 12y^2 + 12y$

Your solution

Solutions on pp. S12–S13

Objective B   **To factor a trinomial of the form $ax^2 + bx + c$ by grouping**

In the preceding objective, trinomials of the form $ax^2 + bx + c$ were factored by using trial factors. In this objective, these trinomials will be factored by grouping.

To factor $ax^2 + bx + c$, first find two factors of $a \cdot c$ whose sum is b. Then use factoring by grouping to write the factorization of the trinomial.

HOW TO   Factor: $2x^2 + 13x + 15$

Find two positive factors of 30 ($2 \cdot 15$) whose sum is 13.

Positive Factors of 30	Sum
1, 30	31
2, 15	17
3, 10	13

• Once the required sum has been found, the remaining factors need not be checked.

$2x^2 + 13x + 15 = 2x^2 + 3x + 10x + 15$

        $= (2x^2 + 3x) + (10x + 15)$

        $= x(2x + 3) + 5(2x + 3)$

        $= (2x + 3)(x + 5)$

• Use the factors of 30 whose sum is 13 to write $13x$ as $3x + 10x$.
• Factor by grouping.

Check: $(2x + 3)(x + 5) = 2x^2 + 10x + 3x + 15$

                        $= 2x^2 + 13x + 15$

HOW TO   Factor: $6x^2 - 11x - 10$

Find two factors of -60 [$6(-10)$] whose sum is -11.

Factors of -60	Sum
1, -60	-59
-1, 60	59
2, -30	-28
-2, 30	28
3, -20	-17
-3, 20	17
4, -15	-11

$6x^2 - 11x - 10 = 6x^2 + 4x - 15x - 10$

        $= (6x^2 + 4x) - (15x + 10)$

        $= 2x(3x + 2) - 5(3x + 2)$

        $= (3x + 2)(2x - 5)$

• Use the factors of -60 whose sum is -11 to write $-11x$ as $4x - 15x$.
• Factor by grouping. Recall that $-15x - 10 = -(15x + 10)$.

Check: $(3x + 2)(2x - 5) = 6x^2 - 15x + 4x - 10$

                        $= 6x^2 - 11x - 10$

HOW TO   Factor: $3x^2 - 2x - 4$

Find two factors of -12 [$3(-4)$] whose sum is -2.

Factors of −12	Sum
1, −12	−11
−1, 12	11
2, −6	−4
−2, 6	4
3, −4	−1
−3, 4	1

TAKE NOTE
$3x^2 - 2x - 4$ is a prime polynomial.

Because no integer factors of -12 have a sum of -2, $3x^2 - 2x - 4$ is nonfactorable over the integers.

Example 3

Factor: $2x^2 + 19x - 10$

Solution

Factors of −20 [2(−10)]	Sum
−1, 20	19

$$2x^2 + 19x - 10 = 2x^2 - x + 20x - 10$$
$$= (2x^2 - x) + (20x - 10)$$
$$= x(2x - 1) + 10(2x - 1)$$
$$= (2x - 1)(x + 10)$$

You Try It 3

Factor: $2a^2 + 13a - 7$

Your solution

Example 4

Factor: $24x^2y - 76xy + 40y$

Solution
The GCF is $4y$.
$24x^2y - 76xy + 40y = 4y(6x^2 - 19x + 10)$

Negative Factors of 60 [6(10)]	Sum
−1, −60	−61
−2, −30	−32
−3, −20	−23
−4, −15	−19

$$6x^2 - 19x + 10 = 6x^2 - 4x - 15x + 10$$
$$= (6x^2 - 4x) - (15x - 10)$$
$$= 2x(3x - 2) - 5(3x - 2)$$
$$= (3x - 2)(2x - 5)$$

$$24x^2y - 76xy + 40y = 4y(6x^2 - 19x + 10)$$
$$= 4y(3x - 2)(2x - 5)$$

You Try It 4

Factor: $15x^3 + 40x^2 - 80x$

Your solution

Solutions on p. S13

5.3 Exercises

Objective A   **To factor a trinomial of the form $ax^2 + bx + c$ by using trial factors**

For Exercises 1 to 70, factor by using trial factors.

1. $2x^2 + 3x + 1$ **2.** $5x^2 + 6x + 1$ **3.** $2y^2 + 7y + 3$ **4.** $3y^2 + 7y + 2$

5. $2a^2 - 3a + 1$ **6.** $3a^2 - 4a + 1$ **7.** $2b^2 - 11b + 5$ **8.** $3b^2 - 13b + 4$

9. $2x^2 + x - 1$ **10.** $4x^2 - 3x - 1$ **11.** $2x^2 - 5x - 3$ **12.** $3x^2 + 5x - 2$

13. $2t^2 - t - 10$ **14.** $2t^2 + 5t - 12$ **15.** $3p^2 - 16p + 5$ **16.** $6p^2 + 5p + 1$

17. $12y^2 - 7y + 1$ **18.** $6y^2 - 5y + 1$ **19.** $6z^2 - 7z + 3$ **20.** $9z^2 + 3z + 2$

21. $6t^2 - 11t + 4$ **22.** $10t^2 + 11t + 3$ **23.** $8x^2 + 33x + 4$ **24.** $7x^2 + 50x + 7$

25. $5x^2 - 62x - 7$ **26.** $9x^2 - 13x - 4$ **27.** $12y^2 + 19y + 5$ **28.** $5y^2 - 22y + 8$

29. $7a^2 + 47a - 14$ **30.** $11a^2 - 54a - 5$ **31.** $3b^2 - 16b + 16$ **32.** $6b^2 - 19b + 15$

33. $2z^2 - 27z - 14$ **34.** $4z^2 + 5z - 6$ **35.** $3p^2 + 22p - 16$ **36.** $7p^2 + 19p + 10$

37. $4x^2 + 6x + 2$ **38.** $12x^2 + 33x - 9$ **39.** $15y^2 - 50y + 35$ **40.** $30y^2 + 10y - 20$

41. $2x^3 - 11x^2 + 5x$ **42.** $2x^3 - 3x^2 - 5x$ **43.** $3a^2b - 16ab + 16b$ **44.** $2a^2b - ab - 21b$

45. $3z^2 + 95z + 10$ **46.** $8z^2 - 36z + 1$ **47.** $36x - 3x^2 - 3x^3$ **48.** $-2x^3 + 2x^2 + 4x$

49. $80y^2 - 36y + 4$ **50.** $24y^2 - 24y - 18$ **51.** $8z^3 + 14z^2 + 3z$ **52.** $6z^3 - 23z^2 + 20z$

53. $6x^2y - 11xy - 10y$ **54.** $8x^2y - 27xy + 9y$ **55.** $10t^2 - 5t - 50$

56. $16t^2 + 40t - 96$ **57.** $3p^3 - 16p^2 + 5p$ **58.** $6p^3 + 5p^2 + p$

59. $26z^2 + 98z - 24$ **60.** $30z^2 - 87z + 30$ **61.** $10y^3 - 44y^2 + 16y$

62. $14y^3 + 94y^2 - 28y$ **63.** $4yz^3 + 5yz^2 - 6yz$ **64.** $12a^3 + 14a^2 - 48a$

65. $42a^3 + 45a^2 - 27a$ **66.** $36p^2 - 9p^3 - p^4$ **67.** $9x^2y - 30xy^2 + 25y^3$

68. $8x^2y - 38xy^2 + 35y^3$ **69.** $9x^3y - 24x^2y^2 + 16xy^3$ **70.** $9x^3y + 12x^2y + 4xy$

Objective B   **To factor a trinomial of the form**
$ax^2 + bx + c$ by grouping

For Exercises 71 to 130, factor by grouping.

71. $6x^2 - 17x + 12$ **72.** $15x^2 - 19x + 6$ **73.** $5b^2 + 33b - 14$ **74.** $8x^2 - 30x + 25$

75. $6a^2 + 7a - 24$ **76.** $14a^2 + 15a - 9$ **77.** $4z^2 + 11z + 6$ **78.** $6z^2 - 25z + 14$

79. $22p^2 + 51p - 10$ **80.** $14p^2 - 41p + 15$ **81.** $8y^2 + 17y + 9$ **82.** $12y^2 - 145y + 12$

83. $18t^2 - 9t - 5$ **84.** $12t^2 + 28t - 5$ **85.** $6b^2 + 71b - 12$ **86.** $8b^2 + 65b + 8$

87. $9x^2 + 12x + 4$ **88.** $25x^2 - 30x + 9$ **89.** $6b^2 - 13b + 6$ **90.** $20b^2 + 37b + 15$

91. $33b^2 + 34b - 35$ **92.** $15b^2 - 43b + 22$ **93.** $18y^2 - 39y + 20$ **94.** $24y^2 + 41y + 12$

95. $15a^2 + 26a - 21$ **96.** $6a^2 + 23a + 21$ **97.** $8y^2 - 26y + 15$ **98.** $18y^2 - 27y + 4$

99. $8z^2 + 2z - 15$ **100.** $10z^2 + 3z - 4$ **101.** $15x^2 - 82x + 24$ **102.** $13z^2 + 49z - 8$

103. $10z^2 - 29z + 10$ **104.** $15z^2 - 44z + 32$ **105.** $36z^2 + 72z + 35$ **106.** $16z^2 + 8z - 35$

107. $3x^2 + xy - 2y^2$ **108.** $6x^2 + 10xy + 4y^2$ **109.** $3a^2 + 5ab - 2b^2$ **110.** $2a^2 - 9ab + 9b^2$

111. $4y^2 - 11yz + 6z^2$ **112.** $2y^2 + 7yz + 5z^2$ **113.** $28 + 3z - z^2$ **114.** $15 - 2z - z^2$

115. $8 - 7x - x^2$ **116.** $12 + 11x - x^2$ **117.** $9x^2 + 33x - 60$ **118.** $16x^2 - 16x - 12$

119. $24x^2 - 52x + 24$ **120.** $60x^2 + 95x + 20$ **121.** $35a^4 + 9a^3 - 2a^2$

122. $15a^4 + 26a^3 + 7a^2$ **123.** $15b^2 - 115b + 70$ **124.** $25b^2 + 35b - 30$

125. $3x^2 - 26xy + 35y^2$ **126.** $4x^2 + 16xy + 15y^2$ **127.** $216y^2 - 3y - 3$

128. $360y^2 + 4y - 4$ **129.** $21 - 20x - x^2$ **130.** $18 + 17x - x^2$

APPLYING THE CONCEPTS

131. In your own words, explain how the signs of the last terms of the two binomial factors of a trinomial are determined.

For Exercises 132 to 137, factor.

132. $(x + 1)^2 - (x + 1) - 6$ **133.** $(x - 2)^2 + 3(x - 2) + 2$ **134.** $(y + 3)^2 - 5(y + 3) + 6$

135. $2(y + 2)^2 - (y + 2) - 3$ **136.** $3(a + 2)^2 - (a + 2) - 4$ **137.** $4(y - 1)^2 - 7(y - 1) - 2$

For Exercises 138 to 143, find all integers k such that the trinomial can be factored over the integers.

138. $2x^2 + kx + 3$ **139.** $2x^2 + kx - 3$ **140.** $3x^2 + kx + 2$

141. $3x^2 + kx - 2$ **142.** $2x^2 + kx + 5$ **143.** $2x^2 + kx - 5$

5.4 Special Factoring

Objective A   **To factor the difference of two squares and perfect-square trinomials**

A polynomial in the form $a^2 - b^2$ is called a **difference of two squares.** Recall the following relationship from Objective 4.3D.

Sum and difference of the same terms		**Difference of two squares**
$(a + b)(a - b)$	$=$	$a^2 - b^2$

> *TAKE NOTE*
>
> Note that the polynomial $x^2 + y^2$ is the *sum* of two squares. The sum of two squares is nonfactorable over the integers.

> **Factoring the Difference of Two Squares**
>
> The difference of two squares factors as the sum and difference of the same terms.
> $$a^2 - b^2 = (a + b)(a - b)$$

HOW TO   Factor: $x^2 - 16$

$x^2 - 16 = (x)^2 - (4)^2$ • $x^2 - 16$ is the difference of two squares.

$\quad\quad = (x + 4)(x - 4)$ • Factor the difference of squares.

Check: $(x + 4)(x - 4) = x^2 - 4x + 4x - 16$

$\quad\quad\quad\quad\quad\quad = x^2 - 16$

HOW TO   Factor: $8x^3 - 18x$

$8x^3 - 18x = 2x(4x^2 - 9)$ • The GCF is 2x.

$\quad\quad\quad = 2x[(2x)^2 - 3^2]$ • $4x^2 - 9$ is the difference of two squares.

$\quad\quad\quad = 2x(2x + 3)(2x - 3)$ • Factor the difference of squares.

You should check the factorization.

HOW TO   Factor: $x^2 - 10$

Because 10 cannot be written as the square of an integer, $x^2 - 10$ is nonfactorable over the integers.

A trinomial that can be written as the square of a binomial is called a **perfect-square trinomial.** Recall the pattern for finding the square of a binomial.

$$(a + b)^2 = a^2 + 2ab + b^2$$

Square of the first term ———┘   └——— Square of the last term

Twice the product
of the two terms

> **Factoring a Perfect-Square Trinomial**
>
> A perfect-square trinomial factors as the square of a binomial.
>
> $$a^2 + 2ab + b^2 = (a + b)^2$$
> $$a^2 - 2ab + b^2 = (a - b)^2$$

HOW TO Factor: $4x^2 - 20x + 25$

Because the first and last terms are squares $[(2x)^2 = 4x^2; \, 5^2 = 25]$, try to factor this as the square of a binomial. Check the factorization.

$$4x^2 - 20x + 25 = (2x - 5)^2$$

Check: $(2x - 5)^2 = (2x)^2 + 2(2x)(-5) + 5^2$
$\qquad\qquad\quad = 4x^2 - 20x + 25$ • The factorization is correct.

$$4x^2 - 20x + 25 = (2x - 5)^2$$

HOW TO Factor: $4x^2 + 37x + 9$

Because the first and last terms are squares $[(2x)^2 = 4x^2; \, 3^2 = 9]$, try to factor this as the square of a binomial. Check the proposed factorization.

$$4x^2 + 37x + 9 = (2x + 3)^2$$

Check: $(2x + 3)^2 = (2x)^2 + 2(2x)(3) + 3^2$
$\qquad\qquad\quad = 4x^2 + 12x + 9$

Because $4x^2 + 12x + 9 \neq 4x^2 + 37x + 9$, the proposed factorization is not correct. In this case, the polynomial is not a perfect-square trinomial. It may, however, still factor. In fact, $4x^2 + 37x + 9 = (4x + 1)(x + 9)$.

Example 1

Factor: $16x^2 - y^2$

Solution
$16x^2 - y^2 = (4x)^2 - y^2$ • The difference of two squares

$\qquad\qquad = (4x + y)(4x - y)$ • Factor.

You Try It 1

Factor: $25a^2 - b^2$

Your solution

Example 2

Factor: $z^4 - 16$

Solution
$z^4 - 16 = (z^2)^2 - 4^2$ • The difference of two squares

$\qquad = (z^2 + 4)(z^2 - 4)$ • $z^2 - 4$ is the difference

$\qquad = (z^2 + 4)(z^2 - 2^2)$ of two squares.

$\qquad = (z^2 + 4)(z + 2)(z - 2)$ • Factor.

You Try It 2

Factor: $n^4 - 81$

Your solution

Solutions on p. S13

Example 3

Factor: $9x^2 - 30x + 25$

Solution

$9x^2 = (3x)^2$, $25 = (5)^2$
$9x^2 - 30x + 25 = (3x - 5)^2$

Check: $(3x - 5)^2 = (3x)^2 + 2(3x)(-5) + 5^2$
$= 9x^2 - 30x + 25$

You Try It 3

Factor: $16y^2 + 8y + 1$

Your solution

Example 4

Factor: $9x^2 + 40x + 16$

Solution

Because $9x^2 = (3x)^2$, $16 = 4^2$, and $40x \neq 2(3x)(4)$, the trinomial is not a perfect-square trinomial.

Try to factor by another method.

$9x^2 + 40x + 16 = (9x + 4)(x + 4)$

You Try It 4

Factor: $x^2 + 15x + 36$

Your solution

Solutions on p. S13

Objective B   **To factor completely**

Study Tip

You have now learned to factor many different types of polynomials. You will need to be able to recognize each of the situations described in the box at the right. To test yourself, you might do the exercises in the Chapter Review.

General Factoring Strategy

1. Is there a common factor? If so, factor out the common factor.
2. Is the polynomial the difference of two perfect squares? If so, factor.
3. Is the polynomial a perfect-square trinomial? If so, factor.
4. Is the polynomial a trinomial that is the product of two binomials? If so, factor.
5. Does the polynomial contain four terms? If so, try factoring by grouping.
6. Is each binomial factor nonfactorable over the integers? If not, factor the binomial.

HOW TO   Factor: $z^3 + 4z^2 - 9z - 36$

$z^3 + 4z^2 - 9z - 36 = (z^3 + 4z^2) - (9z + 36)$

$= z^2(z + 4) - 9(z + 4)$

$= (z + 4)(z^2 - 9)$

$= (z + 4)(z + 3)(z - 3)$

- Factor by grouping. Recall that $-9z - 36 = -(9z + 36)$.
- $z^3 + 4z^2 = z^2(z + 4)$
  $9z + 36 = 9(z + 4)$
- Factor out the common binomial factor $(z + 4)$.
- Factor the difference of squares.

Example 5

Factor: $3x^2 - 48$

Solution

The GCF is 3.

$3x^2 - 48 = 3(x^2 - 16)$
$\quad\quad\quad\quad = 3(x + 4)(x - 4)$ • Factor the difference of two squares.

You Try It 5

Factor: $12x^3 - 75x$

Your solution

Example 6

Factor: $x^3 - 3x^2 - 4x + 12$

Solution

Factor by grouping.

$x^3 - 3x^2 - 4x + 12$
$= (x^3 - 3x^2) - (4x - 12)$ • Factor by grouping.
$= x^2(x - 3) - 4(x - 3)$ • **x − 3** is the common factor.
$= (x - 3)(x^2 - 4)$ • $x^2 - 4$ is the difference of two squares.
$= (x - 3)(x + 2)(x - 2)$ • Factor.

You Try It 6

Factor: $a^2b - 7a^2 - b + 7$

Your solution

Example 7

Factor: $4x^2y^2 + 12xy^2 + 9y^2$

Solution

The GCF is y^2.

$4x^2y^2 + 12xy^2 + 9y^2$
$= y^2(4x^2 + 12x + 9)$ • Factor the GCF, **y^2**.
$= y^2(2x + 3)^2$ • Factor the perfect-square trinomial.

You Try It 7

Factor: $4x^3 + 28x^2 - 120x$

Your solution

Solutions on p. S13

5.4 Exercises

Objective A To factor the difference of two squares and perfect-square trinomials

1. **a.** Provide an example of a binomial that is the difference of two squares.

   b. Provide an example of a perfect-square trinomial.

2. Explain why a binomial that is the sum of two squares is nonfactorable over the integers.

For Exercises 3 to 48, factor.

3. $x^2 - 4$

4. $x^2 - 9$

5. $a^2 - 81$

6. $a^2 - 49$

7. $y^2 + 2y + 1$

8. $y^2 + 14y + 49$

9. $a^2 - 2a + 1$

10. $x^2 - 12x + 36$

11. $4x^2 - 1$

12. $9x^2 - 16$

13. $x^6 - 9$

14. $y^{12} - 4$

15. $x^2 + 8x - 16$

16. $z^2 - 18z - 81$

17. $x^2 + 2xy + y^2$

18. $x^2 + 6xy + 9y^2$

19. $4a^2 + 4a + 1$

20. $25x^2 + 10x + 1$

21. $9x^2 - 1$

22. $1 - 49x^2$

23. $1 - 64x^2$

24. $t^2 + 36$

25. $x^2 + 64$

26. $64a^2 - 16a + 1$

27. $9a^2 + 6a + 1$

28. $x^4 - y^2$

29. $b^4 - 16a^2$

30. $16b^2 + 8b + 1$

31. $4a^2 - 20a + 25$

32. $4b^2 + 28b + 49$

33. $9a^2 - 42a + 49$

34. $9x^2 - 16y^2$

35. $25z^2 - y^2$

36. $x^2y^2 - 4$

37. $a^2b^2 - 25$

38. $16 - x^2y^2$

39. $25x^2 - 1$

40. $25a^2 + 30ab + 9b^2$

41. $4a^2 - 12ab + 9b^2$

42. $49x^2 + 28xy + 4y^2$

43. $4y^2 - 36yz + 81z^2$

44. $64y^2 - 48yz + 9z^2$

45. $\dfrac{1}{x^2} - 4$

46. $\dfrac{9}{a^2} - 16$

47. $9a^2b^2 - 6ab + 1$

48. $16x^2y^2 - 24xy + 9$

Objective B   **To factor completely**

For Exercises 49 to 126, factor.

49. $8y^2 - 2$

50. $12n^2 - 48$

51. $3a^3 + 6a^2 + 3a$

52. $4rs^2 - 4rs + r$

53. $m^4 - 256$

54. $81 - t^4$

55. $9x^2 + 13x + 4$

56. $x^2 + 10x + 16$

57. $16y^4 + 48y^3 + 36y^2$

58. $36c^4 - 48c^3 + 16c^2$

59. $y^8 - 81$

60. $32s^4 - 2$

61. $25 - 20p + 4p^2$

62. $9 + 24a + 16a^2$

63. $(4x - 3)^2 - y^2$

64. $(2x + 5)^2 - 25$

65. $(x^2 - 4x + 4) - y^2$

66. $(4x^2 + 12x + 9) - 4y^2$

67. $5x^2 - 5$

68. $2x^2 - 18$

69. $x^3 + 4x^2 + 4x$

70. $y^3 - 10y^2 + 25y$

71. $x^4 + 2x^3 - 35x^2$

72. $a^4 - 11a^3 + 24a^2$

73. $5b^2 + 75b + 180$

74. $6y^2 - 48y + 72$

75. $3a^2 + 36a + 10$

76. $5a^2 - 30a + 4$

77. $2x^2y + 16xy - 66y$

78. $3a^2b + 21ab - 54b$

79. $x^3 - 6x^2 - 5x$

80. $b^3 - 8b^2 - 7b$

81. $3y^2 - 36$

82. $3y^2 - 147$

83. $20a^2 + 12a + 1$

84. $12a^2 - 36a + 27$

85. $x^2y^2 - 7xy^2 - 8y^2$

86. $a^2b^2 + 3a^2b - 88a^2$

87. $10a^2 - 5ab - 15b^2$

88. $16x^2 - 32xy + 12y^2$

89. $50 - 2x^2$

90. $72 - 2x^2$

91. $a^2b^2 - 10ab^2 + 25b^2$

92. $a^2b^2 + 6ab^2 + 9b^2$

93. $12a^3b - a^2b^2 - ab^3$

94. $2x^3y - 7x^2y^2 + 6xy^3$

95. $12a^3 - 12a^2 + 3a$

96. $18a^3 + 24a^2 + 8a$

97. $243 + 3a^2$

98. $75 + 27y^2$

99. $12a^3 - 46a^2 + 40a$

100. $24x^3 - 66x^2 + 15x$

101. $4a^3 + 20a^2 + 25a$

102. $2a^3 - 8a^2b + 8ab^2$

103. $27a^2b - 18ab + 3b$

104. $a^2b^2 - 6ab^2 + 9b^2$

105. $48 - 12x - 6x^2$

106. $21x^2 - 11x^3 - 2x^4$

107. $x^4 - x^2y^2$

108. $b^4 - a^2b^2$

109. $18a^3 + 24a^2 + 8a$

110. $32xy^2 - 48xy + 18x$

111. $2b + ab - 6a^2b$

112. $15y^2 - 2xy^2 - x^2y^2$

113. $4x^4 - 38x^3 + 48x^2$

114. $3x^2 - 27y^2$

115. $x^4 - 25x^2$

116. $y^3 - 9y$

117. $a^4 - 16$

118. $15x^4y^2 - 13x^3y^3 - 20x^2y^4$

119. $45y^2 - 42y^3 - 24y^4$

120. $a(2x - 2) + b(2x - 2)$

121. $4a(x - 3) - 2b(x - 3)$

122. $x^2(x - 2) - (x - 2)$

123. $y^2(a - b) - (a - b)$

124. $a(x^2 - 4) + b(x^2 - 4)$

125. $x(a^2 - b^2) - y(a^2 - b^2)$

126. $4(x - 5) - x^2(x - 5)$

APPLYING THE CONCEPTS

For Exercises 127 to 132, find all integers k such that the trinomial is a perfect-square trinomial.

127. $4x^2 - kx + 9$

128. $x^2 + 6x + k$

129. $64x^2 + kxy + y^2$

130. $x^2 - 2x + k$

131. $25x^2 - kx + 1$

132. $x^2 + 10x + k$

133. **Number Sense** Select any odd integer greater than 1, square it, and then subtract 1. Is the result evenly divisible by 8? Prove that this procedure always produces a number divisible by 8. (*Suggestion:* Any odd integer greater than 1 can be expressed as $2n + 1$, where n is a natural number.)

5.5 Solving Equations

Objective A   **To solve equations by factoring**

The Multiplication Property of Zero states that the product of a number and zero is zero. This property is stated below.

$$\text{If } a \text{ is a real number, then } a \cdot 0 = 0 \cdot a = 0.$$

Now consider $x \cdot y = 0$. For this to be a true equation, then either $x = 0$ or $y = 0$.

Principle of Zero Products

If the product of two factors is zero, then at least one of the factors must be zero.

If $a \cdot b = 0$, then $a = 0$ or $b = 0$.

The Principle of Zero Products is used to solve some equations.

HOW TO   Solve: $(x - 2)(x - 3) = 0$

$(x - 2)(x - 3) = 0$

$x - 2 = 0 \qquad x - 3 = 0$ • Let each factor equal zero (the Principle of Zero Products).

$\qquad x = 2 \qquad\qquad x = 3$ • Solve each equation for x.

Check:

$$\begin{array}{c|c} (x - 2)(x - 3) = 0 \\ \hline (2 - 2)(2 - 3) & 0 \\ 0(-1) & 0 \\ 0 = 0 \end{array}$$ • A true equation

$$\begin{array}{c|c} (x - 2)(x - 3) = 0 \\ \hline (3 - 2)(3 - 3) & 0 \\ (1)(0) & 0 \\ 0 = 0 \end{array}$$ • A true equation

The solutions are 2 and 3.

An equation that can be written in the form $ax^2 + bx + c = 0$, $a \neq 0$, is a **quadratic equation.** A quadratic equation is in **standard form** when the polynomial is in descending order and equal to zero. The quadratic equations at the right are in standard form.

$3x^2 + 2x + 1 = 0$
$a = 3, b = 2, c = 1$

$4x^2 - 3x + 2 = 0$
$a = 4, b = -3, c = 2$

HOW TO Solve: $2x^2 + x = 6$

$$2x^2 + x = 6$$

$$2x^2 + x - 6 = 0$$ • Write the equation in standard form.

$$(2x - 3)(x + 2) = 0$$ • Factor.

$$2x - 3 = 0 \qquad x + 2 = 0$$ • Use the Principle of Zero Products.

$$2x = 3 \qquad\qquad x = -2$$ • Solve each equation for *x*.

$$x = \frac{3}{2}$$

Check: $\frac{3}{2}$ and -2 check as solutions.

The solutions are $\frac{3}{2}$ and -2.

Example 1

Solve: $x(x - 3) = 0$

Solution

$x(x - 3) = 0$

$x = 0 \qquad\qquad x - 3 = 0$ • Use the Principle
$\qquad\qquad\qquad x = 3$ of Zero Products.

The solutions are 0 and 3.

You Try It 1

Solve: $2x(x + 7) = 0$

Your solution

Example 2

Solve: $2x^2 - 50 = 0$

Solution

$\qquad 2x^2 - 50 = 0$

$\qquad 2(x^2 - 25) = 0$ • Factor the GCF, **2**.

$2(x + 5)(x - 5) = 0$ • Factor the difference of two squares.

$x + 5 = 0 \qquad x - 5 = 0$ • Use the Principle of
$\quad x = -5 \qquad\quad x = 5$ Zero Products.

The solutions are -5 and 5.

You Try It 2

Solve: $4x^2 - 9 = 0$

Your solution

Example 3

Solve: $(x - 3)(x - 10) = -10$

Solution

$(x - 3)(x - 10) = -10$

$\quad x^2 - 13x + 30 = -10$ • Multiply $(x - 3)(x - 10)$.

$\quad x^2 - 13x + 40 = 0$ • Add 10 to each side of
$\quad (x - 8)(x - 5) = 0$ the equation. The equation
$\qquad\qquad\qquad\qquad$ is now in standard form.

$x - 8 = 0 \qquad x - 5 = 0$

$\quad x = 8 \qquad\qquad x = 5$

The solutions are 8 and 5.

You Try It 3

Solve: $(x + 2)(x - 7) = 52$

Your solution

Solutions on pp. S13–S14

Objective B   **To solve application problems**

Example 4

The sum of the squares of two consecutive positive even integers is equal to 100. Find the two integers.

You Try It 4

The sum of the squares of two consecutive positive integers is 61. Find the two integers.

Strategy

First positive even integer: n
Second positive even integer: $n + 2$

The sum of the square of the first positive even integer and the square of the second positive even integer is 100.

Your strategy

Solution

$$n^2 + (n + 2)^2 = 100$$
$$n^2 + n^2 + 4n + 4 = 100$$
$$2n^2 + 4n + 4 = 100$$
$$2n^2 + 4n - 96 = 0$$
$$2(n^2 + 2n - 48) = 0$$
$$2(n - 6)(n + 8) = 0$$

$$n - 6 = 0 \qquad n + 8 = 0 \qquad \bullet \text{ Principle of}$$
$$n = 6 \qquad\qquad n = -8 \qquad \text{ Zero Products.}$$

Because -8 is not a positive even integer, it is not a solution.

$$n = 6$$
$$n + 2 = 6 + 2 = 8$$

The two integers are 6 and 8.

Your solution

Solution on p. S14

Example 5

A stone is thrown into a well with an initial speed of 4 ft/s. The well is 420 ft deep. How many seconds later will the stone hit the bottom of the well? Use the equation $d = vt + 16t^2$, where d is the distance in feet that the stone travels in t seconds when its initial speed is v feet per second.

Strategy

To find the time for the stone to drop to the bottom of the well, replace the variables d and v by their given values and solve for t.

Solution

$$d = vt + 16t^2$$
$$420 = 4t + 16t^2$$
$$0 = -420 + 4t + 16t^2$$
$$0 = 16t^2 + 4t - 420$$
$$0 = 4(4t^2 + t - 105)$$
$$0 = 4(4t + 21)(t - 5)$$

$4t + 21 = 0$ $t - 5 = 0$ • **Principle of**
      $4t = -21$ $t = 5$ **Zero Products.**
        $t = -\dfrac{21}{4}$

Because the time cannot be a negative number, $-\dfrac{21}{4}$ is not a solution.

The stone will hit the bottom of the well 5 s later.

You Try It 5

The length of a rectangle is 4 in. longer than twice the width. The area of the rectangle is 96 in². Find the length and width of the rectangle.

Your strategy

Your solution

Solution on p. S14

5.5 Exercises

Objective A   **To solve equations by factoring**

1. ✏ In your own words, explain the Principle of Zero Products.

2. Fill in the blanks. If $(x + 5)(2x - 7) = 0$, then _____ = 0 or _____ = 0.

For Exercises 3 to 60, solve.

3. $(y + 3)(y + 2) = 0$ **4.** $(y - 3)(y - 5) = 0$ **5.** $(z - 7)(z - 3) = 0$ **6.** $(z + 8)(z - 9) = 0$

7. $x(x - 5) = 0$ **8.** $x(x + 2) = 0$ **9.** $a(a - 9) = 0$ **10.** $a(a + 12) = 0$

11. $y(2y + 3) = 0$ **12.** $t(4t - 7) = 0$ **13.** $2a(3a - 2) = 0$ **14.** $4b(2b + 5) = 0$

15. $(b + 2)(b - 5) = 0$ **16.** $(b - 8)(b + 3) = 0$ **17.** $x^2 - 81 = 0$ **18.** $x^2 - 121 = 0$

19. $4x^2 - 49 = 0$ **20.** $16x^2 - 1 = 0$ **21.** $9x^2 - 1 = 0$ **22.** $16x^2 - 49 = 0$

23. $x^2 + 6x + 8 = 0$ **24.** $x^2 - 8x + 15 = 0$ **25.** $z^2 + 5z - 14 = 0$ **26.** $z^2 + z - 72 = 0$

27. $2a^2 - 9a - 5 = 0$ **28.** $3a^2 + 14a + 8 = 0$ **29.** $6z^2 + 5z + 1 = 0$ **30.** $6y^2 - 19y + 15 = 0$

31. $x^2 - 3x = 0$ **32.** $a^2 - 5a = 0$ **33.** $x^2 - 7x = 0$ **34.** $2a^2 - 8a = 0$

35. $a^2 + 5a = -4$ **36.** $a^2 - 5a = 24$ **37.** $y^2 - 5y = -6$ **38.** $y^2 - 7y = 8$

39. $2t^2 + 7t = 4$ **40.** $3t^2 + t = 10$ **41.** $3t^2 - 13t = -4$ **42.** $5t^2 - 16t = -12$

43. $x(x - 12) = -27$ **44.** $x(x - 11) = 12$ **45.** $y(y - 7) = 18$ **46.** $y(y + 8) = -15$

47. $p(p + 3) = -2$ **48.** $p(p - 1) = 20$ **49.** $y(y + 4) = 45$ **50.** $y(y - 8) = -15$

51. $x(x + 3) = 28$ **52.** $p(p - 14) = 15$ **53.** $(x + 8)(x - 3) = -30$ **54.** $(x + 4)(x - 1) = 14$

55. $(z - 5)(z + 4) = 52$ **56.** $(z - 8)(z + 4) = -35$ **57.** $(z - 6)(z + 1) = -10$

58. $(a + 3)(a + 4) = 72$ **59.** $(a - 4)(a + 7) = -18$ **60.** $(2x + 5)(x + 1) = -1$

Objective B   **To solve application problems**

61. **Number Sense**   The square of a positive number is six more than five times the positive number. Find the number.

62. **Number Sense**   The square of a negative number is fifteen more than twice the negative number. Find the number.

63. **Number Sense**   The sum of two numbers is six. The sum of the squares of the two numbers is twenty. Find the two numbers.

64. **Number Sense**   The sum of two numbers is eight. The sum of the squares of the two numbers is thirty-four. Find the two numbers.

65. **Number Sense**   The sum of the squares of two consecutive positive integers is forty-one. Find the two integers.

66. **Number Sense**   The sum of the squares of two consecutive positive even integers is one hundred. Find the two integers.

67. **Number Sense**   The sum of two numbers is ten. The product of the two numbers is twenty-one. Find the two numbers.

68. **Number Sense**   The sum of two numbers is thirteen. The product of the two numbers is forty. Find the two numbers.

Sum of Natural Numbers   The formula $S = \frac{n^2 + n}{2}$ gives the sum, S, of the first n natural numbers. Use this formula for Exercises 69 and 70.

69. How many consecutive natural numbers beginning with 1 will give a sum of 78?

70. How many consecutive natural numbers beginning with 1 will give a sum of 171?

Sports   The formula $N = \frac{t^2 - t}{2}$ gives the number, N, of football games that must be scheduled in a league with t teams if each team is to play every other team once. Use this formula for Exercises 71 and 72.

71. How many teams are in a league that schedules 15 games in such a way that each team plays every other team once?

72. How many teams are in a league that schedules 45 games in such a way that each team plays every other team once?

Physics   The distance, s, in feet, that an object will fall (neglecting air resistance) in t seconds is given by $s = vt + 16t^2$, where v is the initial velocity of the object in feet per second. Use this formula for Exercises 73 and 74.

73. An object is released from the top of a building 192 ft high. The initial velocity is 16 ft/s, and air resistance is neglected. How many seconds later will the object hit the ground?

74. In October 2003, the world's tallest building, Taipei 101, was completed. The top of the spire is 1667 ft above ground. If an object is released from this building at a point 640 ft above the ground at an initial velocity of 48 ft/s, assuming no air resistance, how many seconds later will the object reach the ground?

Sports   The height, h, in feet, an object will attain (neglecting air resistance) in t seconds is given by $h = vt - 16t^2$, where v is the initial velocity of the object in feet per second. Use this formula for Exercises 75 and 76.

75. A golf ball is thrown onto a cement surface and rebounds straight up. The initial velocity of the rebound is 60 ft/s. How many seconds later will the golf ball return to the ground?

76. A foul ball leaves a bat, hits home plate, and travels straight up with an initial velocity of 64 ft/s. How many seconds later will the ball be 64 ft above the ground?

77. Geometry   The length of a rectangle is 5 in. more than twice its width. Its area is 75 in². Find the length and width of the rectangle.

78. **Geometry**   The width of a rectangle is 5 ft less than the length. The area of the rectangle is 176 ft². Find the length and width of the rectangle.

79. **Geometry**   The height of a triangle is 4 m more than twice the length of the base. The area of the triangle is 35 m². Find the height of the triangle.

80. **Geometry**   The length of each side of a square is extended 5 in. The area of the resulting square is 64 in². Find the length of a side of the original square.

81. **Publishing**   The page of a book measures 6 in. by 9 in. A uniform border around the page leaves 28 in² for type. What are the dimensions of the type area?

82. **Gardening**   A small garden measures 8 ft by 10 ft. A uniform border around the garden increases the total area to 143 ft². What is the width of the border?

83. **Landscaping**   A landscape designer decides to increase the radius of a circular lawn by 3 ft. This increases the area of the lawn by 100 ft². Find the radius of the original circular lawn. Round to the nearest hundredth.

84. **Geometry**   A circle has a radius of 10 in. Find the increase in area that occurs when the radius is increased by 2 in. Round to the nearest hundredth.

APPLYING THE CONCEPTS

85. Find $3n^2$ if $n(n + 5) = -4$.

86. Find $2n^2$ if $n(n + 3) = 4$.

For Exercises 87 to 90, solve.

87. $2y(y + 4) = -5(y + 3)$

88. $(b + 5)^2 = 16$

89. $p^3 = 9p^2$

90. $(x + 3)(2x - 1) = (3 - x)(5 - 3x)$

91. Explain the error made in solving the equation at the right. Solve the equation correctly.

$$(x + 2)(x - 3) = 6$$
$$x + 2 = 6 \quad x - 3 = 6$$
$$x = 4 \qquad x = 9$$

92. Explain the error made in solving the equation at the right. Solve the equation correctly.

$$x^2 = x$$
$$\frac{x^2}{x} = \frac{x}{x}$$
$$x = 1$$

Focus on Problem Solving

Making a Table

There are six students using a gym. The wall on the gym has six lockers that are numbered 1, 2, 3, 4, 5, and 6. After a practice, the first student goes by and opens all the lockers. The second student shuts every second locker, the third student changes every third locker (opens a locker if it is shut, shuts a locker if it is open), the fourth student changes every fourth locker, the fifth student changes every fifth locker, and the sixth student changes every sixth locker. After the sixth student makes changes, which lockers are open?

One method of solving this problem would be to create a table as shown below.

Student Locker	1	2	3	4	5	6
1	O	O	O	O	O	O
2	O	C	C	C	C	C
3	O	O	C	C	C	C
4	O	C	C	O	O	O
5	O	O	O	O	C	C
6	O	C	O	O	O	C

From this table, lockers 1 and 4 are open after the sixth student passes through.

Now extend this to more lockers and students. In each case, the nth student changes multiples of the nth locker. For instance, the 8th student would change the 8th, 16th, 24th, . . .

1. Suppose there were 10 lockers and 10 students. Which lockers would remain open?

2. Suppose there were 16 lockers and 16 students. Which lockers would remain open?

3. Suppose there are 25 lockers and 25 students. Which lockers would remain open?

4. Suppose there are 40 lockers and 40 students. Which lockers would remain open?

5. Suppose there are 50 lockers and 50 students. Which lockers would remain open?

6. Make a conjecture as to which lockers would be open if there were 100 lockers and 100 students.

7. Give a reason why your conjecture should be true. [*Hint:* Consider how many factors there are for the door numbers that remain open and those that remain closed. For instance, with 40 lockers and 40 students, locker 36 (which remains open) has factors 1, 2, 3, 4, 6, 9, 12, 18, 36—an odd number of factors. Locker 28, a closed locker, has factors 1, 2, 4, 7, 14, 28—an even number of factors.]

Projects and Group Activities

Evaluating Polynomials Using a Graphing Calculator

A graphing calculator can be used to evaluate a polynomial. To illustrate the method, consider the polynomial $2x^3 - 3x^2 + 4x - 7$. The keystrokes below are for a TI-83 Plus calculator, but the keystrokes for other calculators will closely follow these keystrokes.

Press the `Y=` key. You will see a screen similar to the one below. Press `CLEAR` to erase any expression next to Y1.

Enter the polynomial as follows. The `^` key is used to enter an exponent.

> **TAKE NOTE**
> Once the polynomial has been entered in Y1, there are several methods to evaluate it. We will show just one option.

```
Plot1 Plot2 Plot3
\Y1 =
\Y2 =
\Y3 =
\Y4 =
\Y5 =
\Y6 =
\Y7 =
```

```
Plot1 Plot2 Plot3
\Y1 ■ 2X^3-3X^2+4X-7
\Y2 =
\Y3 =
\Y4 =
\Y5 =
\Y6 =
\Y7 =
```

To evaluate the polynomial when $x = 3$, first return to what is called the HOME screen by pressing `2nd` QUIT.

Enter the following keystrokes. Sample screens are shown at the right.

(1) 3 `STO▸` `X,T,θ,n` `ENTER`

(2) `VARS` `▸` `ENTER`

(3) `ENTER`

(4) `ENTER`

The value of the polynomial when $x = 3$ is 32.

To evaluate the polynomial at a different value of x, repeat Steps 1 through 4. For instance, to evaluate the polynomial when $x = -4$, we would have

(1) `(-)` 4 `STO▸` `X,T,θ,n` `ENTER`

(2) `VARS` `▸` `ENTER`

(3) `ENTER`

(4) `ENTER`

The value of the polynomial when $x = -4$ is -199.

Here are some practice exercises. Evaluate for the given value.

1. $2x^2 - 3x + 7; x = 4$

2. $3x^2 + 7x - 12; x = -3$

3. $3x^3 - 2x^2 + 6x - 8; x = 3$

4. $2x^3 + 4x^2 - x - 2; x = 2$

5. $x^4 - 3x^3 + 6x^2 + 5x - 1; x = 2$

6. $x^5 - x^3 + 2x - 7; x = -4$

Exploring Integers   *Number theory* is a branch of mathematics that focuses on integers and the relationships that exist among the integers. Some of the results from this field of study have important, practical applications for sending sensitive information such as credit card numbers over the Internet. In this project, you will be asked to discover some of those relationships.

1. If n is an integer, explain why the product $n(n + 1)$ is always an even number.

2. If n is an integer, explain why $2n$ is always an even integer.

3. If n is an integer, explain why $2n + 1$ is always an odd integer.

4. Select any odd integer greater than 1, square it, and then subtract 1. Try this for various odd integers greater than 1. Is the result always evenly divisible by 8?

5. Prove the assertion in Exercise 4. [*Suggestion:* From Exercise 2, an odd integer can be represented as $2n + 1$. Therefore, the assertion in Exercise 4 can be stated "$(2n + 1)^2 - 1$ is evenly divisible by 8." Expand this expression and explain why the result must be divisible by 8. You will need to use the result from Exercise 1.

6. The integers 2 and 3 are consecutive prime numbers. Are there any other consecutive prime numbers? Why?

7. If n is a positive integer, for what values of n is $n^2 - 1$ a prime number?

8. A Mersenne prime number is a prime that can be written in the form $2^n - 1$, where n is also a prime number. For instance, $2^5 - 1 = 32 - 1 = 31$. Because 5 and 31 are prime numbers, 31 is a Mersenne prime number. On the other hand, $2^{11} - 1 = 2048 - 1 = 2047$. In this case, although 11 is a prime number, $2047 = 23 \cdot 89$, which is not a prime number. Find two Mersenne prime numbers other than 31.

Using the World Wide Web

At the address **http://www.utm.edu/research/primes/mersenne/,** you can find more information on Mersenne prime numbers. By searching other web sites, you can also find information about various topics in math.

The web site **http://mathforum.org/dr.math/** is an especially rich source of information. You can even submit math questions to this site and get an answer from Dr. Math. One student posed the question, "What is the purpose of the number zero?" Here is the reply from Dr. Math.

> The invention of zero was one of the most important breakthroughs in the history of civilization. More important, in my opinion, than the invention of the wheel. I think that it's a fairly deep concept.
>
> One crucial purpose that zero holds is as a placeholder in our system of notation. When we write the number 408, we're really using a shorthand notation. What we really mean by 408 is "4 times 100, plus 0 times 10, plus 8 times 1." Without the number zero, we wouldn't be able to tell the numbers 408, 48, 480, 408,000, and 4800 apart. So yes, zero is important.
>
> Another crucial role that zero plays in mathematics is that of an "additive identity element." What this means is that when you add zero to any number, you get the number that you started with. For instance, $5 + 0 = 5$. That may seem obvious and trivial, but

it's actually quite important to have such a number. For instance, when you're manipulating some numerical quantity and you want to change its form but not its value, you might add some fancy version of zero to it, like this:

$$
\begin{aligned}
x^2 + y^2 &= x^2 + y^2 + 2xy - 2xy \\
&= x^2 + 2xy + y^2 - 2xy \\
&= (x + y)^2 - 2xy
\end{aligned}
$$

Now if we wanted to, we could use this as a proof that $(x + y)^2$ is always greater than $2xy$; the expression we started with was positive, so the one we ended up with must be positive, too. Therefore, subtracting $2xy$ from $(x + y)^2$ must leave us with a positive number. Neat stuff.*

*Ask Dr. Math. Copyright © 1994–1997 *The Math Forum.* Used by permission.

Chapter 5 Summary

Key Words	Examples
The *greatest common factor (GCF) of two or more monomials* is the product of the GCF of the coefficients and the common variable factors. [5.1A, p. 237]	The GCF of $8x^2y$ and $12xyz$ is $4xy$.
To *factor a polynomial* means to write the polynomial as a product of other polynomials. [5.1A, p. 237]	To factor $x^2 + 3x + 2$ means to write it as the product $(x + 1)(x + 2)$.
A factor that has two terms is called a *binomial factor.* [5.1B, p. 239]	$(x + 1)$ is a binomial factor of $3x(x + 1)$.
A polynomial that does not factor using only integers is *nonfactorable over the integers.* [5.2A, p. 244]	The trinomial $x^2 + x + 4$ is nonfactorable over the integers. There are no integers whose product is 4 and whose sum is 1.
A polynomial is *factored completely* if it is written as a product of factors that are nonfactorable over the integers. [5.2B, p. 245]	The polynomial $3y^3 + 9y^2 - 12y$ is factored completely as $3y(y + 4)(y - 1)$.

An equation that can be written in the form $ax^2 + bx + c = 0$, $a \neq 0$, is a *quadratic equation*. A quadratic equation is in *standard form* when the polynomial is written in descending order and equal to zero. [5.5A, p. 267]

The equation $2x^2 - 3x + 7 = 0$ is a quadratic equation in standard form.

Essential Rules and Procedures

Examples

Factoring by Grouping [5.1B, p. 239]

A polynomial can be factored by grouping if its terms can be grouped and factored in such a way that a common binomial factor is found.

$3a^2 - a - 15ab + 5b$
$= (3a^2 - a) - (15ab - 5b)$
$= a(3a - 1) - 5b(3a - 1)$
$= (3a - 1)(a - 5b)$

Factoring $x^2 + bx + c$: IMPORTANT RELATIONSHIPS
[5.2A, p. 243]

1. When the constant term of the trinomial is positive, the constant terms of the binomials have the same sign. They are both positive when the coefficient of the x term in the trinomial is positive. They are both negative when the coefficient of the x term in the trinomial is negative.

$x^2 + 6x + 8 = (x + 4)(x + 2)$

$x^2 - 6x + 5 = (x - 5)(x - 1)$

2. When the constant term of the trinomial is negative, the constant terms of the binomials have opposite signs.

$x^2 - 4x - 21 = (x + 3)(x - 7)$

3. In the trinomial, the coefficient of x is the sum of the constant terms of the binomials.

In the three examples above, note that $6 = 4 + 2$, $-6 = -5 + (-1)$, and $-4 = 3 + (-7)$.

4. In the trinomial, the constant term is the product of the constant terms of the binomials.

In the three examples above, note that $8 = 4 \cdot 2$, $5 = -5(-1)$, and $-21 = 3(-7)$.

To factor $ax^2 + bx + c$ by grouping [5.3B, p. 253]

First find two factors of $a \cdot c$ whose sum is b. Then use factoring by grouping to write the factorization of the trinomial.

$3x^2 - 11x - 20$
$a \cdot c = 3(-20) = -60$
The product of 4 and -15 is -60.
The sum of 4 and -15 is -11.
$3x^2 + 4x - 15x - 20$
$= (3x^2 + 4x) - (15x + 20)$
$= x(3x + 4) - 5(3x + 4)$
$= (3x + 4)(x - 5)$

Factoring the Difference of Two Squares [5.4A, p. 259]
The difference of two squares factors as the sum and difference of the same terms.

$a^2 - b^2 = (a + b)(a - b)$

$x^2 - 64 = (x + 8)(x - 8)$
$4x^2 - 81 = (2x)^2 - 9^2$
$\qquad = (2x + 9)(2x - 9)$

Factoring a Perfect-Square Trinomial [5.4A, p. 260]
A perfect-square trinomial is the square of a binomial.

$a^2 + 2ab + b^2 = (a + b)^2$
$a^2 - 2ab + b^2 = (a - b)^2$

$x^2 + 14x + 49 = (x + 7)^2$
$x^2 - 10x + 25 = (x - 5)^2$

General Factoring Strategy [5.4B, p. 261]

1. Is there a common factor? Is so, factor out the common factor.

   $6x^2 - 8x = 2x(3x - 4)$

2. Is the polynomial the difference of two perfect squares? If so, factor.

   $9x^2 - 25 = (3x + 5)(3x - 5)$

3. Is the polynomial a perfect-square trinomial? If so, factor.

   $9x^2 + 6x + 1 = (3x + 1)^2$

4. Is the polynomial a trinomial that is the product of two binomials? If so, factor.

   $6x^2 + 5x - 6 = (3x - 2)(2x + 3)$

5. Does the polynomial contain four terms? If so, try factoring by grouping.

   $x^3 - 3x^2 + 2x - 6$
   $= (x^3 - 3x^2) + (2x - 6)$
   $= x^2(x - 3) + 2(x - 3)$
   $= (x - 3)(x^2 + 2)$

6. Is each binomial factor nonfactorable over the integers? If not, factor the binomial.

   $x^4 - 16 = (x^2 + 4)(x^2 - 4)$
   $\qquad = (x^2 + 4)(x + 2)(x - 2)$

Principle of Zero Products [5.5A, p. 267]
If the product of two factors is zero, then at least one of the factors must be zero.

If $a \cdot b = 0$, then $a = 0$ or $b = 0$.

The Principle of Zero Products is used to solve a quadratic equation by factoring.

$x^2 + x = 12$
$x^2 + x - 12 = 0$
$(x - 3)(x + 4) = 0$

$x - 3 = 0 \qquad x + 4 = 0$
$\quad x = 3 \qquad\qquad x = -4$

Chapter 5 Review Exercises

1. Factor: $b^2 - 13b + 30$

2. Factor: $4x(x - 3) - 5(3 - x)$

3. Factor $2x^2 - 5x + 6$ by using trial factors.

4. Factor: $5x^3 + 10x^2 + 35x$

5. Factor: $14y^9 - 49y^6 + 7y^3$

6. Factor: $y^2 + 5y - 36$

7. Factor $6x^2 - 29x + 28$ by using trial factors.

8. Factor: $12a^2b + 3ab^2$

9. Factor: $a^6 - 100$

10. Factor: $n^4 - 2n^3 - 3n^2$

11. Factor $12y^2 + 16y - 3$ by using trial factors.

12. Factor: $12b^3 - 58b^2 + 56b$

13. Factor: $9y^4 - 25z^2$

14. Factor: $c^2 + 8c + 12$

15. Factor $18a^2 - 3a - 10$ by grouping.

16. Solve: $4x^2 + 27x = 7$

17. Factor: $4x^3 - 20x^2 - 24x$

18. Factor: $3a^2 - 15a - 42$

19. Factor $2a^2 - 19a - 60$ by grouping.

20. Solve: $(x + 1)(x - 5) = 16$

21. Factor: $21ax - 35bx - 10by + 6ay$

22. Factor: $a^2b^2 - 1$

23. Factor: $10x^2 + 25x + 4xy + 10y$

24. Factor: $5x^2 - 5x - 30$

25. Factor: $3x^2 + 36x + 108$

26. Factor $3x^2 - 17x + 10$ by grouping.

27. **Sports**   The length of the field in field hockey is 20 yd less than twice the width of the field. The area of the field in field hockey is 6000 yd². Find the length and width of the field.

28. **Image Projection**   The size, S, of an image from a slide projector depends on the distance, d, of the screen from the projector and is given by $S = d^2$. Find the distance between the projector and the screen when the size of the picture is 400 ft².

29. **Photography**   A rectangular photograph has dimensions 15 in. by 12 in. A picture frame around the photograph increases the total area to 270 in². What is the width of the frame?

30. **Gardening**   The length of each side of a square garden plot is extended 4 ft. The area of the resulting square is 576 ft². Find the length of a side of the original garden plot.

Chapter 5 Test

1. Factor: $ab + 6a - 3b - 18$

2. Factor: $2y^4 - 14y^3 - 16y^2$

3. Factor $8x^2 + 20x - 48$ by grouping.

4. Factor $6x^2 + 19x + 8$ by using trial factors.

5. Factor: $a^2 - 19a + 48$

6. Factor: $6x^3 - 8x^2 + 10x$

7. Factor: $x^2 + 2x - 15$

8. Solve: $4x^2 - 1 = 0$

9. Factor: $5x^2 - 45x - 15$

10. Factor: $p^2 + 12p + 36$

11. Solve: $x(x - 8) = -15$

12. Factor: $3x^2 + 12xy + 12y^2$

13. Factor: $b^2 - 16$

14. Factor $6x^2y^2 + 9xy^2 + 3y^2$ by grouping.

15.   Factor: $p^2 + 5p + 6$

16.   Factor: $a(x - 2) + b(x - 2)$

17.   Factor: $x(p + 1) - (p + 1)$

18.   Factor: $3a^2 - 75$

19.   Factor $2x^2 + 4x - 5$ by using trial factors.

20.   Factor: $x^2 - 9x - 36$

21.   Factor: $4a^2 - 12ab + 9b^2$

22.   Factor: $4x^2 - 49y^2$

23.   Solve: $(2a - 3)(a + 7) = 0$

24.   **Number Sense**   The sum of two numbers is ten. The sum of the squares of the two numbers is fifty-eight. Find the two numbers.

25.   **Geometry**   The length of a rectangle is 3 cm longer than twice its width. The area of the rectangle is 90 cm². Find the length and width of the rectangle.

Cumulative Review Exercises

1. Subtract: $-2 - (-3) - 5 - (-11)$

2. Simplify: $(3 - 7)^2 \div (-2) - 3 \cdot (-4)$

3. Evaluate $-2a^2 \div (2b) - c$ when $a = -4$, $b = 2$, and $c = -1$.

4. Simplify: $-\dfrac{3}{4}(-20x^2)$

5. Simplify: $-2[4x - 2(3 - 2x) - 8x]$

6. Solve: $-\dfrac{5}{7}x = -\dfrac{10}{21}$

7. Solve: $3x - 2 = 12 - 5x$

8. Solve: $-2 + 4[3x - 2(4 - x) - 3] = 4x + 2$

9. 120% of what number is 54?

10. Simplify: $(-3a^3b^2)^2$

11. Multiply: $(x + 2)(x^2 - 5x + 4)$

12. Divide: $(8x^2 + 4x - 3) \div (2x - 3)$

13. Simplify: $(x^{-4}y^3)^2$

14. Factor: $3a - 3b - ax + bx$

15. Factor: $15xy^2 - 20xy^4$

16. Factor: $x^2 - 5xy - 14y^2$

17. Factor: $p^2 - 9p - 10$

18. Factor: $18a^3 + 57a^2 + 30a$

19. Factor: $36a^2 - 49b^2$

20. Factor: $4x^2 + 28xy + 49y^2$

21. Factor: $9x^2 + 15x - 14$

22. Factor: $18x^2 - 48xy + 32y^2$

23. Factor: $3y(x - 3) - 2(x - 3)$

24. Solve: $3x^2 + 19x - 14 = 0$

25. Carpentry   A board 10 ft long is cut into two pieces. Four times the length of the shorter piece is 2 ft less than three times the length of the longer piece. Find the length of each piece.

26. Business   A portable MP3 player that regularly sells for $165 is on sale for $99. Find the discount rate. Use the formula $S = R - rR$.

27. Geometry   Given that lines ℓ_1 and ℓ_2 are parallel, find the measures of angles a and b.

28. Travel   A family drove to a resort at an average speed of 42 mph and later returned over the same road at an average speed of 56 mph. Find the distance to the resort if the total driving time was 7 h.

29. Consecutive Integers   Find three consecutive even integers such that five times the middle integer is twelve more than twice the sum of the first and third integers.

30. Geometry   The length of the base of a triangle is three times the height. The area of the triangle is 24 in^2. Find the length of the base of the triangle.

chapter

6 Rational Expressions

Deep-sea diving is not only a recreational sport enjoyed by tourists and adventure-seekers. Commercial deep-sea divers work on a variety of projects, including pipeline systems, offshore oilfield sites, and search-and-recovery missions. For any deep-sea diver, whether diving for recreation or for business, underwater survival depends on an oxygen tank. On land, the average adult at rest inhales and exhales about one cubic foot of air every four minutes. This air contains approximately 21% oxygen. The percent of oxygen needed by a person changes when he or she is underwater. **Exercise 82 on page 309** uses a variable expression that gives the recommended percent of oxygen for a diver as a function of the diver's depth.

 Need help? For online student resources, such as section quizzes, visit this textbook's website at **math.college.hmco.com/students.**

OBJECTIVES

Section 6.1
A To simplify a rational expression
B To multiply rational expressions
C To divide rational expressions

Section 6.2
A To find the least common multiple (LCM) of two or more polynomials
B To express two fractions in terms of the LCM of their denominators

Section 6.3
A To add or subtract rational expressions with the same denominators
B To add or subtract rational expressions with different denominators

Section 6.4
A To simplify a complex fraction

Section 6.5
A To solve an equation containing fractions

Section 6.6
A To solve a proportion
B To solve application problems
C To solve problems involving similar triangles

Section 6.7
A To solve a literal equation for one of the variables

Section 6.8
A To solve work problems
B To use rational expressions to solve uniform motion problems

Do these exercises to prepare for Chapter 6.

1. Find the least common multiple (LCM) of 12 and 18.

2. Simplify: $\dfrac{9x^3y^4}{3x^2y^7}$

3. Subtract: $\dfrac{3}{4} - \dfrac{8}{9}$

4. Divide: $\left(-\dfrac{8}{11}\right) \div \dfrac{4}{5}$

5. If a is a nonzero number, are the following two quantities equal: $\dfrac{0}{a}$ and $\dfrac{a}{0}$?

6. Solve: $\dfrac{2}{3}x - \dfrac{3}{4} = \dfrac{5}{6}$

7. Line l_1 is parallel to line l_2. Find the measure of angle a.

8. Factor: $x^2 - 4x - 12$

9. Factor: $2x^2 - x - 3$

10. At 9:00 AM, Anthony begins walking on a park trail at a rate of 9 m/min. Ten minutes later his sister Jean begins walking the same trail in pursuit of her brother at a rate of 12 m/min. At what time will Jean catch up to Anthony?

GO FIGURE · · ·

A mouse begins at corner A of a 12 ft by 12 ft square maze traveling clockwise at a constant speed of 2 feet per second. Six seconds later a second mouse starts from the same corner A and travels clockwise at a constant speed of 3 feet per second. How far apart are the two mice 18 seconds after the second mouse begins?

6.1 Multiplication and Division of Rational Expressions

Objective A   **To simplify a rational expression**

A fraction in which the numerator and denominator is a polynomial is called a **rational expression.** Examples of rational expressions are shown at the right.

$$\frac{5}{z}, \quad \frac{x^2 + 1}{2x - 1}, \quad \frac{y^2 + y - 1}{4y^2 + 1}$$

Care must be exercised with a rational expression to ensure that when the variables are replaced with numbers, the resulting denominator is not zero. Consider the rational expression at the right. The value of x cannot be 3 because the denominator would then be zero.

$$\frac{4x^2 - 9}{2x - 6}$$

$$\frac{4(3)^2 - 9}{2(3) - 6} = \frac{27}{0} \quad \text{Not a real number}$$

In the **simplest form of a rational expression,** the numerator and denominator have no common factors. The Multiplication Property of One is used to write a rational expression in simplest form.

HOW TO   Simplify: $\dfrac{x^2 - 4}{x^2 - 2x - 8}$

$$\frac{x^2 - 4}{x^2 - 2x - 8} = \frac{(x - 2)(x + 2)}{(x - 4)(x + 2)}$$

● Factor the numerator and denominator.

$$= \frac{x - 2}{x - 4} \cdot \boxed{\frac{x + 2}{x + 2}} = \frac{x - 2}{x - 4} \cdot 1$$

$$= \frac{x - 2}{x - 4}, \, x \neq -2, 4$$

● The restrictions, $x \neq -2$ or 4, are necessary to prevent division by zero.

This simplification is usually shown with slashes through the common factors:

$$\frac{x^2 - 4}{x^2 - 2x - 8} = \frac{(x - 2)\overset{1}{\cancel{(x + 2)}}}{(x - 4)\underset{1}{\cancel{(x + 2)}}}$$

● Factor the numerator and denominator.

$$= \frac{x - 2}{x - 4}, \, x \neq -2, 4$$

● Divide by the common factors. The restrictions, $x \neq -2$ or 4, are necessary to prevent division by zero.

In summary, to simplify a rational expression, factor the numerator and denominator. Then divide the numerator and denominator by the common factors.

HOW TO   Simplify: $\dfrac{10 + 3x - x^2}{x^2 - 4x - 5}$

$$\frac{10 + 3x - x^2}{x^2 - 4x - 5} = \frac{-(x^2 - 3x - 10)}{x^2 - 4x - 5}$$

● Because the coefficient of x^2 in the numerator is -1, factor -1 from the numerator.

$$= \frac{-\overset{1}{\cancel{(x - 5)}}(x + 2)}{\underset{1}{\cancel{(x - 5)}}(x + 1)}$$

● Factor the numerator and denominator. Divide by the common factors.

$$= -\frac{x + 2}{x + 1}, \, x \neq -1, 5$$

For the remaining examples, we will omit the restrictions on the variables that prevent division by zero and assume the values of the variables are such that division by zero is not possible.

Example 1

Simplify: $\dfrac{4x^3y^4}{6x^4y}$

Solution

$\dfrac{4x^3y^4}{6x^4y} = \dfrac{2y^3}{3x}$ • Use rules of exponents.

You Try It 1

Simplify: $\dfrac{6x^5y}{12x^2y^3}$

Your solution

Example 2

Simplify: $\dfrac{9 - x^2}{x^2 + x - 12}$

Solution

$\dfrac{9 - x^2}{x^2 + x - 12} = \dfrac{\overset{-1}{\cancel{(3 - x)}}(3 + x)}{\underset{1}{\cancel{(x - 3)}}(x + 4)}$ • $(3 - x) = -1(x - 3)$

$= -\dfrac{x + 3}{x + 4}$

You Try It 2

Simplify: $\dfrac{x^2 + 2x - 24}{16 - x^2}$

Your solution

Example 3

Simplify: $\dfrac{x^2 + 2x - 15}{x^2 - 7x + 12}$

Solution

$\dfrac{x^2 + 2x - 15}{x^2 - 7x + 12} = \dfrac{(x + 5)\overset{1}{\cancel{(x - 3)}}}{\underset{1}{\cancel{(x - 3)}}(x - 4)} = \dfrac{x + 5}{x - 4}$

You Try It 3

Simplify: $\dfrac{x^2 + 4x - 12}{x^2 - 3x + 2}$

Your solution

Solutions on p. S14

Objective B   **To multiply rational expressions**

The product of two fractions is a fraction whose numerator is the product of the numerators of the two fractions and whose denominator is the product of the denominators of the two fractions.

> **Multiplying Rational Expressions**
>
> Multiply the numerators. $\dfrac{a}{b} \cdot \dfrac{c}{d} = \dfrac{ac}{bd}$
> Multiply the denominators.

$$\dfrac{2}{3} \cdot \dfrac{4}{5} = \dfrac{8}{15} \qquad \dfrac{3x}{y} \cdot \dfrac{2}{z} = \dfrac{6x}{yz} \qquad \dfrac{x + 2}{x} \cdot \dfrac{3}{x - 2} = \dfrac{3x + 6}{x^2 - 2x}$$

HOW TO Multiply: $\dfrac{x^2 + 3x}{x^2 - 3x - 4} \cdot \dfrac{x^2 - 5x + 4}{x^2 + 2x - 3}$

$\dfrac{x^2 + 3x}{x^2 - 3x - 4} \cdot \dfrac{x^2 - 5x + 4}{x^2 + 2x - 3}$

$= \dfrac{x(x + 3)}{(x - 4)(x + 1)} \cdot \dfrac{(x - 4)(x - 1)}{(x + 3)(x - 1)}$

- Factor the numerator and denominator of each fraction.

$= \dfrac{x \cancel{(x + 3)} \cancel{(x - 4)} \cancel{(x - 1)}}{\cancel{(x - 4)}(x + 1)\cancel{(x + 3)}\cancel{(x - 1)}}$

- Multiply. Then divide by the common factors.

$= \dfrac{x}{x + 1}$

- Write the answer in simplest form.

Example 4

Multiply: $\dfrac{10x^2 - 15x}{12x - 8} \cdot \dfrac{3x - 2}{20x - 25}$

Solution

$\dfrac{10x^2 - 15x}{12x - 8} \cdot \dfrac{3x - 2}{20x - 25}$

$= \dfrac{5x(2x - 3)}{4(3x - 2)} \cdot \dfrac{(3x - 2)}{5(4x - 5)}$ • Factor.

$= \dfrac{\cancel{5}x(2x - 3)\cancel{(3x - 2)}}{4\cancel{(3x - 2)}\cancel{5}(4x - 5)}$ • Divide by common factors.

$= \dfrac{x(2x - 3)}{4(4x - 5)}$

You Try It 4

Multiply: $\dfrac{12x^2 + 3x}{10x - 15} \cdot \dfrac{8x - 12}{9x + 18}$

Your solution

Example 5

Multiply: $\dfrac{x^2 + x - 6}{x^2 + 7x + 12} \cdot \dfrac{x^2 + 3x - 4}{4 - x^2}$

Solution

$\dfrac{x^2 + x - 6}{x^2 + 7x + 12} \cdot \dfrac{x^2 + 3x - 4}{4 - x^2}$

$= \dfrac{(x + 3)(x - 2)}{(x + 3)(x + 4)} \cdot \dfrac{(x + 4)(x - 1)}{(2 - x)(2 + x)}$ • Factor.

$= \dfrac{\cancel{(x + 3)}\overset{-1}{\cancel{(x - 2)}}\cancel{(x + 4)}(x - 1)}{\cancel{(x + 3)}\cancel{(x + 4)}\cancel{(2 - x)}(2 + x)}$ • Divide by common factors.

$= -\dfrac{x - 1}{x + 2}$

You Try It 5

Multiply: $\dfrac{x^2 + 2x - 15}{9 - x^2} \cdot \dfrac{x^2 - 3x - 18}{x^2 - 7x + 6}$

Your solution

Solutions on pp. S14–S15

Objective C   **To divide rational expressions**

The **reciprocal of a rational expression** is the rational expression with the numerator and denominator interchanged.

$$\text{Fraction} \left\{ \begin{array}{ccc} \dfrac{a}{b} & & \dfrac{b}{a} \\[2mm] x^2 = \dfrac{x^2}{1} & & \dfrac{1}{x^2} \\[2mm] \dfrac{x+2}{x} & & \dfrac{x}{x+2} \end{array} \right\} \text{Reciprocal}$$

Dividing Rational Expressions

Multiply the dividend by the reciprocal of the divisor.

$$\frac{a}{b} \div \frac{c}{d} = \frac{a}{b} \cdot \frac{d}{c} = \frac{ad}{bc}$$

$$\frac{4}{x} \div \frac{y}{5} = \frac{4}{x} \cdot \frac{5}{y} = \frac{20}{xy}$$

$$\frac{x+4}{x} \div \frac{x-2}{4} = \frac{x+4}{x} \cdot \frac{4}{x-2} = \frac{4(x+4)}{x(x-2)}$$

The basis for the division rule is shown at the right.

$$\frac{a}{b} \div \frac{c}{d} = \frac{\dfrac{a}{b}}{\dfrac{c}{d}} = \frac{\dfrac{a}{b} \cdot \dfrac{d}{c}}{\dfrac{c}{d} \cdot \dfrac{d}{c}} = \frac{\dfrac{a}{b} \cdot \dfrac{d}{c}}{1} = \frac{a}{b} \cdot \frac{d}{c}$$

Example 6

Divide: $\dfrac{xy^2 - 3x^2y}{z^2} \div \dfrac{6x^2 - 2xy}{z^3}$

Solution

$\dfrac{xy^2 - 3x^2y}{z^2} \div \dfrac{6x^2 - 2xy}{z^3}$

$= \dfrac{xy^2 - 3x^2y}{z^2} \cdot \dfrac{z^3}{6x^2 - 2xy}$ • Multiply by the reciprocal.

$= \dfrac{xy\overset{-1}{(y - 3x)} \cdot z^3}{\underset{1}{z^2 \cdot 2x(3x - y)}} = -\dfrac{yz}{2}$

You Try It 6

Divide: $\dfrac{a^2}{4bc^2 - 2b^2c} \div \dfrac{a}{6bc - 3b^2}$

Your solution

Example 7

Divide: $\dfrac{2x^2 + 5x + 2}{2x^2 + 3x - 2} \div \dfrac{3x^2 + 13x + 4}{2x^2 + 7x - 4}$

Solution

$\dfrac{2x^2 + 5x + 2}{2x^2 + 3x - 2} \div \dfrac{3x^2 + 13x + 4}{2x^2 + 7x - 4}$

$= \dfrac{2x^2 + 5x + 2}{2x^2 + 3x - 2} \cdot \dfrac{2x^2 + 7x - 4}{3x^2 + 13x + 4}$ • Multiply by the reciprocal.

$= \dfrac{(2x + 1)\overset{1}{(x + 2)} \cdot \overset{1}{(2x - 1)}\overset{1}{(x + 4)}}{\underset{1}{(2x - 1)}\underset{1}{(x + 2)} \cdot (3x + 1)\underset{1}{(x + 4)}} = \dfrac{2x + 1}{3x + 1}$

You Try It 7

Divide: $\dfrac{3x^2 + 26x + 16}{3x^2 - 7x - 6} \div \dfrac{2x^2 + 9x - 5}{x^2 + 2x - 15}$

Your solution

Solutions on p. S15

6.1 Exercises

Objective A   **To simplify a rational expression**

1. ✎ Explain the procedure for writing a rational expression in simplest form.

2. ✎ Explain why the following simplification is incorrect.

$$\frac{x + 3}{x} = \frac{\overset{1}{\cancel{x}} + 3}{\underset{1}{\cancel{x}}} = 4$$

For Exercises 3 to 29, simplify.

3. $\dfrac{9x^3}{12x^4}$

4. $\dfrac{16x^2y}{24xy^3}$

5. $\dfrac{(x + 3)^2}{(x + 3)^3}$

6. $\dfrac{(2x - 1)^5}{(2x - 1)^4}$

7. $\dfrac{3n - 4}{4 - 3n}$

8. $\dfrac{5 - 2x}{2x - 5}$

9. $\dfrac{6y(y + 2)}{9y^2(y + 2)}$

10. $\dfrac{12x^2(3 - x)}{18x(3 - x)}$

11. $\dfrac{6x(x - 5)}{8x^2(5 - x)}$

12. $\dfrac{14x^3(7 - 3x)}{21x(3x - 7)}$

13. $\dfrac{a^2 + 4a}{ab + 4b}$

14. $\dfrac{x^2 - 3x}{2x - 6}$

15. $\dfrac{4 - 6x}{3x^2 - 2x}$

16. $\dfrac{5xy - 3y}{9 - 15x}$

17. $\dfrac{y^2 - 3y + 2}{y^2 - 4y + 3}$

18. $\dfrac{x^2 + 5x + 6}{x^2 + 8x + 15}$

19. $\dfrac{x^2 + 3x - 10}{x^2 + 2x - 8}$

20. $\dfrac{a^2 + 7a - 8}{a^2 + 6a - 7}$

21. $\dfrac{x^2 + x - 12}{x^2 - 6x + 9}$

22. $\dfrac{x^2 + 8x + 16}{x^2 - 2x - 24}$

23. $\dfrac{x^2 - 3x - 10}{25 - x^2}$

24. $\dfrac{4 - y^2}{y^2 - 3y - 10}$

25. $\dfrac{2x^3 + 2x^2 - 4x}{x^3 + 2x^2 - 3x}$

26. $\dfrac{3x^3 - 12x}{6x^3 - 24x^2 + 24x}$

27. $\dfrac{6x^2 - 7x + 2}{6x^2 + 5x - 6}$

28. $\dfrac{2n^2 - 9n + 4}{2n^2 - 5n - 12}$

29. $\dfrac{x^2 + 3x - 28}{24 - 2x - x^2}$

Objective B **To multiply rational expressions**

For Exercises 30 to 55, multiply.

30. $\dfrac{8x^2}{9y^3} \cdot \dfrac{3y^2}{4x^3}$

31. $\dfrac{14a^2b^3}{15x^5y^2} \cdot \dfrac{25x^3y}{16ab}$

32. $\dfrac{12x^3y^4}{7a^2b^3} \cdot \dfrac{14a^3b^4}{9x^2y^2}$

33. $\dfrac{18a^4b^2}{25x^2y^3} \cdot \dfrac{50x^5y^6}{27a^6b^2}$

34. $\dfrac{3x - 6}{5x - 20} \cdot \dfrac{10x - 40}{27x - 54}$

35. $\dfrac{8x - 12}{14x + 7} \cdot \dfrac{42x + 21}{32x - 48}$

36. $\dfrac{3x^2 + 2x}{2xy - 3y} \cdot \dfrac{2xy^3 - 3y^3}{3x^3 + 2x^2}$

37. $\dfrac{4a^2x - 3a^2}{2by + 5b} \cdot \dfrac{2b^3y + 5b^3}{4ax - 3a}$

38. $\dfrac{x^2 + 5x + 4}{x^3y^2} \cdot \dfrac{x^2y^3}{x^2 + 2x + 1}$

39. $\dfrac{x^2 + x - 2}{xy^2} \cdot \dfrac{x^3y}{x^2 + 5x + 6}$

40. $\dfrac{x^4y^2}{x^2 + 3x - 28} \cdot \dfrac{x^2 - 49}{xy^4}$

41. $\dfrac{x^5y^3}{x^2 + 13x + 30} \cdot \dfrac{x^2 + 2x - 3}{x^7y^2}$

42. $\dfrac{2x^2 - 5x}{2xy + y} \cdot \dfrac{2xy^2 + y^2}{5x^2 - 2x^3}$

43. $\dfrac{3a^3 + 4a^2}{5ab - 3b} \cdot \dfrac{3b^3 - 5ab^3}{3a^2 + 4a}$

44. $\dfrac{x^2 - 2x - 24}{x^2 - 5x - 6} \cdot \dfrac{x^2 + 5x + 6}{x^2 + 6x + 8}$

45. $\dfrac{x^2 - 8x + 7}{x^2 + 3x - 4} \cdot \dfrac{x^2 + 3x - 10}{x^2 - 9x + 14}$

46. $\dfrac{x^2 + 2x - 35}{x^2 + 4x - 21} \cdot \dfrac{x^2 + 3x - 18}{x^2 + 9x + 18}$

47. $\dfrac{y^2 + y - 20}{y^2 + 2y - 15} \cdot \dfrac{y^2 + 4y - 21}{y^2 + 3y - 28}$

48. $\dfrac{x^2 - 3x - 4}{x^2 + 6x + 5} \cdot \dfrac{x^2 + 5x + 6}{8 + 2x - x^2}$

49. $\dfrac{25 - n^2}{n^2 - 2n - 35} \cdot \dfrac{n^2 - 8n - 20}{n^2 - 3n - 10}$

50. $\dfrac{12x^2 - 6x}{x^2 + 6x + 5} \cdot \dfrac{2x^4 + 10x^3}{4x^2 - 1}$

51. $\dfrac{8x^3 + 4x^2}{x^2 - 3x + 2} \cdot \dfrac{x^2 - 4}{16x^2 + 8x}$

52. $\dfrac{16 + 6x - x^2}{x^2 - 10x - 24} \cdot \dfrac{x^2 - 6x - 27}{x^2 - 17x + 72}$

53. $\dfrac{x^2 - 11x + 28}{x^2 - 13x + 42} \cdot \dfrac{x^2 + 7x + 10}{20 - x - x^2}$

54. $\dfrac{2x^2 + 5x + 2}{2x^2 + 7x + 3} \cdot \dfrac{x^2 - 7x - 30}{x^2 - 6x - 40}$

55. $\dfrac{x^2 - 4x - 32}{x^2 - 8x - 48} \cdot \dfrac{3x^2 + 17x + 10}{3x^2 - 22x - 16}$

Objective C   **To divide rational expressions**

56. ✎   What is the reciprocal of a rational expression?

57. ✎   Explain how to divide rational expressions.

For Exercises 58 to 77, divide.

58. $\dfrac{4x^2y^3}{15a^2b^3} \div \dfrac{6xy}{5a^3b^5}$

59. $\dfrac{9x^3y^4}{16a^4b^2} \div \dfrac{45x^4y^2}{14a^7b}$

60. $\dfrac{6x - 12}{8x + 32} \div \dfrac{18x - 36}{10x + 40}$

61. $\dfrac{28x + 14}{45x - 30} \div \dfrac{14x + 7}{30x - 20}$

62. $\dfrac{6x^3 + 7x^2}{12x - 3} \div \dfrac{6x^2 + 7x}{36x - 9}$

63. $\dfrac{5a^2y + 3a^2}{2x^3 + 5x^2} \div \dfrac{10ay + 6a}{6x^3 + 15x^2}$

64. $\dfrac{x^2 + 4x + 3}{x^2y} \div \dfrac{x^2 + 2x + 1}{xy^2}$

65. $\dfrac{x^3y^2}{x^2 - 3x - 10} \div \dfrac{xy^4}{x^2 - x - 20}$

66. $\dfrac{x^2 - 49}{x^4y^3} \div \dfrac{x^2 - 14x + 49}{x^4y^3}$

67. $\dfrac{x^2y^5}{x^2 - 11x + 30} \div \dfrac{xy^6}{x^2 - 7x + 10}$

68. $\dfrac{4ax - 8a}{c^2} \div \dfrac{2y - xy}{c^3}$

69. $\dfrac{3x^2y - 9xy}{a^2b} \div \dfrac{3x^2 - x^3}{ab^2}$

70. $\dfrac{x^2 - 5x + 6}{x^2 - 9x + 18} \div \dfrac{x^2 - 6x + 8}{x^2 - 9x + 20}$

71. $\dfrac{x^2 + 3x - 40}{x^2 + 2x - 35} \div \dfrac{x^2 + 2x - 48}{x^2 + 3x - 18}$

72. $\dfrac{x^2 + 2x - 15}{x^2 - 4x - 45} \div \dfrac{x^2 + x - 12}{x^2 - 5x - 36}$

73. $\dfrac{y^2 - y - 56}{y^2 + 8y + 7} \div \dfrac{y^2 - 13y + 40}{y^2 - 4y - 5}$

74. $\dfrac{8 + 2x - x^2}{x^2 + 7x + 10} \div \dfrac{x^2 - 11x + 28}{x^2 - x - 42}$

75. $\dfrac{x^2 - x - 2}{x^2 - 7x + 10} \div \dfrac{x^2 - 3x - 4}{40 - 3x - x^2}$

76. $\dfrac{2x^2 - 3x - 20}{2x^2 - 7x - 30} \div \dfrac{2x^2 - 5x - 12}{4x^2 + 12x + 9}$

77. $\dfrac{6n^2 + 13n + 6}{4n^2 - 9} \div \dfrac{6n^2 + n - 2}{4n^2 - 1}$

APPLYING THE CONCEPTS

78. Given the expression $\dfrac{9}{x^2 + 1}$, choose some values of x and evaluate the expression for those values. Is it possible to choose a value of x for which the value of the expression is greater than 10? If so, what is that value of x? If not, explain why it is not possible.

79. Given the expression $\dfrac{1}{y - 3}$, choose some values of y and evaluate the expression for those values. Is it possible to choose a value of y for which the value of the expression is greater than 10,000,000? If so, what is that value of y? If not, explain why it is not possible.

Geometry For Exercises 80 and 81, write in simplest form the ratio of the shaded area of the figure to the total area of the figure.

80.

81.

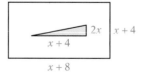

For Exercises 82 to 84, complete the simplification.

82. $\dfrac{8x}{9y} \div \underline{} = \dfrac{10y}{3}$

83. $\dfrac{n}{n + 3} \div \underline{} = \dfrac{n}{n - 2}$

84. $\underline{} \div \dfrac{n - 1}{4n^3} = 2n^2(n + 1)$

6.2 Expressing Fractions in Terms of the Least Common Multiple (LCM)

Objective A   **To find the least common multiple (LCM) of two or more polynomials**

Recall that the least common multiple (LCM) of two or more numbers is the smallest number that contains the prime factorization of each number.

The LCM of 12 and 18 is 36 because 36 contains the prime factors of 12 and the prime factors of 18.

$$12 = 2 \cdot 2 \cdot 3$$
$$18 = 2 \cdot 3 \cdot 3$$

$$\text{Factors of 12}$$
$$\text{LCM} = 36 = \overbrace{2 \cdot \underbrace{2 \cdot 3 \cdot 3}}$$
$$\text{Factors of 18}$$

The **least common multiple (LCM) of two or more polynomials** is the polynomial of least degree that contains all the factors of each polynomial.

To find the LCM of two or more polynomials, first factor each polynomial completely. The LCM is the product of each factor the greatest number of times it occurs in any one factorization.

> **TAKE NOTE**
> The LCM must contain the factors of each polynomial. As shown with the braces at the right, the LCM contains the factors of $4x^2 + 4x$ and the factors of $x^2 + 2x + 1$.

HOW TO   Find the LCM of $4x^2 + 4x$ and $x^2 + 2x + 1$.

The LCM of the polynomials is the product of the LCM of the numerical coefficients and each variable factor the greatest number of times it occurs in any one factorization.

$$4x^2 + 4x = 4x(x + 1) = 2 \cdot 2 \cdot x(x + 1)$$
$$x^2 + 2x + 1 = (x + 1)(x + 1)$$

$$\text{Factors of } 4x^2 + 4x$$
$$\text{LCM} = \overbrace{2 \cdot 2 \cdot x(x + 1)}(x + 1) = 4x(x + 1)(x + 1)$$
$$\underbrace{\text{Factors of } x^2 + 2x + 1}$$

Example 1

Find the LCM of $4x^2y$ and $6xy^2$.

Solution
$4x^2y = 2 \cdot 2 \cdot x \cdot x \cdot y$
$6xy^2 = 2 \cdot 3 \cdot x \cdot y \cdot y$
$\text{LCM} = 2 \cdot 2 \cdot 3 \cdot x \cdot x \cdot y \cdot y = 12x^2y^2$

You Try It 1

Find the LCM of $8uv^2$ and $12uw$.

Your solution

Example 2

Find the LCM of $x^2 - x - 6$ and $9 - x^2$.

Solution
$x^2 - x - 6 = (x - 3)(x + 2)$
$9 - x^2 = -(x^2 - 9) = -(x + 3)(x - 3)$
$\text{LCM} = (x - 3)(x + 2)(x + 3)$

You Try It 2

Find the LCM of $m^2 - 6m + 9$ and $m^2 - 2m - 3$.

Your solution

Solutions on p. S15

Objective B   **To express two fractions in terms of the LCM of their denominators**

When adding and subtracting fractions, it is frequently necessary to express two or more fractions in terms of a common denominator. This common denominator is the LCM of the denominators of the fractions.

HOW TO Write the fractions $\dfrac{x + 1}{4x^2}$ and $\dfrac{x - 3}{6x^2 - 12x}$ in terms of the LCM of the denominators.

Find the LCM of the denominators.

The LCM is $12x^2(x - 2)$.

For each fraction, multiply the numerator and the denominator by the factors whose product with the denominator is the LCM.

$$\frac{x + 1}{4x^2} = \frac{x + 1}{4x^2} \cdot \frac{3(x - 2)}{3(x - 2)} = \frac{3x^2 - 3x - 6}{12x^2(x - 2)} \longleftarrow$$

$$\frac{x - 3}{6x^2 - 12x} = \frac{x - 3}{6x(x - 2)} \cdot \frac{2x}{2x} = \frac{2x^2 - 6x}{12x^2(x - 2)} \longleftarrow$$

$\left. \right\rbrace$ LCM

Example 3

Write the fractions $\dfrac{x + 2}{3x^2}$ and $\dfrac{x - 1}{8xy}$ in terms of the LCM of the denominators.

Solution
The LCM is $24x^2y$.

$$\frac{x + 2}{3x^2} = \frac{x + 2}{3x^2} \cdot \frac{8y}{8y} = \frac{8xy + 16y}{24x^2y}$$

$$\frac{x - 1}{8xy} = \frac{x - 1}{8xy} \cdot \frac{3x}{3x} = \frac{3x^2 - 3x}{24x^2y}$$

You Try It 3

Write the fractions $\dfrac{x - 3}{4xy^2}$ and $\dfrac{2x + 1}{9y^2z}$ in terms of the LCM of the denominators.

Your solution

Example 4

Write the fractions $\dfrac{2x - 1}{2x - x^2}$ and $\dfrac{x}{x^2 + x - 6}$ in terms of the LCM of the denominators.

Solution
$$\frac{2x - 1}{2x - x^2} = \frac{2x - 1}{-(x^2 - 2x)} = -\frac{2x - 1}{x^2 - 2x}$$

The LCM is $x(x - 2)(x + 3)$.

$$\frac{2x - 1}{2x - x^2} = -\frac{2x - 1}{x(x - 2)} \cdot \frac{x + 3}{x + 3} = -\frac{2x^2 + 5x - 3}{x(x - 2)(x + 3)}$$

$$\frac{x}{x^2 + x - 6} = \frac{x}{(x - 2)(x + 3)} \cdot \frac{x}{x} = \frac{x^2}{x(x - 2)(x + 3)}$$

You Try It 4

Write the fractions $\dfrac{x + 4}{x^2 - 3x - 10}$ and $\dfrac{2x}{25 - x^2}$ in terms of the LCM of the denominators.

Your solution

Solutions on p. S15

6.2 Exercises

Objective A   **To find the least common multiple (LCM) of two or more polynomials**

For Exercises 1 to 33, find the LCM of the polynomials.

1. $8x^3y$
$12xy^2$

2. $6ab^2$
$18ab^3$

3. $10x^4y^2$
$15x^3y$

4. $12a^2b$
$18ab^3$

5. $8x^2$
$4x^2 + 8x$

6. $6y^2$
$4y + 12$

7. $2x^2y$
$3x^2 + 12x$

8. $4xy^2$
$6xy^2 + 12y^2$

9. $9x(x + 2)$
$12(x + 2)^2$

10. $8x^2(x - 1)^2$
$10x^3(x - 1)$

11. $3x + 3$
$2x^2 + 4x + 2$

12. $4x - 12$
$2x^2 - 12x + 18$

13. $(x - 1)(x + 2)$
$(x - 1)(x + 3)$

14. $(2x - 1)(x + 4)$
$(2x + 1)(x + 4)$

15. $(2x + 3)^2$
$(2x + 3)(x - 5)$

16. $(x - 7)(x + 2)$
$(x - 7)^2$

17. $x - 1$
$x - 2$
$(x - 1)(x - 2)$

18. $(x + 4)(x - 3)$
$x + 4$
$x - 3$

19. $x^2 - x - 6$
$x^2 + x - 12$

20. $x^2 + 3x - 10$
$x^2 + 5x - 14$

21. $x^2 + 5x + 4$
$x^2 - 3x - 28$

22. $x^2 - 10x + 21$
$x^2 - 8x + 15$

23. $x^2 - 2x - 24$
$x^2 - 36$

24. $x^2 + 7x + 10$
$x^2 - 25$

25. $x^2 - 7x - 30$
$x^2 - 5x - 24$

26. $2x^2 - 7x + 3$
$2x^2 + x - 1$

27. $3x^2 - 11x + 6$
$3x^2 + 4x - 4$

28. $2x^2 - 9x + 10$
$2x^2 + x - 15$

29. $6 + x - x^2$
$x + 2$
$x - 3$

30. $15 + 2x - x^2$
$x - 5$
$x + 3$

31. $5 + 4x - x^2$
$x - 5$
$x + 1$

32. $x^2 + 3x - 18$
$3 - x$
$x + 6$

33. $x^2 - 5x + 6$
$1 - x$
$x - 6$

> **Objective B** To express two fractions in terms
> of the LCM of their denominators

For Exercises 34 to 53, write the fraction in terms of the LCM of the denominators.

34. $\dfrac{4}{x}, \dfrac{3}{x^2}$

35. $\dfrac{5}{ab^2}, \dfrac{6}{ab}$

36. $\dfrac{x}{3y^2}, \dfrac{z}{4y}$

37. $\dfrac{5y}{6x^2}, \dfrac{7}{9xy}$

38. $\dfrac{y}{x(x-3)}, \dfrac{6}{x^2}$

39. $\dfrac{a}{y^2}, \dfrac{6}{y(y+5)}$

40. $\dfrac{9}{(x-1)^2}, \dfrac{6}{x(x-1)}$

41. $\dfrac{a^2}{y(y+7)}, \dfrac{a}{(y+7)^2}$

42. $\dfrac{3}{x-3}, \dfrac{5}{x(3-x)}$

43. $\dfrac{b}{y(y-4)}, \dfrac{b^2}{4-y}$

44. $\dfrac{3}{(x-5)^2}, \dfrac{2}{5-x}$

45. $\dfrac{3}{7-y}, \dfrac{2}{(y-7)^2}$

46. $\dfrac{3}{x^2+2x}, \dfrac{4}{x^2}$

47. $\dfrac{2}{y-3}, \dfrac{3}{y^3-3y^2}$

48. $\dfrac{x-2}{x+3}, \dfrac{x}{x-4}$

49. $\dfrac{x^2}{2x-1}, \dfrac{x+1}{x+4}$

50. $\dfrac{3}{x^2+x-2}, \dfrac{x}{x+2}$

51. $\dfrac{3x}{x-5}, \dfrac{4}{x^2-25}$

52. $\dfrac{x}{x^2+x-6}, \dfrac{2x}{x^2-9}$

53. $\dfrac{x-1}{x^2+2x-15}, \dfrac{x}{x^2+6x+5}$

APPLYING THE CONCEPTS

54. ✎ When is the LCM of two polynomials equal to their product?

For Exercises 55 to 60, write each fraction in terms of the LCM of the denominators.

55. $\dfrac{8}{10^3}, \dfrac{9}{10^5}$

56. $3, \dfrac{2}{n}$

57. $x, \dfrac{x}{x^2-1}$

58. $\dfrac{x^2+1}{(x-1)^3}, \dfrac{x+1}{(x-1)^2}, \dfrac{1}{x-1}$

59. $\dfrac{c}{6c^2+7cd+d^2}, \dfrac{d}{3c^2-3d^2}$

60. $\dfrac{1}{ab+3a-3b-b^2}, \dfrac{1}{ab+3a+3b+b^2}$

6.3 Addition and Subtraction of Rational Expressions

Objective A   **To add or subtract rational expressions with the same denominators**

When adding rational expressions in which the denominators are the same, add the numerators. The denominator of the sum is the common denominator.

$$\frac{5x}{18} + \frac{7x}{18} = \frac{5x + 7x}{18} = \frac{12x}{18} = \frac{2x}{3}$$

$$\frac{x}{x^2 - 1} + \frac{1}{x^2 - 1} = \frac{x + 1}{x^2 - 1} = \frac{\overset{1}{\cancel{(x + 1)}}}{(x - 1)\underset{1}{\cancel{(x + 1)}}} = \frac{1}{x - 1}$$

Note that the sum is written in simplest form.

When subtracting rational expressions with like denominators, subtract the numerators. The denominator of the difference is the common denominator. Write the answer in simplest form.

$$\frac{2x}{x - 2} - \frac{4}{x - 2} = \frac{2x - 4}{x - 2} = \frac{2\overset{1}{\cancel{(x - 2)}}}{\underset{1}{\cancel{x - 2}}} = 2$$

$$\frac{3x - 1}{x^2 - 5x + 4} - \frac{2x + 3}{x^2 - 5x + 4} = \frac{(3x - 1) - (2x + 3)}{x^2 - 5x + 4} = \frac{3x - 1 - 2x - 3}{x^2 - 5x + 4}$$

$$= \frac{x - 4}{x^2 - 5x + 4} = \frac{\overset{1}{\cancel{(x - 4)}}}{\underset{1}{\cancel{(x - 4)}}(x - 1)} = \frac{1}{x - 1}$$

Adding and Subtracting Rational Expressions with the Same Denominator

Add or subtract the numerators. Place the result over the common denominator. $\dfrac{a}{b} + \dfrac{c}{b} = \dfrac{a + c}{b}$ $\dfrac{a}{b} - \dfrac{c}{b} = \dfrac{a - c}{b}$

Example 1

Subtract: $\dfrac{3x^2}{x^2 - 1} - \dfrac{x + 4}{x^2 - 1}$

Solution

$$\frac{3x^2}{x^2 - 1} - \frac{x + 4}{x^2 - 1} = \frac{3x^2 - (x + 4)}{x^2 - 1}$$

$$= \frac{3x^2 - x - 4}{x^2 - 1}$$

$$= \frac{(3x - 4)\overset{1}{\cancel{(x + 1)}}}{(x - 1)\underset{1}{\cancel{(x + 1)}}} = \frac{3x - 4}{x - 1}$$

You Try It 1

Subtract: $\dfrac{2x^2}{x^2 - x - 12} - \dfrac{7x + 4}{x^2 - x - 12}$

Your solution

Solution on p. S15

Example 2

Simplify:

$$\frac{2x^2 + 5}{x^2 + 2x - 3} - \frac{x^2 - 3x}{x^2 + 2x - 3} + \frac{x - 2}{x^2 + 2x - 3}$$

Solution

$$\frac{2x^2 + 5}{x^2 + 2x - 3} - \frac{x^2 - 3x}{x^2 + 2x - 3} + \frac{x - 2}{x^2 + 2x - 3}$$

$$= \frac{(2x^2 + 5) - (x^2 - 3x) + (x - 2)}{x^2 + 2x - 3}$$

$$= \frac{2x^2 + 5 - x^2 + 3x + x - 2}{x^2 + 2x - 3}$$

$$= \frac{x^2 + 4x + 3}{x^2 + 2x - 3}$$

$$= \frac{\overset{1}{\cancel{(x + 3)}}(x + 1)}{\underset{1}{\cancel{(x + 3)}}(x - 1)} = \frac{x + 1}{x - 1}$$

You Try It 2

Simplify:

$$\frac{x^2 - 1}{x^2 - 8x + 12} - \frac{2x + 1}{x^2 - 8x + 12} + \frac{x}{x^2 - 8x + 12}$$

Your solution

Solution on p. S15

Objective B   **To add or subtract rational expressions with different denominators**

Before two fractions with unlike denominators can be added or subtracted, each fraction must be expressed in terms of a common denominator. This common denominator is the LCM of the denominators of the fractions.

HOW TO   Add: $\dfrac{x - 3}{x^2 - 2x} + \dfrac{6}{x^2 - 4}$

The LCM is $x(x - 2)(x + 2)$.   • Find the LCM of the denominators.

$$\frac{x - 3}{x^2 - 2x} + \frac{6}{x^2 - 4}$$

$$= \frac{x - 3}{x(x - 2)} \cdot \frac{x + 2}{x + 2} + \frac{6}{(x - 2)(x + 2)} \cdot \frac{x}{x}$$ • Write each fraction in terms of the LCM.

$$= \frac{x^2 - x - 6}{x(x - 2)(x + 2)} + \frac{6x}{x(x - 2)(x + 2)}$$ • Multiply the factors in the numerators.

$$= \frac{(x^2 - x - 6) + 6x}{x(x - 2)(x + 2)}$$ • Add the fractions.

$$= \frac{x^2 + 5x - 6}{x(x - 2)(x + 2)}$$ • Simplify.

$$= \frac{(x + 6)(x - 1)}{x(x - 2)(x + 2)}$$ • Factor.

After combining the numerators over the common denominator, the last step is to factor the numerator to determine whether there are common factors in the numerator and denominator. For the previous example, there are no common factors, so the answer is in simplest form.

The process of adding and subtracting rational expressions is summarized below.

Adding and Subtracting Rational Expressions

1. Find the LCM of the denominators.
2. Write each fraction as an equivalent fraction using the LCM as the denominator.
3. Add or subtract the numerators and place the result over the common denominator.
4. Write the answer in simplest form.

Example 3

Simplify: $\dfrac{y}{x} - \dfrac{4y}{3x} + \dfrac{3y}{4x}$

Solution

The LCM of the denominators is $12x$.

$\dfrac{y}{x} - \dfrac{4y}{3x} + \dfrac{3y}{4x}$

$= \dfrac{y}{x} \cdot \dfrac{12}{12} - \dfrac{4y}{3x} \cdot \dfrac{4}{4} + \dfrac{3y}{4x} \cdot \dfrac{3}{3}$ • Write each fraction using the LCM.

$= \dfrac{12y}{12x} - \dfrac{16y}{12x} + \dfrac{9y}{12x}$

$= \dfrac{12y - 16y + 9y}{12x} = \dfrac{5y}{12x}$ • Combine the numerators.

You Try It 3

Simplify: $\dfrac{z}{8y} - \dfrac{4z}{3y} + \dfrac{5z}{4y}$

Your solution

Solution on p. S15

Example 4

Subtract: $\dfrac{2x}{x-3} - \dfrac{5}{3-x}$

Solution

Remember $3 - x = -(x-3)$.

Therefore, $\dfrac{5}{3-x} = \dfrac{5}{-(x-3)} = \dfrac{-5}{x-3}$.

$\dfrac{2x}{x-3} - \dfrac{5}{3-x}$

$= \dfrac{2x}{x-3} - \dfrac{-5}{x-3}$ • The LCM is $x - 3$.

$= \dfrac{2x - (-5)}{x-3} = \dfrac{2x+5}{x-3}$ • Combine the numerators.

You Try It 4

Add: $\dfrac{5x}{x-2} + \dfrac{3}{2-x}$

Your solution

Example 5

Subtract: $\dfrac{2x}{2x-3} - \dfrac{1}{x+1}$

Solution

The LCM is $(2x-3)(x+1)$.

$\dfrac{2x}{2x-3} - \dfrac{1}{x+1}$

$= \dfrac{2x}{2x-3} \cdot \dfrac{x+1}{x+1} - \dfrac{1}{x+1} \cdot \dfrac{2x-3}{2x-3}$

$= \dfrac{2x^2+2x}{(2x-3)(x+1)} - \dfrac{2x-3}{(2x-3)(x+1)}$

$= \dfrac{(2x^2+2x)-(2x-3)}{(2x-3)(x+1)}$

$= \dfrac{2x^2+2x-2x+3}{(2x-3)(x+1)} = \dfrac{2x^2+3}{(2x-3)(x+1)}$

You Try It 5

Add: $\dfrac{4x}{3x-1} + \dfrac{9}{x+4}$

Your solution

Example 6

Add: $1 + \dfrac{3}{x^2}$

Solution

The LCM is x^2.

$1 + \dfrac{3}{x^2} = 1 \cdot \dfrac{x^2}{x^2} + \dfrac{3}{x^2} = \dfrac{x^2}{x^2} + \dfrac{3}{x^2} = \dfrac{x^2+3}{x^2}$

You Try It 6

Subtract: $2 - \dfrac{1}{x-3}$

Your solution

Solutions on pp. S15–S16

Example 7

Add: $\dfrac{x + 3}{x^2 - 2x - 8} + \dfrac{3}{4 - x}$

Solution

Recall: $\dfrac{3}{4 - x} = \dfrac{-3}{x - 4}$

The LCM is $(x - 4)(x + 2)$.

$\dfrac{x + 3}{x^2 - 2x - 8} + \dfrac{3}{4 - x}$

$= \dfrac{x + 3}{(x - 4)(x + 2)} + \dfrac{(-3)}{x - 4}$

$= \dfrac{x + 3}{(x - 4)(x + 2)} + \dfrac{(-3)}{x - 4} \cdot \dfrac{x + 2}{x + 2}$

$= \dfrac{x + 3}{(x - 4)(x + 2)} + \dfrac{(-3)(x + 2)}{(x - 4)(x + 2)}$

$= \dfrac{(x + 3) + (-3)(x + 2)}{(x - 4)(x + 2)}$

$= \dfrac{x + 3 - 3x - 6}{(x - 4)(x + 2)}$

$= \dfrac{-2x - 3}{(x - 4)(x + 2)}$

You Try It 7

Add: $\dfrac{2x - 1}{x^2 - 25} + \dfrac{2}{5 - x}$

Your solution

Example 8

Simplify: $\dfrac{3x + 2}{2x^2 - x - 1} - \dfrac{3}{2x + 1} + \dfrac{4}{x - 1}$

Solution

The LCM is $(2x + 1)(x - 1)$.

$\dfrac{3x + 2}{2x^2 - x - 1} - \dfrac{3}{2x + 1} + \dfrac{4}{x - 1}$

$= \dfrac{3x + 2}{(2x + 1)(x - 1)} - \dfrac{3}{2x + 1} \cdot \dfrac{x - 1}{x - 1} + \dfrac{4}{x - 1} \cdot \dfrac{2x + 1}{2x + 1}$

$= \dfrac{3x + 2}{(2x + 1)(x - 1)} - \dfrac{3x - 3}{(2x + 1)(x - 1)} + \dfrac{8x + 4}{(2x + 1)(x - 1)}$

$= \dfrac{(3x + 2) - (3x - 3) + (8x + 4)}{(2x + 1)(x - 1)}$

$= \dfrac{3x + 2 - 3x + 3 + 8x + 4}{(2x + 1)(x - 1)}$

$= \dfrac{8x + 9}{(2x + 1)(x - 1)}$

You Try It 8

Simplify: $\dfrac{2x - 3}{3x^2 - x - 2} + \dfrac{5}{3x + 2} - \dfrac{1}{x - 1}$

Your solution

Solutions on p. S16

6.3 Exercises

> **Objective A**   **To add or subtract rational expressions with the same denominators**

For Exercises 1 to 20, simplify.

1. $\dfrac{3}{y^2} + \dfrac{8}{y^2}$

2. $\dfrac{6}{ab} - \dfrac{2}{ab}$

3. $\dfrac{3}{x+4} - \dfrac{10}{x+4}$

4. $\dfrac{x}{x+6} - \dfrac{2}{x+6}$

5. $\dfrac{3x}{2x+3} + \dfrac{5x}{2x+3}$

6. $\dfrac{6y}{4y+1} - \dfrac{11y}{4y+1}$

7. $\dfrac{2x+1}{x-3} + \dfrac{3x+6}{x-3}$

8. $\dfrac{4x+3}{2x-7} + \dfrac{3x-8}{2x-7}$

9. $\dfrac{5x-1}{x+9} - \dfrac{3x+4}{x+9}$

10. $\dfrac{6x-5}{x-10} - \dfrac{3x-4}{x-10}$

11. $\dfrac{x-7}{2x+7} - \dfrac{4x-3}{2x+7}$

12. $\dfrac{2n}{3n+4} - \dfrac{5n-3}{3n+4}$

13. $\dfrac{x}{x^2+2x-15} - \dfrac{3}{x^2+2x-15}$

14. $\dfrac{3x}{x^2+3x-10} - \dfrac{6}{x^2+3x-10}$

15. $\dfrac{2x+3}{x^2-x-30} - \dfrac{x-2}{x^2-x-30}$

16. $\dfrac{3x-1}{x^2+5x-6} - \dfrac{2x-7}{x^2+5x-6}$

17. $\dfrac{4y+7}{2y^2+7y-4} - \dfrac{y-5}{2y^2+7y-4}$

18. $\dfrac{x+1}{2x^2-5x-12} + \dfrac{x+2}{2x^2-5x-12}$

19. $\dfrac{2x^2+3x}{x^2-9x+20} + \dfrac{2x^2-3}{x^2-9x+20} - \dfrac{4x^2+2x+1}{x^2-9x+20}$

20. $\dfrac{2x^2+3x}{x^2-2x-63} - \dfrac{x^2-3x+21}{x^2-2x-63} - \dfrac{x-7}{x^2-2x-63}$

Objective B **To add or subtract rational expressions with different denominators**

21. Explain the process of writing equivalent rational expressions with the LCM of the denominators of the rational expressions as the new denominator.

22. Explain the process of adding rational expressions with different denominators.

For Exercises 23 to 80, simplify.

23. $\dfrac{4}{x} + \dfrac{5}{y}$

24. $\dfrac{7}{a} + \dfrac{5}{b}$

25. $\dfrac{12}{x} - \dfrac{5}{2x}$

26. $\dfrac{5}{3a} - \dfrac{3}{4a}$

27. $\dfrac{1}{2x} - \dfrac{5}{4x} + \dfrac{7}{6x}$

28. $\dfrac{7}{4y} + \dfrac{11}{6y} - \dfrac{8}{3y}$

29. $\dfrac{5}{3x} - \dfrac{2}{x^2} + \dfrac{3}{2x}$

30. $\dfrac{6}{y^2} + \dfrac{3}{4y} - \dfrac{2}{5y}$

31. $\dfrac{2}{x} - \dfrac{3}{2y} + \dfrac{3}{5x} - \dfrac{1}{4y}$

32. $\dfrac{5}{2a} + \dfrac{7}{3b} - \dfrac{2}{b} - \dfrac{3}{4a}$

33. $\dfrac{2x+1}{3x} + \dfrac{x-1}{5x}$

34. $\dfrac{4x-3}{6x} + \dfrac{2x+3}{4x}$

35. $\dfrac{x-3}{6x} + \dfrac{x+4}{8x}$

36. $\dfrac{2x-3}{2x} + \dfrac{x+3}{3x}$

37. $\dfrac{2x+9}{9x} - \dfrac{x-5}{5x}$

38. $\dfrac{3y-2}{12y} - \dfrac{y-3}{18y}$

39. $\dfrac{x+4}{2x} - \dfrac{x-1}{x^2}$

40. $\dfrac{x-2}{3x^2} - \dfrac{x+4}{x}$

41. $\dfrac{x-10}{4x^2} + \dfrac{x+1}{2x}$

42. $\dfrac{x+5}{3x^2} + \dfrac{2x+1}{2x}$

43. $\dfrac{4}{x+4} - x$

44. $2x + \dfrac{1}{x}$

45. $5 - \dfrac{x-2}{x+1}$

46. $3 + \dfrac{x-1}{x+1}$

47. $\dfrac{x+3}{6x} - \dfrac{x-3}{8x^2}$

48. $\dfrac{x+2}{xy} - \dfrac{3x-2}{x^2y}$

49. $\dfrac{3x-1}{xy^2} - \dfrac{2x+3}{xy}$

50. $\dfrac{4x-3}{3x^2y} + \dfrac{2x+1}{4xy^2}$

51. $\dfrac{5x+7}{6xy^2} - \dfrac{4x-3}{8x^2y}$

52. $\dfrac{x-2}{8x^2} - \dfrac{x+7}{12xy}$

53. $\dfrac{3x-1}{6y^2} - \dfrac{x+5}{9xy}$

54. $\dfrac{4}{x-2} + \dfrac{5}{x+3}$

55. $\dfrac{2}{x-3} + \dfrac{5}{x-4}$

56. $\dfrac{6}{x-7} - \dfrac{4}{x+3}$

57. $\dfrac{3}{y+6} - \dfrac{4}{y-3}$

58. $\dfrac{2x}{x+1} + \dfrac{1}{x-3}$

59. $\dfrac{3x}{x-4} + \dfrac{2}{x+6}$

60. $\dfrac{4x}{2x-1} - \dfrac{5}{x-6}$

61. $\dfrac{6x}{x+5} - \dfrac{3}{2x+3}$

62. $\dfrac{2a}{a-7} + \dfrac{5}{7-a}$

63. $\dfrac{4x}{6-x} + \dfrac{5}{x-6}$

64. $\dfrac{x}{x^2-9} + \dfrac{3}{x-3}$

65. $\dfrac{y}{y^2-16} + \dfrac{1}{y-4}$

66. $\dfrac{2x}{x^2-x-6} - \dfrac{3}{x+2}$

67. $\dfrac{(x-1)^2}{(x+1)^2} - 1$

68. $1 - \dfrac{(y-2)^2}{(y+2)^2}$

69. $\dfrac{x}{1-x^2} - 1 + \dfrac{x}{1+x}$

70. $\dfrac{y}{x-y} + 2 - \dfrac{x}{y-x}$

71. $\dfrac{3x-1}{x^2-10x+25} - \dfrac{3}{x-5}$

72. $\dfrac{2a+3}{a^2-7a+12} - \dfrac{2}{a-3}$

73. $\dfrac{x+4}{x^2-x-42} + \dfrac{3}{7-x}$

74. $\dfrac{x+3}{x^2-3x-10} + \dfrac{2}{5-x}$

75. $\dfrac{1}{x+1} + \dfrac{x}{x-6} - \dfrac{5x-2}{x^2-5x-6}$

76. $\dfrac{x}{x-4} + \dfrac{5}{x+5} - \dfrac{11x-8}{x^2+x-20}$

77. $\dfrac{3x+1}{x-1} - \dfrac{x-1}{x-3} + \dfrac{x+1}{x^2-4x+3}$

78. $\dfrac{4x+1}{x-8} - \dfrac{3x+2}{x+4} - \dfrac{49x+4}{x^2-4x-32}$

79. $\dfrac{2x+9}{3-x} + \dfrac{x+5}{x+7} - \dfrac{2x^2+3x-3}{x^2+4x-21}$

80. $\dfrac{3x+5}{x+5} - \dfrac{x+1}{2-x} - \dfrac{4x^2-3x-1}{x^2+3x-10}$

APPLYING THE CONCEPTS

81. Transportation Suppose that you drive about 12,000 mi per year and that the cost of gasoline averages $1.70 per gallon.

  a. Let x represent the number of miles per gallon your car gets. Write a variable expression for the amount you spend on gasoline in 1 year.

  b. Write and simplify a variable expression for the amount of money you will save each year if you can increase your gas mileage by 5 miles per gallon.

  c. If you currently get 25 miles per gallon and you increase your gas mileage by 5 miles per gallon, how much will you save in 1 year?

82. Deep-Sea Diving A recommended percent of oxygen (by volume) in the air that a deep-sea diver breathes is given by $\dfrac{660}{d+33}$, where d is the depth, in feet, at which the diver is working.

  a. What is the recommended percent (to the nearest percent) of oxygen for a diver working at a depth of 50 ft?

  b. As the depth of the diver increases, does the recommended amount of oxygen increase or decrease?

  c. At sea level, the oxygen content of air is approximately 21%. Is this less than or more than the recommended amount of oxygen for a diver working at the water's surface?

6.4 Complex Fractions

Objective A **To simplify a complex fraction**

Point of Interest

There are many instances of complex fractions in application problems. The fraction $\dfrac{1}{\dfrac{1}{r_1} + \dfrac{1}{r_2}}$ is used to determine the total resistance in certain electric circuits.

TAKE NOTE

You may use either method to simplify a complex fraction. The result will be the same.

A **complex fraction** is a fraction whose numerator or denominator contains one or more fractions. Examples of complex fractions are shown at the right.

$$\dfrac{3}{2 - \dfrac{1}{2}}, \quad \dfrac{4 + \dfrac{1}{x}}{3 + \dfrac{2}{x}}, \quad \dfrac{\dfrac{1}{x - 1} + x + 3}{x - 3 + \dfrac{1}{x + 4}}$$

To simplify a complex fraction, use one of the following methods.

Simplifying Complex Fractions

Method 1: Multiply by 1 in the form $\dfrac{\text{LCM}}{\text{LCM}}$.

1. Determine the LCM of the denominators of the fractions in the numerator and denominator of the complex fraction.
2. Multiply the numerator and denominator of the complex fraction by the LCM.
3. Simplify.

Method 2: Multiply the numerator by the reciprocal of the denominator.

1. Simplify the numerator to a single fraction and simplify the denominator to a single fraction.
2. Using the definition for dividing fractions, multiply the numerator by the reciprocal of the denominator.
3. Simplify.

Here is an example of using Method 1.

HOW TO Simplify: $\dfrac{9 - \dfrac{4}{x^2}}{3 + \dfrac{2}{x}}$

The LCM of x and x^2 is x^2.

• Find the **LCM** of the denominators of the fractions in the numerator and the denominator.

$$\dfrac{9 - \dfrac{4}{x^2}}{3 + \dfrac{2}{x}} = \dfrac{9 - \dfrac{4}{x^2}}{3 + \dfrac{2}{x}} \cdot \dfrac{x^2}{x^2}$$

• Multiply the numerator and denominator by the **LCM**.

$$= \dfrac{9 \cdot x^2 - \dfrac{4}{x^2} \cdot x^2}{3 \cdot x^2 + \dfrac{2}{x} \cdot x^2} = \dfrac{9x^2 - 4}{3x^2 + 2x}$$

• Use the **Distributive Property**.

$$= \dfrac{(3x - 2)\cancel{(3x + 2)}}{x\cancel{(3x + 2)}} = \dfrac{3x - 2}{x}$$

• Simplify.

Here is the same example using the second method.

HOW TO   Simplify: $\dfrac{9 - \dfrac{4}{x^2}}{3 + \dfrac{2}{x}}$

$\dfrac{9 - \dfrac{4}{x^2}}{3 + \dfrac{2}{x}} = \dfrac{\dfrac{9x^2}{x^2} - \dfrac{4}{x^2}}{\dfrac{3x}{x} + \dfrac{2}{x}} = \dfrac{\dfrac{9x^2 - 4}{x^2}}{\dfrac{3x + 2}{x}}$

- Simplify the numerator to a single fraction and simplify the denominator to a single fraction.

$= \dfrac{9x^2 - 4}{x^2} \cdot \dfrac{x}{3x + 2}$

- Multiply the numerator by the reciprocal of the denominator.

$= \dfrac{x(3x - 2)\cancel{(3x + 2)}^{1}}{x^2\cancel{(3x + 2)}_{1}}$

- Simplify.

$= \dfrac{3x - 2}{x}$

For the examples below, we will use the first method.

Example 1

Simplify: $\dfrac{\dfrac{1}{x} + \dfrac{1}{2}}{\dfrac{1}{x^2} - \dfrac{1}{4}}$

Solution

The LCM of x, 2, x^2, and 4 is $4x^2$.

$\dfrac{\dfrac{1}{x} + \dfrac{1}{2}}{\dfrac{1}{x^2} - \dfrac{1}{4}} = \dfrac{\dfrac{1}{x} + \dfrac{1}{2}}{\dfrac{1}{x^2} - \dfrac{1}{4}} \cdot \dfrac{4x^2}{4x^2}$

- Multiply by the LCM.

$= \dfrac{\dfrac{1}{x} \cdot 4x^2 + \dfrac{1}{2} \cdot 4x^2}{\dfrac{1}{x^2} \cdot 4x^2 - \dfrac{1}{4} \cdot 4x^2}$

- Distributive Property

$= \dfrac{4x + 2x^2}{4 - x^2}$

- Simplify.

$= \dfrac{2x\cancel{(2 + x)}^{1}}{(2 - x)\cancel{(2 + x)}_{1}}$

$= \dfrac{2x}{2 - x}$

You Try It 1

Simplify: $\dfrac{\dfrac{1}{3} - \dfrac{1}{x}}{\dfrac{1}{9} - \dfrac{1}{x^2}}$

Your solution

Solution on p. S16

Example 2

Simplify: $\dfrac{1 - \dfrac{2}{x} - \dfrac{15}{x^2}}{1 - \dfrac{11}{x} + \dfrac{30}{x^2}}$

Solution

The LCM of x and x^2 is x^2.

$\dfrac{1 - \dfrac{2}{x} - \dfrac{15}{x^2}}{1 - \dfrac{11}{x} + \dfrac{30}{x^2}} = \dfrac{1 - \dfrac{2}{x} - \dfrac{15}{x^2}}{1 - \dfrac{11}{x} + \dfrac{30}{x^2}} \cdot \dfrac{x^2}{x^2}$

• Multiply by the LCM.

$= \dfrac{1 \cdot x^2 - \dfrac{2}{x} \cdot x^2 - \dfrac{15}{x^2} \cdot x^2}{1 \cdot x^2 - \dfrac{11}{x} \cdot x^2 + \dfrac{30}{x^2} \cdot x^2}$

• Distributive Property

$= \dfrac{x^2 - 2x - 15}{x^2 - 11x + 30}$

$= \dfrac{\cancel{(x - 5)}(x + 3)}{\cancel{(x - 5)}(x - 6)} = \dfrac{x + 3}{x - 6}$

• Simplify.

You Try It 2

Simplify: $\dfrac{1 + \dfrac{4}{x} + \dfrac{3}{x^2}}{1 + \dfrac{10}{x} + \dfrac{21}{x^2}}$

Your solution

Example 3

Simplify: $\dfrac{x - 8 + \dfrac{20}{x + 4}}{x - 10 + \dfrac{24}{x + 4}}$

Solution

The LCM is $x + 4$.

$\dfrac{x - 8 + \dfrac{20}{x + 4}}{x - 10 + \dfrac{24}{x + 4}}$

$= \dfrac{x - 8 + \dfrac{20}{x + 4}}{x - 10 + \dfrac{24}{x + 4}} \cdot \dfrac{x + 4}{x + 4}$

• Multiply by the LCM.

$= \dfrac{(x - 8)(x + 4) + \dfrac{20}{x + 4} \cdot (x + 4)}{(x - 10)(x + 4) + \dfrac{24}{x + 4} \cdot (x + 4)}$

• Distributive Property

$= \dfrac{x^2 - 4x - 32 + 20}{x^2 - 6x - 40 + 24} = \dfrac{x^2 - 4x - 12}{x^2 - 6x - 16}$

• Simplify.

$= \dfrac{(x - 6)\cancel{(x + 2)}}{(x - 8)\cancel{(x + 2)}} = \dfrac{x - 6}{x - 8}$

You Try It 3

Simplify: $\dfrac{x + 3 - \dfrac{20}{x - 5}}{x + 8 + \dfrac{30}{x - 5}}$

Your solution

Solutions on p. S16

6.4 Exercises

Objective A To simplify a complex fraction

For Exercises 1 to 30, simplify.

1. $\dfrac{1 + \dfrac{3}{x}}{1 - \dfrac{9}{x^2}}$

2. $\dfrac{1 + \dfrac{4}{x}}{1 - \dfrac{16}{x^2}}$

3. $\dfrac{2 - \dfrac{8}{x + 4}}{3 - \dfrac{12}{x + 4}}$

4. $\dfrac{5 - \dfrac{25}{x + 5}}{1 - \dfrac{3}{x + 5}}$

5. $\dfrac{1 + \dfrac{5}{y - 2}}{1 - \dfrac{2}{y - 2}}$

6. $\dfrac{2 - \dfrac{11}{2x - 1}}{3 - \dfrac{17}{2x - 1}}$

7. $\dfrac{4 - \dfrac{2}{x + 7}}{5 + \dfrac{1}{x + 7}}$

8. $\dfrac{5 + \dfrac{3}{x - 8}}{2 - \dfrac{1}{x - 8}}$

9. $\dfrac{1 - \dfrac{1}{x} - \dfrac{6}{x^2}}{1 - \dfrac{9}{x^2}}$

10. $\dfrac{1 + \dfrac{4}{x} + \dfrac{4}{x^2}}{1 - \dfrac{2}{x} - \dfrac{8}{x^2}}$

11. $\dfrac{1 - \dfrac{5}{x} - \dfrac{6}{x^2}}{1 + \dfrac{6}{x} + \dfrac{5}{x^2}}$

12. $\dfrac{1 - \dfrac{7}{a} + \dfrac{12}{a^2}}{1 + \dfrac{1}{a} - \dfrac{20}{a^2}}$

13. $\dfrac{1 - \dfrac{6}{x} + \dfrac{8}{x^2}}{\dfrac{4}{x^2} + \dfrac{3}{x} - 1}$

14. $\dfrac{1 + \dfrac{3}{x} - \dfrac{18}{x^2}}{\dfrac{21}{x^2} - \dfrac{4}{x} - 1}$

15. $\dfrac{x - \dfrac{4}{x + 3}}{1 + \dfrac{1}{x + 3}}$

16. $\dfrac{y + \dfrac{1}{y - 2}}{1 + \dfrac{1}{y - 2}}$

17. $\dfrac{1 - \dfrac{x}{2x + 1}}{x - \dfrac{1}{2x + 1}}$

18. $\dfrac{1 - \dfrac{2x - 2}{3x - 1}}{x - \dfrac{4}{3x - 1}}$

19. $\dfrac{x - 5 + \dfrac{14}{x + 4}}{x + 3 - \dfrac{2}{x + 4}}$

20. $\dfrac{a + 4 + \dfrac{5}{a - 2}}{a + 6 + \dfrac{15}{a - 2}}$

21. $\dfrac{x + 3 - \dfrac{10}{x - 6}}{x + 2 - \dfrac{20}{x - 6}}$

22. $\dfrac{x - 7 + \dfrac{5}{x - 1}}{x - 3 + \dfrac{1}{x - 1}}$

23. $\dfrac{y - 6 + \dfrac{22}{2y + 3}}{y - 5 + \dfrac{11}{2y + 3}}$

24. $\dfrac{x + 2 - \dfrac{12}{2x - 1}}{x + 1 - \dfrac{9}{2x - 1}}$

25. $\dfrac{x - \dfrac{2}{2x - 3}}{2x - 1 - \dfrac{8}{2x - 3}}$

26. $\dfrac{x + 3 - \dfrac{18}{2x + 1}}{x - \dfrac{6}{2x + 1}}$

27. $\dfrac{\dfrac{1}{x} - \dfrac{2}{x - 1}}{\dfrac{3}{x} + \dfrac{1}{x - 1}}$

28. $\dfrac{\dfrac{3}{n + 1} + \dfrac{1}{n}}{\dfrac{2}{n + 1} + \dfrac{3}{n}}$

29. $\dfrac{\dfrac{3}{2x - 1} - \dfrac{1}{x}}{\dfrac{4}{x} + \dfrac{2}{2x - 1}}$

30. $\dfrac{\dfrac{4}{3x + 1} + \dfrac{3}{x}}{\dfrac{6}{x} - \dfrac{2}{3x + 1}}$

APPLYING THE CONCEPTS

For Exercises 31 to 39, simplify.

31. $1 + \dfrac{1}{1 + \dfrac{1}{2}}$

32. $1 + \dfrac{1}{1 + \dfrac{1}{1 + \dfrac{1}{2}}}$

33. $1 - \dfrac{1}{1 - \dfrac{1}{x}}$

34. $\dfrac{a^{-1} - b^{-1}}{a^{-2} - b^{-2}}$

35. $\left(\dfrac{y}{4} - \dfrac{4}{y}\right) \div \left(\dfrac{4}{y} - 3 + \dfrac{y}{2}\right)$

36. $\left(\dfrac{b}{8} - \dfrac{8}{b}\right) \div \left(\dfrac{8}{b} - 5 + \dfrac{b}{2}\right)$

37. $\dfrac{1 + x^{-1}}{1 - x^{-1}}$

38. $\dfrac{x + x^{-1}}{x - x^{-1}}$

39. $\dfrac{x^{-1}}{y^{-1}} + \dfrac{x}{y}$

6.5 Solving Equations Containing Fractions

Objective A **To solve an equation containing fractions**

Recall that to solve an equation containing fractions, clear denominators by multiplying each side of the equation by the LCM of the denominators. Then solve for the variable.

> **HOW TO** Solve: $\dfrac{3x-1}{4}+\dfrac{2}{3}=\dfrac{7}{6}$
>
> $$\frac{3x-1}{4}+\frac{2}{3}=\frac{7}{6}$$
>
> $$12\left(\frac{3x-1}{4}+\frac{2}{3}\right)=12\cdot\frac{7}{6}$$
>
> $$12\left(\frac{3x-1}{4}\right)+12\cdot\frac{2}{3}=12\cdot\frac{7}{6}$$
>
> $$\overset{3}{\cancel{12}}\left(\frac{3x-1}{\underset{1}{\cancel{4}}}\right)+\frac{\overset{4}{\cancel{12}}}{1}\cdot\frac{2}{\underset{1}{\cancel{3}}}=\frac{\overset{2}{\cancel{12}}}{1}\cdot\frac{7}{\underset{1}{\cancel{6}}}$$
>
> $$9x-3+8=14$$
>
> $$9x+5=14$$
>
> $$9x=9$$
>
> $$x=1$$

- The LCM is **12**. To clear denominators, multiply each side of the equation by the LCM.
- Simplify using the **Distributive Property** and the Properties of Fractions.
- Solve for *x*.

1 checks as a solution. The solution is 1.

Occasionally, a value of the variable that appears to be a solution of an equation will make one of the denominators zero. In this case, that value is not a solution of the equation.

> **HOW TO** Solve: $\dfrac{2x}{x-2}=1+\dfrac{4}{x-2}$
>
> $$\frac{2x}{x-2}=1+\frac{4}{x-2}$$
>
> $$(x-2)\frac{2x}{x-2}=(x-2)\left(1+\frac{4}{x-2}\right)$$
>
> $$(x-2)\frac{2x}{x-2}=(x-2)\cdot1+(x-2)\frac{4}{x-2}$$
>
> $$\frac{\overset{1}{(\cancel{x-2})}}{1}\cdot\frac{2x}{\underset{1}{\cancel{x-2}}}=(x-2)\cdot1+\frac{\overset{1}{(\cancel{x-2})}}{1}\cdot\frac{4}{\underset{1}{\cancel{x-2}}}$$
>
> $$2x=x-2+4$$
>
> $$2x=x+2$$
>
> $$x=2$$

- The LCM is *x* − **2**. Multiply each side of the equation by the LCM.
- Simplify using the **Distributive Property** and the Properties of Fractions.
- Solve for *x*.

When *x* is replaced by 2, the denominators of $\dfrac{2x}{x-2}$ and $\dfrac{4}{x-2}$ are zero.

Therefore, the equation has no solution.

Example 1

Solve: $\dfrac{x}{x+4} = \dfrac{2}{x}$

Solution

The LCM is $x(x+4)$.

$$\dfrac{x}{x+4} = \dfrac{2}{x}$$

$$x(x+4)\left(\dfrac{x}{x+4}\right) = x(x+4)\left(\dfrac{2}{x}\right)$$ • Multiply by the LCM.

$$\dfrac{x(\cancel{x+4})}{1} \cdot \dfrac{x}{\cancel{x+4}} = \dfrac{\cancel{x}(x+4)}{1} \cdot \dfrac{2}{\cancel{x}}$$ • Divide by common factors.

$$x^2 = (x+4)2$$ • Simplify.

$$x^2 = 2x + 8$$

Solve the quadratic equation by factoring.

$$x^2 - 2x - 8 = 0$$ • Write in standard form.

$$(x-4)(x+2) = 0$$ • Factor.

$$x - 4 = 0 \qquad x + 2 = 0$$ • Principle of Zero Products

$$x = 4 \qquad\quad x = -2$$

Both 4 and -2 check as solutions.
The solutions are 4 and -2.

You Try It 1

Solve: $\dfrac{x}{x+6} = \dfrac{3}{x}$

Your solution

Example 2

Solve: $\dfrac{3x}{x-4} = 5 + \dfrac{12}{x-4}$

Solution

The LCM is $x - 4$.

$$\dfrac{3x}{x-4} = 5 + \dfrac{12}{x-4}$$

$$(x-4)\left(\dfrac{3x}{x-4}\right) = (x-4)\left(5 + \dfrac{12}{x-4}\right)$$ • Clear denominators.

$$\dfrac{(\cancel{x-4})}{1} \cdot \dfrac{3x}{\cancel{x-4}} = (x-4)5 + \dfrac{(\cancel{x-4})}{1} \cdot \dfrac{12}{\cancel{x-4}}$$

$$3x = (x-4)5 + 12$$ • Solve for x.

$$3x = 5x - 20 + 12$$

$$3x = 5x - 8$$

$$-2x = -8$$

$$x = 4$$

4 does not check as a solution.
The equation has no solution.

You Try It 2

Solve: $\dfrac{5x}{x+2} = 3 - \dfrac{10}{x+2}$

Your solution

Solutions on p. S17

6.5 Exercises

Objective A To solve an equation containing fractions

1. Can 2 be a solution of the equation $\dfrac{6x}{x+1} - \dfrac{x}{x-2} = 4$? Explain your answer.

2. After multiplying each side of an equation by a variable expression, why must we check the solution?

For Exercises 3 to 35, solve.

3. $\dfrac{2x}{3} - \dfrac{5}{2} = -\dfrac{1}{2}$

4. $\dfrac{x}{3} - \dfrac{1}{4} = \dfrac{1}{12}$

5. $\dfrac{x}{3} - \dfrac{1}{4} = \dfrac{x}{4} - \dfrac{1}{6}$

6. $\dfrac{2y}{9} - \dfrac{1}{6} = \dfrac{y}{9} + \dfrac{1}{6}$

7. $\dfrac{2x-5}{8} + \dfrac{1}{4} = \dfrac{x}{8} + \dfrac{3}{4}$

8. $\dfrac{3x+4}{12} - \dfrac{1}{3} = \dfrac{5x+2}{12} - \dfrac{1}{2}$

9. $\dfrac{6}{2a+1} = 2$

10. $\dfrac{12}{3x-2} = 3$

11. $\dfrac{9}{2x-5} = -2$

12. $\dfrac{6}{4-3x} = 3$

13. $2 + \dfrac{5}{x} = 7$

14. $3 + \dfrac{8}{n} = 5$

15. $1 - \dfrac{9}{x} = 4$

16. $3 - \dfrac{12}{x} = 7$

17. $\dfrac{2}{y} + 5 = 9$

18. $\dfrac{6}{x} + 3 = 11$

19. $\dfrac{3}{x-2} = \dfrac{4}{x}$

20. $\dfrac{5}{x+3} = \dfrac{3}{x-1}$

21. $\dfrac{2}{3x-1} = \dfrac{3}{4x+1}$

22. $\dfrac{5}{3x-4} = \dfrac{-3}{1-2x}$

23. $\dfrac{-3}{2x+5} = \dfrac{2}{x-1}$

24. $\dfrac{4}{5y - 1} = \dfrac{2}{2y - 1}$

25. $\dfrac{4x}{x - 4} + 5 = \dfrac{5x}{x - 4}$

26. $\dfrac{2x}{x + 2} - 5 = \dfrac{7x}{x + 2}$

27. $2 + \dfrac{3}{a - 3} = \dfrac{a}{a - 3}$

28. $\dfrac{x}{x + 4} = 3 - \dfrac{4}{x + 4}$

29. $\dfrac{x}{x - 1} = \dfrac{8}{x + 2}$

30. $\dfrac{x}{x + 12} = \dfrac{1}{x + 5}$

31. $\dfrac{2x}{x + 4} = \dfrac{3}{x - 1}$

32. $\dfrac{5}{3n - 8} = \dfrac{n}{n + 2}$

33. $x + \dfrac{6}{x - 2} = \dfrac{3x}{x - 2}$

34. $x - \dfrac{6}{x - 3} = \dfrac{2x}{x - 3}$

35. $\dfrac{8}{y} = \dfrac{2}{y - 2} + 1$

APPLYING THE CONCEPTS

36. Explain the procedure for solving an equation containing fractions. Include in your discussion how the LCM of the denominators is used to eliminate fractions in the equation.

For Exercises 37 to 42, solve.

37. $\dfrac{3}{5}y - \dfrac{1}{3}(1 - y) = \dfrac{2y - 5}{15}$

38. $\dfrac{3}{4}a = \dfrac{1}{2}(3 - a) + \dfrac{a - 2}{4}$

39. $\dfrac{b + 2}{5} = \dfrac{1}{4}b - \dfrac{3}{10}(b - 1)$

40. $\dfrac{x}{2x^2 - x - 1} = \dfrac{3}{x^2 - 1} + \dfrac{3}{2x + 1}$

41. $\dfrac{x + 1}{x^2 + x - 2} = \dfrac{x + 2}{x^2 - 1} + \dfrac{3}{x + 2}$

42. $\dfrac{y + 2}{y^2 - y - 2} + \dfrac{y + 1}{y^2 - 4} = \dfrac{1}{y + 1}$

6.6 Ratio and Proportion

To solve a proportion

Quantities such as 4 meters, 15 seconds, and 8 gallons are number quantities written with units. In these examples the units are meters, seconds, and gallons.

A **ratio** is the quotient of two quantities that have the same unit.

The length of a living room is 16 ft and the width is 12 ft. The ratio of the length to the width is written

$$\frac{16 \text{ ft}}{12 \text{ ft}} = \frac{16}{12} = \frac{4}{3}$$ A ratio is in simplest form when the two numbers do not have a common factor. Note that the units are not written.

A **rate** is the quotient of two quantities that have different units.

There are 2 lb of salt in 8 gal of water. The salt-to-water rate is

$$\frac{2 \text{ lb}}{8 \text{ gal}} = \frac{1 \text{ lb}}{4 \text{ gal}}$$ A rate is in simplest form when the two numbers do not have a common factor. The units are written as part of the rate.

A **proportion** is an equation that states the equality of two ratios or rates. Examples of proportions are shown at the right.

$$\frac{30 \text{ mi}}{4 \text{ h}} = \frac{15 \text{ mi}}{2 \text{ h}} \qquad \frac{4}{6} = \frac{8}{12} \qquad \frac{3}{4} = \frac{x}{8}$$

HOW TO Solve the proportion $\dfrac{4}{x} = \dfrac{2}{3}$.

$$\frac{4}{x} = \frac{2}{3}$$

$$3x\left(\frac{4}{x}\right) = 3x\left(\frac{2}{3}\right)$$ • The LCM of the denominators is **3x**. To clear denominators, multiply each side of the proportion by the LCM.

$$12 = 2x$$ • Solve the equation.

$$6 = x$$

The solution is 6.

Example 1 Solve: $\dfrac{8}{x + 3} = \dfrac{4}{x}$

Solution

$$\frac{8}{x + 3} = \frac{4}{x}$$

$$x(x + 3)\frac{8}{x + 3} = x(x + 3)\frac{4}{x}$$ • Clear denominators.

$$8x = 4(x + 3)$$ • Solve for *x*.

$$8x = 4x + 12$$

$$4x = 12$$

$$x = 3$$

The solution is 3.

You Try It 1 Solve: $\dfrac{2}{x + 3} = \dfrac{6}{5x + 5}$

Your solution

Solution on p. S17

Objective B   To solve application problems

Example 2

The monthly loan payment for a car is $28.35 for each $1000 borrowed. At this rate, find the monthly payment for a $6000 car loan.

Strategy

To find the monthly payment, write and solve a proportion, using *P* to represent the monthly car payment.

Solution

$$\frac{28.35}{1000} = \frac{P}{6000}$$ • Write a proportion.

$$6000\left(\frac{28.35}{1000}\right) = 6000\left(\frac{P}{6000}\right)$$ • Clear denominators.

$$170.10 = P$$

The monthly payment is $170.10.

You Try It 2

Sixteen ceramic tiles are needed to tile a 9-square-foot area. At this rate, how many square feet can be tiled using 256 ceramic tiles?

Your strategy

Your solution

Solution on p. S17

Objective C   To solve problems involving similar triangles

Similar objects have the same shape but not necessarily the same size. A tennis ball is similar to a basketball. A model ship is similar to an actual ship.

Similar objects have corresponding parts; for example, the rudder on the model ship corresponds to the rudder on the actual ship. The relationship between the sizes of each of the corresponding parts can be written as a ratio, and each ratio will be the same. If the rudder on the model ship is $\frac{1}{100}$ the size of the rudder on the actual ship, then the model wheelhouse is $\frac{1}{100}$ the size of the actual wheelhouse, the width of the model is $\frac{1}{100}$ the width of the actual ship, and so on.

The two triangles *ABC* and *DEF* shown at the right are similar. Side *AB* corresponds to *DE*, side *BC* corresponds to *EF*, and side *AC* corresponds to *DF*. The height *CH* corresponds to the height *FK*. The ratios of corresponding parts of similar triangles are equal.

$$\frac{AB}{DE} = \frac{4}{8} = \frac{1}{2}, \quad \frac{AC}{DF} = \frac{3}{6} = \frac{1}{2}, \quad \frac{BC}{EF} = \frac{2}{4} = \frac{1}{2}, \quad \text{and} \quad \frac{CH}{FK} = \frac{1.5}{3} = \frac{1}{2}$$

Because the ratios of corresponding parts are equal, three proportions can be formed using the sides of the triangles.

$$\frac{AB}{DE} = \frac{AC}{DF}, \quad \frac{AB}{DE} = \frac{BC}{EF}, \quad \text{and} \quad \frac{AC}{DF} = \frac{BC}{EF}$$

Three proportions can also be formed by using the sides and height of the triangles.

$$\frac{AB}{DE} = \frac{CH}{FK}, \qquad \frac{AC}{DF} = \frac{CH}{FK}, \qquad \text{and} \qquad \frac{BC}{EF} = \frac{CH}{FK}$$

The measures of the corresponding angles in similar triangles are equal. Therefore,

$$m\angle A = m\angle D, \qquad m\angle B = m\angle E, \qquad \text{and} \qquad m\angle C = m\angle F$$

It is also true that if the measures of the three angles of one triangle are equal, respectively, to the measures of the three angles of another triangle, then the two triangles are similar.

T A K E   N O T E

Vertical angles of intersecting lines, corresponding angles of parallel lines, and angles of a triangle are discussed in Section 3.5.

A line *DE* is drawn parallel to the base *AB* in the triangle at the right. $m\angle x = m\angle m$ and $m\angle y = m\angle n$ because corresponding angles are equal. $m\angle C = m\angle C$; thus the measures of the three angles of triangle *DEC* are equal, respectively, to the measures of the three angles of triangle *ABC*. Triangle *DEC* is similar to triangle *ABC*.

The sum of the measures of the three angles of a triangle is 180°. If two angles of one triangle are equal in measure to two angles of another triangle, then the third angles must be equal in measure. Thus we can say that if two angles of one triangle are equal in measure to two angles of another triangle, then the two triangles are similar.

HOW TO   The line segments *AB* and *CD* intersect at point *O* in the figure at the right. Angles *C* and *D* are right angles. Find the length of *DO*.

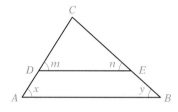

First we must determine whether triangle *AOC* is similar to triangle *BOD*.

$m\angle C = m\angle D$ because they are right angles.

$m\angle x = m\angle y$ because they are vertical angles.

Triangle *AOC* is similar to triangle *BOD* because two angles of one triangle are equal in measure to two angles of the other triangle.

$$\frac{AC}{DB} = \frac{CO}{DO}$$ • Use a proportion to find the length of the unknown side.

$$\frac{4}{7} = \frac{3}{DO}$$ • $AC = 4$, $CO = 3$, and $DB = 7$.

$$7(DO)\frac{4}{7} = 7(DO)\frac{3}{DO}$$ • To clear denominators, multiply each side of the proportion by **7(DO)**.

$$4(DO) = 7(3)$$ • Solve for **DO**.

$$4(DO) = 21$$

$$DO = 5.25$$

HOW TO Triangles *ABC* and *DEF* at the right are similar. Find the area of triangle *ABC*.

$$\frac{AB}{DE} = \frac{CH}{FG}$$ • Solve a proportion to find the height of triangle **ABC**.

$$\frac{5}{12} = \frac{CH}{3}$$ • **AB** = 5, **DE** = 12, and **FG** = 3.

$$12 \cdot \frac{5}{12} = 12 \cdot \frac{CH}{3}$$ • To clear denominators, multiply each side of the proportion by **12**.

$$5 = 4(CH)$$ • Solve for **CH**.

$$1.25 = CH$$ • The height is 1.25 in. The base is 5 in.

$$A = \frac{1}{2}bh = \frac{1}{2}(5)(1.25) = 3.125$$ • Use the formula for the area of a triangle.

The area of triangle *ABC* is 3.125 in².

Example 3

In the figure below, *AB* is parallel to *DC*, and angles *B* and *D* are right angles. *AB* = 12 m, *DC* = 4 m, and *AC* = 18 m. Find the length of *CO*.

Strategy
Triangle *AOB* is similar to triangle *COD*. Solve a proportion to find the length of *CO*. Let *x* represent the length of *CO* and 18 − *x* represent the length of *AO*.

Solution

$$\frac{DC}{AB} = \frac{CO}{AO}$$ • Write a proportion.

$$\frac{4}{12} = \frac{x}{18 - x}$$ • Substitute.

$$12(18 - x) \cdot \frac{4}{12} = 12(18 - x) \cdot \frac{x}{18 - x}$$ • Clear denominators.

$$4(18 - x) = 12x$$ • Solve for *x*.
$$72 - 4x = 12x$$
$$72 = 16x$$
$$4.5 = x$$ • *x* is the length of **CO**.

The length of *CO* is 4.5 m.

You Try It 3

In the figure below, *AB* is parallel to *DC*, and angles *A* and *D* are right angles. *AB* = 10 cm, *CD* = 4 cm, and *DO* = 3 cm. Find the area of triangle *AOB*.

Your strategy

Your solution

Solution on p. S17

6.6 Exercises

Objective A   To solve a proportion

1. What is a proportion?

2. Explain a method for solving a proportion.

For Exercises 3 to 17, solve.

3. $\dfrac{x}{12} = \dfrac{3}{4}$

4. $\dfrac{6}{x} = \dfrac{2}{3}$

5. $\dfrac{4}{9} = \dfrac{x}{27}$

6. $\dfrac{16}{9} = \dfrac{64}{x}$

7. $\dfrac{x + 3}{12} = \dfrac{5}{6}$

8. $\dfrac{3}{5} = \dfrac{x - 4}{10}$

9. $\dfrac{18}{x + 4} = \dfrac{9}{5}$

10. $\dfrac{2}{11} = \dfrac{20}{x - 3}$

11. $\dfrac{2}{x} = \dfrac{4}{x + 1}$

12. $\dfrac{16}{x - 2} = \dfrac{8}{x}$

13. $\dfrac{x + 3}{4} = \dfrac{x}{8}$

14. $\dfrac{x - 6}{3} = \dfrac{x}{5}$

15. $\dfrac{2}{x - 1} = \dfrac{6}{2x + 1}$

16. $\dfrac{9}{x + 2} = \dfrac{3}{x - 2}$

17. $\dfrac{2x}{7} = \dfrac{x - 2}{14}$

Objective B   To solve application problems

18. **Cooking**   Simple syrup used in making some desserts requires 2 c of sugar for every $\frac{2}{3}$ c of boiling water. At this rate, how many cups of sugar are required for 2 c boiling water?

19. **Surveys**   An exit poll survey showed that 4 out of every 7 voters cast a ballot in favor of an amendment to a city charter. At this rate, how many voters voted in favor of the amendment if 35,000 people voted?

20. **Surveys**   In a city of 25,000 homes, a survey was taken to determine the number with cable television. Of the 300 homes surveyed, 210 had cable television. Estimate the number of homes in the city that have cable television.

21. **Cartography**   On a map, two cities are $2\frac{5}{8}$ in. apart. If $\frac{3}{8}$ in. on the map represents 25 mi, find the number of miles between the two cities.

22. **Business**   A company decides to accept a large shipment of 10,000 computer chips if there are 2 or fewer defects in a sample of 100 randomly chosen chips. Assuming that there are 300 defective chips in the shipment and that the rate of defective chips in the sample is the same as the rate in the shipment, will the shipment be accepted?

23. **Taxes**   The sales tax on a car that sold for $12,000 is $780. At this rate, how much higher is the sales tax on a car that sells for $13,500?

24. **Art**   Leonardo da Vinci measured various distances on the human body in order to make accurate drawings. He determined that generally the ratio of the kneeling height of a person to the standing height of that person was $\frac{3}{4}$. Using this ratio, determine how tall a person is who has a kneeling height of 48 in.

25. **Art**   In one of Leonardo da Vinci's notebooks, he wrote that "…from the top to the bottom of the chin is the sixth part of a face, and it is the fifty-fourth part of the man." Suppose the distance from the top to the bottom of the chin of a person is 1.25 in. Using da Vinci's measurements, find the height of this person.

26. **Conservation**   As part of a conservation effort for a lake, 40 fish are caught, tagged, and then released. Later 80 fish are caught. Four of the 80 fish are found to have tags. Estimate the number of fish in the lake.

27. **Conservation**   In a wildlife preserve, 10 elk are captured, tagged, and then released. Later 15 elk are captured and 2 are found to have tags. Estimate the number of elk in the preserve.

28. **Rocketry**   The engine of a small rocket burns 170,000 lb of fuel in 1 min. At this rate, how many pounds of fuel does the rocket burn in 45 s?

Objective C   **To solve problems involving similar triangles**

Triangles *ABC* and *DEF* in Exercises 29 to 36 are similar. Round answers to the nearest tenth.

29. Find side *AC*.

30. Find side *DE*.

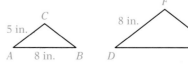

31. Find the height of triangle *ABC*.

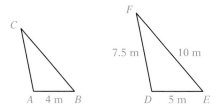

32. Find the height of triangle *DEF*.

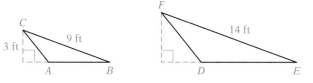

33. Find the perimeter of triangle *DEF*.

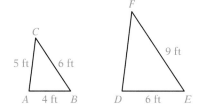

34. Find the perimeter of triangle *ABC*.

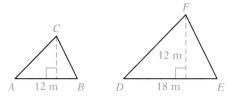

35. Find the area of triangle *ABC*.

36. Find the area of triangle *ABC*.

37. Given *BD ∥ AE*, *BD* measures 5 cm, *AE* measures 8 cm, and *AC* measures 10 cm, find the length of *BC*.

38. Given *AC ∥ DE*, *BD* measures 8 m, *AD* measures 12 m, and *BE* measures 6 m, find the length of *BC*.

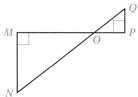

39. Given *DE ∥ AC*, *DE* measures 6 in., *AC* measures 10 in., and *AB* measures 15 in., find the length of *DA*.

40. Given *MP* and *NQ* intersect at *O*, *NO* measures 25 ft, *MO* measures 20 ft, and *PO* measures 8 ft, find the length of *QO*.

41. Given *MP* and *NQ* intersect at *O*, *NO* measures 24 cm, *MN* measures 10 cm, *MP* measures 39 cm, and *QO* measures 12 cm, find the length of *OP*.

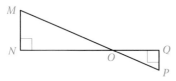

42. Given *MQ* and *NP* intersect at *O*, *NO* measures 12 m, *MN* measures 9 m, *PQ* measures 3 m, and *MQ* measures 20 m, find the perimeter of triangle *OPQ*.

43. Indirect Measurement Similar triangles can be used as an indirect way to measure inaccessible distances. The diagram at the right represents a river of width *DC*. The triangles *AOB* and *DOC* are similar. The distances *AB*, *BO*, and *OC* can be measured. Find the width of the river.

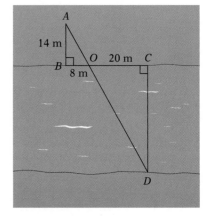

44. Indirect Measurement The sun's rays cast a shadow as shown in the diagram at the right. Find the height of the flagpole. Write the answer in terms of feet.

APPLYING THE CONCEPTS

45. Lottery Tickets Three people put their money together to buy lottery tickets. The first person put in $25, the second person put in $30, and the third person put in $35. One of their tickets was a winning ticket. If they won $4.5 million, what was the first person's share of the winnings?

46. Clubs No one belongs to both the Math Club and the Photography Club, but the two clubs join to hold a car wash. Ten members of the Math Club and 6 members of the Photography Club participate. The profits from the car wash are $120. If each club's profits are proportional to the number of members participating, what share of the profits does the Math Club receive?

47. Sports A basketball player has made 5 out of every 6 foul shots attempted in one year of play. If 42 foul shots were missed that year, how many foul shots did the basketball player make?

6.7 Literal Equations

Objective A   To solve a literal equation for one of the variables

A **literal equation** is an equation that contains more than one variable. Examples of literal equations are shown at the right.

$$2x + 3y = 6$$
$$4w - 2x + z = 0$$

Formulas are used to express a relationship among physical quantities. A **formula** is a literal equation that states a rule about measurements. Examples of formulas are shown at the right.

$$\frac{1}{R_1} + \frac{1}{R_2} = \frac{1}{R} \qquad \text{(Physics)}$$
$$s = a + (n - 1)d \qquad \text{(Mathematics)}$$
$$A = P + Prt \qquad \text{(Business)}$$

The Addition and Multiplication Properties can be used to solve a literal equation for one of the variables. The goal is to rewrite the equation so that the variable being solved for is alone on one side of the equation and all the other numbers and variables are on the other side.

> **HOW TO**   Solve $A = P(1 + i)$ for i.
>
> The goal is to rewrite the equation so that i is on one side of the equation and all other variables are on the other side.
>
> $$A = P(1 + i)$$
> $$A = P + Pi \qquad \bullet \text{ Use the \textbf{Distributive Property} to remove parentheses.}$$
> $$A - P = P - P + Pi \qquad \bullet \text{ Subtract } \textbf{\textit{P}} \text{ from each side of the equation.}$$
> $$A - P = Pi$$
> $$\frac{A - P}{P} = \frac{Pi}{P} \qquad \bullet \text{ Divide each side of the equation by } \textbf{\textit{P}}.$$
> $$\frac{A - P}{P} = i$$

Example 1

Solve $3x - 4y = 12$ for y.

Solution

$$3x - 4y = 12$$
$$3x - 3x - 4y = -3x + 12 \quad \bullet \text{ Subtract } 3x.$$
$$-4y = -3x + 12$$
$$\frac{-4y}{-4} = \frac{-3x + 12}{-4} \quad \bullet \text{ Divide by } -4.$$
$$y = \frac{3}{4}x - 3$$

You Try It 1

Solve $5x - 2y = 10$ for y.

Your solution

Solution on p. S17

Example 2

Solve $I = \dfrac{E}{R + r}$ for R.

Solution

$$I = \frac{E}{R + r}$$

$$(R + r)I = (R + r)\frac{E}{R + r}$$ • Multiply by $(R + r)$.

$$RI + rI = E$$
$$RI + rI - rI = E - rI$$ • Subtract rI.
$$RI = E - rI$$
$$\frac{RI}{I} = \frac{E - rI}{I}$$ • Divide by I.

$$R = \frac{E - rI}{I}$$

You Try It 2

Solve $s = \dfrac{A + L}{2}$ for L.

Your solution

Example 3

Solve $L = a(1 + ct)$ for c.

Solution

$$L = a(1 + ct)$$
$$L = a + act$$ • Distributive Property
$$L - a = a - a + act$$ • Subtract a.
$$L - a = act$$
$$\frac{L - a}{at} = \frac{act}{at}$$ • Divide by at.

$$\frac{L - a}{at} = c$$

You Try It 3

Solve $S = a + (n - 1)d$ for n.

Your solution

Example 4

Solve $S = C - rC$ for C.

Solution

$$S = C - rC$$
$$S = (1 - r)C$$ • Factor.
$$\frac{S}{1 - r} = \frac{(1 - r)C}{1 - r}$$ • Divide by $(1 - r)$.

$$\frac{S}{1 - r} = C$$

You Try It 4

Solve $S = rS + C$ for S.

Your solution

Solutions on p. S18

6.7 Exercises

To solve a literal equation for one of the variables

For Exercises 1 to 15, solve for y.

1. $3x + y = 10$ **2.** $2x + y = 5$ **3.** $4x - y = 3$ **4.** $5x - y = 7$

5. $3x + 2y = 6$ **6.** $2x + 3y = 9$ **7.** $2x - 5y = 10$ **8.** $5x - 2y = 4$

9. $2x + 7y = 14$ **10.** $6x - 5y = 10$ **11.** $x + 3y = 6$ **12.** $x + 2y = 8$

13. $y - 2 = 3(x + 2)$ **14.** $y + 4 = -2(x - 3)$ **15.** $y - 1 = -\dfrac{2}{3}(x + 6)$

For Exercises 16 to 24, solve for x.

16. $x + 3y = 6$ **17.** $x + 6y = 10$ **18.** $3x - y = 3$

19. $2x - y = 6$ **20.** $2x + 5y = 10$ **21.** $4x + 3y = 12$

22. $x - 2y + 1 = 0$ **23.** $x - 4y - 3 = 0$ **24.** $5x + 4y + 20 = 0$

For Exercises 25 to 40, solve the formula for the given variable.

25. $d = rt; t$ (Physics)   **26.** $E = IR; R$ (Physics)

27. $PV = nRT; T$ (Chemistry)   **28.** $A = bh; h$ (Geometry)

29. $P = 2l + 2w; l$ (Geometry)   **30.** $F = \dfrac{9}{5}C + 32; C$ (Temperature conversion)

31. $A = \dfrac{1}{2}h(b_1 + b_2); b_1$ (Geometry)   **32.** $C = \dfrac{5}{9}(F - 32); F$ (Temperature conversion)

33. $V = \dfrac{1}{3}Ah; h$ (Geometry)   **34.** $P = R - C; C$ (Business)

35. $R = \dfrac{C - S}{t}; S$ (Business)   **36.** $P = \dfrac{R - C}{n}; R$ (Business)

37. $A = P + Prt; P$ (Business)   **38.** $T = fm - gm; m$ (Engineering)

39. $A = Sw + w; w$ (Physics)   **40.** $a = S - Sr; S$ (Mathematics)

APPLYING THE CONCEPTS

Business   Break-even analysis is a method used to determine the sales volume required for a company to break even, or experience neither a profit nor a loss on the sale of a product. The break-even point represents the number of units that must be made and sold for income from sales to equal the cost of the product. The break-even point can be calculated using the formula $B = \dfrac{F}{S - V}$, where F is the fixed costs, S is the selling price per unit, and V is the variable costs per unit. Use this information for Exercise 41.

41. a. Solve the formula $B = \dfrac{F}{S - V}$ for S.

   b. Use your answer to part **a** to find the selling price per unit required for a company to break even. The fixed costs are $20,000, the variable costs per unit are $80, and the company plans to make and sell 200 desks.

   c. Use your answer to part **a** to find the selling price per unit required for a company to break even. The fixed costs are $15,000, the variable costs per unit are $50, and the company plans to make and sell 600 cameras.

6.8 Application Problems

Objective A   **To solve work problems**

If a painter can paint a room in 4 h, then in 1 h the painter can paint $\frac{1}{4}$ of the room. The painter's rate of work is $\frac{1}{4}$ of the room each hour. The **rate of work** is the part of a task that is completed in 1 unit of time.

A pipe can fill a tank in 30 min. This pipe can fill $\frac{1}{30}$ of the tank in 1 min. The rate of work is $\frac{1}{30}$ of the tank each minute. If a second pipe can fill the tank in x min, the rate of work for the second pipe is $\frac{1}{x}$ of the tank each minute.

In solving a work problem, the goal is to determine the time it takes to complete a task. The basic equation that is used to solve work problems is

Rate of work × time worked = part of task completed

For example, if a faucet can fill a sink in 6 min, then in 5 min the faucet will fill $\frac{1}{6} \times 5 = \frac{5}{6}$ of the sink. In 5 min the faucet completes $\frac{5}{6}$ of the task.

HOW TO   A painter can paint a wall in 20 min. The painter's apprentice can paint the same wall in 30 min. How long will it take them to paint the wall when they work together?

> **Strategy for Solving a Work Problem**
>
> 1. For each person or machine, write a numerical or variable expression for the rate of work, the time worked, and the part of the task completed. The results can be recorded in a table.

Unknown time to paint the wall working together: t

	Rate of Work	·	Time Worked	=	Part of Task Completed
Painter	$\frac{1}{20}$	·	t	=	$\frac{t}{20}$
Apprentice	$\frac{1}{30}$	·	t	=	$\frac{t}{30}$

> 2. Determine how the parts of the task completed are related. Use the fact that the sum of the parts of the task completed must equal 1, the complete task.

$$\frac{t}{20} + \frac{t}{30} = 1$$

- The sum of the part of the task completed by the painter and the part of the task completed by the apprentice is 1.

$$60\left(\frac{t}{20} + \frac{t}{30}\right) = 60 \cdot 1$$

- Multiply by the **LCM of 20 and 30**.

$$3t + 2t = 60$$

- Distributive Property

$$5t = 60$$

$$t = 12$$

Working together, they will paint the wall in 12 min.

Example 1

A small water pipe takes three times longer to fill a tank than does a large water pipe. With both pipes open it takes 4 h to fill the tank. Find the time it would take the small pipe, working alone, to fill the tank.

You Try It 1

Two computer printers that work at the same rate are working together to print the payroll checks for a large corporation. After they work together for 2 h, one of the printers quits. The second requires 3 h more to complete the payroll checks. Find the time it would take one printer, working alone, to print the payroll.

Strategy

• Time for large pipe to fill the tank: t
  Time for small pipe to fill the tank: $3t$

Your strategy

Fills tank in $3t$ hours   Fills tank in t hours

Fills $\frac{4}{3t}$ of the tank in 4 hours   Fills $\frac{4}{t}$ of the tank in 4 hours

	Rate	*Time*	*Part*
Small pipe	$\frac{1}{3t}$	4	$\frac{4}{3t}$
Large pipe	$\frac{1}{t}$	4	$\frac{4}{t}$

• The sum of the parts of the task completed by each pipe must equal 1.

Solution

$$\frac{4}{3t} + \frac{4}{t} = 1$$

$$3t\left(\frac{4}{3t} + \frac{4}{t}\right) = 3t \cdot 1$$ • **Multiply by the LCM of 3t and t.**

$$4 + 12 = 3t$$ • **Distributive Property**

$$16 = 3t$$

$$\frac{16}{3} = t$$

$$3t = 3\left(\frac{16}{3}\right) = 16$$

The small pipe working alone takes 16 h to fill the tank.

Your solution

Solution on p. S18

Objective B   **To use rational expressions to solve uniform motion problems**

A car that travels constantly in a straight line at 30 mph is in uniform motion. **Uniform motion** means that the speed or direction of an object does not change.

The basic equation used to solve uniform motion problems is

Distance = rate × time

An alternative form of this equation can be written by solving the equation for time.

$$\frac{\text{Distance}}{\text{Rate}} = \text{time}$$

This form of the equation is useful when the total time of travel for two objects or the time of travel between two points is known.

HOW TO   The speed of a boat in still water is 20 mph. The boat traveled 75 mi down a river in the same amount of time it took to travel 45 mi up the river. Find the rate of the river's current.

> **Strategy for Solving a Uniform Motion Problem**
>
> **1.** For each object, write a numerical or variable expression for the distance, rate, and time. The results can be recorded in a table.

The unknown rate of the river's current: r

	Distance	÷	*Rate*	=	*Time*
Down river	75	÷	$20 + r$	=	$\frac{75}{20 + r}$
Up river	45	÷	$20 - r$	=	$\frac{45}{20 - r}$

> **2.** Determine how the times traveled by each object are related. For example, it may be known that the times are equal, or the total time may be known.

$$\frac{75}{20 + r} = \frac{45}{20 - r}$$ • The time down the river is equal to the time up the river.

$$(20 + r)(20 - r)\frac{75}{20 + r} = (20 + r)(20 - r)\frac{45}{20 - r}$$ • Multiply by the **LCM**.

$$(20 - r)75 = (20 + r)45$$
$$1500 - 75r = 900 + 45r$$ • Distributive Property
$$-120r = -600$$
$$r = 5$$

The rate of the river's current is 5 mph.

Example 2

A cyclist rode the first 20 mi of a trip at a constant rate. For the next 16 mi, the cyclist reduced the speed by 2 mph. The total time for the 36 mi was 4 h. Find the rate of the cyclist for each leg of the trip.

Strategy

• Rate for the first 20 mi: r
  Rate for the next 16 mi: $r - 2$

	Distance	*Rate*	*Time*
First 20 mi	20	r	$\dfrac{20}{r}$
Next 16 mi	16	$r - 2$	$\dfrac{16}{r - 2}$

• The total time for the trip was 4 h.

Solution

$$\frac{20}{r} + \frac{16}{r - 2} = 4$$

$$r(r - 2)\left[\frac{20}{r} + \frac{16}{r - 2}\right] = r(r - 2) \cdot 4$$

$$(r - 2)20 + 16r = 4r^2 - 8r$$

$$20r - 40 + 16r = 4r^2 - 8r$$

$$36r - 40 = 4r^2 - 8r$$

• The total time was 4 h.

• Multiply by the LCM.

• Distributive Property

Solve the quadratic equation by factoring.

$$0 = 4r^2 - 44r + 40$$

$$0 = 4(r^2 - 11r + 10)$$

$$0 = 4(r - 10)(r - 1)$$

$$r - 10 = 0 \qquad r - 1 = 0$$

$$r = 10 \qquad r = 1$$

• Standard form

• Factor.

• Principle of Zero Products

The solution $r = 1$ mph is not possible, because the rate on the last 16 mi would then be -1 mph.

10 mph was the rate for the first 20 mi.
8 mph was the rate for the next 16 mi.

You Try It 2

The total time it took for a sailboat to sail back and forth across a lake 6 km wide was 2 h. The rate sailing back was three times the rate sailing across. Find the rate sailing out across the lake.

Your strategy

Your solution

Solution on p. S18

6.8 Exercises

Objective A   **To solve work problems**

1. ✏️ Explain the meaning of the phrase "rate of work."

2. If $\frac{2}{5}$ of a room can be painted in 1 h, what is the rate of work? At the same rate, how long will it take to paint the entire room?

3. A park has two sprinklers that are used to fill a fountain. One sprinkler can fill the fountain in 3 h, whereas the second sprinkler can fill the fountain in 6 h. How long will it take to fill the fountain with both sprinklers operating?

4. One grocery clerk can stock a shelf in 20 min, whereas a second clerk requires 30 min to stock the same shelf. How long would it take to stock the shelf if the two clerks worked together?

5. One person with a skiploader requires 12 h to remove a large quantity of earth. A second, larger skiploader can remove the same amount of earth in 4 h. How long would it take to remove the earth with both skiploaders working together?

6. An experienced painter can paint a fence twice as fast as an inexperienced painter. Working together, the painters require 4 h to paint the fence. How long would it take the experienced painter, working alone, to paint the fence?

7. One computer can solve a complex prime factorization problem in 75 h. A second computer can solve the same problem in 50 h. How long would it take both computers, working together, to solve the problem?

8. A new machine can make 10,000 aluminum cans three times faster than an older machine. With both machines working, 10,000 cans can be made in 9 h. How long would it take the new machine, working alone, to make the 10,000 cans?

9. A small air conditioner can cool a room 5° in 75 min. A larger air conditioner can cool the room 5° in 50 min. How long would it take to cool the room 5° with both air conditioners working?

10. One printing press can print the first edition of a book in 55 min, whereas a second printing press requires 66 min to print the same number of copies. How long would it take to print the first edition with both presses operating?

11. Two oil pipelines can fill a small tank in 30 min. Using one of the pipelines would require 45 min to fill the tank. How long would it take the second pipeline, working alone, to fill the tank?

12. Working together, two dock workers can load a crate in 6 min. One dock worker, working alone, can load the crate in 15 min. How long would it take the second dock worker, working alone, to load the crate?

13. A mason can construct a retaining wall in 10 h. With the mason's apprentice assisting, the task takes 6 h. How long would it take the apprentice, working alone, to construct the wall?

14. A mechanic requires 2 h to repair a transmission, whereas an apprentice requires 6 h to make the same repairs. The mechanic worked alone for 1 h and then stopped. How long will it take the apprentice, working alone, to complete the repairs?

15. One computer technician can wire a modem in 4 h, whereas it takes 6 h for a second technician to do the same job. After working alone for 2 h, the first technician quit. How long will it take the second technician to complete the wiring?

16. A wallpaper hanger requires 2 h to hang the wallpaper on one wall of a room. A second wallpaper hanger requires 4 h to hang the same amount of paper. The first wallpaper hanger worked alone for 1 h and then quit. How long will it take the second wallpaper hanger, working alone, to complete the wall?

17. Two welders who work at the same rate are welding the girders of a building. After they work together for 10 h, one of the welders quits. The second welder requires 20 more hours to complete the welds. Find the time it would have taken one of the welders, working alone, to complete the welds.

18. A large and a small heating unit are being used to heat the water of a pool. The larger unit, working alone, requires 8 h to heat the pool. After both units have been operating for 2 h, the larger unit is turned off. The small unit requires 9 h more to heat the pool. How long would it take the small unit, working alone, to heat the pool?

19. Two machines that fill cereal boxes work at the same rate. After they work together for 7 h, one machine breaks down. The second machine requires 14 h more to finish filling the boxes. How long would it have taken one of the machines, working alone, to fill the boxes?

20. A large and a small drain are opened to drain a pool. The large drain can empty the pool in 6 h. After both drains have been open for 1 h, the large drain becomes clogged and is closed. The smaller drain remains open and requires 9 h more to empty the pool. How long would it have taken the small drain, working alone, to empty the pool?

Objective B   **To use rational expressions to solve uniform motion problems**

21. Running at a constant speed, a jogger ran 24 mi in 3 h. How far did the jogger run in 2 h?

22. For uniform motion, distance = rate · time. How is time related to distance and rate? How is rate related to distance and time?

23. Commuting from work to home, a lab technician traveled 10 mi at a constant rate through congested traffic. On reaching the express-way, the technician increased the speed by 20 mph. An additional 20 mi was traveled at the increased speed. The total time for the trip was 1 h. Find the rate of travel through the congested traffic.

24. The president of a company traveled 1800 mi by jet and 300 mi on a prop plane. The rate of the jet was four times the rate of the prop plane. The entire trip took a total of 5 h. Find the rate of the jet plane.

25. As part of a conditioning program, a jogger ran 8 mi in the same amount of time a cyclist rode 20 mi. The rate of the cyclist was 12 mph faster than the rate of the jogger. Find the rate of the jogger and that of the cyclist.

26. An express train travels 600 mi in the same amount of time it takes a freight train to travel 360 mi. The rate of the express train is 20 mph faster than that of the freight train. Find the rate of each train.

27. To assess the damage done by a fire, a forest ranger traveled 1080 mi by jet and then an additional 180 mi by helicopter. The rate of the jet was four times the rate of the helicopter. The entire trip took a total of 5 h. Find the rate of the jet.

28. A twin-engine plane can fly 800 mi in the same time that it takes a single-engine plane to fly 600 mi. The rate of the twin-engine plane is 50 mph faster than that of the single-engine plane. Find the rate of the twin-engine plane.

29. As part of an exercise plan, Camille Ellison walked for 40 min and then ran for 20 min. If Camille runs 3 mph faster than she walks and covered 5 mi during the 1-hour exercise period, what is her walking speed?

30. A car and a bus leave a town at 1 P.M. and head for a town 300 mi away. The rate of the car is twice the rate of the bus. The car arrives 5 h ahead of the bus. Find the rate of the car.

31. A car is traveling at a rate that is 36 mph faster than the rate of a cyclist. The car travels 384 mi in the same time it takes the cyclist to travel 96 mi. Find the rate of the car.

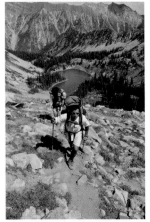

32. A backpacker hiking into a wilderness area walked 9 mi at a constant rate and then reduced this rate by 1 mph. Another 4 mi was hiked at this reduced rate. The time required to hike the 4 mi was 1 h less than the time required to walk the 9 mi. Find the rate at which the hiker walked the first 9 mi.

33. A plane can fly 180 mph in calm air. Flying with the wind, the plane can fly 600 mi in the same amount of time it takes to fly 480 mi against the wind. Find the rate of the wind.

34. A commercial jet can fly 550 mph in calm air. Traveling with the jet stream, the plane flew 2400 mi in the same amount of time it takes to fly 2000 mi against the jet stream. Find the rate of the jet stream.

35. A cruise ship can sail at 28 mph in calm water. Sailing with the gulf current, the ship can sail 170 mi in the same amount of time that it can sail 110 mi against the gulf current. Find the rate of the gulf current.

36. Rowing with the current of a river, a rowing team can row 25 mi in the same amount of time it takes to row 15 mi against the current. The rate of the rowing team in calm water is 20 mph. Find the rate of the current.

37. On a recent trip, a trucker traveled 330 mi at a constant rate. Because of road construction, the trucker then had to reduce the speed by 25 mph. An additional 30 mi was traveled at the reduced rate. The total time for the entire trip was 7 h. Find the rate of the trucker for the first 330 mi.

APPLYING THE CONCEPTS

38. **Work**   One pipe can fill a tank in 2 h, a second pipe can fill the tank in 4 h, and a third pipe can fill the tank in 5 h. How long will it take to fill the tank with all three pipes working?

39. **Transportation**   Because of bad weather, a bus driver reduced the usual speed along a 150-mile bus route by 10 mph. The bus arrived only 30 min later than its usual arrival time. How fast does the bus usually travel?

Focus on Problem Solving

Negations and If ... then Sentences

The sentence "George Washington was the first president of the United States" is a true sentence. The **negation** of that sentence is "George Washington was **not** the first president of the United States." That sentence is false. In general, the negation of a true sentence is a false statement.

The negation of a false sentence is a true sentence. For instance, the sentence "The moon is made of green cheese" is a false statement. The negation of that sentence, "The moon is **not** made of green cheese," is true.

The words *all, no* (or *none*), and *some* are called **quantifiers.** Writing the negation of a sentence that contains these words requires special attention. Consider the sentence "All pets are dogs." This sentence is not true because there are pets that are not dogs; cats, for example, are pets. Because the sentence is false, its negation must be true. You might be tempted to write "All pets are not dogs," but that sentence is not true because some pets are dogs. The correct negation of "All pets are dogs" is "Some pets are not dogs." Note the use of the word *some* in the negation.

Now consider the sentence "Some computers are portable." Because that sentence is true, its negation must be false. Writing "Some computers are not portable" as the negation is not correct, because that sentence is true. The negation of "Some computers are portable" is "No computers are portable."

The sentence "No flowers have red blooms" is false, because there is at least one flower (some roses, for example) that has red blooms. Because the sentence is false, its negation must be true. The negation is "Some flowers have red blooms."

Statement	*Negation*
All *A* are *B*.	Some *A* are not *B*.
No *A* are *B*.	Some *A* are *B*.
Some *A* are *B*.	No *A* are *B*.
Some *A* are not *B*.	All *A* are *B*.

Write the negation of the sentence.

1. All cats like milk.

2. All computers need people.

3. Some trees are tall.

4. No politicians are honest.

5. No houses have kitchens.

6. All police officers are tall.

7. All lakes are not polluted.

8. Some drivers are unsafe.

9. Some speeches are interesting.

10. All laws are good.

11. All businesses are not profitable.

12. All motorcycles are not large.

13. Some vegetables are good for you to eat.

14. Some banks are not open on Sunday.

A **premise** is a known or assumed fact. A premise can be stated using one of the quantifiers (*all, no, none,* or *some*) or using an *If … then* sentence. For instance, the sentence "All triangles have three sides" can be written "*If a figure is a triangle, then it has three sides.*"

We can write the sentence "No whole numbers are negative numbers" as an *If … then* sentence: If a number is a whole number, then it is not a negative number.

Write the sentence as an *If … then* sentence.

15. All students at Barlock College must take a life science course.

16. All baseballs are round.

17. All computers need people.

18. All cats like milk.

19. No odd number is evenly divisible by 2.

20. No prime number greater than 2 is an even number.

21. No rectangles have five sides.

22. All roads lead to Rome.

23. All dogs have fleas.

24. No triangle has four angles.

Projects and Group Activities

Intensity of Illumination

You are already aware that the standard unit of length in the metric system is the meter (m) and that the standard unit of mass in the metric system is the gram (g). You may not know that the standard unit of light intensity is the **candela (cd).**

The rate at which light falls on a 1-square-unit area of surface is called the **intensity of illumination.** Intensity of illumination is measured in **lumens (lm).** A lumen is defined in the following illustration.

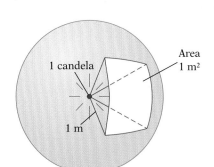

Picture a source of light equal to 1 cd positioned at the center of a hollow sphere that has a radius of 1 m. The rate at which light falls on 1 m² of the inner surface of the sphere is equal to 1 lm. If a light source equal to 4 cd is positioned at the center of the sphere, each square meter of the inner surface receives four times as much illumination, or 4 lm.

Light rays diverge as they leave a light source. The light that falls on an area of 1 m² at a distance of 1 m from the source of light spreads out over an area of 4 m² when it is 2 m from the source. The same light spreads out over an area of 9 m² when it is 3 m from the light source and over an area of 16 m² when it is 4 m from the light source. Therefore, as a surface moves farther away from the source of light, the intensity of illumination on the surface decreases from its value at 1 m to $\left(\frac{1}{2}\right)^2$, or $\frac{1}{4}$, that value at 2 m; to $\left(\frac{1}{3}\right)^2$, or $\frac{1}{9}$, that value at 3 m; and to $\left(\frac{1}{4}\right)^2$, or $\frac{1}{16}$, that value at 4 m.

The formula for the intensity of illumination is

$$I = \frac{s}{r^2}$$

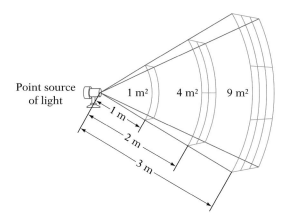

where I is the intensity of illumination in lumens, s is the strength of the light source in candelas, and r is the distance in meters between the light source and the illuminated surface.

A 30-candela lamp is 0.5 m above a desk. Find the illumination on the desk.

$$I = \frac{s}{r^2}$$

$$I = \frac{30}{(0.5)^2} = 120$$

The illumination on the desk is 120 lm.

1. A 100-candela light is hanging 5 m above a floor. What is the intensity of the illumination on the floor beneath it?

2. A 25-candela source of light is 2 m above a desk. Find the intensity of illumination on the desk.

3. How strong a light source is needed to cast 20 lm of light on a surface 4 m from the source?

4. How strong a light source is needed to cast 80 lm of light on a surface 5 m from the source?

5. How far from the desk surface must a 40-candela light source be positioned if the desired intensity of illumination is 10 lm?

6. Find the distance between a 36-candela light source and a surface if the intensity of illumination on the surface is 0.01 lm.

7. Two lights cast the same intensity of illumination on a wall. One light is 6 m from the wall and has a rating of 36 cd. The second light is 8 m from the wall. Find the candela rating of the second light.

8. A 40-candela light source and a 10-candela light source both throw the same intensity of illumination on a wall. The 10-candela light is 6 m from the wall. Find the distance from the 40-candela light to the wall.

Chapter 6 Summary

Key Words	**Examples**
A *rational expression* is a fraction in which the numerator and denominator are polynomials. A rational expression is in *simplest form* when the numerator and denominator have no common factors. [6.1A, p. 289]	$\frac{2x + 1}{x^2 + 4}$ is a rational expression in simplest form.
The *reciprocal of a rational expression* is the rational expression with the numerator and denominator interchanged. [6.1C, p. 292]	The reciprocal of $\frac{3x - y}{x + 4}$ is $\frac{x + 4}{3x - y}$.
The *least common multiple (LCM) of two or more polynomials* is the polynomial of least degree that contains all the factors of each polynomial. [6.2A, p. 297]	The LCM of $3x^2 - 6x$ and $x^2 - 4$ is $3x(x - 2)(x + 2)$, because it contains the factors of $3x^2 - 6x = 3x(x - 2)$ and the factors of $x^2 - 4 = (x - 2)(x + 2)$.
A *complex fraction* is a fraction whose numerator or denominator contains one or more fractions. [6.4A, p. 310]	$\dfrac{x - \dfrac{2}{x + 1}}{1 - \dfrac{4}{x}}$ is a complex fraction.
A *ratio* is the quotient of two quantities that have the same unit. A *rate* is the ratio of two quantities that have different units. [6.6A, p. 319]	$\frac{9}{4}$ is a ratio. $\frac{60 \text{ m}}{12 \text{ s}}$ is a rate.
A *proportion* is an equation that states the equality of two ratios or rates. [6.6A, p. 319]	$\frac{3}{8} = \frac{12}{32}$ and $\frac{x \text{ ft}}{12 \text{ s}} = \frac{15 \text{ ft}}{160 \text{ s}}$ are proportions.
A *literal equation* is an equation that contains more than one variable. A *formula* is a literal equation that states a rule about measurements. [6.7A, p. 327]	$3x - 4y = 12$ is a literal equation. $A = LW$ is a literal equation that is also the formula for the area of a rectangle.

Essential Rules and Procedures   Examples

Simplifying Rational Expressions [6.1A, p. 289]
Factor the numerator and denominator. Divide the numerator and denominator by the common factors.

$$\frac{x^2 - 3x - 10}{x^2 - 5} = \frac{(x + 2)(x - 5)}{(x + 5)(x - 5)}$$
$$= \frac{x + 2}{x + 5}$$

Multiplying Rational Expressions [6.1B, p. 290]
Multiply the numerators. Multiply the denominators. Write the answer in simplest form.

$$\frac{a}{b} \cdot \frac{c}{d} = \frac{ac}{bd}$$

$$\frac{x^2 - 3x}{x^2 + x} \cdot \frac{x^2 + 5x + 4}{x^2 - 4x + 3}$$
$$= \frac{x(x - 3)}{x(x + 1)} \cdot \frac{(x + 1)(x + 4)}{(x - 3)(x - 1)}$$
$$= \frac{x(x - 3)(x + 1)(x + 4)}{x(x + 1)(x - 3)(x - 1)}$$
$$= \frac{x + 4}{x - 1}$$

Dividing Rational Expressions [6.1C, p. 292]
Multiply the dividend by the reciprocal of the divisor. Write the answer in simplest form.

$$\frac{a}{b} \div \frac{c}{d} = \frac{a}{b} \cdot \frac{d}{c} = \frac{ad}{bc}$$

$$\frac{4x + 16}{3x - 6} \div \frac{x^2 + 6x + 8}{x^2 - 4}$$

$$= \frac{4x + 16}{3x - 6} \cdot \frac{x^2 - 4}{x^2 + 6x + 8}$$

$$= \frac{4(x + 4)}{3(x - 2)} \cdot \frac{(x - 2)(x + 2)}{(x + 4)(x + 2)}$$

$$= \frac{4}{3}$$

Adding and Subtracting Rational Expressions [6.3B, p. 303]

1. Find the LCM of the denominators.

2. Write each fraction as an equivalent fraction using the LCM as the denominator.

3. Add or subtract the numerators and place the result over the common denominator.

4. Write the answer in simplest form.

$$\frac{a}{b} + \frac{c}{b} = \frac{a + c}{b} \qquad \frac{a}{b} - \frac{c}{b} = \frac{a - c}{b}$$

$$\frac{x}{x + 1} - \frac{x + 3}{x - 2}$$

$$= \frac{x}{x + 1} \cdot \frac{x - 2}{x - 2} - \frac{x + 3}{x - 2} \cdot \frac{x + 1}{x + 1}$$

$$= \frac{x(x - 2)}{(x + 1)(x - 2)} - \frac{(x + 3)(x + 1)}{(x + 1)(x - 2)}$$

$$= \frac{x(x - 2) - (x + 3)(x + 1)}{(x + 1)(x - 2)}$$

$$= \frac{(x^2 - 2x) - (x^2 + 4x + 3)}{(x + 1)(x - 2)}$$

$$= \frac{-6x - 3}{(x + 1)(x - 2)}$$

Simplifying Complex Fractions [6.4A, p. 310]

Method 1: Multiply by 1 in the form $\dfrac{\text{LCM}}{\text{LCM}}$.

1. Determine the LCM of the denominators of the fractions in the numerator and denominator of the complex fraction.

2. Multiply the numerator and denominator of the complex fraction by the LCM.

3. Simplify.

Method 1: $\dfrac{\dfrac{1}{x} + \dfrac{1}{y}}{\dfrac{1}{x} - \dfrac{1}{y}} = \dfrac{\dfrac{1}{x} + \dfrac{1}{y}}{\dfrac{1}{x} - \dfrac{1}{y}} \cdot \dfrac{xy}{xy}$

$$= \frac{\dfrac{1}{x} \cdot xy + \dfrac{1}{y} \cdot xy}{\dfrac{1}{x} \cdot xy - \dfrac{1}{y} \cdot xy}$$

$$= \frac{y + x}{y - x}$$

Method 2: Multiply the numerator by the reciprocal of the denominator.

1. Simplify the numerator to a single fraction and simplify the denominator to a single fraction.

2. Using the definition for dividing fractions, multiply the numerator by the reciprocal of the denominator.

3. Simplify.

Method 2: $\dfrac{\dfrac{1}{x} + \dfrac{1}{y}}{\dfrac{1}{x} - \dfrac{1}{y}} = \dfrac{\dfrac{y + x}{xy}}{\dfrac{y - x}{xy}}$

$$= \frac{y + x}{xy} \cdot \frac{xy}{y - x}$$

$$= \frac{y + x}{y - x}$$

Solving Equations Containing Fractions [6.5A, p. 315]
Clear denominators by multiplying each side of the equation by
the LCM of the denominators. Then solve for the variable.

$$\frac{1}{2a} = \frac{2}{a} - \frac{3}{8}$$

$$8a\left(\frac{1}{2a}\right) = 8a\left(\frac{2}{a}\right) - 8a\left(\frac{3}{8}\right)$$

$$4 = 16 - 3a$$

$$-12 = -3a$$

$$4 = a$$

Similar Triangles [6.6C, pp. 320–321]
Similar triangles have the same shape but not necessarily the
same size. The ratios of corresponding parts of similar triangles
are equal. The measures of the corresponding angles of similar
triangles are equal.

Triangles *ABC* and *DFE* are similar
triangles. The ratios of corresponding
parts are equal to $\frac{2}{3}$.

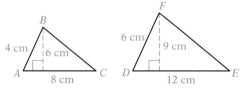

If two angles of one triangle are equal in measure to two angles
of another triangle, then the two triangles are similar.

Triangles *AOB* and *COD* are similar
because $m\angle AOB = m\angle COD$ and
$m\angle B = m\angle D$.

Solving Literal Equations [6.7A, p. 327]
Rewrite the equation so that the letter being solved for is alone
on one side of the equation and all numbers and other variables
are on the other side.

Solve $2x + ax = 5$ for x.

$$2x + ax = 5$$

$$x(2 + a) = 5$$

$$\frac{x(2 + a)}{2 + a} = \frac{5}{2 + a}$$

$$x = \frac{5}{2 + a}$$

Work Problems [6.8A, p. 331]
Rate of work × time worked = part of task completed

Pat can do a certain job in 3 h. Chris
can do the same job in 5 h. How
long would it take them, working
together, to get the job done?

$$\frac{t}{3} + \frac{t}{5} = 1$$

Uniform Motion Problems with Rational Expressions
[6.8B, p. 333]
$$\frac{\text{Distance}}{\text{Rate}} = \text{time}$$

Train A's speed is 15 mph faster than
train B's speed. Train A travels 150 mi
in the same amount of time it takes
train B to travel 120 mi. Find the
rate of train B.

$$\frac{120}{r} = \frac{150}{r + 15}$$

Chapter 6 Review Exercises

1. Divide: $\dfrac{6a^2b^7}{25x^3y} \div \dfrac{12a^3b^4}{5x^2y^2}$

2. Add: $\dfrac{x+7}{15x} + \dfrac{x-2}{20x}$

3. Multiply: $\dfrac{3x^3 + 9x^2}{6xy^2 - 18y^2} \cdot \dfrac{4xy^3 - 12y^3}{5x^2 + 15x}$

4. Divide: $\dfrac{2x(x-y)}{x^2y(x+y)} \div \dfrac{3(x-y)}{x^2y^2}$

5. Simplify: $\dfrac{x - \dfrac{16}{5x-2}}{3x - 4 - \dfrac{88}{5x-2}}$

6. Simplify: $\dfrac{x^2 + x - 30}{15 + 2x - x^2}$

7. Simplify: $\dfrac{16x^5y^3}{24xy^{10}}$

8. Solve: $\dfrac{20}{x+2} = \dfrac{5}{16}$

9. Divide: $\dfrac{10 - 23y + 12y^2}{6y^2 - y - 5} \div \dfrac{4y^2 - 13y + 10}{18y^2 + 3y - 10}$

10. Solve $3ax - x = 5$ for x.

11. Solve: $\dfrac{2}{x} + \dfrac{3}{4} = 1$

12. Add: $\dfrac{x}{y} + \dfrac{3}{x}$

13. Solve $5x + 4y = 20$ for y.

14. Multiply: $\dfrac{8ab^2}{15x^3y} \cdot \dfrac{5xy^4}{16a^2b}$

15. Simplify: $\dfrac{1 - \dfrac{1}{x}}{1 - \dfrac{8x-7}{x^2}}$

16. Write each fraction in terms of the LCM of the denominators.

$$\dfrac{x}{12x^2 + 16x - 3}, \dfrac{4x^2}{6x^2 + 7x - 3}$$

17. Solve $T = 2(ab + bc + ca)$ for a.

18. Solve: $\dfrac{5}{7} + \dfrac{x}{2} = 2 - \dfrac{x}{7}$

19. Simplify: $\dfrac{2 + \dfrac{1}{x}}{3 - \dfrac{2}{x}}$

20. Subtract: $\dfrac{2x}{x-5} - \dfrac{x+1}{x-2}$

21. Solve $i = \dfrac{100m}{c}$ for c.

22. Solve: $\dfrac{x+8}{x+4} = 1 + \dfrac{5}{x+4}$

23. Divide: $\dfrac{20x^2 - 45x}{6x^3 + 4x^2} \div \dfrac{40x^3 - 90x^2}{12x^2 + 8x}$

24. Add: $\dfrac{2y}{5y-7} + \dfrac{3}{7-5y}$

25. Subtract: $\dfrac{5x + 3}{2x^2 + 5x - 3} - \dfrac{3x + 4}{2x^2 + 5x - 3}$

26. Find the LCM of $10x^2 - 11x + 3$ and $20x^2 - 17x + 3$.

27. Solve $4x + 9y = 18$ for y.

28. Multiply: $\dfrac{2x^2 - 5x - 3}{3x^2 - 7x - 6} \cdot \dfrac{3x^2 + 8x + 4}{x^2 + 4x + 4}$

29. Solve: $\dfrac{20}{2x + 3} = \dfrac{17x}{2x + 3} - 5$

30. Add: $\dfrac{x - 1}{x + 2} + \dfrac{3x - 2}{5 - x} + \dfrac{5x^2 + 15x - 11}{x^2 - 3x - 10}$

31. Solve: $\dfrac{6}{x - 7} = \dfrac{8}{x - 6}$

32. Solve: $\dfrac{3}{20} = \dfrac{x}{80}$

33. **Geometry**   Given that *MP* and *NQ* intersect at *O*, *NQ* measures 25 cm, *MO* measures 6 cm, and *PO* measures 9 cm, find the length of *QO*.

34. **Geometry**   Triangles *ABC* and *DEF* are similar triangles. Find the area of triangle *DEF*.

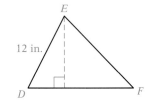

35. **Work**   One hose can fill a pool in 15 h. The second hose can fill the pool in 10 h. How long would it take to fill the pool using both hoses?

36. **Travel**   A car travels 315 mi in the same amount of time in which a bus travels 245 mi. The rate of the car is 10 mph faster than that of the bus. Find the rate of the car.

37. **Travel**   The rate of a jet is 400 mph in calm air. Traveling with the wind, the jet can fly 2100 mi in the same amount of time it takes to fly 1900 mi against the wind. Find the rate of the wind.

38. **Sports**   A pitcher's earned run average (ERA) is the average number of runs allowed in 9 innings of pitching. If a pitcher allows 15 runs in 100 innings, find the pitcher's ERA.

Chapter 6 Test

1. Subtract: $\dfrac{x}{x + 3} - \dfrac{2x - 5}{x^2 + x - 6}$

2. Solve: $\dfrac{3}{x + 4} = \dfrac{5}{x + 6}$

3. Multiply: $\dfrac{x^2 + 2x - 3}{x^2 + 6x + 9} \cdot \dfrac{2x^2 - 11x + 5}{2x^2 + 3x - 5}$

4. Simplify: $\dfrac{16x^5y}{24x^2y^4}$

5. Solve $d = s + rt$ for t.

6. Solve: $\dfrac{6}{x} - 2 = 1$

7. Simplify: $\dfrac{x^2 + 4x - 5}{1 - x^2}$

8. Find the LCM of $6x - 3$ and $2x^2 + x - 1$.

9. Subtract: $\dfrac{2}{2x - 1} - \dfrac{3}{3x + 1}$

10. Divide: $\dfrac{x^2 + 3x + 2}{x^2 + 5x + 4} \div \dfrac{x^2 - x - 6}{x^2 + 2x - 15}$

11. Simplify: $\dfrac{1 + \dfrac{1}{x} - \dfrac{12}{x^2}}{1 + \dfrac{2}{x} - \dfrac{8}{x^2}}$

12. Write each fraction in terms of the LCM of the denominators.
$$\dfrac{3}{x^2 - 2x}, \dfrac{x}{x^2 - 4}$$

13. Subtract: $\dfrac{2x}{x^2 + 3x - 10} - \dfrac{4}{x^2 + 3x - 10}$

14. Solve $3x - 8y = 16$ for y.

15. Solve: $\dfrac{2x}{x + 1} - 3 = \dfrac{-2}{x + 1}$

16. Multiply: $\dfrac{x^3 y^4}{x^2 - 4x + 4} \cdot \dfrac{x^2 - x - 2}{x^6 y^4}$

17. **Geometry**   Given $AE \parallel BD$, AB measures 5 ft, ED measures 8 ft, and BC measures 3 ft, find the length of CE.

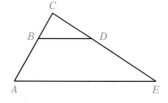

18. **Chemistry**   A saltwater solution is formed by mixing 4 lb of salt with 10 gal of water. At this rate, how many additional pounds of salt are required for 15 gal of water?

19. **Work**   A pool can be filled with one pipe in 6 h, whereas a second pipe requires 12 h to fill the pool. How long would it take to fill the pool with both pipes turned on?

20. **Travel**   A small plane can fly at 110 mph in calm air. Flying with the wind, the plane can fly 260 mi in the same amount of time it takes to fly 180 mi against the wind. Find the rate of the wind.

21. **Landscaping**   A landscape architect uses three sprinklers for each 200 ft² of lawn. At this rate, how many sprinklers are needed for a 3600-square-foot lawn?

Cumulative Review Exercises

1. Evaluate: $\left(\dfrac{2}{3}\right)^2 \div \left(\dfrac{3}{2} - \dfrac{2}{3}\right) + \dfrac{1}{2}$

2. Evaluate $-a^2 + (a - b)^2$ when $a = -2$ and $b = 3$.

3. Simplify: $-2x - (-3y) + 7x - 5y$

4. Simplify: $2[3x - 7(x - 3) - 8]$

5. Solve: $4 - \dfrac{2}{3}x = 7$

6. Solve: $3[x - 2(x - 3)] = 2(3 - 2x)$

7. Find $16\dfrac{2}{3}\%$ of 60.

8. Simplify: $(a^2 b^5)(ab^2)$

9. Multiply: $(a - 3b)(a + 4b)$

10. Divide: $\dfrac{15b^4 - 5b^2 + 10b}{5b}$

11. Divide: $(x^3 - 8) \div (x - 2)$

12. Factor: $12x^2 - x - 1$

13. Factor: $y^2 - 7y + 6$

14. Factor: $2a^3 + 7a^2 - 15a$

15. Factor: $4b^2 - 100$

16. Solve: $(x + 3)(2x - 5) = 0$

17. Simplify: $\dfrac{12x^4 y^2}{18xy^7}$

18. Simplify: $\dfrac{x^2 - 7x + 10}{25 - x^2}$

19. Divide: $\dfrac{x^2 - x - 56}{x^2 + 8x + 7} \div \dfrac{x^2 - 13x + 40}{x^2 - 4x - 5}$

20. Subtract: $\dfrac{2}{2x - 1} - \dfrac{1}{x + 1}$

21. Simplify: $\dfrac{1 - \dfrac{2}{x} - \dfrac{15}{x^2}}{1 - \dfrac{25}{x^2}}$

22. Solve: $\dfrac{3x}{x - 3} - 2 = \dfrac{10}{x - 3}$

23. Solve: $\dfrac{2}{x - 2} = \dfrac{12}{x + 3}$

24. Solve $f = v + at$ for t.

25. **Number Sense** Translate "the difference between five times a number and thirteen is the opposite of eight" into an equation and solve.

26. **Metallurgy** A silversmith mixes 60 g of an alloy that is 40% silver with 120 g of another silver alloy. The resulting alloy is 60% silver. Find the percent of silver in the 120-gram alloy.

27. **Geometry** The length of the base of a triangle is 2 in. less than twice the height. The area of the triangle is 30 in². Find the base and height of the triangle.

28. **Insurance** A life insurance policy costs $16 for every $1000 of coverage. At this rate, how much money would a policy of $5000 cost?

29. **Work** One water pipe can fill a tank in 9 min, whereas a second pipe requires 18 min to fill the tank. How long would it take both pipes, working together, to fill the tank?

30. **Travel** The rower of a boat can row at a rate of 5 mph in calm water. Rowing with the current, the boat travels 14 mi in the same amount of time it takes to travel 6 mi against the current. Find the rate of the current.

chapter

7 Linear Equations in Two Variables

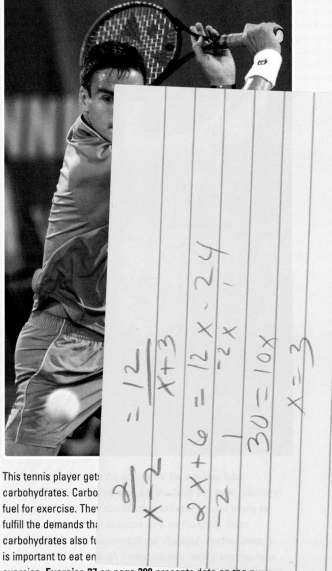

This tennis player gets ... ... the ... ... carbohydrates. Carbo... ... ... fuel for exercise. The ... ... fulfill the demands tha... ... ... carbohydrates also fu... ... ... is important to eat en... ... ... exercise. **Exercise 27 on page 390** presents data on the number of grams of carbohydrates burned as a strenuous tennis workout progresses.

OBJECTIVES

Section 7.1

To graph points in a rectangular coordinate system

To determine ordered-pair solutions of an equation in two variables

To determine whether a set of ordered pairs is a function

To evaluate a function written in functional notation

Section 7.2

To graph an equation of the form $y = mx + b$

To graph an equation of the form $Ax + By = C$

To solve application problems

Section 7.3

A To find the x- and y-intercepts of a straight line

B To find the slope of a straight line

C To graph a line using the slope and the y-intercept

Section 7.4

A To find the equation of a line given a point and the slope

B To find the equation of a line given two points

C To solve application problems

Need help? For online student resources, such as section quizzes, visit this textbook's website at **math.college.hmco.com/students**.

PREP TEST • • •

Do these exercises to prepare for Chapter 7.

1. Simplify: $-\dfrac{5 - (-7)}{4 - 8}$

2. Evaluate $\dfrac{a - b}{c - d}$ when $a = 3$, $b = -2$, $c = -3$, and $d = 2$.

3. Simplify: $-3(x - 4)$

4. Solve: $3x + 6 = 0$

5. Solve $4x + 5y = 20$ when $y = 0$.

6. Solve $3x - 7y = 11$ when $x = -1$.

7. Divide: $\dfrac{12x - 15}{-3}$

8. Solve: $\dfrac{2x + 1}{3} = \dfrac{3x}{4}$

9. Solve $3x - 5y = 15$ for y.

10. Solve $y + 3 = -\dfrac{1}{2}(x + 4)$ for y.

GO FIGURE • • •

Points A, B, C, and D lie on the same line and in that order. The ratio of AB to AC is $\dfrac{1}{4}$ and the ratio of BC to CD is $\dfrac{1}{2}$. Find the ratio of AB to CD.

7.1 The Rectangular Coordinate System

Objective A   **To graph points in a rectangular coordinate system**

Before the 15th century, geometry and algebra were considered separate branches of mathematics. That all changed when René Descartes, a French mathematician who lived from 1596 to 1650, founded **analytic geometry.** In this geometry, a *coordinate system* is used to study relationships between variables.

A **rectangular coordinate system** is formed by two number lines, one horizontal and one vertical, that intersect at the zero point of each line. The point of intersection is called the **origin.** The two lines are called **coordinate axes,** or simply **axes.** The axes determine a **plane,** which can be thought of as a large, flat sheet of paper. The two axes divide the plane into four regions called **quadrants,** which are numbered counterclockwise from I to IV.

Each point in the plane can be identified by a pair of numbers called an **ordered pair.** The first number of the pair measures a horizontal distance and is called the **abscissa.** The second number of the pair measures a vertical distance and is called the **ordinate.** The **coordinates of a point** are the numbers in the ordered pair associated with the point. The abscissa is also called the **first coordinate** of the ordered pair, and the ordinate is also called the **second coordinate** of the ordered pair.

When drawing a rectangular coordinate system, we often label the horizontal axis x and the vertical axis y. In this case, the coordinate system is called an ***xy*-coordinate system.** The coordinates of the points are given by ordered pairs (x, y), where the abscissa is called the ***x*-coordinate** and the ordinate is called the ***y*-coordinate.**

To **graph or plot a point in the plane,** place a dot at the location given by the ordered pair. The **graph of an ordered pair** (x, y) is the dot drawn at the coordinates of the point in the plane. The points whose coordinates are $(3, 4)$ and $(-2.5, -3)$ are graphed in the figures below.

 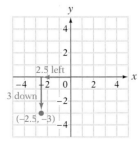

The points whose coordinates are $(3, -1)$ and $(-1, 3)$ are graphed at the right. Note that the graphed points are in different locations. *The order of the coordinates of an ordered pair is important.*

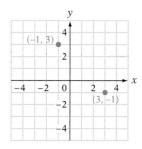

Each point in the plane is associated with an ordered pair, and each ordered pair is associated with a point in the plane. Although only the labels for integers are given on a coordinate grid, the graph of any ordered pair can be approximated. For example, the points whose coordinates are $(-2.3, 4.1)$ and $(\pi, 1)$ are shown on the graph at the right.

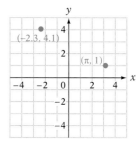

Example 1 Graph the ordered pairs $(-2, -3)$, $(3, -2)$, $(0, -2)$, and $(3, 0)$.

Solution

You Try It 1 Graph the ordered pairs $(-4, 1)$, $(3, -3)$, $(0, 4)$, and $(-3, 0)$.

Your solution

Example 2 Give the coordinates of the points labeled A and B. Give the abscissa of point C and the ordinate of point D.

Solution The coordinates of A are $(-4, 2)$.
The coordinates of B are $(4, 4)$.
The abscissa of C is -1.
The ordinate of D is 1.

You Try It 2 Give the coordinates of the points labeled A and B. Give the abscissa of point D and the ordinate of point C.

Your solution

Solutions on p. S18

Objective B **To determine ordered-pair solutions of an equation in two variables**

An *xy*-coordinate system is used to study the relationship between two variables. Frequently this relationship is given by an equation. Examples of equations in two variables include

$$y = 2x - 3 \qquad 3x + 2y = 6 \qquad x^2 - y = 0$$

A **solution of an equation in two variables** is an ordered pair (x, y) whose coordinates make the equation a true statement.

> **HOW TO**   Is $(-3, 7)$ a solution of $y = -2x + 1$?
>
> $y = -2x + 1$
>
> | 7 | $-2(-3) + 1$ | • Replace *x* by **−3**; replace *y* by **7**. |
> | | $6 + 1$ | |
> | $7 = 7$ | | • The results are equal. |
>
> $(-3, 7)$ is a solution of the equation $y = -2x + 1$.

Besides $(-3, 7)$, there are many other ordered-pair solutions of $y = -2x + 1$. For example, $(0, 1)$, $\left(-\dfrac{3}{2}, 4\right)$, and $(4, -7)$ are also solutions. In general, an equation in two variables has an infinite number of solutions. By choosing any value of x and substituting that value into the equation, we can calculate a corresponding value of y.

> **HOW TO**   Find the ordered-pair solution of $y = \dfrac{2}{3}x - 3$ that corresponds to $x = 6$.
>
> $y = \dfrac{2}{3}x - 3$
>
> $\quad = \dfrac{2}{3}(6) - 3$ • Replace *x* by **6**.
>
> $\quad = 4 - 3 = 1$ • Simplify.
>
> The ordered-pair solution is $(6, 1)$.

The solutions of an equation in two variables can be graphed in an *xy*-coordinate system.

> **HOW TO**   Graph the ordered-pair solutions of $y = -2x + 1$ when $x = -2, -1, 0, 1,$ and 2.

Use the values of x to determine ordered-pair solutions of the equation. It is convenient to record these in a table.

x	$y = -2x + 1$	y	(x, y)
-2	$-2(-2) + 1$	5	$(-2, 5)$
-1	$-2(-1) + 1$	3	$(-1, 3)$
0	$-2(0) + 1$	1	$(0, 1)$
1	$-2(1) + 1$	-1	$(1, -1)$
2	$-2(2) + 1$	-3	$(2, -3)$

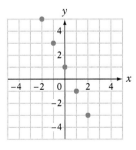

Example 3

Is $(3, -2)$ a solution of $3x - 4y = 15$?

Solution

$$3x - 4y = 15$$

$$\begin{array}{c|c} 3(3) - 4(-2) & 15 \\ 9 + 8 & \\ & 17 \neq 15 \end{array}$$ • Replace *x* by **3** and *y* by **−2**.

No. $(3, -2)$ is not a solution of $3x - 4y = 15$.

You Try It 3

Is $(-2, 4)$ a solution of $x - 3y = -14$?

Your solution

Example 4

Graph the ordered-pair solutions of $2x - 3y = 6$ when $x = -3, 0, 3,$ and 6.

Solution

$$2x - 3y = 6$$
$$-3y = -2x + 6$$ • Solve $2x - 3y = 6$ for *y*.
$$y = \frac{2}{3}x - 2$$

Replace *x* in $y = \frac{2}{3}x - 2$ by $-3, 0, 3,$ and 6. For each value of *x*, determine the value of *y*.

x	$y = \frac{2}{3}x = 2$	y	(x, y)
-3	$\frac{2}{3}(-3) - 2$	-4	$(-3, -4)$
0	$\frac{2}{3}(0) - 2$	-2	$(0, -2)$
3	$\frac{2}{3}(3) - 2$	0	$(3, 0)$
6	$\frac{2}{3}(6) - 2$	2	$(6, 2)$

You Try It 4

Graph the ordered-pair solutions of $x + 2y = 4$ when $x = -4, -2, 0,$ and 2.

Your solution

Solutions on pp. S18–S19

Objective C   **To determine whether a set of ordered pairs is a function**

Discovering a relationship between two variables is an important task in the application of mathematics. Here are some examples.

- Botanists study the relationship between the number of bushels of wheat yielded per acre and the amount of watering per acre.
- Environmental scientists study the relationship between the incidents of skin cancer and the amount of ozone in the atmosphere.
- Business analysts study the relationship between the price of a product and the number of products that are sold at that price.

Each of these relationships can be described by a set of ordered pairs.

> **Definition of a Relation**
>
> A **relation** is any set of ordered pairs.

The following table shows the number of hours that each of 9 students spent studying for a midterm exam and the grade that each of these 9 students received.

Hours	3	3.5	2.75	2	4	4.5	3	2.5	5
Grade	78	75	70	65	85	85	80	75	90

This information can be written as the relation

$$\{(3, 78), (3.5, 75), (2.75, 70), (2, 65), (4, 85), (4.5, 85), (3, 80), (2.5, 75), (5, 90)\}$$

where the first coordinate of the ordered pair is the hours spent studying and the second coordinate is the score on the midterm.

The **domain** of a relation is the set of first coordinates of the ordered pairs; the **range** is the set of second coordinates. For the relation above,

Domain = {2, 2.5, 2.75, 3, 3.5, 4, 4.5, 5}   Range = {65, 70, 75, 78, 80, 85, 90}

The **graph of a relation** is the graph of the ordered pairs that belong to the relation. The graph of the relation given above is shown at the right. The horizontal axis represents the hours spent studying (the domain); the vertical axis represents the test score (the range). The axes could be labeled H for hours studied and S for test score.

A *function* is a special type of relation in which no two ordered pairs have the same first coordinate.

> **Definition of a Function**
>
> A **function** is a relation in which no two ordered pairs have the same first coordinate.

The table at the right is the grading scale for a 100-point test. This table defines a relationship between the *score* on the test and a *letter grade*. Some of the ordered pairs of this function are (78, C), (97, A), (84, B), and (82, B).

Score	Grade
90–100	A
80–89	B
70–79	C
60–69	D
0–59	F

The grading-scale table defines a function, because no two ordered pairs can have the *same* first coordinate and *different* second coordinates. For instance, it is not possible to have the ordered pairs (72, C), and (72, B)—same first coordinate (test score) but different second coordinates (test grade). The domain of this function is {0, 1, 2,…, 99, 100}. The range is {A, B, C, D, F }.

The example of hours spent studying and test score given earlier is *not* a function, because (3, 78) and (3, 80) are ordered pairs of the relation that have the *same* first coordinate but *different* second coordinates.

Consider, again, the grading-scale example. Note that (84, B) and (82, B) are ordered pairs of the function. Ordered pairs of a function may have the same *second* coordinates but not the same first coordinates.

Although relations and functions can be given by tables, they are frequently given by an equation in two variables.

The equation $y = 2x$ expresses the relationship between a number, x, and twice the number, y. For instance, if $x = 3$, then $y = 6$, which is twice 3. To indicate exactly which ordered pairs are determined by the equation, the domain (values of x) is specified. If $x \in \{-2, -1, 0, 1, 2\}$, then the ordered pairs determined by the equation are $\{(-2, -4), (-1, -2), (0, 0), (1, 2), (2, 4)\}$. This relation is a function because no two ordered pairs have the same first coordinate.

The graph of the function $y = 2x$ with domain $\{-2, -1, 0, 1, 2\}$ is shown at the right. The horizontal axis (domain) is labeled x; the vertical axis (range) is labeled y.

The domain $\{-2, -1, 0, 1, 2\}$ was chosen arbitrarily. Other domains could have been selected. The type of application usually influences the choice of the domain.

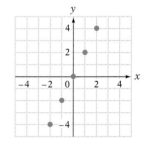

For the equation $y = 2x$, we say that "y is a function of x" because the set of ordered pairs is a function.

Not all equations, however, define a function. For instance, the equation $|y| = x + 2$ does not define y as a function of x. The ordered pairs (2, 4) and (2, −4) both satisfy the equation. Thus there are two ordered pairs with the same first coordinate but different second coordinates.

Example 5

The number of tournaments and the total earnings of the top five Ladies Professional Golf Association (LPGA) players is given in the following table.

Player	Tournaments	Winnings
A. Sorenstam	16	$1,914,506
Se Ri Pak	25	$1,561,928
Grace Park	25	$1,374,702
Hee-Won Han	26	$1,101,060
Juli Inkster	20	$1,012,455

Write a relation where the first coordinate is the number of tournaments played and the second coordinate is the winnings per tournament rounded to the nearest dollar. Is the relation a function?

Solution

Find the winnings per tournament for each player by dividing the player's winnings by the number of tournaments played.

Sorenstam: $1,914,506 \div 16 \approx 119,657$
Pak: $1,561,928 \div 25 \approx 62,477$
Park: $1,374,702 \div 25 \approx 54,988$
Han: $1,101,060 \div 26 \approx 42,348$
Inkster: $1,012,455 \div 20 \approx 50,623$

The relation is {(16, 119,657), (25, 62,477), (25, 54,988), (26, 42,348), (20, 50,623)}. The relation is not a function. Two ordered pairs have the same first coordinate but different second coordinates.

Example 6

Does $y = x^2 + 3$, where $x \in \{-2, -1, 1, 3\}$, define y as a function of x?

Solution
Determine the ordered pairs defined by the equation. Replace x in $y = x^2 + 3$ by the given values and solve for y.

{(−2, 7), (−1, 4), (1, 4), (3, 12)}

No two ordered pairs have the same first coordinate. Therefore, the relation is a function and the equation $y = x^2 + 3$ defines y as a function of x.

Note that (−1, 4) and (1, 4) are ordered pairs that belong to this function. Ordered pairs of a function may have the same *second* coordinate but not the same *first* coordinate.

You Try It 5

Six students decided to go on a diet and fitness program over the summer. Their weights (in pounds) at the beginning and end of the program are given in the table below.

Beginning	End
145	140
140	125
150	130
165	150
140	130
165	160

Write a relation wherein the first coordinate is the weight at the beginning of the summer and the second coordinate is the weight at the end of the summer. Is the relation a function?

Your solution

You Try It 6

Does $y = \frac{1}{2}x + 1$, where $x \in \{-4, 0, 2\}$, define y as a function of x?

Your solution

Solutions on p. S19

Objective D   **To evaluate a function written in functional notation**

When an equation defines y as a function of x, **functional notation** is frequently used to emphasize that the relation is a function. In this case, it is common to replace y in the function's equation with the symbol $f(x)$, where

$f(x)$ is read "f of x" or "the value of f at x."

For instance, the equation $y = x^2 + 3$ from Example 6 defined y as a function of x. The equation can also be written in functional notation as

$$f(x) = x^2 + 3$$

where y has been replaced by $f(x)$.

The symbol $f(x)$ is called the **value of a function at x** because it is the result of evaluating a variable expression. For instance, $f(4)$ means to replace x by 4 and then simplify the resulting numerical expression.

$$f(x) = x^2 + 3$$
$$f(4) = 4^2 + 3 \qquad \bullet \text{ Replace } x \text{ by } \mathbf{4}.$$
$$= 16 + 3 = 19$$

This process is called **evaluating a function.**

> **HOW TO**   Given $f(x) = x^2 + x - 3$, find $f(-2)$.
>
> $$f(x) = x^2 + x - 3$$
> $$f(-2) = (-2)^2 + (-2) - 3 \qquad \bullet \text{ Replace } x \text{ by } \mathbf{-2}.$$
> $$= 4 - 2 - 3 = -1$$
> $$f(-2) = -1$$

In this example, $f(-2)$ is the second coordinate of an ordered pair of the function; the first coordinate is -2. Therefore, an ordered pair of this function is $(-2, f(-2))$, or, because $f(-2) = -1$, $(-2, -1)$.

For the function given by $y = f(x) = x^2 + x - 3$, y is called the **dependent variable** because its value depends on the value of x. The **independent variable** is x.

Functions can be written using other letters or even combinations of letters. For instance, some calculators use $ABS(x)$ for the absolute-value function. Thus the equation $y = |x|$ would be written $ABS(x) = |x|$, where $ABS(x)$ replaces y.

Example 7

Given $G(t) = \dfrac{3t}{t + 4}$, find $G(1)$.

Solution

$$G(t) = \frac{3t}{t + 4}$$

$$G(1) = \frac{3(1)}{1 + 4} \qquad \bullet \text{ Replace } t \text{ by } \mathbf{1}. \text{ Then simplify.}$$

$$G(1) = \frac{3}{5}$$

You Try It 7

Given $H(x) = \dfrac{x}{x - 4}$, find $H(8)$.

Your solution

Solution on p. S19

7.1 Exercises

To graph points in a rectangular coordinate system

1. Graph $(-2, 1)$, $(3, -5)$, $(-2, 4)$, and $(0, 3)$.

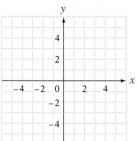

2. Graph $(5, -1)$, $(-3, -3)$, $(-1, 0)$, and $(1, -1)$.

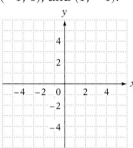

3. Graph $(0, 0)$, $(0, -5)$, $(-3, 0)$, and $(0, 2)$.

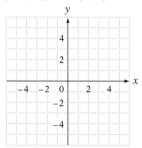

4. Graph $(-4, 5)$, $(-3, 1)$, $(3, -4)$, and $(5, 0)$.

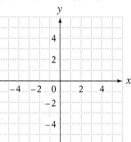

5. Graph $(-1, 4)$, $(-2, -3)$, $(0, 2)$, and $(4, 0)$.

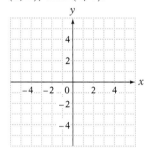

6. Graph $(5, 2)$, $(-4, -1)$, $(0, 0)$, and $(0, 3)$.

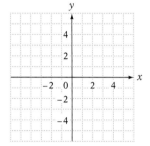

7. Find the coordinates of each of the points.

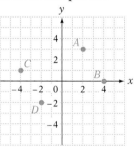

8. Find the coordinates of each of the points.

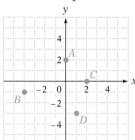

9. Find the coordinates of each of the points.

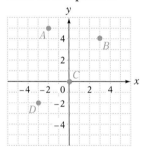

10. Find the coordinates of each of the points.

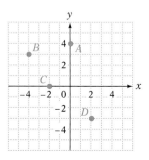

11. **a.** Name the abscissas of points A and C.
    b. Name the ordinates of points B and D.

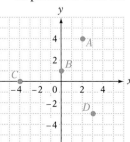

12. **a.** Name the abscissas of points A and C.
    b. Name the ordinates of points B and D.

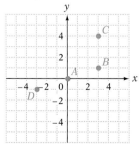

13. ✏️ Suppose you are helping a student who is having trouble graphing ordered pairs. The work of the student is at the right. What can you say to this student to correct the error that is being made?

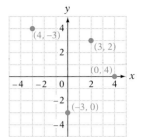

14. a. What are the signs of the coordinates of a point in the third quadrant?

   b. What are the signs of the coordinates of a point in the fourth quadrant?

   c. On an *xy*-coordinate system, what is the name of the axis for which all the *x*-coordinates are zero?

   d. On an *xy*-coordinate system, what is the name of the axis for which all the *y*-coordinates are zero?

Objective B   **To determine ordered-pair solutions of an equation in two variables**

15. Is $(3, 4)$ a solution of $y = -x + 7$?

16. Is $(2, -3)$ a solution of $y = x + 5$?

17. Is $(-1, 2)$ a solution of $y = \frac{1}{2}x - 1$?

18. Is $(1, -3)$ a solution of $y = -2x - 1$?

19. Is $(4, 1)$ a solution of $2x - 5y = 4$?

20. Is $(-5, 3)$ a solution of $3x - 2y = 9$?

21. Is $(0, 4)$ a solution of $3x - 4y = -4$?

22. Is $(-2, 0)$ a solution of $x + 2y = -1$?

For Exercises 23 to 28, graph the ordered-pair solutions of the equation for the given values of x.

23. $y = 2x; x = -2, -1, 0, 2$

24. $y = -2x; x = -2, -1, 0, 2$

25. $y = \frac{2}{3}x + 1; x = -3, 0, 3$

26. $y = -\frac{1}{3}x - 2; x = -3, 0, 3$

27. $2x + 3y = 6; x = -3, 0, 3$

28. $x - 2y = 4; x = -2, 0, 2$

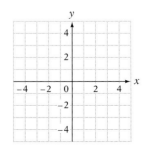

Objective C   **To determine whether a set of ordered pairs is a function**

29. **Biology**   The table below shows the length, in centimeters, of the humerus (the long bone of the forelimb, from shoulder to elbow) and the total wingspan, in centimeters, of several pterosaurs, which are extinct flying reptiles of the order Pterosauria. Write a relation where the first coordinate is the length of the humerus and the second is the wingspan. Is the relation a function?

Humerus (in centimeters)	24	32	22	15	4.4	17	15	4.4
Wingspan (in centimeters)	600	750	430	300	68	370	310	55

30. **Environmental Science**   The table below, based in part on data from the National Oceanic and Atmospheric Administration, shows the average annual concentration of atmospheric carbon dioxide (in parts per million) and the average sea surface temperature (in degrees Celsius) for eight consecutive years. Write a relation wherein the first coordinate is the carbon dioxide concentration and the second coordinate is the average sea surface temperature. Is the relation a function?

Carbon dioxide concentration (in parts per million)	352	353	354	355	356	358	360	361
Surface sea temperature (in degrees Celsius)	15.4	15.4	15.1	15.1	15.2	15.4	15.3	15.5

31. **Sports**   The table at the right shows the number of home runs and the number of at-bats for the top five home runs leaders in the National League for the 2003 season. Write a relation where the first coordinate is the number of at-bats and the second coordinate is the number of home runs per at-bats rounded to the nearest thousand. Is the relation a function?

Player	At-bats	Home runs
Barry Bonds	390	45
Albert Pujois	591	43
Sammy Sosa	517	40
Gary Sheffield	576	39
Jeff Bagwell	605	39

32. **Nielsen Ratings**   The ratings (each rating point is 1,055,000 households) and share (the percentage of television sets in use tuned to a specific program) for selected television shows for a week in November 2003 are shown in the table at the right. Write a relation where the first coordinate is the ratings and the second coordinate is the share. Is the relation a function?

Television Show	Rating	Share
CSI	18.1	27.0
E.R.	13.6	22.0
Friends	13.4	21.0
CSI: Miami	13.2	21.0
60 Minutes	11.3	18.0

33. Does $y = -2x - 3$, where $x \in \{-2, -1, 0, 3\}$, define y as a function of x?

34. Does $y = 2x + 3$, where $x \in \{-2, -1, 1, 4\}$, define y as a function of x?

35. Does $|y| = x - 1$, where $x \in \{1, 2, 3, 4\}$, define y as a function of x?

36. Does $|y| = x + 2$, where $x \in \{-2, -1, 0, 3\}$, define y as a function of x?

37. Does $y = x^2$, where $x \in \{-2, -1, 0, 1, 2\}$, define y as a function of x?

38. Does $y = x^2 - 1$, where $x \in \{-2, -1, 0, 1, 2\}$, define y as a function of x?

Objective D **To evaluate a function written in functional notation**

39. Given $f(x) = 3x - 4$, find $f(4)$.

40. Given $f(x) = 5x + 1$, find $f(2)$.

41. Given $f(x) = x^2$, find $f(3)$.

42. Given $f(x) = x^2 - 1$, find $f(1)$.

43. Given $G(x) = x^2 + x$, find $G(-2)$.

44. Given $H(x) = x^2 - x$, find $H(-2)$.

45. Given $s(t) = \dfrac{3}{t-1}$, find $s(-2)$.

46. Given $P(x) = \dfrac{4}{2x+1}$, find $P(-2)$.

47. Given $h(x) = 3x^2 - 2x + 1$, find $h(3)$.

48. Given $Q(r) = 4r^2 - r - 3$, find $Q(2)$.

49. Given $f(x) = \dfrac{x}{x+5}$, find $f(-3)$.

50. Given $v(t) = \dfrac{2t}{2t+1}$, find $v(3)$.

51. Given $g(x) = x^3 - x^2 + 2x - 7$, find $g(0)$.

52. Given $F(z) = \dfrac{z}{z^2+1}$, find $F(0)$.

APPLYING THE CONCEPTS

53. ✎ Write a few sentences that describe the similarities and differences between relations and functions.

54. ✎ The graph of $y^2 = x$, where $x \in \{0, 1, 4, 9\}$, is shown at the right. Is this the graph of a function? Explain your answer.

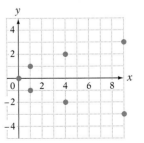

55. ✎ Is it possible to evaluate $f(x) = \dfrac{5}{x-1}$ when $x = 1$? If so, what is $f(1)$? If not, explain why not.

7.2 Linear Equations in Two Variables

Objective A   **To graph an equation of the form $y = mx + b$**

The **graph of an equation in two variables** is a graph of the ordered-pair solutions of the equation.

Consider $y = 2x + 1$. Choosing $x = -2, -1, 0, 1,$ and 2 and determining the corresponding values of y produces some of the ordered pairs of the equation. These are recorded in the table at the right. See the graph of the ordered pairs in Figure 1.

x	$y = 2x + 1$	y	(x, y)
-2	$2(-2) + 1$	-3	$(-2, -3)$
-1	$2(-1) + 1$	-1	$(-1, -1)$
0	$2(0) + 1$	1	$(0, 1)$
1	$2(1) + 1$	3	$(1, 3)$
2	$2(2) + 1$	5	$(2, 5)$

Choosing values of x that are not integers produces more ordered pairs to graph, such as $\left(-\frac{5}{2}, -4\right)$ and $\left(\frac{3}{2}, 4\right)$, as shown in Figure 2. Choosing still other values of x would result in more and more ordered pairs being graphed. The result would be so many dots that the graph would appear as the straight line shown in Figure 3, which is the graph of $y = 2x + 1$.

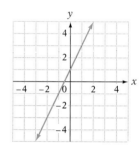

Figure 1   **Figure 2**   **Figure 3**

Equations in two variables have characteristic graphs. The equation $y = 2x + 1$ is an example of a *linear equation*, or *linear function*, because its graph is a straight line. It is also called a *first-degree equation* in two variables because the exponent on each variable is the first power.

> **Linear Equation in Two Variables**
>
> Any equation of the form $y = mx + b$, where m is the coefficient of x and b is a constant, is a **linear equation in two variables**, or a **first-degree equation in two variables** or a **linear function**. The graph of a linear equation in two variables is a straight line.

Examples of linear equations are shown at the right. These equations represent linear functions because there is only one possible y for each x. Note that for $y = 3 - 2x$, m is the coefficient of x and b is the constant.

$$y = 2x + 1 \qquad (m = 2, b = 1)$$
$$y = x - 4 \qquad (m = 1, b = -4)$$
$$y = -\frac{3}{4}x \qquad \left(m = -\frac{3}{4}, b = 0\right)$$
$$y = 3 - 2x \qquad (m = -2, b = 3)$$

The equation $y = x^2 + 4x + 3$ is not a linear equation in two variables because there is a term with a variable squared. The equation $y = \frac{3}{x - 4}$ is not a linear equation because a variable occurs in the denominator of a fraction.

To graph a linear equation, choose some values of x and then find the corresponding values of y. Because a straight line is determined by two points, it is sufficient to find only two ordered-pair solutions. However, it is recommended that at least three ordered-pair solutions be used to ensure accuracy.

HOW TO Graph $y = -\dfrac{3}{2}x + 2$.

This is a linear equation with $m = -\dfrac{3}{2}$ and $b = 2$. Find at least three solutions. Because m is a fraction, choose values of x that will simplify the calculations. We have chosen $-2, 0,$ and 4 for x. (Any values of x could have been selected.)

x	$y = -\dfrac{3}{2}x = 2$	y	(x, y)
-2	$-\dfrac{3}{2}(-2) + 2$	5	$(-2, 5)$
0	$-\dfrac{3}{2}(0) + 2$	2	$(0, 2)$
4	$-\dfrac{3}{2}(4) + 2$	-4	$(4, -4)$

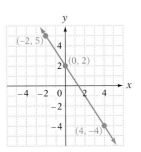

The graph of $y = -\dfrac{3}{2}x + 2$ is shown at the right.

Remember that a graph is a drawing of the ordered-pair solutions of the equation. Therefore, every point on the graph is a solution of the equation, and every solution of the equation is a point on the graph.

The graph at the right is the graph of $y = x + 2$. Note that $(-4, -2)$ and $(1, 3)$ are points on the graph and that these points are solutions of $y = x + 2$. The point whose coordinates are $(4, 1)$ is not a point on the graph and is not a solution of the equation.

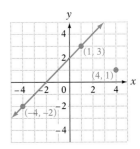

Example 1 Graph $y = 3x - 2$.

Solution

x	y
0	-2
-1	-5
2	4

You Try It 1 Graph $y = 3x + 1$.

Your solution

Solution on p. S19

Example 2 Graph $y = 2x$.

Solution

x	y
0	0
2	4
−2	−4

You Try It 2 Graph $y = -2x$.

Your solution

Example 3 Graph $y = \dfrac{1}{2}x - 1$.

Solution

x	y
0	−1
2	0
−2	−2

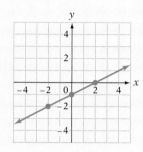

You Try It 3 Graph $y = \dfrac{1}{3}x - 3$.

Your solution

Solutions on p. S19

Objective B

To graph an equation of the form $Ax + By = C$

The equation $Ax + By = C$, where A and B are coefficients and C is a constant, is called the **standard form of a linear equation in two variables.** Examples are shown at the right.

$2x + 3y = 6$	$(A = 2, B = 3, C = 6)$
$x - 2y = -4$	$(A = 1, B = -2, C = -4)$
$2x + y = 0$	$(A = 2, B = 1, C = 0)$
$4x - 5y = 2$	$(A = 4, B = -5, C = 2)$

To graph an equation of the form $Ax + By = C$, first solve the equation for y. Then follow the same procedure used for graphing $y = mx + b$.

Study Tip

Remember that a How To example indicates a worked-out example. Using paper and pencil, work through the example. See *AIM for Success*, page xxv.

HOW TO Graph $3x + 4y = 12$.

$3x + 4y = 12$
$4y = -3x + 12$

$y = -\dfrac{3}{4}x + 3$

x	y
0	3
4	0
−4	6

- Solve for y.
- Subtract $3x$ from each side of the equation.

- Divide each side of the equation by 4.

- Find three ordered-pair solutions of the equation.

- Graph the ordered pairs and then draw a line through the points.

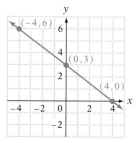

The graph of a linear equation with one of the variables missing is either a horizontal or a vertical line.

The equation $y = 2$ could be written $0 \cdot x + y = 2$. Because $0 \cdot x = 0$ for any value of x, the value of y is always 2 no matter what value of x is chosen. For instance, replace x by -4, by -1, by 0, and by 3. In each case, $y = 2$.

$$0x + y = 2$$
$$0(-4) + y = 2 \qquad (-4, 2) \text{ is a solution.}$$
$$0(-1) + y = 2 \qquad (-1, 2) \text{ is a solution.}$$
$$0(0) + y = 2 \qquad (0, 2) \text{ is a solution.}$$
$$0(3) + y = 2 \qquad (3, 2) \text{ is a solution.}$$

The solutions are plotted in the graph at the right, and a line is drawn through the plotted points. Note that the line is horizontal.

> **Graph of a Horizontal Line**
>
> The graph of $y = b$ is a horizontal line passing through $(0, b)$.

The equation $x = -2$ could be written $x + 0 \cdot y = -2$. Because $0 \cdot y = 0$ for any value of y, the value of x is always -2 no matter what value of y is chosen. For instance, replace y by -2, by 0, by 2, and by 3. In each case, $x = -2$.

$$x + 0y = -2$$
$$x + 0(-2) = -2 \qquad (-2, -2) \text{ is a solution.}$$
$$x + 0(0) = -2 \qquad (-2, 0) \text{ is a solution.}$$
$$x + 0(2) = -2 \qquad (-2, 2) \text{ is a solution.}$$
$$x + 0(3) = -2 \qquad (-2, 3) \text{ is a solution.}$$

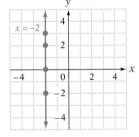

The solutions are plotted in the graph at the right, and a line is drawn through the plotted points. Note that the line is vertical.

> **Graph of a Vertical Line**
>
> The graph of $x = a$ is a vertical line passing through $(a, 0)$.

HOW TO Graph $x = -3$ and $y = 1$ on the same coordinate grid.

- The graph of $x = -3$ is a vertical line passing through $(-3, 0)$.

- The graph of $y = 1$ is a horizontal line passing through $(0, 1)$.

Example 4   Graph $2x - 5y = 10$.

Solution   Solve $2x - 5y = 10$ for y.

$2x - 5y = 10$

$-5y = -2x + 10$

$y = \dfrac{2}{5}x - 2$

x	y
0	−2
5	0
−5	−4

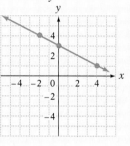

You Try It 4   Graph $5x - 2y = 10$.

Your solution

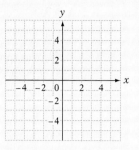

Example 5   Graph $x + 2y = 6$.

Solution   Solve $x + 2y = 6$ for y.

$x + 2y = 6$

$2y = -x + 6$

$y = -\dfrac{1}{2}x + 3$

x	y
0	3
−2	4
4	1

You Try It 5   Graph $x - 3y = 9$.

Your solution

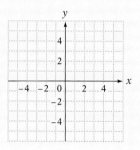

Example 6   Graph $y = -2$.

Solution

The graph of an equation of the form $y = b$ is a horizontal line passing through the point $(0, b)$.

You Try It 6   Graph $y = 3$.

Your solution

Example 7   Graph $x = 3$.

Solution

The graph of an equation of the form $x = a$ is a vertical line passing through the point $(a, 0)$.

You Try It 7   Graph $x = -4$.

Your solution

Solutions on p. S19

Objective C **To solve application problems**

There are a variety of applications of linear functions.

HOW TO Solve: The temperature of a cup of water that has been placed in a microwave oven to be heated can be approximated by the equation $T = 0.7s + 65$, where T is the temperature of the water s seconds after the microwave oven is turned on.

a. Graph this equation for $0 \leq s \leq 220$. (Note: In many applications, the domain of the variable is given so that the equation makes sense. For instance, it would not be sensible to have values of s that are less than 0. This would correspond to negative time. The choice of 220 is somewhat arbitrary and was chosen so that the water would not boil over.)

b. The point whose coordinates are (120, 149) is on the graph of this equation. Write a sentence that describes the meaning of this ordered pair.

Solution

a.

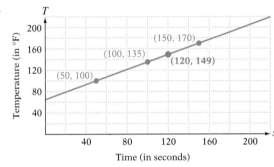

• Choosing $s = 50, 100,$ and 150, you find the corresponding ordered pairs (50, 100), (100, 135), and (150, 170). Plot these points and draw a line through the points.

b. The point whose coordinates are (120, 149) means that 120 s (2 min) after the oven is turned on, the water temperature is 149°F.

Example 8

The number of kilobytes, K, of an MP3 file that remain to be downloaded t seconds after starting the download is given by $K = 935 - 5.5t$. Graph this equation for $0 \leq t \leq 170$. The point whose coordinates are (50, 660) are on this graph. Write a sentence that describes the meaning of this ordered pair.

Solution

The ordered pair (50, 660) means that after 50 s, there are 660 K remaining to be downloaded.

You Try It 8

A car is traveling at a uniform speed of 40 mph. The distance, d, the car travels in t hours is given by $d = 40t$. Graph this equation for $0 \leq t \leq 5$. The point whose coordinates are (3, 120) is on the graph. Write a sentence that describes the meaning of this ordered pair.

Your solution

Solution on p. S19

7.2 Exercises

Objective A To graph an equation of the form $y = mx + b$

For Exercises 1 to 18, graph.

1. $y = 2x - 3$

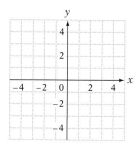

2. $y = -2x + 2$

3. $y = \frac{1}{3}x$

4. $y = -3x$

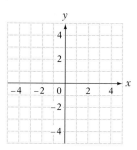

5. $y = \frac{2}{3}x - 1$

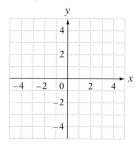

6. $y = \frac{3}{4}x + 2$

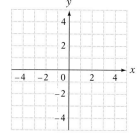

7. $y = -\frac{1}{4}x + 2$

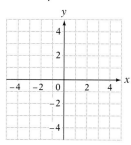

8. $y = -\frac{1}{3}x + 1$

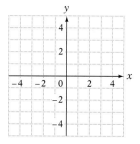

9. $y = -\frac{2}{5}x + 1$

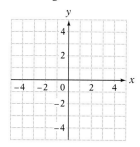

10. $y = -\frac{1}{2}x + 3$

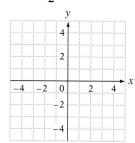

11. $y = 2x - 4$

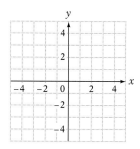

12. $y = 3x - 4$

13. $y = x - 3$

14. $y = x + 2$

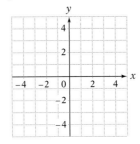

15. $y = -x + 2$

16. $y = -x - 1$

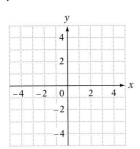

17. $y = -\dfrac{2}{3}x + 1$

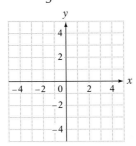

18. $y = 5x - 4$

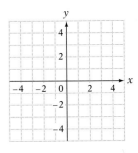

Objective B **To graph an equation of the form $Ax = By + C$**

For Exercises 19 to 36, graph.

19. $3x + y = 3$

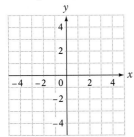

20. $2x + y = 4$

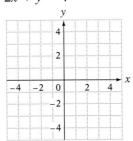

21. $2x + 3y = 6$

22. $3x + 2y = 4$

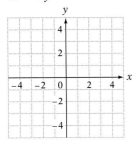

23. $x - 2y = 4$

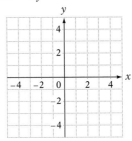

24. $x - 3y = 6$

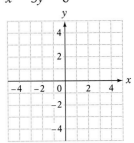

25. $2x - 3y = 6$

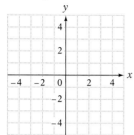

26. $3x - 2y = 8$

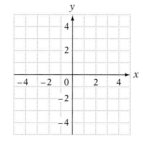

27. $2x + 5y = 10$

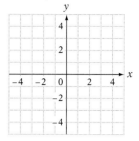

28. $3x + 4y = 12$

29. $x = 3$

30. $y = -4$

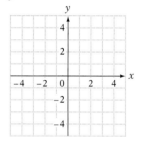

31. $x + 4y = 4$

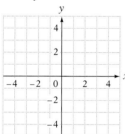

32. $4x - 3y = 12$

33. $y = 4$

34. $x = -2$

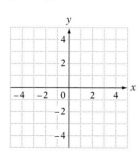

35. $\dfrac{x}{5} + \dfrac{y}{4} = 1$

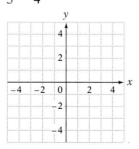

36. $\dfrac{x}{4} - \dfrac{y}{3} = 1$

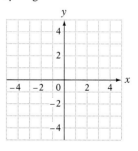

Objective C **To solve application problems**

37. **Emergency Response** A rescue helicopter is rushing at a constant speed of 150 mph to reach several people stranded in the ocean 11 mi away after their boat sank. The rescuers can determine how far they are from the victims using the equation $D = 11 - 2.5t$, where D is the distance in miles and t is the time elapsed in minutes. Graph this equation for $0 \leq t \leq 4$. The point (3, 3.5) is on the graph. Write a sentence that describes the meaning of this ordered pair.

38. **Business** A custom-illustrated sign or banner can be commissioned for a cost of $25 for the material and $10.50 per square foot for the artwork. The equation that represents this cost is given by $y = 10.50x + 25$, where y is the cost and x is the number of square feet in the sign. Graph this equation for $0 \leq x \leq 20$. The point (15, 182.5) is on the graph. Write a sentence that describes the meaning of this ordered pair.

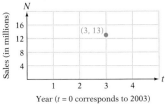

39. **Veterinary Science** According to some veterinarians, the age, x, of a dog can be translated to "human years" by using the equation $H = 4x + 16$, where H is the human equivalent age for the dog. Graph this equation for $2 \leq x \leq 21$. The point whose coordinates are (6, 40) is on this graph. Write a sentence that explains the meaning of this ordered pair.

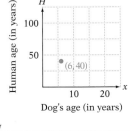

40. **Business** Judging on the basis of data from the Consumer Electronics Association, the projected number, N (in millions), of sales of high-definition televisions (HDTVs) can be approximated by $N = 3t + 4$, where $0 \leq t \leq 4$ and $t = 0$ corresponds to the year 2003. Graph this equation. The point whose coordinates are (3, 13) is on this graph. Write a sentence that explains the meaning of this ordered pair in the context of the problem.

APPLYING THE CONCEPTS

41. Graph $y = 2x - 2$, $y = 2x$, and $y = 2x + 3$. What observation can you make about the graphs?

42. Graph $y = x + 3$, $y = 2x + 3$, and $y = -\frac{1}{2}x + 3$. What observation can you make about the graphs?

43. For the equation $y = 3x + 2$, when the value of x changes from 1 to 2, does the value of y increase or decrease? What is the change in y? Suppose that the value of x changes from 13 to 14. What is the change in y?

44. For the equation $y = -2x + 1$, when the value of x changes from 1 to 2, does the value of y increase or decrease? What is the change in y? Suppose the value of x changes from 13 to 14. What is the change in y?

45. **Telecommunications** A long-distance telephone company offers a flat rate of $.99 for the first 15 minutes of a phone call and then $.15 for each additional minute. The graph of this situation is a combination of the graphs of two linear equations: $C = 0.99$ when $0 < t \leq 15$ and $C = 0.15(t - 15) + 0.99$ when $t > 15$. The graph is shown at the right.
a. What is the cost of a telephone call that lasts 5 minutes?
b. What is the cost of a telephone call that lasts 20 minutes?

7.3 Intercepts and Slopes of Straight Lines

Objective A To find the *x*- and *y*-intercepts of a straight line

The graph of the equation $2x + 3y = 6$ is shown at the right. The graph crosses the *x*-axis at the point (3, 0) and crosses the *y*-axis at the point (0, 2). The point at which a graph crosses the *x*-axis is called the **x-intercept.** At the *x*-intercept, the *y*-coordinate is 0. The point at which a graph crosses the *y*-axis is called the **y-intercept.** At the *y*-intercept, the *x*-coordinate is 0.

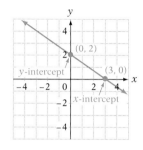

HOW TO   Find the *x*- and *y*-intercepts of the graph of the equation $2x - 3y = 12$.

TAKE NOTE

To find the *x*-intercept, let $y = 0$ and solve for *x*.
To find the *y*-intercept, let $x = 0$ and solve for *y*.

To find the *x*-intercept, let $y = 0$.
(Any point on the *x*-axis has *y*-coordinate 0.)

$$2x - 3y = 12$$
$$2x - 3(0) = 12$$
$$2x = 12$$
$$x = 6$$

The *x*-intercept is (6, 0).

To find the *y*-intercept, let $x = 0$.
(Any point on the *y*-axis has *x*-coordinate 0.)

$$2x - 3y = 12$$
$$2(0) - 3y = 12$$
$$-3y = 12$$
$$y = -4$$

The *y*-intercept is (0, −4).

HOW TO   Find the *y*-intercept of $y = 3x + 4$.

$$y = 3x + 4 = 3(0) + 4 = 4 \qquad \bullet \text{ Let } x = 0.$$

The *y*-intercept is (0, 4).

For any equation of the form $y = mx + b$, the *y*-intercept is (0, *b*).

Some linear equations can be graphed by finding the *x*- and *y*-intercepts and then drawing a line through these two points.

Example 1   Find the *x*- and *y*-intercepts for $x - 2y = 4$. Graph the line.

Solution
To find the *x*-intercept, let $y = 0$ and solve for *x*.
$$x - 2y = 4$$
$$x - 2(0) = 4$$
$$x = 4 \qquad (4, 0)$$
To find the *y*-intercept, let $x = 0$ and solve for *y*.
$$x - 2y = 4$$
$$0 - 2y = 4$$
$$-2y = 4$$
$$y = -2 \qquad (0, -2)$$

You Try It 1   Find the *x*- and *y*-intercepts for $y = 2x - 4$. Graph the line.

Your solution

Solution on p. S20

Objective B   **To find the slope of a straight line**

The graphs of $y = \frac{2}{3}x + 1$ and $y = 2x + 1$ are shown in Figure 1. Each graph crosses the y-axis at the point $(0, 1)$, but the graphs have different slants. The **slope** of a line is a measure of the slant of the line. The symbol for slope is m.

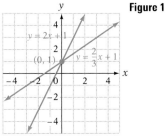

Figure 1

TAKE NOTE
The change in the y values can be thought of as the *rise* of the line, and the change in the x values can be thought of as the *run*. Then

$$\text{Slope} = m = \frac{\text{rise}}{\text{run}}$$

$$m = \frac{\text{rise}}{\text{run}}$$

The slope of a line containing two points is the ratio of the change in the y values of the two points to the change in the x values. The line containing the points $(-2, -3)$ and $(6, 1)$ is graphed in Figure 2. The change in the y values is the difference between the two ordinates.

$$\text{Change in } y = 1 - (-3) = 4$$

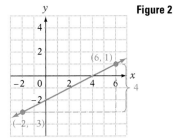

Figure 2

The change in the x values is the difference between the two abscissas (Figure 3).

$$\text{Change in } x = 6 - (-2) = 8$$

$$\text{Slope} = m = \frac{\text{change in } y}{\text{change in } x} = \frac{4}{8} = \frac{1}{2}$$

Figure 3

Slope Formula

If $P_1(x_1, y_1)$ and $P_2(x_2, y_2)$ are two points on a line and $x_1 \neq x_2$, then $m = \dfrac{y_2 - y_1}{x_2 - x_1}$ (Figure 4). If $x_1 = x_2$, the slope is undefined.

Figure 4

HOW TO Find the slope of the line containing the points $(-1, 1)$ and $(2, 3)$.

Let P_1 be $(-1, 1)$ and P_2 be $(2, 3)$. Then $x_1 = -1, y_1 = 1, x_2 = 2$, and $y_2 = 3$.

$$m = \frac{y_2 - y_1}{x_2 - x_1} = \frac{3 - 1}{2 - (-1)} = \frac{2}{3}$$

The slope is $\frac{2}{3}$.

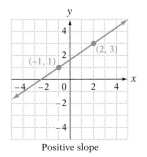

Positive slope

A line that slants upward to the right always has a **positive slope**.

TAKE NOTE
Positive slope means that the value of y increases as the value of x increases.

Note that you obtain the same results if the points are named oppositely. Let P_1 be $(2, 3)$ and P_2 be $(-1, 1)$. Then $x_1 = 2, y_1 = 3, x_2 = -1$, and $y_2 = 1$.

$$m = \frac{y_2 - y_1}{x_2 - x_1} = \frac{1 - 3}{-1 - 2} = \frac{-2}{-3} = \frac{2}{3}$$

The slope is $\frac{2}{3}$. Therefore, it does not matter which point is named P_1 and which is named P_2; the slope remains the same.

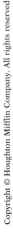

HOW TO   Find the slope of the line containing the points $(-3, 4)$ and $(2, -2)$.

Let P_1 be $(-3, 4)$ and P_2 be $(2, -2)$.

$$m = \frac{y_2 - y_1}{x_2 - x_1} = \frac{-2 - 4}{2 - (-3)} = \frac{-6}{5} = -\frac{6}{5}$$

The slope is $-\frac{6}{5}$.

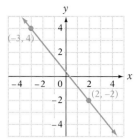

Negative slope

A line that slants downward to the right always has a **negative slope.**

HOW TO   Find the slope of the line containing the points $(-1, 3)$ and $(4, 3)$.

Let P_1 be $(-1, 3)$ and P_2 be $(4, 3)$.

$$m = \frac{y_2 - y_1}{x_2 - x_1} = \frac{3 - 3}{4 - (-1)} = \frac{0}{5} = 0$$

The slope is 0.

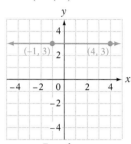

Zero slope

A horizontal line has **zero slope.**

HOW TO   Find the slope of the line containing the points $(2, -2)$ and $(2, 4)$.

Let P_1 be $(2, -2)$ and P_2 be $(2, 4)$.

$$m = \frac{y_2 - y_1}{x_2 - x_1} = \frac{4 - (-2)}{2 - 2} = \frac{6}{0}$$ Division by zero is not defined.

A vertical line has undefined slope.

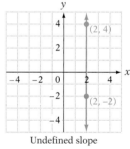

Undefined slope

Two lines in the plane that never intersect are called parallel lines. The lines l_1 and l_2 in the figure at the right are parallel. Calculating the slope of each line, we have

Slope of l_1: $m_1 = \dfrac{y_2 - y_1}{x_2 - x_1} = \dfrac{5 - 1}{3 - (-3)} = \dfrac{4}{6} = \dfrac{2}{3}$

Slope of l_2: $m_2 = \dfrac{y_2 - y_1}{x_2 - x_1} = \dfrac{-1 - (-5)}{3 - (-3)} = \dfrac{4}{6} = \dfrac{2}{3}$

Note that these parallel lines have the same slope. This is always true.

Parallel Lines

Two nonvertical lines in the plane are parallel if and only if they have the same slope. Vertical lines in the plane are parallel.

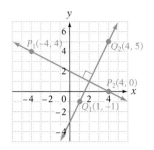

Two lines that intersect at a 90° angle (right angle) are perpendicular lines. The lines at the left are perpendicular.

> **Perpendicular Lines**
>
> Two nonvertical lines in the plane are perpendicular if and only if the product of their slopes is −1. A vertical and horizontal line are perpendicular.

The slope of the line between P_1 and P_2 is $m_1 = \dfrac{0 - 4}{4 - (-4)} = -\dfrac{4}{8} = -\dfrac{1}{2}$. The slope of the line between Q_1 and Q_2 is $m_2 = \dfrac{5 - (-1)}{4 - 1} = \dfrac{6}{3} = 2$. The product of the slopes is $\left(-\dfrac{1}{2}\right)2 = -1$. Because the product of the slopes is −1, the graphs are perpendicular.

There are many applications of the concept of slope. Here is a possibility.

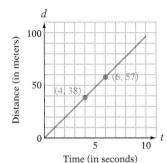

When Florence Griffith-Joyner set the world record for the 100-meter dash, her average rate of speed was approximately 9.5 meters per second. The graph at the right shows the distance she ran during her record-setting run. From the graph, note that after 4 s she had traveled 38 m and that after 6 s she had traveled 57 m. The slope of the line between these two points is

$$m = \frac{57 - 38}{6 - 4} = \frac{19}{2} = 9.5$$

Note that the slope of the line is the same as the rate she was running, 9.5 meters per second. The average speed of an object is related to slope.

Example 2

Find the slope of the line containing the points with coordinates $(-2, -3)$ and $(3, 4)$.

Solution

Let $P_1 = (-2, -3)$ and $P_2 = (3, 4)$.

$m = \dfrac{y_2 - y_1}{x_2 - x_1} = \dfrac{4 - (-3)}{3 - (-2)}$ • $y_2 = 4$, $y_1 = -3$, $x_2 = 3$, $x_1 = -2$

$= \dfrac{7}{5}$

The slope is $\dfrac{7}{5}$.

Example 3

Find the slope of the line containing the points with coordinates $(-1, 4)$ and $(-1, 0)$.

Solution

Let $P_1 = (-1, 4)$ and $P_2 = (-1, 0)$.

$m = \dfrac{y_2 - y_1}{x_2 - x_1} = \dfrac{0 - 4}{-1 - (-1)}$ • $y_2 = 0$, $y_1 = 4$, $x_2 = -1$, $x_1 = -1$

$= \dfrac{-4}{0}$

The slope is undefined.

You Try It 2

Find the slope of the line containing the points with coordinates $(1, 4)$ and $(-3, 8)$.

Your solution

You Try It 3

Find the slope of the line containing the points with coordinates $(-1, 2)$ and $(4, 2)$.

Your solution

Solutions on p. S20

Example 4

The graph below shows the height of a plane above an airport during its 30-minute descent from cruising altitude to landing. Find the slope of the line. Write a sentence that explains the meaning of the slope.

Solution

$$m = \frac{5000 - 20{,}000}{25 - 10} = \frac{-15{,}000}{15}$$

$$= -1000$$

A slope of -1000 means that the height of the plane is *decreasing* at the rate of 1000 ft/min.

You Try It 4

The graph below shows the approximate decline in the value of a used car over a 5-year period. Find the slope of the line. Write a sentence that states the meaning of the slope.

Your solution

Solution on p. S20

Objective C

To graph a line using the slope and the *y*-intercept

The graph of the equation $y = \frac{2}{3}x + 1$ is shown at the right. The points $(-3, -1)$ and $(3, 3)$ are on the graph. The slope of the line between the two points is

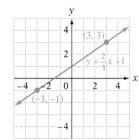

$$m = \frac{3 - (-1)}{3 - (-3)} = \frac{4}{6} = \frac{2}{3}$$

Observe that the slope of the line is the coefficient of x in the equation $y = \frac{2}{3}x + 1$. Also recall that the y-intercept is $(0, 1)$, where 1 is the constant term of the equation.

TAKE NOTE

Here are some equations in slope-intercept form.

$y = 2x - 3$: Slope is 2; y-intercept is $(0, -3)$.

$y = -x + 2$: Slope is -1 (recall that $-x = -1x$); y-intercept is $(0, 2)$.

$y = \frac{x}{2}$: Because $\frac{x}{2} = \frac{1}{2}x$, slope is $\frac{1}{2}$; y-intercept is $(0, 0)$.

Slope-Intercept Form of a Linear Equation

An equation of the form $y = mx + b$ is called the **slope-intercept form** of a straight line. The slope of the line is m, the coefficient of x. The y-intercept is $(0, b)$, where b is the constant term of the equation.

If a linear equation is not in slope-intercept form, solve the equation for y.

HOW TO Find the slope and y-intercept for the graph of $3x + 2y = 12$.

$3x + 2y = 12$

$2y = -3x + 12$ • Write the equation in slope-intercept form by solving for y.

$y = -\frac{3}{2}x + 6$

The slope is $-\frac{3}{2}$; the y-intercept is $(0, 6)$.

When an equation of a line is in slope-intercept form, the graph can be drawn using the slope and the y-intercept. First locate the y-intercept. Use the slope to find a second point on the line. Then draw a line through the two points.

HOW TO   Graph $y = 2x - 3$.

y-intercept $= (0, b) = (0, -3)$

$m = 2 = \dfrac{2}{1} = \dfrac{\text{change in } y}{\text{change in } x}$

Beginning at the y-intercept, move right 1 unit (change in x) and then up 2 units (change in y).

$(1, -1)$ is a second point on the graph.

Draw a line through the two points $(0, -3)$ and $(1, -1)$.

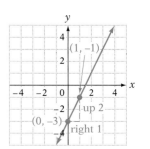

Example 5   Graph $y = -\dfrac{2}{3}x + 1$ by using the slope and y-intercept.

Solution y-intercept $= (0, b) = (0, 1)$

$m = -\dfrac{2}{3} = \dfrac{-2}{3} = \dfrac{\text{change in } y}{\text{change in } x}$

You Try It 5   Graph $y = -\dfrac{1}{4}x - 1$ by using the slope and y-intercept.

Your solution

Example 6   Graph $2x - 3y = 6$ by using the slope and y-intercept.

Solution   Solve the equation for y.

$2x - 3y = 6$

$-3y = -2x + 6$

$y = \dfrac{2}{3}x - 2$

y-intercept $= (0, -2);\ m = \dfrac{2}{3}$

You Try It 6   Graph $x - 2y = 4$ by using the slope and y-intercept.

Your solution

Solutions on p. S20

7.3 Exercises

Objective A   To find the *x*- and *y*-intercepts of a straight line

For Exercises 1 to 12, find the *x*- and *y*-intercepts.

1. $x - y = 3$ 　　　　 **2.** $3x + 4y = 12$ 　　　　 **3.** $y = 3x - 6$ 　　　　 **4.** $y = 2x + 10$

5. $x - 5y = 10$ 　　　 **6.** $3x + 2y = 12$ 　　　 **7.** $y = 3x + 12$ 　　　 **8.** $y = 5x + 10$

9. $2x - 3y = 0$ 　　　 **10.** $3x + 4y = 0$ 　　　 **11.** $y = -\dfrac{1}{2}x + 3$ 　　 **12.** $y = \dfrac{2}{3}x - 4$

For Exercises 13 to 18, find the *x*- and *y*-intercepts and then graph.

13. $5x + 2y = 10$ 　　　　　　 **14.** $x - 3y = 6$ 　　　　　　 **15.** $y = \dfrac{3}{4}x - 3$

 　　　 　　　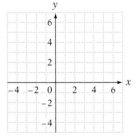

16. $y = \dfrac{2}{5}x - 2$ 　　　　　　 **17.** $5y - 3x = 15$ 　　　　　　 **18.** $9y - 4x = 18$

 　　　 　　　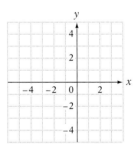

Objective B   To find the slope of a straight line

19. Explain how to find the slope of a line given two points on the line.

20. What is the difference between a line that has zero slope and one that has undefined slope?

For Exercises 21 to 32, find the slope of the line containing the given points.

21. $P_1(4, 2), P_2(3, 4)$ **22.** $P_1(2, 1), P_2(3, 4)$ **23.** $P_1(-1, 3), P_2(2, 4)$ **24.** $P_1(-2, 1), P_2(2, 2)$

25. $P_1(2, 4), P_2(4, -1)$ **26.** $P_1(1, 3), P_2(5, -3)$ **27.** $P_1(3, -4), P_2(3, 5)$ **28.** $P_1(-1, 2), P_2(-1, 3)$

29. $P_1(4, -2), P_2(3, -2)$ **30.** $P_1(5, 1), P_2(-2, 1)$ **31.** $P_1(0, -1), P_2(3, -2)$ **32.** $P_1(3, 0), P_2(2, -1)$

For Exercises 33 to 40, determine whether the line through P_1 and P_2 is parallel, perpendicular, or neither parallel nor perpendicular to the line through Q_1 and Q_2.

33. $P_1(-3, 4), P_2(2, -5); Q_1(3, 6), Q_2(-2, -3)$ **34.** $P_1(4, -5), P_2(6, -9); Q_1(5, -4), Q_2(1, 4)$

35. $P_1(0, 1), P_2(2, 4); Q_1(-4, -7), Q_2(2, 5)$ **36.** $P_1(5, 1), P_2(3, -2); Q_1(0, -2), Q_2(3, -4)$

37. $P_1(-2, 4), P_2(2, 4); Q_1(-3, 6), Q_2(4, 6)$ **38.** $P_1(1, -1), P_2(3, -2); Q_1(-4, 1), Q_2(2, -5)$

39. $P_1(7, -1), P_2(-4, 6); Q_1(3, 0), Q_2(-5, 3)$ **40.** $P_1(5, -2), P_2(-1, 3); Q_1(3, 4), Q_2(-2, -2)$

41. **Business** The graph at the right is based on data from *InfoSync World*. It shows the projected camera-phone sales worldwide through 2008. Find the slope of the line. Write a sentence that states the meaning of the slope in the context of this problem.

42. **Deep-Sea Diving** The pressure, in pounds per square inch, on a diver is shown in the graph at the right. Find the slope of the line. Write a sentence that explains the meaning of the slope.

43. **Depreciation** The graph at the right, based on data from *Kelley Blue Book*, shows the decline in value of a 2002 Porsche Boxster as the number of miles the car is driven increases. Find the slope of the line. Write a sentence that states the meaning of the slope in the context of this problem.

44. **Environmental Science** The stratosphere extends from approximately 11 km to 50 km above Earth. The graph at the right shows how the temperature, in degrees Celsius, changes in the stratosphere. Explain the meaning of the horizontal line segment from A to B. Find the slope of the line from B to C and explain its meaning.

Objective C   **To graph a line using the slope and the _y_-intercept**

For Exercises 45 to 50, find the slope and y-intercept for the graph of the equation.

45. $2x - 3y = 6$

46. $4x + 3y = 12$

47. $2x + 5y = 10$

48. $2x + y = 0$

49. $x - 4y = 0$

50. $2x + 3y = 8$

For Exercises 51 to 65, graph by using the slope and y-intercept.

51. $y = 3x + 1$

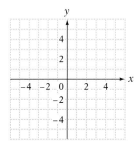

52. $y = -2x - 1$

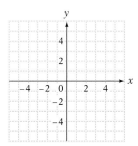

53. $y = \dfrac{2}{5}x - 2$

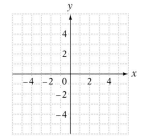

54. $y = \dfrac{3}{4}x + 1$

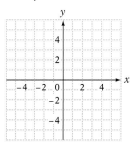

55. $2x + y = 3$

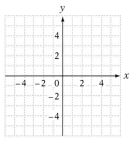

56. $3x - y = 1$

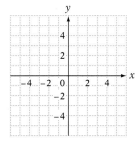

57. $x - 2y = 4$

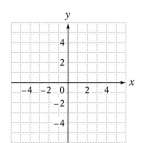

58. $x + 3y = 6$

59. $y = \dfrac{2}{3}x$

60. $y = \dfrac{1}{2}x$

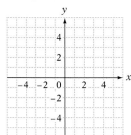

61. $y = -x + 1$

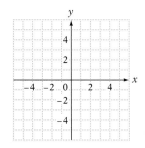

62. $y = -x - 3$

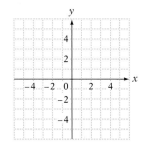

63. $3x - 4y = 12$

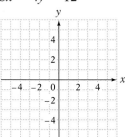

64. $5x - 2y = 10$

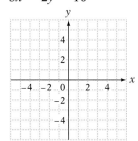

65. $y = -4x + 2$

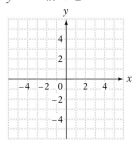

APPLYING THE CONCEPTS

66. Do all straight lines have a *y*-intercept? If not, give an example of one that does not.

67. If two lines have the same slope and the same *y*-intercept, must the graphs of the lines be the same? If not, give an example.

68. **a.** Graph: $\dfrac{x}{3} + \dfrac{y}{4} = 1$ **b.** Graph: $\dfrac{x}{2} - \dfrac{y}{3} = 1$ **c.** Graph: $-\dfrac{x}{4} + \dfrac{y}{2} = 1$

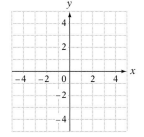

d. What observation can you make about the *x*- and *y*-intercepts of these graphs and the coefficients of *x* and *y*? Use this observation to draw the graph of $\dfrac{x}{4} - \dfrac{y}{3} = 1$.

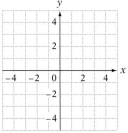

69. **Safety** What does the highway sign at the right have to do with slope?

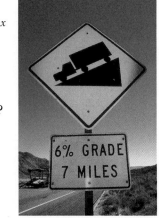

7.4 Equations of Straight Lines

Objective A To find the equation of a line given a point and the slope

In earlier sections, the equation of a line was given and you were asked to determine some properties of the line, such as its intercepts and slope. Here, the process is reversed. Given properties of a line, you will determine its equation.

If the slope and y-intercept of a line are known, the equation of the line can be determined by using the slope-intercept form of a straight line.

> **HOW TO** Find the equation of the line with slope $-\frac{1}{2}$ and y-intercept $(0, 3)$.
>
> $y = mx + b$ • Use the slope-intercept formula.
>
> $y = -\dfrac{1}{2}x + 3$ • $m = -\dfrac{1}{2}$; $(0, b) = (0, 3)$, so $b = 3$.
>
> The equation of the line is $y = -\dfrac{1}{2}x + 3$.

When the slope and the coordinates of a point other than the y-intercept are known, the equation of the line can be found by using the formula for slope.

Suppose a line passes through the point $(3, 1)$ and has a slope of $\frac{2}{3}$. The equation of the line with these properties is determined by letting (x, y) be the coordinates of an unknown point on the line. Because the slope of the line is known, use the slope formula to write an equation. Then solve for y.

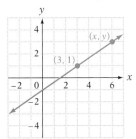

$$\frac{y - 1}{x - 3} = \frac{2}{3}$$ • $\dfrac{y_2 - y_1}{x_2 - x_1} = m$; $m = \dfrac{2}{3}$; $(x_2, y_2) = (x, y)$; $(x_1, y_1) = (3, 1)$

$$\frac{y - 1}{x - 3}(x - 3) = \frac{2}{3}(x - 3)$$ • Multiply each side by $(x - 3)$.

$$y - 1 = \frac{2}{3}x - 2$$ • Simplify.

$$y = \frac{2}{3}x - 1$$ • Solve for y. Add 1 to each side.

The equation of the line is $y = \frac{2}{3}x - 1$.

The same procedure that was used above is used to derive the *point-slope formula*. We use this formula to determine the equation of a line when we are given the coordinates of a point on the line and the slope of the line.

Let (x_1, y_1) be the given coordinates of a point on a line, m the given slope of the line, and (x, y) the coordinates of an unknown point on the line. Then

$$\frac{y - y_1}{x - x_1} = m$$ • Formula for slope.

$$\frac{y - y_1}{x - x_1}(x - x_1) = m(x - x_1)$$ • Multiply each side by $x - x_1$.

$$y - y_1 = m(x - x_1)$$ • Simplify.

> **Point-Slope Formula**
>
> If (x_1, y_1) is a point on a line with slope m, then $y - y_1 = m(x - x_1)$.

HOW TO Find the equation of the line that passes through the point $(2, 3)$ and has slope -2.

$y - y_1 = m(x - x_1)$ • Use the point-slope formula.

$y - 3 = -2(x - 2)$ • $m = -2$; $(x_1, y_1) = (2, 3)$

$y - 3 = -2x + 4$ • Solve for y.

$y = -2x + 7$

The equation of the line is $y = -2x + 7$.

Example 1

Find the equation of the line whose slope is $-\dfrac{2}{3}$ and whose y-intercept is $(0, -1)$.

Solution

Because the slope and y-intercept are known, use the slope-intercept formula, $y = mx + b$.

$y = -\dfrac{2}{3}x - 1$ • $m = -\dfrac{2}{3}$; $b = -1$

You Try It 1

Find the equation of the line whose slope is $\dfrac{5}{3}$ and whose y-intercept is $(0, 2)$.

Your solution

Example 2

Use the point-slope formula to find the equation of the line that passes through the point $(-2, -1)$ and has slope $\dfrac{3}{2}$.

Solution

$y - y_1 = m(x - x_1)$

$y - (-1) = \dfrac{3}{2}[x - (-2)]$ • $m = \dfrac{3}{2}$;

$\qquad\qquad\qquad\qquad (x_1, y_1) = (-2, -1)$

$y + 1 = \dfrac{3}{2}(x + 2)$

$y + 1 = \dfrac{3}{2}x + 3$

$y = \dfrac{3}{2}x + 2$

You Try It 2

Use the point-slope formula to find the equation of the line that passes through the point $(4, -2)$ and has slope $\dfrac{3}{4}$.

Your solution

Solutions on p. S20

Objective B   **To find the equation of a line given two points**

The point-slope formula is used to find the equation of a line when a point on the line and the slope of the line are known. But this formula can also be used to find the equation of a line given two points on the line. In this case,

1. Use the slope formula to determine the slope of the line between the points.

2. Use the point-slope formula, the slope you just calculated, and one of the given points to find the equation of the line.

HOW TO   Find the equation of the line that passes through the points whose coordinates are $(-3, -1)$ and $(3, 3)$.

Use the slope formula to determine the slope of the line between the points.

$$m = \frac{y_2 - y_1}{x_2 - x_1} = \frac{3 - (-1)}{3 - (-3)} = \frac{4}{6} = \frac{2}{3} \qquad \bullet \ (x_1, y_1) = (-3, -1); (x_2, y_2) = (3, 3)$$

Use the point-slope formula, the slope you just calculated, and one of the given points to find the equation of the line.

$$y - y_1 = m(x - x_1) \qquad \bullet \text{ Point-slope formula}$$

$$y - (-1) = \frac{2}{3}[x - (-3)] \qquad \bullet \ m = \frac{2}{3}; (x_1, y_1) = (-3, -1)$$

$$y + 1 = \frac{2}{3}(x + 3)$$

$$y + 1 = \frac{2}{3}x + 2$$

$$y = \frac{2}{3}x + 1$$

Check:

<table>
<tr><td>$y = \frac{2}{3}x + 1$</td><td></td><td></td><td>$y = \frac{2}{3}x + 1$</td><td></td><td></td></tr>
<tr><td>-1</td><td>$\frac{2}{3}(-3) + 1$</td><td>$\bullet \ (x, y) = (-3, -1)$</td><td>$3$</td><td>$\frac{2}{3}(3) + 1$</td><td>$\bullet \ (x, y) = (3, 3)$</td></tr>
<tr><td>-1</td><td>$-2 + 1$</td><td></td><td>3</td><td>$2 + 1$</td><td></td></tr>
<tr><td>$-1 = -1$</td><td></td><td></td><td>$3 = 3$</td><td></td><td></td></tr>
</table>

The equation of the line that passes through the two points is $y = \frac{2}{3}x + 1$.

TAKE NOTE

You can verify that the equation $y = \frac{2}{3}x + 1$ passes through the points $(-3, -1)$ and $(3, 3)$ by substituting the coordinates of these points into the equation.

Example 3

Find the equation of the line that passes through the points $(-4, 0)$ and $(2, -3)$.

Solution

Find the slope of the line between the two points.

$$m = \frac{y_2 - y_1}{x_2 - x_1} = \frac{-3 - 0}{2 - (-4)} = \frac{-3}{6} = -\frac{1}{2}$$

Use the point-slope formula.

$$y - y_1 = m(x - x_1) \qquad \bullet \text{ Point-slope formula}$$

$$y - 0 = -\frac{1}{2}[x - (-4)] \qquad \bullet \ m = -\frac{1}{2}; (x_1, y_1) = (-4, 0)$$

$$y = -\frac{1}{2}(x + 4)$$

$$y = -\frac{1}{2}x - 2$$

The equation of the line is $y = -\frac{1}{2}x - 2$.

You Try It 3

Find the equation of the line that passes through the points $(-6, -2)$ and $(3, 1)$.

Your solution

Solution on p. S20

Objective C   **To solve application problems**

A **linear model** is a first-degree equation that is used to describe a relationship between quantities. In many cases, a linear model is used to approximate collected data. The data are graphed as points in a coordinate system, and then a line is drawn that approximates the data. The graph of the points is called a **scatter diagram;** the line is called the **line of best fit.**

Consider an experiment to determine the weight required to stretch a spring a certain distance. Data from such an experiment are shown in the table below.

Distance (in inches)	2.5	4	2	3.5	1	4.5
Weight (in pounds)	63	104	47	85	27	115

The accompanying graph shows the scatter diagram, which is the plotted points, and the line of fit, which is the line that approximately goes through the plotted points. The equation of the line of best fit is $y = 25.6x - 1.3$, where x is the number of inches the spring is stretched and y is the weight in pounds.

The table below shows the values that the model would predict to the nearest tenth. Good linear models should predict values that are close to the actual values. A more thorough analysis of lines of best fit is undertaken in statistics courses.

Distance, x	2.5	4	2	3.5	1	4.5
Weight predicted using $y = 25.6x - 1.3$	62.7	101.1	49.9	88.3	24.3	113.9

Example 4

The data in the table below show the percent of people in the U.S. that purchase their music from the Internet. (*Source*: The Recording Industry Association of America) The line of best fit is $y = 0.62x + 0.667$, where x is the year (with 1997 corresponding to $x = 0$) and y the percent of people in the U.S. purchasing music from the Internet.

Year	0	1	2	3	4	5
Percent	0.3	1.1	2.4	3.2	2.9	3.4

Graph the data and the line of best fit in the coordinate system below. Write a sentence that describes the meaning of the slope of the line.

Solution

Year ($x = 0$ corresponds to 1997)

The slope of the line means that the percent of people purchasing music from the Internet is increasing 0.62% per year.

You Try It 4

The data in the table below show a reading test grade and the final exam grade in a history class. The line of best fit is $y = 8.3x - 7.8$, where x is the reading test score and y is the history test score.

Reading	8.5	9.4	10.0	11.4	12.0
History	64	68	76	87	92

Graph the data and the line of best fit in the coordinate system below. Write a sentence that describes the meaning of the slope of the line of best fit.

Your solution

Reading score

Solution on p. S20

7.4 Exercises

Objective A   **To find the equation of a line given a point and the slope**

1. What is the point-slope formula and how is it used?

2. Can the point-slope formula be used to find the equation of any line? If not, equations for which types of lines cannot be found using this formula?

3. Find the equation of the line that contains the point $(0, 2)$ and has slope 2.

4. Find the equation of the line that contains the point $(0, -1)$ and has slope -2.

5. Find the equation of the line that contains the point $(-1, 2)$ and has slope -3.

6. Find the equation of the line that contains the point $(2, -3)$ and has slope 3.

7. Find the equation of the line that contains the point $(3, 1)$ and has slope $\frac{1}{3}$.

8. Find the equation of the line that contains the point $(-2, 3)$ and has slope $\frac{1}{2}$.

9. Find the equation of the line that contains the point $(4, -2)$ and has slope $\frac{3}{4}$.

10. Find the equation of the line that contains the point $(2, 3)$ and has slope $-\frac{1}{2}$.

11. Find the equation of the line that contains the point $(5, -3)$ and has slope $-\frac{3}{5}$.

12. Find the equation of the line that contains the point $(5, -1)$ and has slope $\frac{1}{5}$.

13. Find the equation of the line that contains the point $(2, 3)$ and has slope $\frac{1}{4}$.

14. Find the equation of the line that contains the point $(-1, 2)$ and has slope $-\frac{1}{2}$.

Objective B   **To find the equation of a line given two points**

15. Find the equation of the line that passes through the points $(1, -1)$ and $(-2, -7)$.

16. Find the equation of the line that passes through the points $(2, 3)$ and $(3, 2)$.

17. Find the equation of the line that passes through the points $(-2, 1)$ and $(1, -5)$.

18. Find the equation of the line that passes through the points $(-1, -3)$ and $(2, -12)$.

19. Find the equation of the line that passes through the points $(0, 0)$ and $(-3, -2)$.

20. Find the equation of the line that passes through the points $(0, 0)$ and $(-5, 1)$.

21. Find the equation of the line that passes through the points (2, 3) and (−4, 0).

22. Find the equation of the line that passes through the points (3, −1) and (0, −3).

23. Find the equation of the line that passes through the points (−4, 1) and (4, −5).

24. Find the equation of the line that passes through the points (−5, 0) and (10, −3).

25. Find the equation of the line that passes through the points (−2, 1) and (2, 4).

26. Find the equation of the line that passes through the points (3, −2) and (−3, −3).

Objective C **To solve application problems**

27. **Sports** The data in the table below show the number of carbohydrates used for various amounts of time during a strenuous tennis workout. The line of best fit is $y = 1.55x + 1.45$, where x is the time of the workout in minutes and y is the number of carbohydrates used in grams.

Time of workout, x (in minutes)	5	10	20	30	60
Carbohydrates used, y (in grams)	10	15	33	49	94

Graph the data and the line of best fit in the coordinate system at the right. Write a sentence that describes the meaning of the slope of the line of best fit in the context of this problem.

28. **Sports** The data in the table below show the amount of water a professional tennis player loses for various times during a tennis match. The line of best fit is $y = 34.6x + 207$, where x is the time of the workout in minutes and y is the milliliters of water lost during the match.

Time of workout, x (in minutes)	10	20	30	40	50	60
Water lost, y (in milliliters)	600	900	1200	1500	2000	2300

Graph the data and the line of best fit in the coordinate system at the right. Write a sentence that describes the meaning of the slope of the line of best fit in the context of this problem.

29. **Entertainment** The data in the table below show the decline in the percent of music purchased in stores in the U.S. (*Source:* RIAA) The line of best fit is $y = -3x + 55$, where x is the year (with $x = 0$ corresponding to 1997) and y is the percent of music purchased in stores in the U.S.

Year, x	1	2	3	4	5	6
Percent, y	52	51	45	42	42	37

Graph the data and the line of best fit in the coordinate system at the right. Write a sentence that describes the meaning of the slope of the line of best fit in the context of this problem.

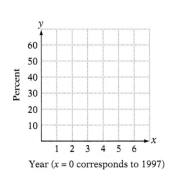

30. Evaporation   The data in the table below show the amount of water that evaporates from swimming pools of various surface areas. The line of best fit is $y = 0.17x - 1$, where x is the surface area of the swimming pool and y is the number of gallons of water that evaporates in one day.

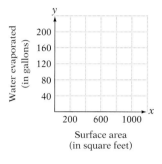

Surface area, x (in square feet)	100	200	300	400	600	1000
Water evaporated, y (in gallons)	25	30	45	60	100	170

Graph the data and the line of best fit in the coordinate system at the right. Write a sentence that describes the meaning of the slope of the line of best fit in the context of this problem.

APPLYING THE CONCEPTS

In Exercises 31 to 34, the first two given points are on a line. Determine whether the third point is on the line.

31. $(-3, 2), (4, 1); (-1, 0)$

32. $(2, -2), (3, 4); (-1, 5)$

33. $(-3, -5), (1, 3); (4, 9)$

34. $(-3, 7), (0, -2); (1, -5)$

35. If $(-2, 4)$ are the coordinates of a point on the line whose equation is $y = mx + 1$, what is the slope of the line?

36. If $(3, 1)$ are the coordinates of a point on the line whose equation is $y = mx - 3$, what is the slope of the line?

37. If $(0, -3), (6, -7),$ and $(3, n)$ are coordinates of points on the same line, determine n.

38. If $(-4, 11), (2, -4),$ and $(6, n)$ are coordinates of points on the same line, determine n.

The formula $y - y_1 = \frac{y_2 - y_1}{x_2 - x_1}(x - x_1)$, where $x_1 \neq x_2$, is called the **two-point formula** for a straight line. This formula can be used to find the equation of a line given two points. Use this formula for Exercises 39 and 40.

39. Find the equation of the line passing through $(-2, 3)$ and $(4, -1)$.

40. Find the equation of the line passing through $(3, -1)$ and $(4, -3)$.

41. Explain why the condition $x_1 \neq x_2$ is placed on the two-point formula given above.

42. Explain how the two-point formula given above can be derived from the point-slope formula.

Focus on Problem Solving

Counterexamples   Some of the exercises in this text ask you to determine whether a statement is true or false. For instance, the statement "Every real number has a reciprocal" is false because 0 is a real number and 0 does not have a reciprocal.

Finding an example, such as "0 has no reciprocal," to show that a statement is not always true is called finding a counterexample. A **counterexample** is an example that shows that a statement is not always true.

Here are some counterexamples to the statement "The square of a number is always larger than the number."

$$\left(\frac{1}{2}\right)^2 = \frac{1}{4} \quad \text{but} \quad \frac{1}{4} < \frac{1}{2} \qquad 1^2 = 1 \quad \text{but} \quad 1 = 1$$

For Exercises 1 to 9, answer true if the statement is always true. If there is an instance when the statement is false, give a counterexample.

1. The product of two integers is always a positive number.

2. The sum of two prime numbers is never a prime number.

3. For all real numbers, $|x + y| = |x| + |y|$.

4. If x and y are nonzero real numbers and $x > y$, then $x^2 > y^2$.

5. The quotient of any two nonzero real numbers is less than either one of the numbers.

6. The reciprocal of a positive number is always smaller than the number.

7. If $x < 0$, then $|x| = -x$.

8. For any two real numbers x and y, $x + y > x - y$.

9. The list of numbers 1, 11, 111, 1111, 11111, ... contains infinitely many composite numbers. (*Hint:* A number is divisible by 3 if the sum of the digits of the number is divisible by 3.)

Projects and Group Activities

Graphing Linear Equations with a Graphing Utility

A computer or graphing calculator screen is divided into *pixels*. There are approximately 6000 to 790,000 pixels available on the screen (depending on the computer or calculator). The greater the number of pixels, the smoother a graph will appear. A portion of a screen is shown at the left. Each little rectangle represents one pixel.

The graphing utilities that are used by computers or calculators to graph an equation do basically what we have shown in the text: They choose values of x and, for each, calculate the corresponding value of y. The pixel corresponding to the ordered pair is then turned on. The graph is jagged because pixels are much larger than the dots we draw on paper.

The graph of $y = 0.45x$ is shown at the left as the calculator drew it (jagged). The x- and y-axes have been chosen so that each pixel represents $\frac{1}{10}$ of a unit. Consider the region of the graph where $x = 1$, 1.1, and 1.2.

The corresponding values of y are 0.45, 0.495, and 0.54. Because the y-axis is in tenths, the numbers 0.45, 0.495, and 0.54 are rounded to the nearest tenth before plotting. Rounding 0.45, 0.495, and 0.54 to the nearest tenth results in 0.5 for each number. Thus the ordered pairs (1, 0.45), (1.1, 0.495), and (1.2, 0.54) are graphed as (1, 0.5), (1.1, 0.5), and (1.2, 0.5). These points appear as three illuminated horizontal pixels. However, if you use the TRACE feature of the calculator (see the Appendix), the actual y-coordinate for each value of x is displayed.

Here are the keystrokes for a TI-83 calculator to graph $y = \frac{2}{3}x + 1$. First the equation is entered. Then the domain (Xmin to Xmax) and the range (Ymin to Ymax) are entered. This is called the **viewing window.**

Y= CLEAR 2 X,T,θ,n ÷ 3 + 1 WINDOW (-) 10 ENTER 10 ENTER 1

ENTER (-) 10 ENTER 10 ENTER 1 ENTER GRAPH

By changing the keystrokes 2 X,T,θ,n ÷ 3 + 1, you can graph different equations.

Integrating Technology

See the Keystroke Guide: Y= and WINDOW for assistance.

For Exercises 1 to 4, graph on a graphing calculator.

1. $y = 2x + 1$ **2.** $y = -\frac{1}{2}x - 2$ **3.** $3x + 2y = 6$ **4.** $4x + 3y = 75$

Graphs of Motion   A graph can be useful in analyzing the motion of a body. For example, consider an airplane in uniform motion traveling at 100 m/s. The table at the right shows the distance, in meters, traveled by the plane at the end of each of five one-second intervals.

Time (in seconds)	Distance (in meters)
0	0
1	100
2	200
3	300
4	400
5	500

These data can be graphed on a rectangular coordinate system and a straight line drawn through the points plotted. The travel time is shown along the horizontal axis, and the distance traveled by the plane is shown along the vertical axis. (Note that the units along the two axes are not the same length.)

To write the equation for the line just graphed, use the coordinates of any two points on the line to find the slope. The y-intercept is (0, 0).

Let $(x_1, y_1) = (1, 100)$ and $(x_2, y_2) = (2, 200)$.

$$m = \frac{y_2 - y_1}{x_2 - x_1} = \frac{200 - 100}{2 - 1} = 100$$

$y = mx + b$
$y = 100x + 0$
$y = 100x$

Note that the slope of the line, 100, is equal to the speed, 100 m/s. *The slope of a distance-time graph represents the speed of the object.*

The distance-time graphs for two planes are shown at the left. One plane is traveling at 100 m/s, and the other at 200 m/s. The slope of the line representing the faster plane is greater than the slope of the line representing the slower plane.

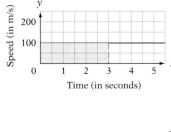

In the speed-time graph at the left, the time a plane has been flying at 100 m/s is shown along the horizontal axis, and its speed is shown along the vertical axis. Because the speed is constant, the graph is a horizontal line.

The area between the horizontal line graphed and the horizontal axis is equal to the distance traveled by the plane up to that time. For example, the area of the shaded region on the graph is

$$\text{Length} \cdot \text{width} = (3\text{ s})(100\text{ m/s}) = 300\text{ m}$$

The distance traveled by the plane in 3 s is equal to 300 m.

1. A car in uniform motion is traveling at 20 m/s.
   a. Prepare a distance-time graph for the car for 0 s to 5 s.
   b. Find the slope of the line.
   c. Find the equation of the line.
   d. Prepare a speed-time graph for the car for 0 s to 5 s.
   e. Find the distance traveled by the car after 3 s.

2. One car in uniform motion is traveling at 10 m/s. A second car in uniform motion is traveling at 15 m/s.
   a. Prepare one distance-time graph for both cars for 0 s to 5 s.
   b. Find the slope of each line.
   c. Find the equation of each line graphed.
   d. Assuming that the cars started at the same point at 0 s, find the distance between the cars at the end of 5 s.

Chapter 7 Summary

Key Words

A *rectangular coordinate system* is formed by two number lines, one horizontal and one vertical, that intersect at the zero point of each line. The number lines that make up a rectangular coordinate system are called the *coordinate axes*, or simply *axes*. The *origin* is the point of intersection of the two coordinate axes. Generally, the horizontal axis is labeled the *x*-axis and the vertical axis is labeled the *y*-axis. The coordinate system divides the plane into four regions called *quadrants*. The *coordinates of a point* in the plane are given by an *ordered pair* (x, y). The first number in the ordered pair is called the *abscissa* or *x*-coordinate. The second number in the ordered pair is the *ordinate* or *y*-coordinate. The *graph of an ordered pair* (x, y) is the dot drawn at the coordinates of the point in the plane. [7.1A, p. 353]

Examples

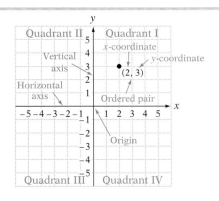

A *solution of an equation in two variables* is an ordered pair (x, y) that makes the equation a true statement. [7.1B, p. 355]

The ordered pair $(-1, 1)$ is a solution of the equation $y = 2x + 3$ because when -1 is substituted for x and 1 is substituted for y, the result is a true equation.

A *relation* is any set of ordered pairs. The *domain* of a relation is the set of first coordinates of the ordered pairs. The *range* is the set of second coordinates of the ordered pairs. [7.1C, p. 357]

For the relation $\{(-1, 2), (2, 4), (3, 5), (3, 7)\}$, the domain is $\{-1, 2, 3\}$; the range is $\{2, 4, 5, 7\}$.

A *function* is a relation in which no two ordered pairs have the same first coordinate. [7.1C, p. 357]

The relation $\{(-2, -3), (0, 4), (1, 5)\}$ is a function. No two ordered pairs have the same first coordinate.

The *graph of an equation in two variables* is a graph of the ordered-pair solutions of the equation. An equation of the form $y = mx + b$ is a *linear equation in two variables*. [7.2A, p. 365]

$y = 2x + 3$ is a linear equation in two variables. Its graph is shown at the right.

An equation written in the form $Ax + By = C$ is the *standard form of a linear equation in two variables*. [7.2B, p. 367]

$2x + 7y = 10$ is an example of a linear equation in two variables written in standard form.

The point at which a graph crosses the *x*-axis is called the *x-intercept*. At the *x*-intercept, the *y*-coordinate is 0. The point at which a graph crosses the *y*-axis is called the *y-intercept*. At the *y*-intercept, the *x*-coordinate is 0. [7.3A, p. 375]

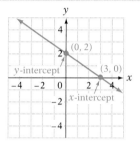

The *slope* of a line is a measure of the slant of the line. The symbol for slope is *m*. A line with *positive slope* slants upward to the right. A line with *negative slope* slants downward to the right. A horizontal line has *zero slope*. A vertical line has an *undefined slope*. [7.3A, pp. 376–377]

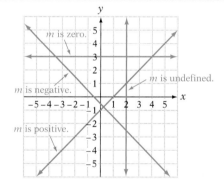

When data are graphed as points in a coordinate system, the graph is called a *scatter diagram*. A line drawn to approximate the data is called the *line of best fit*. [7.4C, p. 388]

The graph shown at the right is the scatter diagram and line of best fit for the spring data on page 388.

Essential Rules and Procedures

Examples

Functional Notation [7.1D, p. 360]
The equation of a function is written in functional notation when *y* is replaced by the symbol $f(x)$, where $f(x)$ is read "*f* of *x*" or "the value of *f* at *x*." To evaluate a function at a given value of *x*, replace *x* by the given value and then simplify the resulting numerical expression to find the value of $f(x)$.

$y = x^2 + 2x - 1$ is written in functional notation as $f(x) = x^2 + 2x - 1$. To evaluate $f(x) = x^2 + 2x - 1$ at $x = -3$, find $f(-3)$.

$$f(-3) = (-3)^2 + 2(-3) - 1$$
$$= 9 - 6 - 1 = 2$$

Horizontal and Vertical Lines [7.2B, p. 368]

The graph of $y = b$ is a horizontal line passing through $(0, b)$.
The graph of $x = a$ is a vertical line passing through $(a, 0)$.

The graph of $y = -2$ is a horizontal line passing through $(0, -2)$. The graph of $x = 3$ is a vertical line passing through $(3, 0)$.

To find the x-intercept, let $y = 0$ and solve for x.
To find the y-intercept, let $x = 0$ and solve for y.
[7.3A, p. 375]

To find the x-intercept of $4x - 5y = 20$, let $y = 0$ and solve for x. To find the y-intercept, let $x = 0$ and solve for y.

$$4x - 5y = 20 \qquad\qquad 4x - 5y = 20$$
$$4x - 5(0) = 20 \qquad\qquad 4(0) - 5y = 20$$
$$4x = 20 \qquad\qquad\qquad -5y = 20$$
$$x = 5 \qquad\qquad\qquad\quad y = -4$$

The x-intercept is $(5, 0)$. The y-intercept is $(0, -4)$.

Slope Formula [7.3B, p. 376]

If $P_1(x_1, y_1)$ and $P_2(x_2, y_2)$ are two points on a line and $x_1 \neq x_2$, then

$$m = \frac{y_2 - y_1}{x_2 - x_1}$$

To find the slope of the line between the points $(1, -2)$ and $(-3, -1)$, let $P_1 = (1, -2)$ and $P_2 = (-3, -1)$. Then

$$m = \frac{y_2 - y_1}{x_2 - x_1} = \frac{-1 - (-2)}{-3 - 1} = \frac{1}{-4} = -\frac{1}{4}.$$

Parallel Lines [7.3B, p. 377]

Two nonvertical lines in the plane are parallel if and only if they have the same slope. Vertical lines in the plane are parallel.

The slope of the line through $P_1(3, -6)$ and $P_2(5, -10)$ is $m_1 = \frac{-10 - (-6)}{5 - 3} = -2$.
The slope of the line through $Q_1(4, -5)$ and $Q_2(0, 3)$ is $m_2 = \frac{3 - (-5)}{0 - 4} = -2$.
Because $m_1 = m_2$, the lines are parallel.

Perpendicular Lines [7.3B, p. 378]

Two nonvertical lines in the plane are perpendicular if and only if the product of their slopes is -1. A vertical and horizontal line are perpendicular.

The slope of the line through $P_1(5, -3)$ and $P_2(2, -1)$ is $m_1 = \frac{-1 - (-3)}{2 - 5} = -\frac{2}{3}$.
The slope of the line through $Q_1(1, -4)$ and $Q_2(3, -1)$ is $m_2 = \frac{-1 - (-4)}{3 - 1} = \frac{3}{2}$.
Because $m_1 m_2 = \left(-\frac{2}{3}\right)\left(\frac{3}{2}\right) = -1$, the lines are perpendicular.

Slope-Intercept Form of a Linear Equation [7.3C, p. 379]

An equation of the form $y = mx + b$ is called the slope-intercept form of a straight line. The slope of the line is m, the coefficient of x. The y-intercept is $(0, b)$, where b is the constant term of the equation.

For the line with equation $y = -3x + 2$, the slope is -3 and the y-intercept is $(0, 2)$.

Point-Slope Formula [7.4A, p. 386]

If (x_1, y_1) is a point on a line with slope m, then

$$y - y_1 = m(x - x_1)$$

The equation of the line that passes through the point $(5, -3)$ and has slope -2 is:

$$y - y_1 = m(x - x_1)$$
$$y - (-3) = -2(x - 5)$$
$$y + 3 = -2x + 10$$
$$y = -2x + 7$$

Chapter 7 Review Exercises

1. **a.** Graph the ordered pairs $(-2, 4)$ and $(3, -2)$.
    b. Name the abscissa of point A.
    c. Name the ordinate of point B.

    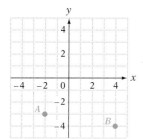

2. Graph the ordered-pair solutions of
    $y = -\frac{1}{2}x - 2$ when $x \in \{-4, -2, 0, 2\}$.

    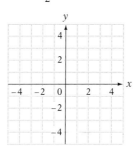

3. Determine the equation of the line that passes through the points $(-1, 3)$ and $(2, -5)$.

4. Determine the equation of the line that passes through the point $(6, 1)$ and has slope $-\frac{5}{2}$.

5. Graph $y = \frac{1}{4}x + 3$.

    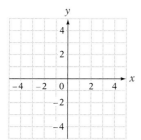

6. Graph $5x + 3y = 15$.

    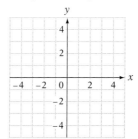

7. Is the line that passes through $(7, -5)$ and $(6, -1)$ parallel, perpendicular, or neither parallel nor perpendicular to the line that passes through $(4, 5)$ and $(2, -3)$?

8. Given $f(x) = x^2 - 2$, find $f(-1)$.

9. Determine the equation of the line that passes through the points $(-2, 5)$ and $(4, 1)$.

10. Does $y = -x + 3$, where $x \in \{-2, 0, 3, 5\}$, define y as a function of x?

11. Find the slope of the line containing the points $(9, 8)$ and $(-2, 1)$.

12. Find the x- and y-intercepts of $3x - 2y = 24$.

13. Find the slope of the line containing the points $(-2, -3)$ and $(4, -3)$.

14. Graph the line that has slope $\frac{1}{2}$ and y-intercept $(0, -1)$.

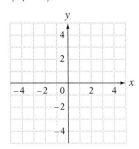

15. Graph $x = -3$.

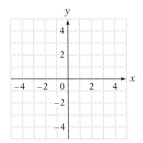

16. Graph the line that has slope $-\frac{2}{3}$ and y-intercept $(0, 2)$.

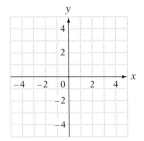

17. Graph $y = -2x - 1$.

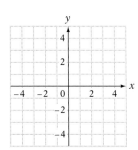

18. Graph the line that has slope 2 and y-intercept $(0, -4)$.

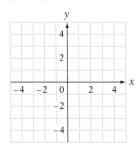

19. Graph $3x - 2y = -6$.

20. **Health** The height and weight of 8 seventh-grade students are shown in the following table. Write a relation where the first coordinate is height, in inches, and the second coordinate is weight, in pounds. Is the relation a function?

Height (in inches)	55	57	53	57	60	61	58	54
Weight (in pounds)	95	101	94	98	100	105	97	95

21. **Business** An online research service charges a monthly access fee of $75 plus $.45 per minute to use the service. An equation that represents the monthly cost to use this service is $C = 0.45x + 75$, where C is the monthly cost and x is the number of minutes of access used. Graph this equation for $0 \leq x \leq 100$. The point (50, 97.5) is on the graph. Write a sentence that describes the meaning of this ordered pair.

22. **Telecommunications** The data in the table below are estimates of the projected annual increase in average telephone bills for a family. The line of best fit is $y = 34x + 657$, where x is the year (with $x = 0$ corresponding to 1999) and y is the annual cost, in dollars, of telephone bills.

Year, x	0	1	2	3	4	5	6
Telephone bills, y (in dollars)	658	690	708	772	809	830	849

Graph the data and the line of best fit in the coordinate system at the right. Write a sentence that describes the meaning of the slope of the line of best fit.

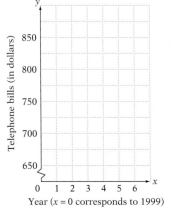

Chapter 7 Test

1. Find the ordered-pair solution of $2x - 3y = 15$ corresponding to $x = 3$.

2. Graph the ordered-pair solutions of $y = -\frac{3}{2}x + 1$ for $x \in \{-2, 0, 4\}$.

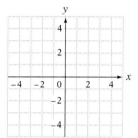

3. Does $y = \frac{1}{2}x - 3$ define y as a function of x for $x \in \{-2, 0, 4\}$?

4. Given $f(t) = t^2 + t$, find $f(2)$.

5. Given $f(x) = x^2 - 2x$, find $f(-1)$.

6. **Emergency Response**   The distance a house is from a fire station and the amount of damage that the house sustained in a fire are given in the following table. Write a relation wherein the first coordinate of the ordered pair is the distance, in miles, from the fire station and the second coordinate is the amount of damage in thousands of dollars. Is the relation a function?

| Distance (in miles) | 3.5 | 4.0 | 5.2 | 5.0 | 4.0 | 6.3 | 5.4 |
| Damage (in thousands of dollars) | 25 | 30 | 45 | 38 | 42 | 12 | 34 |

7. Graph $y = 3x + 1$.

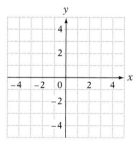

8. Graph $y = -\frac{3}{4}x + 3$.

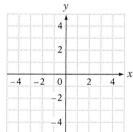

9. Graph $3x - 2y = 6$.

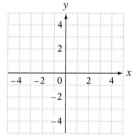

10. Graph $x + 3 = 0$.

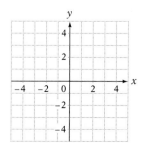

11. Graph the line that has slope $-\frac{2}{3}$ and y-intercept $(0, 4)$.

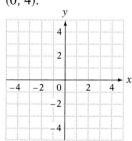

12. Graph the line that has slope 2 and y-intercept -2.

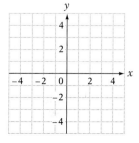

13. **Sports**   The equation for the speed of a ball that is thrown straight up with an initial speed of 128 ft/s is $v = 128 - 32t$, where v is the speed of the ball after t seconds. Graph this equation for $0 \leq t \leq 4$. The point whose coordinates are $(1, 96)$ is on the graph. Write a sentence that describes this ordered pair.

14. **Wages**   The graph at the right shows the projected increase in the average hourly wage of a U.S. worker for the years 2000 through 2006 (with $x = 0$ corresponding to 2000). Find the slope of the line. Write a sentence that states the meaning of the slope.

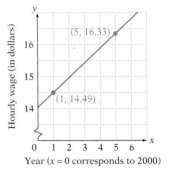

15. **Tuition**   The data in the table below are the average annual tuition costs for 4-year private colleges in the United States. The line of best fit is $y = 809x + 12{,}195$, where x is the year (with $x = 0$ corresponding to 1995) and y is the annual tuition cost in dollars rounded to the nearest 100.

Year, x	0	1	2	3	4	5
Tuition Costs, y (in dollars)	12,400	12,800	13,700	14,700	15,400	16,300

Graph the data and the line of best fit in the coordinate system at the right. Write a sentence that describes the meaning of the slope of the line of best fit.

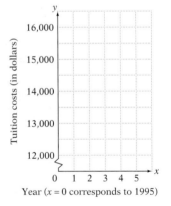

16. Find the x- and y-intercepts for $6x - 4y = 12$.

17. Find the x- and y-intercepts for $y = \frac{1}{2}x + 1$.

18. Find the slope of the line containing the points $(2, -3)$ and $(4, 1)$.

19. Is the line that passes through $(2, 5)$ and $(-1, 1)$ parallel, perpendicular, or neither parallel nor perpendicular to the line that passes through $(-2, 3)$ and $(4, 11)$?

20. Find the slope of the line containing the points $(-5, 2)$ and $(-5, 7)$.

21. Find the slope of the line whose equation is $2x + 3y = 6$.

22. Find the equation of the line that contains the point $(0, -1)$ and has slope 3.

23. Find the equation of the line that contains the point $(-3, 1)$ and has slope $\frac{2}{3}$.

24. Find the equation of the line that passes through the points $(5, -4)$ and $(-3, 1)$.

25. Find the equation of the line that passes through the points $(-2, 0)$ and $(5, -2)$.

Cumulative Review Exercises

1. Simplify: $12 - 18 \div 3 \cdot (-2)^2$

2. Evaluate $\dfrac{a - b}{a^2 - c}$ when $a = -2$, $b = 3$, and $c = -4$.

3. Given $f(x) = \dfrac{2}{x - 1}$, find $f(-2)$.

4. Solve: $2x - \dfrac{2}{3} = \dfrac{7}{3}$

5. Solve: $3x - 2[x - 3(2 - 3x)] = x - 7$

6. Write $6\dfrac{2}{3}\%$ as a fraction.

7. Simplify: $(-2x^2y)^3(2xy^2)^2$

8. Simplify: $\dfrac{-15x^7}{5x^5}$

9. Divide: $(x^2 - 4x - 21) \div (x - 7)$

10. Factor: $5x^2 + 15x + 10$

11. Factor: $x(a + 2) + y(a + 2)$

12. Solve: $x(x - 2) = 8$

13. Multiply: $\dfrac{x^5y^3}{x^2 - x - 6} \cdot \dfrac{x^2 - 9}{x^2y^4}$

14. Subtract: $\dfrac{3x}{x^2 + 5x - 24} - \dfrac{9}{x^2 + 5x - 24}$

15. Solve: $3 - \dfrac{1}{x} = \dfrac{5}{x}$

16. Solve $4x - 5y = 15$ for y.

17. Find the ordered-pair solution of $y = 2x - 1$ corresponding to $x = -2$.

18. Find the slope of the line that contains the points $(2, 3)$ and $(-2, 3)$.

19. Find the equation of the line that contains the point $(2, -1)$ and has slope $\frac{1}{2}$.

20. Find the equation of the line that contains the point $(0, 2)$ and has slope -3.

21. Find the equation of the line that contains the point $(-1, 0)$ and has slope 2.

22. Find the equation of the line that contains the point $(6, 1)$ and has slope $\frac{2}{3}$.

23. Business A suit that regularly sells for $89 is on sale for 30% off the regular price. Find the sale price.

24. Geometry The measure of the first angle of a triangle is 3° more than the measure of the second angle. The measure of the third angle is 5° more than twice the measure of the second angle. Find the measure of each angle.

25. Taxes The real estate tax for a home that costs $50,000 is $625. At this rate, what is the value of a home for which the real estate tax is $1375?

26. Business An electrician requires 6 h to wire a garage. An apprentice can do the same job in 10 h. How long would it take to wire the garage if both the electrician and the apprentice were working?

27. Graph $y = \frac{1}{2}x - 1$.

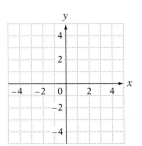

28. Graph the line that has slope $-\frac{2}{3}$ and y-intercept 2.

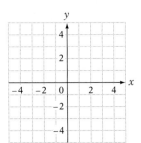

8 Systems of Linear Equations

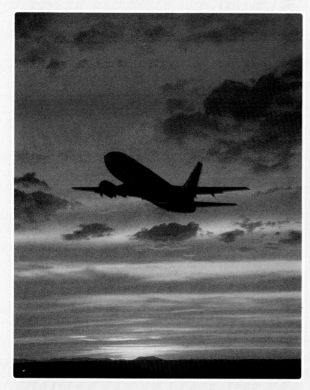

Recall from Section 3.1 that if a boat is traveling with the current, the effective speed of the boat is increased. If the boat is traveling against the current, the current slows the boat down. The same is true for airplanes and the wind. If the plane in the photo above is flying with a tailwind, its effective speed is increased. If the plane is flying against a headwind, the wind will slow the plane down. In **Exercises 2, 5, 6, and 8 on page 433,** you will use systems of equations to determine the rates of planes in calm air and rates of the wind.

OBJECTIVES

Section 8.1

A To solve a system of linear equations by graphing

Section 8.2

A To solve a system of linear equations by the substitution method

B To solve investment problems

Section 8.3

A To solve a system of linear equations by the addition method

Section 8.4

A To solve rate-of-wind or rate-of-current problems

B To solve application problems using two variables

 Need help? For online student resources, such as section quizzes, visit this textbook's website at **math.college.hmco.com/students.**

Do these exercises to prepare for Chapter 8.

1. Solve $3x - 4y = 24$ for y.

2. Solve: $50 + 0.07x = 0.05(x + 1400)$

3. Simplify: $-3(2x - 7y) + 3(2x + 4y)$

4. Simplify: $4x + 2(3x - 5)$

5. Is $(-4, 2)$ a solution of $3x - 5y = -22$?

6. Find the x- and y-intercepts for $3x - 4y = 12$.

7. Are the graphs of $3x + y = 6$ and $y = -3x - 4$ parallel?

8. Graph: $y = \dfrac{5}{4}x - 2$

9. **Pharmacology**   A pharmacist has 20 ml of an 80% acetic acid solution. How many milliliters of a 55% acetic acid solution should be mixed with the 20-milliliter solution to produce a solution that is 75% acetic acid?

10. **Hiking**   One hiker starts along a trail walking at 3 mph. One-half hour later, another hiker starts on the same walking trail at a speed of 4 mph. How long after the second hiker started will the two hikers be side-by-side?

GO FIGURE • • •

Two children are running from school to home. Carla runs half the time and walks half the time. James runs half the distance and walks half the distance. Assuming the children run and walk at the same rate, which child arrives home first?

8.1 Solving Systems of Linear Equations by Graphing

Objective A To solve a system of linear equations by graphing

Two or more equations considered together are called a **system of equations.** Three examples of *linear* systems of equations in *two* variables are shown below, along with the graphs of the equations of each system.

System I	**System II**	**System III**
$x - 2y = -8$	$4x + 2y = 6$	$4x + 6y = 12$
$2x + 5y = 11$	$y = -2x + 3$	$6x + 9y = -9$

  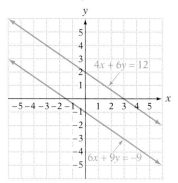

For system I, the two lines intersect at a single point, $(-2, 3)$. Because this point lies on both lines, it is a solution of each equation of the system of equations. We can check this by replacing x by -2 and y by 3. The check is shown below.

$$\begin{array}{c|c}
x - 2y = -8 & \\
\hline
-2 - 2(3) & -8 \\
-2 - 6 & -8 \\
-8 = -8 \ \checkmark &
\end{array}
\qquad
\begin{array}{c|c}
2x + 5y = 11 & \\
\hline
2(-2) + 5(3) & 11 \\
-4 + 15 & 11 \\
11 = 11 \ \checkmark &
\end{array}$$

● Replace x by **−2** and replace y by **3**.

A **solution of a system of equations in two variables** is an ordered pair that is a solution of each equation of the system. The ordered pair $(-2, 3)$ is a solution of system I.

HOW TO Is $(-1, 4)$ a solution of the system of equations? $7x + 3y = 5$
   $3x - 2y = 12$

$$\begin{array}{c|c}
7x + 3y = 5 & \\
\hline
7(-1) + 3(4) & 5 \\
-7 + 12 & 5 \\
5 = 5 \ \checkmark &
\end{array}
\qquad
\begin{array}{c|c}
3x - 2y = 12 & \\
\hline
3(-1) - 2(4) & 12 \\
-3 - 8 & 12 \\
-11 \ne 12 &
\end{array}$$

● Replace x by **−1** and replace y by **4**.

● Does not check

Because $(-1, 4)$ is not a solution of both equations, $(-1, 4)$ is not a solution of the system of equations.

Using the system of equations above and the graph at the right, note that the graph of the ordered pair $(-1, 4)$ lies on the graph of $7x + 3y = 5$ but not on *both* lines. The ordered pair $(-1, 4)$ is *not* a solution of the system of equations. The graph of the ordered pair $(2, -3)$ does lie on both lines and therefore the ordered pair $(2, -3)$ is a solution of the system of equations.

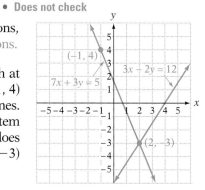

System II from the preceding page and the graph of the equations of that system are shown again at the right. Note that the graph of $y = -2x + 3$ lies directly on top of the graph of $4x + 2y = 6$. Thus the two lines intersect at an infinite number of points. The graphs intersect at an infinite number of points, so there are an infinite number of solutions of this system of equations. Because each equation represents the same set of points, the solutions of the system of equations can be stated by using the ordered pairs of either one of the equations. Therefore, we can say, "The solutions are the ordered pairs that satisfy $4x + 2y = 6$," or we can say "The solutions are the ordered pairs that satisfy $y = -2x + 3$."

$$4x + 2y = 6$$
$$y = -2x + 3$$

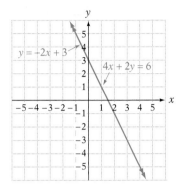

System III from the preceding page and the graph of the equations of that system are shown again at the right. Note that in this case, the graphs of the lines are parallel and do not intersect. Because the graphs do not intersect, there is no point that is on both lines. Therefore, the system of equations has no solution.

$$4x + 6y = 12$$
$$6x + 9y = -9$$

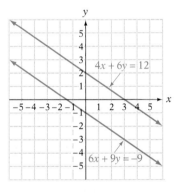

The preceding examples illustrate three types of systems of linear equations. An **independent system** has exactly one solution—the graphs intersect at one point. A **dependent system** has an infinite number of solutions—the graphs are the same line. An **inconsistent system** has no solution—the graphs are parallel lines.

Independent:
one solution

Dependent:
infinitely many solutions

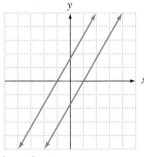

Inconsistent:
no solutions

HOW TO The graphs of the equations for the system of equations below are shown at the right. What is the solution of the system of equations?

$$2x + 3y = 6$$
$$2x + y = -2$$

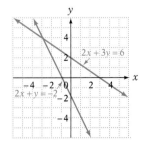

The graphs intersect at $(-3, 4)$. This is an *independent* system of equations. The solution of the system of equations is $(-3, 4)$.

TAKE NOTE

Because both equations represent the same ordered pairs, we can also say that the solutions of the system of equations are the ordered pairs that satisfy $x = \frac{1}{2}y + 1$. Either answer is correct.

HOW TO   The graphs of the equations for the system of equations at the right are shown below. What is the solution of the system of equations?

$$y = 2x - 2$$
$$x = \frac{1}{2}y + 1$$

The two graphs lie directly on top of one another. Thus the two lines intersect at an infinite number of points, and the system of equations has an infinite number of solutions. This is a *dependent* system of equations. The solutions of the system of equations are the ordered pairs that satisfy $y = 2x - 2$.

Integrating Technology

The Projects and Group Activities at the end of this chapter discusses using a calculator to approximate the solution of an independent system of equations. Also see the Keystroke Guide: *Intersect*.

Solving a system of equations means finding the ordered-pair solutions of the system. One way to do this is to draw the graphs of the equations in the system of equations and determine where the graphs intersect.

To solve a system of linear equations in two variables by graphing, graph each equation on the same coordinate system, and then determine the points of intersection.

HOW TO   Solve by graphing: $2x - y = -1$
$x + 2y = 7$

Graph each line.

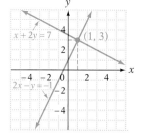

The point of intersection of the two graphs lies on both lines and is therefore the solution of the system of equations.

The system of equations is independent. (1, 3) is a solution of each equation.

The solution is (1, 3).

HOW TO   Solve by graphing: $y = 2x + 2$
$4x - 2y = 4$

Graph each line.

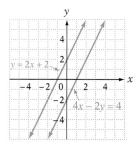

The graphs do not intersect. The system of equations is inconsistent.

The system of equations has no solution.

Example 1

Is (1, −3) a solution of the following system?
$3x + 2y = -3$
$x - 3y = 6$

Solution
Replace x by 1 and y by −3.

$3x + 2y = -3$		$x - 3y = 6$	
$3 \cdot 1 + 2(-3)$	-3	$1 - 3(-3)$	6
$3 + (-6)$	-3	$1 - (-9)$	6
$-3 = -3$		$10 \neq 6$	

No, (1, −3) is not a solution of the system of equations.

You Try It 1

Is (−1, −2) a solution of the following system?
$2x - 5y = 8$
$-x + 3y = -5$

Your solution

Example 2

Solve by graphing:
$x - 2y = 2$
$x + y = 5$

Solution

The solution is (4, 1).

You Try It 2

Solve by graphing:
$x + 3y = 3$
$-x + y = 5$

Your solution

Example 3

Solve by graphing:
$4x - 2y = 6$
$y = 2x - 3$

Solution

The solutions are the ordered pairs that satisfy the equation $y = 2x - 3$.

You Try It 3

Solve by graphing:
$y = 3x - 1$
$6x - 2y = -6$

Your solution

Solutions on p. S21

8.1 Exercises

Objective A   **To solve a system of linear equations by graphing**

For Exercises 1 and 2, match each system of equations (systems I, II, and III) with (a) independent, (b) dependent, or (c) inconsistent.

1. I

II

III

2. I

II

III

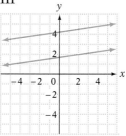

For Exercises 3 to 10, use the graphs of the equations of the system of equations to find the solution of the system of equations.

3.

4.

5.

6.

7.

8.

9.

10.

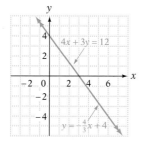

11. Is $(2, 3)$ a solution of $\begin{array}{l} 3x + 4y = 18 \\ 2x - y = 1 \end{array}$?

12. Is $(2, -1)$ a solution of $\begin{array}{l} x - 2y = 4 \\ 2x + y = 3 \end{array}$?

13. Is $(4, 3)$ a solution of $\begin{array}{l} 5x - 2y = 14 \\ x + y = 8 \end{array}$?

14. Is $(2, 5)$ a solution of $\begin{array}{l} 3x + 2y = 16 \\ 2x - 3y = 4 \end{array}$?

15. Is $(2, -3)$ a solution of $\begin{array}{l} y = 2x - 7 \\ 3x - y = 9 \end{array}$?

16. Is $(-1, -2)$ a solution of $\begin{array}{l} 3x - 4y = 5 \\ y = x - 1 \end{array}$?

17. Is $(0, 0)$ a solution of $\begin{array}{l} 3x + 4y = 0 \\ y = x \end{array}$?

18. Is $(3, -4)$ a solution of $\begin{array}{l} 5x - 2y = 23 \\ 2x - 5y = 25 \end{array}$?

For Exercises 19 to 38, solve by graphing.

19. $x - y = 3$
$x + y = 5$

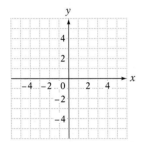

20. $2x - y = 4$
$x + y = 5$

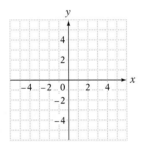

21. $x + 2y = 6$
$x - y = 3$

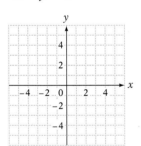

22. $3x - y = 3$
$2x + y = 2$

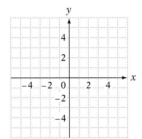

23. $3x - 2y = 6$
$y = 3$

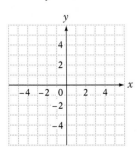

24. $x = 2$
$3x + 2y = 4$

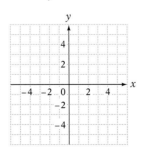

25. $x = 3$
      $y = -2$

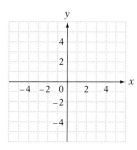

26. $x + 1 = 0$
      $y - 3 = 0$

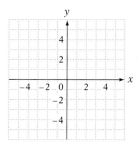

27. $y = 2x - 6$
    $x + y = 0$

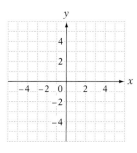

28. $5x - 2y = 11$
       $y = 2x - 5$

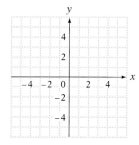

29. $2x + y = -2$
    $6x + 3y = 6$

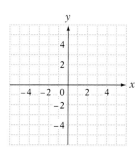

30. $x + y = 5$
    $3x + 3y = 6$

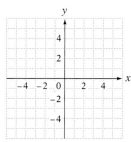

31. $y = 2x - 2$
   $4x - 2y = 4$

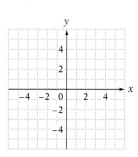

32. $y = -\dfrac{1}{3}x + 1$
   $2x + 6y = 6$

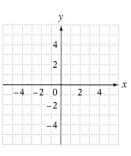

33. $x - y = 5$
   $2x - y = 6$

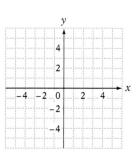

34. $5x - 2y = 10$
    $3x + 2y = 6$

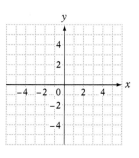

35. $3x + 4y = 0$
$2x - 5y = 0$

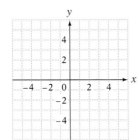

36. $2x - 3y = 0$
$y = -\dfrac{1}{3}x$

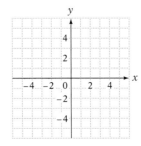

37. $x - 3y = 3$
$2x - 6y = 12$

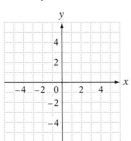

38. $4x + 6y = 12$
$6x + 9y = 18$

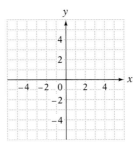

APPLYING THE CONCEPTS

39.   Determine whether the statement is always true, sometimes true, or never true.

   a. A solution of a system of two equations in two variables is a point in the plane.

   b. Two parallel lines have the same slope.

   c. Two different lines with the same y-intercept are parallel.

   d. Two different lines with the same slope are parallel.

40.   Explain how you can determine from the graph of a system of two linear equations in two variables whether it is an independent system of equations.

41.   Explain how you can determine from the graph of a system of two linear equations in two variables whether it is an inconsistent system of equations.

42.   Write a system of equations that has $(-2, 4)$ as its only solution.

43.   Write a system of equations for which there is no solution.

44.   Write a system of equations that is a dependent system of equations.

8.2 Solving Systems of Linear Equations by the Substitution Method

Objective A **To solve a system of linear equations by the substitution method**

A graphical solution of a system of equations is based on approximating the coordinates of a point of intersection. Algebraic methods can be used to find an exact solution of a system of equations. To solve a system of equations by the **substitution method,** one variable must be written in terms of the other variable.

HOW TO Solve by the substitution method:
$$(1) \quad 2x + 5y = -11$$
$$(2) \quad \quad y = 3x - 9$$

Equation (2) states that $y = 3x - 9$. Substitute $3x - 9$ for y in Equation (1). Then solve for x.

$$2x + 5y = -11 \qquad \bullet \text{ This is Equation (1).}$$
$$2x + 5(3x - 9) = -11 \qquad \bullet \text{ From Equation (2), substitute } 3x - 9 \text{ for } y.$$
$$2x + 15x - 45 = -11 \qquad \bullet \text{ Solve for } x.$$
$$17x - 45 = -11$$
$$17x = 34$$
$$x = 2$$

Now substitute the value of x into Equation (2) and solve for y.

$$y = 3x - 9 \qquad \bullet \text{ This is Equation (2).}$$
$$y = 3(2) - 9 \qquad \bullet \text{ Substitute } 2 \text{ for } x.$$
$$y = 6 - 9 = -3$$

The solution is the ordered pair $(2, -3)$.

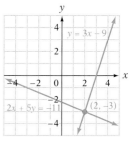

The graph of the equations in this system of equations is shown at the right. Note that the lines intersect at the point whose coordinates are $(2, -3)$, which is the algebraic solution we determined by the substitution method.

To solve a system of equations by the substitution method, we may need to solve one of the equations in the system of equations for one of its variables. For instance, the first step in solving the system of equations

$$(1) \quad x + 2y = -3$$
$$(2) \quad 2x - 3y = 5$$

is to solve an equation of the system for one of its variables. Either equation can be used.

Solving Equation (1) for x:

$$x + 2y = -3$$
$$x = -2y - 3$$

Solving Equation (2) for x:

$$2x - 3y = 5$$
$$2x = 3y + 5$$
$$x = \frac{3y + 5}{2} = \frac{3}{2}y + \frac{5}{2}$$

Because solving Equation (1) for x does not result in fractions, it is the easier of the two equations to use.

Here is the solution of the system of equations given on the preceding page.

HOW TO Solve by the substitution method: (1) $x + 2y = -3$
(2) $2x - 3y = 5$

To use the substitution method, we must solve an equation for one of its variables. Equation (1) is used here because solving it for x does not result in fractions.

$x + 2y = -3$

(3) $\qquad x = -2y - 3$ • Solve for x. This is Equation (3).

Now substitute $-2y - 3$ for x in Equation (2) and solve for y.

$2x - 3y = 5$ • This is Equation (2).

$2(-2y - 3) - 3y = 5$ • From Equation (3), substitute **−2y − 3** for x.

$-4y - 6 - 3y = 5$ • Solve for y.

$-7y - 6 = 5$

$-7y = 11$

$y = -\dfrac{11}{7}$

Substitute the value of y into Equation (3) and solve for x.

$x = -2y - 3$ • This is Equation (3).

$= -2\left(-\dfrac{11}{7}\right) - 3$ • Substitute $-\dfrac{11}{7}$ for y.

$= \dfrac{22}{7} - 3 = \dfrac{22}{7} - \dfrac{21}{7} = \dfrac{1}{7}$

The solution is $\left(\dfrac{1}{7}, -\dfrac{11}{7}\right)$.

The graph of the system of equations given above is shown at the right. It would be difficult to determine the exact solution of this system of equations from the graphs of the equations.

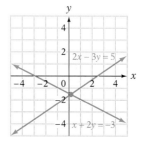

HOW TO Solve by the substitution method: (1) $y = 3x - 1$
(2) $y = -2x - 6$

$y = -2x - 6$

$3x - 1 = -2x - 6$ • Substitute **3x − 1** for y in Equation (2).

$5x = -5$ • Solve for x.

$x = -1$

Substitute this value of x into Equation (1) or Equation (2) and solve for y. Equation (1) is used here.

$y = 3x - 1$

$y = 3(-1) - 1 = -4$ • Substitute **−1** for x.

The solution is $(-1, -4)$.

The substitution method can be used to analyze inconsistent and dependent systems of equations. If, when solving a system of equations algebraically, the variable is eliminated and the result is a false equation, such as $0 = 4$, the system of equations is inconsistent. If the variable is eliminated and the result is a true equation, such as $12 = 12$, the system of equations is dependent.

HOW TO   Solve by the substitution method: (1) $2x + 3y = 3$

(2) $y = -\dfrac{2}{3}x + 3$

$$2x + 3y = 3$$ • This is Equation (1).

$$2x + 3\left(-\dfrac{2}{3}x + 3\right) = 3$$ • From Equation (2), replace y with $-\dfrac{2}{3}x + 3$.

$$2x - 2x + 9 = 3$$ • Solve for x.

$$9 = 3$$ • This is not a true equation.

Because $9 = 3$ is not a true equation, the system of equations has no solution.

Solving Equation (1) above for y, we have $y = -\dfrac{2}{3}x + 1$.

Comparing this with Equation (2) reveals that the slopes are equal and the y-intercepts are different. The graphs of the equations that make up this system of equations are parallel and thus never intersect. Because the graphs do not intersect, there are no solutions of the system of equations. The system of equations is inconsistent.

HOW TO   Solve by the substitution method: (1) $x = 2y + 3$

(2) $4x - 8y = 12$

$$4x - 8y = 12$$ • This is Equation (2).

$$4(2y + 3) - 8y = 12$$ • From Equation (1), replace x by $2y + 3$.

$$8y + 12 - 8y = 12$$ • Solve for y.

$$12 = 12$$ • This is a true equation.

The true equation $12 = 12$ indicates that any ordered pair (x, y) that satisfies one equation of the system satisfies the other equation. Therefore, the system of equations has an infinite number of solutions. The solutions are the ordered pairs (x, y) that are solutions of $x = 2y + 3$.

TAKE NOTE

As we mentioned in the previous section, when the system of equations is dependent, either equation can be used to write the ordered-pair solutions. Thus we could have said, "The solutions are the ordered pairs (x, y) that are solutions of $4x - 8y = 12$." Also note that, as we show at the right, if we solve each equation for y, the equations have the same slope-intercept form. This means we could also say, "The solutions are the ordered pairs (x, y) that are solutions of $y = \dfrac{1}{2}x - \dfrac{3}{2}$." When a system of equations is dependent, there are many ways in which the solutions can be stated.

If we write Equation (1) and Equation (2) in slope-intercept form, we have

$$x = 2y + 3 \qquad\qquad 4x - 8y = 12$$
$$-2y = -x + 3 \qquad\qquad -8y = -4x + 12$$
$$y = \dfrac{1}{2}x - \dfrac{3}{2} \qquad\qquad y = \dfrac{1}{2}x - \dfrac{3}{2}$$

The slope-intercept forms of the equations are the same, and therefore the graphs are the same. If we graph these two equations, we essentially graph one over the other. Accordingly, the graphs intersect at an infinite number of points.

Example 1 Solve by substitution:
(1) $3x + 4y = -2$
(2) $-x + 2y = 4$

Solution

$-x + 2y = 4$ • Solve Equation (2) for x.
$-x = -2y + 4$
$x = 2y - 4$

Substitute in Equation (1).
(1) $3x + 4y = -2$
$3(2y - 4) + 4y = -2$ • $x = 2y - 4$
$6y - 12 + 4y = -2$ • Solve for y.
$10y - 12 = -2$
$10y = 10$
$y = 1$

Substitute in $x = 2y - 4$.
$x = 2y - 4$
$x = 2(1) - 4$ • $y = 1$
$x = 2 - 4$
$x = -2$

The solution is $(-2, 1)$.

You Try It 1 Solve by substitution:
(1) $7x - y = 4$
(2) $3x + 2y = 9$

Your solution

Example 2 Solve by substitution:
$4x + 2y = 5$
$y = -2x + 1$

Solution

$4x + 2y = 5$
$4x + 2(-2x + 1) = 5$ • $y = -2x + 1$
$4x - 4x + 2 = 5$ • Solve for x.
$2 = 5$ • Not a true equation

The system of equations is inconsistent and therefore does not have a solution.

You Try It 2 Solve by substitution:
$3x - y = 4$
$y = 3x + 2$

Your solution

Example 3 Solve by substitution:
$y = 3x - 2$
$6x - 2y = 4$

Solution

$6x - 2y = 4$
$6x - 2(3x - 2) = 4$ • $y = 3x - 2$
$6x - 6x + 4 = 4$ • Solve for x.
$4 = 4$ • A true equation

The system of equations is dependent. The solutions are the ordered pairs that satisfy the equation $y = 3x - 2$.

You Try It 3 Solve by substitution:
$y = -2x + 1$
$6x + 3y = 3$

Your solution

Solutions on p. S21

Objective B   **To solve investment problems**

The annual simple interest that an investment earns is given by the equation $Pr = I$, where P is the principal, or the amount invested, r is the simple interest rate, and I is the simple interest.

For instance, if you invest $750 at a simple interest rate of 6%, then the interest earned after 1 year is calculated as follows:

$$Pr = I$$
$$750(0.06) = I \qquad \bullet \text{ Replace } P \text{ by } \mathbf{750} \text{ and } r \text{ by } \mathbf{0.06} \ (6\%).$$
$$45 = I \qquad \bullet \text{ Simplify.}$$

The amount of interest earned is $45.

HOW TO   A medical lab technician decided to open an Individual Retirement Account (IRA) by placing $2000 in two simple interest accounts. On one account, a corporate bond fund, the annual simple interest rate is 7.5%. On the second account, a real estate investment trust, the annual simple interest rate is 9%. If the technician wants to have annual earnings of $168 from these two investments, how much must be invested in each account?

Strategy for Solving Simple-Interest Investment Problems

1. For each amount invested, use the equation $Pr = I$. Write a numerical or variable expression for the principal, the interest rate, and the interest earned.

Amount invested at 7.5%: x
Amount invested at 9%: y

	Principal, P	$\cdot$	Interest rate, r	$=$	Interest earned, I
Amount at 7.5%	x	$\cdot$	0.075	$=$	$0.075x$
Amount at 9%	y	$\cdot$	0.09	$=$	$0.09y$

2. Write a system of equations. One equation will express the relationship among the amounts invested. The second equation will express the relationship among the amounts of interest earned by the investments.

The total amount invested is $2000: $x + y = 2000$
The total annual interest earned is $168: $0.075x + 0.09y = 168$

Solve the system of equations.
(1) $x + y = 2000$
(2) $0.075x + 0.09y = 168$

Solve Equation (1) for y and substitute into Equation (2).
(3) $y = -x + 2000$

$$0.075x + 0.09(-x + 2000) = 168 \qquad \bullet \text{ Substitute } -\mathbf{x} + \mathbf{2000} \text{ for } y.$$
$$0.075x - 0.09x + 180 = 168$$
$$-0.015x = -12$$
$$x = 800$$

Substitute the value of x into Equation (3) and solve for y.

$$y = -x + 2000$$
$$y = -800 + 2000 = 1200 \qquad \bullet \text{ Substitute } -\mathbf{800} \text{ for } x.$$

The amount invested at 7.5% is $800. The amount invested at 9% is $1200.

Example 4

A hair stylist invested some money at an annual simple interest rate of 5.2%. A second investment, $1000 more than the first, was invested at an annual simple interest rate of 7.2% so that the total interest earned was $320. How much was invested in each account?

Strategy

• Amount invested at 5.2%: x
  Amount invested at 7.2%: y

	Principal	*Rate*	*Interest*
Amount at 5.2%	x	0.052	$0.052x$
Amount at 7.2%	y	0.072	$0.072y$

• The second investment is $1000 more than the first investment:

$y = x + 1000$

The sum of the interest earned at 5.2% and the interest earned at 7.2% equals $320.

$0.052x + 0.072y = 320$

Solution

(1) $\qquad\qquad\qquad y = x + 1000$

(2) $\quad 0.052x + 0.072y = 320$

Replace y in Equation (2) by $x + 1000$ from Equation (1). Then solve for x.

$$0.052x + 0.072y = 320$$
$$0.052x + 0.072(x + 1000) = 320 \quad \bullet \; y = x + 1000$$
$$0.052x + 0.072x + 72 = 320 \quad \bullet \; \text{Solve for } x.$$
$$0.124x + 72 = 320$$
$$0.124x = 248$$
$$x = 2000$$

$y = x + 1000$

$\quad = 2000 + 1000 \quad \bullet \; x = 2000$

$\quad = 3000$

$2000 was invested at an annual simple interest rate of 5.2%; $3000 was invested at 7.2%.

You Try It 4

The manager of a city's investment income wished to place $330,000 in two simple interest accounts. The first account earns 6.5% annual interest and the second account earns 4.5%. How much should be invested in each account so that both accounts earn the same annual interest?

Your strategy

Your solution

Solution on pp. S21–S22

8.2 Exercises

Objective A   **To solve a system of linear equations by the substitution method**

1. ✎ Describe in your own words the process of solving a system of equations by the substitution method.

2. ✎ When you solve a system of equations by the substitution method, how do you determine whether the system of equations is dependent?

For Exercises 3 to 32, solve by substitution.

3. $2x + 3y = 7$
   $x = 2$

4. $y = 3$
   $3x - 2y = 6$

5. $y = x - 3$
   $x + y = 5$

6. $y = x + 2$
   $x + y = 6$

7. $x = y - 2$
   $x + 3y = 2$

8. $x = y + 1$
   $x + 2y = 7$

9. $y = 4 - 3x$
   $3x + y = 5$

10. $y = 2 - 3x$
    $6x + 2y = 7$

11. $x = 3y + 3$
    $2x - 6y = 12$

12. $x = 2 - y$
    $3x + 3y = 6$

13. $3x + 5y = -6$
    $x = 5y + 3$

14. $y = 2x + 3$
    $4x - 3y = 1$

15. $3x + y = 4$
    $4x - 3y = 1$

16. $x - 4y = 9$
    $2x - 3y = 11$

17. $3x - y = 6$
    $x + 3y = 2$

18. $4x - y = -5$
    $2x + 5y = 13$

19. $3x - y = 5$
    $2x + 5y = -8$

20. $3x + 4y = 18$
    $2x - y = 1$

21. $4x + 3y = 0$
    $2x - y = 0$

22. $5x + 2y = 0$
    $x - 3y = 0$

23. $2x - y = 2$
    $6x - 3y = 6$

24. $3x + y = 4$
$9x + 3y = 12$

25. $x = 3y + 2$
$y = 2x + 6$

26. $x = 4 - 2y$
$y = 2x - 13$

27. $y = 2x + 11$
$y = 5x - 19$

28. $y = 2x - 8$
$y = 3x - 13$

29. $y = -4x + 2$
$y = -3x - 1$

30. $x = 3y + 7$
$x = 2y - 1$

31. $x = 4y - 2$
$x = 6y + 8$

32. $x = 3 - 2y$
$x = 5y - 10$

> **Objective B**   **To solve investment problems**

33. An investment of $3500 is divided between two simple interest accounts. On one account, the annual simple interest rate is 5%, and on the second account, the annual simple interest rate is 7.5%. How much should be invested in each account so that the total interest from the two accounts is $215?

34. A mortgage broker purchased two trust deeds for a total of $250,000. One trust deed earns 7% simple annual interest, and the second one earns 8% simple annual interest. If the total annual interest from the two trust deeds is $18,500, what was the purchase price of each trust deed?

35. When Sara Whitehorse changed jobs, she rolled over the $6000 in her retirement account into two simple interest accounts. On one account, the annual simple interest rate is 9%; on the second account, the annual simple interest rate is 6%. How much must be invested in each account if the accounts earn the same amount of annual interest?

36. An animal trainer decided to take the $15,000 won on a game show and deposit it in two simple interest accounts. Part of the winnings were placed in an account paying 7% annual simple interest, and the remainder was used to purchase a government bond that earns 6.5% annual simple interest. The amount of interest earned for 1 year was $1020. How much was invested in each account?

37. A police officer has chosen a high-yield stock fund that earns 8% annual simple interest for part of a $6000 investment. The remaining portion is used to purchase a preferred stock that earns 11% annual simple interest. How much should be invested in each account so that the amount earned on the 8% account is twice the amount earned on the 11% account?

38. To plan for the purchase of a new car, a deposit was made into an account that earns 7% annual simple interest. Another deposit, $1500 less than the first deposit, was placed in a certificate of deposit earning 9% annual simple interest. The total interest earned on both accounts for 1 year was $505. How much money was deposited in the certificate of deposit?

39. The Pacific Investment Group invested some money in a certificate of deposit (CD) that earns 6.5% annual simple interest. Twice the amount invested at 6.5% was invested in a second CD that earns 8.5% annual simple interest. If the total annual interest earned from the two investments was $4935, how much was invested at 6.5%?

40. A corporation gave a university $300,000 to support product safety research. The university deposited some of the money in a 10% simple interest account and the remainder in an 8.5% simple interest account. How much should be deposited in each account so that the annual interest earned is $28,500?

41. Ten co-workers formed an investment club, and each deposited $2000 in the club's account. They decided to take the total amount and invest some of it in preferred stock that pays 8% annual simple interest and the remainder in a municipal bond that pays 7% annual simple interest. The amount of interest earned each year from the investments was $1520. How much was invested in each?

42. A financial consultant advises a client to invest part of $30,000 in municipal bonds that earn 6.5% annual simple interest and the remainder of the money in 8.5% corporate bonds. How much should be invested in each so that the total interest earned each year is $2190?

43. Alisa Rhodes placed some money in a real estate investment trust that earns 7.5% annual simple interest. A second investment, which was one-half the amount placed in the real estate investment trust, was used to purchase a trust deed that earns 9% annual simple interest. If the total annual interest earned from the two investments was $900, how much was invested in the trust deed?

APPLYING THE CONCEPTS

44. Business   Suppose a breadmaker costs $180 and that the ingredients and electricity needed to make one loaf of bread cost $.95. If a comparable loaf of bread at a grocery store costs $1.55, how many loaves of bread must you make before the breadmaker pays for itself?

45. Business   Suppose a natural gas clothes dryer costs $240 and uses $.45 worth of gas to dry a load of clothes for one hour. If a laundromat charges $1.75 to use a dryer for one hour, how many loads of clothes must you dry before purchasing a gas dryer is more economical?

For Exercises 46 to 48, find the value of k for which the system of equations has no solution.

46. $2x - 3y = 7$
$kx - 3y = 4$

47. $8x - 4y = 1$
$2x - ky = 3$

48. $x = 4y + 4$
$kx - 8y = 4$

49. The following was offered as a solution of the system of equations

(1) $\quad y = \dfrac{1}{2}x + 2$

(2) $\quad 2x + 5y = 10$

$\begin{array}{ll} 2x + 5y = 10 & \bullet \text{ Equation (2)} \\[2mm] 2x + 5\left(\dfrac{1}{2}x + 2\right) = 10 & \bullet \text{ Substitute } \dfrac{1}{2}x + 2 \text{ for } y. \\[2mm] 2x + \dfrac{5}{2}x + 10 = 10 & \bullet \text{ Solve for } x. \\[2mm] \dfrac{9}{2}x = 0 & \\[2mm] x = 0 & \end{array}$

At this point the student stated that because $x = 0$, the system of equations has no solution. If this assertion is correct, is the system of equations independent, dependent, or inconsistent? If the assertion is not correct, what is the correct solution?

50. When you solve a system of equations by the substitution method, how do you determine whether the system of equations is inconsistent?

51. Investments A sales representative invests in a stock paying a 9% dividend. A research consultant invests $5000 more than the sales representative in bonds paying 8% annual simple interest. The research consultant's income from the investment is equal to the sales representative's. Find the amount of the research consultant's investment.

52. Investments A plant manager invested $3000 more in stocks than in bonds. The stocks paid 8% annual simple interest, and the bonds paid 9.5% annual simple interest. Both investments yielded the same income. Find the total annual interest received on both investments.

53. Compound Interest The exercises in this objective were based on annual *simple* interest, r, which means that the amount of interest earned after 1 year is given by $I = Pr$. For **compound interest,** the interest earned for a certain period of time (usually daily or monthly) is added to the principal before the interest for the next period is calculated. The compound interest earned in 1 year is given by the formula $I = P\left[\left(1 + \dfrac{r}{n}\right)^n - 1\right]$, where n is the number of times per year that interest is compounded. For instance, if interest is compounded daily, then $n = 365$; if interest is compounded monthly, then $n = 12$. Suppose an investment of $5000 is made into three different accounts. The first account earns 8% annual simple interest, the second earns 8% compounded monthly ($n = 12$), and the third earns 8% compounded daily ($n = 365$). Find the amount of interest earned from each account.

8.3 Solving Systems of Equations by the Addition Method

Objective A **To solve a system of linear equations by the addition method**

Another method of solving a system of equations is called the **addition method.** This method is based on the Addition Property of Equations.

Note, for the system of equations at the right, the effect of adding Equation (2) to Equation (1). Because $2y$ and $-2y$ are opposites, adding the equations results in an equation with only one variable.

$$\begin{array}{ll}(1) & 5x + 2y = 11 \\ (2) & 3x - 2y = 13 \\ \hline & 8x + 0y = 24 \\ & 8x = 24\end{array}$$

Solving $8x = 24$ for x gives the first coordinate of the ordered-pair solution of the system of equations.

$$\frac{8x}{8} = \frac{24}{8}$$
$$x = 3$$

The second coordinate is found by substituting the value of x into Equation (1) or Equation (2) and then solving for y. Equation (1) is used here.

$$\begin{array}{ll}(1) & 5x + 2y = 11 \\ & 5(3) + 2y = 11 \\ & 15 + 2y = 11 \\ & 2y = -4 \\ & y = -2\end{array}$$

The solution is $(3, -2)$.

Sometimes adding the two equations does not eliminate one of the variables. In this case, use the Multiplication Property of Equations to rewrite one or both of the equations so that the coefficients of one variable are opposites. Then add the equations and solve for the variables.

HOW TO Solve by the addition method:
$$\begin{array}{ll}(1) & 4x + y = 5 \\ (2) & 2x - 5y = 19\end{array}$$

Multiply Equation (2) by -2. The coefficients of x will then be opposites.

$$\begin{array}{ll} & -2(2x - 5y) = -2 \cdot 19 \\ (3) & -4x + 10y = -38\end{array}$$
- Multiply Equation (2) by **−2**.
- Simplify. This is Equation (3).

Add Equation (1) to Equation (3). Then solve for y.

$$\begin{array}{ll}(1) & 4x + y = 5 \\ (3) & -4x + 10y = -38 \\ \hline & 11y = -33 \\ & y = -3\end{array}$$
- Note that the coefficients of x are opposites.
- Add the two equations.
- Solve for y.

Substitute the value of y into Equation (1) or Equation (2) and solve for x. Equation (1) is used here.

$$\begin{array}{ll}(1) & 4x + y = 5 \\ & 4x + (-3) = 5 \\ & 4x - 3 = 5 \\ & 4x = 8 \\ & x = 2\end{array}$$
- Substitute **−3** for y.
- Solve for x.

The solution is $(2, -3)$.

Sometimes each equation of a system of equations must be multiplied by a constant so that the coefficients of one of the variable terms are opposites.

HOW TO Solve by the addition method:
(1) $3x + 7y = 2$
(2) $5x - 3y = -26$

To eliminate x, multiply Equation (1) by 5 and Equation (2) by -3. Note at the right how the constants are chosen.

$5(3x + 7y) = 5 \cdot 2$

$-3(5x - 3y) = -3(-26)$

• The negative is used so that the coefficients will be opposites.

$$\begin{array}{r} 15x + 35y = 10 \\ -15x + 9y = 78 \\ \hline 44y = 88 \\ y = 2 \end{array}$$

• 5 times Equation (1)
• -3 times Equation (2)
• Add the equations.
• Solve for y.

Substitute the value of y into Equation (1) or Equation (2) and solve for x. Equation (1) is used here.

(1) $\quad 3x + 7y = 2$
$\quad\quad 3x + 7(2) = 2$ • Substitute **2** for y.
$\quad\quad 3x + 14 = 2$ • Solve for x.
$\quad\quad 3x = -12$
$\quad\quad x = -4$

The solution is $(-4, 2)$.

For the above system of equations, the value of x was determined by substitution. This value can also be determined by eliminating y from the system.

$$\begin{array}{r} 9x + 21y = 6 \\ 35x - 21y = -182 \\ \hline 44x = -176 \\ x = -4 \end{array}$$

• 3 times Equation (1)
• 7 times Equation (2)
• Add the equations.
• Solve for x.

Note that this is the same value of x as was determined by using substitution.

HOW TO Solve by the addition method:
(1) $2x - 5y = 4$
(2) $4x - 10y = 8$

Eliminate x. Multiply Equation (1) by -2.

$\quad -2(2x - 5y) = -2(4)$ • -2 times Equation (1)
(3) $\quad -4x + 10y = -8$ • This is Equation (3).

Add Equation (3) to Equation (2) and solve for y.

(2) $\quad 4x - 10y = 8$
(3) $\quad -4x + 10y = -8$
$\quad\quad\quad 0x + 0y = 0$
$\quad\quad\quad\quad\quad 0 = 0$

The equation $0 = 0$ means that the system of equations is dependent. Therefore, the solutions of the system of equations are the ordered pairs that satisfy $2x - 5y = 4$.

The graphs of the two equations in the system of equations above are shown at the left. Note that one line is on top of the other line and therefore they intersect infinitely often. The system of equations is dependent and the solutions are the ordered pairs that belong to either one of the equations.

TAKE NOTE

When you use the addition method to solve a system of equations and the result is an equation that is always true (like the one at the right), the system of equations is dependent. Compare this with the following example.

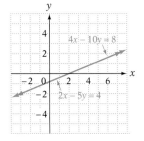

HOW TO Solve by the addition method: (1) $2x + y = 2$
(2) $4x + 2y = -5$

Eliminate y. Multiply Equation (1) by -2.

(1) $-2(2x + y) = -2 \cdot 2$ • **−2** times Equation (1)
(3) $-4x - 2y = -4$ • This is Equation (3).

Add Equation (2) to Equation (3) and solve for x.

(3) $-4x - 2y = -4$
(2) $\underline{4x + 2y = -5}$
     $0x + 0y = -9$ • Add Equation (2) to Equation (3).
        $0 = -9$ • This is not a true equation.

The system of equations is inconsistent and therefore does not have a solution.

The graphs of the two equations in the system of equations above are shown at the left. Note that the graphs are parallel and therefore do not intersect. Thus the system of equations has no solution.

Example 1

Solve by the addition method:
(1) $2x + 4y = 7$
(2) $5x - 3y = -2$

Solution
Eliminate x.
$5(2x + 4y) = 5 \cdot 7$ • **5** times Equation (1)
$-2(5x - 3y) = -2(-2)$ • **−2** times Equation (2)

$10x + 20y = 35$
$\underline{-10x + 6y = 4}$
    $26y = 39$ • Add the equations.

$y = \dfrac{39}{26} = \dfrac{3}{2}$ • Solve for y.

Substitute $\dfrac{3}{2}$ for y in Equation (1).

(1) $2x + 4y = 7$

$2x + 4\left(\dfrac{3}{2}\right) = 7$ • Replace y by $\dfrac{3}{2}$.

$2x + 6 = 7$ • Solve for x.

$2x = 1$

$x = \dfrac{1}{2}$

The solution is $\left(\dfrac{1}{2}, \dfrac{3}{2}\right)$.

You Try It 1

Solve by the addition method:
(1) $x - 2y = 1$
(2) $2x + 4y = 0$

Your solution

Solution on p. S22

Example 2

Solve by the addition method:
(1) $\quad 6x + 9y = 15$
(2) $\quad 4x + 6y = 10$

Solution

Eliminate x.

$\quad 4(6x + 9y) = 4 \cdot 15$ • **4 times Equation (1)**
$-6(4x + 6y) = -6 \cdot 10$ • **−6 times Equation (2)**

$\quad\quad 24x + 36y = 60$
$\underline{-24x - 36y = -60}$

$\quad\quad\quad 0x + 0y = 0$ • **Add the equations.**
$\quad\quad\quad\quad\quad 0 = 0$

The system of equations is dependent.
The solutions are the ordered pairs that
satisfy the equation $6x + 9y = 15$.

You Try It 2

Solve by the addition method:
$\quad 2x - 3y = 4$
$-4x + 6y = -8$

Your solution

Example 3

Solve by the addition method:
(1) $\quad\quad 2x = y + 8$
(2) $\quad 3x + 2y = 5$

Solution

Write Equation (1) in the form
$Ax + By = C$.

$\quad\quad\quad\quad 2x = y + 8$
(3) $\quad\quad 2x - y = 8$ • **This is Equation (3).**

Eliminate y.

$2(2x - y) = 2 \cdot 8$ • **2 times Equation (3).**
$\quad 3x + 2y = 5$ • **This is Equation (2).**

$\quad 4x - 2y = 16$
$\underline{\quad 3x + 2y = 5}$

$\quad\quad\quad 7x = 21$ • **Add the equations.**
$\quad\quad\quad\quad x = 3$

Replace x in Equation (1).
(1) $\quad 2x = y + 8$
$\quad 2 \cdot 3 = y + 8$ • **Replace x by 3.**
$\quad\quad 6 = y + 8$
$\quad\quad -2 = y$

The solution is $(3, -2)$.

You Try It 3

Solve by the addition method:
$4x + 5y = 11$
$\quad 3y = x + 10$

Your solution

Solutions on p. S22

8.3 Exercises

Objective A   **To solve a system of linear equations by the addition method**

For Exercises 1 to 36, solve by the addition method.

1. $x + y = 4$
$\quad x - y = 6$

2. $2x + y = 3$
$\quad x - y = 3$

3. $x + y = 4$
$\quad 2x + y = 5$

4. $x - 3y = 2$
$\quad x + 2y = -3$

5. $2x - y = 1$
$\quad x + 3y = 4$

6. $x - 2y = 4$
$\quad 3x + 4y = 2$

7. $4x - 5y = 22$
$\quad x + 2y = -1$

8. $3x - y = 11$
$\quad 2x + 5y = 13$

9. $2x - y = 1$
$\quad 4x - 2y = 2$

10. $x + 3y = 2$
$\quad 3x + 9y = 6$

11. $4x + 3y = 15$
$\quad 2x - 5y = 1$

12. $3x - 7y = 13$
$\quad 6x + 5y = 7$

13. $2x - 3y = 1$
$\quad 4x - 6y = 2$

14. $2x + 4y = 6$
$\quad 3x + 6y = 9$

15. $3x - 6y = -1$
$\quad 6x - 4y = 2$

16. $5x + 2y = 3$
$\quad 3x - 10y = -1$

17. $5x + 7y = 10$
$\quad 3x - 14y = 6$

18. $7x + 10y = 13$
$\quad 4x + 5y = 6$

19. $3x - 2y = 0$
$\quad 6x + 5y = 0$

20. $5x + 2y = 0$
$\quad 3x + 5y = 0$

21. $2x - 3y = 16$
$\quad 3x + 4y = 7$

22. $3x + 4y = 10$
$\quad 4x + 3y = 11$

23. $5x + 3y = 7$
$\quad 2x + 5y = 1$

24. $-2x + 7y = 9$
$\quad 3x + 2y = -1$

25. $3x + 4y = 4$
$5x + 12y = 5$

26. $2x + 5y = 2$
$3x + 3y = 1$

27. $8x - 3y = 11$
$6x - 5y = 11$

28. $4x - 8y = 36$
$3x - 6y = 15$

29. $5x + 15y = 20$
$2x + 6y = 12$

30. $y = 2x - 3$
$3x + 4y = -1$

31. $3x = 2y + 7$
$5x - 2y = 13$

32. $2y = 4 - 9x$
$9x - y = 25$

33. $2x + 9y = 16$
$5x = 1 - 3y$

34. $3x - 4 = y + 18$
$4x + 5y = -21$

35. $2x + 3y = 7 - 2x$
$7x + 2y = 9$

36. $5x - 3y = 3y + 4$
$4x + 3y = 11$

APPLYING THE CONCEPTS

37. Describe in your own words the process of solving a system of equations by the addition method.

38. The point of intersection of the graphs of the equations $Ax + 2y = 2$ and $2x + By = 10$ is $(2, -2)$. Find A and B.

39. The point of intersection of the graphs of the equations $Ax - 4y = 9$ and $4x + By = -1$ is $(-1, -3)$. Find A and B.

40. For what value of k is the system of equations dependent?

a. $2x + 3y = 7$
$4x + 6y = k$

b. $y = \dfrac{2}{3}x - 3$
$y = kx - 3$

c. $x = ky - 1$
$y = 2x + 2$

41. For what value of k is the system of equations inconsistent?

a. $x + y = 7$
$kx + y = 3$

b. $x + 2y = 4$
$kx + 3y = 2$

c. $2x + ky = 1$
$x + 2y = 2$

8.4 Application Problems in Two Variables

Objective A   **To solve rate-of-wind or rate-of-current problems**

We normally need two variables to solve motion problems that involve an object moving with or against a wind or current.

HOW TO   Flying with the wind, a small plane can fly 600 mi in 3 h. Against the wind, the plane can fly the same distance in 4 h. Find the rate of the plane in calm air and the rate of the wind.

> **Strategy for Solving Rate-of-Wind or Rate-of-Current Problems**
>
> Choose one variable to represent the rate of the object in calm conditions and a second variable to represent the rate of the wind or current. Using these variables, express the rate of the object with and against the wind or current. Use the equation $rt = d$ to write expressions for the distance traveled by the object. The results can be recorded in a table.

Rate of plane in calm air: p
Rate of wind: w

	Rate	·	*Time*	=	*Distance*
With the wind	$p + w$	·	3	=	$3(p + w)$
Against the wind	$p - w$	·	4	=	$4(p - w)$

> Determine how the expressions for distance are related.

The distance traveled with the wind is 600 mi. $3(p + w) = 600$
The distance traveled against the wind is 600 mi. $4(p - w) = 600$

Solve the system of equations.

$$3(p + w) = 600 \qquad \rightarrow \qquad \frac{1}{3} \cdot 3(p + w) = \frac{1}{3} \cdot 600 \qquad \rightarrow \qquad p + w = 200$$

$$4(p - w) = 600 \qquad \qquad \frac{1}{4} \cdot 4(p - w) = \frac{1}{4} \cdot 600 \qquad \qquad p - w = 150$$

$$2p = 350$$
$$p = 175$$

$$p + w = 200$$
$$175 + w = 200 \qquad \bullet\ p = \mathbf{175}$$
$$w = 25$$

The rate of the plane in calm air is 175 mph.
The rate of the wind is 25 mph.

Example 1

A 450-mile trip from one city to another takes 3 h when a plane is flying with the wind. The return trip, against the wind, takes 5 h. Find the rate of the plane in still air and the rate of the wind.

Strategy

• Rate of the plane in still air: p
  Rate of the wind: w

	Rate	Time	Distance
With wind	$p + w$	3	$3(p + w)$
Against wind	$p - w$	5	$5(p - w)$

• The distance traveled with the wind is 450 mi. The distance traveled against the wind is 450 mi.

Solution

$$3(p + w) = 450 \qquad \frac{1}{3} \cdot 3(p + w) = \frac{1}{3} \cdot 450$$

$$5(p - w) = 450 \qquad \frac{1}{5} \cdot 5(p - w) = \frac{1}{5} \cdot 450$$

$$p + w = 150$$
$$p - w = 90$$

$$2p = 240$$
$$p = 120$$

$$p + w = 150$$
$$120 + w = 150 \qquad \bullet \ p = 120$$
$$w = 30$$

The rate of the plane in still air is 120 mph.
The rate of the wind is 30 mph.

You Try It 1

A canoeist paddling with the current can travel 15 mi in 3 h. Against the current, it takes the canoeist 5 h to travel the same distance. Find the rate of the current and the rate of the canoeist in calm water.

Your strategy

Your solution

Solution on pp. S22–S23

Objective B   **To solve application problems using two variables**

The application problems in this section are varieties of those problems solved earlier in the text. Each of the strategies for the problems in this section will result in a system of equations.

HOW TO A jeweler purchased 5 oz of a gold alloy and 20 oz of a silver alloy for a total cost of $540. The next day, at the same prices per ounce, the jeweler purchased 4 oz of the gold alloy and 25 oz of the silver alloy for a total cost of $450. Find the cost per ounce of the gold and silver alloys.

> **Strategy for Solving an Application Problem in Two Variables**
>
> Choose one variable to represent one of the unknown quantities and a second variable to represent the other unknown quantity. Write numerical or variable expressions for all of the remaining quantities. These results can be recorded in two tables, one for each of the conditions.

Cost per ounce of gold: g
Cost per ounce of silver: s

First day:

	Amount	·	*Unit Cost*	=	*Value*
Gold	5	·	g	=	$5g$
Silver	20	·	s	=	$20s$

Second day:

	Amount	·	*Unit Cost*	=	*Value*
Gold	4	·	g	=	$4g$
Silver	25	·	s	=	$25s$

> Determine a system of equations. Each table will give one equation of the system.

The total value of the purchase on the first day was $540. $5g + 20s = 540$

The total value of the purchase on the second day was $450. $4g + 25s = 450$

Solve the system of equations.

$5g + 20s = 540$ $4(5g + 20s) = 4 \cdot 540$ $20g + 80s = 2160$
$4g + 25s = 450$ $-5(4g + 25s) = -5 \cdot 450$ $\underline{-20g - 125s = -2250}$
                                                          $-45s = -90$
                                                          $s = 2$

$5g + 20s = 540$
$5g + 20(2) = 540$ • $s = 2$
$5g + 40 = 540$
$5g = 500$
$g = 100$

The cost per ounce of the gold alloy was $100.
The cost per ounce of the silver alloy was $2.

Example 2

A store owner purchased 20 incandescent light bulbs and 30 fluorescent bulbs for a total cost of $40. A second purchase, at the same prices, included 30 incandescent bulbs and 10 fluorescent bulbs for a total cost of $25. Find the cost of an incandescent bulb and of a fluorescent bulb.

Strategy

Cost of an incandescent bulb: b
Cost of a fluorescent bulb: f

First purchase:

	Amount	Unit Cost	Value
Incandescent	20	b	$20b$
Fluorescent	30	f	$30f$

Second purchase:

	Amount	Unit Cost	Value
Incandescent	30	b	$30b$
Fluorescent	10	f	$10f$

The total cost of the first purchase was $40.
The total cost of the second purchase was $25.

Solution

$$20b + 30f = 40 \qquad 3(20b + 30f) = 3 \cdot 40$$
$$30b + 10f = 25 \qquad -2(30b + 10f) = -2 \cdot 25$$

$$60b + 90f = 120$$
$$\underline{-60b - 20f = -50}$$
$$70f = 70$$
$$f = 1$$

$$20b + 30f = 40$$
$$20b + 30(1) = 40 \qquad \bullet \; f = 1$$
$$20b = 10$$
$$b = \frac{1}{2}$$

The cost of an incandescent bulb was $.50.
The cost of a fluorescent bulb was $1.00.

You Try It 2

A citrus grower purchased 25 orange trees and 20 grapefruit trees for $290. The next week, at the same prices, the grower bought 20 orange trees and 30 grapefruit trees for $330. Find the cost of an orange tree and the cost of a grapefruit tree.

Your strategy

Your solution

Solution on p. S23

8.4 Exercises

Objective A   To solve rate-of-wind or rate-of-current problems

1. A whale swimming against an ocean current traveled 60 mi in 2 h. Swimming in the opposite direction, with the current, the whale was able to travel the same distance in 1.5 h. Find the speed of the whale in calm water and the rate of the ocean current.

2. A plane flying with the jet stream flew from Los Angeles to Chicago, a distance of 2250 mi, in 5 h. Flying against the jet stream, the plane could fly only 1750 mi in the same amount of time. Find the rate of the plane in calm air and the rate of the wind.

3. A rowing team rowing with the current traveled 40 km in 2 h. Rowing against the current, the team could travel only 16 km in 2 h. Find the rowing rate in calm water and the rate of the current.

4. The bird capable of the fastest flying speed is the swift. A swift flying with the wind to a favorite feeding spot traveled 26 mi in 0.2 h. On returning, now against the wind, the swift was able to travel only 16 mi in the same amount of time. What is the rate of the swift in calm air and what was the rate of the wind?

5. A private Learjet 31A was flying with a tailwind and traveled 1120 mi in 2 h. Flying against the wind on the return trip, the jet was able to travel only 980 mi in 2 h. Find the speed of the jet in calm air and the rate of the wind.

6. A plane flying with a tailwind flew 300 mi in 2 h. Against the wind, it took 3 h to travel the same distance. Find the rate of the plane in calm air and the rate of the wind.

7. A Boeing Apache Longbow military helicopter traveling directly into a strong headwind was able to travel 450 mi in 2.5 h. The return trip, now with a tailwind, took 1 h 40 min. Find the speed of the helicopter in calm air and the rate of the wind.

8. A seaplane pilot flying with the wind flew from an ocean port to a lake, a distance of 240 mi, in 2 h. Flying against the wind, the pilot flew from the lake to the ocean port in 2 h 40 min. Find the rate of the plane in calm air and the rate of the wind.

9. Rowing with the current, a canoeist paddled 14 mi in 2 h. Against the current, the canoeist could paddle only 10 mi in the same amount of time. Find the rate of the canoeist in calm water and the rate of the current.

Objective B   To solve application problems using two variables

10. **Internet Services**   A computer online service charges one hourly price for regular use but a higher hourly rate for designated "premium" areas. One customer was charged $28 after spending 2 h in premium areas and 9 regular hours; another spent 3 h in the premium areas and 6 regular hours and was charged $27. What does the online service charge per hour for regular and premium services?

11. **Flour Mixtures**   A baker purchased 12 lb of wheat flour and 15 lb of rye flour for a total cost of $18.30. A second purchase, at the same prices, included 15 lb of wheat flour and 10 lb of rye flour. The cost of the second purchase was $16.75. Find the cost per pound of the wheat flour and of the rye flour.

12. Investments An investor owned 300 shares of an oil company and 200 shares of a movie company. The quarterly dividend from the two stocks was $165. After the investor sold 100 shares of the oil company and bought an additional 100 shares of the movie company, the quarterly dividend became $185. Find the dividend per share for each stock.

13. Chemistry A lab technician mixed 40 ml of reagent I with 60 ml of reagent II to produce a solution that was 31% hydrochloric acid. A second solution, which was 31.5% hydrochloric acid, was created by mixing 35 ml of reagent I with 65 ml of reagent II. What is the percent hydrochloric acid in the two reagents?

14. Food Mixtures A pastry chef created a 50-ounce sugar solution that was 34% sugar from a 20% sugar solution and a 40% sugar solution. How much of the 20% sugar solution and how much of the 40% sugar solution were used?

15. Fuel Mixtures The octane number of 87 on gasoline means that it will fight engine "knock" as effectively as a reference fuel that is 87% isooctane, a type of gas. Suppose you want to fill an empty 18-gallon tank with some 87-octane gasoline and some 93-octane fuel to produce a mixture that is 89-octane. How much of each type of gasoline must you use?

APPLYING THE CONCEPTS

16. Geometry Two angles are supplementary. The measure of the larger angle is 15° more than twice the measure of the smaller angle. Find the measures of the two angles. (Supplementary angles are two angles whose sum is 180°.)

17. Coin Problem The value of the nickels and dimes in a coin bank is $.25. If the number of nickels and the number of dimes were doubled, the value of the coins would be $.50. How many nickels and how many dimes are in the bank?

18. Investments An investor has $5000 to invest in two accounts. The first account earns 8% annual simple interest, and the second account earns 10% annual simple interest. How much money should be invested in each account so that the annual simple interest earned is $600?

19. Ancient Problem Solve the following problem, which dates from a Chinese manuscript called the Jinzhang that is approximately 2100 years old. "The price of 1 acre of good land is 300 pieces of gold; the price of 7 acres of bad land is 500 pieces of gold. One has purchased altogether 100 acres. The price was 10,000 pieces of gold. How much good land and how much bad land was bought?" Adapted from Victor J. Katz, *A History of Mathematics, An Introduction* (New York: Harper-Collins, 1993), p. 15.

20. Coin Problem A coin bank contains only nickels or dimes, but there are no more than 27 coins. The value of the coins is $2.10. How many different combinations of nickels and dimes could be in the coin bank?

Focus on Problem Solving

Using a Table and Searching for a Pattern

Consider the numbers 10, 12, and 28 and the sum of the proper factors (the natural number factors less than the number) of those numbers.

$$10: 1 + 2 + 5 = 8 \qquad 12: 1 + 2 + 3 + 4 + 6 = 16 \qquad 28: 1 + 2 + 4 + 7 + 14 = 28$$

10 is called a **deficient number** because the sum of its proper factors is less than the number ($8 < 10$). 12 is called an **abundant number** because the sum of its proper factors is greater than the number ($16 > 12$), and 28 is called a **perfect number** because the sum of its proper divisors equals the number ($28 = 28$).

Our goal for this Focus on Problem Solving is to try to find a method that will determine whether a number is deficient, abundant, or perfect without having to first find all the factors and then add them up. We will use a table and search for a pattern.

Before we begin, recall that a prime number is a number greater than 1 whose only factors are itself and 1, and that each natural number greater than 1 has a unique prime factorization. For instance, the prime factorization of 36 is given by $36 = 2^2 \cdot 3^2$. Note that the proper factors of 36 (1, 2, 3, 4, 6, 9, 12, 18) can be represented in terms of the same prime numbers.

$$1 = 2^0, \quad 2 = 2^1, \quad 3 = 3^1, \quad 4 = 2^2, \quad 6 = 2 \cdot 3, \quad 9 = 3^2, \quad 12 = 2^2 \cdot 3, \quad 18 = 2 \cdot 3^2$$

Now let us consider a trial problem of determining whether 432 is deficient, abundant, or perfect.

TAKE NOTE

This table contains the factor $432 = 2^4 \cdot 3^3$, which is not a proper factor.

We write the prime factorization of 432 as $2^4 \cdot 3^3$ and place the factors of 432 in a table as shown at the right. This table contains *all* the factors of 432 represented in terms of the prime number factors. The sum of each column is shown at the bottom.

	1	$1 \cdot 3$	$1 \cdot 3^2$	$1 \cdot 3^3$
	2	$2 \cdot 3$	$2 \cdot 3^2$	$2 \cdot 3^3$
	2^2	$2^2 \cdot 3$	$2^2 \cdot 3^2$	$2^2 \cdot 3^3$
	2^3	$2^3 \cdot 3$	$2^3 \cdot 3^2$	$2^3 \cdot 3^3$
	2^4	$2^4 \cdot 3$	$2^4 \cdot 3^2$	$2^4 \cdot 3^3$
Sum	31	$31 \cdot 3$	$31 \cdot 3^2$	$31 \cdot 3^3$

Here is the calculation of the sum for the column headed by $1 \cdot 3$.

$$1 \cdot 3 + 2 \cdot 3 + 2^2 \cdot 3 + 2^3 \cdot 3 + 2^4 \cdot 3 = (1 + 2 + 2^2 + 2^3 + 2^4)3 = 31(3)$$

For the column headed by $1 \cdot 3^2$ there is a similar situation.

$$1 \cdot 3^2 + 2 \cdot 3^2 + 2^2 \cdot 3^2 + 2^3 \cdot 3^2 + 2^4 \cdot 3^2 = (1 + 2 + 2^2 + 2^3 + 2^4)3^2 = 31(3^2)$$

The sum of *all* the factors (including 432) is the sum of the last row.

$$\text{Sum of all factors} = 31 + 31 \cdot 3 + 31 \cdot 3^2 + 31 \cdot 3^3 = 31(1 + 3 + 3^2 + 3^3) = 31(40) = 1240$$

To find the sum of the *proper* factors, we must subtract 432 from 1240; we get 808 (see Take Note). Thus 432 is abundant.

We now look for some pattern for the sum of all the factors. Note that

$$\text{Sum of left column} \quad 1 + 2 + 2^2 + 2^3 + 2^4 = \underbrace{\qquad}\underbrace{\qquad} = 1 + 3 + 3^2 + 3^3 \quad \text{Sum of top row}$$

$$31(40) = 1240$$

This suggests that the sum of the proper factors can be found by finding the sum of all the prime power factors for each prime, multiplying those numbers, and then subtracting the original number. Although we have not proved this for all cases, it is a true statement.

For instance, to find the sum of the proper factors of 3240, first find the prime factorization.

$$3240 = 2^3 \cdot 3^4 \cdot 5$$

Now find the following sums:

$$1 + 2 + 2^2 + 2^3 = 15 \qquad 1 + 3 + 3^2 + 3^3 + 3^4 = 121 \qquad 1 + 5 = 6$$

The sum of the proper factors = $(15)(121)6 - 3240 = 7650$. Thus 3240 is abundant.

Determine whether the number is deficient, abundant, or perfect.

1. 200   **2.** 3125   **3.** 8128   **4.** 10,000

5. Is a prime number deficient, abundant, or perfect?

Projects and Group Activities

Finding a Pattern   The Focus on Problem Solving involved finding the sum of $1 + 3 + 3^2$ and $1 + 2 + 2^2 + 2^3$. For sums that contain a larger number of terms, it may be difficult or time consuming to try to evaluate the sum. Perhaps there is a pattern for these sums that we can use to calculate them without having to evaluate each exponential expression and then add the results.

Look at the following from the calculation of the sum of the factors of 3240.

$$1 + 2 + 2^2 + 2^3 = 1 + 2 + 4 + 8 = 15$$

$$\frac{2^{3+1} - 1}{2 - 1} = \frac{2^4 - 1}{1} = 16 - 1 = 15$$

$$1 + 3 + 3^2 + 3^3 + 3^4 = 1 + 3 + 9 + 27 + 81 = 121$$

$$\frac{3^{4+1} - 1}{3 - 1} = \frac{3^5 - 1}{2} = \frac{243 - 1}{2} = \frac{242}{2} = 121$$

Consider another sum of this type.

$$1 + 7 + 7^2 + 7^3 + 7^4 + 7^5 = 1 + 7 + 49 + 343 + 2401 + 16{,}807 = 19{,}608$$

$$\frac{7^{5+1} - 1}{7 - 1} = \frac{7^6 - 1}{6} = \frac{117{,}649 - 1}{6} = \frac{117{,}648}{6} = 19{,}608$$

On the basis of the examples shown above, make a conjecture about the value of $1 + n + n^2 + n^3 + n^4 + \cdots + n^k$, where n and k are natural numbers greater than 1.

Solving a System of Equations with a Graphing Calulator

A graphing calculator can be used to approximate the solution of a system of equations in two variables. Graph each equation of the system of equations, and then approximate the coordinates of the point of intersection. The process by which you approximate the solution depends on what model of calculator you have. In all cases, however, you must first solve each equation in the system of equations for y.

Solve: $2x - 5y = 9$
$4x + 3y = 2$

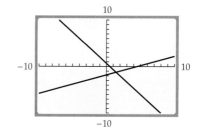

• Solve each equation for *y*.

$$2x - 5y = 9 \qquad\qquad 4x + 3y = 2$$
$$-5y = -2x + 9 \qquad\quad 3y = -4x + 2$$
$$y = \frac{2}{5}x - \frac{9}{5} \qquad\qquad y = -\frac{4}{3}x + \frac{2}{3}$$

TAKE NOTE

The graphing calculator screens shown here are taken from a TI-83 Plus. Similar screens would appear if we used a different model graphing calculator.

For the TI-83 and TI-83 Plus, press ⬛Y= . Enter one equation as Y1 and the other as Y2. The result should be similar to the screen at the left below. Press ⬛GRAPH . The graphs of the two equations should appear on the screen, as shown at the right below. If the point of intersection is not on the screen, adjust the viewing window by pressing the ⬛WINDOW key.

```
Plot1 Plot2 Plot3
\Y1 ▤ 2X/5−9/5
\Y2 ▤ -4X/3+2/3
\Y3 =
\Y4 =
\Y5 =
\Y6 =
\Y7 =
```

Integrating *Technology*

See the Keystroke Guide: *Intersect* for instructions on using a graphing calculator to solve systems of equations.

Press ⬛2nd CALC 5 ⬛ENTER ⬛ENTER ⬛ENTER . After a few seconds, the point of intersection will show on the bottom of the screen as X = 1.4230769, Y = −1.230769.

For Exercises 1 to 4, solve by using a graphing calculator.

1. $4x - 5y = 8$
$ 5x + 7y = 7$

2. $3x + 2y = 11$
$ 7x - 6y = 13$

3. $x = 3y + 2$
$ y = 4x - 2$

4. $x = 2y - 5$
$ x = 3y + 2$

Chapter 8 Summary

Key Words

Two or more equations considered together are called a *system of equations.* [8.1A, p. 405]

A *solution of a system of equations in two variables* is an ordered pair that is a solution of each equation of the system. [8.1A, p. 405]

An *independent system* of linear equations has exactly one solution. The graphs of the equations in an independent system of linear equations intersect at one point. [8.1A, p. 406]

Examples

An example of a system of equations is
$$2x - 3y = 9$$
$$3x + 4y = 5$$

The solution of the system of equations shown above is the ordered pair $(3, -1)$ because it is a solution of each equation of the system of equations.

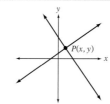

A *dependent system* of linear equations has an infinite number of solutions. The graphs of the equations in a dependent system of linear equations are the same line. [8.1A, p. 406]

If, when solving a system of equations algebraically, the variable is eliminated and the result is a true equation, such as $5 = 5$, the system of equations is dependent. [8.2A, p. 415]

An *inconsistent system* of linear equations has no solution. The graphs of the equations of an inconsistent system of linear equations are parallel lines. [8.1A, p. 406]

If, when solving a system of equations algebraically, the variable is eliminated and the result is a false equation, such as $0 = 4$, the system of equations is inconsistent. [8.2A, p. 415]

Essential Rules and Procedures

Examples

To solve a system of linear equations in two variables by graphing, graph each equation on the same coordinate system, and then determine the points of intersection. [8.1A, p. 407]

Solve by graphing: $x + 2y = 4$
$$2x + y = -1$$

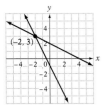

The solution is $(-2, 3)$.

To solve a system of equations by the substitution method, one variable must be written in terms of the other variable. [8.2A, p. 413]

Solve by substitution: $2x + y = 5$ (1)
$$3x - 2y = 11 \quad (2)$$
$2x + y = 5$
$\quad y = -2x + 5$ • Solve Equation (1) for y.

$3x - 2y = 11$
$3x - 2(-2x + 5) = 11$ • Substitute for y
$\quad 3x + 4x - 10 = 11$ in Equation (2).
$\quad\quad 7x - 10 = 11$
$\quad\quad\quad\quad 7x = 21$
$\quad\quad\quad\quad\quad x = 3$
$y = -2x + 5$
$y = -2(3) + 5$
$y = -1$
The solution is $(3, -1)$.

To solve a system of linear equations by the addition method, use the Multiplication Property of Equations to rewrite one or both of the equations so that the coefficients of one variable are opposites. Then add the equations and solve for the variables. [8.3A, p. 423]

Solve by the addition method:
$2x + 5y = 8$ (1)
$3x - 4y = -11$ (2)
$\quad 6x + 15y = 24$ • 3 times Equation (1)
$\underline{-6x + 8y = 22}$ • −2 times Equation (2)
$\quad\quad\quad 23y = 46$ • Add the equations.
$\quad\quad\quad\quad y = 2$ • Solve for y.
$2x + 5y = 8$
$2x + 5(2) = 8$ • Replace y by 2 in Equation (1).
$2x + 10 = 8$ • Solve for x.
$\quad\quad 2x = -2$
$\quad\quad\quad x = -1$
The solution is $(-1, 2)$.

Chapter 8 Review Exercises

1. Is $(-1, -3)$ a solution of this system of equations? $5x + 4y = -17$
$$2x - y = 1$$

2. Is $(-2, 0)$ a solution of this system of equations? $-x + 9y = 2$
$$6x - 4y = 12$$

3. Solve by graphing:
$3x - y = 6$
$$y = -3$$

4. Solve by graphing:
$4x - 2y = 8$
$$y = 2x - 4$$

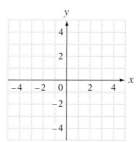

5. Solve by graphing:
$x + 2y = 3$
$$y = -\frac{1}{2}x + 1$$

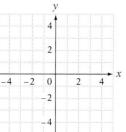

6. Solve by substitution:
$4x + 7y = 3$
$$x = y - 2$$

7. Solve by substitution:
$6x - y = 0$
$$7x - y = 1$$

8. Solve by the addition method:
$3x + 8y = -1$
$$x - 2y = -5$$

9. Solve by the addition method:
$6x + 4y = -3$
$$12x - 10y = -15$$

10. Solve by substitution:
$12x - 9y = 18$
$$y = \frac{4}{3}x - 3$$

11. Solve by substitution:
$8x - y = 2$
$$y = 5x + 1$$

12. Solve by the addition method:
$4x - y = 9$
$$2x + 3y = -13$$

13. Solve by the addition method:
$5x + 7y = 21$
$$20x + 28y = 63$$

14. Solve by substitution:
$4x + 3y = 12$
$$y = -\frac{4}{3}x + 4$$

15. Solve by substitution:
$$7x + 3y = -16$$
$$x - 2y - 5$$

16. Solve by the addition method:
$$3x + y = -2$$
$$-9x - 3y = 6$$

17. Solve by the addition method:
$$6x - 18y = 7$$
$$9x + 24y = 2$$

18. **Sculling**   A sculling team rowing with the current went 24 mi in 2 h. Rowing against the current, the sculling team went 18 mi in 3 h. Find the rate of the sculling team in calm water and the rate of the current.

19. **Investments**   An investor bought 1500 shares of stock, some at $6 per share and the rest at $25 per share. If $12,800 worth of stock was purchased, how many shares of each kind did the investor buy?

20. **Travel**   A flight crew flew 420 km in 3 h with a tailwind. Flying against the wind, the flight crew flew 440 km in 4 h. Find the rate of the flight crew in calm air and the rate of the wind.

21. **Travel**   A small plane flying with the wind flew 360 mi in 3 h. Against a headwind, the plane took 4 h to fly the same distance. Find the rate of the plane in calm air and the rate of the wind.

22. **Postage**   A small wood-carving company mailed 190 advertisements, some requiring $.25 postage and others requiring $.45. The total cost for mailing was $59.50. Find the number of advertisements mailed at each rate.

23. **Investments**   Terra Cotta Art Center receives an annual income of $915 from two simple interest investments. One investment, in a corporate bond fund, earns 8.5% annual simple interest. The second investment, in a real estate investment trust, earns 7% annual simple interest. If the total amount invested in the two accounts is $12,000, how much is invested in each account?

24. **Grain Mixtures**   A silo contains a mixture of lentils and corn. If 50 bushels of lentils were added, there would be twice as many bushels of lentils as of corn. If 150 bushels of corn were added instead, there would be the same amount of corn as of lentils. How many bushels of each were originally in the silo?

25. **Investments**   Mosher Children's Hospital received a $300,000 donation that it invested in two simple interest accounts, one earning 5.4% and the other earning 6.6%. If each account earned the same amount of annual interest, how much was invested in each account?

Chapter 8 Test

1. Is $(-2, 3)$ a solution of this system?
$2x + 5y = 11$
$x + 3y = 7$

2. Is $(1, -3)$ a solution of this system?
$3x - 2y = 9$
$4x + y = 1$

3. Solve by graphing: $3x + 2y = 6$
$5x + 2y = 2$

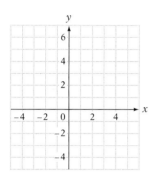

4. Solve by substitution:
$4x - y = 11$
$y = 2x - 5$

5. Solve by substitution:
$x = 2y + 3$
$3x - 2y = 5$

6. Solve by substitution:
$3x + 5y = 1$
$2x - y = 5$

7. Solve by substitution:
$3x - 5y = 13$
$x + 3y = 1$

8. Solve by substitution:
$2x - 4y = 1$
$y = \dfrac{1}{2}x + 3$

9. Solve by the addition method:
$4x + 3y = 11$
$5x - 3y = 7$

10. Solve by the addition method:
$2x - 5y = 6$
$4x + 3y = -1$

11. Solve by the addition method:

$x + 2y = 8$

$3x + 6y = 24$

12. Solve by the addition method:

$7x + 3y = 11$

$2x - 5y = 9$

13. Solve by the addition method:

$5x + 6y = -7$

$3x + 4y = -5$

14. Travel   With the wind, a plane flies 240 mi in 2 h. Against the wind, the plane requires 3 h to fly the same distance. Find the rate of the plane in calm air and the rate of the wind.

15. Entertainment   For the first performance of a play in a community theater, 50 reserved-seat tickets and 80 general-admission tickets were sold. The total receipts were $980. For the second performance, 60 reserved-seat tickets and 90 general-admission tickets were sold. The total receipts were $1140. Find the price of a reserved-seat ticket and the price of a general-admission ticket.

16. Investments   Bernardo Community Library received a $28,000 donation that it invested in two accounts, one earning 7.6% simple interest and the other earning 6.4% simple interest. If both accounts earned the same amount of annual interest, how much was invested in each account?

Cumulative Review Exercises

1. Evaluate $\dfrac{a^2 - b^2}{2a}$ when $a = 4$ and $b = -2$.

2. Solve: $-\dfrac{3}{4}x = \dfrac{9}{8}$

3. Given $f(x) = x^2 + 2x - 1$, find $f(2)$.

4. Multiply: $(2a^2 - 3a + 1)(2 - 3a)$

5. Simplify: $\dfrac{(-2x^2y)^4}{-8x^3y^2}$

6. Divide: $(4b^2 - 8b + 4) \div (2b - 3)$

7. Simplify: $\dfrac{8x^{-2}y^5}{-2xy^4}$

8. Factor: $4x^2y^4 - 64y^2$

9. Solve: $(x - 5)(x + 2) = -6$

10. Divide: $\dfrac{x^2 - 6x + 8}{2x^3 + 6x^2} \div \dfrac{2x - 8}{4x^3 + 12x^2}$

11. Add: $\dfrac{x - 1}{x + 2} + \dfrac{2x + 1}{x^2 + x - 2}$

12. Simplify: $\dfrac{x + 4 - \dfrac{7}{x - 2}}{x + 8 + \dfrac{21}{x - 2}}$

13. Solve: $\dfrac{x}{2x - 3} + 2 = \dfrac{-7}{2x - 3}$

14. Solve $A = P + Prt$ for r.

15. Find the x- and y-intercepts for $2x - 3y = 12$.

16. Find the slope of the line that passes through the points $(2, -3)$ and $(-3, 4)$.

17. Find the equation of the line that passes through the point $(-2, 3)$ and has slope $-\dfrac{3}{2}$.

18. Is $(2, 0)$ a solution of this system?
$$5x - 3y = 10$$
$$4x + 7y = 8$$

19. Solve by substitution:
$$3x - 5y = -23$$
$$x + 2y = -4$$

20. Solve by the addition method:
$$5x - 3y = 29$$
$$4x + 7y = -5$$

21. **Investments**   A total of $8750 is invested in two accounts. On one account, the annual simple interest rate is 9.6%; on the second account, the annual simple interest rate is 7.2%. How much should be invested in each account so that both accounts earn the same interest?

22. **Travel**   A passenger train leaves a train depot $\frac{1}{2}$ h after a freight train leaves the same depot. The freight train is traveling 8 mph slower than the passenger train. Find the rate of each train if the passenger train overtakes the freight train in 3 h.

23. **Geometry**   The length of each side of a square is extended 4 in. The area of the resulting square is 144 in². Find the length of a side of the original square.

24. **Travel**   A plane can travel 160 mph in calm air. Flying with the wind, the plane can fly 570 mi in the same amount of time as it takes to fly 390 mi against the wind. Find the rate of the wind.

25. Graph $2x - 3y = 6$.

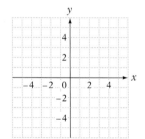

26. Solve by graphing: $3x + 2y = 6$
$3x - 2y = 6$

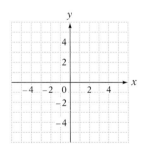

27. **Travel**   With the current, a motorboat can travel 48 mi in 3 h. Against the current, the boat requires 4 h to travel the same distance. Find the rate of the boat in calm water.

28. **Food Mixtures**   A child adds 8 g of sugar to a 50-gram serving of a breakfast cereal that is 25% sugar. What is now the percent concentration of sugar in the mixture? Round to the nearest tenth of a percent.

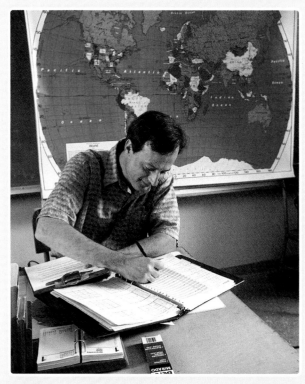

You know how to determine the average of four exam scores when each exam has the same weight: add the four scores and divide the sum by 4. But do you know how to determine what score you must receive on a fifth exam to earn a specific average on the five exams? In this chapter you will learn how to use an inequality to determine that fifth and final score. **Exercises 86 and 87 on page 460** give you different scores for the first four tests, and ask you to determine the score needed for the fifth test in order to achieve a specific average.

OBJECTIVES

Section 9.1

A To write a set using the roster method
B To write a set using set-builder notation
C To graph an inequality on the number line

Section 9.2

A To solve an inequality using the Addition Property of Inequalities
B To solve an inequality using the Multiplication Property of Inequalities
C To solve application problems

Section 9.3

A To solve general inequalities
B To solve application problems

Section 9.4

A To graph an inequality in two variables

Need help? For online student resources, such as section quizzes, visit this textbook's website at **math.college.hmco.com/students.**

Do these exercises to prepare for Chapter 9.

1. Place the correct symbol, $<$ or $>$, between the two numbers.

$-45 \quad -27$

2. Simplify: $3x - 5(2x - 3)$

3. State the Addition Property of Equations.

4. State the Multiplication Property of Equations.

5. **Nutrition** A certain grade of hamburger contains 15% fat. How many pounds of fat are in 3 lb of this hamburger?

6. Solve: $4x - 5 = -7$

7. Solve: $4 = 2 - \dfrac{3}{4}x$

8. Solve: $7 - 2(2x - 3) = 3x - 1$

9. Graph: $y = \dfrac{2}{3}x - 3$

10. Graph: $3x + 4y = 12$

GO FIGURE • • •

Without using a calculator, which is the largest number: 2^{150}, 3^{100}, or 5^{50}?

9.1 Sets

Objective A   **To write a set using the roster method**

Recall that a *set* is a collection of objects, which are called the *elements* of the set. The roster method of writing a set encloses a list of the elements in braces.

The set of the last three letters of the alphabet is written {x, y, z}.

The set of the positive integers less than 5 is written {1, 2, 3, 4}.

HOW TO   Use the roster method to write the set of integers between 0 and 10.

$A = \{1, 2, 3, 4, 5, 6, 7, 8, 9\}$ • **A set can be designated by a capital letter. Note that 0 and 10 are not elements of the set.**

HOW TO   Use the roster method to write the set of natural numbers.

$A = \{1, 2, 3, 4, \ldots\}$ • **The three dots mean that the pattern of numbers continues without end.**

The **empty set,** or **null set,** is the set that contains no elements. The symbol ∅ or { } is used to represent the empty set.

The set of people who have run a 2-minute mile is an empty set.

Union of Two Sets

The **union** of two sets, written $A \cup B$, is the set that contains the elements of A and the elements of B.

HOW TO   Find $A \cup B$, given $A = \{1, 2, 3, 4\}$ and $B = \{3, 4, 5, 6\}$.

$A \cup B = \{1, 2, 3, 4, 5, 6\}$ • **The union of A and B contains all the elements of A and all the elements of B. Elements in both sets are listed only once.**

Intersection of Two Sets

The **intersection** of two sets, written $A \cap B$, is the set that contains the elements that are common to both A and B.

HOW TO   Find $A \cap B$, given $A = \{1, 2, 3, 4\}$ and $B = \{3, 4, 5, 6\}$.

$A \cap B = \{3, 4\}$ • **The intersection of A and B contains the elements common to A and B.**

Example 1

Use the roster method to write the set of the odd positive integers less than 12.

Solution
$A = \{1, 3, 5, 7, 9, 11\}$

You Try It 1

Use the roster method to write the set of the odd negative integers greater than -10.

Your solution

Solution on p. S23

Example 2

Use the roster method to write the set of the even positive integers.

Solution
$A = \{2, 4, 6, \ldots\}$

You Try It 2

Use the roster method to write the set of the odd positive integers.

Your solution

Example 3

Find $D \cup E$, given
$D = \{6, 8, 10, 12\}$ and
$E = \{-8, -6, 10, 12\}$.

Solution
$D \cup E = \{-8, -6, 6, 8, 10, 12\}$

You Try It 3

Find $A \cup B$, given
$A = \{-2, -1, 0, 1, 2\}$ and
$B = \{0, 1, 2, 3, 4\}$.

Your solution

Example 4

Find $A \cap B$, given
$A = \{5, 6, 9, 11\}$ and
$B = \{5, 9, 13, 15\}$.

Solution
$A \cap B = \{5, 9\}$

You Try It 4

Find $C \cap D$, given
$C = \{10, 12, 14, 16\}$ and
$D = \{10, 16, 20, 26\}$.

Your solution

Example 5

Find $A \cap B$, given
$A = \{1, 2, 3, 4\}$ and
$B = \{8, 9, 10, 11\}$.

Solution
$A \cap B = \varnothing$

You Try It 5

Find $A \cap B$, given
$A = \{-5, -4, -3, -2\}$ and
$B = \{2, 3, 4, 5\}$.

Your solution

Solutions on p. S23

Objective B   **To write a set using set-builder notation**

Point of Interest

The symbol $\in$ was first used in the book *Arithmeticae Principia*, published in 1889. It is the first letter of the Greek word $\varepsilon\sigma\tau\iota$ which means "is." The symbols for union and intersection were also introduced at that time.

Another method of representing sets is called **set-builder notation.** This method of writing a set uses a rule to describe the elements of the set. Using set-builder notation, we represent the set of all positive integers less than 10 as

$\{x \,|\, x < 10, x \in \text{positive integers}\}$, which is read "the set of all x such that x is less than 10 and x is an element of the positive integers."

HOW TO   Use set-builder notation to write the set of real numbers greater than 4.

$\{x \,|\, x > 4, x \in \text{real numbers}\}$ • "$x \in$ real numbers" is read "x is an element of the real numbers."

Example 6

Use set-builder notation to write the set of negative integers greater than -100.

Solution

$\{x \mid x > -100, x \in \text{negative integers}\}$

You Try It 6

Use set-builder notation to write the set of positive even integers less than 59.

Your solution

Example 7

Use set-builder notation to write the set of real numbers less than 60.

Solution

$\{x \mid x < 60, x \in \text{real numbers}\}$

You Try It 7

Use set-builder notation to write the set of real numbers greater than -3.

Your solution

Solutions on p. S23

Objective C **To graph an inequality on the number line**

An expression that contains the symbol $>$, $<$, $\geq$ (is greater than or equal to), or $\leq$ (is less than or equal to) is called an **inequality.** An inequality expresses the relative order of two mathematical expressions. The expressions can be either numerical or variable.

$$4 > 2$$
$$3x \leq 7$$
$$x^2 - 2x > y + 4$$
Inequalities

The **solution set of an inequality** is the set of numbers each element of which, when substituted for the variable, results in a true inequality. The solution set of an inequality can be graphed on the number line.

TAKE NOTE

For the remainder of this section, all variables will represent real numbers. Given this convention, the expression $\{x \mid x > 1, x \in \text{real numbers}\}$ will be written as $\{x \mid x > 1\}$, as shown in the example at the right.

HOW TO Graph: $\{x \mid x > 1\}$

The graph is the real numbers greater than 1. The parenthesis at 1 indicates that 1 is not included in the graph.

HOW TO Graph: $\{x \mid x \geq 1\}$

The bracket at 1 indicates that 1 is included in the solution set.

HOW TO Graph: $\{x \mid -1 > x\}$

$-1 > x$ is equivalent to $x < -1$. The numbers less than -1 are to the left of -1 on the number line.

The union of two sets is the set that contains all the elements of each set.

HOW TO Graph: $\{x \mid x > 4\} \cup \{x \mid x < 1\}$

The graph is the numbers greater than 4 and the numbers less than 1.

The intersection of two sets is the set that contains the elements common to both sets.

> **HOW TO** Graph: $\{x | x > -1\} \cap \{x | x < 2\}$
>
> The graphs of $\{x | x > -1\}$ and $\{x | x < 2\}$ are shown at the right.
>
>
>
>
> The graph of $\{x | x > -1\} \cap \{x | x < 2\}$ is the numbers between -1 and 2.
>
>

Example 8

Graph: $\{x | x < 5\}$

Solution
The solution set is the numbers less than 5.

You Try It 8

Graph: $\{x | -2 < x\}$

Your solution

Example 9

Graph: $\{x | x > -2\} \cap \{x | x < 1\}$

Solution
The solution set is the numbers between -2 and 1.

You Try It 9

Graph: $\{x | x > -1\} \cup \{x | x < -3\}$

Your solution

Example 10

Graph: $\{x | x > 3\} \cup \{x | x < 1\}$

Solution
The solution set is the numbers greater than 3 and the numbers less than 1.

You Try It 10

Graph: $\{x | x \leq 4\} \cap \{x | x \geq -4\}$

Your solution

Example 11

Graph: $\{x | x \leq 5\} \cup \{x | x \geq -3\}$

Solution
The solution set is the real numbers.

You Try It 11

Graph: $\{x | x < 2\} \cup \{x | x \geq -2\}$

Your solution

Solutions on pp. S23–S24

9.1 Exercises

Objective A   **To write a set using the roster method**

1. ✎ Explain how to find the union of two sets.

2. ✎ Explain how to find the intersection of two sets.

For Exercises 3 to 8, use the roster method to write the set.

3. The integers between 15 and 22

4. The integers between -10 and -4

5. The odd integers between 8 and 18

6. The even integers between -11 and -1

7. The letters of the alphabet between a and d

8. The letters of the alphabet between p and v

For Exercises 9 to 16, find $A \cup B$.

9. $A = \{3, 4, 5\}$ $B = \{4, 5, 6\}$

10. $A = \{-3, -2, -1\}$ $B = \{-2, -1, 0\}$

11. $A = \{-10, -9, -8\}$ $B = \{8, 9, 10\}$

12. $A = \{a, b, c\}$ $B = \{x, y, z\}$

13. $A = \{a, b, d, e\}$ $B = \{c, d, e, f\}$

14. $A = \{m, n, p, q\}$ $B = \{m, n, o\}$

15. $A = \{1, 3, 7, 9\}$ $B = \{7, 9, 11, 13\}$

16. $A = \{-3, -2, -1\}$ $B = \{-1, 1, 2\}$

For Exercises 17 to 22, find $A \cap B$.

17. $A = \{3, 4, 5\}$ $B = \{4, 5, 6\}$

18. $A = \{-4, -3, -2\}$ $B = \{-6, -5, -4\}$

19. $A = \{-4, -3, -2\}$ $B = \{2, 3, 4\}$

20. $A = \{1, 2, 3, 4\}$ $B = \{1, 2, 3, 4\}$

21. $A = \{a, b, c, d, e\}$ $B = \{c, d, e, f, g\}$

22. $A = \{m, n, o, p\}$ $B = \{k, l, m, n\}$

Objective B   **To write a set using set-builder notation**

For Exercises 23 to 30, use set-builder notation to write the set.

23. The negative integers greater than -5

24. The positive integers less than 5

25. The integers greater than 30

26. The integers less than -70

27. The even integers greater than 5

28. The odd integers less than -2

29. The real numbers greater than 8

30. The real numbers less than 57

To graph an inequality on the number line

For Exercises 31 to 40, graph.

31. $\{x\,|\,x > 2\}$

<+++++++++++++>
-5 -4 -3 -2 -1 0 1 2 3 4 5

32. $\{x\,|\,x \geq -1\}$

<+++++++++++++>
-5 -4 -3 -2 -1 0 1 2 3 4 5

33. $\{x\,|\,x \leq 0\}$

<+++++++++++++>
-5 -4 -3 -2 -1 0 1 2 3 4 5

34. $\{x\,|\,x < 4\}$

<+++++++++++++>
-5 -4 -3 -2 -1 0 1 2 3 4 5

35. $\{x\,|\,x > -2\} \cup \{x\,|\,x < -4\}$

<+++++++++++++>
-5 -4 -3 -2 -1 0 1 2 3 4 5

36. $\{x\,|\,x > 4\} \cup \{x\,|\,x < -2\}$

<+++++++++++++>
-5 -4 -3 -2 -1 0 1 2 3 4 5

37. $\{x\,|\,x > -2\} \cap \{x\,|\,x < 4\}$

<+++++++++++++>
-5 -4 -3 -2 -1 0 1 2 3 4 5

38. $\{x\,|\,x > -3\} \cap \{x\,|\,x < 3\}$

<+++++++++++++>
-5 -4 -3 -2 -1 0 1 2 3 4 5

39. $\{x\,|\,x \geq -2\} \cup \{x\,|\,x < 4\}$

<+++++++++++++>
-5 -4 -3 -2 -1 0 1 2 3 4 5

40. $\{x\,|\,x > 0\} \cup \{x\,|\,x \leq 4\}$

<+++++++++++++>
-5 -4 -3 -2 -1 0 1 2 3 4 5

APPLYING THE CONCEPTS

41. Determine whether the statement is always true, sometimes true, or never true.
   a. Given that $a > 0$ and $b < 0$, then $ab > 0$.
   b. Given that $a < 0$, then $a^2 > 0$.
   c. Given that $a > 0$ and $b < 0$, then $a^2 > b$.

42. **a.** By trying various sets, make a conjecture about whether or not finding the union of two sets is a commutative operation.
   b. By trying various sets, make a conjecture about whether or not finding the union of two sets is an associative operation.

43. **a.** By trying various sets, make a conjecture about whether or not finding the intersection of two sets is a commutative operation.
   b. By trying various sets, make a conjecture about whether or not finding the intersection of two sets is an associative operation.

9.2 The Addition and Multiplication Properties of Inequalities

Objective A To solve an inequality using the Addition Property of Inequalities

Recall that the solution set of an inequality is the set of numbers each element of which, when substituted for the variable, results in a true inequality.

The inequality at the right is true if the variable is replaced by 7, 9.3, or $\frac{15}{2}$.

$$x + 5 > 8$$

$$\left.\begin{array}{l} 7 + 5 > 8 \\ 9.3 + 5 > 8 \\ \dfrac{15}{2} + 5 > 8 \end{array}\right\} \text{True inequalities}$$

The inequality $x + 5 > 8$ is false if the variable is replaced by 2, 1.5, or $-\frac{1}{2}$.

$$\left.\begin{array}{l} 2 + 5 > 8 \\ 1.5 + 5 > 8 \\ -\dfrac{1}{2} + 5 > 8 \end{array}\right\} \text{False inequalities}$$

There are many values of the variable x that will make the inequality $x + 5 > 8$ true. The solution set of $x + 5 > 8$ is any number greater than 3.

At the right is the graph of the solution set of $x + 5 > 8$.

In solving an inequality, the goal is to rewrite the given inequality in the form *variable > constant* or *variable < constant*. The Addition Property of Inequalities is used to rewrite an inequality in this form.

Addition Property of Inequalities

The same term can be added to each side of an inequality without changing the solution set of the inequality.

If $a > b$, then $a + c > b + c$.
If $a < b$, then $a + c < b + c$.

The Addition Property of Inequalities also holds true for an inequality containing the symbol $\geq$ or $\leq$.

The Addition Property of Inequalities is used when, in order to rewrite an inequality in the form *variable > constant* or *variable < constant*, we must remove a term from one side of the inequality. Add the opposite of that term to each side of the inequality.

HOW TO Solve: $x - 4 < -3$

$$x - 4 < -3$$
$$x - 4 + 4 < -3 + 4 \qquad \bullet \textbf{ Add 4 } \text{to each side of the inequality.}$$
$$x < 1 \qquad \bullet \textbf{ Simplify.}$$

At the right is the graph of the solution set of $x - 4 < -3$.

Because subtraction is defined in terms of addition, the Addition Property of Inequalities allows the same term to be subtracted from each side of an inequality.

> **HOW TO** Solve: $5x - 6 \leq 4x - 4$
>
> $$5x - 6 \leq 4x - 4$$
> $$5x - 4x - 6 \leq 4x - 4x - 4 \qquad \bullet \textbf{ Subtract } \textbf{4x} \text{ from each side of the inequality.}$$
> $$x - 6 \leq -4 \qquad \bullet \textbf{ Simplify.}$$
> $$x - 6 + 6 \leq -4 + 6 \qquad \bullet \textbf{ Add 6 } \text{to each side of the inequality.}$$
> $$x \leq 2 \qquad \bullet \textbf{ Simplify.}$$

Example 1

Solve $3 < x + 5$ and graph the solution set.

Solution
$$3 < x + 5$$
$$3 - 5 < x + 5 - 5 \qquad \bullet \textbf{ Subtract 5.}$$
$$-2 < x$$

You Try It 1

Solve $x + 2 < -2$ and graph the solution set.

Your solution

Example 2

Solve: $7x - 14 \leq 6x - 16$

Solution
$$7x - 14 \leq 6x - 16$$
$$7x - 6x - 14 \leq 6x - 6x - 16 \qquad \bullet \textbf{ Subtract 6x.}$$
$$x - 14 \leq -16$$
$$x - 14 + 14 \leq -16 + 14 \qquad \bullet \textbf{ Add 14.}$$
$$x \leq -2$$

You Try It 2

Solve: $5x + 3 > 4x + 5$

Your solution

Solutions on p. S24

Objective B   **To solve an inequality using the Multiplication Property of Inequalities**

In solving an inequality, the goal is to rewrite the given inequality in the form *variable* > *constant* or *variable* < *constant*. The Multiplication Property of Inequalities is used when, in order to rewrite an inequality in this form, we must remove a coefficient from one side of the inequality.

Multiplication Property of Inequalities

Each side of an inequality can be multiplied by the same positive number without changing the solution set of the inequality.

$$\text{If } a > b \text{ and } c > 0, \text{ then } ac > bc.$$
$$\text{If } a < b \text{ and } c > 0, \text{ then } ac < bc.$$

If each side of an inequality is multiplied by the same negative number and the inequality symbol is reversed, then the solution set of the inequality is not changed.

$$\text{If } a > b \text{ and } c < 0, \text{ then } ac < bc.$$
$$\text{If } a < b \text{ and } c < 0, \text{ then } ac > bc.$$

TAKE NOTE

Any time an inequality is multiplied or divided by a negative number, the inequality symbol must be reversed. Compare the next two examples.

$2x < -4$	Divide each
$\dfrac{2x}{2} < \dfrac{-4}{2}$	side by *positive* 2.
$x < -2$	Inequality *is not* reversed.
$-2x < 4$	Divide each
$\dfrac{-2x}{-2} > \dfrac{4}{-2}$	side by *negative* 2.
$x > -2$	Inequality *is* reversed.

$5 > 4$	• A true inequality
$5(2) > 4(2)$	• Multiply by *positive* 2.
$10 > 8$	• Still a true inequality
$6 < 9$	• A true inequality
$6(-3) > 9(-3)$	• Multiply by *negative* 3 and *reverse* the inequality.
$-18 > -27$	• Still a true inequality

The Multiplication Property of Inequalities also holds true for an inequality containing the symbol $\geq$ or $\leq$.

HOW TO   Solve $-\dfrac{3}{2}x \leq 6$ and graph the solution set.

$$-\dfrac{3}{2}x \leq 6$$

• Multiply each side of the inequality by $-\dfrac{2}{3}$. Because $-\dfrac{2}{3}$ is a negative number, the inequality symbol must be reversed.

$$-\dfrac{2}{3}\left(-\dfrac{3}{2}x\right) \geq -\dfrac{2}{3}(6)$$

$$x \geq -4$$

• Graph $\{x \mid x \geq -4\}$.

Because division is defined in terms of multiplication, the Multiplication Property of Inequalities allows each side of an inequality to be divided by a nonzero constant.

TAKE NOTE

As shown in the example at the right, the goal in solving an inequality can be *constant < variable* or *constant > variable*. We could have written the answer to this example as $x > -\dfrac{2}{3}$.

HOW TO   Solve: $-4 < 6x$

$$-4 < 6x$$

$$\dfrac{-4}{6} < \dfrac{6x}{6}$$

• Divide each side of the inequality by 6.

$$-\dfrac{2}{3} < x$$

• Simplify: $\dfrac{-4}{6} = -\dfrac{2}{3}$.

Example 3   Solve $-7x > 14$ and graph the solution set.

Solution

$$-7x > 14$$

$$\frac{-7x}{-7} < \frac{14}{-7} \qquad \bullet \text{ Divide by } -7.$$

$$x < -2$$

$$
\begin{array}{c}
\longleftrightarrow \\
-5\ -4\ -3\ -2\ -1\ \ 0\ \ 1\ \ 2\ \ 3\ \ 4\ \ 5
\end{array}
$$

You Try It 3   Solve $-3x > -9$ and graph the solution set.

Your solution

$$
\begin{array}{c}
\longleftrightarrow \\
-5\ -4\ -3\ -2\ -1\ \ 0\ \ 1\ \ 2\ \ 3\ \ 4\ \ 5
\end{array}
$$

Example 4   Solve: $-\dfrac{5}{8}x \le \dfrac{5}{12}$

Solution

$$-\frac{5}{8}x \le \frac{5}{12}$$

$$-\frac{8}{5}\left(-\frac{5}{8}x\right) \ge -\frac{8}{5}\left(\frac{5}{12}\right) \qquad \bullet \text{ Multiply by } -\frac{8}{5}.$$

$$x \ge -\frac{2}{3}$$

You Try It 4   Solve: $-\dfrac{3}{4}x \ge 18$

Your solution

Solutions on p. S24

Objective C   **To solve application problems**

Example 5

A student must have at least 450 points out of 500 points on five tests to receive an A in a course. One student's results on the first four tests were 94, 87, 77, and 95. What scores on the last test will enable this student to receive an A in the course?

Strategy

To find the scores, write and solve an inequality using N to represent the possible scores on the last test.

Solution

Total number of points on the five tests	is greater than or equal to	450

$$94 + 87 + 77 + 95 + N \ge 450$$
$$353 + N \ge 450 \qquad \bullet \text{ Simplify.}$$
$$353 - 353 + N \ge 450 - 353 \qquad \bullet \text{ Subtract 353.}$$
$$N \ge 97$$

The student's score on the last test must be greater than or equal to 97.

You Try It 5

An appliance dealer will make a profit on the sale of a television set if the cost of the new set is less than 70% of the selling price. What selling prices will enable the dealer to make a profit on a television set that costs the dealer $314?

Your strategy

Your solution

Solution on p. S24

9.2 Exercises

 To solve an inequality using the Addition Property of Inequalities

For Exercises 1 to 8, solve the inequality and graph the solution set.

1. $x + 1 < 3$

2. $y + 2 < 2$

3. $x - 5 > -2$

4. $x - 3 > -2$

5. $7 \le n + 4$

6. $3 \le 5 + x$

7. $x - 6 \le -10$

8. $y - 8 \le -11$

For Exercises 9 to 12, write an inequality that represents the set of numbers shown in the graph.

9.

10.

11.

12.

For Exercises 13 to 42, solve.

13. $y - 3 \ge -12$

14. $x + 8 \ge -14$

15. $3x - 5 < 2x + 7$

16. $5x + 4 < 4x - 10$

17. $8x - 7 \ge 7x - 2$

18. $3n - 9 \ge 2n - 8$

19. $2x + 4 < x - 7$

20. $9x + 7 < 8x - 7$

21. $4x - 8 \le 2 + 3x$

22. $5b - 9 < 3 + 4b$

23. $6x + 4 \ge 5x - 2$

24. $7x - 3 \ge 6x - 2$

25. $2x - 12 > x - 10$

26. $3x + 9 > 2x + 7$

27. $d + \dfrac{1}{2} < \dfrac{1}{3}$

28. $x - \dfrac{3}{8} < \dfrac{5}{6}$

29. $x + \dfrac{5}{8} \geq -\dfrac{2}{3}$

30. $y + \dfrac{5}{12} \geq -\dfrac{3}{4}$

31. $x - \dfrac{3}{8} < \dfrac{1}{4}$

32. $y + \dfrac{5}{9} \leq \dfrac{5}{6}$

33. $2x - \dfrac{1}{2} < x + \dfrac{3}{4}$

34. $6x - \dfrac{1}{3} \leq 5x - \dfrac{1}{2}$

35. $3x + \dfrac{5}{8} > 2x + \dfrac{5}{6}$

36. $4b - \dfrac{7}{12} \geq 3b - \dfrac{9}{16}$

37. $3.8x < 2.8x - 3.8$

38. $1.2x < 0.2x - 7.3$

39. $x + 5.8 \leq 4.6$

40. $n - 3.82 \leq 3.95$

41. $x - 3.5 < 2.1$

42. $x - 0.23 \leq 0.47$

Objective B **To solve an inequality using the Multiplication Property of Inequalities**

For Exercises 43 to 52, solve the inequality and graph the solution set.

43. $3x < 12$

44. $8x \leq -24$

45. $15 \leq 5y$

46. $-48 < 24x$

47. $16x \leq 16$

48. $3x > 0$

49. $-8x > 8$

50. $-2n \leq -8$

51. $-6b > 24$

52. $-4x < 8$

For Exercises 53 to 79, solve.

53. $-5y \geq 0$

54. $-3z < 0$

55. $7x > 2$

56. $6x \leq -1$

57. $2x \leq -5$

58. $\dfrac{5}{6}n < 15$

59. $\dfrac{3}{4}x < 12$

60. $\dfrac{2}{3}y \geq 4$

61. $10 \leq \dfrac{5}{8}x$

62. $4 \geq \dfrac{2}{3}x$

63. $-\dfrac{3}{7}x \leq 6$

64. $-\dfrac{2}{11}b \geq -6$

65. $-\dfrac{4}{7}x \geq -12$

66. $\dfrac{2}{3}n < \dfrac{1}{2}$

67. $-\dfrac{3}{5}x < 0$

68. $-\dfrac{2}{3}x \geq 0$

69. $-\dfrac{3}{8}x \geq \dfrac{9}{14}$

70. $-\dfrac{3}{5}x < -\dfrac{6}{7}$

71. $-\dfrac{4}{5}x < -\dfrac{8}{15}$

72. $-\dfrac{3}{4}y \geq -\dfrac{5}{8}$

73. $-\dfrac{8}{9}x \geq -\dfrac{16}{27}$

74. $1.5x \leq 6.30$

75. $2.3x \leq 5.29$

76. $-3.5d > 7.35$

77. $-0.24x > 0.768$

78. $4.25m > -34$

79. $-3.9x \geq -19.5$

Objective C   **To solve application problems**

80. **Number Sense**   Three-fifths of a number is greater than two-thirds. Find the smallest integer that satisfies this inequality.

81. **Sports**   To be eligible for a basketball tournament, a basketball team must win at least 60% of its remaining games. If the team has 17 games remaining, how many games must the team win to qualify for the tournament?

82. **Taxes**   To avoid a tax penalty, at least 90% of a self-employed person's total annual income tax liability must be paid in estimated tax payments during the year. What amount of income tax must a person with an annual income tax liability of $3500 pay in estimated tax payments?

83. **Recycling** A service organization will receive a bonus of $200 for collecting more than 1850 lb of aluminum cans during its four collection drives. On the first three drives, the organization collected 505 lb, 493 lb, and 412 lb. How many pounds of cans must the organization collect on the fourth drive to receive the bonus?

84. **Software Development** Computer software engineers are fond of saying that software takes at least twice as long to develop as they think it will. According to that saying, how many hours will it take to develop a software product that an engineer thinks can be finished in 50 h?

85. **Health** A government agency recommends a minimum daily allowance of vitamin C of 60 mg. How many additional milligrams of vitamin C does a person who has already drunk a glass of orange juice with 10 mg of vitamin C need in order to satisfy the recommended daily allowance?

86. **Grading** To pass a course with a B grade, a student must have an average of 80 points on five tests. The student's grades on the first four tests were 75, 83, 86, and 78. What scores can the student receive on the fifth test to earn a B grade?

87. **Grading** A professor scores all tests with a maximum of 100 points. To earn an A grade in this course, a student must have an average of 92 on four tests. One student's grades on the first three tests were 89, 86, and 90. Can this student earn an A grade?

88. **Health** A health official recommends a maximum cholesterol level of 200 units. How many units must a patient with a cholesterol level of 275 units reduce her cholesterol level to satisfy the recommended maximum level?

APPLYING THE CONCEPTS

For Exercises 89 to 94, given that $a > b$ and that a and b are real numbers, determine for which real numbers c the statement is true. Use set-builder notation to write the answer.

89. $ac > bc$

90. $ac < bc$

91. $a + c > b + c$

92. $a + c < b + c$

93. $\dfrac{a}{c} > \dfrac{b}{c}$

94. $\dfrac{a}{c} < \dfrac{b}{c}$

95. ✎ In your own words, state the Addition Property of Inequalities.

96. ✎ In your own words, state the Multiplication Property of Inequalities.

9.3 General Inequalities

Objective A **To solve general inequalities**

Solving an inequality frequently requires application of both the Addition and the Multiplication Properties of Inequalities.

HOW TO Solve: $4y - 3 \geq 6y + 5$

$$4y - 3 \geq 6y + 5$$

$$4y - 6y - 3 \geq 6y - 6y + 5$$ • **Subtract 6y** from each side of the inequality.

$$-2y - 3 \geq 5$$ • **Simplify.**

$$-2y - 3 + 3 \geq 5 + 3$$ • **Add 3** to each side of the inequality.

$$-2y \geq 8$$ • **Simplify.**

$$\frac{-2y}{-2} \leq \frac{8}{-2}$$ • **Divide** each side of the inequality by **−2**. Because −2 is a negative number, the inequality symbol must be reversed.

$$y \leq -4$$

When an inequality contains parentheses, one of the steps in solving the inequality requires the use of the Distributive Property.

HOW TO Solve: $-2(x - 7) > 3 - 4(2x - 3)$

$$-2(x - 7) > 3 - 4(2x - 3)$$

$$-2x + 14 > 3 - 8x + 12$$ • Use the **Distributive Property** to remove parentheses.

$$-2x + 14 > -8x + 15$$ • **Simplify.**

$$-2x + 8x + 14 > -8x + 8x + 15$$ • **Add 8x** to each side of the inequality.

$$6x + 14 > 15$$ • **Simplify.**

$$6x + 14 - 14 > 15 - 14$$ • **Subtract 14** from each side of the inequality.

$$6x > 1$$ • **Simplify.**

$$\frac{6x}{6} > \frac{1}{6}$$ • **Divide** each side of the inequality by **6**.

$$x > \frac{1}{6}$$

Example 1 Solve: $7x - 3 \leq 3x + 17$

Solution

$$7x - 3 \leq 3x + 17$$
$$7x - 3x - 3 \leq 3x - 3x + 17$$ • Subtract 3x.
$$4x - 3 \leq 17$$
$$4x - 3 + 3 \leq 17 + 3$$ • Add 3.
$$4x \leq 20$$
$$\frac{4x}{4} \leq \frac{20}{4}$$ • Divide by 4.
$$x \leq 5$$

You Try It 1 Solve: $5 - 4x > 9 - 8x$

Your solution

Solution on p. S24

Example 2

Solve: $3(3 - 2x) \geq -5x - 2(3 - x)$

Solution

$$
\begin{aligned}
3(3 - 2x) &\geq -5x - 2(3 - x) \\
9 - 6x &\geq -5x - 6 + 2x &&\bullet \text{ Distributive Property} \\
9 - 6x &\geq -3x - 6 \\
9 - 6x + 3x &\geq -3x + 3x - 6 &&\bullet \text{ Add } 3x. \\
9 - 3x &\geq -6 \\
9 - 9 - 3x &\geq -6 - 9 &&\bullet \text{ Subtract 9.} \\
-3x &\geq -15 \\
\frac{-3x}{-3} &\leq \frac{-15}{-3} &&\bullet \text{ Divide by } -3. \\
x &\leq 5
\end{aligned}
$$

You Try It 2

Solve: $8 - 4(3x + 5) \leq 6(x - 8)$

Your solution

Solution on p. S24

Objective B **To solve application problems**

Example 3

A rectangle is 10 ft wide and $(2x + 4)$ ft long. Express as an integer the maximum length of the rectangle when the area is less than 200 ft². (The area of a rectangle is equal to its length times its width.)

Strategy

To find the maximum length:

- Replace the variables in the area formula by the given values and solve for x.
- Replace the variable in the expression $2x + 4$ with the value found for x.

Solution

Length times width	is less than	200 ft²

$$
\begin{aligned}
10(2x + 4) &< 200 \\
20x + 40 &< 200 &&\bullet \text{ Distributive Property} \\
20x + 40 - 40 &< 200 - 40 &&\bullet \text{ Subtract 40.} \\
20x &< 160 \\
\frac{20x}{20} &< \frac{160}{20} &&\bullet \text{ Divide by 20.} \\
x &< 8
\end{aligned}
$$

The length is $(2x + 4)$ ft. Because $x < 8$, $2x + 4 < 2(8) + 4 = 20$. Therefore, the length is less than 20 ft. The maximum length is 19 ft.

You Try It 3

Company A rents cars for $8 a day and $.10 for every mile driven. Company B rents cars for $10 a day and $.08 per mile driven. You want to rent a car for 1 week. What is the maximum number of miles you can drive a Company A car if it is to cost you less than a Company B car?

Your strategy

Your solution

Solution on p. S24

9.3 Exercises

Objective A   To solve general inequalities

For Exercises 1 to 20, solve.

1. $4x - 8 < 2x$

2. $7x - 4 < 3x$

3. $2x - 8 > 4x$

4. $3y + 2 > 7y$

5. $8 - 3x \le 5x$

6. $10 - 3x \le 7x$

7. $3x + 2 > 5x - 8$

8. $2n - 9 \ge 5n + 4$

9. $5x - 2 < 3x - 2$

10. $8x - 9 > 3x - 9$

11. $0.1(180 + x) > x$

12. $x > 0.2(50 + x)$

13. $2(2y - 5) \le 3(5 - 2y)$

14. $2(5x - 8) \le 7(x - 3)$

15. $5(2 - x) > 3(2x - 5)$

16. $4(3d - 1) > 3(2 - 5d)$

17. $4 - 3(3 - n) \le 3(2 - 5n)$

18. $15 - 5(3 - 2x) \le 4(x - 3)$

19. $2x - 3(x - 4) \ge 4 - 2(x - 7)$

20. $4 + 2(3 - 2y) \le 4(3y - 5) - 6y$

Objective B   To solve application problems

21. **Wages**   The sales agent for a jewelry company is offered a flat monthly salary of $3200 or a salary of $1000 plus an 11% commission on the selling price of each item sold by the agent. If the agent chooses the $3200, what dollar amount does the agent expect to sell in 1 month?

22. **Sports**   A baseball player is offered an annual salary of $200,000 or a base salary of $100,000 plus a bonus of $1000 for each hit over 100 hits. How many hits must the baseball player make to earn more than $200,000?

23. **Comparing Services**   A computer bulletin board service charges a flat fee of $10 per month or a fee of $4 per month plus $.10 for each minute the service is used. How many minutes must a person use this service to exceed $10?

24. **Comparing Services**   A site licensing fee for a computer program is $1500. Paying this fee allows the company to use the program at any computer terminal within the company. Alternatively, the company can choose to pay $200 for each individual computer it has. How many individual computers must a company have for the site license to be more economical for the company?

25. **Health**   For a product to be labeled orange juice, a state agency requires that at least 80% of the drink be real orange juice. How many ounces of artificial flavors can be added to 32 oz of real orange juice and have it still be legal to label the drink orange juice?

26. **Health**   Grade A hamburger cannot contain more than 20% fat. How much fat can a butcher mix with 300 lb of lean meat to meet the 20% requirement?

27. **Transportation**   A shuttle service taking skiers to a ski area charges $8 per person each way. Four skiers are debating whether to take the shuttle bus or rent a car for $45 plus $.25 per mile. Assuming that the skiers will share the cost of the car and that they want the least expensive method of transportation, find how far away the ski area is if they choose the shuttle service.

APPLYING THE CONCEPTS

28. Determine whether the statement is always true, sometimes true, or never true, given that a, b, and c are real numbers.
    a. If $a > b$, then $-a > -b$.
    b. If $a < b$, then $ac < bc$.
    c. If $a > b$, then $a + c > b + c$.
    d. If $a \neq 0$, $b \neq 0$, and $a > b$, then $\frac{1}{a} > \frac{1}{b}$.

For Exercises 29 and 30, use the roster method to list the set of positive integers that are solutions of the inequality.

29. $7 - 2b \leq 15 - 5b$ 30. $-6(2 - d) \geq 4d - 9$

For Exercises 31 and 32, use the roster method to list the set of integers that are common to the solution sets of the two inequalities.

31. $5x - 12 \leq x + 8$ 32. $3(x + 2) > 9x - 2$
    $3x - 4 \geq 2 + x$ $4(x + 5) > 3(x + 6)$

33. Determine the solution set of $2 - 3(x + 4) < 5 - 3x$.

34. Determine the solution set of $3x + 2(x - 1) > 5(x + 1)$.

9.4 Graphing Linear Inequalities

Objective A   To graph an inequality in two variables

Point of Interest

Linear inequalities play an important role in applied mathematics. They are used in a branch of mathematics called *linear programming*, which was developed during World War II to solve problems in supplying the Air Force with the machine parts necessary to keep planes flying. Today, its applications extend to many other disciplines.

The graph of the linear equation $y = x - 2$ separates a plane into three sets:

The set of points on the line
The set of points above the line
The set of points below the line

The point $(3, 1)$ is a solution of $y = x - 2$.

$$y = x - 2$$
$$1 \mid 3 - 2$$
$$1 = 1$$

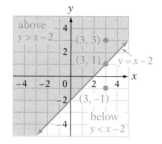

The point $(3, 3)$ is a solution of $y > x - 2$.

$$y > x - 2$$
$$3 \mid 3 - 2$$
$$3 > 1$$

Any point above the line is a solution of $y > x - 2$.

Study Tip

Be sure to do all you need to do in order to be successful at graphing linear inequalities: Read through the introductory material, work through the How To examples, study the paired examples, do the You Try Its, and check your solutions against those in the back of the book. See *AIM for Success*, page xxv.

The point $(3, -1)$ is a solution of $y < x - 2$.

$$y < x - 2$$
$$-1 \mid 3 - 2$$
$$-1 < 1$$

Any point below the line is a solution of $y < x - 2$.

The solution set of $y = x - 2$ is all points on the line. The solution set of $y > x - 2$ is all points above the line. The solution set of $y < x - 2$ is all points below the line. The solution set of an inequality in two variables is a **half-plane.**

The following illustrates the procedure for graphing a linear inequality.

HOW TO   Graph the solution set of $2x + 3y \leq 6$.

Solve the inequality for y.
$$2x + 3y \leq 6$$
$$2x - 2x + 3y \leq -2x + 6$$
$$3y \leq -2x + 6$$
$$\frac{3y}{3} \leq \frac{-2x + 6}{3}$$
$$y \leq -\frac{2}{3}x + 2$$

• **Subtract 2x** from each side.
• **Simplify.**

• **Divide each side by 3**.

• **Simplify.**

Change the inequality to an equality and graph $y = -\frac{2}{3}x + 2$. If the inequality is $\geq$ or $\leq$, the line is in the solution set and is shown by a solid line. If the inequality is $>$ or $<$, the line is not a part of the solution set and is shown by a dotted line.

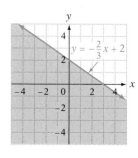

If the inequality is $>$ or $\geq$, shade the upper half-plane. If the inequality is $<$ or $\leq$, shade the lower half-plane.

Example 1

Graph the solution set of $3x + y > -2$.

Solution

$$3x + y > -2$$
$$3x - 3x + y > -3x - 2 \quad \bullet \text{ Subtract } 3x.$$
$$y > -3x - 2$$

Graph $y = -3x - 2$ as a dotted line.
Shade the upper half-plane.

Example 2

Graph the solution set of $2x - y \geq 2$.

Solution

$$2x - y \geq 2$$
$$2x - 2x - y \geq -2x + 2 \quad \bullet \text{ Subtract } 2x.$$
$$-y \geq -2x + 2$$
$$-1(-y) \leq -1(-2x + 2) \quad \bullet \text{ Multiply by } -1.$$
$$y \leq 2x - 2$$

Graph $y = 2x - 2$ as a solid line.
Shade the lower half-plane.

Example 3

Graph the solution set of $y > -1$.

Solution
Graph $y = -1$ as a dotted line.
Shade the upper half-plane.

You Try It 1

Graph the solution set of $x - 3y < 2$.

Your solution

You Try It 2

Graph the solution set of $2x - 4y \leq 8$.

Your solution

You Try It 3

Graph the solution set of $x < 3$.

Your solution

Solutions on pp. S24–S25

9.4 Exercises

Objective A To graph an inequality in two variables

For Exercises 1 to 18, graph the solution set of the inequality.

1. $x + y > 4$

2. $x - y > -3$

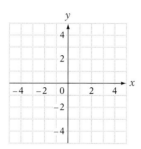

3. $2x - y < -3$

4. $3x - y < 9$

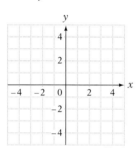

5. $2x + y \geq 4$

6. $3x + y \geq 6$

7. $y \leq -2$

8. $y > 3$

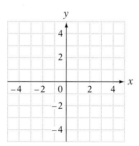

9. $3x - 2y < 8$

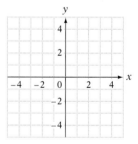

10. $5x + 4y > 4$

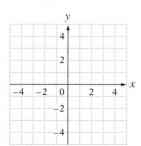

11. $-3x - 4y \geq 4$

12. $-5x - 2y \geq 8$

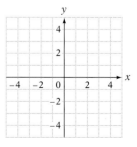

13. $6x + 5y \le -10$

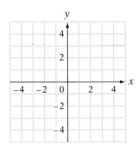

14. $2x + 2y \le -4$

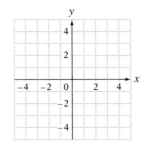

15. $-4x + 3y < -12$

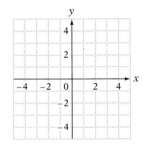

16. $-4x + 5y < 15$

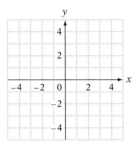

17. $-2x + 3y \le 6$

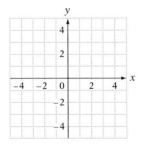

18. $3x - 4y > 12$

APPLYING THE CONCEPTS

For Exercises 19 to 21, graph the solution set of the inequality.

19. $\dfrac{x}{4} + \dfrac{y}{2} > 1$

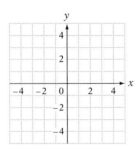

20. $2x - 3(y + 1) > y - (4 - x)$

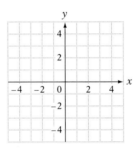

21. $4y - 2(x + 1) \ge 3(y - 1) + 3$

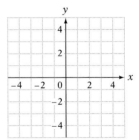

For Exercises 22 to 24, write the inequality given its graph.

22.

23.

24.

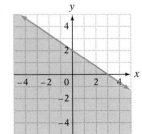

Focus on Problem Solving

Graphing Data Graphs are very useful in displaying data. By studying a graph, we can reach various conclusions about the data.

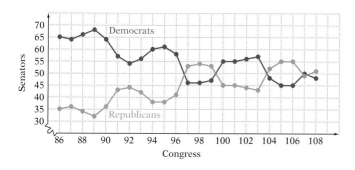

The double-line graph at the left shows the number of Democrats and the number of Republicans in the U.S. Senate for the 86th Congress (1959–1961) through the 108th Congress (2003–2005).

1. How many Democratic and how many Republican senators were in the 90th Congress?

2. In which Congress was the difference between the numbers of Democrats and Republicans the greatest?

3. In which Congress did the majority first change from Democratic to Republican?

4. Between which two Congresses did the number of Republican senators increase but the number of Democratic senators remain the same?

5. In what percent of the Congresses did the number of Democrats exceed the number of Republicans? Round to the nearest tenth.

6. In which Congresses were there a greater number of Republican senators than Democratic senators?

Year	DVD Recorders	DVD Players
2003	1	59
2004	2	62
2005	9	60
2006	18	54
2007	28	44
2008	44	34

The table at the left, based on data from Allied Business Intelligence, Inc., shows the actual and projected number (in millions) of worldwide shipments of DVD players and DVD recorders for the years 2003 through 2008.

7. Make a double-line graph of the data.

8. In which year will the number of DVD recorders shipped first exceed the number of DVD players shipped?

9. Based on this data, for the year 2009, would you expect the difference between the number of DVD recorders shipped and the number of DVD players shipped to be less than or greater than the same number in 2008?

Projects and Group Activities

Mean and Standard Deviation An automotive engineer tests the miles-per-gallon ratings of 15 cars and records the results as follows:

25   22   21   27   25   35   29   31   25   26   21   39   34   32   28

The **mean** of the data is the sum of the measurements divided by the number of measurements. The symbol for the mean is $\bar{x}$.

$$\text{Mean} = \bar{x} = \frac{\text{sum of all data values}}{\text{number of data values}}$$

To find the mean for the data on p. 469, add the numbers and then divide by 15.

$$\bar{x} = \frac{25 + 22 + 21 + 27 + 25 + 35 + 29 + 31 + 25 + 26 + 21 + 39 + 34 + 32 + 28}{15}$$

$$= \frac{420}{15} = 28$$

The mean number of miles per gallon for the 15 cars tested was 28 mi/gal.

The mean is one of the most frequently computed averages. It is the one that is commonly used to calculate a student's performance in a class.

The scores for a history student on five tests were 78, 82, 91, 87, and 93. What was the mean score for this student?

To find the mean, add the numbers. Then divide by 5.

$$\bar{x} = \frac{78 + 82 + 91 + 87 + 93}{5}$$

$$= \frac{431}{5} = 86.2$$

The mean score for the history student was 86.2.

Consider two students, each of whom has taken five exams.

Scores for student A

| 84 | 86 | 83 | 85 | 87 |

$$\bar{x} = \frac{84 + 86 + 83 + 85 + 87}{5} = \frac{425}{5} = 85$$

The mean for student A is 85.

Scores for student B

| 90 | 75 | 94 | 68 | 98 |

$$\bar{x} = \frac{90 + 75 + 94 + 68 + 98}{5} = \frac{425}{5} = 85$$

The mean for student B is 85.

For each of these students, the mean (average) for the five exams is 85. However, student A has a more consistent record of scores than student B. One way to measure the consistency, or "clustering" near the mean, of data is to use the **standard deviation.**

To calculate the standard deviation:

Step 1. Sum the squares of the differences between each value of the data and the mean.

Step 2. Divide the result in Step 1 by the number of items in the set of data.

Step 3. Take the square root of the result in Step 2.

The calculation for student A is shown at the right.

Step 1:

x	$x - \bar{x}$	$(x - \bar{x})^2$
84	$84 - 85$	$(-1)^2 = 1$
86	$86 - 85$	$1^2 = 1$
83	$83 - 85$	$(-2)^2 = 4$
85	$85 - 85$	$0^2 = 0$
87	$87 - 85$	$2^2 = 4$
		Total $= 10$

Step 2: $\frac{10}{5} = 2$

The symbol for standard deviation is the lowercase Greek letter *sigma*, σ.

Step 3: $\sigma = \sqrt{2} \approx 1.414$

The standard deviation for student A's scores is approximately 1.414.

Following a similar procedure for student B shows that the standard deviation for student B's scores is approximately 11.524. Because the standard deviation of student B's scores is greater than that of student A's (11.524 > 1.414), student B's scores are not as consistent as those of student A.

1. The weights in ounces of 6 newborn infants were recorded by a hospital. The weights were 96, 105, 84, 90, 102, and 99. Find the standard deviation of the weights. Round to the nearest hundredth.

2. The numbers of rooms occupied in a hotel on 6 consecutive days were 234, 321, 222, 246, 312, and 396. Find the standard deviation for the number of rooms occupied. Round to the nearest hundredth.

3. Seven coins were tossed 100 times. The numbers of heads recorded for each coin were 56, 63, 49, 50, 48, 53, and 52. Find the standard deviation of the number of heads. Round to the nearest hundredth.

4. The temperatures, in degrees Fahrenheit, for 11 consecutive days at a desert resort were 95°, 98°, 98°, 104°, 97°, 100°, 96°, 97°, 108°, 93°, and 104°. For the same days, temperatures in Antarctica were 27°, 28°, 28°, 30°, 28°, 27°, 30°, 25°, 24°, 26°, and 21°. Which location has the greater standard deviation of temperatures?

5. The scores for 5 college basketball games were 56, 68, 60, 72, and 64. The scores for 5 professional basketball games were 106, 118, 110, 122, and 114. Which scores have the greater standard deviation?

6. The weights in pounds of the 5-man front line of a college football team are 210, 245, 220, 230, and 225. Find the standard deviation of the weights. Round to the nearest hundredth.

7. One student received test scores of 85, 92, 86, and 89. A second student received scores of 90, 97, 91, and 94 (exactly 5 points more on each test). Are the means of the two students the same? If not, what is the relationship between the means of the two students? Are the standard deviations of the scores of the two students the same? If not, what is the relationship between the standard deviations of the scores of the two students?

8. Grade-point average (GPA) is a *weighted* mean. It is called a weighted mean because a grade in a 5-unit course has more influence on your GPA than a grade in a 2-unit course. GPA is calculated by multiplying the numerical equivalent of each grade by the number of units, adding those products, and then dividing by the total number of units. Calculate your GPA for the last quarter or semester.

9. If you average 40 mph for 1 h and then 50 mph for 1 h, is your average speed $\frac{40 + 50}{2} = 45$ mph? Why or why not?

10. A company is negotiating with its employees the terms of a raise in salary. One proposal would add $500 a year to each employee's salary. The second proposal would give each employee a 4% raise. Explain how each of these proposals would affect the current mean and standard deviation of salaries for the company.

Chapter 9 Summary

Key Words	**Examples**
The *empty set* or *null set,* written ∅, is the set that contains no elements. [9.1A, p. 447]	The set of cars that can travel faster than 1000 mph is an empty set.
The *union* of two sets, written $A \cup B$, is the set that contains the elements of A and the elements of B. [9.1A, p. 447]	Let $A = \{2, 4, 6, 8\}$ and $B = \{0, 1, 2, 3, 4\}$. Then $A \cup B = \{0, 1, 2, 3, 4, 6, 8\}$.
The *intersection* of two sets, written $A \cap B$, is the set that contains the elements that are common to both A and B. [9.1A, p. 447]	Let $A = \{2, 4, 6, 8\}$ and $B = \{0, 1, 2, 3, 4\}$. Then $A \cap B = \{2, 4\}$.
Set-builder notation uses a rule to describe the elements of a set. [9.1B, p. 448]	Using set-builder notation, the set of real numbers greater than 2 is written $\{x \mid x > 2, x \in \text{real numbers}\}$.
An *inequality* is an expression that contains the symbol $<$, $>$, $\leq$, or $\geq$. The *solution set of an inequality* is a set of numbers each element of which, when substituted for the variable, results in a true inequality. The solution set of an inequality can be graphed on a number line. [9.1C, p. 449]	$3x - 1 < 5$ is an inequality. The solution set of $3x - 1 < 5$ is $\{x \mid x < 2\}$. The graph of the solution set is ⟵++++++++++)++++⟶ . $\quad$ –5 –4 –3 –2 –1 0 1 2 3 4 5
The solution set of a linear inequality in two variables is a *half-plane.* [9.4A, p. 465]	The solution set of $3x + 4y \geq 12$ is the half-plane shown at the right. 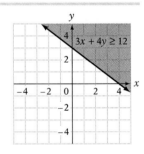

Essential Rules and Procedures   **Examples**

Addition Property of Inequalities   [9.2A, p. 453]
The same term can be added to each side of an inequality without changing the solution set of the inequality.

$$\text{If } a > b, \text{ then } a + c > b + c.$$
$$\text{If } a < b, \text{ then } a + c < b + c.$$

$$x - 3 < -7$$
$$x - 3 + 3 < -7 + 3$$
$$x < -4$$

Multiplication Property of Inequalities   [9.2B, p. 455]
Each side of an inequality can be multiplied by the same positive number without changing the solution set of the inequality.

$$\text{If } a > b \text{ and } c > 0, \text{ then } ac > bc.$$
$$\text{If } a < b \text{ and } c > 0, \text{ then } ac < bc.$$

$$4x > -8$$
$$\frac{4x}{4} > \frac{-8}{4}$$
$$x > -2$$

If each side of an inequality is multiplied by the same negative number and the inequality symbol is reversed, then the solution set of the inequality is not changed.

$$\text{If } a > b \text{ and } c < 0, \text{ then } ac < bc.$$
$$\text{If } a < b \text{ and } c < 0, \text{ then } ac > bc.$$

$$-2x < 6$$
$$\frac{-2x}{-2} > \frac{6}{-2}$$
$$x > -3$$

Chapter 9 Review Exercises

1. Solve: $2x - 3 > x + 15$

2. Find $A \cap B$, given $A = \{0, 2, 4, 6, 8\}$ and $B = \{-2, -4\}$.

3. Use set-builder notation to write the set of odd integers greater than -8.

4. Find $A \cup B$, given $A = \{6, 8, 10\}$ and $B = \{2, 4, 6\}$.

5. Use the roster method to write the set of odd positive integers less than 8.

6. Solve: $12 - 4(x - 1) \leq 5(x - 4)$

7. Graph: $\{x \mid x > 3\}$

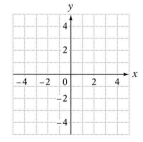

8. Solve: $3x + 4 \geq -8$

9. Graph: $3x + 2y \leq 12$

10. Graph: $5x + 2y < 6$

11. Use set-builder notation to write the set of real numbers greater than 3.

12. Solve and graph the solution set of $x - 3 > -1$.

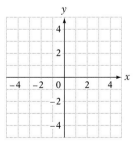

13. Find $A \cap B$, given $A = \{1, 5, 9, 13\}$ and $B = \{1, 3, 5, 7, 9\}$.

14. Graph: $\{x \mid x < 2\} \cup \{x \mid x > 5\}$

15. Graph: $\{x|x > -1\} \cap \{x|x \le 2\}$

<-+--+--+--+--+--+--+--+--+--+--+->
-5 -4 -3 -2 -1 0 1 2 3 4 5

16. Solve: $-15x \le 45$

17. Solve: $6x - 9 < 4x + 3(x + 3)$

18. Solve: $5 - 4(x + 9) > 11(12x - 9)$

19. Solve: $-\dfrac{3}{4}x > \dfrac{2}{3}$

20. Graph: $2x - 3y < 9$

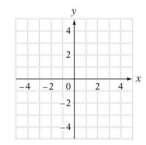

21. Solve: $7x - 2(x + 3) \ge x + 10$

22. **Floral Delivery**   Florist A charges a \$3 delivery fee plus \$21 per bouquet delivered. Florist B charges a \$15 delivery fee plus \$18 per bouquet delivered. A church wants to supply each resident of a small nursing home with a bouquet for Grandparents Day. Find the number of residents of the nursing home if using florist B is more economical than using florist A.

23. **Landscaping**   The width of a rectangular garden is 12 ft. The length of the garden is $(3x + 5)$ ft. Express as an integer the minimum length of the garden when the area is greater than 276 ft². (The area of a rectangle is equal to its length times its width.)

12 ft

$(3x + 5)$ ft

24. **Number Sense**   Six less than a number is greater than twenty-five. Find the smallest integer that will satisfy the inequality.

25. **Grading**   A student's grades on five sociology tests were 68, 82, 90, 73, and 95. What is the lowest score the student can receive on the next test and still be able to attain a minimum of 480 points?

Chapter 9 Test

1. Graph: $\{x \mid x < 5\} \cap \{x \mid x > 0\}$

$$\text{-5 -4 -3 -2 -1 0 1 2 3 4 5}$$

2. Use set-builder notation to write the set of the positive integers less than 50.

3. Use the roster method to write the set of the even positive integers between 3 and 9.

4. Solve: $3(2x - 5) \geq 8x - 9$

5. Solve: $x + \dfrac{1}{2} > \dfrac{5}{8}$

6. Graph: $\{x \mid x > -2\}$

<!-- number line -->
$$\text{-5 -4 -3 -2 -1 0 1 2 3 4 5}$$

7. Solve: $5 - 3x > 8$

8. Use set-builder notation to write the set of the real numbers greater than -23.

9. Graph the solution set of $3x + y > 4$.

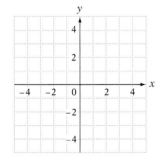

10. Graph the solution set of $4x - 5y \geq 15$.

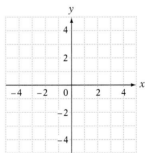

11. Find $A \cap B$, given $A = \{6, 8, 10, 12\}$ and $B = \{12, 14, 16\}$.

12. Solve and graph the solution set of $4 + x < 1$.

13. Solve: $-\dfrac{3}{8}x \leq 5$

14. Solve: $6x - 3(2 - 3x) < 4(2x - 7)$

15. Solve and graph the solution set of $\frac{2}{3}x \geq 2$.

16. Solve: $2x - 7 \leq 6x + 9$

17. Safety   To ride a certain roller coaster at an amusement park, a person must be at least 48 in. tall. How many inches must a child who is 43 in. tall grow to be eligible to ride the roller coaster?

18. Geometry   A rectangle is 15 ft long and $(2x - 4)$ ft wide. Express as an integer the maximum width of the rectangle if the area is less than 180 ft². (The area of a rectangle is equal to its length times its width.)

$(2x - 4)$ ft

15 ft

19. Machining   A ball bearing for a rotary engine must have a circumference between 0.1220 in. and 0.1240 in. What are the allowable diameters for the bearing? Round to the nearest ten-thousandth. Recall that $C = \pi d$.

20. Wages   A stockbroker receives a monthly salary that is the greater of $2500 or $1000 plus 2% of the total value of all stock transactions the broker processes during the month. What dollar amounts of transactions did the broker process in a month for which the broker's salary was $2500?

Cumulative Review Exercises

1. Simplify: $2[5a - 3(2 - 5a) - 8]$

2. Solve: $\dfrac{5}{8} - 4x = \dfrac{1}{8}$

3. Solve: $2x - 3[x - 2(x - 3)] = 2$

4. Simplify: $(-3a)(-2a^3b^2)^2$

5. Simplify: $\dfrac{27a^3b^2}{(-3ab^2)^3}$

6. Divide: $(16x^2 - 12x - 2) \div (4x - 1)$

7. Given $f(x) = x^2 - 4x - 5$, find $f(-1)$.

8. Factor: $27a^2x^2 - 3a^2$

9. Divide: $\dfrac{x^2 - 2x}{x^2 - 2x - 8} \div \dfrac{x^3 - 5x^2 + 6x}{x^2 - 7x + 12}$

10. Subtract: $\dfrac{4a}{2a - 3} - \dfrac{2a}{a + 3}$

11. Solve: $\dfrac{5y}{6} - \dfrac{5}{9} = \dfrac{y}{3} - \dfrac{5}{6}$

12. Solve $R = \dfrac{C - S}{t}$ for C.

13. Find the slope of the line that passes through the points $(2, -3)$ and $(-1, 4)$.

14. Find the equation of the line that passes through the point $(1, -3)$ and has slope $-\dfrac{3}{2}$.

15. Solve by substitution.
$$x = 3y + 1$$
$$2x + 5y = 13$$

16. Solve by the addition method.
$$9x - 2y = 17$$
$$5x + 3y = -7$$

17. Find $A \cup B$, given $A = \{0, 1, 2\}$ and $B = \{-10, -2\}$.

18. Use set-builder notation to write the set of the real numbers less than 48.

19. Graph: $\{x | x > 1\} \cup \{x | x < -1\}$

← —+—+—+—+—+—+—+—+—+—+—+—→
-5 -4 -3 -2 -1 0 1 2 3 4 5

20. Graph the solution set of $\frac{3}{8}x > -\frac{3}{4}$.

← —+—+—+—+—+—+—+—+—+—+—+—→
-5 -4 -3 -2 -1 0 1 2 3 4 5

21. Solve: $-\frac{4}{5}x > 12$

22. Solve: $15 - 3(5x - 7) < 2(7 - 2x)$

23. **Number Sense** Three-fifths of a number is less than negative fifteen. What integers satisfy this inequality? Write the answer in set-builder notation.

24. **Rental Agencies** Company A rents cars for $6 a day and $.25 for every mile driven. Company B rents cars for $15 a day and $.10 per mile. You want to rent a car for 6 days. What is the maximum number of miles you can drive a Company A car if it is to cost you less than a Company B car?

25. **Conservation** In a lake, 100 fish are caught, tagged, and then released. Later, 150 fish are caught. Three of these 150 fish are found to have tags. Estimate the number of fish in the lake.

26. **Geometry** The measure of the first angle of a triangle is 30° more than the measure of the second angle. The measure of the third angle is 10° more than twice the measure of the second angle. Find the measure of each angle.

27. Graph: $y = 2x - 1$

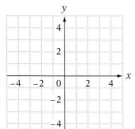

28. Graph the solution set of $6x - 3y \geq 6$.

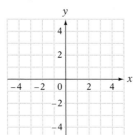

10

Radical Expressions

OBJECTIVES

Section 10.1

A To simplify numerical radical expressions
B To simplify variable radical expressions

Section 10.2

A To add and subtract radical expressions

Section 10.3

A To multiply radical expressions
B To divide radical expressions

Section 10.4

A To solve an equation containing a radical expression
B To solve application problems

Think back to some of the carnival rides you enjoyed as a child. Was your favorite ride the Ferris wheel, or the tilt-a-whirl? Was it the scrambler, or the carousel? All of these rides involve mathematics and physics. **Exercise 29 on page 502** uses a formula to describe the speed of a child riding a merry-go-round. The formula, which uses a radical expression, is $v = \sqrt{12r}$, where v is the speed in feet per second and r is the distance in feet from the center of the merry-go-round to the rider. You will learn more about radical expressions in this chapter.

Need help? For online student resources, such as section quizzes, visit this textbook's website at **math.college.hmco.com/students.**

Do these exercises to prepare for Chapter 10.

1. Evaluate: $-|-14|$

2. Simplify: $3x^2y - 4xy^2 - 5x^2y$

3. Solve: $1.5h = 21$

4. Solve: $3x - 2 = 5 - 2x$

5. Simplify: $x^3 \cdot x^3$

6. Expand: $(x + y)^2$

7. Expand: $(2x - 3)^2$

8. Multiply: $(2 - 3v)(2 + 3v)$

9. Multiply: $(a - 5)(a + 5)$

10. Simplify: $\dfrac{2x^4y^3}{18x^2y}$

GO FIGURE • • •

Two pieces of lumber are used to support two walls as shown in the figure. How high above the ground do the pieces of lumber cross? (Hint: Triangles *ABC* and *EBD* are similar triangles. Therefore, the ratio of the corresponding heights equals the ratio of corresponding sides. That is, $\dfrac{y}{8 - y} = \dfrac{10}{12}$. Solve for y and use this value to write another equation based on different similar triangles.)

10.1 Introduction to Radical Expressions

Objective A   **To simplify numerical radical expressions**

Point of Interest

The radical symbol was first used in 1525 but was written as √. Some historians suggest that the radical symbol also developed into the symbols for "less than" and "greater than." Because typesetters of that time did not want to make additional symbols, the radical was rotated to the position ⟩ and used as a "greater than" symbol and rotated to ⟨ and used for the "less than" symbol. Other evidence, however, suggests that the "less than" and "greater than" symbols were developed independently of the radical symbol.

A **square root** of a positive number a is a number whose square is a.

A square root of 16 is 4 because $4^2 = 16$.
A square root of 16 is -4 because $(-4)^2 = 16$.

Every positive number has two square roots, one a positive and one a negative number. The symbol $\sqrt{}$, called a **radical sign,** is used to indicate the positive or **principal square root** of a number. For example, $\sqrt{16} = 4$ and $\sqrt{25} = 5$. The number under the radical sign is called the **radicand.**

When the negative square root of a number is to be found, a negative sign is placed in front of the radical. For example, $-\sqrt{16} = -4$ and $-\sqrt{25} = -5$.

The square of an integer is a **perfect square.** For instance, 49, 81, and 144 are perfect squares. The principal square root of a perfect-square integer is a positive integer.

$$7^2 = 49 \qquad \sqrt{49} = 7$$
$$9^2 = 81 \qquad \sqrt{81} = 9$$
$$12^2 = 144 \qquad \sqrt{144} = 12$$

If a number is not a perfect square, its square root can only be approximated. For example, 2 and 7 are not perfect squares. The square roots of these numbers are *irrational numbers.* Their decimal approximations never terminate or repeat.

$$\sqrt{2} \approx 1.4142135\ldots \qquad \sqrt{7} \approx 2.6457513\ldots$$

TAKE NOTE

Recall that a factor of a number divides the number evenly. For instance, 6 is a factor of 18. The perfect square 9 is also a factor of 18. It is a *perfect-square factor* of 18, whereas 6 is not a perfect-square factor of 18.

Radical expressions that contain radicands that are not perfect squares are frequently written in simplest form. A radical expression is in *simplest form* when the radicand contains no factor greater than 1 that is a perfect square. For instance, $\sqrt{50}$ is not in simplest form because 25 is a perfect-square factor of 50. The radical expression $\sqrt{15}$ is in simplest form because there are no perfect-square factors of 15 that are greater than 1.

The Product Property of Square Roots and a knowledge of perfect squares is used to simplify radicands that are not perfect squares.

> **The Product Property of Square Roots**
>
> If a and b are positive real numbers, then $\sqrt{ab} = \sqrt{a} \cdot \sqrt{b}$.

The chart below shows the square roots of some perfect squares.

Square Roots of Perfect Squares

$\sqrt{1} = 1$	$\sqrt{16} = 4$	$\sqrt{49} = 7$	$\sqrt{100} = 10$
$\sqrt{4} = 2$	$\sqrt{25} = 5$	$\sqrt{64} = 8$	$\sqrt{121} = 11$
$\sqrt{9} = 3$	$\sqrt{36} = 6$	$\sqrt{81} = 9$	$\sqrt{144} = 12$

HOW TO   Simplify: $\sqrt{72}$

$$\sqrt{72} = \sqrt{36 \cdot 2}$$
• Write the radicand as the product of **a perfect square** and a factor that does not contain a perfect square.

$$= \sqrt{36}\sqrt{2}$$
• Use the Product Property of Square Roots to write the expression as a product.

$$= 6\sqrt{2}$$
• Simplify.

Note that 72 must be written as the product of a perfect square and a *factor that does not contain a perfect square*. Therefore, it would not be correct to simplify $\sqrt{72}$ as $\sqrt{9 \cdot 8}$. Although 9 is a perfect-square factor of 72, 8 also contains a perfect-square factor ($8 = 4 \cdot 2$). Therefore, $\sqrt{8}$ is not in simplest form. Remember to find the largest perfect-square factor of the radicand.

$$\sqrt{72} = \sqrt{9 \cdot 8}$$
$$= \sqrt{9} \cdot \sqrt{8}$$
$$= 3\sqrt{8}$$

Not in simplest form

HOW TO   Simplify: $\sqrt{147}$

$\sqrt{147} = \sqrt{49 \cdot 3}$ • Write the radicand as the product of **a perfect square** and a factor that does not contain a perfect square.

$= \sqrt{49}\sqrt{3}$ • Use the Product Property of Square Roots to write the expression as a product.

$= 7\sqrt{3}$ • Simplify.

HOW TO   Simplify: $\sqrt{360}$

$\sqrt{360} = \sqrt{36 \cdot 10}$ • Write the radicand as the product of **a perfect square** and a factor that does not contain a perfect square.

$= \sqrt{36}\sqrt{10}$ • Use the Product Property of Square Roots to write the expression as a product.

$= 6\sqrt{10}$ • Simplify.

From the last example, note that $\sqrt{360} = 6\sqrt{10}$. The two expressions are different representations of the same number. Using a calculator, we find that $\sqrt{360} \approx 18.973666$ and $6\sqrt{10} \approx 6(3.1622777) = 18.9736662$.

HOW TO   Simplify: $\sqrt{-16}$

Because the square of any real number is positive, there is no real number whose square is -16. $\sqrt{-16}$ is not a real number.

Example 1   Simplify: $3\sqrt{90}$

Solution
$3\sqrt{90} = 3\sqrt{9 \cdot 10}$ • **9** is a perfect-square factor.
$= 3\sqrt{9}\sqrt{10}$ • Product Property of Square Roots
$= 3 \cdot 3\sqrt{10}$
$= 9\sqrt{10}$

You Try It 1   Simplify: $-5\sqrt{32}$

Your solution

Example 2   Simplify: $\sqrt{252}$

Solution
$\sqrt{252} = \sqrt{36 \cdot 7}$ • **36** is a perfect-square factor.
$= \sqrt{36}\sqrt{7}$ • Product Property of Square Roots
$= 6\sqrt{7}$

You Try It 2   Simplify: $\sqrt{216}$

Your solution

Solutions on p. S25

Objective B   **To simplify variable radical expressions**

Variable expressions that contain radicals do not always represent real numbers. For example, if $a = -4$, then

$$\sqrt{a^3} = \sqrt{(-4)^3} = \sqrt{-64}$$

and $\sqrt{-64}$ is not a real number.

Now consider the expression $\sqrt{x^2}$. Evaluate this expression for $x = -2$ and $x = 2$.

$$\sqrt{x^2} \qquad\qquad\qquad\qquad \sqrt{x^2}$$
$$\sqrt{(-2)^2} = \sqrt{4} = 2 = |-2| \qquad\qquad \sqrt{2^2} = \sqrt{4} = 2 = |2|$$

This suggests the following:

For any real number a, $\sqrt{a^2} = |a|$. If $a \geq 0$, then $\sqrt{a^2} = a$.

In order to avoid variable expressions that do not represent real numbers, and so that absolute value signs are not needed for certain expressions, the variables in this chapter will represent *positive* numbers unless otherwise stated.

A variable or a product of variables written in exponential form is a perfect square when each exponent is an even number.

To find the square root of a perfect square, remove the radical sign and multiply each exponent by $\frac{1}{2}$.

HOW TO   Simplify: $\sqrt{a^6}$

$\sqrt{a^6} = a^3$ • Remove the radical sign and multiply the exponent by $\frac{1}{2}$.

A variable radical expression is in simplest form when the radicand contains no factor greater than 1 that is a perfect square.

HOW TO   Simplify: $\sqrt{x^7}$

$\sqrt{x^7} = \sqrt{x^6 \cdot x}$ • Write x^7 as the product of **a perfect square** and x.
$\quad = \sqrt{x^6}\sqrt{x}$ • Use the Product Property of Square Roots.
$\quad = x^3\sqrt{x}$ • Simplify the perfect square.

HOW TO   Simplify: $3x\sqrt{8x^3y^{13}}$

$3x\sqrt{8x^3y^{13}} = 3x\sqrt{4x^2y^{12}(2xy)}$ • Write the radicand as the product of **perfect squares** and factors that do not contain a perfect square.
$\quad = 3x\sqrt{4x^2y^{12}}\sqrt{2xy}$ • Use the Product Property of Square Roots.
$\quad = 3x \cdot 2xy^6\sqrt{2xy}$ • Simplify.
$\quad = 6x^2y^6\sqrt{2xy}$

HOW TO   Simplify: $\sqrt{25(x + 2)^2}$

$$\sqrt{25(x + 2)^2} = 5(x + 2)$$
$$= 5x + 10$$

Example 3

Simplify: $\sqrt{b^{15}}$

Solution
$\sqrt{b^{15}} = \sqrt{b^{14} \cdot b}$ • b^{14} is a perfect square.
$\quad\quad = \sqrt{b^{14}} \cdot \sqrt{b} = b^7\sqrt{b}$

You Try It 3

Simplify: $\sqrt{y^{19}}$

Your solution

Example 4

Simplify: $\sqrt{24x^5}$

Solution
$\sqrt{24x^5} = \sqrt{4x^4(6x)}$ • 4 and x^4 are perfect squares.
$\quad\quad = \sqrt{4x^4}\sqrt{6x}$
$\quad\quad = 2x^2\sqrt{6x}$

You Try It 4

Simplify: $\sqrt{45b^7}$

Your solution

Example 5

Simplify: $2a\sqrt{18a^3b^{10}}$

Solution
$2a\sqrt{18a^3b^{10}}$
$\quad = 2a\sqrt{9a^2b^{10}(2a)}$ • 9, a^2, and b^{10} are perfect squares.
$\quad = 2a\sqrt{9a^2b^{10}}\sqrt{2a}$
$\quad = 2a \cdot 3ab^5\sqrt{2a}$
$\quad = 6a^2b^5\sqrt{2a}$

You Try It 5

Simplify: $3a\sqrt{28a^9b^{18}}$

Your solution

Example 6

Simplify: $\sqrt{16(x + 5)^2}$

Solution
$\sqrt{16(x + 5)^2} = 4(x + 5) = 4x + 20$

You Try It 6

Simplify: $\sqrt{25(a + 3)^2}$

Your solution

Example 7

Simplify: $\sqrt{x^2 + 10x + 25}$

Solution
$\sqrt{x^2 + 10x + 25} = \sqrt{(x + 5)^2} = x + 5$

You Try It 7

Simplify: $\sqrt{x^2 + 14x + 49}$

Your solution

Solutions on p. S25

10.1 Exercises

Objective A   **To simplify numerical radical expressions**

1. ✏️ Describe in your own words how to simplify a radical expression.

2. ✏️ Explain why $2\sqrt{2}$ is in simplest form and $\sqrt{8}$ is not in simplest form.

For Exercises 3 to 26, simplify.

3. $\sqrt{16}$ 4. $\sqrt{64}$ 5. $\sqrt{49}$ 6. $\sqrt{144}$ 7. $\sqrt{32}$ 8. $\sqrt{50}$

9. $\sqrt{8}$ 10. $\sqrt{12}$ 11. $6\sqrt{18}$ 12. $-3\sqrt{48}$ 13. $5\sqrt{40}$ 14. $2\sqrt{28}$

15. $\sqrt{15}$ 16. $\sqrt{21}$ 17. $\sqrt{29}$ 18. $\sqrt{13}$ 19. $-9\sqrt{72}$ 20. $11\sqrt{80}$

21. $\sqrt{45}$ 22. $\sqrt{225}$ 23. $\sqrt{0}$ 24. $\sqrt{210}$ 25. $6\sqrt{128}$ 26. $9\sqrt{288}$

For Exercises 27 to 32, find the decimal approximation rounded to the nearest thousandth.

27. $\sqrt{240}$ 28. $\sqrt{300}$ 29. $\sqrt{288}$ 30. $\sqrt{600}$ 31. $\sqrt{256}$ 32. $\sqrt{324}$

Objective B   **To simplify variable radical expressions**

For Exercises 33 to 72, simplify.

33. $\sqrt{x^6}$ 34. $\sqrt{x^{12}}$ 35. $\sqrt{y^{15}}$ 36. $\sqrt{y^{11}}$

37. $\sqrt{a^{20}}$ 38. $\sqrt{a^{16}}$ 39. $\sqrt{x^4y^4}$ 40. $\sqrt{x^{12}y^8}$

41. $\sqrt{4x^4}$ 42. $\sqrt{25y^8}$ 43. $\sqrt{24x^2}$ 44. $\sqrt{x^3y^{15}}$

45. $\sqrt{60x^5}$ 46. $\sqrt{72y^7}$ 47. $\sqrt{49a^4b^8}$ 48. $\sqrt{144x^2y^8}$

49. $\sqrt{18x^5y^7}$ 50. $\sqrt{32a^5b^{15}}$ 51. $\sqrt{40x^{11}y^7}$ 52. $\sqrt{72x^9y^3}$

53. $\sqrt{80a^9b^{10}}$ 54. $\sqrt{96a^5b^7}$ 55. $2\sqrt{16a^2b^3}$ 56. $5\sqrt{25a^4b^7}$

57. $x\sqrt{x^4y^2}$

58. $y\sqrt{x^3y^6}$

59. $4\sqrt{20a^4b^7}$

60. $5\sqrt{12a^3b^4}$

61. $3x\sqrt{12x^2y^7}$

62. $4y\sqrt{18x^5y^4}$

63. $2x^2\sqrt{8x^2y^3}$

64. $3y^2\sqrt{27x^4y^3}$

65. $\sqrt{25(a+4)^2}$

66. $\sqrt{81(x+y)^4}$

67. $\sqrt{4(x+2)^4}$

68. $\sqrt{9(x+2)^8}$

69. $\sqrt{x^2+4x+4}$

70. $\sqrt{b^2+8b+16}$

71. $\sqrt{y^2+2y+1}$

72. $\sqrt{a^2+6a+9}$

APPLYING THE CONCEPTS

73. Automotive Safety Traffic accident investigators can estimate the speed S, in miles per hour, of a car from the length of its skid mark by using the formula $S = \sqrt{30fl}$, where f is the coefficient of friction (which depends on the type of road surface) and l is the length, in feet, of the skid mark. Say the coefficient of friction is 1.2 and the length of a skid mark is 60 ft.
a. Determine the speed of the car as a radical expression in simplest form.
b. Write the answer to part **a** as a decimal rounded to the nearest integer.

74. Travel The distance a passenger in an airplane can see to the horizon can be approximated by $d = 1.2\sqrt{h}$, where d is the distance to the horizon in miles and h is the height of the plane in feet. To the nearest tenth of a mile, what is the distance to the horizon of a passenger who is flying at an altitude of 5000 ft?

75. If a and b are positive real numbers, does $\sqrt{a+b} = \sqrt{a} + \sqrt{b}$? If not, give an example in which the expressions are not equal.

76. a. Find the two-digit perfect square that has exactly nine factors.
b. Find two whole numbers such that their difference is 10, the smaller number is a perfect square, and the larger number is two less than a perfect square.

77. You are to grade this solution to the problem "Write $\sqrt{72}$ in simplest form." Is the solution correct? If not, what error was made? What is the correct solution?

$\sqrt{72} = \sqrt{4}\sqrt{18}$
$= 2\sqrt{18}$

78. Simplify: **a.** $\sqrt{\sqrt{16}}$ **b.** $\sqrt{\sqrt{81}}$

79. Given $f(x) = \sqrt{2x-1}$, find each of the following. Write your answer in simplest form.
a. $f(1)$ **b.** $f(5)$ **c.** $f(14)$

10.2 Addition and Subtraction of Radical Expressions

Objective A   **To add and subtract radical expressions**

The Distributive Property is used to simplify the sum or difference of radical expressions with like radicands.

$$5\sqrt{2} + 3\sqrt{2} = (5 + 3)\sqrt{2} = 8\sqrt{2}$$

$$6\sqrt{2x} - 4\sqrt{2x} = (6 - 4)\sqrt{2x} = 2\sqrt{2x}$$

Radical expressions that are in simplest form and have unlike radicands cannot be simplified by the Distributive Property.

$2\sqrt{3} + 4\sqrt{2}$ cannot be simplified by the Distributive Property.

HOW TO   Simplify: $4\sqrt{8} - 10\sqrt{2}$

$$
\begin{aligned}
4\sqrt{8} - 10\sqrt{2} &= 4\sqrt{4 \cdot 2} - 10\sqrt{2} \\
&= 4\sqrt{4}\sqrt{2} - 10\sqrt{2} \\
&= 4 \cdot 2\sqrt{2} - 10\sqrt{2} \\
&= 8\sqrt{2} - 10\sqrt{2} \\
&= (8 - 10)\sqrt{2} \\
&= -2\sqrt{2}
\end{aligned}
$$

- Use the Product Property of Square Roots.

- Simplify the expression by using the Distributive Property.

HOW TO   Simplify: $8\sqrt{18x} - 2\sqrt{32x}$

$$
\begin{aligned}
8\sqrt{18x} - 2\sqrt{32x} &= 8\sqrt{9 \cdot 2x} - 2\sqrt{16 \cdot 2x} \\
&= 8\sqrt{9}\sqrt{2x} - 2\sqrt{16}\sqrt{2x} \\
&= 8 \cdot 3\sqrt{2x} - 2 \cdot 4\sqrt{2x} \\
&= 24\sqrt{2x} - 8\sqrt{2x} \\
&= (24 - 8)\sqrt{2x} \\
&= 16\sqrt{2x}
\end{aligned}
$$

- Use the Product Property of Square Roots.

- Simplify the expression by using the Distributive Property.

Example 1

Simplify: $5\sqrt{2} - 3\sqrt{2} + 12\sqrt{2}$

Solution

$$
\begin{aligned}
&5\sqrt{2} - 3\sqrt{2} + 12\sqrt{2} \\
&\quad = (5 - 3 + 12)\sqrt{2} \quad \text{• Distributive Property} \\
&\quad = 14\sqrt{2}
\end{aligned}
$$

You Try It 1

Simplify: $9\sqrt{3} + 3\sqrt{3} - 18\sqrt{3}$

Your solution

Solution on p. S25

Example 2

Simplify: $3\sqrt{12} - 5\sqrt{27}$

Solution

$3\sqrt{12} - 5\sqrt{27}$

$\quad = 3\sqrt{4 \cdot 3} - 5\sqrt{9 \cdot 3}$ • Simplify $\sqrt{12}$ and $\sqrt{27}$.

$\quad = 3\sqrt{4}\,\sqrt{3} - 5\sqrt{9}\,\sqrt{3}$

$\quad = 3 \cdot 2\sqrt{3} - 5 \cdot 3\sqrt{3}$

$\quad = 6\sqrt{3} - 15\sqrt{3}$

$\quad = (6 - 15)\sqrt{3}$ • Distributive Property

$\quad = -9\sqrt{3}$

You Try It 2

Simplify: $2\sqrt{50} - 5\sqrt{32}$

Your solution

Example 3

Simplify: $3\sqrt{12x^3} - 2x\sqrt{3x}$

Solution

$3\sqrt{12x^3} - 2x\sqrt{3x}$

$\quad = 3\sqrt{4x^2 \cdot 3x} - 2x\sqrt{3x}$ • Simplify $\sqrt{12x^3}$.

$\quad = 3\sqrt{4x^2}\sqrt{3x} - 2x\sqrt{3x}$

$\quad = 3 \cdot 2x\sqrt{3x} - 2x\sqrt{3x}$

$\quad = 6x\sqrt{3x} - 2x\sqrt{3x}$

$\quad = (6x - 2x)\sqrt{3x}$ • Distributive Property

$\quad = 4x\sqrt{3x}$

You Try It 3

Simplify: $y\sqrt{28y} + 7\sqrt{63y^3}$

Your solution

Example 4

Simplify: $2x\sqrt{8y} - 3\sqrt{2x^2y} + 2\sqrt{32x^2y}$

Solution

$2x\sqrt{8y} - 3\sqrt{2x^2y} + 2\sqrt{32x^2y}$

$\quad = 2x\sqrt{4 \cdot 2y} - 3\sqrt{x^2 \cdot 2y} + 2\sqrt{16x^2 \cdot 2y}$

$\quad = 2x\sqrt{4}\sqrt{2y} - 3\sqrt{x^2}\sqrt{2y} + 2\sqrt{16x^2}\sqrt{2y}$

$\quad = 2x \cdot 2\sqrt{2y} - 3 \cdot x\sqrt{2y} + 2 \cdot 4x\sqrt{2y}$

$\quad = 4x\sqrt{2y} - 3x\sqrt{2y} + 8x\sqrt{2y}$

$\quad = 9x\sqrt{2y}$

You Try It 4

Simplify: $2\sqrt{27a^5} - 4a\sqrt{12a^3} + a^2\sqrt{75a}$

Your solution

Solutions on p. S25

10.2 Exercises

To add and subtract radical expressions

1. Which of the numbers 2, 9, 20, 25, 50, 81, and 100 are *not* perfect squares?

2. Write down a number that has a perfect-square factor that is greater than 1.

3. ✏ Write a sentence or two that you could email to a friend to explain the concept of a perfect-square factor.

4. Name the perfect-square factors of 540. What number is the largest perfect-square factor of 540?

For Exercises 5 to 60, simplify.

5. $2\sqrt{2} + \sqrt{2}$

6. $3\sqrt{5} + 8\sqrt{5}$

7. $-3\sqrt{7} + 2\sqrt{7}$

8. $4\sqrt{5} - 10\sqrt{5}$

9. $-3\sqrt{11} - 8\sqrt{11}$

10. $-3\sqrt{3} - 5\sqrt{3}$

11. $2\sqrt{x} + 8\sqrt{x}$

12. $3\sqrt{y} + 2\sqrt{y}$

13. $8\sqrt{y} - 10\sqrt{y}$

14. $-5\sqrt{2a} + 2\sqrt{2a}$

15. $-2\sqrt{3b} - 9\sqrt{3b}$

16. $-7\sqrt{5a} - 5\sqrt{5a}$

17. $3x\sqrt{2} - x\sqrt{2}$

18. $2y\sqrt{3} - 9y\sqrt{3}$

19. $2a\sqrt{3a} - 5a\sqrt{3a}$

20. $-5b\sqrt{3x} - 2b\sqrt{3x}$

21. $3\sqrt{xy} - 8\sqrt{xy}$

22. $-4\sqrt{xy} + 6\sqrt{xy}$

23. $\sqrt{45} + \sqrt{125}$

24. $\sqrt{32} - \sqrt{98}$

25. $2\sqrt{2} + 3\sqrt{8}$

26. $4\sqrt{128} - 3\sqrt{32}$

27. $5\sqrt{18} - 2\sqrt{75}$

28. $5\sqrt{75} - 2\sqrt{18}$

29. $5\sqrt{4x} - 3\sqrt{9x}$

30. $-3\sqrt{25y} + 8\sqrt{49y}$

31. $3\sqrt{3x^2} - 5\sqrt{27x^2}$

32. $-2\sqrt{8y^2} + 5\sqrt{32y^2}$

33. $2x\sqrt{xy^2} - 3y\sqrt{x^2y}$

34. $4a\sqrt{b^2a} - 3b\sqrt{a^2b}$

35. $3x\sqrt{12x} - 5\sqrt{27x^3}$

36. $2a\sqrt{50a} + 7\sqrt{32a^3}$

37. $4y\sqrt{8y^3} - 7\sqrt{18y^5}$

38. $2a\sqrt{8ab^2} - 2b\sqrt{2a^3}$

39. $b^2\sqrt{a^5b} + 3a^2\sqrt{ab^5}$

40. $y^2\sqrt{x^5y} + x\sqrt{x^3y^5}$

41. $4\sqrt{2} - 5\sqrt{2} + 8\sqrt{2}$

42. $3\sqrt{3} + 8\sqrt{3} - 16\sqrt{3}$

43. $5\sqrt{x} - 8\sqrt{x} + 9\sqrt{x}$

44. $\sqrt{x} - 7\sqrt{x} + 6\sqrt{x}$

45. $8\sqrt{2} - 3\sqrt{y} - 8\sqrt{2}$

46. $8\sqrt{3} - 5\sqrt{2} - 5\sqrt{3}$

47. $8\sqrt{8} - 4\sqrt{32} - 9\sqrt{50}$

48. $2\sqrt{12} - 4\sqrt{27} + \sqrt{75}$

49. $-2\sqrt{3} + 5\sqrt{27} - 4\sqrt{45}$

50. $-2\sqrt{8} - 3\sqrt{27} + 3\sqrt{50}$

51. $4\sqrt{75} + 3\sqrt{48} - \sqrt{99}$

52. $2\sqrt{75} - 5\sqrt{20} + 2\sqrt{45}$

53. $\sqrt{25x} - \sqrt{9x} + \sqrt{16x}$

54. $\sqrt{4x} - \sqrt{100x} - \sqrt{49x}$

55. $3\sqrt{3x} + \sqrt{27x} - 8\sqrt{75x}$

56. $5\sqrt{5x} + 2\sqrt{45x} - 3\sqrt{80x}$

57. $2a\sqrt{75b} - a\sqrt{20b} + 4a\sqrt{45b}$

58. $2b\sqrt{75a} - 5b\sqrt{27a} + 2b\sqrt{20a}$

59. $x\sqrt{3y^2} - 2y\sqrt{12x^2} + xy\sqrt{3}$

60. $a\sqrt{27b^2} + 3b\sqrt{147a^2} - ab\sqrt{3}$

APPLYING THE CONCEPTS

61. Given $G(x) = \sqrt{x + 5} + \sqrt{5x + 3}$, write $G(3)$ in simplest form.

62. Is the equation $\sqrt{a^2 + b^2} = \sqrt{a} + \sqrt{b}$ true for all real numbers a and b?

63. Explain the steps in simplifying $4\sqrt{2a^3b} + 5\sqrt{8a^3b}$.

64. For each equation, write "ok" if the equation is correct. If the equation is incorrect, correct the right-hand side.
  a. $3\sqrt{ab} + 5\sqrt{ab} = 8\sqrt{2ab}$
  b. $7\sqrt{x^3} - 3x\sqrt{x} - x\sqrt{16x} = 0$
  c. $5 - 2\sqrt{y} = 3\sqrt{y}$

10.3 Multiplication and Division of Radical Expressions

Objective A   To multiply radical expressions

The Product Property of Square Roots is used to multiply radical expressions.

$$\sqrt{2x}\,\sqrt{3y} = \sqrt{2x \cdot 3y} = \sqrt{6xy}$$

HOW TO   Simplify: $\sqrt{2x^2}\sqrt{32x^5}$

$$
\begin{aligned}
\sqrt{2x^2}\,\sqrt{32x^5} &= \sqrt{2x^2 \cdot 32x^5} \\
&= \sqrt{64x^7} \\
&= \sqrt{64x^6 \cdot x} \\
&= \sqrt{64x^6}\,\sqrt{x} = 8x^3\sqrt{x}
\end{aligned}
$$

- Use the Product Property of Square Roots.
- Multiply the radicands.
- Simplify.

HOW TO   Simplify: $\sqrt{2x}(x + \sqrt{2x})$

$$
\begin{aligned}
\sqrt{2x}(x + \sqrt{2x}) &= \sqrt{2x}(x) + \sqrt{2x}\sqrt{2x} \\
&= x\sqrt{2x} + \sqrt{4x^2} \\
&= x\sqrt{2x} + 2x
\end{aligned}
$$

- Use the Distributive Property to remove parentheses.
- Simplify.

Use FOIL to multiply radical expressions with two terms.

HOW TO   Simplify: $(\sqrt{2} - 3x)(\sqrt{2} + x)$

$$
\begin{aligned}
(\sqrt{2} - 3x)(\sqrt{2} + x) &= \sqrt{2 \cdot 2} + x\sqrt{2} - 3x\sqrt{2} - 3x^2 \\
&= \sqrt{4} + (x - 3x)\sqrt{2} - 3x^2 \\
&= 2 - 2x\sqrt{2} - 3x^2
\end{aligned}
$$

- Use the FOIL method to remove parentheses.

The expressions $a + b$ and $a - b$, which differ only in the sign of one term, are called **conjugates.** Recall that $(a + b)(a - b) = a^2 - b^2$.

HOW TO   Simplify: $(2 + \sqrt{7})(2 - \sqrt{7})$

$$
\begin{aligned}
(2 + \sqrt{7})(2 - \sqrt{7}) &= 2^2 - (\sqrt{7})^2 \\
&= 4 - 7 = -3
\end{aligned}
$$

- $(2 + \sqrt{7})(2 - \sqrt{7})$ is the product of conjugates.

> **TAKE NOTE**
> For $x > 0$,
> $(\sqrt{x})^2 = x$ because
> $(\sqrt{x})^2 = \sqrt{x} \cdot \sqrt{x} = \sqrt{x^2} = x$.

HOW TO   Simplify: $(3 + \sqrt{y})(3 - \sqrt{y})$

$$
\begin{aligned}
(3 + \sqrt{y})(3 - \sqrt{y}) &= 3^2 - (\sqrt{y})^2 \\
&= 9 - y
\end{aligned}
$$

- $(3 + \sqrt{y})(3 - \sqrt{y})$ is the product of conjugates.

Example 1   Simplify: $\sqrt{3x^4}\sqrt{2x^2y}\sqrt{6xy^2}$

Solution

$$
\begin{aligned}
&\sqrt{3x^4}\sqrt{2x^2y}\sqrt{6xy^2} \\
&= \sqrt{36x^7y^3} \\
&= \sqrt{36x^6y^2 \cdot xy} \\
&= \sqrt{36x^6y^2}\sqrt{xy} \\
&= 6x^3y\sqrt{xy}
\end{aligned}
$$

- Product Property of Square Roots
- Simplify.

You Try It 1   Simplify: $\sqrt{5a}\sqrt{15a^3b^4}\sqrt{20b^5}$

Your solution

Solution on p. S26

Example 2

Simplify: $\sqrt{3ab}(\sqrt{3a} + \sqrt{9b})$

Solution

$\sqrt{3ab}(\sqrt{3a} + \sqrt{9b})$

$= \sqrt{9a^2b} + \sqrt{27ab^2}$ • **Distributive Property**

$= \sqrt{9a^2 \cdot b} + \sqrt{9b^2 \cdot 3a}$ • **Simplify.**

$= \sqrt{9a^2}\sqrt{b} + \sqrt{9b^2}\sqrt{3a}$

$= 3a\sqrt{b} + 3b\sqrt{3a}$

You Try It 2

Simplify: $\sqrt{5x}(\sqrt{5x} - \sqrt{25y})$

Your solution

Example 3

Simplify: $(\sqrt{a} - \sqrt{b})(\sqrt{a} + \sqrt{b})$

Solution

$(\sqrt{a} - \sqrt{b})(\sqrt{a} + \sqrt{b})$

$= (\sqrt{a})^2 - (\sqrt{b})^2$ • **Product of conjugates**

$= a - b$

You Try It 3

Simplify: $(2\sqrt{x} + 7)(2\sqrt{x} - 7)$

Your solution

Example 4

Simplify: $(2\sqrt{x} - \sqrt{y})(5\sqrt{x} - 2\sqrt{y})$

Solution

$(2\sqrt{x} - \sqrt{y})(5\sqrt{x} - 2\sqrt{y})$

$= 10(\sqrt{x})^2 - 4\sqrt{xy} - 5\sqrt{xy} + 2(\sqrt{y})^2$ • **FOIL**

$= 10x - 9\sqrt{xy} + 2y$

You Try It 4

Simplify: $(3\sqrt{x} - \sqrt{y})(5\sqrt{x} - 2\sqrt{y})$

Your solution

Solutions on p. S26

Objective B **To divide radical expressions**

> **The Quotient Property of Square Roots**
>
> If a and b are positive real numbers, then $\sqrt{\dfrac{a}{b}} = \dfrac{\sqrt{a}}{\sqrt{b}}$ and $\dfrac{\sqrt{a}}{\sqrt{b}} = \sqrt{\dfrac{a}{b}}$.

This property states that the square root of a quotient is equal to the quotient of the square roots.

HOW TO Simplify: $\sqrt{\dfrac{4x^2}{z^6}}$

$\sqrt{\dfrac{4x^2}{z^6}} = \dfrac{\sqrt{4x^2}}{\sqrt{z^6}}$ • **Rewrite the radical expression as the quotient of the square roots.**

$= \dfrac{2x}{z^3}$ • **Simplify.**

Point of Interest

A radical expression that occurs in Einstein's Theory of Relativity is

$$\dfrac{1}{\sqrt{1 - \dfrac{v^2}{c^2}}}$$

where v is the velocity of an object and c is the speed of light.

HOW TO Simplify: $\sqrt{\dfrac{24x^3y^7}{3x^7y^2}}$

$$\sqrt{\dfrac{24x^3y^7}{3x^7y^2}} = \sqrt{\dfrac{8y^5}{x^4}}$$

- Simplify the radicand.

$$= \dfrac{\sqrt{8y^5}}{\sqrt{x^4}}$$

- Rewrite the radical expression as the quotient of the square roots.

$$= \dfrac{\sqrt{4y^4 \cdot 2y}}{\sqrt{x^4}}$$

- Simplify.

$$= \dfrac{\sqrt{4y^4}\sqrt{2y}}{\sqrt{x^4}}$$

$$= \dfrac{2y^2\sqrt{2y}}{x^2}$$

The Quotient Property of Square Roots is used to divide radical expressions.

HOW TO Simplify: $\dfrac{\sqrt{4x^2y}}{\sqrt{xy}}$

$$\dfrac{\sqrt{4x^2y}}{\sqrt{xy}} = \sqrt{\dfrac{4x^2y}{xy}}$$

- Use the Quotient Property of Square Roots.

$$= \sqrt{4x}$$

- Simplify the radicand.

$$= \sqrt{4}\sqrt{x}$$

- Simplify the radical expression.

$$= 2\sqrt{x}$$

The previous examples all result in radical expressions written in simplest form.

> **Simplest Form of a Radical Expression**
>
> For a radical expression to be in simplest form, three conditions must be met:
>
> 1. The radicand contains no factor greater than 1 that is a perfect square.
> 2. There is no fraction under the radical sign.
> 3. There is no radical in the denominator of a fraction.

The procedure used to remove a radical from the denominator is called **rationalizing the denominator.**

HOW TO Simplify: $\dfrac{2}{\sqrt{3}}$

$$\dfrac{2}{\sqrt{3}} = \dfrac{2}{\sqrt{3}} \cdot \boxed{\dfrac{\sqrt{3}}{\sqrt{3}}}$$

- To rationalize the denominator, multiply the expression by $\dfrac{\sqrt{3}}{\sqrt{3}}$, which equals 1.

$$= \dfrac{2\sqrt{3}}{(\sqrt{3})^2}$$

- The radicand in the denominator is a perfect square.

$$= \dfrac{2\sqrt{3}}{3}$$

- Simplify.

When the denominator contains a radical expression with two terms, rationalize the denominator by multiplying the numerator and denominator by the conjugate of the denominator.

HOW TO Simplify: $\dfrac{\sqrt{2y}}{\sqrt{y}+3}$

$$\dfrac{\sqrt{2y}}{\sqrt{y}+3} = \dfrac{\sqrt{2y}}{\sqrt{y}+3} \cdot \dfrac{\sqrt{y}-3}{\sqrt{y}-3}$$

- Multiply the numerator and denominator by $\sqrt{y}-3$, the conjugate of $\sqrt{y}+3$.

$$= \dfrac{\sqrt{2y^2}-3\sqrt{2y}}{(\sqrt{y})^2-3^2} = \dfrac{y\sqrt{2}-3\sqrt{2y}}{y-9}$$

Example 5 Simplify: $\dfrac{\sqrt{4x^2y^5}}{\sqrt{3x^4y}}$

Solution

$$\dfrac{\sqrt{4x^2y^5}}{\sqrt{3x^4y}} = \sqrt{\dfrac{4x^2y^5}{3x^4y}} = \sqrt{\dfrac{4y^4}{3x^2}} = \dfrac{\sqrt{4y^4}}{\sqrt{3x^2}}$$

$$= \dfrac{2y^2}{x\sqrt{3}} = \dfrac{2y^2}{x\sqrt{3}} \cdot \dfrac{\sqrt{3}}{\sqrt{3}}$$

- Rationalize the denominator.

$$= \dfrac{2y^2\sqrt{3}}{3x}$$

You Try It 5 Simplify: $\dfrac{\sqrt{15x^6y^7}}{\sqrt{3x^7y^9}}$

Your solution

Example 6 Simplify: $\dfrac{\sqrt{2}}{\sqrt{2}+\sqrt{6}}$

Solution

$$\dfrac{\sqrt{2}}{\sqrt{2}+\sqrt{6}}$$

- Multiply the numerator and denominator by the conjugate of the denominator.

$$= \dfrac{\sqrt{2}}{\sqrt{2}+\sqrt{6}} \cdot \dfrac{\sqrt{2}-\sqrt{6}}{\sqrt{2}-\sqrt{6}}$$

$$= \dfrac{(\sqrt{2})^2-\sqrt{12}}{2-6} = \dfrac{2-2\sqrt{3}}{-4}$$

$$= \dfrac{2(1-\sqrt{3})}{-4} = \dfrac{1-\sqrt{3}}{-2} = -\dfrac{1-\sqrt{3}}{2}$$

You Try It 6 Simplify: $\dfrac{\sqrt{3}}{\sqrt{3}-\sqrt{6}}$

Your solution

Example 7 Simplify: $\dfrac{3-\sqrt{5}}{2+3\sqrt{5}}$

Solution

$$\dfrac{3-\sqrt{5}}{2+3\sqrt{5}} = \dfrac{3-\sqrt{5}}{2+3\sqrt{5}} \cdot \dfrac{2-3\sqrt{5}}{2-3\sqrt{5}}$$

- Rationalize the denominator.

$$= \dfrac{6-9\sqrt{5}-2\sqrt{5}+3(\sqrt{5})^2}{4-9\cdot 5}$$

$$= \dfrac{6-11\sqrt{5}+15}{4-45}$$

$$= \dfrac{21-11\sqrt{5}}{-41} = -\dfrac{21-11\sqrt{5}}{41}$$

You Try It 7 Simplify: $\dfrac{5+\sqrt{y}}{1-2\sqrt{y}}$

Your solution

Solutions on p. S26

10.3 Exercises

Objective A   **To multiply radical expressions**

1. ✏️ Explain in words and then write in symbols the Product Property of Square Roots.

2. ✏️ Give an example to show that $\sqrt{a^2} \neq a$.

For Exercises 3 to 37, simplify.

3. $\sqrt{5} \cdot \sqrt{5}$

4. $\sqrt{11} \cdot \sqrt{11}$

5. $\sqrt{3} \cdot \sqrt{12}$

6. $\sqrt{2} \cdot \sqrt{8}$

7. $\sqrt{x} \cdot \sqrt{x}$

8. $\sqrt{y} \cdot \sqrt{y}$

9. $\sqrt{xy^3} \cdot \sqrt{x^5y}$

10. $\sqrt{a^3b^5} \cdot \sqrt{ab^5}$

11. $\sqrt{3a^2b^5} \cdot \sqrt{6ab^7}$

12. $\sqrt{5x^3y} \cdot \sqrt{10x^2y}$

13. $\sqrt{6a^3b^2} \cdot \sqrt{24a^5b}$

14. $\sqrt{8ab^5} \cdot \sqrt{12a^7b}$

15. $\sqrt{2}(\sqrt{2} - \sqrt{3})$

16. $3(\sqrt{12} - \sqrt{3})$

17. $\sqrt{x}(\sqrt{x} - \sqrt{y})$

18. $\sqrt{b}(\sqrt{a} - \sqrt{b})$

19. $\sqrt{5}(\sqrt{10} - \sqrt{x})$

20. $\sqrt{6}(\sqrt{y} - \sqrt{18})$

21. $\sqrt{8}(\sqrt{2} - \sqrt{5})$

22. $\sqrt{10}(\sqrt{20} - \sqrt{a})$

23. $(\sqrt{x} - 3)^2$

24. $(2\sqrt{a} - y)^2$

25. $\sqrt{3a}(\sqrt{3a} - \sqrt{3b})$

26. $\sqrt{5x}(\sqrt{10x} - \sqrt{x})$

27. $\sqrt{2ac} \cdot \sqrt{5ab} \cdot \sqrt{10cb}$

28. $\sqrt{3xy} \cdot \sqrt{6x^3y} \cdot \sqrt{2y^2}$

29. $(\sqrt{5} + 3)(2\sqrt{5} - 4)$

30. $(2 - 3\sqrt{7})(5 + 2\sqrt{7})$

31. $(4 + \sqrt{8})(3 + \sqrt{2})$

32. $(6 - \sqrt{27})(2 + \sqrt{3})$

33. $(2\sqrt{x} + 4)(3\sqrt{x} - 1)$

34. $(5 + \sqrt{y})(6 - 3\sqrt{y})$

35. $(3\sqrt{x} - 2y)(5\sqrt{x} - 4y)$

36. $(5\sqrt{x} + 2\sqrt{y})(3\sqrt{x} - \sqrt{y})$

37. $(\sqrt{x} - \sqrt{y})(\sqrt{x} + \sqrt{y})$

Objective B   **To divide radical expressions**

38. ✎ Why is $\dfrac{\sqrt{3}}{3}$ in simplest form but $\dfrac{1}{\sqrt{3}}$ not in simplest form?

39. ✎ Why can we multiply $\dfrac{2}{\sqrt{5}}$ by $\dfrac{\sqrt{5}}{\sqrt{5}}$ without changing the value of $\dfrac{2}{\sqrt{5}}$?

For Exercises 40 to 69, simplify.

40. $\dfrac{\sqrt{32}}{\sqrt{2}}$

41. $\dfrac{\sqrt{45}}{\sqrt{5}}$

42. $\dfrac{\sqrt{98}}{\sqrt{2}}$

43. $\dfrac{\sqrt{48}}{\sqrt{3}}$

44. $\dfrac{\sqrt{27a}}{\sqrt{3a}}$

45. $\dfrac{\sqrt{72x^5}}{\sqrt{2x}}$

46. $\dfrac{\sqrt{15x^3y}}{\sqrt{3xy}}$

47. $\dfrac{\sqrt{40x^5y^2}}{\sqrt{5xy}}$

48. $\dfrac{\sqrt{2a^5b^4}}{\sqrt{98ab^4}}$

49. $\dfrac{\sqrt{48x^5y^2}}{\sqrt{3x^3y}}$

50. $\dfrac{\sqrt{9xy^2}}{\sqrt{27x}}$

51. $\dfrac{\sqrt{4x^2y}}{\sqrt{3xy^3}}$

52. $\dfrac{\sqrt{16x^3y^2}}{\sqrt{8x^3y}}$

53. $\dfrac{\sqrt{2}}{\sqrt{8}+4}$

54. $\dfrac{1}{\sqrt{2}-3}$

55. $\dfrac{5}{\sqrt{7}-3}$

56. $\dfrac{3}{5+\sqrt{5}}$

57. $\dfrac{\sqrt{3}}{5-\sqrt{27}}$

58. $\dfrac{7}{\sqrt{2}-7}$

59. $\dfrac{3-\sqrt{6}}{5-2\sqrt{6}}$

60. $\dfrac{6-2\sqrt{3}}{4+3\sqrt{3}}$

61. $\dfrac{-6}{4+\sqrt{2}}$

62. $\dfrac{\sqrt{2}+2\sqrt{6}}{2\sqrt{2}-3\sqrt{6}}$

63. $\dfrac{2\sqrt{3}-\sqrt{6}}{5\sqrt{3}+2\sqrt{6}}$

64. $\dfrac{3+\sqrt{x}}{2-\sqrt{x}}$

65. $\dfrac{-\sqrt{15}}{3-\sqrt{12}}$

66. $\dfrac{\sqrt{a}-4}{2\sqrt{a}+2}$

67. $\dfrac{\sqrt{xy}}{\sqrt{x}-\sqrt{y}}$

68. $\dfrac{\sqrt{x}}{\sqrt{x}-\sqrt{y}}$

69. $\dfrac{-12}{\sqrt{6}-3}$

APPLYING THE CONCEPTS

70. ✎ In your own words, describe the process of rationalizing the denominator.

71. Show that $1+\sqrt{6}$ and $1-\sqrt{6}$ are solutions of the equation $x^2-2x-5=0$.

72. Answer true or false. If the equation is false, correct it.

  a. $(\sqrt{y})^4 = y^2$ **b.** $(2\sqrt{x})^3 = 8x\sqrt{x}$ **c.** $(\sqrt{x}+1)^2 = x+1$ **d.** $\dfrac{1}{2-\sqrt{3}} = 2+\sqrt{3}$

10.4 Solving Equations Containing Radical Expressions

Objective A
To solve an equation containing a radical expression

An equation that contains a variable expression in a radicand is a **radical equation.**

$$\sqrt{x} = 4$$
$$\sqrt{x + 2} = \sqrt{x - 7}$$ } Radical equations

The following property of equality, which states that if two numbers are equal, the squares of the numbers are equal, is used to solve radical equations.

> **Property of Squaring Both Sides of an Equation**
>
> If a and b are real numbers and $a = b$, then $a^2 = b^2$.

To solve a radical equation with one radical, use the following procedure.

> **Solving a Radical Equation**
>
> **1.** Write the equation with the radical alone on one side.
> **2.** Square both sides of the equation.
> **3.** Solve for the variable.
> **4.** Check the solution(s) in the original equation.

HOW TO   Solve: $\sqrt{x - 2} - 7 = 0$

$$\sqrt{x - 2} - 7 = 0$$
$$\sqrt{x - 2} = 7$$
$$(\sqrt{x - 2})^2 = 7^2$$
$$x - 2 = 49$$
$$x = 51$$

- Isolate the radical by adding 7 to both sides of the equation.
- Square both sides of the equation.
- Solve the resulting equation.

Study Tip

When we suggest that you check a solution, substitute the solution into the original equation.

Check:
$$\sqrt{x - 2} - 7 = 0$$
$$\overline{\sqrt{51 - 2} - 7} \,\Big|\, 0$$
$$\sqrt{49} - 7 \,\Big|\, 0$$
$$7 - 7 \,\Big|\, 0$$
$$0 = 0 \quad \text{A true equation}$$

The solution is 51.

When both sides of an equation are squared, the resulting equation may have a solution that is not a solution of the original equation. Checking a proposed solution of a radical equation, as we did above, is a necessary step.

HOW TO   Solve: $\sqrt{2x - 5} + 3 = 0$

$$\sqrt{2x - 5} + 3 = 0$$
$$\sqrt{2x - 5} = -3$$
$$(\sqrt{2x - 5})^2 = (-3)^2$$
$$2x - 5 = 9$$
$$2x = 14$$
$$x = 7$$

- Isolate the radical by subtracting 3 from both sides of the equation.
- Square both sides of the equation.
- Solve for x.

Here is the check for the equation on the preceding page.

$Check:$ $\dfrac{\sqrt{2x-5}+3=0}{}$

$$\begin{array}{c|c} \sqrt{2 \cdot 7 - 5} + 3 & 0 \\ \sqrt{14 - 5} + 3 & 0 \\ \sqrt{9} + 3 & 0 \\ 3 + 3 & 0 \\ 6 & \neq 0 \end{array}$$

7 does not check as a solution. The equation has no solution.

Example 1

Solve: $\sqrt{3x} + 2 = 5$

Solution

$\sqrt{3x} + 2 = 5$

$\sqrt{3x} = 3$ • Isolate $\sqrt{3x}$.

$(\sqrt{3x})^2 = 3^2$ • Square both

$3x = 9$ sides.

$x = 3$ • Solve for x.

The solution is 3.

$Check:$

$\dfrac{\sqrt{3x}+2=5}{}$

$$\begin{array}{c|c} \sqrt{3 \cdot 3} + 2 & 5 \\ \sqrt{9} + 2 & 5 \\ 3 + 2 & 5 \\ 5 & = 5 \end{array}$$

You Try It 1

Solve: $\sqrt{4x} + 3 = 7$

Your solution

Example 2

Solve: $1 = \sqrt{x} - \sqrt{x - 5}$

Solution

When an equation contains two radicals, isolate the radicals one at a time.

$1 = \sqrt{x} - \sqrt{x - 5}$

$1 + \sqrt{x - 5} = \sqrt{x}$ • Isolate $\sqrt{x}$.

$(1 + \sqrt{x - 5})^2 = (\sqrt{x})^2$ • Square both sides.

$1 + 2\sqrt{x - 5} + (x - 5) = x$ • Square the binomial.

$2\sqrt{x - 5} = 4$ • Simplify.

$\sqrt{x - 5} = 2$ • Isolate $\sqrt{x - 5}$.

$(\sqrt{x - 5})^2 = 2^2$ • Square both sides.

$x - 5 = 4$

$x = 9$ • Solve for x.

$Check:$

$\dfrac{1 = \sqrt{x} - \sqrt{x - 5}}{}$

$$\begin{array}{c|c} 1 & \sqrt{9} - \sqrt{9 - 5} \\ 1 & \sqrt{9} - \sqrt{4} \\ 1 & 3 - 2 \\ 1 & = 1 \end{array}$$

The solution is 9.

You Try It 2

Solve: $\sqrt{x} + \sqrt{x + 9} = 9$

Your solution

Solutions on p. S26

Objective B   **To solve application problems**

A **right triangle** is a triangle that contains a 90° angle. The side opposite the 90° angle is called the **hypotenuse.** The other two sides are called **legs.**

Pythagoras (c. 580 B.C.–520 B.C.)

Pythagoras, a Greek mathematician who lived around 550 B.C., is given credit for the Pythagorean Theorem. It states that the square of the hypotenuse of a right triangle is equal to the sum of the squares of the two legs. Actually, this theorem was known to the Babylonians around 1200 B.C.

Point of Interest

The first known proof of this theorem occurs in a Chinese text, *Arithmetic Classic*, which was first written around 600 B.C. (but there are no existing copies) and revised over a period of 500 years. The earliest known copy of this text dates from approximately 100 B.C.

> **Pythagorean Theorem**
>
> If a and b are the lengths of the legs of a right triangle and c is the length of the hypotenuse, then $c^2 = a^2 + b^2$.

Using this theorem, we can find the hypotenuse of a right triangle when we know the two legs. Use the formula

$$
\begin{aligned}
\text{Hypotenuse} &= \sqrt{(\text{leg})^2 + (\text{leg})^2} \\
c &= \sqrt{a^2 + b^2} \\
&= \sqrt{(5)^2 + (12)^2} \\
&= \sqrt{25 + 144} \\
&= \sqrt{169} \\
&= 13
\end{aligned}
$$

The leg of a right triangle can be found when one leg and the hypotenuse are known. Use the formula

$$
\begin{aligned}
\text{Leg} &= \sqrt{(\text{hypotenuse})^2 - (\text{leg})^2} \\
a &= \sqrt{c^2 - b^2} \\
&= \sqrt{(25)^2 - (20)^2} \\
&= \sqrt{625 - 400} \\
&= \sqrt{225} \\
&= 15
\end{aligned}
$$

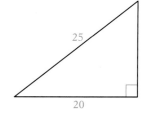

Example 3 and You Try It 3 on the following page illustrate the use of the Pythagorean Theorem. Example 4 and You Try It 4 illustrate other applications of radical equations.

Example 3	You Try It 3

A guy wire is attached to a point 20 m above the ground on a telephone pole. The wire is anchored to the ground at a point 8 m from the base of the pole. Find the length of the guy wire. Round to the nearest tenth.

A ladder 8 ft long is resting against a building. How high on the building will the ladder reach when the bottom of the ladder is 3 ft from the building? Round to the nearest hundredth.

Strategy

To find the length of the guy wire, use the Pythagorean Theorem. One leg is 20 m. The other leg is 8 m. The guy wire is the hypotenuse. Solve the Pythagorean Theorem for the hypotenuse.

Your strategy

Solution
$$c = \sqrt{a^2 + b^2}$$
$$= \sqrt{(20)^2 + (8)^2} \qquad \bullet \ a = 20, \ b = 8$$
$$= \sqrt{400 + 64} = \sqrt{464} \approx 21.5$$

The guy wire has a length of approximately 21.5 m.

Your solution

Example 4	You Try It 4

How far would a submarine periscope have to be above the water to locate a ship 4 mi away? The equation for the distance in miles that the lookout can see is $d = \sqrt{1.5h}$, where h is the height in feet above the surface of the water. Round to the nearest hundredth.

Find the length of a pendulum that makes one swing in 2.5 s. The equation for the time for one swing is $T = 2\pi\sqrt{\frac{L}{32}}$, where T is the time in seconds and L is the length in feet. Use 3.14 for π. Round to the nearest hundredth.

Strategy

To find the height above the water, replace d in the equation with the given value and solve for h.

Your strategy

Solution
$$d = \sqrt{1.5h}$$
$$4 = \sqrt{1.5h} \qquad \bullet \ d = 4$$
$$4^2 = (\sqrt{1.5h})^2$$
$$16 = 1.5h$$
$$10.67 \approx h$$

The periscope must be approximately 10.67 ft above the water.

Your solution

Solutions on pp. S26–S27

10.4 Exercises

> **Objective A**   **To solve an equation containing a radical expression**

For Exercises 1 to 22, solve and check.

1. $\sqrt{x} = 5$ **2.** $\sqrt{y} = 7$ **3.** $\sqrt{a} = 12$ **4.** $\sqrt{a} = 9$ **5.** $\sqrt{5x} = 5$

6. $\sqrt{4x} + 5 = 2$ **7.** $\sqrt{3x} + 9 = 4$ **8.** $\sqrt{3x - 2} = 4$ **9.** $\sqrt{5x + 6} = 1$

10. $\sqrt{2x + 1} = 7$ **11.** $\sqrt{5x + 4} = 3$ **12.** $0 = 2 - \sqrt{3 - x}$ **13.** $0 = 5 - \sqrt{10 + x}$

14. $\sqrt{5x + 2} = 0$ **15.** $\sqrt{3x - 7} = 0$ **16.** $\sqrt{3x} - 6 = -4$

17. $\sqrt{x^2 + 5} = x + 1$ **18.** $\sqrt{x^2 - 5} = 5 - x$ **19.** $\sqrt{x + 4} - \sqrt{x - 1} = 1$

20. $\sqrt{x} + \sqrt{x - 12} = 2$ **21.** $\sqrt{2x + 1} - \sqrt{2x - 4} = 1$ **22.** $\sqrt{3x + 1} - \sqrt{3x - 2} = 1$

> **Objective B**   **To solve application problems**

23. **Sports**   The infield of a baseball diamond is a square. The distance between successive bases is 90 ft. The pitcher's mound is on the diagonal between home plate and second base at a distance of 60.5 ft from home plate. (See the figure at the right.) Is the pitcher's mound more or less than halfway between home plate and second base?

24. **Sports**   The infield of a softball diamond is a square. The distance between successive bases is 60 ft. The pitcher's mound is on the diagonal between home plate and second base at a distance of 46 ft from home plate. Is the pitcher's mound more or less than halfway between home plate and second base?

25. **Periscopes**   How far would a submarine periscope have to be above the water to locate a ship 5 mi away? The equation for the distance in miles that the lookout can see is $d = \sqrt{1.5h}$, where h is the height in feet above the surface of the water. Round to the nearest hundredth.

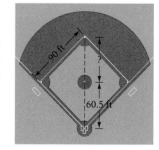

26. **Building Maintenance**   A 16-foot ladder is leaning against a building. How high on the building will the ladder reach when the bottom of the ladder is 5 ft from the building? (See the figure at the right.) Round to the nearest tenth.

28.8 in.

36 in.

27. **Home Entertainment**   The measure of a big-screen television is given by the length of a diagonal across the screen. A 36-inch television has a width of 28.8 in. Find the height of the screen to the nearest tenth of an inch.

28. **Home Entertainment**   The measure of a television screen is given by the length of a diagonal across the screen. A 33-inch television has a width of 26.4 in. Find the height of the screen to the nearest tenth of an inch.

29. **Recreation**   The speed of a child riding a merry-go-round at a carnival is given by the equation $v = \sqrt{12r}$, where v is the speed in feet per second and r is the distance in feet from the center of the merry-go-round to the rider. If a child is moving at 15 ft/s, how far is the child from the center of the merry-go-round?

30. **Time**   Find the length of a pendulum that makes one swing in 1.5 s. The equation for the time of one swing of a pendulum is $T = 2\pi\sqrt{\dfrac{L}{32}}$, where T is the time in seconds and L is the length in feet. Use 3.14 for π. Round to the nearest hundredth.

APPLYING THE CONCEPTS

31. **Geometry**   In the coordinate plane, a triangle is formed by drawing lines between the points (0, 0) and (5, 0), (5, 0) and (5, 12), and (5, 12) and (0, 0). Find the perimeter of the triangle.

32. **Geometry**   The hypotenuse of a right triangle is $5\sqrt{2}$ cm, and one leg is $4\sqrt{2}$ cm.
    a. Find the perimeter of the triangle.
    b. Find the area of the triangle.

33. If a and b are real numbers and $a^2 = b^2$, does $a = b$? Explain your answer.

34. **Geometry**   Can the Pythagorean Theorem be used to find the length of side c of the triangle at the right? If so, determine c. If not, explain why the theorem cannot be used.

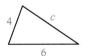

35. **Fountain Design**   A circular fountain is being designed for a triangular plaza in a cultural center. The fountain is placed so that each side of the triangle touches the fountain, as shown in the diagram at the right. Find the area of the fountain. The formula for the radius of the circle is given by

$$r = \sqrt{\frac{(s-a)(s-b)(s-c)}{s}}$$

28 ft   37 ft

31 ft

where $s = \dfrac{1}{2}(a + b + c)$ and a, b, and c are the lengths of the sides of the triangle. Round to the nearest hundredth.

36. Complete each statement using $<$, $=$, or $>$.
    a. For an acute triangle with longest side c, $a^2 + b^2$ ____ c^2.
    b. For a right triangle with longest side c, $a^2 + b^2$ ____ c^2.
    c. For an obtuse triangle with longest side c, $a^2 + b^2$ ____ c^2.

Focus on Problem Solving

Deductive Reasoning

Deductive reasoning uses a rule or statement of fact to reach a conclusion. For instance, if two angles of one triangle are equal to two angles of another triangle, then the two triangles are similar. Thus any time we establish this fact about two triangles, we know that the triangles are similar. Below are two examples of deductive reasoning.

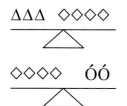

Given that △△△ = ◇◇◇◇ and ◇◇◇◇ = ÓÓ, then △△△△△△ is equivalent to how many Ós?

Because 3 △s = 4 ◇s and 4 ◇s = 2 Ós, 3 △s = 2 Ós.

6 △s is twice 3 △s. We need to find twice 2 Ós, which is 4 Ós.

Therefore, △△△△△△ = ÓÓÓÓ.

Lomax, Parish, Thorpe, and Wong are neighbors. Each drives a different type of vehicle: a compact car, a sedan, a sports car, or a station wagon. From the following statements, determine which type of vehicle each of the neighbors drives.

1. Although the vehicle owned by Lomax has more mileage on it than does either the sedan or the sports car, it does not have the highest mileage of all four cars. (Use X1 in the chart below to eliminate possibilities that this statement rules out.)

2. Wong and the owner of the sports car live on one side of the street, and Thorpe and the owner of the compact car live on the other side of the street. (Use X2 to eliminate possibilities that this statement rules out.)

3. Thorpe owns the vehicle with the most mileage on it. (Use X3 to eliminate possibilities that this statement rules out.)

> **TAKE NOTE**
> To use the chart to solve this problem, write an X in a box to indicate that a possibility has been eliminated. Write a √ to show that a match has been found. When a row or column has 3 X's, a √ is written in the remaining open box in that row or column of the chart.

	Compact	*Sedan*	*Sports Car*	*Wagon*
Lomax	√	X1	X1	X2
Parish	X2	X2	√	X2
Thorpe	X2	X3	X2	√
Wong	X2		X2	

Lomax drives the compact car, Parish drives the sports car, Thorpe drives the station wagon, and Wong drives the sedan.

1. Given that ‡‡ = ••••• and ••••• = ΛΛ, then ‡‡‡‡‡ = how many Λs?

2. Given that □□□□□□ = ÓÓÓÓ and ÓÓÓÓ = ÎÎ, then □□□ = how many Îs?

3. Given that ¤¤¤¤ = ΩΩΩ and ΩΩΩ = ΔΔ, then ΔΔΔΔ = how many ¤s?

4. Given that ¥¥¥¥¥ = §§ and §§ = ÂÂÂ, then ÂÂÂÂÂÂ = how many ¥s?

5. Anna, Kay, Megan, and Nicole decide to travel together during spring break, but they need to find a destination where each of them will be able to participate in her favorite sport (golf, horseback riding, sailing, or tennis). From the following statements, determine the favorite sport of each student.

    a. Anna and the student whose favorite sport is sailing both like to swim, whereas Nicole and the student whose favorite sport is tennis would prefer to scuba dive.

    b. Megan and the student whose favorite sport is sailing are roommates. Nicole and the student whose favorite sport is golf live by themselves in singles.

6. Chang, Nick, Pablo, and Saul each take a different form of transportation (bus, car, subway, or taxi) from the office to the airport. From the following statements, determine which form of transportation each takes.

    a. Chang spent more on transportation than the fellow who took the bus but less than the fellow who took the taxi.

    b. Pablo, who did not travel by bus and who spent the least on transportation, arrived at the airport after Nick but before the fellow who took the subway.

    c. Saul spent less on transportation than either Chang or Nick.

Projects and Group Activities

Distance to the Horizon

The formula $d = \sqrt{1.5h}$ can be used to calculate the approximate distance d (in miles) that a person can see who uses a periscope h feet above the water. That formula is derived by using the Pythagorean Theorem.

Consider the diagram (not to scale) at the right, which shows Earth as a sphere and the periscope as extending h feet above its surface. From geometry, because AB is tangent to the circle and OA is a radius, triangle AOB is a right triangle. Therefore,

$$(OA)^2 + (AB)^2 = (OB)^2$$

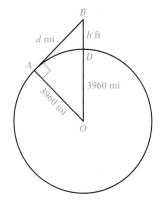

Substituting into this formula, we have

$$3960^2 + d^2 = \left(3960 + \frac{h}{5280}\right)^2$$

• Because h is in feet, $\frac{h}{5280}$ is in miles.

$$3960^2 + d^2 = 3960^2 + \frac{2 \cdot 3960}{5280}h + \left(\frac{h}{5280}\right)^2$$

$$d^2 = \frac{3}{2}h + \left(\frac{h}{5280}\right)^2$$

$$d = \sqrt{\frac{3}{2}h + \left(\frac{h}{5280}\right)^2}$$

At this point, an assumption is made that $\sqrt{\frac{3}{2}h + \left(\frac{h}{5280}\right)^2} \approx \sqrt{1.5h}$, where we have written $\frac{3}{2}$ as 1.5. Thus $d \approx \sqrt{1.5h}$ is used to approximate the distance that can be seen using a periscope h feet above the water.

1. Write a paragraph that justifies the assumption that

$$\sqrt{\frac{3}{2}h + \left(\frac{h}{5280}\right)^2} \approx \sqrt{1.5h}$$

(*Suggestion:* Evaluate each expression for various values of h. Because h is the height of a periscope above water, it is unlikely that $h > 25$ ft.)

2. The distance d is the distance from the top of the periscope to A. The distance along the surface of the water is given by arc AD. This distance can be approximated by the equation

$$L \approx \sqrt{1.5h} + 0.306186\left(\sqrt{\frac{h}{5280}}\right)^3$$

Using this formula, calculate L when $h = 10$.

Chapter 10 Summary

Key Words

Key Words	Examples
A *square root* of a positive number a is a number whose square is a. Every positive number has two square roots, one a positive and one a negative number. The square root of a negative number is not a real number. [10.1A, p. 481]	A square root of 49 is 7 because $7^2 = 49$. A square root of 49 is -7 because $(-7)^2 = 49$. $\sqrt{-9}$ is not a real number.
The symbol $\sqrt{\ }$ is called a *radical sign* and is used to indicate the positive or *principal square root* of a number. The negative square root of a number is indicated by placing a negative sign in front of the radical. The *radicand* is the expression under the radical sign. [10.1A, p. 481]	$\sqrt{49} = 7$ $-\sqrt{49} = -7$ In the expression $\sqrt{49xy}$, $49xy$ is the radicand.
The square of an integer is a *perfect square*. If a number is not a perfect square, its square root can only be approximated. Such square roots are *irrational numbers*. Their decimal representations never terminate or repeat. [10.1A, p. 481]	1, 4, 9, 16, 25, 36, 49, 64,... are examples of perfect squares. 7 is not a perfect square. $\sqrt{7}$ is an irrational number.
Conjugates are expressions with two terms that differ only in the sign of one term. The expressions $a + b$ and $a - b$ are conjugates. [10.3A, p. 491]	$-5 + \sqrt{11}$ and $-5 - \sqrt{11}$ are conjugates. $\sqrt{x} - 3$ and $\sqrt{x} + 3$ are conjugates.
A *radical equation* is an equation that contains a variable expression in a radicand. [10.4A, p. 497]	$\sqrt{2x} + 5 = 9$ is a radical equation. $2x + \sqrt{5} = 9$ is not a radical equation.
A *right triangle* is a triangle that contains a 90° angle. The side opposite the 90° angle is the *hypotenuse*. The other two sides are called *legs*. [10.4B, p. 499]	

Essential Rules and Procedures

Essential Rules and Procedures	Examples
The Product Property of Square Roots [10.1A, p. 481] If a and b are positive real numbers, then $\sqrt{ab} = \sqrt{a} \cdot \sqrt{b}$. Use the Product Property of Square Roots and a knowledge of perfect squares to simplify radicands that are not perfect squares.	$\sqrt{28} = \sqrt{4 \cdot 7} = \sqrt{4} \cdot \sqrt{7} = 2\sqrt{7}$ $\sqrt{9x^7} = \sqrt{9x^6 \cdot x} = \sqrt{9x^6}\sqrt{x} = 3x^3\sqrt{x}$

Adding or Subtracting Radical Expressions [10.2A, p. 487]
The Distributive Property is used to simplify the sum or
difference of radical expressions with like radicands.

$8\sqrt{2x} - 3\sqrt{2x} = (8 - 3)\sqrt{2x} = 5\sqrt{2x}$

Multiplying Radical Expressions [10.3A, p. 491]
The Product Property of Square Roots is used to multiply
radical expressions.

$\sqrt{2y}(\sqrt{3} - \sqrt{x}) = \sqrt{6y} - \sqrt{2xy}$

Use FOIL to multiply radical expressions with two terms.

$(3 - \sqrt{x})(5 + \sqrt{x})$
$= 15 + 3\sqrt{x} - 5\sqrt{x} - (\sqrt{x})^2$
$= 15 - 2\sqrt{x} - x$

The Quotient Property of Square Roots [10.3B, p. 492]

If a and b are positive real numbers, then $\sqrt{\dfrac{a}{b}} = \dfrac{\sqrt{a}}{\sqrt{b}}$ and

$\dfrac{\sqrt{a}}{\sqrt{b}} = \sqrt{\dfrac{a}{b}}$.

$\dfrac{\sqrt{27}}{\sqrt{3}} = \sqrt{\dfrac{27}{3}} = \sqrt{9} = 3$

The Quotient Property of Square Roots is used to divide radical
expressions.

$\dfrac{\sqrt{3x^5y}}{\sqrt{75xy^3}} = \sqrt{\dfrac{3x^5y}{75xy^3}} = \sqrt{\dfrac{x^4}{25y^2}} = \dfrac{x^2}{5y}$

Simplest Form of a Radical Expression [10.3B, p. 493]
For a radical expression to be in simplest form, three conditions
must be met:

1. The radicand contains no factor greater than 1 that is a
   perfect square.

2. There is no fraction under the radical sign.

3. There is no radical in the denominator of a fraction.

$\sqrt{12}$, $\sqrt{\dfrac{3}{4}}$, and $\dfrac{1}{\sqrt{3}}$ are not in
simplest form.

$5\sqrt{3}$ and $\dfrac{\sqrt{3}}{3}$ are in simplest form.

Rationalizing the Denominator [10.3B, p. 493]
The procedure used to remove a radical from the denominator
is called **rationalizing the denominator.**

$\dfrac{5}{\sqrt{7}} = \dfrac{5}{\sqrt{7}} \cdot \dfrac{\sqrt{7}}{\sqrt{7}} = \dfrac{5\sqrt{7}}{7}$

Property of Squaring Both Sides of an Equation
[10.4A, p. 497]
If a and b are real numbers and $a = b$, then $a^2 = b^2$.

$\sqrt{x} = 5$
$(\sqrt{x})^2 = 5^2$
$x = 25$

Solving a Radical Equation Containing One Radical
[10.4A, p. 497]

1. Write the equation with the radical alone on one side.

2. Square both sides of the equation.

3. Solve for the variable.

4. Check the solution(s) in the original equation.

$\sqrt{2x} - 1 = 5$
$\sqrt{2x} = 6$ • Isolate the radical.
$(\sqrt{2x})^2 = 6^2$ • Square both sides.
$2x = 36$
$x = 18$ • Solve for x.
The solution checks.

Pythagorean Theorem [10.4B, p. 499]
If a and b are lengths of the legs of a right triangle and c is the
length of the hypotenuse, then $c^2 = a^2 + b^2$.

Two legs of a right triangle measure
4 cm and 7 cm. Find the length of
the hypotenuse.
$c = \sqrt{a^2 + b^2}$
$c = \sqrt{4^2 + 7^2}$ • $a = 4$, $b = 7$
$c = \sqrt{16 + 49}$
$c = \sqrt{65}$
The length of the hypotenuse is $\sqrt{65}$ cm.

Chapter 10 Review Exercises

1. Simplify: $\sqrt{3}(\sqrt{12} - \sqrt{3})$

2. Simplify: $3\sqrt{18a^5b}$

3. Simplify: $2\sqrt{36}$

4. Simplify: $\sqrt{6a}(\sqrt{3a} + \sqrt{2a})$

5. Simplify: $\dfrac{12}{\sqrt{6}}$

6. Simplify: $2\sqrt{8} - 3\sqrt{32}$

7. Simplify: $(3 - \sqrt{7})(3 + \sqrt{7})$

8. Solve: $\sqrt{x + 3} - \sqrt{x} = 1$

9. Simplify: $\dfrac{2x}{\sqrt{3} - \sqrt{5}}$

10. Simplify: $-3\sqrt{120}$

11. Solve: $\sqrt{5x} = 10$

12. Simplify: $5\sqrt{48}$

13. Simplify: $\dfrac{\sqrt{98x^7y^9}}{\sqrt{2x^3y}}$

14. Solve: $3 - \sqrt{7x} = 5$

15. Simplify: $6a\sqrt{80b} - \sqrt{180a^2b} + 5a\sqrt{b}$

16. Simplify: $4\sqrt{250}$

17. Simplify: $2x\sqrt{60x^3y^3} + 3x^2y\sqrt{15xy}$

18. Simplify: $(4\sqrt{y} - \sqrt{5})(2\sqrt{y} + 3\sqrt{5})$

19. Simplify: $3\sqrt{12x} + 5\sqrt{48x}$

20. Solve: $\sqrt{2x - 3} + 4 = 0$

21. Simplify: $\dfrac{8}{\sqrt{x} - 3}$

22. Simplify: $4y\sqrt{243x^{17}y^9}$

23. Simplify: $y\sqrt{24y^6}$

24. Solve: $2x + 4 = \sqrt{x^2 + 3}$

25. Simplify:
$2x^2\sqrt{18x^2y^5} + 6y\sqrt{2x^6y^3} - 9xy^2\sqrt{8x^4y}$

26. Simplify: $\dfrac{16}{\sqrt{a}}$

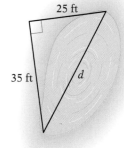

25 ft

35 ft d

27. Surveying To find the distance across a pond, a surveyor constructs a right triangle as shown at the right. Find the distance d across the pond. Round to the nearest foot.

28. Physics The weight of an object is related to the distance the object is above the surface of Earth. An equation for this relationship is

$d = 4000\sqrt{\dfrac{W_0}{W_d}} - 4000$, where W_0 is an object's weight on the surface of Earth and W_d is the object's weight at a distance of d miles above Earth's surface. If a space explorer weighs 36 lb at a distance of 4000 mi above the surface of Earth, how much does the explorer weigh on the surface of Earth?

29. Tsunamis A tsunami is a great sea wave produced by underwater earthquakes or volcanic eruption. The velocity of a tsunami as it approaches land depends on the depth of the water and can be approximated by the equation $v = 3\sqrt{d}$, where d is the depth of the water in feet and v is the velocity of the tsunami in feet per second. Find the depth of the water if the velocity is 30 ft/s.

30. Bicycle Safety A bicycle will overturn if it rounds a corner too sharply or too fast. An equation for the maximum velocity at which a cyclist can turn a corner without tipping over is $v = 4\sqrt{r}$, where v is the velocity of the bicycle in miles per hour and r is the radius of the corner in feet. What is the radius of the sharpest corner that a cyclist can safely turn while riding at 20 mph?

Chapter 10 Test

1. Simplify: $\sqrt{121x^8y^2}$

2. Simplify: $\sqrt{3x^2y}\sqrt{6xy^2}\sqrt{2x}$

3. Simplify: $5\sqrt{8} - 3\sqrt{50}$

4. Simplify: $\sqrt{45}$

5. Simplify: $\dfrac{\sqrt{162}}{\sqrt{2}}$

6. Solve: $\sqrt{9x} + 3 = 18$

7. Simplify: $\sqrt{32a^5b^{11}}$

8. Simplify: $\dfrac{\sqrt{98a^6b^4}}{\sqrt{2a^3b^2}}$

9. Simplify: $\dfrac{2}{\sqrt{3} - 1}$

10. Simplify: $\sqrt{8x^3y}\sqrt{10xy^4}$

11. Solve: $\sqrt{x-5} + \sqrt{x} = 5$

12. Simplify: $3\sqrt{8y} - 2\sqrt{72x} + 5\sqrt{18y}$

13. Simplify: $\sqrt{72x^7y^2}$

14. Simplify: $(\sqrt{y} - 3)(\sqrt{y} + 5)$

15. Simplify: $2x\sqrt{3xy^3} - 2y\sqrt{12x^3y} - 3xy\sqrt{xy}$

16. Simplify: $\dfrac{2 - \sqrt{5}}{6 + \sqrt{5}}$

17. Simplify: $\sqrt{a}(\sqrt{a} - \sqrt{b})$

18. Simplify: $\sqrt{75}$

19. **Time**   Find the length of a pendulum that makes one swing in 3 s. The equation for the time of one swing of a pendulum is $T = 2\pi\sqrt{\dfrac{L}{32}}$, where T is the time in seconds and L is the length in feet. Use 3.14 for π. Round to the nearest hundredth.

20. **Camping**   A support rope for a tent is attached to the top of a pole and then secured to the ground as shown in the figure at the right. If the rope is 8 ft long and the pole is 4 ft high, how far, x, from the base of the pole should the rope be secured? Round to the nearest foot.

Cumulative Review Exercises

1. Simplify:

$$\left(\frac{2}{3}\right)^2 \cdot \left(\frac{3}{4} - \frac{3}{2}\right) + \left(\frac{1}{2}\right)^2$$

2. Simplify:

$$-3[x - 2(3 - 2x) - 5x] + 2x$$

3. Solve:
$$2x - 4[3x - 2(1 - 3x)] = 2(3 - 4x)$$

4. Simplify: $(-3x^2y)(-2x^3y^4)$

5. Simplify: $\dfrac{12b^4 - 6b^2 + 2}{-6b^2}$

6. Given $f(x) = \dfrac{2x}{x - 3}$, find $f(-3)$.

7. Factor: $2a^3 - 16a^2 + 30a$

8. Multiply: $\dfrac{3x^3 - 6x^2}{4x^2 + 4x} \cdot \dfrac{3x - 9}{9x^3 - 45x^2 + 54x}$

9. Subtract: $\dfrac{x + 2}{x - 4} - \dfrac{6}{(x - 4)(x - 3)}$

10. Solve: $\dfrac{x}{2x - 5} - 2 = \dfrac{3x}{2x - 5}$

11. Find the equation of the line that contains the point $(-2, -3)$ and has slope $\frac{1}{2}$.

12. Solve by substitution:
$$4x - 3y = 1$$
$$2x + y = 3$$

13. Solve by the addition method:
$$5x + 4y = 7$$
$$3x - 2y = 13$$

14. Solve: $3(x - 7) \geq 5x - 12$

15. Simplify: $\sqrt{108}$

16. Simplify: $3\sqrt{32} - 2\sqrt{128}$

17. Simplify: $2a\sqrt{2ab^3} + b\sqrt{8a^3b} - 5ab\sqrt{ab}$

18. Simplify: $\sqrt{2a^9b}\sqrt{98ab^3}\sqrt{2a}$

19. Simplify: $\sqrt{3}(\sqrt{6} - \sqrt{x^2})$

20. Simplify: $\dfrac{\sqrt{320}}{\sqrt{5}}$

21. Simplify: $\dfrac{3}{2 - \sqrt{5}}$

22. Solve: $\sqrt{3x - 2} - 4 = 0$

23. Business   The selling price of a book is $29.40. The markup rate used by the bookstore is 20%. Find the cost of the book. Use the formula $S = C + rC$, where S is the selling price, C is the cost, and r is the markup rate.

24. Chemistry   How many ounces of pure water must be added to 40 oz of a 12% salt solution to make a salt solution that is 5% salt?

25. Number Sense   The sum of two numbers is twenty-one. The product of the two numbers is one hundred four. Find the two numbers.

26. Work   A small water pipe takes twice as long to fill a tank as does a larger water pipe. With both pipes open, it takes 16 h to fill the tank. Find the time it would take the small pipe working alone to fill the tank.

27. Solve by graphing: $3x - 2y = 8$
$4x + 5y = 3$

28. Graph the solution set of $3x + y \le 2$.

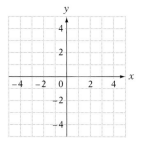

29. Number Sense   The square root of the sum of two consecutive integers is equal to 9. Find the smaller integer.

30. Physics   A stone is dropped from a building and hits the ground 5 s later. How high is the building? The equation for the distance an object falls in T seconds is $T = \sqrt{\dfrac{d}{16}}$, where d is the distance in feet.

chapter

11 Quadratic Equations

This photo shows employees at the Fender factory in Corona, California, in the process of making guitars. During this stage, the Fender Stratocaster bodies are hand-sanded. The objective of the Fender Musical Instruments Corporation, as with any business, is to earn a profit. Profit is the difference between a company's revenue (the total amount of money the company earns by selling its products or services) and its costs (the total amount of money the company spends in doing business). The **Focus on Problem Solving on page 540** uses quadratic equations to determine the maximum profit a company can earn.

OBJECTIVES

Section 11.1
A To solve a quadratic equation by factoring

B To solve a quadratic equation by taking square roots

Section 11.2
A To solve a quadratic equation by completing the square

Section 11.3
A To solve a quadratic equation by using the quadratic formula

Section 11.4
A To graph a quadratic equation of the form $y = ax^2 + bx + c$

Section 11.5
A To solve application problems

Need help? For online student resources, such as section quizzes, visit this textbook's website at **math.college.hmco.com/students**.

Do these exercises to prepare for Chapter 11.

1. Evaluate $b^2 - 4ac$ when $a = 2$, $b = -3$, and $c = -4$.

2. Solve: $5x + 4 = 3$

3. Factor: $x^2 + x - 12$

4. Factor: $4x^2 - 12x + 9$

5. Is $x^2 - 10x + 25$ a perfect square trinomial?

6. Solve: $\dfrac{5}{x - 2} = \dfrac{15}{x}$

7. Graph: $y = -2x + 3$

8. Simplify: $\sqrt{28}$

9. If a is *any* real number, simplify $\sqrt{a^2}$.

10. Exercising Walking at a constant speed of 4.5 mph, Lucy and Sam walked from the beginning to the end of a hiking trail. When they reached the end, they immediately started back along the same path at a constant speed of 3 mph. If the round-trip took 2 h, what is the length of the hiking trail?

GO FIGURE ••• •

Find the value of $\dfrac{1}{\sqrt{1} + \sqrt{2}} + \dfrac{1}{\sqrt{2} + \sqrt{3}} + \dfrac{1}{\sqrt{3} + \sqrt{4}} + \cdots + \dfrac{1}{\sqrt{8} + \sqrt{9}}$.

11.1 Solving Quadratic Equations by Factoring or by Taking Square Roots

Objective A   **To solve a quadratic equation by factoring**

An equation of the form $ax^2 + bx + c = 0$, where $a, b,$ and c are real numbers and $a \neq 0$, is a **quadratic equation.**

$4x^2 - 3x + 1 = 0$, $a = 4, b = -3, c = 1$
$3x^2 - 4 = 0$, $a = 3, b = 0, c = -4$
$\dfrac{x^2}{2} - 2x + 4 = 0$, $a = \dfrac{1}{2}, b = -2, c = 4$

A quadratic equation is also called a **second-degree equation.**

A quadratic equation is in **standard form** when the polynomial is in descending order and equal to zero.

Recall that the Principle of Zero Products states that if the product of two factors is zero, then at least one of the factors must be zero.

If $a \cdot b = 0$,
then $a = 0$ or $b = 0$.

The Principle of Zero Products can be used to solve quadratic equations by factoring. Write the equation in standard form, factor the polynomial, apply the Principle of Zero Products, and solve for the variable.

HOW TO   Solve by factoring: $2x^2 - x = 1$

$$2x^2 - x = 1$$
$$2x^2 - x - 1 = 0$$ • Write the equation in standard form.
$$(2x + 1)(x - 1) = 0$$ • Factor.
$$2x + 1 = 0 \qquad x - 1 = 0$$ • Use the Principle of Zero Products to set each factor equal to zero.
$$2x = -1 \qquad\qquad x = 1$$ • Rewrite each equation in the form *variable* = *constant*.
$$x = -\frac{1}{2}$$

TAKE NOTE
You should always check your solutions by substituting the proposed solutions back into the *original* equation.

Check:

$$
\begin{array}{c|c}
2x^2 - x = 1 & 2x^2 - x = 1 \\
\hline
2\left(-\dfrac{1}{2}\right)^2 - \left(-\dfrac{1}{2}\right) \;\Big|\; 1 & 2(1)^2 - 1 \;\Big|\; 1 \\
2 \cdot \dfrac{1}{4} + \dfrac{1}{2} \;\Big|\; 1 & 2 \cdot 1 - 1 \;\Big|\; 1 \\
\dfrac{1}{2} + \dfrac{1}{2} \;\Big|\; 1 & 2 - 1 \;\Big|\; 1 \\
1 = 1 & 1 = 1
\end{array}
$$

The solutions are $-\dfrac{1}{2}$ and 1.

HOW TO Solve by factoring: $3x^2 - 4x + 8 = (4x + 1)(x - 2)$

$3x^2 - 4x + 8 = (4x + 1)(x - 2)$

$3x^2 - 4x + 8 = 4x^2 - 7x - 2$

$0 = x^2 - 3x - 10$
$0 = (x - 5)(x + 2)$
$x - 5 = 0 \qquad x + 2 = 0$

$x = 5 \qquad\qquad x = -2$

- Multiply the factors on the right side of the equation.
- Write the equation in standard form.
- Factor.
- Use the Principle of Zero Products to set each factor equal to zero.
- Rewrite each equation in the form *variable* = *constant*.

Check:

$3x^2 - 4x + 8 = (4x + 1)(x - 2)$	
$3(5)^2 - 4(5) + 8$	$(4[5] + 1)(5 - 2)$
$3(25) - 4(5) + 8$	$(20 + 1)(3)$
$75 - 20 + 8$	$(21)(3)$
$63 = 63$	

$3x^2 - 4x + 8 = (4x + 1)(x - 2)$	
$3(-2)^2 - 4(-2) + 8$	$(4[-2] + 1)(-2 - 2)$
$3(4) - 4(-2) + 8$	$(-8 + 1)(-4)$
$12 + 8 + 8$	$(-7)(-4)$
$28 = 28$	

The solutions are 5 and -2.

HOW TO Solve by factoring: $x^2 - 10x + 25 = 0$

$x^2 - 10x + 25 = 0$
$(x - 5)(x - 5) = 0$
$x - 5 = 0 \qquad x - 5 = 0$
$x = 5 \qquad\quad x = 5$

- Factor.
- Use the Principle of Zero Products.
- Solve each equation for *x*.

The solution is 5.

In this last example, 5 is called a **double root** of the quadratic equation.

Example 1

Solve by factoring: $\dfrac{z^2}{2} - \dfrac{z}{4} - \dfrac{1}{4} = 0$

Solution

$\dfrac{z^2}{2} - \dfrac{z}{4} - \dfrac{1}{4} = 0$

$4\left(\dfrac{z^2}{2} - \dfrac{z}{4} - \dfrac{1}{4}\right) = 4(0)$ • Multiply each side by 4.

$2z^2 - z - 1 = 0$
$(2z + 1)(z - 1) = 0$ • Factor.

$2z + 1 = 0 \qquad z - 1 = 0$ • Principle of
$2z = -1 \qquad\quad z = 1$ Zero Products
$z = -\dfrac{1}{2}$

The solutions are $-\dfrac{1}{2}$ and 1.

You Try It 1

Solve by factoring: $\dfrac{3y^2}{2} + y - \dfrac{1}{2} = 0$

Your solution

Solution on p. S27

Objective B   **To solve a quadratic equation by taking square roots**

Consider a quadratic equation of the form $x^2 = a$. This equation can be solved by factoring.

$$x^2 = 25$$
$$x^2 - 25 = 0$$
$$(x - 5)(x + 5) = 0$$
$$x - 5 = 0 \qquad x + 5 = 0$$
$$x = 5 \qquad\quad x = -5$$

The solutions are 5 and -5. The solutions are plus or minus the same number, which is frequently written by using $\pm$; for example, "the solutions are ± 5." An alternative method of solving this equation is suggested by the fact that ± 5 can be written as $\pm\sqrt{25}$.

> **Principle of Taking the Square Root of Each Side of an Equation**
>
> If $x^2 = a$, then $x = \pm\sqrt{a}$.

HOW TO   Solve by taking square roots: $x^2 = 25$

$$x^2 = 25$$
$$\sqrt{x^2} = \sqrt{25}$$
$$x = \pm\sqrt{25} = \pm 5$$

• Take the square root of each side of the equation. Then simplify.

The solutions are 5 and -5.

HOW TO   Solve by taking square roots: $3x^2 = 36$

$$3x^2 = 36$$
$$x^2 = 12$$
$$\sqrt{x^2} = \sqrt{12}$$
$$x = \pm\sqrt{12}$$
$$x = \pm 2\sqrt{3}$$

• Solve for x^2.
• Take the square root of each side.
• Simplify.

The solutions are $2\sqrt{3}$ and $-2\sqrt{3}$.

HOW TO   Solve by taking square roots: $49y^2 - 25 = 0$

$$49y^2 - 25 = 0$$
$$49y^2 = 25$$
$$y^2 = \frac{25}{49}$$
$$\sqrt{y^2} = \sqrt{\frac{25}{49}}$$
$$y = \pm\frac{5}{7}$$

• Solve for y^2.

• Take the square root of each side.

• Simplify.

The solutions are $\frac{5}{7}$ and $-\frac{5}{7}$.

An equation that contains the square of a binomial can be solved by taking square roots.

> **HOW TO**   Solve by taking square roots: $2(x - 1)^2 - 36 = 0$
>
> $$2(x - 1)^2 - 36 = 0$$
> $$2(x - 1)^2 = 36$$ • Solve for $(x - 1)^2$.
> $$(x - 1)^2 = 18$$
> $$\sqrt{(x - 1)^2} = \sqrt{18}$$ • Take the square root of each side of the equation.
> $$x - 1 = \pm\sqrt{18}$$
> $$x - 1 = \pm 3\sqrt{2}$$ • Simplify.
> $$x = 1 \pm 3\sqrt{2}$$ • Solve for x.
>
> The solutions are $1 + 3\sqrt{2}$ and $1 - 3\sqrt{2}$.

Example 2

Solve by taking square roots:
$x^2 + 16 = 0$

Solution
$x^2 + 16 = 0$
$\quad\; x^2 = -16$ • Solve for x^2.
$\sqrt{x^2} = \sqrt{-16}$ • Take square roots.

$\sqrt{-16}$ is not a real number.

The equation has no real number solution.

You Try It 2

Solve by taking square roots:
$x^2 + 81 = 0$

Your solution

Example 3

Solve by taking square roots:
$5(y - 4)^2 = 25$

Solution
$5(y - 4)^2 = 25$
$\quad (y - 4)^2 = 5$ • Solve for $(y - 4)^2$.
$\sqrt{(y - 4)^2} = \sqrt{5}$ • Take square roots.
$\quad\; y - 4 = \pm\sqrt{5}$ • Simplify.
$\quad\quad\quad y = 4 \pm\sqrt{5}$ • Solve for y.

The solutions are $4 + \sqrt{5}$ and $4 - \sqrt{5}$.

You Try It 3

Solve by taking square roots:
$7(z + 2)^2 = 21$

Your solution

Solutions on p. S27

11.1 Exercises

Objective A   **To solve a quadratic equation by factoring**

For Exercises 1 to 4, solve for x.

1. $(x + 3)(x - 5) = 0$ **2.** $x(x - 7) = 0$

3. $(2x + 5)(3x - 1) = 0$ **4.** $(x - 4)(2x - 7) = 0$

For Exercises 5 to 34, solve by factoring.

5. $x^2 + 2x - 15 = 0$ **6.** $t^2 + 3t - 10 = 0$ **7.** $z^2 - 4z + 3 = 0$ **8.** $s^2 - 5s + 4 = 0$

9. $p^2 + 3p + 2 = 0$ **10.** $v^2 + 6v + 5 = 0$ **11.** $x^2 - 6x + 9 = 0$ **12.** $y^2 - 8y + 16 = 0$

13. $12y^2 + 8y = 0$ **14.** $6x^2 - 9x = 0$ **15.** $r^2 - 10 = 3r$ **16.** $t^2 - 12 = 4t$

17. $3v^2 - 5v + 2 = 0$ **18.** $2p^2 - 3p - 2 = 0$ **19.** $3s^2 + 8s = 3$

20. $3x^2 + 5x = 12$ **21.** $\dfrac{3}{4}z^2 - z = -\dfrac{1}{3}$ **22.** $\dfrac{r^2}{2} = 1 - \dfrac{r}{12}$

23. $4t^2 = 4t + 3$ **24.** $5y^2 + 11y = 12$ **25.** $4v^2 - 4v + 1 = 0$

26. $9s^2 - 6s + 1 = 0$ **27.** $x^2 - 9 = 0$ **28.** $t^2 - 16 = 0$

29. $4y^2 - 1 = 0$ **30.** $9z^2 - 4 = 0$ **31.** $x + 15 = x(x - 1)$

32. $p + 18 = p(p - 2)$ **33.** $r^2 - r - 2 = (2r - 1)(r - 3)$ **34.** $s^2 + 5s - 4 = (2s + 1)(s - 4)$

Objective B   **To solve a quadratic equation by taking square roots**

For Exercises 35 to 61, solve by taking square roots.

35. $x^2 = 36$ **36.** $y^2 = 49$ **37.** $v^2 - 1 = 0$

38. $z^2 - 64 = 0$ **39.** $4x^2 - 49 = 0$ **40.** $9w^2 - 64 = 0$

41. $9y^2 = 4$

42. $4z^2 = 25$

43. $16v^2 - 9 = 0$

44. $25x^2 - 64 = 0$

45. $y^2 + 81 = 0$

46. $z^2 + 49 = 0$

47. $w^2 - 24 = 0$

48. $v^2 - 48 = 0$

49. $(x - 1)^2 = 36$

50. $(y + 2)^2 = 49$

51. $2(x + 5)^2 = 8$

52. $4(z - 3)^2 = 100$

53. $9(x - 1)^2 - 16 = 0$

54. $4(y + 3)^2 - 81 = 0$

55. $49(v + 1)^2 - 25 = 0$

56. $81(y - 2)^2 - 64 = 0$

57. $(x - 4)^2 - 20 = 0$

58. $(y + 5)^2 - 50 = 0$

59. $(x + 1)^2 + 36 = 0$

60. $2\left(z - \dfrac{1}{2}\right)^2 = 12$

61. $3\left(v + \dfrac{3}{4}\right)^2 = 36$

APPLYING THE CONCEPTS

For Exercises 62 and 63, solve for x.

62. $(3x^2 - 13)^2 = 4$

63. $(6x^2 - 5)^2 = 1$

64. Investments The value A of an initial investment of P dollars after 2 years is given by $A = P(1 + r)^2$, where r is the annual percentage rate earned by the investment. If an initial investment of \$1500 grew to a value of \$1782.15 in 2 years, what was the annual percentage rate?

65. Investments An initial investment of \$5000 grew to a value of \$5832 in 2 years. Use the formula in Exercise 64 to find the annual percentage rate.

66. Physics The kinetic energy of a moving body is given by $E = \dfrac{1}{2}mv^2$, where E is the kinetic energy, m is the mass, and v is the velocity in meters per second. What is the velocity of a moving body whose mass is 5 kg and whose kinetic energy is 250 newton-meters?

67. Automotive Safety On a certain type of street surface, the equation $d = 0.0074v^2$ can be used to approximate the distance d, in feet, a car traveling v miles per hour will slide when its brakes are applied. After applying the brakes, the owner of a car involved in an accident skidded 40 ft. Did the traffic officer investigating the accident issue the car owner a ticket for speeding if the speed limit is 65 mph?

11.2 Solving Quadratic Equations by Completing the Square

Objective A   **To solve a quadratic equation by completing the square**

Recall that a perfect-square trinomial is the square of a binomial.

Perfect-Square Trinomial		Square of a Binomial
$x^2 + 6x + 9$	$=$	$(x + 3)^2$
$x^2 - 10x + 25$	$=$	$(x - 5)^2$
$x^2 + 8x + 16$	$=$	$(x + 4)^2$

For each perfect-square trinomial, the square of $\frac{1}{2}$ of the coefficient of x equals the constant term.

$$x^2 + 6x + 9, \qquad \left(\frac{1}{2} \cdot 6\right)^2 = 9$$

$$x^2 - 10x + 25, \qquad \left[\frac{1}{2}(-10)\right]^2 = 25$$

$$x^2 + 8x + 16, \qquad \left(\frac{1}{2} \cdot 8\right)^2 = 16$$

Adding to a binomial the constant term that makes it a perfect-square trinomial is called **completing the square.**

HOW TO   Complete the square of $x^2 - 8x$. Write the resulting perfect-square trinomial as the square of a binomial.

$$\left[\frac{1}{2}(-8)\right]^2 = 16$$
  • Find the constant term.

$$x^2 - 8x + 16$$
  • Complete the square of $x^2 - 8x$ by adding the constant term.

$$x^2 - 8x + 16 = (x - 4)^2$$
  • Write the resulting perfect-square trinomial as the square of a binomial.

HOW TO   Complete the square of $y^2 + 5y$. Write the resulting perfect-square trinomial as the square of a binomial.

$$\left(\frac{1}{2} \cdot 5\right)^2 = \left(\frac{5}{2}\right)^2 = \frac{25}{4}$$
  • Find the constant term.

$$y^2 + 5y + \frac{25}{4}$$
  • Complete the square of $y^2 + 5y$ by adding the constant term.

$$y^2 + 5y + \frac{25}{4} = \left(y + \frac{5}{2}\right)^2$$
  • Write the resulting perfect-square trinomial as the square of a binomial.

Point of Interest

Early mathematicians solved quadratic equations by literally *completing the square*. For these mathematicians, all equations had geometric interpretations. They found that a quadratic equation could be solved by making certain figures into squares. See the second of the Projects and Group Activities at the end of this chapter for an idea of how this was done.

A quadratic equation that cannot be solved by factoring can be solved by completing the square. When the quadratic equation is in the form $x^2 + bx = c$, add to each side of the equation the term that completes the square on $x^2 + bx$. Factor the perfect-square trinomial, and write it as the square of a binomial. Take the square root of each side of the equation and then solve for x.

S t u d y   T i p

This is a new skill and one that is difficult for many students. Be sure to do all you need to do in order to be successful at solving quadratic equations by completing the square: Read through the introductory material, work through the How To examples, study the paired examples, and do the You Try Its and check your solutions against the ones in the back of the book. See *AIM for Success*, page xxv.

HOW TO Solve by completing the square: $x^2 + 8x - 2 = 0$

$$x^2 + 8x - 2 = 0$$
$$x^2 + 8x = 2$$

• Add 2 to each side of the equation.

$$x^2 + 8x + \left(\frac{1}{2} \cdot 8\right)^2 = 2 + \left(\frac{1}{2} \cdot 8\right)^2$$

• Complete the square of $x^2 + 8x$. Add $\left(\frac{1}{2} \cdot 8\right)^2$ to each side of the equation.

$$x^2 + 8x + 16 = 2 + 16$$

• Simplify.

$$(x + 4)^2 = 18$$

• Factor the perfect-square trinomial.

$$\sqrt{(x + 4)^2} = \sqrt{18}$$

• Take the square root of each side of the equation.

$$x + 4 = \pm\sqrt{18}$$
$$x + 4 = \pm 3\sqrt{2}$$
$$x = -4 \pm 3\sqrt{2}$$

• Solve for x.

Check:

$x^2 + 8x - 2 = 0$	
$(-4 + 3\sqrt{2})^2 + 8(-4 + 3\sqrt{2}) - 2$	0
$16 - 24\sqrt{2} + 18 - 32 + 24\sqrt{2} - 2$	0
	$0 = 0$

$x^2 + 8x - 2 = 0$	
$(-4 - 3\sqrt{2})^2 + 8(-4 - 3\sqrt{2}) - 2$	0
$16 + 24\sqrt{2} + 18 - 32 - 24\sqrt{2} - 2$	0
	$0 = 0$

The solutions are $-4 + 3\sqrt{2}$ and $-4 - 3\sqrt{2}$.

If the coefficient of the second-degree term is not 1, a step in completing the square is to multiply each side of the equation by the reciprocal of that coefficient.

HOW TO Solve by completing the square: $2x^2 - 3x + 1 = 0$

$$2x^2 - 3x + 1 = 0$$
$$2x^2 - 3x = -1$$

• Subtract 1 from each side of the equation.

$$\frac{1}{2}(2x^2 - 3x) = \frac{1}{2} \cdot (-1)$$

• For us to complete the square, the coefficient of x^2 must be 1. Multiply each side of the equation by $\frac{1}{2}$.

$$x^2 - \frac{3}{2}x = -\frac{1}{2}$$

$$x^2 - \frac{3}{2}x + \left[\frac{1}{2}\left(-\frac{3}{2}\right)\right]^2 = -\frac{1}{2} + \left[\frac{1}{2}\left(-\frac{3}{2}\right)\right]^2$$

• Complete the square. Add $\left[\frac{1}{2}\left(-\frac{3}{2}\right)\right]^2$ to each side of the equation.

$$x^2 - \frac{3}{2}x + \frac{9}{16} = -\frac{1}{2} + \frac{9}{16}$$

• Simplify.

$$\left(x - \frac{3}{4}\right)^2 = \frac{1}{16}$$

• Factor the perfect-square trinomial.

$$\sqrt{\left(x - \frac{3}{4}\right)^2} = \sqrt{\frac{1}{16}}$$

• Take the square root of each side of the equation.

$$x - \frac{3}{4} = \pm\frac{1}{4}$$

$$x = \frac{3}{4} \pm \frac{1}{4}$$

• Solve for x.

$$x = \frac{3}{4} + \frac{1}{4} = 1 \qquad x = \frac{3}{4} - \frac{1}{4} = \frac{1}{2}$$

The solutions are $\frac{1}{2}$ and 1.

Example 1

Solve by completing the square:
$2x^2 - 4x - 1 = 0$

Solution

$2x^2 - 4x - 1 = 0$

$2x^2 - 4x = 1$ • Add 1.

$\dfrac{1}{2}(2x^2 - 4x) = \dfrac{1}{2} \cdot 1$ • Multiply by $\dfrac{1}{2}$.

$x^2 - 2x = \dfrac{1}{2}$ • The coefficient of x^2 is 1.

Complete the square.

$x^2 - 2x + 1 = \dfrac{1}{2} + 1$ • $\left[\dfrac{1}{2} \cdot (-2)\right]^2 = [-1]^2 = 1$

$(x - 1)^2 = \dfrac{3}{2}$ • Factor.

$\sqrt{(x - 1)^2} = \sqrt{\dfrac{3}{2}}$ • Take square roots.

$x - 1 = \pm\dfrac{\sqrt{6}}{2}$ • Simplify.

$x = 1 \pm \dfrac{\sqrt{6}}{2}$

$x = 1 + \dfrac{\sqrt{6}}{2} \qquad x = 1 - \dfrac{\sqrt{6}}{2}$

$= \dfrac{2 + \sqrt{6}}{2} \qquad\quad = \dfrac{2 - \sqrt{6}}{2}$

Check:

$$2x^2 - 4x - 1 = 0$$

$$\begin{array}{c|c} 2\left(\dfrac{2 + \sqrt{6}}{2}\right)^2 - 4\left(\dfrac{2 + \sqrt{6}}{2}\right) - 1 & 0 \\ 2\left(\dfrac{4 + 4\sqrt{6} + 6}{4}\right) - 2(2 + \sqrt{6}) - 1 & 0 \\ 2 + 2\sqrt{6} + 3 - 4 - 2\sqrt{6} - 1 & 0 \\ 0 = 0 \end{array}$$

$$2x^2 - 4x - 1 = 0$$

$$\begin{array}{c|c} 2\left(\dfrac{2 - \sqrt{6}}{2}\right)^2 - 4\left(\dfrac{2 - \sqrt{6}}{2}\right) - 1 & 0 \\ 2\left(\dfrac{4 - 4\sqrt{6} + 6}{4}\right) - 2(2 - \sqrt{6}) - 1 & 0 \\ 2 - 2\sqrt{6} + 3 - 4 + 2\sqrt{6} - 1 & 0 \\ 0 = 0 \end{array}$$

The solutions are $\dfrac{2 + \sqrt{6}}{2}$ and $\dfrac{2 - \sqrt{6}}{2}$.

You Try It 1

Solve by completing the square:
$3x^2 - 6x - 2 = 0$

Your solution

Solution on pp. S27–S28

Example 2

Solve by completing the square:
$x^2 + 4x + 5 = 0$

Solution

$x^2 + 4x + 5 = 0$

$x^2 + 4x = -5$ • Subtract 5.

Complete the square.

$x^2 + 4x + 4 = -5 + 4$ • $\left(\frac{1}{2} \cdot 4\right)^2 = 2^2 = \mathbf{4}$

$(x + 2)^2 = -1$ • Factor.

$\sqrt{(x + 2)^2} = \sqrt{-1}$ • Take square roots.

$\sqrt{-1}$ is not a real number.

The quadratic equation has no real number solution.

You Try It 2

Solve by completing the square:
$x^2 + 6x + 12 = 0$

Your solution

Example 3

Solve $\frac{x^2}{4} + \frac{3x}{2} + 1 = 0$ by completing the square. Approximate the solutions to the nearest thousandth.

Solution

$\frac{x^2}{4} + \frac{3x}{2} + 1 = 0$

$4\left(\frac{x^2}{4} + \frac{3x}{2} + 1\right) = 4(0)$ • Multiply by **4**.

$x^2 + 6x + 4 = 0$

$x^2 + 6x = -4$ • Subtract 4.

Complete the square.

$x^2 + 6x + 9 = -4 + 9$ • $\left(\frac{1}{2} \cdot 6\right)^2 = 3^2 = \mathbf{9}$

$(x + 3)^2 = 5$ • Factor.

$\sqrt{(x + 3)^2} = \sqrt{5}$ • Take square roots.

$x + 3 = \pm\sqrt{5}$

$x + 3 = \sqrt{5}$ $x + 3 = -\sqrt{5}$

$x = -3 + \sqrt{5}$ $x = -3 - \sqrt{5}$

$\approx -3 + 2.236$ $\approx -3 - 2.236$

≈ -0.764 ≈ -5.236

The solutions are approximately -0.764 and -5.236.

You Try It 3

Solve $\frac{x^2}{8} + x + 1 = 0$ by completing the square. Approximate the solutions to the nearest thousandth.

Your solution

Solutions on p. S28

11.2 Exercises

Objective A   **To solve a quadratic equation by completing the square**

For Exercises 1 to 4, complete the square of each binomial. Write the resulting trinomial as the square of a binomial.

1. $x^2 - 8x$

2. $x^2 + 6x$

3. $x^2 + 5x$

4. $x^2 - 3x$

For Exercises 5 to 53, solve by completing the square.

5. $x^2 + 2x - 3 = 0$

6. $y^2 + 4y - 5 = 0$

7. $z^2 - 6z - 16 = 0$

8. $w^2 + 8w - 9 = 0$

9. $x^2 = 4x - 4$

10. $z^2 = 8z - 16$

11. $v^2 - 6v + 13 = 0$

12. $x^2 + 4x + 13 = 0$

13. $y^2 + 5y + 4 = 0$

14. $v^2 - 5v - 6 = 0$

15. $w^2 + 7w = 8$

16. $y^2 + 5y = -4$

17. $v^2 + 4v + 1 = 0$

18. $y^2 - 2y - 5 = 0$

19. $x^2 + 6x = 5$

20. $w^2 - 8w = 3$

21. $\dfrac{z^2}{2} = z + \dfrac{1}{2}$

22. $\dfrac{y^2}{10} = y - 2$

23. $p^2 + 3p = 1$

24. $r^2 + 5r = 2$

25. $t^2 - 3t = -2$

26. $z^2 - 5z = -3$

27. $v^2 + v - 3 = 0$

28. $x^2 - x = 1$

29. $y^2 = 7 - 10y$

30. $v^2 = 14 + 16v$

31. $r^2 - 3r = 5$

32. $s^2 + 3s = -1$

33. $t^2 - t = 4$

34. $y^2 + y - 4 = 0$

35. $x^2 - 3x + 5 = 0$

36. $z^2 + 5z + 7 = 0$

37. $2t^2 - 3t + 1 = 0$

38. $2x^2 - 7x + 3 = 0$ **39.** $2r^2 + 5r = 3$ **40.** $2y^2 - 3y = 9$ **41.** $2s^2 = 7s - 6$

42. $2x^2 = 3x + 20$ **43.** $2v^2 = v + 1$ **44.** $2z^2 = z + 3$ **45.** $3r^2 + 5r = 2$

46. $3t^2 - 8t = 3$ **47.** $3y^2 + 8y + 4 = 0$ **48.** $3z^2 - 10z - 8 = 0$ **49.** $4x^2 + 4x - 3 = 0$

50. $4v^2 + 4v - 15 = 0$ **51.** $6s^2 + 7s = 3$ **52.** $6z^2 = z + 2$ **53.** $6p^2 = 5p + 4$

For Exercises 54 to 59, solve by completing the square. Approximate the solutions to the nearest thousandth.

54. $y^2 + 3y = 5$ **55.** $w^2 + 5w = 2$ **56.** $2z^2 - 3z = 7$

57. $2x^2 + 3x = 11$ **58.** $4x^2 + 6x - 1 = 0$ **59.** $4x^2 + 2x - 3 = 0$

APPLYING THE CONCEPTS

60. Explain why the equation $(x - 2)^2 = -4$ does not have a real number solution.

For Exercises 61 to 69, solve.

61. $\dfrac{x^2}{6} - \dfrac{x}{3} = 1$ **62.** $\sqrt{x + 2} = x - 4$ **63.** $\sqrt{3x + 4} - x = 2$

64. $\dfrac{x}{3} + \dfrac{3}{x} = \dfrac{8}{3}$ **65.** $\dfrac{x + 1}{2} + \dfrac{3}{x - 1} = 4$ **66.** $\dfrac{x - 2}{3} + \dfrac{2}{x + 2} = 4$

67. $4\sqrt{x + 1} - x = 4$ **68.** $\sqrt{2x^2 + 7} = x + 2$ **69.** $3\sqrt{x - 1} + 3 = x$

70. **Sports** A basketball player shoots at a basket 25 ft away. The height of the ball above the ground at time t is given by $h = -16t^2 + 32t + 6.5$. How many seconds after the ball is released does it hit the basket? *Hint:* When it hits the basket, $h = 10$ ft.

71. **Sports** A ball player hits a ball. The height of the ball above the ground can be approximated by the equation $h = -16t^2 + 76t + 5$. When will the ball hit the ground? *Hint:* The ball strikes the ground when $h = 0$ ft.

5 ft

11.3 Solving Quadratic Equations by Using the Quadratic Formula

Objective A   **To solve a quadratic equation by using the quadratic formula**

Any quadratic equation can be solved by completing the square. Applying this method to the standard form of a quadratic equation produces a formula that can be used to solve any quadratic equation.

Solve $ax^2 + bx + c = 0$ by completing the square.

$$ax^2 + bx + c = 0$$

Add the opposite of the constant term to each side of the equation.

$$ax^2 + bx + c + (-c) = 0 + (-c)$$
$$ax^2 + bx = -c$$

Multiply each side of the equation by the reciprocal of a, the coefficient of x^2.

$$\frac{1}{a}(ax^2 + bx) = \frac{1}{a}(-c)$$
$$x^2 + \frac{b}{a}x = -\frac{c}{a}$$

Complete the square by adding $\left(\frac{1}{2} \cdot \frac{b}{a}\right)^2$ to each side of the equation.

$$x^2 + \frac{b}{a}x + \left(\frac{1}{2} \cdot \frac{b}{a}\right)^2 = \left(\frac{1}{2} \cdot \frac{b}{a}\right)^2 - \frac{c}{a}$$
$$x^2 + \frac{b}{a}x + \frac{b^2}{4a^2} = \frac{b^2}{4a^2} - \frac{c}{a}$$

Simplify the right side of the equation.

$$x^2 + \frac{b}{a}x + \frac{b^2}{4a^2} = \frac{b^2}{4a^2} - \left(\frac{c}{a} \cdot \frac{4a}{4a}\right)$$
$$x^2 + \frac{b}{a}x + \frac{b^2}{4a^2} = \frac{b^2}{4a^2} - \frac{4ac}{4a^2}$$
$$x^2 + \frac{b}{a}x + \frac{b^2}{4a^2} = \frac{b^2 - 4ac}{4a^2}$$

Factor the perfect-square trinomial on the left side of the equation.

$$\left(x + \frac{b}{2a}\right)^2 = \frac{b^2 - 4ac}{4a^2}$$

Take the square root of each side of the equation.

$$\sqrt{\left(x + \frac{b}{2a}\right)^2} = \sqrt{\frac{b^2 - 4ac}{4a^2}}$$
$$x + \frac{b}{2a} = \pm\frac{\sqrt{b^2 - 4ac}}{2a}$$

Solve for x.

$$x + \frac{b}{2a} = \frac{\sqrt{b^2 - 4ac}}{2a}$$
$$x = -\frac{b}{2a} + \frac{\sqrt{b^2 - 4ac}}{2a}$$
$$= \frac{-b + \sqrt{b^2 - 4ac}}{2a}$$

$$x + \frac{b}{2a} = -\frac{\sqrt{b^2 - 4ac}}{2a}$$
$$x = -\frac{b}{2a} - \frac{\sqrt{b^2 - 4ac}}{2a}$$
$$= \frac{-b - \sqrt{b^2 - 4ac}}{2a}$$

The Quadratic Formula

The solutions of $ax^2 + bx + c = 0$, $a \neq 0$, are

$$x = \frac{-b \pm \sqrt{b^2 - 4ac}}{2a}$$

HOW TO Solve by using the quadratic formula: $2x^2 = 4x - 1$

$$2x^2 = 4x - 1$$
$$2x^2 - 4x + 1 = 0$$

- Write the equation in standard form.
- Subtract $4x$ from and add 1 to each side of the equation.

$$x = \frac{-b \pm \sqrt{b^2 - 4ac}}{2a}$$

- The quadratic formula

$$= \frac{-(-4) \pm \sqrt{(-4)^2 - (4 \cdot 2 \cdot 1)}}{2 \cdot 2}$$

- $a = 2$, $b = -4$, $c = 1$. Replace a, b, and c by their values.

$$= \frac{4 \pm \sqrt{16 - 8}}{4} = \frac{4 \pm \sqrt{8}}{4}$$

- Simplify.

$$= \frac{4 \pm 2\sqrt{2}}{4} = \frac{2 \pm \sqrt{2}}{2}$$

The solutions are $\dfrac{2 + \sqrt{2}}{2}$ and $\dfrac{2 - \sqrt{2}}{2}$.

> **TAKE NOTE**
> $$\frac{4 \pm 2\sqrt{2}}{4} = \frac{2(2 \pm \sqrt{2})}{2 \cdot 2}$$
> $$= \frac{2 \pm \sqrt{2}}{2}$$

Example 1

Solve by using the quadratic formula:
$2x^2 - 3x + 1 = 0$

Solution
$2x^2 - 3x + 1 = 0$
- Standard form

$$x = \frac{-(-3) \pm \sqrt{(-3)^2 - 4(2)(1)}}{2 \cdot 2}$$
- $a = 2$, $b = -3$, $c = 1$

$$= \frac{3 \pm \sqrt{9 - 8}}{4} = \frac{3 \pm \sqrt{1}}{4} = \frac{3 \pm 1}{4}$$

$$x = \frac{3 + 1}{4} \qquad x = \frac{3 - 1}{4}$$

$$= \frac{4}{4} = 1 \qquad = \frac{2}{4} = \frac{1}{2}$$

The solutions are 1 and $\dfrac{1}{2}$.

You Try It 1

Solve by using the quadratic formula:
$3x^2 + 4x - 4 = 0$

Your solution

Example 2

Solve by using the quadratic formula:
$\dfrac{x^2}{2} = 2x - \dfrac{5}{4}$

Solution $\dfrac{x^2}{2} = 2x - \dfrac{5}{4}$

$$4\left(\frac{x^2}{2}\right) = 4\left(2x - \frac{5}{4}\right)$$
- Multiply by **4**.

$$2x^2 = 8x - 5$$

$$2x^2 - 8x + 5 = 0$$
- Standard form

$$x = \frac{-(-8) \pm \sqrt{(-8)^2 - 4(2)(5)}}{2 \cdot 2}$$
- $a = 2$, $b = -8$, $c = 5$

$$= \frac{8 \pm \sqrt{64 - 40}}{4} = \frac{8 \pm \sqrt{24}}{4}$$

$$= \frac{8 \pm 2\sqrt{6}}{4} = \frac{4 \pm \sqrt{6}}{2}$$

The solutions are $\dfrac{4 + \sqrt{6}}{2}$ and $\dfrac{4 - \sqrt{6}}{2}$.

You Try It 2

Solve by using the quadratic formula:
$\dfrac{x^2}{4} + \dfrac{x}{2} = \dfrac{1}{4}$

Your solution

Solutions on p. S28

11.3 Exercises

Objective A   **To solve a quadratic equation by using the quadratic formula**

For Exercises 1 to 38, solve by using the quadratic formula.

1. $x^2 - 4x - 5 = 0$ **2.** $y^2 + 3y + 2 = 0$ **3.** $z^2 - 2z - 15 = 0$ **4.** $v^2 + 5v + 4 = 0$

5. $y^2 = 2y + 3$ **6.** $w^2 = 3w + 18$ **7.** $r^2 = 5 - 4r$ **8.** $z^2 = 3 - 2z$

9. $2y^2 - y - 1 = 0$ **10.** $2t^2 - 5t + 3 = 0$ **11.** $w^2 + 3w + 5 = 0$

12. $x^2 - 2x + 6 = 0$ **13.** $p^2 - p = 0$ **14.** $2v^2 + v = 0$

15. $4t^2 - 9 = 0$ **16.** $4s^2 - 25 = 0$ **17.** $4y^2 + 4y = 15$

18. $6y^2 + 5y - 4 = 0$ **19.** $2x^2 + x + 1 = 0$ **20.** $3r^2 - r + 2 = 0$

21. $\frac{1}{2}t^2 - t = \frac{5}{2}$ **22.** $y^2 - 4y = 6$ **23.** $\frac{1}{3}t^2 + 2t - \frac{1}{3} = 0$

24. $z^2 + 4z + 1 = 0$ **25.** $w^2 = 4w + 9$ **26.** $y^2 = 8y + 3$

27. $9y^2 + 6y - 1 = 0$ **28.** $9s^2 - 6s - 2 = 0$ **29.** $4p^2 + 4p + 1 = 0$

30. $9z^2 + 12z + 4 = 0$ **31.** $\frac{x^2}{2} = x - \frac{5}{4}$ **32.** $r^2 = \frac{5}{3}r - 2$

33. $4p^2 + 16p = -11$ **34.** $4y^2 - 12y = -1$ **35.** $4x^2 = 4x + 11$

36. $4s^2 + 12s = 3$

37. $9v^2 = -30v - 23$

38. $9t^2 = 30t + 17$

For Exercises 39 to 47, solve by using the quadratic formula. Approximate the solutions to the nearest thousandth.

39. $x^2 - 2x - 21 = 0$

40. $y^2 + 4y - 11 = 0$

41. $s^2 - 6s - 13 = 0$

42. $w^2 + 8w - 15 = 0$

43. $2p^2 - 7p - 10 = 0$

44. $3t^2 - 8t - 1 = 0$

45. $4z^2 + 8z - 1 = 0$

46. $4x^2 + 7x + 1 = 0$

47. $5v^2 - v - 5 = 0$

APPLYING THE CONCEPTS

48. Factoring, completing the square, and using the quadratic formula are three methods of solving quadratic equations. Describe each method, and cite the advantages and disadvantages of using each.

49. Explain why the equation $0x^2 + 3x + 4 = 0$ cannot be solved by the quadratic formula.

50. Solve $x^2 + ax + b = 0$ for x.

51. True or False?
 a. The equations $x = \sqrt{12 - x}$ and $x^2 = 12 - x$ have the same solutions.
 b. If $\sqrt{a} + \sqrt{b} = c$, then $a + b = c^2$.
 c. $\sqrt{9} = \pm 3$
 d. $\sqrt{x^2} = |x|$

For Exercises 52 to 57, solve.

52. $\sqrt{x + 3} = x - 3$

53. $\sqrt{x + 4} = x + 4$

54. $\sqrt{x + 1} = x - 1$

55. $\sqrt{x^2 + 2x + 1} = x - 1$

56. $\dfrac{x}{4} + \dfrac{3}{x} = \dfrac{5}{2}$

57. $\dfrac{x + 1}{5} - \dfrac{4}{x - 1} = 2$

58. **Distance** An L-shaped sidewalk from the parking lot to a memorial is shown in the figure at the right. The distance directly across the grass to the memorial is 650 ft. The distance to the corner is 600 ft. Find the distance from the corner to the memorial.

59. **Travel** A commuter plane leaves an airport traveling due south at 400 mph. Another plane leaving at the same time travels due east at 300 mph. Find the distance between the two planes after 2 h.

11.4 Graphing Quadratic Equations in Two Variables

Objective A

To graph a quadratic equation of the form $y = ax^2 + bx + c$

TAKE NOTE

For the equation
$y = 3x^2 - x + 1$, $a = 3$,
$b = -1$, and $c = 1$.

An equation of the form $y = ax^2 + bx + c$, $a \neq 0$, is a **quadratic equation in two variables.** Examples of quadratic equations in two variables are shown at the right.

$$y = 3x^2 - x + 1$$
$$y = -x^2 - 3$$
$$y = 2x^2 - 5x$$

For these equations, y is a function of x, and we can write $f(x) = ax^2 + bx + c$. This equation represents a **quadratic function.**

Point of Interest

Mirrors in some telescopes are ground into the shape of a parabola. The mirror at the Palomar Mountain Observatory is 2 ft thick at the ends and weighs 14.75 tons. The mirror has been ground to a true paraboloid (the three-dimensional version of a parabola) to within 0.0000015 in. A possible equation of the mirror is $y = 2640x^2$.

HOW TO Evaluate $f(x) = 2x^2 - 3x + 4$ when $x = -2$.

$$f(x) = 2x^2 - 3x + 4$$
$$f(-2) = 2(-2)^2 - 3(-2) + 4 \qquad \bullet \text{ Replace } x \text{ by } -2.$$
$$= 2(4) + 6 + 4 = 18 \qquad \bullet \text{ Simplify.}$$

The value of the function when $x = -2$ is 18.

The graph of $y = ax^2 + bx + c$ or $f(x) = ax^2 + bx + c$ is a **parabola.** The graph is ∪-shaped and opens up when a is positive, and down when a is negative. The graphs of two parabolas are shown below.

TAKE NOTE

One of the equations at the right was written as $y = 2x^2 + 3x - 2$ and the other using functional notation as $f(x) = -x^2 + 3x + 2$. Remember that y and $f(x)$ are different symbols for the same quantity.

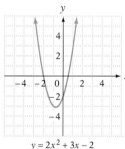

$y = 2x^2 + 3x - 2$
$a = 2$, a positive number
Parabola opens up.

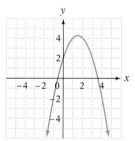

$f(x) = -x^2 + 3x + 2$
$a = -1$, a negative number
Parabola opens down.

HOW TO Graph $y = x^2 - 2x - 3$.

x	y
-2	5
-1	0
0	-3
1	-4
2	-3
3	0
4	5

- Find several solutions of the equation. Because the graph is not a straight line, several solutions must be found in order to determine the ∪-shape. Record the ordered pairs in a table.

Integrating Technology

One of the Projects and Group Activities at the end of this chapter shows how to graph a quadratic equation by using a graphing calculator. You may want to verify the graphs you draw in this section by drawing them on a graphing calculator.

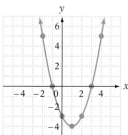

- Graph the ordered-pair solutions on a rectangular coordinate system. Draw a parabola through the points.

For the graph of $y = x^2 - 2x - 3$, shown here again below, note that the graph crosses the x-axis at $(-1, 0)$ and $(3, 0)$. This is also confirmed from the table for the graph (see the preceding page). From the table, note that $y = 0$ when $x = -1$ and when $x = 3$. The x-intercepts of the graph are $(-1, 0)$ and $(3, 0)$.

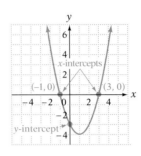

The y-intercept is the point at which the graph crosses the y-axis. At this point, $x = 0$. From the graph, we can see that the y-intercept is $(0, -3)$.

We can find the x-intercepts algebraically by letting $y = 0$ and solving for x.

$y = x^2 - 2x - 3$
$0 = x^2 - 2x - 3$ • Replace y by **0** and solve for x.
$0 = (x + 1)(x - 3)$ • This equation can be solved by factoring. However, it
$x + 1 = 0 \qquad x - 3 = 0$ will be necessary to use the quadratic formula to
$\qquad x = -1 \qquad\quad x = 3$ solve some quadratic equations.

The x-intercepts are $(-1, 0)$ and $(3, 0)$.

We can find the y-intercept algebraically by letting $x = 0$ and solving for y.

$y = x^2 - 2x - 3$
$y = 0^2 - 2(0) - 3$ • Replace x by **0** and simplify.
$\ = -3$

The y-intercept is $(0, -3)$.

Integrating
Technology

The first of the Projects and Group Activities at the end of this chapter shows how to use a graphing calculator to draw the graph of a parabola and how to find the x-intercepts.

> **Graph of a Quadratic Equation in Two Variables**
>
> **To graph a quadratic equation in two variables,** find several solutions of the equation. Graph the ordered-pair solutions on a rectangular coordinate system. Draw a parabola through the points.
>
> **To find the x-intercepts** of the graph of a quadratic equation in two variables, let $y = 0$ and solve for x.
>
> **To find the y-intercept,** let $x = 0$ and solve for y.

Example 1 Graph $y = x^2 - 2x$.

Solution

x	y
-1	3
0	0
1	-1
2	0
3	3

• Find several solutions of the equation.

• Graph the ordered-pair solutions. Draw a parabola through the points.

You Try It 1 Graph $y = x^2 + 2$.

Your solution

Example 2 Find the x- and y-intercepts of the graph of $y = x^2 - 2x - 5$.

Solution

To find the x-intercepts, let $y = 0$ and solve for x. This gives the equation $0 = x^2 - 2x - 5$, which is not factorable over the integers. Use the quadratic formula.

$$x = \frac{-b \pm \sqrt{b^2 - 4ac}}{2a}$$

$$= \frac{-(-2) \pm \sqrt{(-2)^2 - 4(1)(-5)}}{2(1)}$$ • $a = 1, b = -2, c = -5$

$$= \frac{2 \pm \sqrt{24}}{2}$$

$$= \frac{2 \pm 2\sqrt{6}}{2}$$

$$= 1 \pm \sqrt{6}$$

The x-intercepts are $(1 - \sqrt{6}, 0)$ and $(1 + \sqrt{6}, 0)$.

To find the y-intercept, let $x = 0$ and solve for y.

$y = x^2 - 2x - 5$
$= 0^2 - 2(0) - 5$ • Replace x by **0**.
$= -5$

The y-intercept is $(0, -5)$.

You Try It 2 Find the x- and y-intercepts of the graph of $f(x) = x^2 - 6x + 9$.

Your solution

Solutions on pp. S28–S29

11.4 Exercises

Objective A   To graph a quadratic equation of the form $y = ax^2 + bx + c$

For Exercises 1 to 4, determine whether the graph of the equation opens up or down.

1. $y = -\dfrac{1}{3}x^2$

2. $y = x^2 - 2x - 3$

3. $y = 2x^2 - 4$

4. $f(x) = 3 - 2x - x^2$

For Exercises 5 to 10, evaluate the function for the given value of x.

5. $f(x) = x^2 - 2x + 1$; $x = 3$

6. $f(x) = 2x^2 + x - 1$; $x = -2$

7. $f(x) = 4 - x^2$; $x = -3$

8. $f(x) = x^2 + 6x + 9$; $x = -3$

9. $f(x) = -x^2 + 5x - 6$; $x = -4$

10. $f(x) = -2x^2 + 2x - 1$; $x = -3$

For Exercises 11 to 25, graph.

11. $y = x^2$

12. $y = -x^2$

13. $y = -x^2 + 1$

14. $y = x^2 - 1$

15. $y = 2x^2$

16. $y = \dfrac{1}{2}x^2$

17. $y = -\dfrac{1}{2}x^2 + 1$

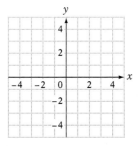

18. $y = 2x^2 - 1$

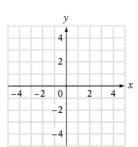

19. $y = x^2 - 4x$

20. $y = x^2 + 4x$

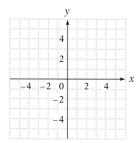

21. $y = x^2 - 2x + 3$

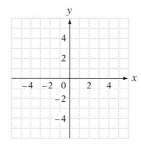

22. $y = x^2 - 4x + 2$

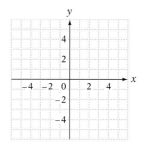

23. $y = -x^2 + 2x + 3$

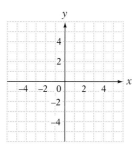

24. $y = -x^2 - 2x + 3$

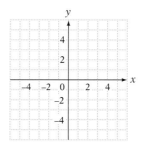

25. $y = -x^2 + 4x - 4$

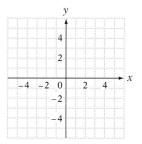

For Exercises 26 to 37, determine the x- and y-intercepts.

26. $y = x^2 - 5x + 6$

27. $y = x^2 + 5x - 6$

28. $f(x) = 9 - x^2$

29. $f(x) = x^2 + 12x + 36$

30. $y = x^2 + 2x - 6$

31. $f(x) = x^2 + 4x - 2$

32. $y = x^2 + 2x + 3$

33. $y = x^2 - x + 1$

34. $f(x) = 2x^2 - x - 3$

35. $f(x) = 2x^2 - 13x + 15$

36. $y = 4 - x - x^2$

37. $y = 2 - 3x - 3x^2$

APPLYING THE CONCEPTS

For Exercises 38 to 41, show that the equation is a quadratic equation in two variables by writing it in the form $y = ax^2 + bx + c$.

38. $y + 1 = (x - 4)^2$ **39.** $y - 2 = 3(x + 1)^2$ **40.** $y - 4 = 2(x - 3)^2$ **41.** $y + 3 = 3(x - 1)^2$

For Exercises 42 to 45, find the x-intercepts.

42. $y = x^3 - x^2 - 6x$

43. $y = x^3 - 4x^2 - 5x$

44. $y = x^3 + x^2 - 4x - 4$
     Hint: Factor by grouping.

45. $y = x^3 + 3x^2 - x - 3$
     Hint: Factor by grouping.

11.5   Application Problems

Objective A   To solve application problems

The application problems in this section are varieties of those problems solved earlier in the text. Each of the strategies for the problems in this section will result in a quadratic equation.

HOW TO   In 5 h, two campers rowed 12 mi down a stream and then rowed back to their campsite. The rate of the stream's current was 1 mph. Find the rate at which the campers rowed.

> **Strategy for Solving an Application Problem**
>
> **1.** Determine the type of problem. For example, is it a distance-rate problem, a geometry problem, or a work problem?

The problem is a distance-rate problem.

> **2.** Choose a variable to represent the unknown quantity. Write numerical or variable expressions for all the remaining quantities. These results can be recorded in a table.

The unknown rate of the campers: r

	Distance	**÷**	**Rate**	**=**	**Time**
Downstream	12	÷	$r + 1$	=	$\dfrac{12}{r + 1}$
Upstream	12	÷	$r - 1$	=	$\dfrac{12}{r - 1}$

> **3.** Determine how the quantities are related.

The total time of the trip was 5 h.

$$\frac{12}{r + 1} + \frac{12}{r - 1} = 5$$

$$(r + 1)(r - 1)\left(\frac{12}{r + 1} + \frac{12}{r - 1}\right) = (r + 1)(r - 1)5$$

$$(r - 1)12 + (r + 1)12 = (r^2 - 1)5$$

$$12r - 12 + 12r + 12 = 5r^2 - 5$$

$$24r = 5r^2 - 5$$

$$0 = 5r^2 - 24r - 5$$

$$0 = (5r + 1)(r - 5)$$

$$5r + 1 = 0 \qquad r - 5 = 0$$

$$5r = -1 \qquad r = 5$$

$$r = -\frac{1}{5}$$

The rowing rate was 5 mph.

TAKE NOTE

The time going downstream plus the time going upstream is equal to the time of the entire trip.

TAKE NOTE

The solution $r = -\dfrac{1}{5}$ is not possible, because the rate cannot be a negative number.

Example 1

A painter and the painter's apprentice working together can paint a room in 2 h. Working alone, the apprentice requires 3 more hours to paint the room than the painter requires working alone. How long does it take the painter working alone to paint the room?

Strategy

• This is a work problem.

• Time for the painter to paint the room: t
  Time for the apprentice to paint the room: $t + 3$

	Rate	*Time*	*Part*
Painter	$\dfrac{1}{t}$	2	$\dfrac{2}{t}$
Apprentice	$\dfrac{1}{t+3}$	2	$\dfrac{2}{t+3}$

• The sum of the parts of the task completed must equal 1.

Solution

$$\frac{2}{t} + \frac{2}{t+3} = 1$$

$$t(t+3)\left(\frac{2}{t} + \frac{2}{t+3}\right) = t(t+3) \cdot 1$$

$$(t+3)2 + t(2) = t(t+3)$$
$$2t + 6 + 2t = t^2 + 3t$$
$$4t + 6 = t^2 + 3t$$
$$0 = t^2 - t - 6$$
$$0 = (t-3)(t+2)$$

$$t - 3 = 0 \qquad t + 2 = 0$$
$$t = 3 \qquad\quad t = -2$$

The solution $t = -2$ is not possible.

The time is 3 h.

You Try It 1

The length of a rectangle is 2 m more than the width. The area is 15 m². Find the width.

Your strategy

Your solution

Solution on p. S29

11.5 Exercises

Objective A   **To solve application problems**

1. **Geometry**   The height of a triangle is 2 m more than twice the length of the base. The area of the triangle is 20 m². Find the height of the triangle and the length of the base.

2. **Geometry**   The length of a rectangle is 4 ft more than twice the width. The area of the rectangle is 160 ft². Find the length and width of the rectangle.

3. **Sports**   The area of the batter's box on a major-league baseball field is 24 ft². The length of the batter's box is 2 ft more than the width. Find the length and width of the batter's box.

4. **Sports**   The length of the batter's box on a softball field is 1 ft less than twice the width. The area of the batter's box is 15 ft². Find the length and width of the batter's box.

5. **Sports**   The length of a swimming pool is twice the width. The area of the pool is 5000 ft². Find the length and width of the pool.

6. **Sports**   The length of a singles tennis court is 24 ft more than twice the width. The area of the tennis court is 2106 ft². Find the length and width of the court.

7. **Sports**   The hang time of a football that is kicked on the opening kick-off is given by $s = -16t^2 + 88t + 4$, where s is the height of the football t seconds after leaving the kicker's foot. What is the hang time of a kickoff that hits the ground without being caught? Round to the nearest tenth.

8. **Manufacturing**   A square piece of cardboard is to be formed into a box to transport pizzas. The box is formed by cutting 2-inch square corners from the cardboard and folding them up as shown in the figure at the right. If the volume of the box is 512 in³, what are the dimensions of the cardboard?

9. **Automotive Safety**   The distance, s, a car needs to come to a stop on a certain surface depends on the velocity, v, in feet per second, of the car when the brakes are applied. The equation is given by $s = 0.0344v^2 - 0.758v$. What is the maximum velocity a car can have when the brakes are applied and stop within 150 ft?

10. **Landscaping**   The perimeter of a rectangular garden is 54 ft. The area of the garden is 180 ft². Find the length and width of the garden.

11. **Food Preparation**   The radius of a large pizza is 1 in. less than twice the radius of a small pizza. The difference between the areas of the two pizzas is 33π in². Find the radius of the large pizza.

12. **Geometry**   The hypotenuse of a right triangle is $\sqrt{13}$ cm. One leg is 1 cm shorter than twice the length of the other leg. Find the lengths of the legs of the right triangle.

13. **Computer Computations**   One computer takes 21 min longer than a second computer to calculate the value of a complex equation. Working together, these computers complete the calculation in 10 min. How long would it take each computer, working separately, to calculate the value?

14. **Plumbing**   A tank has two drains. One drain takes 16 min longer to empty the tank than does a second drain. With both drains open, the tank is emptied in 6 min. How long would it take each drain, working alone, to empty the tank?

15. **Transportation**   Using one engine of a ferryboat, it takes 6 h longer to cross a channel than it does using a second engine alone. With both engines operating, the ferryboat can make the crossing in 4 h. How long would it take each engine, working alone, to power the ferryboat across the channel?

16. **Masonry**   An apprentice mason takes 8 h longer to build a small fireplace than an experienced mason. Working together, they can build the fireplace in 3 h. How long would it take each mason, working alone, to complete the fireplace?

17. **Travel**   It took a small plane 2 h longer to fly 375 mi against the wind than to fly the same distance with the wind. The rate of the wind was 25 mph. Find the rate of the plane in calm air.

18. **Travel**   It took a motorboat 1 h longer to travel 36 mi against the current than to go 36 mi with the current. The rate of the current was 3 mph. Find the rate of the boat in calm water.

APPLYING THE CONCEPTS

19. **Food Preparation**   If a pizza with a diameter of 8 in. costs $6, what should be the cost of a pizza with a diameter of 16 in. if both pizzas cost the same amount per square inch?

20. **Geometry**   A wire 8 ft long is cut into two pieces. A circle is formed from one piece, and a square is formed from the other. The total area of both figures is given by $A = \frac{1}{16}(8 - x)^2 + \frac{x^2}{4\pi}$. What is the length of each piece of wire if the total area is 4.5 ft²?

Focus on Problem Solving

Algebraic Manipulation and Graphing Techniques

Problem solving is often easier when we have both algebraic manipulation and graphing techniques at our disposal. Solving quadratic equations and graphing quadratic equations in two variables are used here to solve problems involving profit.

A company's revenue, R, is the total amount of money the company earned by selling its products. The cost, C, is the total amount of money the company spent to manufacture and sell its products. A company's profit, P, is the difference between the revenue and the cost: $P = R - C$. A company's revenue and cost may be represented by equations.

A company manufactures and sells woodstoves. The total monthly cost, in dollars, to produce n woodstoves is $C = 30n + 2000$. Write a variable expression for the company's monthly profit if the revenue, in dollars, obtained from selling all n woodstoves is $R = 150n - 0.4n^2$.

$$P = R - C$$
$$P = 150n - 0.4n^2 - (30n + 2000)$$
$$P = -0.4n^2 + 120n - 2000$$

• Replace R by $150n - 0.4n^2$ and C by $30n + 2000$. Then simplify.

How many woodstoves must the company manufacture and sell in order to make a profit of $6000 a month?

$$P = -0.4n^2 + 120n - 2000$$
$$6000 = -0.4n^2 + 120n - 2000$$
$$0 = -0.4n^2 + 120n - 8000$$

$$0 = n^2 - 300n + 20{,}000$$

$$0 = (n - 100)(n - 200)$$

$$n - 100 = 0 \qquad n - 200 = 0$$
$$n = 100 \qquad\quad n = 200$$

• Substitute 6000 for P.
• Write the equation in standard form.
• Divide each side of the equation by -0.4.
• Factor.
• Solve for n.

The company will make a monthly profit of $6000 if either 100 or 200 woodstoves are manufactured and sold.

The graph of $P = -0.4n^2 + 120n - 2000$ is shown at the right. Note that when $P = 6000$, the values of n are 100 and 200.

Also note that the coordinates of the highest point on the graph are (150, 7000). This means that the company makes a *maximum* profit of $7000 per month when 150 woodstoves are manufactured and sold.

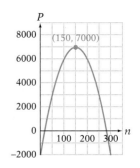

1. The total cost, in dollars, for a company to produce and sell n guitars per month is $C = 240n + 1200$. The company's revenue, in dollars, from selling all n guitars is $R = 400n - 2n^2$.

   a. How many guitars must the company produce and sell each month in order to make a monthly profit of $1200?

   b. Graph the profit equation. What is the maximum monthly profit that the company can make?

Projects and Group Activities

Graphical Solutions of Quadratic Equations

A real number x is called a **zero of a function** if the function evaluated at x is 0. That is, if $f(x) = 0$, then x is called a zero of the function. For instance, evaluating $f(x) = x^2 + x - 6$ when $x = -3$, we have

$$f(x) = x^2 + x - 6$$
$$f(-3) = (-3)^2 + (-3) - 6 \qquad \bullet \text{ Replace } x \text{ by } -3.$$
$$f(-3) = 9 - 3 - 6 = 0$$

For this function $f(-3) = 0$, so -3 is a zero of the function.

Verify that 2 is a zero of $f(x) = x^2 + x - 6$ by showing that $f(2) = 0$.

The graph of $f(x) = x^2 + x - 6$ is shown at the left. Note that the graph crosses the x-axis at -3 and 2, the two zeros of the function. The points $(-3, 0)$ and $(2, 0)$ are x-intercepts of the graph.

Or consider the equation $0 = x^2 + x - 6$, which is $f(x) = x^2 + x - 6$ with $f(x)$ replaced by 0. Solving $0 = x^2 + x - 6$, we have

$$0 = x^2 + x - 6$$
$$0 = (x + 3)(x - 2) \qquad \bullet \text{ Solve by factoring and using the}$$
$$x + 3 = 0 \qquad\qquad x - 2 = 0 \qquad \text{Principle of Zero Products.}$$
$$x = -3 \qquad\qquad x = 2$$

Observe that the solutions of the equation are the zeros of the function. This important connection among the real zeros of a function, the x-intercepts of its graph, and the solutions of the equation is the basis for using a graphing calculator to solve an equation.

The following method of solving a quadratic equation by using a graphing calculator is based on a TI-83 or TI-83 Plus calculator. Other calculators will require a slightly different approach.

> **HOW TO** Approximate the solutions of $x^2 + 4x = 6$ by using a graphing calculator.
>
> Write the equation in standard form. $\qquad\qquad x^2 + 4x = 6$
> $\qquad\qquad\qquad\qquad\qquad\qquad\qquad\qquad\qquad\qquad x^2 + 4x - 6 = 0$
> 1. Press ⬚ **Y=** and enter $x^2 + 4x - 6$ for Y1.
> 2. Press **GRAPH**. If the graph does not appear on the screen, press **ZOOM** 6.
> 3. Press **2nd** CALC 2. Note that the selection for 2 says **zero**. This will begin the calculation of the zeros of the function, which are the solutions of the equation.

Step 1

Step 2

Step 3

Step 4

Step 5

Step 6

Step 7

4. At the bottom of the screen you will see **LeftBound?** This is asking you to move the blinking cursor so that it is to the *left* of the first *x*-intercept. Use the left arrow key to move the cursor to the left of the first *x*-intercept. The values of *x* and *y* that appear on your calculator may be different from the ones shown here. Just be sure that you are to the left of the *x*-intercept. When you are done, press <kbd>ENTER</kbd>.

5. At the bottom of the screen you will see **RightBound?** This is asking you to move the blinking cursor so that it is to the *right* of the *x*-intercept. Use the right arrow key to move the cursor to the right of the *x*-intercept. The values of *x* and *y* that appear on your calculator may be different from the ones shown here. Just be sure that you are to the right of the *x*-intercept. When you are done, press <kbd>ENTER</kbd>.

6. At the bottom of the screen you will see **Guess?** Press <kbd>ENTER</kbd>.

7. The zero of the function is approximately -5.162278. Thus one solution of $x^2 + 4x = 6$ is -5.162278. Also note that the value of *y* is given as $Y1 = {}^-1E^-12$. This is the way the calculator writes a number in scientific notation. We would normally write $Y1 = -1 \times 10^{-12}$. This number is very close to zero.

To find the other solution, we repeat Steps 3 through 7. The screens are shown below.

A second zero of the function is approximately 1.1622777. Thus the two solutions of $x^2 + 4x = 6$ are approximately -5.162278 and 1.1622777.

Use a graphing calculator to approximate the solutions of the following equations.

1. $x^2 + 3x - 4 = 0$ **2.** $x^2 - 4x - 5 = 0$

3. $x^2 + 3.4x = 4.15$ **4.** $2x^2 - \dfrac{5}{9}x = \dfrac{3}{8}$

5. $\pi x^2 - \sqrt{17}x - 2 = 0$ **6.** $\sqrt{2}x^2 + x - \sqrt{7} = 0$

Geometric Construction of Completing the Square

Completing the square as a method for solving a quadratic equation has been known for centuries. The Persian mathematician Al-Khwarismi used this method in a textbook written around 825 A.D. The method was very geometric. That is, Al-Khwarismi literally completed a square. To understand how this method works, consider the following geometric shapes: a square whose area is x^2, a rectangle whose area is x, and another square whose area is 1.

9 squares were added

Now consider the expression $x^2 + 6x$. From our discussion in this chapter, to complete the square, we added $\left(\frac{1}{2} \cdot 6\right)^2 = 3^2 = 9$ to the expression. The geometric construction that Al-Khwarismi used is shown at the left.

Note that it is necessary to add 9 squares to the figure to "complete the square." One of the difficulties of using a geometric method such as this is that it cannot easily be extended to $x^2 - 6x$. There is no way to draw an area of $-6x$! That really did not bother Al-Khwarismi much. Negative numbers were not a significant part of mathematics until well into the 13th century.

1. Show how Al-Khwarismi would have completed the square for $x^2 + 4x$.
2. Show how Al-Khwarismi would have completed the square for $x^2 + 10x$.

Chapter 11 Summary

Key Words

A *quadratic equation* is an equation that can be written in the form $ax^2 + bx + c = 0$, where a, b, and c are real numbers and $a \neq 0$. [11.1A, p. 515]

Examples

$3x^2 - 5x - 3 = 0$ is a quadratic equation. For this equation, $a = 3$, $b = -5$, $c = -3$.

A quadratic equation is in *standard form* when the polynomial is in descending order and equal to zero. [11.1A, p. 515]

$2x - 4 + 5x^2 = 0$ is not in standard form. The same equation in standard form is $5x^2 + 2x - 4 = 0$.

Adding to a binomial the constant term that makes it a perfect-square trinomial is called *completing the square*. [11.2A, p. 521]

Adding to $x^2 - 8x$ the constant term 16 results in a perfect square trinomial: $x^2 - 8x + 16 = (x - 4)^2$.

An equation of the form $y = ax^2 + bx + c$, $a \neq 0$, is a *quadratic equation in two variables*. [11.4A, p. 531]

$y = 2x^2 + 3x - 4$ is a quadratic equation in two variables.

The graph of an equation of the form $y = ax^2 + bx + c$, $a \neq 0$, is a *parabola*. The graph is U-shaped and opens up when $a > 0$ and opens down when $a < 0$. [11.4A, p. 531]

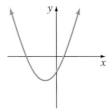

$a > 0$
Parabola opens up

$a < 0$
Parabola opens down

Essential Rules and Procedures

Examples

Solving a Quadratic Equation by Factoring [11.1A, p. 515]
Write the equation in standard form, factor the polynomial, apply the Principle of Zero Products, and solve for the variable.

$$x^2 - 3x = 10$$
$$x^2 - 3x - 10 = 0$$
$$(x + 2)(x - 5) = 0$$
$$x + 2 = 0 \qquad x - 5 = 0$$
$$x = -2 \qquad x = 5$$

Principle of Taking the Square Root of Each Side of an Equation [11.1B, p. 517]
If $x^2 = a$, then $x = \pm\sqrt{a}$.

This principle is used to solve quadratic equations by taking square roots.

$$2x^2 - 36 = 0$$
$$2x^2 = 36$$
$$x^2 = 18$$
$$\sqrt{x^2} = \sqrt{18}$$
$$x = \pm\sqrt{18} = \pm 3\sqrt{2}$$

Solving a Quadratic Equation by Completing the Square [11.2A, p. 521]
When the quadratic equation is in the form $x^2 + bx = c$, add to each side of the equation the term that completes the square on $x^2 + bx$. Factor the perfect-square trinomial, and write it as the square of a binomial. Take the square root of each side of the equation and solve for x.

$$x^2 + 6x = 5$$
$$x^2 + 6x + 9 = 5 + 9$$
$$(x + 3)^2 = 14$$
$$\sqrt{(x + 3)^2} = \sqrt{14}$$
$$x + 3 = \pm\sqrt{14}$$
$$x = -3 \pm \sqrt{14}$$

The Quadratic Formula [11.3A, p. 527]
The solutions of $ax^2 + bx + c = 0$, $a \neq 0$, are

$$x = \frac{-b \pm \sqrt{b^2 - 4ac}}{2a}.$$

$$2x^2 + 3x - 6 = 0$$
$$x = \frac{-b \pm \sqrt{b^2 - 4ac}}{2a}$$
$$= \frac{-3 \pm \sqrt{(3)^2 - 4(2)(-6)}}{2(2)}$$
$$= \frac{-3 \pm \sqrt{9 + 48}}{4} = \frac{-3 \pm \sqrt{57}}{4}$$

Graph of a Quadratic Equation in Two Variables [11.4A, p. 532]
To graph a quadratic equation in two variables, find several solutions of the equation. Graph the ordered-pair solutions on a rectangular coordinate system. Draw a parabola through the points.

To find the x-intercepts of the graph of a quadratic equation in two variables, let $y = 0$ and solve for x.

To find the y-intercept, let $x = 0$ and solve for y.

$y = x^2 - x - 2$

x	y
-2	4
-1	0
0	-2
1	-2
2	0
3	4

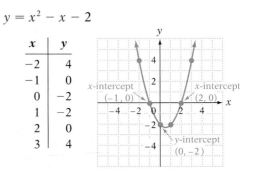

x-intercept $(-1, 0)$

x-intercept $(2, 0)$

y-intercept $(0, -2)$

Chapter 11 Review Exercises

1. Solve by factoring: $6x^2 + 13x - 28 = 0$

2. Solve by taking square roots:
   $49x^2 = 25$

3. Solve by completing the square:
   $x^2 + 2x - 24 = 0$

4. Solve by using the quadratic formula:
   $x^2 + 5x - 6 = 0$

5. Solve by completing the square:
   $2x^2 + 5x = 12$

6. Solve by factoring: $12x^2 + 10 = 29x$

7. Solve by taking square roots:
   $(x + 2)^2 - 24 = 0$

8. Solve by using the quadratic formula:
   $2x^2 + 3 = 5x$

9. Solve by factoring: $6x(x + 1) = x - 1$

10. Solve by taking square roots:
    $4y^2 + 9 = 0$

11. Solve by completing the square:
    $x^2 - 4x + 1 = 0$

12. Solve by using the quadratic formula:
    $x^2 - 3x - 5 = 0$

13. Solve by completing the square:
    $x^2 + 6x + 12 = 0$

14. Solve by factoring: $(x + 9)^2 = x + 11$

15. Solve by taking square roots:
    $\left(x - \dfrac{1}{2}\right)^2 = \dfrac{9}{4}$

16. Solve by completing the square:
    $4x^2 + 16x = 7$

17. Solve by using the quadratic formula: $x^2 - 4x + 8 = 0$

18. Solve by using the quadratic formula: $2x^2 + 5x + 2 = 0$

19. Graph $y = -3x^2$.

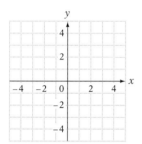

20. Graph $y = -\dfrac{1}{4}x^2$.

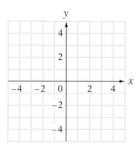

21. Graph $y = 2x^2 + 1$.

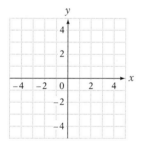

22. Graph $y = x^2 - 4x + 3$.

23. Graph $y = -x^2 + 4x - 5$.

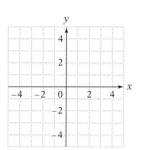

24. Find the x- and y-intercepts of the graph of $y = x^2 - 2x - 15$.

25. **Travel** It took a hawk half an hour longer to fly 70 mi against the wind than to fly 40 mi with the wind. The rate of the wind was 5 mph. Find the rate of the hawk in calm air.

Chapter 11 Test

1. Solve by factoring: $x^2 - 5x - 6 = 0$

2. Solve by factoring: $3x^2 + 7x = 20$

3. Solve by taking square roots:
   $2(x - 5)^2 - 50 = 0$

4. Solve by taking square roots:
   $3(x + 4)^2 - 60 = 0$

5. Solve by completing the square:
   $x^2 + 4x - 16 = 0$

6. Solve by completing the square:
   $x^2 + 3x = 8$

7. Solve by completing the square:
   $2x^2 - 6x + 1 = 0$

8. Solve by completing the square:
   $2x^2 + 8x = 3$

9. Solve by using the quadratic formula:
   $x^2 + 4x + 2 = 0$

10. Solve by using the quadratic formula:
    $x^2 - 3x = 6$

11. Solve by using the quadratic formula:
$2x^2 - 5x - 3 = 0$

12. Solve by using the quadratic formula:
$3x^2 - x = 1$

13. Graph $y = x^2 + 2x - 4$.

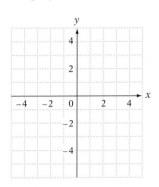

14. Find the x- and y-intercepts of the graph of
$f(x) = x^2 + x - 12$.

15. Geometry   The length of a rectangle is 2 ft less than twice the width. The area of the rectangle is 40 ft². Find the length and width of the rectangle.

16. Travel   It took a motorboat 1 h longer to travel 60 mi against a current than it took the boat to travel 60 mi with the current. The rate of the current was 1 mph. Find the rate of the boat in calm water.

Cumulative Review Exercises

1. Simplify: $2x - 3[2x - 4(3 - 2x) + 2] - 3$

2. Solve: $-\dfrac{3}{5}x = -\dfrac{9}{10}$

3. Solve: $2x - 3(4x - 5) = -3x - 6$

4. Simplify: $(2a^2b)^2(-3a^4b^2)$

5. Divide: $(x^2 - 8) \div (x - 2)$

6. Factor: $3x^3 + 2x^2 - 8x$

7. Divide: $\dfrac{3x^2 - 6x}{4x - 6} \div \dfrac{2x^2 + x - 6}{6x^3 - 24x}$

8. Subtract: $\dfrac{x}{2(x - 1)} - \dfrac{1}{(x - 1)(x + 1)}$

9. Simplify: $\dfrac{1 - \dfrac{7}{x} + \dfrac{12}{x^2}}{2 - \dfrac{1}{x} - \dfrac{15}{x^2}}$

10. Find the x- and y-intercepts for the graph of the line $4x - 3y = 12$.

11. Find the equation of the line that contains the point $(-3, 2)$ and has slope $-\dfrac{4}{3}$.

12. Solve by substitution:
$$3x - y = 5$$
$$y = 2x - 3$$

13. Solve by the addition method:
$$3x + 2y = 2$$
$$5x - 2y = 14$$

14. Solve: $2x - 3(2 - 3x) > 2x - 5$

15. Simplify: $(\sqrt{a} - \sqrt{2})(\sqrt{a} + \sqrt{2})$

16. Simplify: $\dfrac{\sqrt{108a^7b^3}}{\sqrt{3a^4b}}$

17. Simplify: $\dfrac{\sqrt{3}}{5 + 2\sqrt{3}}$

18. Solve: $3 = 8 - \sqrt{5x}$

19. Solve by factoring: $6x^2 - 17x = -5$

20. Solve by taking square roots:
$2(x - 5)^2 = 36$

21. Solve by completing the square:
$3x^2 + 7x = -3$

22. Solve by using the quadratic formula:
$2x^2 - 3x - 2 = 0$

23. **Food Mixtures**   Find the cost per pound of a mixture made from 20 lb of cashews that cost $3.50 per pound and 50 lb of peanuts that cost $1.75 per pound.

24. **Investments**   A stock investment of 100 shares paid a dividend of $215. At this rate, how many additional shares must the investor own to earn a dividend of $752.50?

25. **Travel**   A 720-mile trip from one city to another takes 3 h when a plane is flying with the wind. The return trip, against the wind, takes 4.5 h. Find the rate of the plane in still air and the rate of the wind.

26. **Grading**   A student received a 70, a 91, an 85, and a 77 on four tests in a mathematics class. What scores on the last test will enable the student to receive a minimum of 400 points?

27. **Number Sense**   The sum of the squares of three consecutive odd integers is 83. Find the middle odd integer.

28. **Exercise**   A jogger ran 7 mi at a constant rate and then reduced the rate by 3 mph and ran an additional 8 mi at the reduced rate. The total time spent jogging the 15 mi was 3 h. Find the jogger's rate for the last 8 mi.

29. Graph the solution set of $2x - 3y > 6$.

30. Graph $y = x^2 - 2x - 3$.

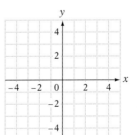

Final Exam

1. Evaluate $-|-3|$.

2. Subtract: $-15 - (-12) - 3$

3. Simplify: $-2^4 \cdot (-2)^4$

4. Simplify: $-7 - \dfrac{12 - 15}{2 - (-1)} \cdot (-4)$

5. Evaluate $\dfrac{a^2 - 3b}{2a - 2b^2}$ when $a = 3$ and $b = -2$.

6. Simplify: $6x - (-4y) - (-3x) + 2y$

7. Simplify: $(-15z)\left(-\dfrac{2}{5}\right)$

8. Simplify: $-2[5 - 3(2x - 7) - 2x]$

9. Solve: $20 = -\dfrac{2}{5}x$

10. Solve: $4 - 2(3x + 1) = 3(2 - x) + 5$

11. Write $\dfrac{1}{8}$ as a percent.

12. Find 19% of 80.

13. Subtract: $(2x^2 - 5x + 1) - (5x^2 - 2x - 7)$

14. Simplify: $(-3xy^3)^4$

15. Multiply: $(3x^2 - x - 2)(2x + 3)$

16. Simplify: $\dfrac{(-2x^2y^3)^3}{(-4xy^4)^2}$

17. Divide: $\dfrac{12x^2y - 16x^3y^2 - 20y^2}{4xy^2}$

18. Divide: $(5x^2 - 2x - 1) \div (x + 2)$

19. Simplify: $(4x^{-2}y)^2 (2xy^{-2})^{-2}$

20. Given $f(t) = \dfrac{t}{t + 1}$, find $f(3)$.

21. Factor: $x^2 - 5x - 6$

22. Factor: $6x^2 - 5x - 6$

23. Factor: $8x^3 - 28x^2 + 12x$

24. Factor: $25x^2 - 16$

25. Factor: $2a(4 - x) - 6(x - 4)$

26. Factor: $75y - 12x^2y$

27. Solve: $2x^2 = 7x - 3$

28. Multiply: $\dfrac{2x^2 - 3x + 1}{4x^2 - 2x} \cdot \dfrac{4x^2 + 4x}{x^2 - 2x + 1}$

29. Subtract: $\dfrac{5}{x + 3} - \dfrac{3x}{2x - 5}$

30. Simplify: $x - \dfrac{1}{1 - \dfrac{1}{x}}$

31. Solve: $\dfrac{5x}{3x - 5} - 3 = \dfrac{7}{3x - 5}$

32. Solve $a = 3a - 2b$ for a.

33. Find the slope of the line that contains the points $(-1, -3)$ and $(2, -1)$.

34. Find the equation of the line that contains the point $(3, -4)$ and has slope $-\dfrac{2}{3}$.

35. Solve by substitution:
$y = 4x - 7$
$y = 2x + 5$

36. Solve by the addition method:
$4x - 3y = 11$
$2x + 5y = -1$

37. Solve: $4 - x \geq 7$

38. Solve: $2 - 2(y - 1) \leq 2y - 6$

39. Simplify: $\sqrt{49x^6}$

40. Simplify: $2\sqrt{27a} + 8\sqrt{48a}$

41. Simplify: $\dfrac{\sqrt{3}}{\sqrt{5} - 2}$

42. Solve: $\sqrt{2x - 3} + 4 = 5$

43. Solve by factoring:
$3x^2 - x = 4$

44. Solve by using the quadratic formula:
$4x^2 - 2x - 1 = 0$

45. Number Sense   Translate and simplify "the sum of twice a number and three times the difference between the number and two."

46. Depreciation   Because of depreciation, the value of an office machine is now $2400. This is 80% of its original value. Find the original value.

47. Business   The manufacturer's cost for a laser printer is $900. The manufacturer then sells the printer for $1485. What is the markup rate?

48. Investment   An investment of $3000 is made at an annual simple interest rate of 8%. How much additional money must be invested at 11% so that the total interest earned is 10% of the total investment?

49. Food Mixtures   A grocer mixes 4 lb of peanuts that cost $2 per pound with 2 lb of walnuts that cost $5 per pound. What is the cost per pound of the resulting mixture?

50. Pharmacology   A pharmacist mixes together 20 L of a solution that is 60% acid and 30 L of a solution that is 20% acid. What is the percent concentration of the acid in the mixture?

51. Travel   At 2 P.M. a small plane had been flying 1 h when a change of wind direction doubled its average ground speed. The complete 860-kilometer trip took 2.5 h. How far did the plane travel in the first hour?

52. Geometry   The angles of a triangle are such that the measure of the second angle is 10° more than the measure of the first angle, and the measure of the third angle is 10° more than the measure of the second angle. Find the measure of each of the three angles.

53. Number Sense The sum of the squares of three consecutive integers is 50. Find the middle integer.

54. Geometry The length of a rectangle is 5 m more than the width. The area of the rectangle is 50 m². Find the dimensions of the rectangle.

55. Paint Mixtures A paint formula requires 2 oz of dye for every 15 oz of base paint. How many ounces of dye are required for 120 oz of base paint?

56. Food Preparation It takes a chef 1 h to prepare a dinner. The chef's apprentice can prepare the dinner in 1.5 h. How long would it take the chef and the apprentice, working together, to prepare the dinner?

57. Travel With the current, a motorboat travels 50 mi in 2.5 h. Against the current, it takes twice as long to travel 50 mi. Find the rate of the boat in calm water and the rate of the current.

58. Travel Flying against the wind, it took a pilot $\frac{1}{2}$ h longer to travel 500 mi than it took flying with the wind. The rate of the plane in calm air is 225 mph. Find the rate of the wind.

59. Graph the line that has slope $-\frac{1}{2}$ and y-intercept $(0, -3)$.

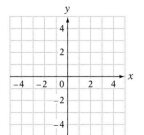

60. Graph $y = x^2 - 4x + 3$.

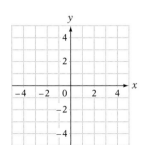

Appendix A

Keystroke Guide for the TI-83 and TI-83 Plus

Basic Operations

Numerical calculations are performed on the **home screen.** You can always return to the home screen by pressing 〔2nd〕 QUIT. Pressing 〔CLEAR〕 erases the home screen.

To evaluate the expression $-2(3 + 5) - 8 \div 4$, use the following keystrokes.

〔(-)〕 2 〔(〕 3 〔+〕 5 〔)〕 〔−〕 8 〔÷〕 4 〔ENTER〕

Note: There is a difference between the key to enter a negative number, 〔(-)〕, and the key for subtraction, 〔−〕. You cannot use these keys interchangeably.

The 〔2nd〕 key is used to access the commands in gold writing above a key. For instance, to evaluate the $\sqrt{49}$, press 〔2nd〕 √ 49 〔)〕 〔ENTER〕.

The 〔ALPHA〕 key is used to place a letter on the screen. One reason to do this is to store a value of a variable. The following keystrokes give A the value of 5.

5 〔STO▸〕 〔ALPHA〕 A 〔ENTER〕

This value is now available in calculations. For instance, we can find the value of $3a^2$ by using the following keystrokes: 3 〔ALPHA〕 A 〔x^2〕. To display the value of the variable on the screen, press 〔2nd〕 RCL 〔ALPHA〕 A.

Note: When you use the 〔ALPHA〕 key, only capital letters are available on the TI-83 calculator.

Evaluating Functions

There are various methods of evaluating a function but all methods require that the expression be entered as one of the ten functions Y_1 to Y_0. To evaluate $f(x) = \frac{x^2}{x - 1}$ when $x = -3$, enter the expression into, for instance, Y_1, and then press 〔VARS〕 ▸ 11 〔(〕 〔(-)〕 3 〔)〕 〔ENTER〕.

Note: If you try to evaluate a function at a number that is not in the domain of the function, you will get an error message. For instance, 1 is not in the domain of $f(x) = \frac{x^2}{x - 1}$. If we try to evaluate the function at 1, the error screen at the right appears.

Evaluating Variable Expressions

To evaluate a variable expression, first store the values of each variable. Then enter the variable expression. For instance, to evaluate $s^2 + 2sl$ when $s = 4$ and $l = 5$, use the following keystrokes.

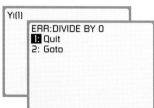

4 〔STO▸〕 〔ALPHA〕 S 〔ENTER〕 5 〔STO▸〕 〔ALPHA〕 L 〔ENTER〕 〔ALPHA〕 S 〔x^2〕 〔+〕 2 〔ALPHA〕 S 〔ALPHA〕 L 〔ENTER〕

Graph To graph a function, use the ⬛ Y= ⬛ key to enter the expression for the function, select a suitable viewing window, and then press ⬛GRAPH⬛ . For instance, to graph $f(x) = 0.1x^3 - 2x - 1$ in the standard viewing window, use the following keystrokes.

⬛ Y= ⬛ 0.1 ⬛X,T,θ,n⬛ ⬛^⬛ 3 ⬛–⬛ 2 ⬛X,T,θ,n⬛ ⬛–⬛ 1 ⬛ZOOM⬛ (scroll to 6) ⬛ENTER⬛

Note: For the keystrokes above, you do not have to scroll to 6. Alternatively, use ⬛ZOOM⬛ 6. This will select the standard viewing window and automatically start the graph. Use the ⬛WINDOW⬛ key to create a custom window for a graph.

Graphing Inequalities To illustrate this feature, we will graph $y \le 2x - 1$. Enter $2x - 1$ into Y_1. Because $y \le 2x - 1$, we want to shade below the graph. Move the cursor to the left of Y_1 and press ⬛ENTER⬛ three times. Press ⬛GRAPH⬛ .

Note: To shade above the graph, move the cursor to the left of Y_1 and press ⬛ENTER⬛ two times. An inequality with the symbol $\le$ or $\ge$ should be graphed with a solid line, and an inequality with the symbol $<$ or $>$ should be graphed with a dashed line. However, the graph of a linear inequality on a graphing calculator does not distinguish between a solid line and a dashed line.

To graph the solution set of a system of inequalities, solve each inequality for y and graph each inequality. The solution set is the intersection of the two inequalities. The solution set of $\begin{matrix} 3x + 2y > 10 \\ 4x - 3y \le 5 \end{matrix}$ is shown at the right.

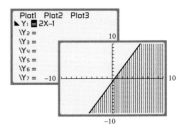

Intersect The INTERSECT feature is used to solve a system of equations. To illustrate this feature, we will use the system of equations $\begin{matrix} 2x - 3y = 13 \\ 3x + 4y = -6 \end{matrix}$.

Note: Some equations can be solved by this method. See the section "Solve an equation" below. Also, this method is used to find a number in the domain of a function for a given number in the range. See the section "Find a domain element."

Solve each of the equations in the system of equations for y. In this case, we have $y = \frac{2}{3}x - \frac{13}{3}$ and $y = -\frac{3}{4}x - \frac{3}{2}$.

Use the Y-editor to enter $\frac{2}{3}x - \frac{13}{3}$ into Y_1 and $-\frac{3}{4}x - \frac{3}{2}$ into Y_2. Graph the two functions in the standard viewing window. (If the window does not show the point of intersection of the two graphs, adjust the window until you can see the point of intersection.)

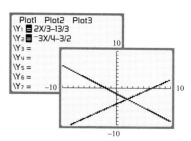

Press (2nd) CALC (scroll to 5, intersect) (ENTER).

Alternatively, you can just press (2nd) CALC 5.

First curve? is shown at the bottom of the screen and identifies one of the two graphs on the screen. Press (ENTER).

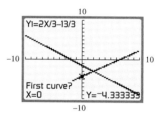

Second curve? is shown at the bottom of the screen and identifies the second of the two graphs on the screen. Press (ENTER).

Guess? shown at the bottom of the screen asks you to use the left or right arrow key to move the cursor to the *approximate* location of the point of intersection. (If there are two or more points of intersection, it does not matter which one you choose first.) Press (ENTER).

The solution of the system of equations is $(2, -3)$.

Solve an equation To illustrate the steps involved, we will solve the equation $2x + 4 = -3x - 1$. The idea is to write the equation as the system of equations $\begin{aligned} y &= 2x + 4 \\ y &= -3x - 1 \end{aligned}$ and then use the steps for solving a system of equations.

Use the Y-editor to enter the left and right sides of the equation into Y_1 and Y_2. Graph the two functions and then follow the steps for Intersect.

The solution is -1, the x-coordinate of the point of intersection.

Find a domain element For this example, we will find a number in the domain of $f(x) = -\frac{2}{3}x + 2$ that corresponds to 4 in the range of the function. This is like solving the system of equations $y = -\frac{2}{3}x + 2$ and $y = 4$.

Use the Y= editor to enter the expression for the function in Y₁ and the desired output, 4, in Y₂. Graph the two functions and then follow the steps for Intersect.

The point of intersection is $(-3, 4)$. The number -3 in the domain of f produces an output of 4 in the range of f.

Math Pressing MATH gives you access to many built-in functions. The following key-strokes will convert 0.125 to a fraction: .125 MATH 1 ENTER.

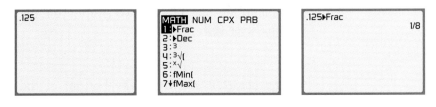

Additional built-in functions under MATH can be found by pressing MATH ▸. For instance, to evaluate $-|-25|$, press (-) MATH ▸ 1 (-) 25) ENTER.

See your owner's manual for assistance with other functions under the MATH key.

Radical Expressions To evaluate a square-root expression, press 2nd √ .

For instance, to evaluate $0.15\sqrt{p^2 + 4p + 10}$ when $p = 100{,}000$, first store 100,000 in P. Then press 0.15 2nd √ ALPHA P x² + 4 ALPHA P + 10) ENTER.

To evaluate a radical expression other than a square root, access $\sqrt[x]{}$ by pressing MATH. For instance, to evaluate $\sqrt[4]{67}$, press 4 (the index of the radical) MATH (scroll to 5) ENTER 67 ENTER.

Scientific Notation To enter a number in scientific notation, use 2nd EE. For instance, to find $\dfrac{3.45 \times 10^{-12}}{1.5 \times 10^{25}}$, press 3.45 2nd EE (-) 12 ÷ 1.5 2nd EE 25 ENTER. The answer is 2.3×10^{-37}.

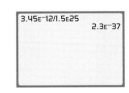

Table There are three steps in creating an input/output table for a function. First use the Y= editor to input the function. The second step is setting up the table, and the third step is displaying the table.

To set up the table, press ⒉ TBLSET. **TblStart** is the first value of the independent variable in the input/output table. △**Tbl** is the difference between successive values. Setting this to 1 means that, for this table, the input values are −2, −1, 0, 1, 2. . . . If △Tbl = 0.5, then the input values are −2, −1.5, −1, −0.5, 0, 0.5, . . .

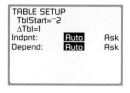

Indpnt is the independent variable. When this is set to **Auto**, values of the independent variable are automatically entered into the table. **Depend** is the dependent variable. When this is set to **Auto**, values of the dependent variable are automatically entered into the table.

To display the table, press ⒉ TABLE. An input/output table for $f(x) = x^2 − 1$ is shown at the right.

Once the table is on the screen, the up and down arrow keys can be used to display more values in the table. For the table at the right, we used the up arrow key to move to $x = −7$.

An input/output table for any given input can be created by selecting **Ask** for the independent variable. The table at the right shows an input/output table for $f(x) = \dfrac{4x}{x − 2}$ for selected values of x. Note the word **ERROR** when 2 was entered. This occurred because f is not defined when $x = 2$.

Note: Using the table feature in **Ask** mode is the same as evaluating a function for given values of the independent variable. For instance, from the table at the right, we have $f(4) = 8$.

Test The TEST feature has many uses, one of which is to graph the solution set of a linear inequality in one variable. To illustrate this feature, we will graph the solution set of $x − 1 < 4$. Press ⦅Y=⦆ ⦅X,T,θ,n⦆ ⦅−⦆ 1 ⒉ TEST (scroll to 5) ⦅ENTER⦆ 4 ⦅GRAPH⦆.

Trace Once a graph is drawn, pressing ⦅TRACE⦆ will place a cursor on the screen, and the coordinates of the point below the cursor are shown at the bottom of the screen. Use the left and right arrow keys to move the cursor along the graph. For the graph at the right, we have $f(4.8) = 3.4592$, where $f(x) = 0.1x^3 − 2x + 2$ is shown at the top left of the screen.

In TRACE mode, you can evaluate a function at any value of the independent variable that is within Xmin and Xmax. To do this, first graph the function. Now press ⦅TRACE⦆ (the value of x) ⦅ENTER⦆. For the graph at the left below, we used $x = −3.5$. If a value of x is chosen outside the window, an error message is displayed.

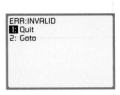

In the example above where we entered -3.5 for x, the value of the function was calculated as 4.7125. This means that $f(-3.5) = 4.7125$. The keystrokes 2nd QUIT VARS ▸ 11 MATH 1 ENTER will convert the decimal value to a fraction.

When the TRACE feature is used with two or more graphs, the up and down arrow keys are used to move between the graphs. The graphs below are for the functions $f(x) = 0.1x^3 - 2x + 2$ and $g(x) = 2x - 3$. By using the up and down arrows, we can place the cursor on either graph. The right and left arrows are used to move along the graph.

Window   The viewing window for a graph is controlled by pressing WINDOW. Xmin and Xmax are the minimum value and maximum value, respectively, of the independent variable shown on the graph. Xscl is the distance between tic marks on the x-axis. Ymin and Ymax are the minimum value and maximum value, respectively, of the dependent variable shown on the graph. Yscl is the distance between tic marks on the y-axis. Leave Xres as 1.

Note: In the standard viewing window, the distance between tic marks on the x-axis is different from the distance between tic marks on the y-axis. This will distort a graph. A more accurate picture of a graph can be created by using a square viewing window. See ZOOM.

Y=   The Y= editor is used to enter the expression for a function. There are ten possible functions, labeled Y_1 to Y_0, that can be active at any one time. For instance, to enter $f(x) = x^2 + 3x - 2$ as Y_1, use the following keystrokes.

Y= X,T,θ,n x^2 + 3 X,T,θ,n − 2

Note: If an expression is already entered for Y_1, place the cursor anywhere on that expression and press CLEAR.

To enter $s = \dfrac{2v - 1}{v^3 - 3}$ into Y_2, place the cursor to the right of the equals sign for Y_2. Then press (2 X,T,θ,n −
1) ÷ (X,T,θ,n ^ 3 − 3) .

Note: When we enter an equation, the independent variable, v in the expression above, is entered using [X,T,θ,n]. The dependent variable, s in the expression above, is one of Y_1 to Y_0. Also note the use of parentheses to ensure the correct order of operations.

Observe the black rectangle that covers the equals sign for the two examples we have shown. This rectangle means that the function is "active." If we were to press [GRAPH], then the graph of both functions would appear. You can make a function inactive by using the arrow keys to move the cursor over the equals sign of that

function and then pressing [ENTER]. This will remove the black rectangle. We have done that for Y_2, as shown at the right. Now if [GRAPH] is pressed, only Y_1 will be graphed.

It is also possible to control the appearance of the graph by moving the cursor on the [Y=] screen to the left of any Y. With the cursor in this position, pressing [ENTER] will change the appearance of the graph. The options are shown at the right.

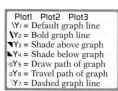

Zero   The ZERO feature of a graphing calculator is used for various calculations: to find the x-intercepts of a function, to solve some equations, and to find the zero of a function.

***x*-intercepts**   To illustrate the procedure for finding x-intercepts, we will use $f(x) = x^2 + x - 2$.

First, use the Y-editor to enter the expression for the function and then graph the function in the standard viewing window. (It may be necessary to adjust this window so that the intercepts are visible.) Once the graph is displayed, use the keystrokes below to find the x-intercepts of the graph of the function.

Press [2nd] CALC (scroll to 2 for **zero** of the function) [ENTER].

Alternatively, you can just press [2nd] CALC 2.

Left Bound? shown at the bottom of the screen asks you to use the left or right arrow key to move the cursor to the *left* of the desired x-intercept. Press [ENTER].

Right Bound? shown at the bottom of the screen asks you to use the left or right arrow key to move the cursor to the *right* of the desired x-intercept. Press [ENTER].

Guess? shown at the bottom of the screen asks you to use the left or right arrow key to move the cursor to the *approximate* location of the desired x-intercept. Press [ENTER].

The x-coordinate of an x-intercept is -2. Therefore, an x-intercept is $(-2, 0)$.

To find the other x-intercept, follow the same steps as above. The screens for this calculation are shown below.

A second x-intercept is $(1, 0)$.

Solve an equation   To use the ZERO feature to solve an equation, first rewrite the equation with all terms on one side. For instance, one way to solve the equation $x^3 - x + 1 = -2x + 3$ is first to rewrite it as $x^3 + x - 2 = 0$. Enter $x^3 + x - 2$ into Y_1 and then follow the steps for finding x-intercepts.

Find the real zeros of a function   To find the real zeros of a function, follow the steps for finding x-intercepts.

Zoom   Pressing ⟨ZOOM⟩ allows you to select some preset viewing windows. This key also gives you access to ZBox, Zoom In, and Zoom Out. These functions enable you to redraw a selected portion of a graph in a new window. Some windows used frequently in this text are shown below.

Appendix B

Tables

Table of Symbols

$+$	add	$<$	is less than
$-$	subtract	$\leq$	is less than or equal to
$\cdot, \times, (a)(b)$	multiply	$>$	is greater than
$\dfrac{a}{b}, \div$	divide	$\geq$	is greater than or equal to
$(\ )$	parentheses, a grouping symbol	(a, b)	an ordered pair whose first component is a and whose second component is b
$[\ ]$	brackets, a grouping symbol	$\circ$	degree (for angles)
π	pi, a number approximately equal to $\dfrac{22}{7}$ or 3.14	$\sqrt{a}$	the principal square root of a
		$\varnothing, \{\}$	the empty set
$-a$	the opposite, or additive inverse, of a	$\lvert a \rvert$	the absolute value of a
$\dfrac{1}{a}$	the reciprocal, or multiplicative inverse, of a	$\cup$	union of two sets
		$\cap$	intersection of two sets
$=$	is equal to	$\in$	is an element of (for sets)
$\approx$	is approximately equal to	$\notin$	is not an element of (for sets)
$\neq$	is not equal to		

Table of Measurement Abbreviations

U.S. Customary System

Length		Capacity		Weight		Area	
in.	inches	oz	fluid ounces	oz	ounces	in^2	square inches
ft	feet	c	cups	lb	pounds	ft^2	square feet
yd	yards	qt	quarts				
mi	miles	gal	gallons				

Metric System

Length		Capacity		Weight/Mass		Area	
mm	millimeter (0.001 m)	ml	milliliter (0.001 L)	mg	milligram (0.001 g)	cm^2	square centimeters
cm	centimeter (0.01 m)	cl	centiliter (0.01 L)	cg	centigram (0.01 g)	m^2	square meters
dm	decimeter (0.1 m)	dl	deciliter (0.1 L)	dg	decigram (0.1 g)		
m	meter	L	liter	g	gram		
dam	decameter (10 m)	dal	decaliter (10 L)	dag	decagram (10 g)		
hm	hectometer (100 m)	hl	hectoliter (100 L)	hg	hectogram (100 g)		
km	kilometer (1000 m)	kl	kiloliter (1000 L)	kg	kilogram (1000 g)		

Time

h	hours	min	minutes	s	seconds

Table of Equations and Formulas

Slope of a line

$$m = \frac{y_2 - y_1}{x_2 - x_1}, \; x_1 \neq x_2$$

Slope-intercept form of a straight line

$$y = mx + b$$

Point-slope formula for a line

$$y - y_1 = m(x - x_1)$$

Quadratic Formula

$$x = \frac{-b \pm \sqrt{b^2 - 4ac}}{2a}$$

$$\text{discriminant} = b^2 - 4ac$$

Perimeter and Area of a Triangle, and Sum of the Measures of the Angles

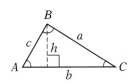

$$P = a + b + c$$

$$A = \frac{1}{2}bh$$

$$A + B + C = 180°$$

Pythagorean Theorem

$$a^2 + b^2 = c^2$$

Perimeter and Area of a Rectangle

$$P = 2L + 2W$$

$$A = LW$$

Perimeter and Area of a Square

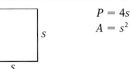

$$P = 4s$$

$$A = s^2$$

Circumference and Area of a Circle

$$C = 2\pi r \quad \text{or} \quad C = \pi d$$

$$A = \pi r^2$$

Volume of a Rectangular Solid

$$V = LWH$$

Volume of a Cube

$$V = s^3$$

Volume of a Sphere

$$V = \frac{4}{3}\pi r^3$$

Volume of a Right Circular Cylinder

$$V = \pi r^2 h$$

Table of Properties

Properties of Real Numbers

The Associative Property of Addition
If a, b, and c are real numbers, then
$(a + b) + c = a + (b + c)$.

The Associative Property of Multiplication
If a, b, and c are real numbers, then
$(a \cdot b) \cdot c = a \cdot (b \cdot c)$.

The Commutative Property of Addition
If a and b are real numbers, then
$a + b = b + a$.

The Commutative Property of Multiplication
If a and b are real numbers, then
$a \cdot b = b \cdot a$.

The Addition Property of Zero
If a is a real number, then
$a + 0 = 0 + a = a$.

The Multiplication Property of One
If a is a real number, then
$a \cdot 1 = 1 \cdot a = a$.

The Multiplication Property of Zero
If a is a real number, then
$a \cdot 0 = 0 \cdot a = 0$.

The Inverse Property of Multiplication
If a is a real number and $a \neq 0$, then
$a \cdot \dfrac{1}{a} = \dfrac{1}{a} \cdot a = 1$.

The Inverse Property of Addition
If a is a real number, then
$a + (-a) = (-a) + a = 0$.

Distributive Property
If a, b, and c are real numbers, then
$a(b + c) = ab + ac$.

Properties of Equations

Addition Property of Equations
If $a = b$, then $a + c = b + c$.

Multiplication Property of Equations
If $a = b$ and $c \neq 0$, then $a \cdot c = b \cdot c$.

Properties of Exponents

If m and n are integers, then $x^m \cdot x^n = x^{m+n}$.
If m and n are integers, then $(x^m)^n = x^{mn}$.

If $x \neq 0$, then $x^0 = 1$.

If m and n are integers and $x \neq 0$, then $\dfrac{x^m}{x^n} = x^{m-n}$.

If m, n, and p are integers, then $(x^m \cdot y^n)^p = x^{mp}y^{np}$.
If n is a positive integer and $x \neq 0$, then

$x^{-n} = \dfrac{1}{x^n}$ and $\dfrac{1}{x^{-n}} = x^n$.

If m, n, and p are integers and $y \neq 0$, then $\left(\dfrac{x^m}{y^n}\right)^p = \dfrac{x^{mp}}{y^{np}}$.

Properties of Inequalities

Addition Property of Inequalities
If $a > b$, then $a + c > b + c$.
If $a < b$, then $a + c < b + c$.

Multiplication Property of Inequalities
If $a > b$ and $c > 0$, then $ac > bc$.
If $a < b$ and $c > 0$, then $ac < bc$.
If $a > b$ and $c < 0$, then $ac < bc$.
If $a < b$ and $c < 0$, then $ac > bc$.

Principle of Zero Products

If $a \cdot b = 0$, then $a = 0$ or $b = 0$.

Properties of Radical Expressions

If a and b are positive real numbers, then $\sqrt{ab} = \sqrt{a}\sqrt{b}$.

If a and b are positive real numbers, then $\sqrt{\dfrac{a}{b}} = \dfrac{\sqrt{a}}{\sqrt{b}}$.

Property of Squaring Both Sides of an Equation

If a and b are real numbers and $a = b$, then $a^2 = b^2$.

Solutions to Chapter 1 "You Try It"

SECTION 1.1

You Try It 1 $A = \{1, 2, 3, 4, 5, 6\}$

You Try It 2 $-5 < -1$
$-1 = -1$
$5 > -1$

The element 5 is greater than -1.

You Try It 3 $|-5| = 5$
$-|-23| = -23$

You Try It 4 $-(-11) = 11$
$-0 = 0$
$-(8) = -8$

You Try It 5 $|-37| = 37$
$|0| = 0$
$|29| = 29$

SECTION 1.2

You Try It 1 $100 + (-43) = 57$

You Try It 2 $(-51) + 42 + 17 + (-102)$
$\quad = -9 + 17 + (-102)$
$\quad = 8 + (-102)$
$\quad = -94$

You Try It 3 $-8 + 7 = -1$

You Try It 4 $19 - (-32) = 19 + 32$
$\quad\quad\quad\quad = 51$

You Try It 5 $-9 - (-12) - 17 - 4$
$\quad = -9 + 12 + (-17) + (-4)$
$\quad = 3 + (-17) + (-4)$
$\quad = -14 + (-4)$
$\quad = -18$

You Try It 6 $-11 - (-12) = -11 + 12 = 1$

You Try It 7

Strategy   To find the difference between the two average temperatures, subtract the smaller number (-130) from the larger number (-17).

Solution $-17 - (-130) = -17 + 130 = 113$

The difference is $113°$F.

SECTION 1.3

You Try It 1 $-7(32) = -224$

You Try It 2 $8(-9)10 = -72(10)$
$\quad\quad\quad\quad = -720$

You Try It 3 $(-2)3(-8)7 = -6(-8)7$
$\quad\quad\quad\quad\quad = 48(7)$
$\quad\quad\quad\quad\quad = 336$

You Try It 4 $-9(34) = -306$

You Try It 5 $(-135) \div (-9) = 15$

You Try It 6 $\dfrac{-72}{4} = -18$

You Try It 7 $-\dfrac{36}{-12} = -(-3)$
$\quad\quad\quad\quad = 3$

You Try It 8 $\dfrac{-72}{-8} = 9$

You Try It 9

Strategy
To find the average daily low temperature:
• Add the seven temperature readings.
• Divide the sum by 7.

Solution
$-6 + (-7) + 0 + (-5) + (-8) + (-1) + (-1) = -28$
$-28 \div 7 = -4$

The average daily low temperature was $-4°$C.

SECTION 1.4

You Try It 1 $-6^3 = -(6 \cdot 6 \cdot 6) = -216$

You Try It 2 $(-3)^4 = (-3)(-3)(-3)(-3) = 81$

You Try It 3 $(3^3)(-2)^3 = (3)(3)(3) \cdot (-2)(-2)(-2)$
$\quad\quad\quad\quad = 27(-8) = -216$

You Try It 4   The product of an odd number of negative factors is odd. Therefore, $(-1)^7 = -1$.

You Try It 5

$-2^2 \cdot (-1)^{12} \cdot (-3)^2 = -(2 \cdot 2) \cdot 1 \cdot (-3) \cdot (-3)$
$$= -4 \cdot 1 \cdot 9 = -36$$

You Try It 6

$7 - 2[2 \cdot 3 - 7 \cdot 2]^2$
$= 7 - 2[6 - 14]^2$ • Perform operations inside
$= 7 - 2[-8]^2$ grouping symbols.
$= 7 - 2[64]$ • Simplify exponential expressions.
$= 7 - 128$ • Do multiplication and division
                  from left to right.
$= -121$ • Do addition and subtraction
                  from left to right.

You Try It 7

$18 - 5[8 - 2(2 - 5)] \div 10$
$= 18 - 5[8 - 2(-3)] \div 10$ • Perform operations
$= 18 - 5[8 + 6] \div 10$ inside grouping symbols.
$= 18 - 5[14] \div 10$
$= 18 - 70 \div 10$ • Do multiplication and
$= 18 - 7$ division from left to right.
$= 11$ • Do addition and
               subtraction from left
               to right.

You Try It 8

$36 \div (8 - 5)^2 - (-3)^2 \cdot 2$
$= 36 \div (3)^2 - (-3)^2 \cdot 2$ • Perform operations inside
               grouping symbols.
$= 36 \div 9 - 9 \cdot 2$ • Simplify exponential expressions.
$= 4 - 9 \cdot 2$ • Do multiplication and division
$= 4 - 18$ from left to right.
$= -14$ • Do addition and subtraction
               from left to right.

SECTION 1.5

You Try It 1

$24 \div 1 = 24$
$24 \div 2 = 12$
$24 \div 3 = 8$
$24 \div 4 = 6$
$24 \div 5$ Remainder is not 0.
$24 \div 6 = 4$ The factors repeat.
The factors of 24 are 1, 2, 3, 4, 6, 8, 12, and 24.

You Try It 2

$$\begin{array}{r|l} & 315 \\ \hline 3 & 105 \\ 3 & 35 \\ 5 & 7 \\ 7 & 1 \end{array}$$
$315 = 3^2 \cdot 5 \cdot 7$

You Try It 3

$$\begin{array}{r|l} & 326 \\ \hline 2 & 163 \\ 163 & 1 \end{array}$$
    • For 163, try prime
$326 = 2 \cdot 163$ numbers up to 13
               because $13^2 > 163$.

You Try It 4

$20 = 2 \cdot 2 \cdot 5$ $21 = 3 \cdot 7$
$LCM = 2 \cdot 2 \cdot 3 \cdot 5 \cdot 7 = 420$

You Try It 5

$42 = 2 \cdot 3 \cdot 7$ $63 = 3 \cdot 3 \cdot 7$
$GCF = 3 \cdot 7 = 21$

SECTION 1.6

You Try It 1

$$\frac{60}{140} = \frac{\overset{1}{\cancel{2}} \cdot \overset{1}{\cancel{2}} \cdot 3 \cdot \overset{1}{\cancel{5}}}{\underset{1}{\cancel{2}} \cdot \underset{1}{\cancel{2}} \cdot \underset{1}{\cancel{5}} \cdot 7} = \frac{3}{7}$$

You Try It 2

$$\frac{4}{9} = 4 \div 9 = 0.\overline{4}$$

You Try It 3

The LCM of 9 and 12 is 36.
$$\frac{5}{9} + \left(-\frac{11}{12}\right) = \frac{20}{36} + \left(-\frac{33}{36}\right) = \frac{20 + (-33)}{36}$$
$$= \frac{-13}{36} = -\frac{13}{36}$$

You Try It 4

The LCM of 8 and 6 is 24.
$$-\frac{5}{6} + \frac{7}{8} = -\frac{20}{24} + \frac{21}{24} = \frac{-20 + 21}{24}$$
$$= \frac{1}{24}$$

You Try It 5

$-6.12 + (-12.881) = -19.001$

You Try It 6

The LCM of 8, 12, and 9 is 72.
$$\frac{7}{8} - \left(-\frac{5}{12}\right) - \frac{1}{9} = \frac{63}{72} - \left(-\frac{30}{72}\right) - \frac{8}{72}$$
$$= \frac{63 - (-30) - 8}{72}$$
$$= \frac{63 + 30 - 8}{72} = \frac{85}{72}$$

You Try It 7

$-12.03 - 19.117 = -12.03 + (-19.117)$
$$= -31.147$$

You Try It 8

Strategy

To find the difference:
• Find the sum of the numbers from the graph that correspond to oil and gas production (8.5, 7.7, 3.6, 3.4, 3.0).
• Find the sum of the numbers from the graph that correspond to oil and gas exports (7.1, 5.1, 2.4, 3.1, 2.5).
• Find the difference between the two sums.

    Solution $8.5 + 7.7 + 3.6 + 3.4 + 3.0 = 26.2$
                      $7.1 + 5.1 + 2.4 + 3.1 + 2.5 = 20.2$
                      $26.2 - 20.2 = 6.0$

The difference between the amount of gas and oil produced and the amount exported is 6 million barrels per day.

You Try It 9

Strategy   To find the fraction of her day spent on these activities, add the fractions $\frac{1}{6}$, $\frac{1}{8}$, and $\frac{1}{4}$.

Solution   The common denominator is 24.

$$\frac{1}{6} + \frac{1}{8} + \frac{1}{4} = \frac{4}{24} + \frac{3}{24} + \frac{6}{24}$$

- Write each fraction in terms of the common denominator.

$$= \frac{4 + 3 + 6}{24} = \frac{13}{24}$$

- Add the fractions.

She spent $\frac{13}{24}$ of her day on these activities.

SECTION 1.7

You Try It 1

$$\frac{5}{8}\left(-\frac{4}{25}\right) = -\frac{5 \cdot 4}{8 \cdot 25}$$

- The signs are different. The product is negative.

$$= -\frac{\overset{1}{\cancel{5}} \cdot \overset{1}{\cancel{2}} \cdot \overset{1}{\cancel{2}}}{\underset{1}{\cancel{2}} \cdot \underset{1}{\cancel{2}} \cdot 2 \cdot \underset{1}{\cancel{5}} \cdot 5}$$

$$= -\frac{1}{10}$$

- Write the answer in simplest form.

You Try It 2

$$-\frac{4}{5} \cdot \left(-\frac{3}{8}\right) \cdot \left(-\frac{10}{27}\right)$$

$$= -\frac{\overset{1}{\cancel{2}} \cdot \overset{1}{\cancel{2}} \cdot \overset{1}{\cancel{3}} \cdot \overset{1}{\cancel{2}} \cdot \overset{1}{\cancel{5}}}{\underset{1}{\cancel{5}} \cdot \underset{1}{\cancel{2}} \cdot \underset{1}{\cancel{2}} \cdot \underset{1}{\cancel{2}} \cdot \underset{1}{\cancel{3}} \cdot 3 \cdot 3}$$

- The product is negative.

$$= -\frac{1}{9}$$

- Write the answer in simplest form.

You Try It 3   The product is negative.
$$0.034(-2.14) = -0.07276$$

You Try It 4   The quotient is negative.

$$\frac{5}{8} \div \left(-\frac{10}{11}\right) = -\left(\frac{5}{8} \cdot \frac{11}{10}\right)$$

- Multiply by the reciprocal of the divisor.

$$= -\frac{\overset{1}{\cancel{5}} \cdot 11}{2 \cdot 2 \cdot 2 \cdot \underset{1}{\cancel{5}}}$$

$$= -\frac{11}{16}$$

- Write the answer in simplest form.

You Try It 5   The quotient is negative.

$$-\frac{1}{3} \div \frac{7}{15} = -\left(\frac{1}{3} \cdot \frac{15}{7}\right)$$

- Multiply by the reciprocal of the divisor.

$$= -\frac{1 \cdot \overset{1}{\cancel{3}} \cdot 5}{\underset{1}{\cancel{3}} \cdot 7}$$

$$= -\frac{5}{7}$$

- Write the answer in simplest form.

You Try It 6   The quotient is positive.
$$-34 \div (-9.02) \approx 3.77$$

You Try It 7 $125\% = 125\left(\frac{1}{100}\right) = \frac{125}{100} = \frac{5}{4}$
$125\% = 125(0.01) = 1.25$

You Try It 8 $16\frac{2}{3}\% = 16\frac{2}{3}\left(\frac{1}{100}\right) = \frac{50}{3}\left(\frac{1}{100}\right) = \frac{1}{6}$

You Try It 9 $\frac{9}{16} = \frac{9}{16}(100\%) = \frac{900}{16}\% = 56.25\%$

or $56\frac{1}{4}\%$

You Try It 10 $0.043 = 0.043(100\%) = 4.3\%$

You Try It 11

Strategy   To find the number of cushions, divide 20 by $1\frac{1}{2}$.

Solution $20 \div 1\frac{1}{2} = 20 \div \frac{3}{2} = \frac{20}{1} \cdot \frac{2}{3}$

$$= \frac{40}{3} = 13\frac{1}{3}$$

The number of cushions must be a whole number. Therefore, the number of cushions is 13.

SECTION 1.8

You Try It 1   To find the complement of 87°, subtract 87° from 90°.

$$90° - 87° = 3°$$

3° is the complement of 87°.

You Try It 2   To find the supplement of 87°, subtract 87° from 180°.

$$180° - 87° = 93°$$

93° is the supplement of 87°.

You Try It 3 $m\angle x$ is the sum of the measures of two angles.
$$m\angle x = 34° + 95° = 129°$$

You Try It 4   Perimeter $= 4 \cdot$ side
$$= 4 \cdot 4.2 \text{ m} = 16.8 \text{ m}$$

You Try It 5   Circumference $= \pi \cdot$ diameter
$$\approx 3.14 \cdot 5 \text{ in.}$$
$$= 15.7 \text{ in.}$$

You Try It 6

 Strategy To find the cost of the metal strip:
- Find the circumference of the table in inches.
- Convert inches to feet.
- Multiply the circumference by the per-foot cost of the metal strip.

 Solution
$$\text{Circumference} = \pi \cdot \text{diameter}$$
$$\approx 3.14 \cdot 36 \text{ in.}$$
$$= 113.04$$

$$\frac{113.04}{12} = 9.42$$

Cost: $9.42(3.21) = 30.2382$

The cost is $30.24.

You Try It 7
$$\text{Area} = \frac{1}{2} \cdot \text{base} \cdot \text{height}$$
$$= \frac{1}{2} \cdot 5 \text{ ft} \cdot 3 \text{ ft} = 7.5 \text{ ft}^2$$

You Try It 8
$$\text{Area} = \pi \cdot (\text{radius})^2$$
$$\approx 3.14 \cdot (6 \text{ in.})^2$$
$$= 113.04 \text{ in}^2$$

You Try It 9
$$\text{Area} = \text{base} \cdot \text{height}$$
$$= 28 \text{ in.} \cdot 15 \text{ in.} = 420 \text{ in}^2$$

You Try It 10

 Strategy To find how much more expensive the wool rug is than the nylon rug:
- Find the area of the rug.
- Multiply the area of the rug by the per-square-foot cost of the wool rug.
- Multiply the area of the rug by the per-square-foot cost of the nylon rug.
- Find the difference between the two costs.

 Solution
$$\text{Area} = \text{length} \cdot \text{width}$$
$$= 15 \text{ ft} \cdot 4 \text{ ft} = 60 \text{ ft}^2$$

Cost of wool rug $= \$1.93 \cdot 60 = \$115.80$

Cost of nylon rug $= \$1.25 \cdot 60 = \$75$

Difference in cost $= \$115.80 - \$75 = \$40.80$

Solutions to Chapter 2 "You Try It"

SECTION 2.1

You Try It 1 -4 is the constant term.

You Try It 2
$$2xy + y^2$$
$$2(-4)(2) + (2)^2$$
$$= 2(-4)(2) + 4$$
$$= (-8)(2) + 4$$
$$= (-16) + 4$$
$$= -12$$

You Try It 3
$$\frac{a^2 + b^2}{a + b}$$
$$\frac{5^2 + (-3)^2}{5 + (-3)} = \frac{25 + 9}{5 + (-3)}$$
$$= \frac{34}{2}$$
$$= 17$$

You Try It 4
$$x^3 - 2(x + y) + z^2$$
$$(2)^3 - 2[2 + (-4)] + (-3)^2$$
$$= 8 - 2(-2) + 9$$
$$= 8 + 4 + 9$$
$$= 12 + 9$$
$$= 21$$

SECTION 2.2

You Try It 1 $3a - 2b - 5a + 6b = -2a + 4b$

You Try It 2 $-3y^2 + 7 + 8y^2 - 14 = 5y^2 - 7$

You Try It 3 $-5(4y^2) = -20y^2$

You Try It 4 $-7(-2a) = 14a$

You Try It 5 $-\dfrac{3}{5}\left(-\dfrac{7}{9}a\right) = \dfrac{7}{15}a$

You Try It 6 $5(3 + 7b) = 15 + 35b$

You Try It 7 $(3a - 1)5 = 15a - 5$

You Try It 8 $-8(-2a + 7b) = 16a - 56b$

You Try It 9 $3(12x^2 - x + 8) = 36x^2 - 3x + 24$

You Try It 10 $3(-a^2 - 6a + 7) = -3a^2 - 18a + 21$

You Try It 11 $3y - 2(y - 7x) = 3y - 2y + 14x$
$$= y + 14x$$

You Try It 12

$$-2(x - 2y) - (-x + 3y) = -2x + 4y + x - 3y$$
$$= -x + y$$

You Try It 13

$$3y - 2[x - 4(2 - 3y)] = 3y - 2[x - 8 + 12y]$$
$$= 3y - 2x + 16 - 24y$$
$$= -2x - 21y + 16$$

SECTION 2.3

You Try It 1   the difference between twice n and one-third of n

$$2n - \frac{1}{3}n$$

You Try It 2   the quotient of 7 less than b and 15

$$\frac{b - 7}{15}$$

You Try It 3   the unknown number: n
the cube of the number: n^3
the total of ten and the cube of the number: $10 + n^3$

$$-4(10 + n^3)$$

You Try It 4   the unknown number: x
the difference between the number and sixty: $x - 60$

$$5(x - 60)$$
$$= 5x - 300$$

You Try It 5   the speed of the older model: s
the speed of the new jet plane is twice the speed of the older model: $2s$

You Try It 6   the length of the longer piece: y
the length of the shorter piece: $6 - y$

Solutions to Chapter 3 "You Try It"

SECTION 3.1

You Try It 1

$$\frac{5 - 4x = 8x + 2}{}$$

$$5 - 4\left(\frac{1}{4}\right) \,\bigg|\, 8\left(\frac{1}{4}\right) + 2$$
$$5 - 1 \,\bigg|\, 2 + 2$$
$$4 = 4$$

Yes, $\frac{1}{4}$ is a solution.

You Try It 2

$$\frac{10x - x^2 = 3x - 10}{}$$

$$10(5) - (5)^2 \,\bigg|\, 3(5) - 10$$
$$50 - 25 \,\bigg|\, 15 - 10$$
$$25 \neq 5$$

No, 5 is not a solution.

You Try It 3

$$\frac{5}{6} = y - \frac{3}{8}$$
$$\frac{5}{6} + \frac{3}{8} = y - \frac{3}{8} + \frac{3}{8}$$
$$\frac{29}{24} = y$$

The solution is $\frac{29}{24}$.

You Try It 4

$$-\frac{2}{5}x = 6$$
$$\left(-\frac{5}{2}\right)\left(-\frac{2}{5}x\right) = \left(-\frac{5}{2}\right)(6)$$
$$x = -15$$

The solution is -15.

You Try It 5

$$4x - 8x = 16$$
$$-4x = 16$$
$$\frac{-4x}{-4} = \frac{16}{-4}$$
$$x = -4$$

The solution is -4.

You Try It 6

$$P \cdot B = A$$
$$\frac{1}{6}B = 18 \qquad \bullet\ 16\frac{2}{3}\% = \frac{1}{6}$$
$$6 \cdot \frac{1}{6}B = 6 \cdot 18$$
$$B = 108$$

18 is $16\frac{2}{3}\%$ of 108.

You Try It 7

Strategy   Use the percent equation. $B = 83.3$, the total revenue received by the BCS; $A = 3.1$, the amount received by the college representing the Pac-10 conference; P is the unknown percent.

Solution
$$P \cdot B = A$$
$$P(83.3) = 3.1 \qquad \bullet\ B = 83.3,\ A = 3.1$$
$$P = \frac{3.1}{83.3} \approx 0.037$$

The college representing the Pac-10 conference received approximately 3.7% of the BCS revenue.

You Try It 8

Strategy To find how much she must deposit into the bank account:
- Find the amount of interest earned on the municipal bond by solving $I = Prt$ for I using $P = 1000$, $r = 6.4\% = 0.064$, and $t = 1$.
- Solve $I = Prt$ for P using the amount of interest earned on the municipal bond as I. $r = 8\% = 0.08$, and $t = 1$.

Solution

$I = Prt$
$= 1000(0.064)(1) = 64$

The interest earned on the municipal bond was $64.

$I = Prt$
$64 = P(0.08)(1)$ • $I = 64$, $r = 0.08$, $t = 1$.
$64 = 0.08P$
$\dfrac{64}{0.08} = \dfrac{0.08P}{0.08}$
$800 = P$

Clarissa must invest $800 in the bank account.

You Try It 9

Strategy To find the number of ounces of cereal in the bowl, solve $Q = Ar$ for A using $Q = 2$ and $r = 25\% = 0.25$.

Solution

$Q = Ar$
$2 = A(0.25)$ • $Q = 2$, $r = 0.25$.
$\dfrac{2}{0.25} = \dfrac{A(0.25)}{0.25}$
$8 = A$

The cereal bowl contains 8 oz of cereal.

You Try It 10

Strategy To find the distance, solve the equation $d = rt$ for d. The time is 3 h. Therefore, $t = 3$. The plane is moving against the wind, which means the headwind is slowing the actual speed of the plane. 250 mph − 25 mph = 225 mph. Thus $r = 225$.

Solution $d = rt$
$d = 225(3)$ • $r = 225$, $t = 3$.
$= 675$

The plane travels 675 mi in 3 h.

SECTION 3.2

You Try It 1

$5x + 7 = 10$
$5x + 7 - 7 = 10 - 7$ • Subtract 7.
$5x = 3$
$\dfrac{5x}{5} = \dfrac{3}{5}$ • Divide by 5.
$x = \dfrac{3}{5}$

The solution is $\dfrac{3}{5}$.

You Try It 2

$2 = 11 + 3x$
$2 - 11 = 11 - 11 + 3x$ • Subtract 11.
$-9 = 3x$
$\dfrac{-9}{3} = \dfrac{3x}{3}$ • Divide by 3.
$-3 = x$

The solution is -3.

You Try It 3

$\dfrac{5}{8} - \dfrac{2x}{3} = \dfrac{5}{4}$
$\dfrac{5}{8} - \dfrac{5}{8} - \dfrac{2}{3}x = \dfrac{5}{4} - \dfrac{5}{8}$ • Recall that $\dfrac{2x}{3} = \dfrac{2}{3}x$.
$-\dfrac{2}{3}x = \dfrac{5}{8}$
$-\dfrac{3}{2}\left(-\dfrac{2}{3}x\right) = -\dfrac{3}{2}\left(\dfrac{5}{8}\right)$ • Multiply by $-\dfrac{3}{2}$.
$x = -\dfrac{15}{16}$

The solution is $-\dfrac{15}{16}$.

You Try It 4

$\dfrac{2}{3}x + 3 = \dfrac{7}{2}$
$6\left(\dfrac{2}{3}x + 3\right) = 6\left(\dfrac{7}{2}\right)$
$6\left(\dfrac{2}{3}x\right) + 6(3) = 6\left(\dfrac{7}{2}\right)$ • Distributive Property
$4x + 18 = 21$
$4x + 18 - 18 = 21 - 18$ • Subtract 18.
$4x = 3$
$\dfrac{4x}{4} = \dfrac{3}{4}$ • Divide by 4.
$x = \dfrac{3}{4}$

The solution is $\dfrac{3}{4}$.

You Try It 5

$$x - 5 + 4x = 25$$
$$5x - 5 = 25$$
$$5x - 5 + 5 = 25 + 5$$
$$5x = 30$$
$$\frac{5x}{5} = \frac{30}{5}$$
$$x = 6$$

The solution is 6.

You Try It 6

Strategy   Given: $S = 986$
  $r = 45\% = 0.45$
Unknown: C

Solution
$$S = C + rC$$
$$986 = C + 0.45C$$
$$986 = 1.45C$$
$$\frac{986}{1.45} = \frac{1.45C}{1.45}$$
$$680 = C$$

The cost of the outboard motor is $680.

You Try It 7

Strategy   Given: $S = 159$
  $r = 25\% = 0.25$
Unknown: R

Solution
$$S = R - rR$$
$$159 = R - 0.25R$$
$$159 = 0.75R$$
$$\frac{159}{0.75} = \frac{0.75R}{0.75}$$
$$212 = R$$

The regular price of the garage door opener is $212.

You Try It 8

Strategy   Given: $P = 45$
Unknown: D

Solution
$$P = 15 + \frac{1}{2}D$$
$$45 = 15 + \frac{1}{2}D$$
$$45 - 15 = 15 - 15 + \frac{1}{2}D$$
$$30 = \frac{1}{2}D$$
$$2(30) = 2 \cdot \frac{1}{2}D$$
$$60 = D$$

The depth is 60 ft.

SECTION 3.3

You Try It 1

$$5x + 4 = 6 + 10x$$
$$5x - 10x + 4 = 6 + 10x - 10x \quad \bullet \text{ Subtract } 10x.$$
$$-5x + 4 = 6$$
$$-5x + 4 - 4 = 6 - 4 \quad \bullet \text{ Subtract } 4.$$
$$-5x = 2$$
$$\frac{-5x}{-5} = \frac{2}{-5} \quad \bullet \text{ Divide by } -5.$$
$$x = -\frac{2}{5}$$

The solution is $-\frac{2}{5}$.

You Try It 2

$$5x - 10 - 3x = 6 - 4x$$
$$2x - 10 = 6 - 4x \quad \bullet \text{ Combine like terms.}$$
$$2x + 4x - 10 = 6 - 4x + 4x \quad \bullet \text{ Add } 4x.$$
$$6x - 10 = 6$$
$$6x - 10 + 10 = 6 + 10 \quad \bullet \text{ Add } 10.$$
$$6x = 16$$
$$\frac{6x}{6} = \frac{16}{6} \quad \bullet \text{ Divide by } 6.$$
$$x = \frac{8}{3}$$

The solution is $\frac{8}{3}$.

You Try It 3

$$5x - 4(3 - 2x) = 2(3x - 2) + 6$$
$$5x - 12 + 8x = 6x - 4 + 6 \quad \bullet \text{ Distributive Property}$$
$$13x - 12 = 6x + 2$$
$$13x - 6x - 12 = 6x - 6x + 2 \quad \bullet \text{ Subtract } 6x.$$
$$7x - 12 = 2$$
$$7x - 12 + 12 = 2 + 12 \quad \bullet \text{ Add } 12.$$
$$7x = 14$$
$$\frac{7x}{7} = \frac{14}{7} \quad \bullet \text{ Divide by } 7.$$
$$x = 2$$

The solution is 2.

You Try It 4

$$-2[3x - 5(2x - 3)] = 3x - 8$$
$$-2[3x - 10x + 15] = 3x - 8 \quad \bullet \text{ Distributive}$$
$$-2[-7x + 15] = 3x - 8 \qquad\quad \text{ Property}$$
$$14x - 30 = 3x - 8$$
$$14x - 3x - 30 = 3x - 3x - 8 \quad \bullet \text{ Subtract } 3x.$$
$$11x - 30 = -8$$
$$11x - 30 + 30 = -8 + 30 \quad \bullet \text{ Add } 30.$$
$$11x = 22$$
$$\frac{11x}{11} = \frac{22}{11} \quad \bullet \text{ Divide by } 11.$$
$$x = 2$$

The solution is 2.

You Try It 5

Strategy   Given: $F_1 = 45$
$F_2 = 80$
$d = 25$
Unknown: x

Solution
$$F_1 x = F_2(d - x)$$
$$45x = 80(25 - x)$$
$$45x = 2000 - 80x$$
$$45x + 80x = 2000 - 80x + 80x$$
$$125x = 2000$$
$$\frac{125x}{125} = \frac{2000}{125}$$
$$x = 16$$

The fulcrum is 16 ft from the 45-pound force.

SECTION 3.4

You Try It 1

The smaller number: n
The larger number: $12 - n$

The total of three times the smaller number and six	amounts to	seven less than the product of four and the larger number

$$3n + 6 = 4(12 - n) - 7$$
$$3n + 6 = 48 - 4n - 7$$
$$3n + 6 = 41 - 4n$$
$$3n + 4n + 6 = 41 - 4n + 4n$$
$$7n + 6 = 41$$
$$7n + 6 - 6 = 41 - 6$$
$$7n = 35$$
$$\frac{7n}{7} = \frac{35}{7}$$
$$n = 5$$

$12 - n = 12 - 5 = 7$

The smaller number is 5.
The larger number is 7.

You Try It 2

Strategy
• First integer: n
Second integer: $n + 1$
Third integer: $n + 2$
• The sum of the three integers is -6.

Solution $n + (n + 1) + (n + 2) = -6$
$$3n + 3 = -6$$
$$3n = -9$$
$$n = -3$$
$$n + 1 = -3 + 1 = -2$$
$$n + 2 = -3 + 2 = -1$$

The three consecutive integers are -3, -2, and -1.

You Try It 3

Strategy
To find the number of tickets that you are purchasing, write and solve an equation using x to represent the number of tickets purchased.

Solution

$3.50 plus $17.50 for each ticket	is	$161

$$3.50 + 17.50x = 161$$
$$3.50 - 3.50 + 17.50x = 161 - 3.50$$
$$17.50x = 157.50$$
$$\frac{17.50x}{17.50} = \frac{157.50}{17.50}$$
$$x = 9$$

You are purchasing 9 tickets.

You Try It 4

Strategy
To find the length, write and solve an equation using x to represent the length of the shorter piece and $22 - x$ to represent the length of the longer piece.

Solution

The length of the longer piece	is	4 in. more than twice the length of the shorter piece

$$22 - x = 2x + 4$$
$$22 - x - 2x = 2x - 2x + 4$$
$$22 - 3x = 4$$
$$22 - 22 - 3x = 4 - 22$$
$$-3x = -18$$
$$\frac{-3x}{-3} = \frac{-18}{-3}$$
$$x = 6$$

$22 - x = 22 - 6 = 16$

The length of the shorter piece is 6 in.
The length of the longer piece is 16 in.

SECTION 3.5

You Try It 1

Strategy The angles labeled are adjacent angles of intersecting lines and are therefore supplementary angles. To find x, write an equation and solve for x.

Solution

$$x + (3x + 20°) = 180°$$
$$4x + 20° = 180°$$
$$4x = 160°$$
$$x = 40°$$

You Try It 2

Strategy $2x = y$ because alternate exterior angles have the same measure. $y + (x + 15°) = 180°$ because adjacent angles of intersecting lines are supplementary angles. Substitute $2x$ for y and solve for x.

Solution

$$y + (x + 15°) = 180°$$
$$2x + (x + 15°) = 180°$$
$$3x + 15° = 180°$$
$$3x = 165°$$
$$x = 55°$$

You Try It 3

Strategy
- To find the measure of angle a, use the fact that $\angle a$ and $\angle y$ are vertical angles.
- To find the measure of angle b, use the fact that the sum of the measures of the interior angles of a triangle is 180°.
- To find the measure of angle d, use the fact that the sum of an interior and an exterior angle is 180°.

Solution $m\angle a = m\angle y = 55°$

$$m\angle a + m\angle b + 90° = 180°$$
$$55° + m\angle b + 90° = 180°$$
$$m\angle b + 145° = 180°$$
$$m\angle b = 35°$$

$$m\angle d + m\angle b = 180°$$
$$m\angle d + 35° = 180°$$
$$m\angle d = 145°$$

You Try It 4

Strategy To find the measure of the third angle, use the fact that the sum of the measures of the interior angles of a triangle is 180°. Write an equation using x to represent the measure of the third angle. Solve the equation for x.

Solution

$$x + 90° + 27° = 180°$$
$$x + 117° = 180°$$
$$x = 63°$$

The measure of the third angle is 63°.

SECTION 3.6

You Try It 1

Strategy • Pounds of $.55 fertilizer: x

	Amount	Cost	Value
$.80 fertilizer	20	.80	0.80(20)
$.55 fertilizer	x	.55	0.55x
$.75 fertilizer	20 + x	.75	0.75(20 + x)

- The sum of the values before mixing equals the value after mixing.

Solution

$$0.80(20) + 0.55x = 0.75(20 + x)$$
$$16 + 0.55x = 15 + 0.75x$$
$$16 - 0.20x = 15$$
$$-0.20x = -1$$
$$x = 5$$

5 lb of the $.55 fertilizer must be added.

You Try It 2

Strategy • Liters of 6% solution: x

	Amount	Percent	Quantity
6% solution	x	0.06	0.06x
12% solution	5	0.12	5(0.12)
8% solution	x + 5	0.08	0.08(x + 5)

- The sum of the quantities before mixing equals the quantity after mixing.

Solution

$$0.06x + 5(0.12) = 0.08(x + 5)$$
$$0.06x + 0.60 = 0.08x + 0.40$$
$$-0.02x + 0.60 = 0.40$$
$$-0.02x = -0.20$$
$$x = 10$$

The pharmacist adds 10 L of the 6% solution to the 12% solution to get an 8% solution.

You Try It 3

Strategy   • Rate of the first train: r
Rate of the second train: $2r$

	Rate	Time	Distance
1st train	r	3	$3r$
2nd train	$2r$	3	$3(2r)$

   • The sum of the distances traveled by the two trains equals 288 mi.

Solution
$$3r + 3(2r) = 288$$
$$3r + 6r = 288$$
$$9r = 288$$
$$r = 32$$

$$2r = 2(32) = 64$$

The first train is traveling at 32 mph. The second train is traveling at 64 mph.

You Try It 4

Strategy   • Time spent flying out: t
Time spent flying back: $5 - t$

	Rate	Time	Distance
Out	150	t	$150t$
Back	100	$5 - t$	$100(5 - t)$

   • The distance out equals the distance back.

Solution
$$150t = 100(5 - t)$$
$$150t = 500 - 100t$$
$$250t = 500$$
$$t = 2 \quad \text{(The time out was 2 h.)}$$

$$\text{The distance out} = 150t = 150(2)$$
$$= 300 \text{ mi}$$

The parcel of land was 300 mi away.

Solutions to Chapter 4 "You Try It"

SECTION 4.1

You Try It 1
$(-4x^3 + 2x^2 - 8) + (4x^3 + 6x^2 - 7x + 5)$
$= (-4x^3 + 4x^3) + (2x^2 + 6x^2) + (-7x) + (-8 + 5)$
$= 8x^2 - 7x - 3$

You Try It 2
$$
\begin{array}{r}
6x^3 \qquad\quad + 2x + 8 \\
-9x^3 + 2x^2 - 12x - 8 \\
\hline
-3x^3 + 2x^2 - 10x
\end{array}
$$

You Try It 3
$(-4w^3 + 8w - 8) - (3w^3 - 4w^2 - 2w - 1)$
$= (-4w^3 + 8w - 8)$
$\quad + (-3w^3 + 4w^2 + 2w + 1)$
$= -7w^3 + 4w^2 + 10w - 7$

You Try It 4
$$
\begin{array}{r}
13y^3 \qquad\quad - 6y - 7 \\
- 4y^2 + 6y + 9 \\
\hline
13y^3 - 4y^2 \qquad\quad + 2
\end{array}
$$

SECTION 4.2

You Try It 1
$(8m^3n)(-3n^5)$
$= [8(-3)](m^3)(n \cdot n^5)$ • **Multiply coefficients. Add**
$= -24m^3n^6$ **exponents with same base.**

You Try It 2
$(12p^4q^3)(-3p^5q^2)$
$= [12(-3)](p^4 \cdot p^5)(q^3 \cdot q^2)$ • **Multiply coefficients. Add**
$= -36p^9q^5$ **exponents with same base.**

You Try It 3
$(-3a^4bc^2)^3 = (-3)^{1 \cdot 3}a^{4 \cdot 3}b^{1 \cdot 3}c^{2 \cdot 3}$ • **Rule for Simplifying**
$= (-3)^3a^{12}b^3c^6$ **the Power of a Product**
$= -27a^{12}b^3c^6$

You Try It 4
$(-xy^4)(-2x^3y^2)^2 = (-xy^4)[(-2)^{1 \cdot 2}x^{3 \cdot 2}y^{2 \cdot 2}]$ • **Rule for**
$= (-xy^4)[(-2)^2x^6y^4]$ **Simplifying**
$= (-xy^4)(4x^6y^4)$ **the Power of**
$= -4x^7y^8$ **a Product**

SECTION 4.3

You Try It 1 $(-2y + 3)(-4y) = 8y^2 - 12y$

You Try It 2
$-a^2(3a^2 + 2a - 7) = -3a^4 - 2a^3 + 7a^2$

You Try It 3
$$
\begin{array}{r}
2y^3 + 2y^2 \qquad\quad - 3 \\
3y - 1 \\
\hline
- 2y^3 - 2y^2 \qquad\quad + 3 = -1(2y^3 + 2y^2 - 3) \\
6y^4 + 6y^3 \qquad\quad - 9y \quad\;\; = 3y(2y^3 + 2y^2 - 3) \\
\hline
6y^4 + 4y^3 - 2y^2 - 9y + 3
\end{array}
$$

You Try It 4

$(4y - 5)(2y - 3) = 8y^2 - 12y - 10y + 15$
$\qquad\qquad\qquad = 8y^2 - 22y + 15$

You Try It 5

$(3b + 2)(3b - 5) = 9b^2 - 15b + 6b - 10$
$\qquad\qquad\qquad = 9b^2 - 9b - 10$

You Try It 6 $(2a + 5c)(2a - 5c) = 4a^2 - 25c^2$

You Try It 7 $(3x + 2y)^2 = 9x^2 + 12xy + 4y^2$

You Try It 8

Strategy   To find the area, replace the variable r in the equation $A = \pi r^2$ by $(x - 4)$ and solve for A.

Solution $A = \pi r^2$
$A = \pi(x - 4)^2$
$A = \pi(x^2 - 8x + 16)$
$A = \pi x^2 - 8\pi x + 16\pi$

The area of the circle is $(\pi x^2 - 8\pi x + 16\pi)$ ft^2.

SECTION 4.4

You Try It 1 $(-2x^2)(x^{-3}y^{-4})^{-2}$
$\qquad = (-2x^2)(x^6 y^8)$ • **Rule for Simplifying**
$\qquad = -2x^8 y^8$ **the Power of a Product**

You Try It 2 $\dfrac{(6a^{-2}b^3)^{-1}}{(4a^3 b^{-2})^{-2}}$

$\qquad = \dfrac{6^{-1}a^2 b^{-3}}{4^{-2}a^{-6}b^4}$ • **Rule for Simplifying the**
$\qquad\qquad\qquad\qquad$ **Power of a Product**
$\qquad = 4^2(6^{-1}a^8 b^{-7})$ • **Rule for Dividing**
$\qquad\qquad\qquad\qquad$ **Exponential Expressions**
$\qquad = \dfrac{16a^8}{6b^7} = \dfrac{8a^8}{3b^7}$

You Try It 3 $\left[\dfrac{6r^3 s^{-3}}{9r^3 s^{-1}}\right]^{-2} = \left[\dfrac{2r^0 s^{-2}}{3}\right]^{-2}$

$\qquad\qquad = \dfrac{2^{-2}s^4}{3^{-2}} = \dfrac{9s^4}{4}$

You Try It 4 $0.000000961 = 9.61 \times 10^{-7}$

You Try It 5 $7.329 \times 10^6 = 7,329,000$

SECTION 4.5

You Try It 1

$\dfrac{24x^2 y^2 - 18xy + 6y}{6xy} = \dfrac{24x^2 y^2}{6xy} - \dfrac{18xy}{6xy} + \dfrac{6y}{6xy}$

$\qquad\qquad = 4xy - 3 + \dfrac{1}{x}$

You Try It 2

$$\begin{array}{r} x^2 + 2x - 1 \\ 2x - 3\overline{)\,2x^3 + x^2 - 8x - 3} \\ \underline{2x^3 - 3x^2} \\ 4x^2 - 8x \\ \underline{4x^2 - 6x} \\ -2x - 3 \\ \underline{-2x + 3} \\ -6 \end{array}$$

$(2x^3 + x^2 - 8x - 3) \div (2x - 3)$

$\qquad = x^2 + 2x - 1 - \dfrac{6}{2x - 3}$

You Try It 3

$$\begin{array}{r} x^2 + x - 1 \\ x - 1\overline{)\,x^3 + 0x^2 - 2x + 1} \\ \underline{x^3 - x^2} \\ x^2 - 2x \\ \underline{x^2 - x} \\ -x + 1 \\ \underline{-x + 1} \\ 0 \end{array}$$

$(x^3 - 2x + 1) \div (x - 1) = x^2 + x - 1$

Solutions to Chapter 5 "You Try It"

SECTION 5.1

You Try It 1   The GCF is $7a^2$.

$14a^2 - 21a^4 b = 7a^2(2) + 7a^2(-3a^2 b)$
$\qquad\qquad\qquad = 7a^2(2 - 3a^2 b)$

You Try It 2   The GCF is 9.

$27b^2 + 18b + 9$
$\qquad = 9(3b^2) + 9(2b) + 9(1)$
$\qquad = 9(3b^2 + 2b + 1)$

You Try It 3
The GCF is $3x^2y^2$.

$6x^4y^2 - 9x^3y^2 + 12x^2y^4$
$= 3x^2y^2(2x^2) + 3x^2y^2(-3x) + 3x^2y^2(4y^2)$
$= 3x^2y^2(2x^2 - 3x + 4y^2)$

You Try It 4
$2y(5x - 2) - 3(2 - 5x)$
$= 2y(5x - 2) + 3(5x - 2)$ • $5x - 2$ is the
$= (5x - 2)(2y + 3)$ common factor.

You Try It 5
$a^2 - 3a + 2ab - 6b$
$= (a^2 - 3a) + (2ab - 6b)$
$= a(a - 3) + 2b(a - 3)$ • $a - 3$ is the common factor.
$= (a - 3)(a + 2b)$

You Try It 6
$2mn^2 - n + 8mn - 4$
$= (2mn^2 - n) + (8mn - 4)$ • $2mn - 1$ is the
$= n(2mn - 1) + 4(2mn - 1)$ common factor.
$= (2mn - 1)(n + 4)$

You Try It 7
$3xy - 9y - 12 + 4x$
$= (3xy - 9y) - (12 - 4x)$ • $-12 + 4x = -(12 - 4x)$
$= 3y(x - 3) - 4(3 - x)$ • $-(3 - x) = (x - 3)$
$= 3y(x - 3) + 4(x - 3)$ • $x - 3$ is the common factor.
$= (x - 3)(3y + 4)$

SECTION 5.2

You Try It 1
Find the positive factors of 20 whose sum is 9.

Factors	Sum
1, 20	21
2, 10	12
4, 5	9

$x^2 + 9x + 20 = (x + 4)(x + 5)$

You Try It 2
Find the factors of -18 whose sum is 7.

Factors	Sum
+1, −18	−17
−1, +18	17
+2, −9	−7
−2, +9	7
+3, −6	−3
−3, +6	3

$x^2 + 7x - 18 = (x + 9)(x - 2)$

You Try It 3
The GCF is $-2x$.

$-2x^3 + 14x^2 - 12x = -2x(x^2 - 7x + 6)$

Factor the trinomial $x^2 - 7x + 6$. Find two negative factors of 6 whose sum is -7.

Factors	Sum
−1, −6	−7
−2, −3	−5

$-2x^3 + 14x^2 - 12x = -2x(x - 6)(x - 1)$

You Try It 4
The GCF is 3.

$3x^2 - 9xy - 12y^2 = 3(x^2 - 3xy - 4y^2)$

Factor the trinomial.

Find the factors of -4 whose sum is -3.

Factors	Sum
+1, −4	−3
−1, +4	3
+2, −2	0

$3x^2 - 9xy - 12y^2 = 3(x + y)(x - 4y)$

SECTION 5.3

You Try It 1
Factor the trinomial $2x^2 - x - 3$.

Positive factors of 2: 1, 2   Factors of -3: $+1, -3$
                                              $-1, +3$

Trial Factors	Middle Term
$(x + 1)(2x - 3)$	$-3x + 2x = -x$
$(x - 3)(2x + 1)$	$x - 6x = -5x$
$(x - 1)(2x + 3)$	$3x - 2x = x$
$(x + 3)(2x - 1)$	$-x + 6x = 5x$

$2x^2 - x - 3 = (x + 1)(2x - 3)$

You Try It 2
The GCF is $-3y$.

$-45y^3 + 12y^2 + 12y = -3y(15y^2 - 4y - 4)$

Factor the trinomial $15y^2 - 4y - 4$.

Positive factors of 15: 1, 15   Factors of -4: 1, −4
                        3, 5   −1, 4
                                                     2, −2

Trial Factors	Middle Term
$(y + 1)(15y - 4)$	$-4y + 15y = 11y$
$(y - 4)(15y + 1)$	$y - 60y = -59y$
$(y - 1)(15y + 4)$	$4y - 15y = -11y$
$(y + 4)(15y - 1)$	$-y + 60y = 59y$
$(y + 2)(15y - 2)$	$-2y + 30y = 28y$
$(y - 2)(15y + 2)$	$2y - 30y = -28y$
$(3y + 1)(5y - 4)$	$-12y + 5y = -7y$
$(3y - 4)(5y + 1)$	$3y - 20y = -17y$
$(3y - 1)(5y + 4)$	$12y - 5y = 7y$
$(3y + 4)(5y - 1)$	$-3y + 20y = 17y$
$(3y + 2)(5y - 2)$	$-6y + 10y = 4y$
$(3y - 2)(5y + 2)$	$6y - 10y = -4y$

$-45y^3 + 12y^2 + 12y = -3y(3y - 2)(5y + 2)$

You Try It 3

Factors of -14 [$2(-7)$]	Sum
$+1, -14$	-13
$-1, +14$	13
$+2, -7$	-5
$-2, +7$	5

$$2a^2 + 13a - 7 = 2a^2 - a + 14a - 7$$
$$= (2a^2 - a) + (14a - 7)$$
$$= a(2a - 1) + 7(2a - 1)$$
$$= (2a - 1)(a + 7)$$

$2a^2 + 13a - 7 = (2a - 1)(a + 7)$

You Try It 4
The GCF is $5x$.

$15x^3 + 40x^2 - 80x = 5x(3x^2 + 8x - 16)$

Factors of -48 [$3(-16)$]	Sum
$+1, -48$	-47
$-1, +48$	47
$+2, -24$	-22
$-2, +24$	22
$+3, -16$	-13
$-3, +16$	13
$+4, -12$	-8
$-4, +12$	8

$$3x^2 + 8x - 16 = 3x^2 - 4x + 12x - 16$$
$$= (3x^2 - 4x) + (12x - 16)$$
$$= x(3x - 4) + 4(3x - 4)$$
$$= (3x - 4)(x + 4)$$

$$15x^3 + 40x^2 - 80x = 5x(3x^2 + 8x - 16)$$
$$= 5x(3x - 4)(x + 4)$$

SECTION 5.4

You Try It 1
$25a^2 - b^2 = (5a)^2 - b^2$ • Difference of
$\qquad\qquad = (5a + b)(5a - b)$ two squares

You Try It 2
$n^4 - 81 = (n^2)^2 - 9^2$ • Difference of
two squares

$\qquad = (n^2 + 9)(n^2 - 9)$ • Difference of
$\qquad = (n^2 + 9)(n + 3)(n - 3)$ two squares

You Try It 3   Because $16y^2 = (4y)^2$, $1 = 1^2$, and
$8y = 2(4y)(1)$, the trinomial is a
perfect-square trinomial.

$\qquad 16y^2 + 8y + 1 = (4y + 1)^2$

You Try It 4   Because $x^2 = (x)^2$, $36 = 6^2$, and
$15x \neq 2(x)(6)$, the trinomial is not a
perfect-square trinomial. Try to factor the trinomial by another method.

$\qquad x^2 + 15x + 36 = (x + 3)(x + 12)$

You Try It 5   The GCF is $3x$.

$\qquad 12x^3 - 75x = 3x(4x^2 - 25)$
$\qquad\qquad\qquad = 3x(2x + 5)(2x - 5)$

You Try It 6
Factor by grouping.

$a^2b - 7a^2 - b + 7$
$\quad = (a^2b - 7a^2) - (b - 7)$
$\quad = a^2(b - 7) - (b - 7)$ • $b - 7$ is the common factor.
$\quad = (b - 7)(a^2 - 1)$ • $a^2 - 1$ is the difference
$\quad = (b - 7)(a + 1)(a - 1)$ of two squares.

You Try It 7
The GCF is $4x$.

$4x^3 + 28x^2 - 120x$
$\quad = 4x(x^2 + 7x - 30)$ • Factor the GCF, $4x$.
$\quad = 4x(x + 10)(x - 3)$ • Factor the trinomial.

SECTION 5.5

You Try It 1
$2x(x + 7) = 0$

$\quad 2x = 0 \qquad x + 7 = 0$ • Principle of
$\quad\ x = 0 \qquad\quad x = -7$ Zero Products

The solutions are 0 and -7.

You Try It 2

$$4x^2 - 9 = 0 \qquad \text{• Difference of two squares}$$
$$(2x - 3)(2x + 3) = 0$$

$$2x - 3 = 0 \qquad 2x + 3 = 0 \qquad \text{• Principle of Zero Products}$$
$$2x = 3 \qquad\qquad 2x = -3$$
$$x = \frac{3}{2} \qquad\qquad x = -\frac{3}{2}$$

The solutions are $\frac{3}{2}$ and $-\frac{3}{2}$.

You Try It 3

$$(x + 2)(x - 7) = 52$$
$$x^2 - 5x - 14 = 52$$
$$x^2 - 5x - 66 = 0$$
$$(x + 6)(x - 11) = 0$$

$$x + 6 = 0 \qquad x - 11 = 0 \qquad \text{• Principle of Zero Products}$$
$$x = -6 \qquad\qquad x = 11$$

The solutions are -6 and 11.

You Try It 4

Strategy   First consecutive positive integer: n
Second consecutive positive integer:
$n + 1$

The sum of the squares of the two consecutive positive integers is 61.

Solution
$$n^2 + (n + 1)^2 = 61$$
$$n^2 + n^2 + 2n + 1 = 61$$
$$2n^2 + 2n + 1 = 61$$
$$2n^2 + 2n - 60 = 0$$
$$2(n^2 + n - 30) = 0$$
$$2(n - 5)(n + 6) = 0$$

$$n - 5 = 0 \qquad n + 6 = 0 \qquad \text{• Principle of Zero Products}$$
$$n = 5 \qquad\qquad n = -6$$

Because -6 is not a positive integer, it is not a solution.

$$n = 5$$
$$n + 1 = 5 + 1 = 6$$

The two integers are 5 and 6.

You Try It 5

Strategy   Width $= x$
Length $= 2x + 4$

The area of the rectangle is 96 in². Use the equation $A = L \cdot W$.

Solution
$$A = L \cdot W$$
$$96 = (2x + 4)x$$
$$96 = 2x^2 + 4x$$
$$0 = 2x^2 + 4x - 96$$
$$0 = 2(x^2 + 2x - 48)$$
$$0 = 2(x + 8)(x - 6)$$

$$x + 8 = 0 \qquad x - 6 = 0 \qquad \text{• Principle of}$$
$$x = -8 \qquad\qquad x = 6 \qquad \text{Zero Products}$$

Because the width cannot be a negative number, -8 is not a solution.

$$x = 6$$
$$2x + 4 = 2(6) + 4 = 12 + 4 = 16$$

The length is 16 in. The width is 6 in.

Solutions to Chapter 6 "You Try It"

SECTION 6.1

You Try It 1

$$\frac{6x^5y}{12x^2y^3} = \frac{\overset{1}{2} \cdot \overset{1}{3} \cdot x^5y}{\underset{1}{2} \cdot 2 \cdot \underset{1}{3} \cdot x^2y^3} = \frac{x^3}{2y^2}$$

You Try It 2

$$\frac{x^2 + 2x - 24}{16 - x^2} = \frac{\overset{-1}{\cancel{(x - 4)}}(x + 6)}{\underset{1}{(4 - x)}(4 + x)} \qquad \text{• } (4 - x) = -1(x - 4)$$

$$= -\frac{x + 6}{x + 4}$$

You Try It 3

$$\frac{x^2 + 4x - 12}{x^2 - 3x + 2} = \frac{\overset{1}{\cancel{(x - 2)}}(x + 6)}{(x - 1)\underset{1}{\cancel{(x - 2)}}} = \frac{x + 6}{x - 1}$$

You Try It 4

$$\frac{12x^2 + 3x}{10x - 15} \cdot \frac{8x - 12}{9x + 18} = \frac{3x(4x + 1)}{5(2x - 3)} \cdot \frac{4(2x - 3)}{9(x + 2)} \qquad \text{• Factor.}$$

$$= \frac{3x(4x + 1) \cdot \overset{1}{2} \cdot 2\underset{1}{\cancel{(2x - 3)}}}{5\underset{1}{\cancel{(2x - 3)}} \cdot \underset{1}{3} \cdot 3(x + 2)}$$

$$= \frac{4x(4x + 1)}{15(x + 2)}$$

You Try It 5

$$\frac{x^2 + 2x - 15}{9 - x^2} \cdot \frac{x^2 - 3x - 18}{x^2 - 7x + 6}$$

$$= \frac{(x - 3)(x + 5)}{(3 - x)(3 + x)} \cdot \frac{(x + 3)(x - 6)}{(x - 1)(x - 6)} \qquad \bullet \text{ Factor.}$$

$$= \frac{\overset{-1}{\cancel{(x - 3)}}(x + 5) \cdot \cancel{(x + 3)}\cancel{(x - 6)}}{\underset{1}{\cancel{(3 - x)}}\underset{1}{\cancel{(3 + x)}} \cdot (x - 1)\underset{1}{\cancel{(x - 6)}}} = -\frac{x + 5}{x - 1}$$

You Try It 6

$$\frac{a^2}{4bc^2 - 2b^2c} \div \frac{a}{6bc - 3b^2}$$

$$= \frac{a^2}{4bc^2 - 2b^2c} \cdot \frac{6bc - 3b^2}{a} \qquad \bullet \text{ Multiply by the reciprocal.}$$

$$= \frac{a^{\cancel{2}} \cdot 3\cancel{b}\cancel{(2c - b)}}{2\cancel{b}c\cancel{(2c - b)} \cdot \cancel{a}} = \frac{3a}{2c}$$

You Try It 7

$$\frac{3x^2 + 26x + 16}{3x^2 - 7x - 6} \div \frac{2x^2 + 9x - 5}{x^2 + 2x - 15}$$

$$= \frac{3x^2 + 26x + 16}{3x^2 - 7x - 6} \cdot \frac{x^2 + 2x - 15}{2x^2 + 9x - 5} \qquad \bullet \text{ Multiply by the reciprocal.}$$

$$= \frac{\cancel{(3x + 2)}(x + 8) \cdot \cancel{(x + 5)}\cancel{(x - 3)}}{\cancel{(3x + 2)}\cancel{(x - 3)} \cdot (2x - 1)\cancel{(x + 5)}} = \frac{x + 8}{2x - 1}$$

SECTION 6.2

You Try It 1

$8uv^2 = 2 \cdot 2 \cdot 2 \cdot u \cdot v \cdot v$

$12uw = 2 \cdot 2 \cdot 3 \cdot u \cdot w$

$\text{LCM} = 2 \cdot 2 \cdot 2 \cdot 3 \cdot u \cdot v \cdot v \cdot w = 24uv^2w$

You Try It 2 $m^2 - 6m + 9 = (m - 3)(m - 3)$
$m^2 - 2m - 3 = (m + 1)(m - 3)$
$\text{LCM} = (m - 3)(m - 3)(m + 1)$

You Try It 3   The LCM is $36xy^2z$.

$$\frac{x - 3}{4xy^2} = \frac{x - 3}{4xy^2} \cdot \frac{9z}{9z} = \frac{9xz - 27z}{36xy^2z}$$

$$\frac{2x + 1}{9y^2z} = \frac{2x + 1}{9y^2z} \cdot \frac{4x}{4x} = \frac{8x^2 + 4x}{36xy^2z}$$

You Try It 4
The LCM is $(x + 2)(x - 5)(x + 5)$.

$$\frac{x + 4}{x^2 - 3x - 10} = \frac{x + 4}{(x + 2)(x - 5)} \cdot \frac{x + 5}{x + 5}$$

$$= \frac{x^2 + 9x + 20}{(x + 2)(x - 5)(x + 5)}$$

$$\frac{2x}{25 - x^2} = \frac{2x}{-(x^2 - 25)} = -\frac{2x}{(x - 5)(x + 5)} \cdot \frac{x + 2}{x + 2}$$

$$= -\frac{2x^2 + 4x}{(x + 2)(x - 5)(x + 5)}$$

SECTION 6.3

You Try It 1

$$\frac{2x^2}{x^2 - x - 12} - \frac{7x + 4}{x^2 - x - 12}$$

$$= \frac{2x^2 - (7x + 4)}{x^2 - x - 12} = \frac{2x^2 - 7x - 4}{x^2 - x - 12}$$

$$= \frac{(2x + 1)\cancel{(x - 4)}}{(x + 3)\underset{1}{\cancel{(x - 4)}}} = \frac{2x + 1}{x + 3}$$

You Try It 2

$$\frac{x^2 - 1}{x^2 - 8x + 12} - \frac{2x + 1}{x^2 - 8x + 12} + \frac{x}{x^2 - 8x + 12}$$

$$= \frac{(x^2 - 1) - (2x + 1) + x}{x^2 - 8x + 12} = \frac{x^2 - 1 - 2x - 1 + x}{x^2 - 8x + 12}$$

$$= \frac{x^2 - x - 2}{x^2 - 8x + 12} = \frac{(x + 1)\cancel{(x - 2)}}{\underset{1}{\cancel{(x - 2)}}(x - 6)} = \frac{x + 1}{x - 6}$$

You Try It 3
The LCM of the denominators is $24y$.

$$\frac{z}{8y} - \frac{4z}{3y} + \frac{5z}{4y}$$

$$= \frac{z}{8y} \cdot \frac{3}{3} - \frac{4z}{3y} \cdot \frac{8}{8} + \frac{5z}{4y} \cdot \frac{6}{6} \qquad \bullet \text{ Write each fraction using the LCM.}$$

$$= \frac{3z}{24y} - \frac{32z}{24y} + \frac{30z}{24y}$$

$$= \frac{3z - 32z + 30z}{24y} = \frac{z}{24y} \qquad \bullet \text{ Combine the numerators.}$$

You Try It 4

$$2 - x = -(x - 2) \qquad \text{Therefore, } \frac{3}{2 - x} = \frac{-3}{x - 2}.$$

$$\frac{5x}{x - 2} + \frac{3}{2 - x} = \frac{5x}{x - 2} + \frac{-3}{x - 2} \qquad \bullet \text{ The LCM is } x - 2.$$

$$= \frac{5x + (-3)}{x - 2} = \frac{5x - 3}{x - 2} \qquad \bullet \text{ Combine the numerators.}$$

You Try It 5
The LCM is $(3x - 1)(x + 4)$.

$$\frac{4x}{3x - 1} + \frac{9}{x + 4} = \frac{4x}{3x - 1} \cdot \frac{x + 4}{x + 4} + \frac{9}{x + 4} \cdot \frac{3x - 1}{3x - 1}$$

$$= \frac{4x^2 + 16x}{(3x - 1)(x + 4)} + \frac{27x - 9}{(3x - 1)(x + 4)}$$

$$= \frac{(4x^2 + 16x) + (27x - 9)}{(3x - 1)(x + 4)}$$

$$= \frac{4x^2 + 16x + 27x - 9}{(3x - 1)(x + 4)}$$

$$= \frac{4x^2 + 43x - 9}{(3x - 1)(x + 4)}$$

You Try It 6   The LCM is $x - 3$.

$$2 - \frac{1}{x - 3} = 2 \cdot \frac{x - 3}{x - 3} - \frac{1}{x - 3}$$

$$= \frac{2x - 6}{x - 3} - \frac{1}{x - 3}$$

$$= \frac{2x - 6 - 1}{x - 3}$$

$$= \frac{2x - 7}{x - 3}$$

You Try It 7

$$\frac{2}{5 - x} = \frac{-2}{x - 5}$$

The LCM is $(x + 5)(x - 5)$.

$$\frac{2x - 1}{x^2 - 25} + \frac{2}{5 - x} = \frac{2x - 1}{(x + 5)(x - 5)} + \frac{-2}{x - 5}$$

$$= \frac{2x - 1}{(x + 5)(x - 5)} + \frac{-2}{x - 5} \cdot \frac{x + 5}{x + 5}$$

$$= \frac{2x - 1}{(x + 5)(x - 5)} + \frac{-2(x + 5)}{(x + 5)(x - 5)}$$

$$= \frac{2x - 1 + (-2)(x + 5)}{(x + 5)(x - 5)}$$

$$= \frac{2x - 1 - 2x - 10}{(x + 5)(x - 5)}$$

$$= \frac{-11}{(x + 5)(x - 5)}$$

$$= -\frac{11}{(x + 5)(x - 5)}$$

You Try It 8
The LCM is $(3x + 2)(x - 1)$.

$$\frac{2x - 3}{3x^2 - x - 2} + \frac{5}{3x + 2} - \frac{1}{x - 1}$$

$$= \frac{2x - 3}{(3x + 2)(x - 1)} + \frac{5}{3x + 2} \cdot \frac{x - 1}{x - 1}$$

$$- \frac{1}{x - 1} \cdot \frac{3x + 2}{3x + 2}$$

$$= \frac{2x - 3}{(3x + 2)(x - 1)} + \frac{5x - 5}{(3x + 2)(x - 1)}$$

$$- \frac{3x + 2}{(3x + 2)(x - 1)}$$

$$= \frac{(2x - 3) + (5x - 5) - (3x + 2)}{(3x + 2)(x - 1)}$$

$$= \frac{2x - 3 + 5x - 5 - 3x - 2}{(3x + 2)(x - 1)}$$

$$= \frac{4x - 10}{(3x + 2)(x - 1)} = \frac{2(2x - 5)}{(3x + 2)(x - 1)}$$

SECTION 6.4

You Try It 1
The LCM of 3, x, 9, and x^2 is $9x^2$.

$$\frac{\dfrac{1}{3} - \dfrac{1}{x}}{\dfrac{1}{9} - \dfrac{1}{x^2}} = \frac{\dfrac{1}{3} - \dfrac{1}{x}}{\dfrac{1}{9} - \dfrac{1}{x^2}} \cdot \frac{9x^2}{9x^2} = \frac{\dfrac{1}{3} \cdot 9x^2 - \dfrac{1}{x} \cdot 9x^2}{\dfrac{1}{9} \cdot 9x^2 - \dfrac{1}{x^2} \cdot 9x^2}$$

   • Multiply by the LCM.

$$= \frac{3x^2 - 9x}{x^2 - 9} = \frac{3x\cancel{(x - 3)}^{1}}{\cancel{(x - 3)}_{1}(x + 3)} = \frac{3x}{x + 3}$$

You Try It 2
The LCM of x and x^2 is x^2.

$$\frac{1 + \dfrac{4}{x} + \dfrac{3}{x^2}}{1 + \dfrac{10}{x} + \dfrac{21}{x^2}} = \frac{1 + \dfrac{4}{x} + \dfrac{3}{x^2}}{1 + \dfrac{10}{x} + \dfrac{21}{x^2}} \cdot \frac{x^2}{x^2}$$

   • Multiply by the LCM.

$$= \frac{1 \cdot x^2 + \dfrac{4}{x} \cdot x^2 + \dfrac{3}{x^2} \cdot x^2}{1 \cdot x^2 + \dfrac{10}{x} \cdot x^2 + \dfrac{21}{x^2} \cdot x^2}$$

   • Distributive Property

$$= \frac{x^2 + 4x + 3}{x^2 + 10x + 21} = \frac{(x + 1)\cancel{(x + 3)}^{1}}{\cancel{(x + 3)}_{1}(x + 7)}$$

$$= \frac{x + 1}{x + 7}$$

You Try It 3
The LCM is $x - 5$.

$$\frac{x + 3 - \dfrac{20}{x - 5}}{x + 8 + \dfrac{30}{x - 5}} = \frac{x + 3 - \dfrac{20}{x - 5}}{x + 8 + \dfrac{30}{x - 5}} \cdot \frac{x - 5}{x - 5}$$

   • Multiply by the LCM.

$$= \frac{(x + 3)(x - 5) - \dfrac{20}{x - 5} \cdot (x - 5)}{(x + 8)(x - 5) + \dfrac{30}{x - 5} \cdot (x - 5)}$$

$$= \frac{x^2 - 2x - 15 - 20}{x^2 + 3x - 40 + 30} = \frac{x^2 - 2x - 35}{x^2 + 3x - 10}$$

$$= \frac{\cancel{(x + 5)}^{1}(x - 7)}{(x - 2)\cancel{(x + 5)}_{1}} = \frac{x - 7}{x - 2}$$

SECTION 6.5

You Try It 1

$$\frac{x}{x+6} = \frac{3}{x}$$

• The LCM is $x(x + 6)$.

$$\frac{x(x+6)}{1} \cdot \frac{x}{x+6} = \frac{x(x+6)}{1} \cdot \frac{3}{x}$$

• Multiply by the LCM.

$$x^2 = (x+6)3$$

• Simplify.

$$x^2 = 3x + 18$$
$$x^2 - 3x - 18 = 0$$
$$(x+3)(x-6) = 0$$

• Factor.

$$x + 3 = 0 \qquad x - 6 = 0$$

• Principle of Zero Products

$$x = -3 \qquad x = 6$$

Both -3 and 6 check as solutions.
The solutions are -3 and 6.

You Try It 2

$$\frac{5x}{x+2} = 3 - \frac{10}{x+2}$$

• The LCM is $x + 2$.

$$\frac{(x+2)}{1} \cdot \frac{5x}{x+2} = \frac{(x+2)}{1}\left(3 - \frac{10}{x+2}\right)$$

• Clear denominators.

$$\frac{x+2}{1} \cdot \frac{5x}{x+2} = \frac{x+2}{1} \cdot 3 - \frac{x+2}{1} \cdot \frac{10}{x+2}$$

$$5x = (x+2)3 - 10$$

• Solve for x.

$$5x = 3x + 6 - 10$$
$$5x = 3x - 4$$
$$2x = -4$$
$$x = -2$$

-2 does not check as a solution.
The equation has no solution.

SECTION 6.6

You Try It 1

$$\frac{2}{x+3} = \frac{6}{5x+5}$$

$$\frac{(x+3)(5x+5)}{1} \cdot \frac{2}{x+3} = \frac{(x+3)(5x+5)}{1} \cdot \frac{6}{5x+5}$$

$$\frac{(x+3)(5x+5)}{1} \cdot \frac{2}{x+3} = \frac{(x+3)(5x+5)}{1} \cdot \frac{6}{5x+5}$$

$$(5x+5)2 = (x+3)6$$

• Solve for x.

$$10x + 10 = 6x + 18$$
$$4x + 10 = 18$$
$$4x = 8$$
$$x = 2$$

The solution is 2.

You Try It 2

Strategy To find the total area that 256 ceramic tiles will cover, write and solve a proportion using x to represent the number of square feet that 256 tiles will cover.

Solution

$$\frac{9}{16} = \frac{x}{256}$$

• Write a proportion.

$$256\left(\frac{9}{16}\right) = 256\left(\frac{x}{256}\right)$$

• Clear denominators.

$$144 = x$$

An area of 144 ft² can be tiled using 256 ceramic tiles.

You Try It 3

Strategy To find the area of triangle AOB:
• Solve a proportion to find the length of AO (the height of triangle AOB).
• Use the formula for the area of a triangle. AB is the base and AO is the height.

Solution

$$\frac{CD}{AB} = \frac{DO}{AO}$$

• Write a proportion.

$$\frac{4}{10} = \frac{3}{AO}$$

• Substitute.

$$10 \cdot AO \cdot \frac{4}{10} = 10 \cdot AO \cdot \frac{3}{AO}$$

$$4(AO) = 30$$
$$AO = 7.5$$

$$A = \frac{1}{2}bh$$

• Area of a triangle

$$= \frac{1}{2}(10)(7.5)$$

• Substitute.

$$= 37.5$$

The area of triangle AOB is 37.5 cm².

SECTION 6.7

You Try It 1

$$5x - 2y = 10$$
$$5x - 5x - 2y = -5x + 10$$

• Subtract $5x$.

$$-2y = -5x + 10$$
$$\frac{-2y}{-2} = \frac{-5x+10}{-2}$$

• Divide by -2.

$$y = \frac{5}{2}x - 5$$

You Try It 2

$$s = \frac{A + L}{2}$$

$$2 \cdot s = 2\left(\frac{A + L}{2}\right) \quad \bullet \text{ Multiply by 2.}$$

$$2s = A + L$$

$$2s - A = A - A + L \quad \bullet \text{ Subtract } A.$$

$$2s - A = L$$

You Try It 3

$$S = a + (n - 1)d$$

$$S = a + nd - d$$

$$S - a = a - a + nd - d \quad \bullet \text{ Subtract } a.$$

$$S - a = nd - d$$

$$S - a + d = nd - d + d \quad \bullet \text{ Add } d.$$

$$S - a + d = nd$$

$$\frac{S - a + d}{d} = \frac{nd}{d} \quad \bullet \text{ Divide by } d.$$

$$\frac{S - a + d}{d} = n$$

You Try It 4

$$S = rS + C$$

$$S - rS = rS - rS + C \quad \bullet \text{ Subtract } rS.$$

$$S - rS = C$$

$$(1 - r)S = C \quad \bullet \text{ Factor.}$$

$$\frac{(1 - r)S}{1 - r} = \frac{C}{1 - r} \quad \bullet \text{ Divide by } 1 - r.$$

$$S = \frac{C}{1 - r}$$

SECTION 6.8

You Try It 1

Strategy   • Time for one printer to complete the job: t

	Rate	*Time*	*Part*
1st printer	$\frac{1}{t}$	2	$\frac{2}{t}$
2nd printer	$\frac{1}{t}$	5	$\frac{5}{t}$

• The sum of the parts of the task completed must equal 1.

Solution

$$\frac{2}{t} + \frac{5}{t} = 1$$

$$t\left(\frac{2}{t} + \frac{5}{t}\right) = t \cdot 1$$

$$2 + 5 = t$$

$$7 = t$$

Working alone, one printer takes 7 h to print the payroll.

You Try It 2

Strategy   • Rate sailing across the lake: r
            Rate sailing back: $3r$

	Distance	*Rate*	*Time*
Across	6	r	$\frac{6}{r}$
Back	6	$3r$	$\frac{6}{3r}$

• The total time for the trip was 2 h.

Solution

$$\frac{6}{r} + \frac{6}{3r} = 2$$

$$3r\left(\frac{6}{r} + \frac{6}{3r}\right) = 3r(2) \quad \bullet \text{ Multiply by the LCM, } 3r.$$

$$3r \cdot \frac{6}{r} + 3r \cdot \frac{6}{3r} = 6r$$

$$18 + 6 = 6r \quad \bullet \text{ Solve for } r.$$

$$24 = 6r$$

$$4 = r$$

The rate across the lake was 4 km/h.

Solutions to Chapter 7 "You Try It"

SECTION 7.1

You Try It 1

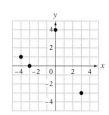

You Try It 2 $A(4, -2)$, $B(-2, 4)$.
The abscissa of D is 0.
The ordinate of C is 0.

You Try It 3

$$\begin{array}{c|c} x - 3y = -14 \\ \hline -2 - 3(4) & -14 \\ -2 - 12 & -14 \\ -14 = -14 \end{array}$$

Yes, $(-2, 4)$ is a solution of
$x - 3y = -14$.

You Try It 4 $x + 2y = 4$
$$2y = -x + 4$$
$$y = -\frac{1}{2}x + 2$$

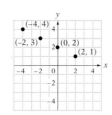

You Try It 5

$\{(145, 140), (140, 125), (150, 130), (165, 150), (140, 130), (165, 160)\}$

No, the relation is not a function. The two ordered
pairs $(140, 125)$ and $(140, 130)$ have the same first
coordinate but different second coordinates.

You Try It 6   Determine the ordered pairs defined
by the equation. Replace x in
$y = \frac{1}{2}x + 1$ by the given values and
solve for y. $\{(-4, -1), (0, 1), (2, 2)\}$
Yes, y is a function of x.

You Try It 7 $H(x) = \dfrac{x}{x - 4}$

$$H(8) = \dfrac{8}{8 - 4}$$ • Replace x by 8.

$$H(8) = \dfrac{8}{4} = 2$$

SECTION 7.2

You Try It 1

You Try It 2

You Try It 3

You Try It 4 $5x - 2y = 10$ • Solve for y.
$$-2y = -5x + 10$$
$$y = \frac{5}{2}x - 5$$

You Try It 5 $x - 3y = 9$ • Solve for y.
$$-3y = -x + 9$$
$$y = \frac{1}{3}x - 3$$

You Try It 6

You Try It 7

You Try It 8

The ordered pair
$(3, 120)$ means that
in 3 h the car will
travel 120 mi.

SECTION 7.3

You Try It 1 *x*-intercept: *y*-intercept:

$$y = 2x - 4 \qquad (0, b)$$
$$0 = 2x - 4 \qquad b = -4$$
$$-2x = -4 \qquad (0, -4)$$
$$x = 2$$
$$(2, 0)$$

You Try It 2 Let $P_1 = (1, 4)$ and $P_2 = (-3, 8)$.

$$m = \frac{y_2 - y_1}{x_2 - x_1} = \frac{8 - 4}{-3 - 1} = \frac{4}{-4} = -1$$

The slope is -1.

You Try It 3 Let $P_1 = (-1, 2)$ and $P_2 = (4, 2)$.

$$m = \frac{y_2 - y_1}{x_2 - x_1} = \frac{2 - 2}{4 - (-1)} = \frac{0}{5} = 0$$

The slope is 0.

You Try It 4 $m = \dfrac{8650 - 6100}{1 - 4} = \dfrac{2550}{-3}$

$m = -850$

A slope of -850 means that the value of the car is decreasing at a rate of $850 per year.

You Try It 5 y-intercept $= (0, b) = (0, -1)$

$$m = -\frac{1}{4}$$

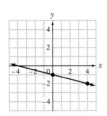

You Try It 6 Solve the equation for y.

$$x - 2y = 4$$
$$-2y = -x + 4$$
$$y = \frac{1}{2}x - 2$$

y-intercept $= (0, b) = (0, -2)$

$$m = \frac{1}{2}$$

SECTION 7.4

You Try It 1 Because the slope and *y*-intercept are known, use the slope-intercept formula, $y = mx + b$.

$$y = mx + b$$
$$y = \frac{5}{3}x + 2 \quad \bullet \; m = \frac{5}{3}; b = 2$$

You Try It 2 $m = \dfrac{3}{4}$ $(x_1, y_1) = (4, -2)$

$$y - y_1 = m(x - x_1)$$
$$y - (-2) = \frac{3}{4}(x - 4)$$
$$y + 2 = \frac{3}{4}x - 3$$
$$y = \frac{3}{4}x - 5$$

The equation of the line is $y = \dfrac{3}{4}x - 5$.

You Try It 3 Find the slope of the line between the two points.

$$m = \frac{y_2 - y_1}{x_2 - x_1} = \frac{1 - (-2)}{3 - (-6)} = \frac{3}{9} = \frac{1}{3}$$

Use the point-slope formula.

$$y - y_1 = m(x - x_1)$$
$$y - (-2) = \frac{1}{3}[x - (-6)] \quad \bullet \; y_1 = -2;$$
$$\qquad\qquad\qquad\qquad\qquad x_1 = -6$$
$$y + 2 = \frac{1}{3}x + 2$$
$$y = \frac{1}{3}x$$

You Try It 4

The slope of the line means that the grade on the history test increases 8.3 points for each 1-point increase in the grade on the reading test.

Solutions to Chapter 8 "You Try It"

SECTION 8.1

You Try It 1

$2x - 5y = 8$		$-x + 3y = -5$	
$2(-1) - 5(-2)$	8	$-(-1) + 3(-2)$	-5
$-2 + 10$	8	$1 + (-6)$	-5
$8 = 8$		$-5 = -5$	

Yes, $(-1, -2)$ is a solution of the system of equations.

You Try It 2

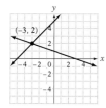

The solution is $(-3, 2)$.

You Try It 3

The lines are parallel. The system of equations is inconsistent and does not have a solution.

SECTION 8.2

You Try It 1

(1) $7x - y = 4$
(2) $3x + 2y = 9$

Solve Equation (1) for y.

$$7x - y = 4$$
$$-y = -7x + 4$$
$$y = 7x - 4$$

Substitute in Equation (2).

$$3x + 2y = 9$$
$$3x + 2(7x - 4) = 9 \qquad \bullet \; y = 7x - 4$$
$$3x + 14x - 8 = 9$$
$$17x - 8 = 9$$
$$17x = 17$$
$$x = 1$$

Substitute in Equation (1).

$$7x - y = 4$$
$$7(1) - y = 4 \qquad \bullet \; x = 1$$
$$7 - y = 4$$
$$-y = -3$$
$$y = 3$$

The solution is $(1, 3)$.

You Try It 2   (1) $3x - y = 4$
(2) $\qquad y = 3x + 2$

$$3x - y = 4$$
$$3x - (3x + 2) = 4 \qquad \bullet \; y = 3x + 2$$
$$3x - 3x - 2 = 4$$
$$-2 = 4$$

This is not a true equation. The system of equations is inconsistent and therefore has no solution.

You Try It 3   (1) $\qquad y = -2x + 1$
(2) $6x + 3y = 3$

$$6x + 3y = 3$$
$$6x + 3(-2x + 1) = 3 \qquad \bullet \; y = -2x + 1$$
$$6x - 6x + 3 = 3$$
$$3 = 3$$

The system of equations is dependent. The solutions are the ordered pairs that satisfy the equation $y = -2x + 1$.

You Try It 4

Strategy   • Amount invested at 6.5%: x
Amount invested at 4.5%: y

	Principal	Rate	Interest
Amount at 6.5%	x	0.065	$0.065x$
Amount at 4.5%	y	0.045	$0.045y$

• The sum of the two investments is $330,000: $x + y = 330{,}000$.
The interest earned at 6.5% equals the interest earned at 4.5%: $0.065x = 0.045y$

Solution
(1) $x + y = 330{,}000$
(2) $0.065x = 0.045y$

Solve Equation (2) for y.

(3) $\qquad y = \dfrac{13}{9}x$

Replace y by $\dfrac{13}{9}x$ in Equation (1) and solve for x.

$$x + y = 330{,}000$$
$$x + \frac{13}{9}x = 330{,}000 \qquad \bullet \ y = \frac{13}{9}x$$
$$\frac{22}{9}x = 330{,}000$$
$$x = 135{,}000$$

Replace x by 135,000 in Equation (3) and solve for y.

$$y = \frac{13}{9}x$$
$$= \frac{13}{9}(135{,}000) = 195{,}000 \qquad \bullet \ x = 135{,}000$$

$135,000 should be invested at 6.5% and $195,000 should be invested at 4.5%.

SECTION 8.3

You Try It 1 (1) $\quad x - 2y = 1$
(2) $\quad 2x + 4y = 0$

Eliminate y.

$$2(x - 2y) = 2 \cdot 1 \qquad \bullet \ \text{Multiply by 2.}$$
$$2x + 4y = 0$$

$$2x - 4y = 2$$
$$2x + 4y = 0$$

Add the equations.

$$4x = 2$$
$$x = \frac{2}{4} = \frac{1}{2}$$

Replace x in Equation (2).

$$2\left(\frac{1}{2}\right) + 4y = 0 \qquad \bullet \ x = \frac{1}{2}$$
$$1 + 4y = 0$$
$$4y = -1$$
$$y = -\frac{1}{4}$$

The solution is $\left(\dfrac{1}{2}, -\dfrac{1}{4}\right)$.

You Try It 2 (1) $\quad 2x - 3y = 4$
(2) $\quad -4x + 6y = -8$

Eliminate y.

$$2(2x - 3y) = 2 \cdot 4 \qquad \bullet \ \text{Multiply by 2.}$$
$$-4x + 6y = -8$$

$$4x - 6y = 8$$
$$-4x + 6y = -8$$

Add the equations.

$$0x + 0y = 0$$
$$0 = 0$$

The system of equations is dependent. The solutions are the ordered pairs that satisfy the equation $2x - 3y = 4$.

You Try It 3 (1) $\quad 4x + 5y = 11$
(2) $\quad 3y = x + 10$

Write equation (2) in the form $Ax + By = C$.

$$3y = x + 10$$
$$-x + 3y = 10$$

Eliminate x.

$$4x + 5y = 11$$
$$4(-x + 3y) = 4 \cdot 10 \qquad \bullet \ \text{Multiply by 4.}$$

$$4x + 5y = 11$$
$$-4x + 12y = 40$$

Add the equations.

$$17y = 51$$
$$y = 3$$

Replace y in Equation (1).

$$4x + 5y = 11$$
$$4x + 5 \cdot 3 = 11 \qquad \bullet \ y = 3$$
$$4x + 15 = 11$$
$$4x = -4$$
$$x = -1$$

The solution is $(-1, 3)$.

SECTION 8.4

You Try It 1

Strategy • Rate of the current: c
Rate of the canoeist in calm water: r

	Rate	Time	Distance
With current	$r + c$	3	$3(r + c)$
Against current	$r - c$	5	$5(r - c)$

- The distance traveled with the current is 15 mi.
  The distance traveled against the current is 15 mi.

Solution

$$3(r + c) = 15 \qquad \frac{1}{3} \cdot 3(r + c) = \frac{1}{3} \cdot 15 \quad \bullet \text{ Multiply by } \frac{1}{3}.$$

$$5(r - c) = 15 \qquad \frac{1}{5} \cdot 5(r - c) = \frac{1}{5} \cdot 15 \quad \bullet \text{ Multiply by } \frac{1}{5}.$$

$$r + c = 5$$
$$r - c = 3$$
$$2r = 8$$
$$r = 4$$

$$r + c = 5$$
$$4 + c = 5 \qquad \bullet \ r = 4$$
$$c = 1$$

The rate of the current: 1 mph.
The rate of the canoeist in calm water: 4 mph.

You Try It 2

Strategy   • Cost of an orange tree: x
Cost of a grapefruit tree: y

First purchase:

	Amount	Unit Cost	Value
Orange trees	25	x	$25x$
Grapefruit trees	20	y	$20y$

Second purchase:

	Amount	Unit Cost	Value
Orange trees	20	x	$20x$
Grapefruit trees	30	y	$30y$

- The total of the first purchase was \$290.
  The total of the second purchase was \$330.

Solution

$$25x + 20y = 290 \qquad 4(25x + 20y) = 4 \cdot 290$$
$$\bullet \text{ Multiply by 4.}$$
$$20x + 30y = 330 \qquad -5(20x + 30y) = -5 \cdot 330$$
$$\bullet \text{ Multiply by } -5.$$

$$100x + 80y = 1160$$
$$-100x - 150y = -1650$$
$$-70y = -490$$
$$y = 7$$

$$25x + 20y = 290$$
$$25x + 20(7) = 290 \qquad \bullet \ y = 7$$
$$25x + 140 = 290$$
$$25x = 150$$
$$x = 6$$

The cost of an orange tree is \$6.
The cost of a grapefruit tree is \$7.

Solutions to Chapter 9 "You Try It"

SECTION 9.1

You Try It 1 $A = \{-9, -7, -5, -3, -1\}$

You Try It 2 $A = \{1, 3, 5, \ldots\}$

You Try It 3 $A \cup B = \{-2, -1, 0, 1, 2, 3, 4\}$

You Try It 4 $C \cap D = \{10, 16\}$

You Try It 5 $A \cap B = \varnothing$

You Try It 6 $\{x \mid x < 59, x \in \text{positive even integers}\}$

You Try It 7 $\{x \mid x > -3, x \in \text{real numbers}\}$

You Try It 8   The solution set is the numbers greater than -2.

You Try It 9   The solution set is the numbers greater than -1 and the numbers less than -3.

You Try It 10   The solution set is the numbers less than or equal to 4 and greater than or equal to -4.

You Try It 11 The solution set is the real numbers.

$$\xleftarrow{\hspace{0.3cm}}\underset{-5\,-4\,-3\,-2\,-1\ \ 0\ \ 1\ \ 2\ \ 3\ \ 4\ \ 5}{\overline{}}\xrightarrow{\hspace{0.3cm}}$$

SECTION 9.2

You Try It 1
$$x + 2 < -2$$
$$x + 2 - 2 < -2 - 2 \quad \text{• Subtract 2.}$$
$$x < -4$$

$$\xleftarrow{\hspace{0.3cm}}\underset{-5\,-4\,-3\,-2\,-1\ \ 0\ \ 1\ \ 2\ \ 3\ \ 4\ \ 5}{\overline{}}\xrightarrow{\hspace{0.3cm}}$$

You Try It 2
$$5x + 3 > 4x + 5$$
$$5x - 4x + 3 > 4x - 4x + 5 \quad \text{• Subtract } 4x.$$
$$x + 3 > 5$$
$$x + 3 - 3 > 5 - 3 \quad \text{• Subtract 3.}$$
$$x > 2$$

You Try It 3
$$-3x > -9$$
$$\frac{-3x}{-3} < \frac{-9}{-3} \quad \text{• Divide by } -3.$$
$$x < 3$$

$$\xleftarrow{\hspace{0.3cm}}\underset{-5\,-4\,-3\,-2\,-1\ \ 0\ \ 1\ \ 2\ \ 3\ \ 4\ \ 5}{\overline{}}\xrightarrow{\hspace{0.3cm}}$$

You Try It 4
$$-\frac{3}{4}x \geq 18$$
$$-\frac{4}{3}\left(-\frac{3}{4}x\right) \leq -\frac{4}{3}(18) \quad \text{• Multiply by } -\frac{4}{3}.$$
$$x \leq -24$$

You Try It 5

Strategy To find the selling prices, write and solve an inequality using p to represent the possible selling prices.

Solution
$$0.70p > 314$$
$$p > 448.571 \quad \text{• Divide by 0.70.}$$

The dealer will make a profit with any selling price greater than or equal to $448.58.

SECTION 9.3

You Try It 1
$$5 - 4x > 9 - 8x$$
$$5 - 4x + 8x > 9 - 8x + 8x \quad \text{• Add } 8x.$$
$$5 + 4x > 9$$
$$5 - 5 + 4x > 9 - 5 \quad \text{• Subtract 5.}$$
$$4x > 4$$
$$\frac{4x}{4} > \frac{4}{4} \quad \text{• Divide by 4.}$$
$$x > 1$$

You Try It 2
$$8 - 4(3x + 5) \leq 6(x - 8)$$
$$8 - 12x - 20 \leq 6x - 48 \quad \text{• Distributive Property}$$
$$-12 - 12x \leq 6x - 48$$
$$-12 - 12x - 6x \leq 6x - 6x - 48 \quad \text{• Subtract } 6x.$$
$$-12 - 18x \leq -48$$
$$-12 + 12 - 18x \leq -48 + 12 \quad \text{• Add 12.}$$
$$-18x \leq -36$$
$$\frac{-18x}{-18} \geq \frac{-36}{-18} \quad \text{• Divide by } -18.$$
$$x \geq 2$$

You Try It 3

Strategy To find the maximum number of miles:
• Write an expression for the cost of each car, using x to represent the number of miles driven during the week.
• Write and solve an inequality.

Solution

Cost of a Company A car	is less than	cost of a Company B car

$$8(7) + 0.10x < 10(7) + 0.08x$$
$$56 + 0.10x < 70 + 0.08x$$
$$56 + 0.10x - 0.08x < 70 + 0.08x - 0.08x \quad \text{• Subtract } 0.08x.$$
$$56 + 0.02x < 70$$
$$56 - 56 + 0.02x < 70 - 56 \quad \text{• Subtract 56.}$$
$$0.02x < 14$$
$$\frac{0.02x}{0.02} < \frac{14}{0.02} \quad \text{• Divide by 0.02.}$$
$$x < 700$$

The maximum number of miles is 699 mi.

SECTION 9.4

You Try It 1
$$x - 3y < 2$$
$$x - x - 3y < -x + 2$$
$$-3y < -x + 2 \quad \text{• Subtract } x.$$
$$\frac{-3y}{-3} > \frac{-x + 2}{-3} \quad \text{• Divide by } -3.$$
$$y > \frac{1}{3}x - \frac{2}{3}$$

You Try It 2

$$2x - 4y \le 8$$
$$2x - 2x - 4y \le -2x + 8 \quad \bullet \text{ Subtract } 2x.$$
$$-4y \le -2x + 8$$
$$\frac{-4y}{-4} \ge \frac{-2x + 8}{-4} \quad \bullet \text{ Divide by } -4.$$
$$y \ge \frac{1}{2}x - 2$$

You Try It 3 $x < 3$

Solutions to Chapter 10 "You Try It"

SECTION 10.1

You Try It 1

$$-5\sqrt{32} = -5\sqrt{16 \cdot 2} \qquad \bullet \text{ 16 is a perfect square.}$$
$$= -5\sqrt{16}\sqrt{2}$$
$$= -5 \cdot 4\sqrt{2} = -20\sqrt{2}$$

You Try It 2

$$\sqrt{216} = \sqrt{36 \cdot 6} \qquad \bullet \text{ 36 is a perfect square.}$$
$$= \sqrt{36}\sqrt{6} = 6\sqrt{6}$$

You Try It 3

$$\sqrt{y^{19}} = \sqrt{y^{18} \cdot y} \qquad \bullet \ y^{18} \text{ is a perfect square.}$$
$$= \sqrt{y^{18}}\sqrt{y} = y^9\sqrt{y}$$

You Try It 4

$$\sqrt{45b^7} = \sqrt{9b^6 \cdot 5b} \qquad \bullet \ 9b^6 \text{ is a perfect square.}$$
$$= \sqrt{9b^6}\sqrt{5b} = 3b^3\sqrt{5b}$$

You Try It 5

$$3a\sqrt{28a^9b^{18}} = 3a\sqrt{4a^8b^{18}(7a)} \qquad \bullet \ 4a^8b^{18} \text{ is a perfect square.}$$
$$= 3a\sqrt{4a^8b^{18}}\sqrt{7a}$$
$$= 3a \cdot 2a^4b^9\sqrt{7a} = 6a^5b^9\sqrt{7a}$$

You Try It 6 $\sqrt{25(a + 3)^2} = 5(a + 3) = 5a + 15$

You Try It 7 $\sqrt{x^2 + 14x + 49} = \sqrt{(x + 7)^2} = x + 7$

SECTION 10.2

You Try It 1

$$9\sqrt{3} + 3\sqrt{3} - 18\sqrt{3} = (9 + 3 - 18)\sqrt{3} = -6\sqrt{3}$$

You Try It 2

$$2\sqrt{50} - 5\sqrt{32}$$
$$= 2\sqrt{25 \cdot 2} - 5\sqrt{16 \cdot 2} \qquad \bullet \text{ Simplify the radicands.}$$
$$= 2\sqrt{25}\sqrt{2} - 5\sqrt{16}\sqrt{2}$$
$$= 2 \cdot 5\sqrt{2} - 5 \cdot 4\sqrt{2}$$
$$= 10\sqrt{2} - 20\sqrt{2}$$
$$= (10 - 20)\sqrt{2} \qquad \bullet \text{ Distributive Property}$$
$$= -10\sqrt{2}$$

You Try It 3

$$y\sqrt{28y} + 7\sqrt{63y^3}$$
$$= y\sqrt{4 \cdot 7y} + 7\sqrt{9y^2 \cdot 7y} \qquad \bullet \text{ Simplify the radicands.}$$
$$= y\sqrt{4}\sqrt{7y} + 7\sqrt{9y^2}\sqrt{7y}$$
$$= y \cdot 2\sqrt{7y} + 7 \cdot 3y\sqrt{7y}$$
$$= 2y\sqrt{7y} + 21y\sqrt{7y}$$
$$= (2y + 21y)\sqrt{7y} \qquad \bullet \text{ Distributive Property}$$
$$= 23y\sqrt{7y}$$

You Try It 4

$$2\sqrt{27a^5} - 4a\sqrt{12a^3} + a^2\sqrt{75a}$$
$$= 2\sqrt{9a^4 \cdot 3a} - 4a\sqrt{4a^2 \cdot 3a} + a^2\sqrt{25 \cdot 3a}$$
$$= 2\sqrt{9a^4}\sqrt{3a} - 4a\sqrt{4a^2}\sqrt{3a}$$
$$\quad + a^2\sqrt{25}\sqrt{3a}$$
$$= 2 \cdot 3a^2\sqrt{3a} - 4a \cdot 2a\sqrt{3a} + a^2 \cdot 5\sqrt{3a}$$
$$= 6a^2\sqrt{3a} - 8a^2\sqrt{3a} + 5a^2\sqrt{3a} = 3a^2\sqrt{3a}$$

SECTION 10.3

You Try It 1 $\sqrt{5a}\sqrt{15a^3b^4}\sqrt{20b^5}$
$$= \sqrt{1500a^4b^9} = \sqrt{100a^4b^8 \cdot 15b}$$
$$= \sqrt{100a^4b^8} \cdot \sqrt{15b}$$
$$= 10a^2b^4\sqrt{15b}$$

You Try It 2
$\sqrt{5x}(\sqrt{5x} - \sqrt{25y})$
$$= \sqrt{25x^2} - \sqrt{125xy} \qquad \bullet \text{ Distributive Property}$$
$$= \sqrt{25x^2} - \sqrt{25 \cdot 5xy} = \sqrt{25x^2} - \sqrt{25}\sqrt{5xy}$$
$$= 5x - 5\sqrt{5xy}$$

You Try It 3
$(2\sqrt{x} + 7)(2\sqrt{x} - 7)$
$$= 4(\sqrt{x})^2 - 7^2 \qquad \bullet \text{ Product of conjugates}$$
$$= 4x - 49$$

You Try It 4
$(3\sqrt{x} - \sqrt{y})(5\sqrt{x} - 2\sqrt{y})$
$$= 15(\sqrt{x})^2 - 6\sqrt{xy} - 5\sqrt{xy} + 2(\sqrt{y})^2 \qquad \bullet \text{ FOIL}$$
$$= 15(\sqrt{x})^2 - 11\sqrt{xy} + 2(\sqrt{y})^2$$
$$= 15x - 11\sqrt{xy} + 2y$$

You Try It 5
$$\frac{\sqrt{15x^6y^7}}{\sqrt{3x^7y^9}} = \sqrt{\frac{15x^6y^7}{3x^7y^9}} = \sqrt{\frac{5}{xy^2}} = \frac{\sqrt{5}}{\sqrt{xy^2}}$$
$$= \frac{\sqrt{5}}{y\sqrt{x}} = \frac{\sqrt{5}}{y\sqrt{x}} \cdot \frac{\sqrt{x}}{\sqrt{x}} \qquad \bullet \text{ Rationalize the denominator.}$$
$$= \frac{\sqrt{5x}}{xy}$$

You Try It 6
$$\frac{\sqrt{3}}{\sqrt{3} - \sqrt{6}} = \frac{\sqrt{3}}{\sqrt{3} - \sqrt{6}} \cdot \frac{\sqrt{3} + \sqrt{6}}{\sqrt{3} + \sqrt{6}} \qquad \bullet \text{ Rationalize the denominator.}$$
$$= \frac{3 + \sqrt{18}}{3 - 6} = \frac{3 + 3\sqrt{2}}{-3}$$
$$= \frac{3(1 + \sqrt{2})}{-3} = -1(1 + \sqrt{2})$$
$$= -1 - \sqrt{2}$$

You Try It 7
$$\frac{5 + \sqrt{y}}{1 - 2\sqrt{y}} = \frac{5 + \sqrt{y}}{1 - 2\sqrt{y}} \cdot \frac{1 + 2\sqrt{y}}{1 + 2\sqrt{y}} \qquad \bullet \text{ Rationalize the denominator.}$$
$$= \frac{5 + 10\sqrt{y} + \sqrt{y} + 2(\sqrt{y})^2}{1 - 4y}$$
$$= \frac{5 + 11\sqrt{y} + 2y}{1 - 4y}$$

SECTION 10.4

You Try It 1 $\sqrt{4x} + 3 = 7$
$$\sqrt{4x} = 4 \qquad \bullet \text{ Isolate } \sqrt{4x}.$$
$$(\sqrt{4x})^2 = 4^2 \qquad \bullet \text{ Square both sides.}$$
$$4x = 16$$
$$x = 4 \qquad \bullet \text{ Solve for } x.$$

Check:
$$\begin{array}{c|c} \sqrt{4x} + 3 = 7 & \\ \hline \sqrt{4 \cdot 4} + 3 & 7 \\ \sqrt{16} + 3 & 7 \\ 4 + 3 & 7 \\ 7 = 7 & \end{array}$$

The solution is 4.

You Try It 2
$\sqrt{x} + \sqrt{x + 9} = 9$
$$\sqrt{x} = 9 - \sqrt{x + 9} \qquad \bullet \text{ Isolate } \sqrt{x}.$$
$$(\sqrt{x})^2 = (9 - \sqrt{x + 9})^2 \qquad \bullet \text{ Square both sides.}$$
$$x = 81 - 18\sqrt{x + 9} + (x + 9)$$
$$-90 = -18\sqrt{x + 9}$$
$$5 = \sqrt{x + 9} \qquad \bullet \text{ Isolate } \sqrt{x + 9}.$$
$$5^2 = (\sqrt{x + 9})^2 \qquad \bullet \text{ Square both sides.}$$
$$25 = x + 9$$
$$16 = x \qquad \bullet \text{ Solve for } x.$$

Check:
$$\begin{array}{c|c} \sqrt{x} + \sqrt{x + 9} = 9 & \\ \hline \sqrt{16} + \sqrt{16 + 9} & 9 \\ \sqrt{16} + \sqrt{25} & 9 \\ 4 + 5 & 9 \\ 9 = 9 & \end{array}$$

The solution is 16.

You Try It 3

Strategy To find the distance, use the Pythagorean Theorem. The hypotenuse is the length of the ladder. One leg is the distance from the bottom of the ladder to the base of the building. The distance along the building from the ground to the top of the ladder is the unknown leg.

Solution $a^2 = \sqrt{c^2 - b^2}$
$\qquad\quad = \sqrt{(8)^2 - (3)^2}$ • $c = 8, b = 3$
$\qquad\quad = \sqrt{64 - 9}$
$\qquad\quad = \sqrt{55}$
$\qquad\quad \approx 7.42$

The distance is approximately 7.42 ft.

You Try It 4

Strategy   To find the length of the pendulum, replace T in the equation with the given value and solve for L.

Solution

$$T = 2\pi\sqrt{\dfrac{L}{32}}$$

$$2.5 = 2(3.14)\sqrt{\dfrac{L}{32}} \quad • \ T = 2.5$$

$$2.5 = 6.28\sqrt{\dfrac{L}{32}}$$

$$\dfrac{2.5}{6.28} = \sqrt{\dfrac{L}{32}}$$

$$\left(\dfrac{2.5}{6.28}\right)^2 = \left(\sqrt{\dfrac{L}{32}}\right)^2$$

$$\dfrac{6.25}{39.4384} = \dfrac{L}{32}$$

$$(32)\left(\dfrac{6.25}{39.4384}\right) = (32)\left(\dfrac{L}{32}\right)$$

$$\dfrac{200}{39.4384} = L$$

$$5.07 \approx L$$

The length of the pendulum is approximately 5.07 ft.

Solutions to Chapter 11 "You Try It"

SECTION 11.1

You Try It 1

$$\dfrac{3y^2}{2} + y - \dfrac{1}{2} = 0$$

$$2\left(\dfrac{3y^2}{2} + y - \dfrac{1}{2}\right) = 2(0) \quad • \ \textbf{Multiply each side by 2.}$$

$$3y^2 + 2y - 1 = 0$$

$$(3y - 1)(y + 1) = 0 \qquad • \ \textbf{Factor.}$$

$3y - 1 = 0 \qquad y + 1 = 0$ • **Principle of**
$\quad 3y = 1 \qquad\qquad y = -1$ **Zero Products**

$$y = \dfrac{1}{3}$$

The solutions are $\dfrac{1}{3}$ and -1.

You Try It 2 $x^2 + 81 = 0$
$\qquad\qquad\qquad x^2 = -81$ • **Solve for** x^2.
$\qquad\qquad\quad \sqrt{x^2} = \sqrt{-81}$ • **Take square roots.**

$\sqrt{-81}$ is not a real number.

The equation has no real number solution.

You Try It 3 $7(z + 2)^2 = 21$
$\qquad\qquad\quad (z + 2)^2 = 3$ • **Solve for** $(z + 2)^2$.
$\qquad\qquad \sqrt{(z + 2)^2} = \sqrt{3}$ • **Take square roots.**
$\qquad\qquad\quad z + 2 = \pm\sqrt{3}$
$\qquad\qquad\qquad\quad z = -2 \pm \sqrt{3}$ • **Solve for** z.

The solutions are $-2 + \sqrt{3}$ and $-2 - \sqrt{3}$.

SECTION 11.2

You Try It 1
$3x^2 - 6x - 2 = 0$
$\quad 3x^2 - 6x = 2$ • **Add 2.**

$$\dfrac{1}{3}(3x^2 - 6x) = \dfrac{1}{3}\cdot 2 \quad • \ \textbf{Multiply by } \tfrac{1}{3}.$$

$$x^2 - 2x = \dfrac{2}{3}$$

Complete the square.

$x^2 - 2x + 1 = \dfrac{2}{3} + 1$

$\quad$ • $\left[\dfrac{1}{2}(-2)\right]^2 = [-1]^2 = 1$

$(x - 1)^2 = \dfrac{5}{3}$

$\quad$ • Factor.

$\sqrt{(x - 1)^2} = \sqrt{\dfrac{5}{3}}$

$\quad$ • Take square roots.

$x - 1 = \pm\sqrt{\dfrac{5}{3}}$

$\quad$ • Simplify.

$x = 1 \pm \sqrt{\dfrac{5}{3}}$

$x = 1 \pm \dfrac{\sqrt{15}}{3}$

$x = \dfrac{3 \pm \sqrt{15}}{3}$

The solutions are $\dfrac{3 + \sqrt{15}}{3}$ and $\dfrac{3 - \sqrt{15}}{3}$.

You Try It 2

$x^2 + 6x + 12 = 0$

$\quad x^2 + 6x = -12$

$\quad$ • Subtract 12.

$x^2 + 6x + 9 = -12 + 9$

$\quad$ • $\left(\dfrac{1}{2} \cdot 6\right)^2 = 3^2 = 9$

$(x + 3)^2 = -3$

$\quad$ • Factor.

$\sqrt{(x + 3)^2} = \sqrt{-3}$

$\quad$ • Take square roots.

$\sqrt{-3}$ is not a real number.

The quadratic equation has no real number solution.

You Try It 3

$\dfrac{x^2}{8} + x + 1 = 0$

$8\left(\dfrac{x^2}{8} + x + 1\right) = 8(0)$

$\quad$ • Multiply by 8.

$\quad x^2 + 8x + 8 = 0$

$\quad\quad x^2 + 8x = -8$

$\quad$ • Subtract 8.

$x^2 + 8x + 16 = -8 + 16$

$\quad$ • $\left(\dfrac{1}{2} \cdot 8\right)^2 = 4^2 = 16$

$(x + 4)^2 = 8$

$\quad$ • Factor.

$\sqrt{(x + 4)^2} = \sqrt{8}$

$\quad$ • Take square roots.

$x + 4 = \pm\sqrt{8}$

$x + 4 = \pm 2\sqrt{2}$

$x = -4 \pm 2\sqrt{2}$

$x = -4 + 2\sqrt{2} \qquad\qquad x = -4 - 2\sqrt{2}$

$\approx -4 + 2(1.414) \qquad \approx -4 - 2(1.414)$

$\approx -4 + 2.828 \qquad\quad \approx -4 - 2.828$

$\approx -1.172 \qquad\qquad\quad \approx -6.828$

The solutions are approximately -1.172 and -6.828.

SECTION 11.3

You Try It 1

$3x^2 + 4x - 4 = 0$

$a = 3, b = 4, c = -4$

$x = \dfrac{-(4) \pm \sqrt{(4)^2 - 4(3)(-4)}}{2 \cdot 3}$

$= \dfrac{-4 \pm \sqrt{16 + 48}}{6}$

$= \dfrac{-4 \pm \sqrt{64}}{6} = \dfrac{-4 \pm 8}{6}$

$x = \dfrac{-4 + 8}{6} \qquad\quad x = \dfrac{-4 - 8}{6}$

$= \dfrac{4}{6} = \dfrac{2}{3} \qquad\qquad = \dfrac{-12}{6} = -2$

The solutions are $\dfrac{2}{3}$ and -2.

You Try It 2

$\dfrac{x^2}{4} + \dfrac{x}{2} = \dfrac{1}{4}$

$4\left(\dfrac{x^2}{4} + \dfrac{x}{2}\right) = 4\left(\dfrac{1}{4}\right)$

$\quad$ • Multiply by 4.

$x^2 + 2x = 1$

$x^2 + 2x - 1 = 0$

$\quad$ • Standard form

$a = 1, b = 2, c = -1$

$x = \dfrac{-(2) \pm \sqrt{(2)^2 - 4(1)(-1)}}{2 \cdot 1}$

$= \dfrac{-2 \pm \sqrt{4 + 4}}{2} = \dfrac{-2 \pm \sqrt{8}}{2}$

$= \dfrac{-2 \pm 2\sqrt{2}}{2} = -1 \pm \sqrt{2}$

The solutions are $-1 + \sqrt{2}$ and $-1 - \sqrt{2}$.

SECTION 11.4

You Try It 1 $\quad y = x^2 + 2$

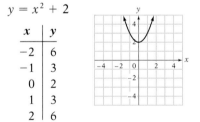

x	y
-2	6
-1	3
0	2
1	3
2	6

You Try It 2 To find the x-intercept, let $f(x) = 0$ and solve for x.

$f(x) = x^2 - 6x + 9$

$0 = x^2 - 6x + 9$

$0 = (x - 3)(x - 3)$

$\quad$ • Factor.

$x - 3 = 0 \qquad x - 3 = 0$

$\quad$ • Principle of

$x = 3 \qquad\qquad x = 3$

$\quad$ Zero Products

The x-intercept: $(3, 0)$.

There is only one x-intercept. The equation has a double root.

To find the y-intercept, evaluate the function at $x = 0$.

$f(x) = x^2 - 6x + 9$
$f(0) = 0^2 - 6(0) + 9 = 9$

The y-intercept: $(0, 9)$.

SECTION 11.5

You Try It 1

Strategy
- This is a geometry problem.
- Width of the rectangle: W
  Length of the rectangle: $W + 2$
- Use the equation $A = L \cdot W$.

Solution

$A = L \cdot W$
$15 = (W + 2)W$ • $A = 15,\ L = W + 2$
$15 = W^2 + 2W$
$0 = W^2 + 2W - 15$
$0 = (W + 5)(W - 3)$ • Factor.

$W + 5 = 0 \qquad W - 3 = 0$ • Principle of
$\qquad W = -5 \qquad\quad W = 3$ Zero Products.

The solution -5 is not possible.
The width is 3 m.

Answers to Chapter 1 Selected Exercises

PREP TEST

1. 127.16 **2.** 46,514 **3.** 4517 **4.** 11,396 **5.** 508 **6.** 24 **7.** 4 **8.** $3 \cdot 7$ **9.** $\frac{2}{5}$ **10.** d

SECTION 1.1

3. $>$ **5.** $<$ **7.** $>$ **9.** $>$ **11.** $>$ **13.** False **15.** True **17.** False **19.** True **21.** True
23. $\{1, 2, 3, 4, 5, 6, 7, 8\}$ **25.** $\{1, 2, 3, 4, 5, 6, 7, 8\}$ **27.** $\{-6, -5, -4, -3, -2, -1\}$ **29.** 5 **31.** $-23, -18$
33. 21, 37 **35.** $-52, -46, 0$ **37.** $-17, 0, 4, 29$ **39.** 5, 6, 7, 8, 9 **41.** $-10, -9, -8, -7, -6, -5$ **43.** -4
45. 9 **47.** 36 **49.** 40 **51.** -39 **53.** 74 **55.** -82 **57.** -81 **59.** $>$ **61.** $<$ **63.** $<$ **65.** $<$
67. a. $11, 7, 3, -1, -5$ **b.** $11, 7, 3, 1, 5$ **69.** Never true

SECTION 1.2

3. -11 **5.** -5 **7.** -83 **9.** -46 **11.** 0 **13.** -5 **15.** 9 **17.** 1 **19.** -10 **21.** -18 **23.** -41
25. -12 **27.** 0 **29.** -34 **31.** 0 **33.** -61 **35.** -65 **37.** -15 **39.** 22 **43.** 8 **45.** -7
47. -9 **49.** 9 **51.** -3 **53.** 18 **55.** -9 **57.** 11 **59.** -18 **61.** 0 **63.** 2 **65.** -138 **67.** -8
69. -12 **71.** -20 **73.** 15 **75.** -39 **77.** 16 **79.** The difference in temperature is 399°C.
81. The difference in elevation is 7046 m. **83.** The continent with the greatest difference between the highest and lowest
elevation is Asia. **85.** The difference in temperature is 396°C. **87.** Yes **89.** The difference between the heights is
20,370 m. **91.** The difference in wind-chill factor is 6°F. **93.** No. For example, the difference between 10 and -8 is
18, which is greater than either 10 or -8.

SECTION 1.3

3. 42 **5.** 60 **7.** -253 **9.** -114 **11.** -105 **13.** -216 **15.** -315 **17.** 336 **19.** -2772 **21.** 0
23. 350 **25.** -336 **27.** -352 **29.** -2 **31.** 8 **33.** -7 **35.** -12 **37.** -6 **39.** 11 **41.** -14
43. 15 **45.** -16 **47.** 0 **49.** -29 **51.** Undefined **53.** -11 **55.** Undefined **57.** -12 **59.** 24
61. 28 **63.** The average daily high temperature was -26°F. **65.** The five-day moving average is $-12, 28, 40, 28,$
48, 52. **67.** The student's score was 74. **69.** The expression $-3x$ is greatest for $x = -6$.

SECTION 1.4

1. 36 **3.** -49 **5.** 9 **7.** 81 **9.** $-4^4 = -(4 \cdot 4 \cdot 4 \cdot 4) = 256$ **11.** $2 \cdot (-3)^2 = 2(-3)(-3) = 2 \cdot 9 = 18$
13. -27 **15.** 216 **17.** -12 **19.** 16 **21.** -864 **23.** -1008 **25.** -81 **27.** $-77,760$ **31.** 9 **33.** 12
35. 1 **37.** 8 **39.** -16 **41.** 12 **43.** 13 **45.** -36 **47.** 13 **49.** 4 **51.** 15 **53.** -1 **55.** 4
57. 1 **59.** 0 **61.** -172

SECTION 1.5

1. 1, 2, 4 **3.** 1, 2, 3, 4, 6, 12 **5.** 1, 2, 4, 8 **7.** 1, 13 **9.** 1, 2, 4, 7, 8, 14, 28, 56 **11.** 1, 3, 5, 9, 15, 45
13. 1, 29 **15.** 1, 2, 4, 13, 26, 52 **17.** 1, 2, 41, 82 **19.** 1, 3, 19, 57 **21.** 1, 2, 3, 4, 6, 8, 12, 16, 24, 48
23. 1, 2, 5, 10, 25, 50 **25.** 1, 7, 11, 77 **27.** 1, 2, 4, 5, 10, 20, 25, 50, 100 **29.** 1, 5, 17, 85 **31.** $2 \cdot 7$ **33.** $2^3 \cdot 3^2$
35. $2^3 \cdot 3$ **37.** $2^2 \cdot 3^2$ **39.** $2 \cdot 13$ **41.** 7^2 **43.** Prime **45.** $2 \cdot 31$ **47.** Prime **49.** $2 \cdot 43$ **51.** $5 \cdot 19$
53. $2 \cdot 3 \cdot 13$ **55.** $2^4 \cdot 3^2$ **57.** $5^2 \cdot 7$ **59.** $2^4 \cdot 5^2$ **61.** 24 **63.** 12 **65.** 36 **67.** 140 **69.** 36 **71.** 240
73. 720 **75.** 216 **77.** 360 **79.** 160 **81.** 24 **83.** 30 **85.** 72 **87.** 150 **89.** 108 **91.** 2 **93.** 1
95. 2 **97.** 6 **99.** 4 **101.** 10 **103.** 8 **105.** 7 **107.** 30 **109.** 1 **111.** 2 **113.** 6 **115.** 12
117. 26 **119.** 18 **121.** Answers may vary. One possibility is 3, 5; 5, 7; 11, 13. **125.** The product of the two
numbers.

SECTION 1.6

1. $\frac{1}{3}$ **3.** $-\frac{4}{11}$ **5.** $-\frac{2}{3}$ **7.** $\frac{3}{2}$ **9.** 0 **11.** $-\frac{3}{5}$ **13.** $-\frac{7}{5}$ **15.** -15 **17.** $\frac{1}{2}$ **19.** 0.8 **21.** $0.1\overline{6}$

23. $-0.\overline{3}$ **25.** $-0.\overline{2}$ **27.** $-0.58\overline{3}$ **29.** $0.91\overline{6}$ **31.** $-0.3\overline{8}$ **33.** 0.5625 **35.** $-0.\overline{857142}$ **37.** $\frac{13}{12}$

39. $\frac{35}{24}$ **41.** $\frac{1}{24}$ **43.** $-\frac{29}{26}$ **45.** $-\frac{19}{12}$ **47.** $\frac{25}{18}$ **49.** $-\frac{7}{48}$ **51.** $-\frac{17}{24}$ **53.** 8.022 **55.** -25.4

57. -9.964 **59.** -2.84 **61.** -6.961 **63.** 0 **65.** -108.677 **67.** $-\frac{1}{12}$ **69.** $-\frac{7}{18}$ **71.** -5.55 **73.** $\frac{2}{27}$

75. $-\frac{1}{8}$ **77.** $-\frac{37}{24}$ **79.** $-\frac{23}{18}$ **81.** $\frac{73}{60}$ **83.** $\frac{29}{16}$ **85.** -1 **87.** $\frac{143}{72}$ **89.** -2.801 **91.** 2.088

93. 21.16 **95.** 0 **97.** 14.452 **99.** -17.606 **101.** 4.001 **103.** $-\frac{29}{24}$ **105.** $-\frac{1}{12}$ **107.** $-\frac{7}{18}$

109. From thinnest to thickest, the coins are dime, penny, quarter, nickel. **111.** The sum of the weights of the four coins is 15.438 g. **113.** The largest difference in diameter between two coins is 6.35 mm. **115.** $\frac{1}{12}$ cup of additional broth is needed. **117.** Apple Computer had negative earnings per share in 1996, 1997, 2001. **119.** Between 1996 and 1997, the decrease in earnings per share was $1.70. **121.** The five countries consume 34.9 million barrels of oil per day. **123.** The difference between the number of barrels of oil per day consumed and the number of imported is 11.5 million barrels. **125.** $\frac{13}{24}$ **127.** $-\frac{5}{24}$ **129.** $-\frac{7}{60}$ **131.** $<$ **133.** $=$ **135.** $\frac{1}{9}$ **137.** $\frac{5}{12}$ **139.** 2, 5 **141.** Larger

SECTION 1.7

1. $\frac{10}{21}$ **3.** $-\frac{3}{16}$ **5.** $-\frac{1}{8}$ **7.** $-\frac{4}{9}$ **9.** $\frac{9}{50}$ **11.** $\frac{9}{16}$ **13.** $\frac{5}{36}$ **15.** $\frac{3}{64}$ **17.** -7 **19.** $-\frac{11}{18}$ **21.** -1.794

23. 0.7407 **25.** -0.408 **27.** -0.22165 **29.** 27.2136 **31.** -0.097722 **33.** $-\frac{3}{5}$ **35.** 1.035 **37.** $-\frac{5}{12}$

39. $\frac{10}{9}$ **41.** $\frac{11}{14}$ **43.** $-\frac{1}{11}$ **45.** $-\frac{11}{20}$ **47.** $-\frac{2}{3}$ **49.** $\frac{14}{5}$ **51.** -2 **53.** -4.925 **55.** 1.05 **57.** -0.09

59. 0.11 **61.** 0.970 **63.** 0.007 **65.** 1.5 **67.** -0.0046 **69.** $\frac{23}{30}$ **71.** $-\frac{1}{8}$ **73.** $-\frac{11}{18}$ **75.** 0 **77.** -2

79. $-\frac{4}{27}$ **81.** -4.09 **83.** -0.1238 **85.** 0 **89.** $\frac{3}{4}$, 0.75 **91.** $\frac{16}{25}$, 0.64 **93.** $\frac{7}{4}$, 1.75 **95.** $\frac{19}{100}$, 0.19

97. $\frac{1}{20}$, 0.05 **99.** $\frac{1}{9}$ **101.** $\frac{1}{8}$ **103.** $\frac{2}{3}$ **105.** $\frac{1}{200}$ **107.** $\frac{5}{6}$ **109.** 0.073 **111.** 0.158 **113.** 0.003

115. 0.099 **117.** 1.212 **119.** 15% **121.** 5% **123.** 17.5% **125.** 115% **127.** 0.8% **129.** 54%

131. $33\frac{1}{3}\%$ **133.** $45\frac{5}{11}\%$ **135.** 87.5% **137.** $166\frac{2}{3}\%$ **139.** The center of the frame is $9\frac{1}{8}$ in. from the left side of the frame and $12\frac{1}{4}$ in. from the bottom of the frame. **141.** The cut should be made $18\frac{1}{4}$ in. from the left side of the board. **143.** The staircase is $87\frac{1}{2}$ in. high. **145.** $\frac{1}{22}$ **147.** $\frac{1}{16}$ ft

SECTION 1.8

1. 90° **3.** 28° **5.** 132° **7.** 83° **9.** 91° **11.** 132° **13.** 51° **15.** 77° **17.** 79° **19.** 292°
21. 9.71 cm **23.** 14 ft 2 in. **25.** 52 in. **27.** 131.88 cm **29.** 3.768 m **31.** The wood framing would cost $76.96. **33.** The cost of the binding is $19.78. **35.** 32 ft² **37.** 378 cm² **39.** 50.24 in² **41.** 16.81 m²
43. 52.5 cm² **45.** 226.865 in² **47.** 138.6 gal of water should be used. **49.** The cost to build the design is $172.
51. The cost to plaster the room is $990.72. **53.** Perimeter: 265.6 m; area: 4056 m² **55.** Perimeter: 340 m; area: 1600 m² **57. a.** 39 in² **b.** 87 in²

CHAPTER 1 REVIEW EXERCISES*

1. -6 [1.2A] **2.** 0.28 [1.6A] **3.** -25 [1.4A] **4.** 10 [1.4B] **5.** 37° [1.8A] **6.** 0.062 [1.7C]
7. -42 [1.3A] **8.** $\frac{7}{12}$ [1.6C] **9.** 34° [1.8A] **10.** -4 [1.1A] **11.** 1, 2, 4, 7, 8, 14, 28, 56 [1.5A]

*Note: The numbers in brackets following the answers in the Chapter Review are a reference to the objective that corresponds to that problem. For example, the reference [1.2A] stands for Section 1.2, Objective A. This notation will be used for all Prep Tests, Chapter Reviews, Chapter Tests, and Cumulative Reviews throughout the text.

12. -1.068 [1.6C] **13.** 62.5% [1.7C] **14.** $0.1\overline{3}$ [1.6A] **15.** -4 [1.2B] **16.** $-\dfrac{2}{15}$ [1.6C] **17.** 4 [1.1B]

18. 18 cm² [1.8C] **19.** -20 [1.3B] **20.** $\dfrac{159}{200}$ [1.7C] **21.** $2 \cdot 2 \cdot 2 \cdot 5 \cdot 7$ [1.5B] **22.** 31 [1.4B]

23. -13 [1.2A] **24.** $\dfrac{17}{40}$ [1.6B] **25.** $54\dfrac{2}{7}\%$ [1.7C] **26.** 28.26 m² [1.8C] **27.** -4.6224 [1.7A]

28. -5 [1.1B] **29.** 1 [1.2B] **30.** $-\dfrac{8}{15}$ [1.7B] **31.** 152° [1.8A] **32.** 44 in. [1.8B]

33. $-|6| < |-10|$ [1.1B] **34.** 1 [1.7B] **35.** The score for the exam was 98. [1.3C]
36. 38.2% of the PDAs were shipped by Palm. [1.7D] **37.** The difference between the boiling point and freezing point of mercury is 396°C. [1.2C] **38.** The sod cost $240.96. [1.8C]

CHAPTER 1 TEST*

1. 17 [1.3B] **2.** $83\dfrac{1}{3}\%$ [1.7C] **3.** 62° [1.8A] **4.** -5.3578 [1.7A] **5.** -14 [1.2B] **6.** $\dfrac{3}{8}$ [1.7C]

7. $\dfrac{1}{24}$ [1.6C] **8.** 8 [1.7B] **9.** 90 [1.3A] **10.** 84.78 in² [1.8B] **11.** -108 [1.4A] **12.** 90 cm² [1.8C]

13. $-2 > -40$ [1.1A] **14.** $-\dfrac{7}{20}$ [1.6B] **15.** -4 [1.1B] **16.** $\dfrac{9}{20}$; 0.45 [1.7C] **17.** -16 [1.2A]

18. -48 [1.3A] **19.** $2 \cdot 3 \cdot 3 \cdot 5 \cdot 11$ [1.5B] **20.** 17 [1.4B] **21.** 4 [1.2B] **22.** $-\dfrac{1}{2}$ [1.7B]

23. 47° [1.8A] **24.** 9 [1.4B] **25.** $0.\overline{7}$ [1.6A] **26. a.** The annual earnings would be $-\$457.6$ million.
b. The average monthly loss would be $-\$67$ thousand. [1.7D] **27.** The cost of the new fencing is $4564. [1.8C]

Answers to Chapter 2 Selected Exercises

PREP TEST

1. 3 [1.2B] **2.** 4 [1.3B] **3.** $\dfrac{1}{12}$ [1.6B] **4.** $-\dfrac{4}{9}$ [1.7B] **5.** $\dfrac{3}{10}$ [1.7B] **6.** -16 [1.4A] **7.** $\dfrac{8}{27}$ [1.7A]

8. 48 [1.4B] **9.** 1 [1.4B] **10.** 12 [1.4B]

SECTION 2.1

1. $2x^2, 5x, \underline{-8}$ **3.** $-a^4, \underline{6}$ **5.** $7\underline{x^2y}, 6xy^2$ **7.** $1, -9$ **9.** $1, -4, -1$ **13.** 10 **15.** 32 **17.** 21 **19.** 16
21. -9 **23.** 41 **25.** -7 **27.** 13 **29.** -15 **31.** 41 **33.** 1 **35.** 5 **37.** 1 **39.** 57 **41.** 5
43. 8 **45.** -3 **47.** -2 **49.** -4 **51.** 225 **53.** 60 **55.** 4 **57.** 81 **59.** $n^x > x^n$ if $x \geq n + 1$

SECTION 2.2

3. $14x$ **5.** $5a$ **7.** $-6y$ **9.** $7 - 3b$ **11.** $5a$ **13.** $-2ab$ **15.** $5xy$ **17.** 0 **19.** $-\dfrac{5}{6}x$ **21.** $6.5x$

23. $0.45x$ **25.** $7a$ **27.** $-14x^2$ **29.** $-\dfrac{11}{24}x$ **31.** $17x - 3y$ **33.** $-2a - 6b$ **35.** $-3x - 8y$ **37.** $-4x^2 - 2x$

39. $12x$ **41.** $-21a$ **43.** $6y$ **45.** $8x$ **47.** $-6a$ **49.** $12b$ **51.** $-15x^2$ **53.** x^2 **55.** a **57.** x **59.** n

61. x **63.** y **65.** $3x$ **67.** $-2x$ **69.** $-8a^2$ **71.** $8y$ **73.** $4y$ **75.** $-2x$ **77.** $6a$ **79.** $-x - 2$

81. $8x - 6$ **83.** $-2a - 14$ **85.** $-6y + 24$ **87.** $35 - 21b$ **89.** $2 - 5y$ **91.** $15x^2 + 6x$ **93.** $2y - 18$

95. $-15x - 30$ **97.** $-6x^2 - 28$ **99.** $-6y^2 + 21$ **101.** $3x^2 - 3y^2$ **103.** $-4x + 12y$ **105.** $-6a^2 + 7b^2$

107. $4x^2 - 12x + 20$ **109.** $\dfrac{3}{2}x - \dfrac{9}{2}y + 6$ **111.** $-12a^2 - 20a + 28$ **113.** $12x^2 - 9x + 12$ **115.** $10x^2 - 20xy - 5y^2$

117. $-8b^2 + 6b - 9$ **119.** $a - 7$ **121.** $-11x + 13$ **123.** $-4y - 4$ **125.** $-2x - 16$ **127.** $14y - 45$

129. $a + 7b$ **131.** $6x + 28$ **133.** $5x - 75$ **135.** $4x - 4$ **137.** $2x - 9$ **139. a.** False. For example, $8 \div 2 \neq 2 \div 8$ **b.** False. For example, $(12 \div 4) \div 2 \neq 12 \div (4 \div 2)$ **c.** False. For example, $(9 - 2) - 3 \neq 9 - (2 - 3)$ **d.** False. For example, $10 - 4 \neq 4 - 10$ **141.** No. 0 does not have a multiplicative inverse.

SECTION 2.3

1. $8 + y$ **3.** $t + 10$ **5.** $z + 14$ **7.** $x^2 - 20$ **9.** $\frac{3}{4}n + 12$ **11.** $8 + \frac{n}{4}$ **13.** $3(y + 7)$ **15.** $t(t + 16)$

17. $\frac{1}{2}x^2 + 15$ **19.** $5n^3 + n^2$ **21.** $r - \frac{r}{3}$ **23.** $x^2 - (x + 17)$ **25.** $9(z + 4)$ **27.** $12 - x$ **29.** $\frac{2}{3}x$ **31.** $\frac{2x}{9}$

33. $11x - 8$ **35.** $(x + 2) - 9; x - 7$ **37.** $\frac{7}{5 + x}$ **39.** $5 + \frac{1}{2}(x + 3); \frac{1}{2}x + \frac{13}{2}$ **41.** $(2x - 4) + x; 3x - 4$

43. $(x - 5)7; 7x - 35$ **45.** $\frac{2x + 5}{x}$ **47.** $x - (3x - 8); -2x + 8$ **49.** $x + 3x; 4x$ **51.** $(x + 6) + 5; x + 11$

53. $x - (x + 10); -10$ **55.** $\frac{1}{6}x + \frac{4}{9}x; \frac{11}{18}x$ **57.** $\frac{x}{3} + x; \frac{4}{3}x$ **59.** Number of emails: A; number of spam emails: $\frac{1}{2}A$

61. Length of one piece: S; length of second piece: $12 - S$ **63.** Distance traveled by the faster car: x; distance traveled by the slower car: $200 - x$ **65.** Number of bones in your body: N; number of bones in your foot: $\frac{1}{4}N$

67. Salary needed in San Francisco: S; salary needed in Daytona Beach: $\frac{1}{2}S$ **69.** $2x$

CHAPTER 2 REVIEW EXERCISES

1. $3x^2 - 24x - 21$ [2.2C]   **2.** $11x$ [2.2A]   **3.** $8a - 4b$ [2.2A]   **4.** $-5n$ [2.2B]   **5.** 79 [2.1A]
6. $10x - 35$ [2.2C]   **7.** $12y^2 + 8y - 10$ [2.2C]   **8.** $-6a$ [2.2B]   **9.** $-42x^2$ [2.2B]   **10.** $-63 - 36x$ [2.2C]
11. $-5y$ [2.2A]   **12.** -4 [2.1A]   **13.** $-6x - 1$ [2.2D]   **14.** $-40a + 40$ [2.2D]   **15.** $24y + 30$ [2.2D]
16. $9c - 5d$ [2.2A]   **17.** $20x$ [2.2B]   **18.** $7x + 46$ [2.2D]   **19.** 29 [2.1A]   **20.** $-9r + 8s$ [2.2A]
21. 50 [2.1A]   **22.** 28 [2.1A]   **23.** $-4x^2 + 6x$ [2.2A]   **24.** $-90x + 25$ [2.2D]   **25.** $-0.2x + 150$ [2.2D]
26. $-\frac{1}{12}x$ [2.2A]   **27.** $28a^2 - 8a + 12$ [2.2C]   **28.** $-4x + 20$ [2.2D]   **29.** -7 [2.1A]   **30.** $36y$ [2.2B]
31. $\frac{2}{3}(x + 10)$ [2.3A]   **32.** $4x$ [2.3A]   **33.** $x - 6$ [2.3A]   **34.** $x + 2x; 3x$ [2.3B]   **35.** $2x - \frac{1}{2}x; \frac{3}{2}x$ [2.3B]
36. $3x + 5(x - 1); 8x - 5$ [2.3B]   **37.** Number of American League cards: A; number of National League cards: $5A$ [2.3C]   **38.** Number of ten-dollar bills: T; number of five-dollar bills: $35 - T$ [2.3C]   **39.** Number of calories in an apple: a; number of calories in a candy bar: $2a + 8$ [2.3C]   **40.** Width of Parthenon: w; length of Parthenon: $1.6w$ [2.3C]   **41.** Kneeling height: h; standing height: $1.3h$ [2.3C]

CHAPTER 2 TEST

1. $5x$ [2.2A]   **2.** $-6x^2 + 21y^2$ [2.2C]   **3.** $-x + 6$ [2.2D]   **4.** $-7x + 33$ [2.2D]   **5.** $-9x - 7y$ [2.2A]
6. 22 [2.1A]   **7.** $2x$ [2.2B]   **8.** $7x + 38$ [2.2D]   **9.** $-10x^2 + 15x - 30$ [2.2C]   **10.** $-2x - 5y$ [2.2A]
11. 3 [2.1A]   **12.** $3x$ [2.2B]   **13.** y^2 [2.2A]   **14.** $-4x + 8$ [2.2C]   **15.** $-10a$ [2.2B]
16. $2x + y$ [2.2D]   **17.** $36y$ [2.2B]   **18.** $15 - 35b$ [2.2C]   **19.** $a^2 - b^2$ [2.3A]
20. $10(x - 3) = 10x - 30$ [2.3B]   **21.** $x + 2x^2$ [2.3B]   **22.** $\frac{6}{x} - 3$ [2.3B]   **23.** $b - 7b$ [2.3A]   **24.** Speed of return throw: s; speed of fastball: $2s$ [2.3C]   **25.** Shorter piece: x; longer piece: $4x - 3$ [2.3C]

CUMULATIVE REVIEW EXERCISES

1. -7 [1.2A]   **2.** 5 [1.2B]   **3.** 24 [1.3A]   **4.** -5 [1.3B]   **5.** $53°$ [1.8A]   **6.** $\frac{11}{48}$ [1.6C]   **7.** $-\frac{1}{6}$ [1.7B]

8. $\frac{1}{4}$ [1.7A]   **9.** 75% [1.7C]   **10.** -5 [1.4B]   **11.** $-\frac{27}{26}$ [1.7B]   **12.** 16 [2.1A]   **13.** $5x^2$ [2.2A]

14. $-7a - 10b$ [2.2A]   **15.** 153.86 cm^2 [1.8C]   **16.** 96 ft [1.8B]   **17.** $24 - 6x$ [2.2C]   **18.** $6y - 18$ [2.2C]

19. $\frac{3}{8}$ [1.7C]   **20.** 0.0105 [1.7C]   **21.** $-8x^2 + 12y^2$ [2.2C]   **22.** $-9y^2 + 9y + 21$ [2.2C]

23. $-7x + 14$ [2.2D]   **24.** $5x - 43$ [2.2D]   **25.** $17x - 24$ [2.2D]   **26.** $-3x + 21y$ [2.2D]

27. $\frac{1}{2}b + b$ [2.3A]   **28.** $\frac{10}{y-2}$ [2.3A]   **29.** $8 - \frac{x}{12}$ [2.3B]   **30.** $x + (x+2); 2x + 2$ [2.3B]

31. 3600 ft^2 [1.8C]   **32.** Speed of dial-up connection: s; speed of DSL connection: $10s$ [2.3C]

Answers to Chapter 3 Selected Exercises

PREP TEST

1. 0.09 [1.7C]   **2.** 75% [1.7C]   **3.** 63 [2.1A]   **4.** $0.65R$ [2.2A]   **5.** $\frac{7}{6}x$ [2.2A]   **6.** $9x - 18$ [2.2C]

7. $1.66x + 1.32$ [2.2C]   **8.** $5 - 2n$ [2.3B]   **9.** Speed of the old card: s; speed of the new card: $5s$ [2.3C]

10. $5 - x$ [2.3C]

SECTION 3.1

3. Yes   **5.** No   **7.** No   **9.** Yes   **11.** No   **13.** Yes   **15.** No   **17.** Yes   **19.** No   **23.** 2   **25.** 15
27. 6   **29.** 3   **31.** 0   **33.** -7 **35.** -7 **37.** -12 **39.** -5 **41.** 15   **43.** 9   **45.** 14   **47.** -1

49. 1   **51.** $-\frac{1}{2}$ **53.** $-\frac{3}{4}$ **55.** $\frac{1}{12}$ **57.** $-\frac{7}{12}$ **59.** 0.6529   **61.** -0.283 **63.** 9.257   **67.** -3 **69.** 0

71. -2 **73.** 9   **75.** 80   **77.** 0   **79.** -7 **81.** 12   **83.** -18 **85.** 15   **87.** -20 **89.** 0   **91.** $\frac{8}{3}$

93. $\frac{1}{3}$ **95.** $-\frac{1}{2}$ **97.** $-\frac{3}{2}$ **99.** $\frac{15}{7}$ **101.** 4   **103.** 3   **105.** 4.745   **107.** 2.06   **109.** -2.13

111. Equal to   **113.** 28   **115.** 0.72   **117.** 64   **119.** 24%   **121.** 7.2   **123.** 400   **125.** 9   **127.** 25%
129. 200%   **131.** 400   **133.** 7.7   **135.** 200   **137.** 400   **139.** 30   **141.** There are 4536 L of oxygen in the
room.   **143.** The median income was $42,428.   **145.** 47.9% of the U.S. population watched Super Bowl XXXVIII.
147. You need to know the number of people three years old and older in the U.S.   **149.** Andrea must invest $1875.
151. Octavia will earn the greater amount of interest.   **153.** $1500 was invested at 8%.   **155.** There are 1.8 g of
platinum in the necklace.   **157.** There are 131.25 lb of wool in the carpet.   **159.** The percent concentration of sugar
is 50%.   **161.** The percent concentration of the resulting mixture is 6%.   **163.** The runner will travel 3 mi.
165. Marcella's average rate of speed is 36 mph.   **167.** It would take Palmer 2.5 h to walk the course.
169. The two joggers will meet 40 min after they start.   **171.** It will take them 0.5 h.   **173.** 40°   **175.** 10°
177. a. Answers will vary.   **b.** Answers will vary.

SECTION 3.2

1. 3   **3.** 6   **5.** -1 **7.** -3 **9.** 2   **11.** 2   **13.** 5   **15.** -3 **17.** 6   **19.** 3   **21.** 1   **23.** 6   **25.** -7

27. 0   **29.** $\frac{3}{4}$ **31.** $\frac{4}{9}$ **33.** $\frac{1}{3}$ **35.** $-\frac{1}{2}$ **37.** $-\frac{3}{4}$ **39.** $\frac{1}{3}$ **41.** $-\frac{1}{6}$ **43.** 1   **45.** 1   **47.** 0   **49.** $\frac{13}{10}$

51. $\frac{2}{5}$ **53.** $-\frac{4}{3}$ **55.** $-\frac{3}{2}$ **57.** 18   **59.** 8   **61.** -16 **63.** 25   **65.** $\frac{3}{4}$ **67.** $\frac{3}{8}$ **69.** $\frac{16}{9}$ **71.** $\frac{1}{18}$

73. $\frac{15}{2}$ **75.** $-\frac{18}{5}$ **77.** 2   **79.** 3   **81.** $x = 7$ **83.** $y = 3$ **85.** 19   **87.** -1 **89.** -11 **91.** The markup
rate is 60%.   **93.** The cost of the basketball is $59.   **95.** The markup rate is 44.4%.   **97.** The cost of the CD is
$8.50.   **99.** The discount rate is 23.2%.   **101.** The regular price of the tool set was $300.   **103.** The markdown rate
is 38%.   **105.** The regular price of the telescope is $275.   **107.** The initial velocity is 8 ft/s.   **109.** The depreciated
value will be $38,000 after 2 years.   **111.** The approximate length is 31.8 in.   **113.** The distance the car will skid is
168 ft.   **115.** The estimated population is 51,000 people.   **117.** $a = 7$ **119.** The regular price is $317.65.
121. 6 m   **123.** 385

SECTION 3.3

1. 2 **3.** 3 **5.** −1 **7.** 2 **9.** −2 **11.** −3 **13.** 0 **15.** −1 **17.** −3 **19.** −1 **21.** 4 **23.** $\frac{2}{3}$

25. $\frac{5}{6}$ **27.** $\frac{3}{4}$ **29.** −17 **31.** 41 **33.** 8 **35.** 1 **37.** 4 **39.** −1 **41.** −1 **43.** 24 **45.** 495

47. $\frac{1}{2}$ **49.** $-\frac{1}{3}$ **51.** $\frac{10}{3}$ **53.** $-\frac{1}{4}$ **55.** 0 **57.** −1 **59.** A force of 25 lb must be applied to the other end.

61. The fulcrum is 6 ft from the 180-pound person. **63.** The fulcrum is 10 ft from the 128-pound acrobat.

65. The minimum force to move the rock is 34.6 lb. **67.** The break-even point is 260 barbecues. **69.** The break-even point is 520 desk lamps. **71.** The break-even point is 3000 softball bats. **73.** No solution **75.** 0

SECTION 3.4

1. $x - 15 = 7$; 22 **3.** $7x = -21$; −3 **5.** $9 - x = 7$; 2 **7.** $5 - 2x = 1$; 2 **9.** $2x + 5 = 15$; 5 **11.** $4x - 6 = 22$; 7
13. $3(4x - 7) = 15$; 3 **15.** $3x = 2(20 - x)$; 8, 12 **17.** $2x - (14 - x) = 1$; 5, 9 **19.** 15, 17, 19 **21.** −1, 1, 3
23. 4, 6 **25.** 5, 7 **27.** The processor speed of the newer personal computer is 4.2$\overline{6}$ GHz. **29.** The lengths of the sides are 6 ft, 6 ft, and 11 ft. **31.** The union member worked 168 h during March. **33.** 37 h of labor was required to paint the house. **35.** There are 1024 vertical pixels. **37.** The length is 13 m; the width is 8 m. **39.** The shorter piece is 3 ft; the longer piece is 9 ft. **41.** The larger scholarship is $5000.

SECTION 3.5

1. 116° **3.** 20° **5.** 20° **7.** 20° **9.** 106° **11.** 11° **13.** $m\angle a$ is 38°; $m\angle b$ is 142° **15.** $m\angle a$ is 47°;
$m\angle b$ is 133° **17.** 20° **19.** 47° **21.** 141° **23.** $m\angle x$ is 155°; $m\angle y$ is 70° **25.** $m\angle a$ is 45°; $m\angle b$ is 135°
27. $90° - x$ **29.** 60° **31.** 35° **33.** 102° **35.** 120° **37.** 360°

SECTION 3.6

1. The amount of $1 herbs is 20 oz. **3.** The mixture will cost $1.84 per pound. **5.** The amount of caramel is 3 lb.
7. The amount of olive oil is 2 c; the amount of vinegar is 8 c. **9.** The cost of the mixture is $3.00 per ounce.
11. To make the mixture, 16 oz of the alloy are needed. **13.** The amount of almonds is 37 lb; the amount of walnuts is 63 lb. **15.** There were 228 adult tickets sold. **17.** The cost per pound of the sugar-coated cereal is $.70. **19.** The percent concentration of gold in the mixture is 24%. **21.** The amount of the 15% acid is 20 gal. **23.** The amount of the 25% wool yarn is 30 lb. **25.** The amount of 9% nitrogen plant food is 6.25 gal. **27.** The percent concentration of sugar in the mixture is 19%. **29.** 20 lb of 40% java bean coffee must be used. **31.** The amount of the 7% solution is 100 ml; the amount of the 4% solution is 200 ml. **33.** 150 oz of pure chocolate must be added. **35.** The percent concentration of the resulting alloy is 50%. **37.** The first plane is traveling at a rate of 105 mph; the second plane is traveling at a rate of 130 mph. **39.** The planes will be 3000 km apart at 11 A.M. **41.** In 2 h the cabin cruiser will be alongside the motorboat. **43.** The corporate offices are 120 mi from the airport. **45.** The rate of the car is 68 mph.
47. The distance between the airports was 300 mi. **49.** The planes will pass each other in 2.5 h after the plane leaves Seattle. **51.** The cyclists will meet after 1.5 h. **53.** The bus overtakes the car in 180 mi. **55.** 75 g of pure water must be added. **57.** 3.75 gal of 20% antifreeze must be drained. **59.** The bicyclist's average speed is $13\frac{1}{3}$ mph.

CHAPTER 3 REVIEW EXERCISES

1. 21 [3.1B] **2.** 10 [3.3B] **3.** 7 [3.2A] **4.** No [3.1A] **5.** 20 [3.1C] **6.** −2 [3.3B]
7. 250% [3.1D] **8.** 4 [3.3A] **9.** −1 [3.3B] **10.** 4 [3.3A] **11.** $671.25 [3.2B] **12.** 35° [3.5A]
13. 26° [3.5A] **14.** The force is 24 lb. [3.3C] **15.** The average speed on the winding road was 32 mph. [3.6C]
16. The discount rate is $33\frac{1}{3}$%. [3.2B] **17.** $m\angle x = 22°$, $m\angle y = 158°$ [3.5B] **18.** The amount of cranberry juice is 7 qt; the amount of apple juice is 3 qt. [3.6A] **19.** The three integers are −1, 0, 1. [3.4A]

20. The angles measure 75°, 60°, and 45°. [3.5B] **21.** $5n - 4 = 16$; 4 [3.4A] **22.** The height of the Eiffel Tower is 1063 ft. [3.4B] **23.** 25° [3.5A] **24.** 60° [3.5A] **25.** The jet overtakes the propeller-driven plane 600 mi from the starting point. [3.6C] **26.** The numbers are 8 and 13. [3.4A] **27.** The mixture is 14% butterfat. [3.6B]

CHAPTER 3 TEST

1. -5 [3.3A] **2.** -5 [3.1B] **3.** -3 [3.2A] **4.** 2 [3.3B] **5.** No [3.1A] **6.** 5 [3.2A] **7.** 0.04 [3.1D]
8. $-\dfrac{1}{3}$ [3.3B] **9.** 2 [3.3A] **10.** -12 [3.1C] **11.** The amount of rye is 10 lb; the amount of wheat is 5 lb. [3.6A]
12. 19° [3.5A] **13.** The discount rate is 20%. [3.2B] **14.** 200 calculators were produced. [3.2B]
15. The measure of one of the equal angles is 70°. [3.5B] **16.** The numbers are 10, 12, and 14. [3.4A] **17.** 1.25 gal of water must be added. [3.6B] **18.** $m\angle a = 138°$, $m\angle b = 42°$ [3.5A] **19.** $3x - 15 = 27$; 14 [3.4A] **20.** The rate of the snowmobile was 6 mph. [3.6C] **21.** The company makes 110 25-inch TVs each day. [3.4B]
22. The smaller number is 8; the larger number is 10. [3.4A] **23.** The distance between the airports is 360 mi. [3.6C]
24. $m\angle x = 138°$; $m\angle y = 130°$ [3.5B] **25.** The final temperature is 60°C. [3.3C]

CUMULATIVE REVIEW EXERCISES

1. 6 [1.2B] **2.** -48 [1.3A] **3.** $-\dfrac{19}{48}$ [1.6C] **4.** -2 [1.7B] **5.** 54 [1.7A] **6.** 24 [1.4B]
7. 6 [2.1A] **8.** $-17x$ [2.2A] **9.** $-5a - 2b$ [2.2A] **10.** $2x$ [2.2B] **11.** $36y$ [2.2B]
12. $2x^2 + 6x - 4$ [2.2C] **13.** $-4x + 14$ [2.2D] **14.** $6x - 34$ [2.2D] **15.** Yes [3.1A] **16.** No [3.1A]
17. 19.2 [3.1D] **18.** -25 [3.1C] **19.** -3 [3.2A] **20.** 3 [3.2A] **21.** 13 [3.3B] **22.** 2 [3.3B]
23. -3 [3.3A] **24.** $\dfrac{1}{2}$ [3.3A] **25.** The final temperature is 60°C. [3.3C] **26.** $12 - 5x = -18$; 6 [3.4A]
27. The area of the garage is 600 ft². [3.4B] **28.** 20 lb of oat flour are needed for the mixture. [3.6A] **29.** 25 g of pure gold must be added. [3.6B] **30.** The length is 12 ft; the width is 10 ft. [1.8B] **31.** 131° [3.5A]
32. The measure of one of the equal angles is 60°. [3.5B] **33.** The length of the track is 120 m. [3.6C]

Answers to Chapter 4 Selected Exercises

PREP TEST

1. 1 [1.2B] **2.** -18 [1.3A] **3.** $\dfrac{2}{3}$ [1.6A] **4.** 48 [2.1A] **5.** 0 [1.6A] **6.** No [2.2A]
7. $5x^2 - 9x - 6$ [2.2A] **8.** 0 [2.2A] **9.** $-6x + 24$ [2.2C] **10.** $-7xy + 10y$ [2.2D]

SECTION 4.1

1. Yes **3.** No **5.** Yes **7.** Yes **9.** Binomial **11.** Trinomial **13.** None of these **15.** Binomial
17. $-2x^2 + 3x$ **19.** $y^2 - 8$ **21.** $5x^2 + 7x + 20$ **23.** $x^3 + 2x^2 - 6x - 6$ **25.** $2a^3 - 3a^2 - 11a + 2$ **27.** $5x^2 + 8x$
29. $7x^2 + xy - 4y^2$ **31.** $3a^2 - 3a + 17$ **33.** $5x^3 + 10x^2 - x - 4$ **35.** $3r^3 + 2r^2 - 11r + 7$ **37.** $4x$
39. $3y^2 - 4y - 2$ **41.** $-7x - 7$ **43.** $4x^3 + 3x^2 + 3x + 1$ **45.** $y^3 + 5y^2 - 2y - 4$ **47.** $-y^2 - 13xy$
49. $2x^2 - 3x - 1$ **51.** $-2x^3 + x^2 + 2$ **53.** $3a^3 - 2$ **55.** $4y^3 + 2y^2 + 2y - 4$ **57.** $x^2 + 9x - 11$

SECTION 4.2

3. $30x^3$ **5.** $-42c^6$ **7.** $9a^7$ **9.** x^3y^4 **11.** $-10x^9y$ **13.** $12x^7y^8$ **15.** $-6x^3y^5$ **17.** x^4y^5z **19.** $a^3b^5c^4$
21. $-30a^5b^8$ **23.** $6a^5b$ **25.** $40y^{10}z^6$ **27.** $x^3y^3z^2$ **29.** $-24a^3b^3c^3$ **31.** $8x^7yz^6$ **33.** $30x^6y^8$ **35.** $-36a^3b^2c^3$
37. x^{15} **39.** x^{14} **41.** x^8 **43.** y^{12} **45.** $-8x^6$ **47.** x^4y^6 **49.** $9x^4y^2$ **51.** $-243x^{15}y^{10}$ **53.** $-8x^7$
55. $24x^8y^7$ **57.** a^4b^6 **59.** $64x^{12}y^3$ **61.** $-18x^3y^4$ **63.** $-8a^7b^5$ **65.** $-54a^9b^3$ **67.** $12x^2$ **69.** $2x^6y^2 + 9x^4y^2$

71. 0   **73.** $17x^4y^8$ **75.** True   **77.** False. $(x^2)^5 = x^{2 \cdot 5} = x^{10}$ **79.** No. $2^{(3^2)}$ is larger.

SECTION 4.3

1. 3, 3, $12x - 15$ **3.** $x^2 - 2x$ **5.** $-x^2 - 7x$ **7.** $3a^3 - 6a^2$ **9.** $-5x^4 + 5x^3$ **11.** $-3x^5 + 7x^3$ **13.** $12x^3 - 6x^2$
15. $6x^2 - 12x$ **17.** $3x^2 + 4x$ **19.** $-x^3y + xy^3$ **21.** $2x^4 - 3x^2 + 2x$ **23.** $2a^3 + 3a^2 + 2a$ **25.** $3x^6 - 3x^4 - 2x^2$
27. $-6y^4 - 12y^3 + 14y^2$ **29.** $-2a^3 - 6a^2 + 8a$ **31.** $6y^4 - 3y^3 + 6y^2$ **33.** $x^3y - 3x^2y^2 + xy^3$ **35.** $x^3 + 4x^2 + 5x + 2$
37. $a^3 - 6a^2 + 13a - 12$ **39.** $-2b^3 + 7b^2 + 19b - 20$ **41.** $-6x^3 + 31x^2 - 41x + 10$ **43.** $x^3 - 3x^2 + 5x - 15$
45. $x^4 - 4x^3 - 3x^2 + 14x - 8$ **47.** $15y^3 - 16y^2 - 70y + 16$ **49.** $5a^4 - 20a^3 - 5a^2 + 22a - 8$
51. $y^4 + 4y^3 + y^2 - 5y + 2$ **53.** $x^2 + 4x + 3$ **55.** $a^2 + a - 12$ **57.** $y^2 - 5y - 24$ **59.** $y^2 - 10y + 21$
61. $2x^2 + 15x + 7$ **63.** $3x^2 + 11x - 4$ **65.** $4x^2 - 31x + 21$ **67.** $3y^2 - 2y - 16$ **69.** $9x^2 + 54x + 77$
71. $21a^2 - 83a + 80$ **73.** $6a^2 - 25ab + 14b^2$ **75.** $2a^2 - 11ab - 63b^2$ **77.** $100a^2 - 100ab + 21b^2$
79. $15x^2 + 56xy + 48y^2$ **81.** $14x^2 - 97xy - 60y^2$ **83.** $56x^2 - 61xy + 15y^2$ **85.** $y^2 - 25$ **87.** $4x^2 - 9$
89. $9x^2 - 49$ **91.** $16 - 9y^2$ **93.** $x^2 + 2x + 1$ **95.** $9a^2 - 30a + 25$ **97.** $4a^2 + 4ab + b^2$ **99.** $x^2 - 4xy + 4y^2$
101. $25x^2 + 20xy + 4y^2$ **103.** The area of the rectangle is $(10x^2 - 35x)$ ft^2.   **105.** The area is $(4x^2 + 4x + 1)$ km^2.
107. The area of the triangle is $(4x^2 + 10x)$ m^2.   **109.** The area is $(60w + 3000)$ yd^2.   **111.** $x^4 + 2x^3 - 5x^2 - 6x + 9$
113. $12x^2 - x - 20$ **115.** $x^3 - 7x^2 - 7$

SECTION 4.4

1. 7, 5, 2   **3.** $\dfrac{1}{25}$ **5.** 64   **7.** $\dfrac{1}{27}$ **9.** 2   **11.** $\dfrac{1}{x^2}$ **13.** a^6 **15.** $\dfrac{4}{x^7}$ **17.** $\dfrac{2}{3z^2}$ **19.** $5b^8$ **21.** $\dfrac{x^2}{3}$ **23.** 1
25. -1 **27.** y^4 **29.** a^3 **31.** p^4 **33.** $2x^3$ **35.** $2k$ **37.** m^5n^2 **39.** $\dfrac{3r^2}{2}$ **41.** $-\dfrac{2a}{3}$ **43.** $\dfrac{1}{y^5}$ **45.** $\dfrac{1}{a^6}$
47. $\dfrac{1}{3x^3}$ **49.** $\dfrac{2}{3x^5}$ **51.** $\dfrac{y^4}{x^2}$ **53.** $\dfrac{2}{5m^3n^8}$ **55.** $\dfrac{1}{p^3q}$ **57.** $\dfrac{1}{2y^3}$ **59.** $\dfrac{7xz}{8y^3}$ **61.** $\dfrac{p^2}{2m^3}$ **63.** $-\dfrac{8x^3}{y^6}$ **65.** $\dfrac{9}{x^2y^4}$
67. $\dfrac{2}{x^4}$ **69.** $-\dfrac{5}{a^8}$ **71.** $-\dfrac{a^5}{8b^4}$ **73.** $\dfrac{10y^3}{x^4}$ **75.** $\dfrac{1}{2x^3}$ **77.** $\dfrac{3}{x^3}$ **79.** $\dfrac{1}{2x^2y^6}$ **81.** $\dfrac{1}{x^6y}$ **83.** $\dfrac{a^4}{y^{10}}$ **85.** $-\dfrac{1}{6x^3}$
87. $-\dfrac{a^2b}{6c^2}$ **89.** $-\dfrac{7b^6}{a^2}$ **91.** $\dfrac{s^8t^4}{4r^{12}}$ **93.** $\dfrac{125p^3}{27m^{15}n^6}$ **97.** 3.24×10^{-9} **99.** 3×10^{-18} **101.** 3.2×10^{16}
103. 1.22×10^{-19} **105.** 5.47×10^8 **107.** 0.000167   **109.** 68,000,000   **111.** 0.0000305   **113.** 0.00000000102
115. 6.023×10^{23} **117.** 3.7×10^{-6} **119.** 1×10^{-9} **121.** 1.6×10^{-19} **123.** $\dfrac{1}{4}, \dfrac{1}{2}, 1, 2, 4$ **125.** $4, 2, 1, \dfrac{1}{2}, \dfrac{1}{4}$
127. False. $(2a)^{-3} = \dfrac{1}{8a^3}$ **129.** False. $(2 + 3)^{-1} = (5)^{-1} = \dfrac{1}{5}$

SECTION 4.5

1. $2a - 5$ **3.** $6y + 4$ **5.** $x - 2$ **7.** $-x + 2$ **9.** $x^2 + 3x - 5$ **11.** $x^4 - 3x^2 - 1$ **13.** $xy + 2$ **15.** $-3y^3 + 5$
17. $3x - 2 + \dfrac{1}{x}$ **19.** $-3x + 7 - \dfrac{6}{x}$ **21.** $4a - 5 + 6b$ **23.** $9x + 6 - 3y$ **25.** $(x + 2), (x - 3)$ **27.** $b - 7$
29. $y - 5$ **31.** $2y - 7$ **33.** $2y + 6 + \dfrac{25}{y - 3}$ **35.** $x - 2 + \dfrac{8}{x + 2}$ **37.** $3y - 5 + \dfrac{20}{2y + 4}$ **39.** $6x - 12 + \dfrac{19}{x + 2}$
41. $b - 5 - \dfrac{24}{b - 3}$ **43.** $3x + 17 + \dfrac{64}{x - 4}$ **45.** $5y + 3 + \dfrac{1}{2y + 3}$ **47.** $4a + 1$ **49.** $2a + 9 + \dfrac{33}{3a - 1}$
51. $x^2 - 5x + 2$ **53.** $x^2 + 5$ **55.** $3ab$

CHAPTER 4 REVIEW EXERCISES

1. $8b^2 - 2b - 15$ [4.3C]   **2.** $21y^2 + 4y - 1$ [4.1A]   **3.** $x^4y^8z^4$ [4.2A]   **4.** $\dfrac{2x^3}{3}$ [4.4A]

5. $-8x^3 - 14x^2 + 18x$ [4.3A]   **6.** $-\dfrac{1}{2a}$ [4.4A]   **7.** $16u^{12}v^{16}$ [4.2B]   **8.** 64 [4.2B]   **9.** $2x^2 + 3x - 8$ [4.1B]

10. $\dfrac{b^6}{a^4}$ [4.4A]   **11.** $-108x^{18}$ [4.2B]   **12.** $25y^2 - 70y + 49$ [4.3D]   **13.** $100a^{15}b^{13}$ [4.2B]

14. $4b^4 + 12b^2 - 1$ [4.5A]   **15.** $-\dfrac{1}{16}$ [4.4A]   **16.** $13y^3 - 12y^2 - 5y - 1$ [4.1B]   **17.** $-x + 2 + \dfrac{1}{x + 3}$ [4.5B]

18. $2ax - 4ay - bx + 2by$ [4.3C]   **19.** $6y^3 + 17y^2 - 2y - 21$ [4.3B]   **20.** $b^2 + 5b + 2 + \dfrac{7}{b - 7}$ [4.5B]

21. $8a^3b^3 - 4a^2b^4 + 6ab^5$ [4.3A]   **22.** $4a^2 - 25b^2$ [4.3D]   **23.** $12b^5 - 4b^4 - 6b^3 - 8b^2 + 5$ [4.3B]

24. $2x^3 + 9x^2 - 3x - 12$ [4.1A]   **25.** $-4y + 8$ [4.5A]   **26.** $a^2 - 49$ [4.3D]   **27.** 3.756×10^{10} [4.4B]

28. $14{,}600{,}000$ [4.4B]   **29.** $-54a^{13}b^5c^7$ [4.2A]   **30.** $2y - 9$ [4.5B]   **31.** $\dfrac{x^4y^6}{9}$ [4.4A]

32. $10a^2 + 31a - 63$ [4.3C]   **33.** 1.27×10^{-7} [4.4B]   **34.** 0.0000000000032 [4.4B]

35. The area is $(2w^2 - w)$ ft². [4.3E]   **36.** The area is $(9x^2 - 12x + 4)$ in². [4.3E]

CHAPTER 4 TEST

1. $4x^3 - 6x^2$ [4.3A]   **2.** $4x - 1 + \dfrac{3}{x^2}$ [4.5A]   **3.** $-\dfrac{4}{x^6}$ [4.4A]   **4.** $-6x^3y^6$ [4.2A]   **5.** $x - 1 + \dfrac{2}{x + 1}$ [4.5B]

6. $x^3 - 7x^2 + 17x - 15$ [4.3B]   **7.** $-8a^6b^3$ [4.2B]   **8.** $\dfrac{9y^{10}}{x^{10}}$ [4.4A]   **9.** $a^2 + 3ab - 10b^2$ [4.3C]

10. $4x^4 - 2x^2 + 5$ [4.5A]   **11.** $x + 7$ [4.5B]   **12.** $6y^4 - 9y^3 + 18y^2$ [4.3A]   **13.** $-4x^4 + 8x^3 - 3x^2 - 14x + 21$
[4.3B]   **14.** $16y^2 - 9$ [4.3D]   **15.** a^4b^7 [4.2A]   **16.** $8ab^4$ [4.4A]   **17.** $4a - 7$ [4.5A]
18. $-5a^3 + 3a^2 - 4a + 3$ [4.1B]   **19.** $4x^2 - 20x + 25$ [4.3D]   **20.** $2x + 3 + \dfrac{2}{2x - 3}$ [4.5B]
21. $-2x^3$ [4.4A]   **22.** $10x^2 - 43xy + 28y^2$ [4.3C]   **23.** $3x^3 + 6x^2 - 8x + 3$ [4.1A]   **24.** 3.02×10^{-9} [4.4B]
25. The area of the circle is $(\pi x^2 - 10\pi x + 25\pi)$ m². [4.3E]

CUMULATIVE REVIEW EXERCISES

1. $\dfrac{5}{144}$ [1.6C]   **2.** $\dfrac{5}{3}$ [1.7A]   **3.** $\dfrac{25}{11}$ [1.7B]   **4.** $-\dfrac{22}{9}$ [2.1A]   **5.** $5x - 3xy$ [2.2A]   **6.** $-9x$ [2.2B]

7. $-18x + 12$ [2.2D]   **8.** -16 [3.1C]   **9.** -16 [3.3A]   **10.** 15 [3.3B]   **11.** 22% [3.1D]

12. $4b^3 - 4b^2 - 8b - 4$ [4.1A]   **13.** $3y^3 + 2y^2 - 10y$ [4.1B]   **14.** a^9b^{15} [4.2B]   **15.** $-8x^3y^6$ [4.2A]

16. $6y^4 + 8y^3 - 16y^2$ [4.3A]   **17.** $10a^3 - 39a^2 + 20a - 21$ [4.3B]   **18.** $15b^2 - 31b + 14$ [4.3C]   **19.** $\dfrac{1}{2b^2}$ [4.4A]

20. $a - 7$ [4.5B]   **21.** 0.0000609 [4.4B]   **22.** $8x - 2x = 18; 3$ [3.4B]   **23.** The percent concentration of orange
juice in the mixture is 28%. [3.6B]   **24.** The car overtakes the cyclist 25 mi from the starting point. [3.6C]
25. The length is 15 m and the width is 6 m. [3.1D]

Answers to Chapter 5 Selected Exercises

PREP TEST

1. $2 \cdot 3 \cdot 5$ [1.5B]   **2.** $-12y + 15$ [2.2C]   **3.** $-a + b$ [2.2C]   **4.** $-3a + 3b$ [2.2D]   **5.** 0 [3.1C]

6. $-\dfrac{1}{2}$ [3.2A]   **7.** $x^2 - 2x - 24$ [4.3C]   **8.** $6x^2 - 11x - 10$ [4.3C]   **9.** x^3 [4.4A]   **10.** $3x^3y$ [4.4A]

SECTION 5.1

3. $5(a + 1)$ **5.** $8(2 - a^2)$ **7.** $4(2x + 3)$ **9.** $6(5a - 1)$ **11.** $x(7x - 3)$ **13.** $a^2(3 + 5a^3)$ **15.** $y(14y + 11)$
17. $2x(x^3 - 2)$ **19.** $2x^2(5x^2 - 6)$ **21.** $4a^5(2a^3 - 1)$ **23.** $xy(xy - 1)$ **25.** $3xy(xy^3 - 2)$ **27.** $xy(x - y^2)$
29. $5y(y^2 - 4y + 1)$ **31.** $3y^2(y^2 - 3y - 2)$ **33.** $3y(y^2 - 3y + 8)$ **35.** $a^2(6a^3 - 3a - 2)$ **37.** $ab(2a - 5ab + 7b)$
39. $2b(2b^4 + 3b^2 - 6)$ **41.** $x^2(8y^2 - 4y + 1)$ **43.** $(a + z)(y + 7)$ **45.** $(a - b)(3r + s)$ **47.** $(m - 7)(t - 7)$
49. $(4a - b)(2y + 1)$ **51.** $(x + 2)(x + 2y)$ **53.** $(p - 2)(p - 3r)$ **55.** $(a + 6)(b - 4)$ **57.** $(2z - 1)(z + y)$

59. $(2v - 3y)(4v + 7)$ **61.** $(2x - 5)(x - 3y)$ **63.** $(y - 2)(3y - a)$ **65.** $(3x - y)(y + 1)$ **67.** $(3s + t)(t - 2)$
69. a. 28 **b.** 496 **71. a.** $r^2(\pi - 2)$ **b.** $2r^2(4 - \pi)$ **c.** $r^2(4 - \pi)$

SECTION 5.2

1. The same   **3.** $(x + 1)(x + 2)$ **5.** $(x + 1)(x - 2)$ **7.** $(a + 4)(a - 3)$ **9.** $(a - 1)(a - 2)$ **11.** $(a + 2)(a - 1)$
13. $(b - 3)(b - 3)$ **15.** $(b + 8)(b - 1)$ **17.** $(y + 11)(y - 5)$ **19.** $(y - 2)(y - 3)$ **21.** $(z - 5)(z - 9)$
23. $(z + 8)(z - 20)$ **25.** $(p + 3)(p + 9)$ **27.** $(x + 10)(x + 10)$ **29.** $(b + 4)(b + 5)$ **31.** $(x + 3)(x - 14)$
33. $(b + 4)(b - 5)$ **35.** $(y + 3)(y - 17)$ **37.** $(p + 3)(p - 7)$ **39.** Nonfactorable over the integers
41. $(x - 5)(x - 15)$ **43.** $(p + 3)(p + 21)$ **45.** $(x + 2)(x + 19)$ **47.** $(x + 9)(x - 4)$ **49.** $(a + 4)(a - 11)$
51. $(a - 3)(a - 18)$ **53.** $(z + 21)(z - 7)$ **55.** $(c + 12)(c - 15)$ **57.** $(p + 9)(p + 15)$ **59.** $(c + 2)(c + 9)$
61. $(x + 15)(x - 5)$ **63.** $(x + 25)(x - 4)$ **65.** $(b - 4)(b - 18)$ **67.** $(a + 45)(a - 3)$ **69.** $(b - 7)(b - 18)$
71. $(z + 12)(z + 12)$ **73.** $(x - 4)(x - 25)$ **75.** $(x + 16)(x - 7)$ **77.** $3(x + 2)(x + 3)$ **79.** $-(x - 2)(x + 6)$
81. $a(b + 8)(b - 1)$ **83.** $x(y + 3)(y + 5)$ **85.** $-2a(a + 1)(a + 2)$ **87.** $4y(y + 6)(y - 3)$ **89.** $2x(x^2 - x + 2)$
91. $6(z + 5)(z - 3)$ **93.** $3a(a + 3)(a - 6)$ **95.** $(x + 7y)(x - 3y)$ **97.** $(a - 5b)(a - 10b)$ **99.** $(s + 8t)(s - 6t)$
101. Nonfactorable over the integers   **103.** $z^2(z + 10)(z - 8)$ **105.** $b^2(b + 2)(b - 5)$ **107.** $3y^2(y + 3)(y + 15)$
109. $-x^2(x + 1)(x - 12)$ **111.** $3y(x + 3)(x - 5)$ **113.** $-3x(x - 3)(x - 9)$ **115.** $(x - 3y)(x - 5y)$
117. $(a - 6b)(a - 7b)$ **119.** $(y + z)(y + 7z)$ **121.** $3y(x + 21)(x - 1)$ **123.** $3x(x + 4)(x - 3)$
125. $4z(z + 11)(z - 3)$ **127.** $4x(x + 3)(x - 1)$ **129.** $5(p + 12)(p - 7)$ **131.** $p^2(p + 12)(p - 3)$
133. $(t - 5s)(t - 7s)$ **135.** $(a + 3b)(a - 11b)$ **137.** $y(x + 6)(x - 9)$ **139.** $-36, 36, -12, 12$
141. $22, -22, 10, -10$ **143.** $6, 10, 12$ **145.** $6, 10, 12$ **147.** $4, 6$

SECTION 5.3

1. $(x + 1)(2x + 1)$ **3.** $(y + 3)(2y + 1)$ **5.** $(a - 1)(2a - 1)$ **7.** $(b - 5)(2b - 1)$ **9.** $(x + 1)(2x - 1)$
11. $(x - 3)(2x + 1)$ **13.** $(t + 2)(2t - 5)$ **15.** $(p - 5)(3p - 1)$ **17.** $(3y - 1)(4y - 1)$ **19.** Nonfactorable over
the integers   **21.** $(2t - 1)(3t - 4)$ **23.** $(x + 4)(8x + 1)$ **25.** Nonfactorable over the integers
27. $(3y + 1)(4y + 5)$ **29.** $(a + 7)(7a - 2)$ **31.** $(b - 4)(3b - 4)$ **33.** $(z - 14)(2z + 1)$ **35.** $(p + 8)(3p - 2)$
37. $2(x + 1)(2x + 1)$ **39.** $5(y - 1)(3y - 7)$ **41.** $x(x - 5)(2x - 1)$ **43.** $b(a - 4)(3a - 4)$ **45.** Nonfactorable
over the integers   **47.** $-3x(x + 4)(x - 3)$ **49.** $4(4y - 1)(5y - 1)$ **51.** $z(2z + 3)(4z + 1)$ **53.** $y(2x - 5)(3x + 2)$
55. $5(t + 2)(2t - 5)$ **57.** $p(p - 5)(3p - 1)$ **59.** $2(z + 4)(13z - 3)$ **61.** $2y(y - 4)(5y - 2)$
63. $yz(z + 2)(4z - 3)$ **65.** $3a(2a + 3)(7a - 3)$ **67.** $y(3x - 5y)(3x - 5y)$ **69.** $xy(3x - 4y)(3x - 4y)$
71. $(2x - 3)(3x - 4)$ **73.** $(b + 7)(5b - 2)$ **75.** $(3a + 8)(2a - 3)$ **77.** $(z + 2)(4z + 3)$ **79.** $(2p + 5)(11p - 2)$
81. $(y + 1)(8y + 9)$ **83.** $(6t - 5)(3t + 1)$ **85.** $(b + 12)(6b - 1)$ **87.** $(3x + 2)(3x + 2)$ **89.** $(2b - 3)(3b - 2)$
91. $(3b + 5)(11b - 7)$ **93.** $(3y - 4)(6y - 5)$ **95.** $(3a + 7)(5a - 3)$ **97.** $(2y - 5)(4y - 3)$ **99.** $(2z + 3)(4z - 5)$
101. Nonfactorable over the integers   **103.** $(2z - 5)(5z - 2)$ **105.** $(6z + 5)(6z + 7)$ **107.** $(x + y)(3x - 2y)$
109. $(a + 2b)(3a - b)$ **111.** $(y - 2z)(4y - 3z)$ **113.** $-(z - 7)(z + 4)$ **115.** $-(x - 1)(x + 8)$
117. $3(x + 5)(3x - 4)$ **119.** $4(2x - 3)(3x - 2)$ **121.** $a^2(5a + 2)(7a - 1)$ **123.** $5(b - 7)(3b - 2)$
125. $(x - 7y)(3x - 5y)$ **127.** $3(8y - 1)(9y + 1)$ **129.** $-(x - 1)(x + 21)$ **133.** $x(x - 1)$ **135.** $(2y + 1)(y + 3)$
137. $(4y - 3)(y - 3)$ **139.** $-5, 5, -1, 1$ **141.** $-5, 5, -1, 1$ **143.** $-9, 9, -3, 3$

SECTION 5.4

1. a. Answers will vary. For instance, $x^2 - 25$.   **b.** Answers will vary. For instance, $x^2 + 6x + 9$.   **3.** $(x + 2)(x - 2)$
5. $(a + 9)(a - 9)$ **7.** $(y + 1)^2$ **9.** $(a - 1)^2$ **11.** $(2x + 1)(2x - 1)$ **13.** $(x^3 + 3)(x^3 - 3)$ **15.** Nonfactorable over
the integers   **17.** $(x + y)^2$ **19.** $(2a + 1)^2$ **21.** $(3x + 1)(3x - 1)$ **23.** $(1 + 8x)(1 - 8x)$ **25.** Nonfactorable over
the integers   **27.** $(3a + 1)^2$ **29.** $(b^2 + 4a)(b^2 - 4a)$ **31.** $(2a - 5)^2$ **33.** $(3a - 7)^2$ **35.** $(5z + y)(5z - y)$
37. $(ab + 5)(ab - 5)$ **39.** $(5x + 1)(5x - 1)$ **41.** $(2a - 3b)^2$ **43.** $(2y - 9z)^2$ **45.** $\left(\dfrac{1}{x} + 2\right)\left(\dfrac{1}{x} - 2\right)$
47. $(3ab - 1)^2$ **49.** $2(2y + 1)(2y - 1)$ **51.** $3a(a + 1)^2$ **53.** $(m^2 + 16)(m + 4)(m - 4)$ **55.** $(x + 1)(9x + 4)$
57. $4y^2(2y + 3)^2$ **59.** $(y^4 + 9)(y^2 + 3)(y^2 - 3)$ **61.** $(5 - 2p)^2$ **63.** $(4x - 3 + y)(4x - 3 - y)$
65. $(x - 2 + y)(x - 2 - y)$ **67.** $5(x + 1)(x - 1)$ **69.** $x(x + 2)^2$ **71.** $x^2(x + 7)(x - 5)$ **73.** $5(b + 3)(b + 12)$
75. Nonfactorable over the integers   **77.** $2y(x + 11)(x - 3)$ **79.** $x(x^2 - 6x - 5)$ **81.** $3(y^2 - 12)$

83. $(2a + 1)(10a + 1)$ **85.** $y^2(x + 1)(x - 8)$ **87.** $5(a + b)(2a - 3b)$ **89.** $-2(x + 5)(x - 5)$ **91.** $b^2(a - 5)^2$
93. $ab(3a - b)(4a + b)$ **95.** $3a(2a - 1)^2$ **97.** $3(81 + a^2)$ **99.** $2a(2a - 5)(3a - 4)$ **101.** $a(2a + 5)^2$
103. $3b(3a - 1)^2$ **105.** $-6(x - 2)(x + 4)$ **107.** $x^2(x + y)(x - y)$ **109.** $2a(3a + 2)^2$ **111.** $-b(3a - 2)(2a + 1)$
113. $2x^2(x - 8)(2x - 3)$ **115.** $x^2(x + 5)(x - 5)$ **117.** $(a^2 + 4)(a + 2)(a - 2)$ **119.** $-3y^2(2y + 5)(4y - 3)$
121. $2(x - 3)(2a - b)$ **123.** $(a - b)(y + 1)(y - 1)$ **125.** $(a + b)(a - b)(x - y)$ **127.** $12, -12$ **129.** $16, -16$
131. $10, -10$

SECTION 5.5

3. $-3, -2$ **5.** $7, 3$ **7.** $0, 5$ **9.** $0, 9$ **11.** $0, -\dfrac{3}{2}$ **13.** $0, \dfrac{2}{3}$ **15.** $-2, 5$ **17.** $-9, 9$ **19.** $-\dfrac{7}{2}, \dfrac{7}{2}$

21. $-\dfrac{1}{3}, \dfrac{1}{3}$ **23.** $-2, -4$ **25.** $-7, 2$ **27.** $-\dfrac{1}{2}, 5$ **29.** $-\dfrac{1}{3}, -\dfrac{1}{2}$ **31.** $0, 3$ **33.** $0, 7$ **35.** $-1, -4$

37. $2, 3$ **39.** $\dfrac{1}{2}, -4$ **41.** $\dfrac{1}{3}, 4$ **43.** $3, 9$ **45.** $-2, 9$ **47.** $-1, -2$ **49.** $-9, 5$ **51.** $-7, 4$ **53.** $-2, -3$

55. $-8, 9$ **57.** $1, 4$ **59.** $-5, 2$ **61.** The number is 6.   **63.** The numbers are 2 and 4.   **65.** The numbers are 4 and 5.   **67.** The numbers are 3 and 7.   **69.** There will be 12 consecutive numbers.   **71.** There are 6 teams in the league.   **73.** The object will hit the ground 3 s later.   **75.** The golf ball will return to the ground 3.75 s later.
77. The length is 15 in. The width is 5 in.   **79.** The height of the triangle is 14 m.   **81.** The dimensions of the type area are 4 in. by 7 in.   **83.** The radius of the original circular lawn was approximately 3.81 ft.   **85.** $3, 48$

87. $-\dfrac{3}{2}, -5$ **89.** $0, 9$

CHAPTER 5 REVIEW EXERCISES

1. $(b - 10)(b - 3)$ [5.2A]   **2.** $(x - 3)(4x + 5)$ [5.1B]   **3.** Nonfactorable over the integers   [5.3A]
4. $5x(x^2 + 2x + 7)$ [5.1A]   **5.** $7y^3(2y^6 - 7y^3 + 1)$ [5.1A]   **6.** $(y + 9)(y - 4)$ [5.2A]   **7.** $(2x - 7)(3x - 4)$ [5.3A]
8. $3ab(4a + b)$ [5.1A]   **9.** $(a^3 + 10)(a^3 - 10)$ [5.4A]   **10.** $n^2(n - 3)(n + 1)$ [5.2B]   **11.** $(6y - 1)(2y + 3)$ [5.3A]
12. $2b(2b - 7)(3b - 4)$ [5.4B]   **13.** $(3y^2 + 5z)(3y^2 - 5z)$ [5.4A]   **14.** $(c + 6)(c + 2)$ [5.2A]

15. $(6a - 5)(3a + 2)$ [5.3B]   **16.** $\dfrac{1}{4}, -7$ [5.5A]   **17.** $4x(x - 6)(x + 1)$ [5.2B]   **18.** $3(a - 7)(a + 2)$ [5.2B]

19. $(a - 12)(2a + 5)$ [5.3B]   **20.** $7, -3$ [5.5A]   **21.** $(3a - 5b)(7x + 2y)$ [5.1B]   **22.** $(ab + 1)(ab - 1)$ [5.4A]
23. $(2x + 5)(5x + 2y)$ [5.1B]   **24.** $5(x - 3)(x + 2)$ [5.2B]   **25.** $3(x + 6)^2$ [5.4B]   **26.** $(x - 5)(3x - 2)$ [5.3B]
27. The length is 100 yd. The width is 60 yd.   [5.5B]   **28.** The distance is 20 ft.   [5.5B]   **29.** The width of the frame is 1.5 in.   [5.5B]   **30.** The side of the original garden plot was 20 ft.   [5.5B]

CHAPTER 5 TEST

1. $(b + 6)(a - 3)$ [5.1B]   **2.** $2y^2(y - 8)(y + 1)$ [5.2B]   **3.** $4(x + 4)(2x - 3)$ [5.3B]   **4.** $(2x + 1)(3x + 8)$ [5.3A]

5. $(a - 16)(a - 3)$ [5.2A]   **6.** $2x(3x^2 - 4x + 5)$ [5.1A]   **7.** $(x + 5)(x - 3)$ [5.2A]   **8.** $-\dfrac{1}{2}, \dfrac{1}{2}$ [5.5A]

9. $5(x^2 - 9x - 3)$ [5.1A]   **10.** $(p + 6)^2$ [5.4A]   **11.** $3, 5$ [5.5A]   **12.** $3(x + 2y)^2$ [5.4B]
13. $(b + 4)(b - 4)$ [5.4A]   **14.** $3y^2(2x + 1)(x + 1)$ [5.3B]   **15.** $(p + 3)(p + 2)$ [5.2A]   **16.** $(x - 2)(a + b)$ [5.1B]
17. $(p + 1)(x - 1)$ [5.1B]   **18.** $3(a + 5)(a - 5)$ [5.4B]   **19.** Nonfactorable over the integers   [5.3A]

20. $(x - 12)(x + 3)$ [5.2A]   **21.** $(2a - 3b)^2$ [5.4A]   **22.** $(2x + 7y)(2x - 7y)$ [5.4A]   **23.** $\dfrac{3}{2}, -7$ [5.5A]

24. The two numbers are 7 and 3.   [5.5B]   **25.** The length is 15 cm. The width is 6 cm.   [5.5B]

CUMULATIVE REVIEW EXERCISES

1. 7 [1.2B]   **2.** 4 [1.4B]   **3.** -7 [2.1A]   **4.** $15x^2$ [2.2B]   **5.** 12 [2.2D]   **6.** $\dfrac{2}{3}$ [3.1C]   **7.** $\dfrac{7}{4}$ [3.3A]

8. 3 [3.3B]   **9.** 45 [3.1D]   **10.** $9a^6b^4$ [4.2B]   **11.** $x^3 - 3x^2 - 6x + 8$ [4.3B]   **12.** $4x + 8 + \dfrac{21}{2x - 3}$ [4.5B]

13. $\dfrac{y^6}{x^8}$ [4.4A]   **14.** $(a - b)(3 - x)$ [5.1B]   **15.** $5xy^2(3 - 4y^2)$ [5.1A]   **16.** $(x - 7y)(x + 2y)$ [5.2A]

17. $(p - 10)(p + 1)$ [5.2A]   **18.** $3a(3a + 2)(2a + 5)$ [5.4B]   **19.** $(6a + 7b)(6a - 7b)$ [5.4A]

20. $(2x + 7y)^2$ [5.4A] **21.** $(3x + 7)(3x - 2)$ [5.3A] **22.** $2(3x - 4y)^2$ [5.4B] **23.** $(x - 3)(3y - 2)$ [5.1B]

24. $\frac{2}{3}, -7$ [5.5A] **25.** The shorter piece is 4 ft long. The longer piece is 6 ft long. [3.4B] **26.** The discount rate is 40%. [3.2B] **27.** $m\angle a = 72°; m\angle b = 108°$ [3.5A] **28.** The distance to the resort is 168 mi. [3.6C] **29.** The integers are 10, 12, and 14. [3.4A] **30.** The length of the base of the triangle is 12 in. [5.5B]

Answers to Chapter 6 Selected Exercises

PREP TEST

1. 36 [1.5C] **2.** $\frac{3x}{y^3}$ [4.4A] **3.** $-\frac{5}{36}$ [1.6C] **4.** $-\frac{10}{11}$ [1.7B] **5.** No [1.7B] **6.** $\frac{19}{8}$ [3.2A]

7. 130° [3.5B] **8.** $(x - 6)(x + 2)$ [5.2A] **9.** $(2x - 3)(x + 1)$ [5.3A] **10.** 9:40 A.M. [3.6C]

SECTION 6.1

3. $\frac{3}{4x}$ **5.** $\frac{1}{x + 3}$ **7.** -1 **9.** $\frac{2}{3y}$ **11.** $-\frac{3}{4x}$ **13.** $\frac{a}{b}$ **15.** $-\frac{2}{x}$ **17.** $\frac{y - 2}{y - 3}$ **19.** $\frac{x + 5}{x + 4}$ **21.** $\frac{x + 4}{x - 3}$

23. $-\frac{x + 2}{x + 5}$ **25.** $\frac{2(x + 2)}{x + 3}$ **27.** $\frac{2x - 1}{2x + 3}$ **29.** $-\frac{x + 7}{x + 6}$ **31.** $\frac{35ab^2}{24x^2y}$ **33.** $\frac{4x^3y^3}{3a^2}$ **35.** $\frac{3}{4}$ **37.** ab^2

39. $\frac{x^2(x - 1)}{y(x + 3)}$ **41.** $\frac{y(x - 1)}{x^2(x + 10)}$ **43.** $-ab^2$ **45.** $\frac{x + 5}{x + 4}$ **47.** 1 **49.** $-\frac{n - 10}{n - 7}$ **51.** $\frac{x(x + 2)}{2(x - 1)}$ **53.** $-\frac{x + 2}{x - 6}$

55. $\frac{x + 5}{x - 12}$ **59.** $\frac{7a^3y^2}{40bx}$ **61.** $\frac{4}{3}$ **63.** $\frac{3a}{2}$ **65.** $\frac{x^2(x + 4)}{y^2(x + 2)}$ **67.** $\frac{x(x - 2)}{y(x - 6)}$ **69.** $-\frac{3by}{ax}$ **71.** $\frac{(x + 6)(x - 3)}{(x + 7)(x - 6)}$

73. 1 **75.** $-\frac{x + 8}{x - 4}$ **77.** $\frac{2n + 1}{2n - 3}$ **81.** $\frac{x}{x + 8}$ **83.** $\frac{n - 2}{n + 3}$

SECTION 6.2

1. $24x^3y^2$ **3.** $30x^4y^2$ **5.** $8x^2(x + 2)$ **7.** $6x^2y(x + 4)$ **9.** $36x(x + 2)^2$ **11.** $6(x + 1)^2$
13. $(x - 1)(x + 2)(x + 3)$ **15.** $(2x + 3)^2(x - 5)$ **17.** $(x - 1)(x - 2)$ **19.** $(x + 2)(x - 3)(x + 4)$
21. $(x + 1)(x + 4)(x - 7)$ **23.** $(x + 4)(x - 6)(x + 6)$ **25.** $(x + 3)(x - 10)(x - 8)$ **27.** $(x - 3)(3x - 2)(x + 2)$
29. $(x + 2)(x - 3)$ **31.** $(x + 1)(x - 5)$ **33.** $(x - 1)(x - 2)(x - 3)(x - 6)$ **35.** $\frac{5}{ab^2}, \frac{6b}{ab^2}$ **37.** $\frac{15y^2}{18x^2y}, \frac{14x}{18x^2y}$

39. $\frac{ay + 5a}{y^2(y + 5)}, \frac{6y}{y^2(y + 5)}$ **41.** $\frac{a^2y + 7a^2}{y(y + 7)^2}, \frac{ay}{y(y + 7)^2}$ **43.** $\frac{b}{y(y - 4)}, -\frac{b^2y}{y(y - 4)}$ **45.** $-\frac{3y - 21}{(y - 7)^2}, \frac{2}{(y - 7)^2}$ **47.** $\frac{2y^2}{y^2(y - 3)},$

$\frac{3}{y^2(y - 3)}$ **49.** $\frac{x^3 + 4x^2}{(2x - 1)(x + 4)}, \frac{2x^2 + x - 1}{(2x - 1)(x + 4)}$ **51.** $\frac{3x^2 + 15x}{(x + 5)(x - 5)}, \frac{4}{(x + 5)(x - 5)}$ **53.** $\frac{x^2 - 1}{(x - 3)(x + 5)(x + 1)},$

$\frac{x^2 - 3x}{(x - 3)(x + 5)(x + 1)}$ **55.** $\frac{800}{10^5}, \frac{9}{10^5}$ **57.** $\frac{x^3 - x}{x^2 - 1}, \frac{x}{x^2 - 1}$ **59.** $\frac{3c^2 - 3cd}{3(6c + d)(c + d)(c - d)}, \frac{6cd + d^2}{3(6c + d)(c + d)(c - d)}$

SECTION 6.3

1. $\frac{11}{y^2}$ **3.** $-\frac{7}{x + 4}$ **5.** $\frac{8x}{2x + 3}$ **7.** $\frac{5x + 7}{x - 3}$ **9.** $\frac{2x - 5}{x + 9}$ **11.** $\frac{-3x - 4}{2x + 7}$ **13.** $\frac{1}{x + 5}$ **15.** $\frac{1}{x - 6}$ **17.** $\frac{3}{2y - 1}$

19. $\frac{1}{x - 5}$ **23.** $\frac{4y + 5x}{xy}$ **25.** $\frac{19}{2x}$ **27.** $\frac{5}{12x}$ **29.** $\frac{19x - 12}{6x^2}$ **31.** $\frac{52y - 35x}{20xy}$ **33.** $\frac{13x + 2}{15x}$ **35.** $\frac{7}{24}$

37. $\frac{x + 90}{45x}$ **39.** $\frac{x^2 + 2x + 2}{2x^2}$ **41.** $\frac{2x^2 + 3x - 10}{4x^2}$ **43.** $\frac{-x^2 - 4x + 4}{x + 4}$ **45.** $\frac{4x + 7}{x + 1}$ **47.** $\frac{4x^2 + 9x + 9}{24x^2}$

49. $\frac{3x - 1 - 2xy - 3y}{xy^2}$ **51.** $\frac{20x^2 + 28x - 12xy + 9y}{24x^2y^2}$ **53.** $\frac{9x^2 - 3x - 2xy - 10y}{18xy^2}$ **55.** $\frac{7x - 23}{(x - 3)(x - 4)}$

57. $\dfrac{-y - 33}{(y + 6)(y - 3)}$ **59.** $\dfrac{3x^2 + 20x - 8}{(x - 4)(x + 6)}$ **61.** $\dfrac{3(4x^2 + 5x - 5)}{(x + 5)(2x + 3)}$ **63.** $\dfrac{-4x + 5}{x - 6}$ **65.** $\dfrac{2(y + 2)}{(y + 4)(y - 4)}$ **67.** $-\dfrac{4x}{(x + 1)^2}$

69. $\dfrac{2x - 1}{(1 + x)(1 - x)}$ **71.** $\dfrac{14}{(x - 5)^2}$ **73.** $\dfrac{-2(x + 7)}{(x + 6)(x - 7)}$ **75.** $\dfrac{x - 4}{x - 6}$ **77.** $\dfrac{2x + 1}{x - 1}$ **79.** $\dfrac{-3(x^2 + 8x + 25)}{(x - 3)(x + 7)}$

81. a. $\dfrac{20{,}400}{x}$ dollars   **b.** $\dfrac{102{,}000}{x(x + 5)}$ dollars   **c.** \$136

SECTION 6.4

1. $\dfrac{x}{x - 3}$ **3.** $\dfrac{2}{3}$ **5.** $\dfrac{y + 3}{y - 4}$ **7.** $\dfrac{2(2x + 13)}{5x + 36}$ **9.** $\dfrac{x + 2}{x + 3}$ **11.** $\dfrac{x - 6}{x + 5}$ **13.** $\dfrac{-x + 2}{x + 1}$ **15.** $x - 1$ **17.** $\dfrac{1}{2x - 1}$

19. $\dfrac{x - 3}{x + 5}$ **21.** $\dfrac{x - 7}{x - 8}$ **23.** $\dfrac{2y - 1}{2y + 1}$ **25.** $\dfrac{x - 2}{2x - 5}$ **27.** $\dfrac{-x - 1}{4x - 3}$ **29.** $\dfrac{x + 1}{2(5x - 2)}$ **31.** $\dfrac{5}{3}$ **33.** $-\dfrac{1}{x - 1}$

35. $\dfrac{y + 4}{2(y - 2)}$ **37.** $\dfrac{x + 1}{x - 1}$ **39.** $\dfrac{y^2 + x^2}{xy}$

SECTION 6.5

3. 3   **5.** 1   **7.** 9   **9.** 1   **11.** $\dfrac{1}{4}$ **13.** 1   **15.** -3 **17.** $\dfrac{1}{2}$ **19.** 8   **21.** 5   **23.** -1 **25.** 5

27. No solution   **29.** 4, 2   **31.** $-\dfrac{3}{2}$, 4   **33.** 3   **35.** 4   **37.** 0   **39.** $-\dfrac{2}{5}$ **41.** 0, $-\dfrac{2}{3}$

SECTION 6.6

3. 9   **5.** 12   **7.** 7   **9.** 6   **11.** 1   **13.** -6 **15.** 4   **17.** $-\dfrac{2}{3}$ **19.** 20,000 voters voted in favor of the amendment.   **21.** The distance between the two cities is 175 mi.   **23.** The sales tax will be \$97.50 higher. **25.** The person is 67.5 in. tall.   **27.** There are approximately 75 elk in the preserve.   **29.** The length of side AC is 6.7 cm.   **31.** The height is 2.9 m.   **33.** The perimeter is 22.5 ft.   **35.** The area is 48 m².   **37.** The length of BC is 6.25 cm.   **39.** The length of DA is 6 in.   **41.** The length of OP is 13 cm.   **43.** The distance across the river is 35 m. **45.** The first person won \$1.25 million.   **47.** The player made 210 foul shots.

SECTION 6.7

1. $y = -3x + 10$ **3.** $y = 4x - 3$ **5.** $y = -\dfrac{3}{2}x + 3$ **7.** $y = \dfrac{2}{5}x - 2$ **9.** $y = -\dfrac{2}{7}x + 2$ **11.** $y = -\dfrac{1}{3}x + 2$

13. $y = 3x + 8$ **15.** $y = -\dfrac{2}{3}x - 3$ **17.** $x = -6y + 10$ **19.** $x = \dfrac{1}{2}y + 3$ **21.** $x = -\dfrac{3}{4}y + 3$ **23.** $x = 4y + 3$

25. $t = \dfrac{d}{r}$ **27.** $T = \dfrac{PV}{nR}$ **29.** $l = \dfrac{P - 2w}{2}$ **31.** $b_1 = \dfrac{2A - hb_2}{h}$ **33.** $h = \dfrac{3V}{A}$ **35.** $S = C - Rt$ **37.** $P = \dfrac{A}{1 + rt}$

39. $w = \dfrac{A}{S + 1}$ **41. a.** $S = \dfrac{F + BV}{B}$ **b.** The required selling price is \$180.   **c.** The required selling price is \$75.

SECTION 6.8

3. It will take 2 h to fill the fountain with both sprinklers working.   **5.** With both skiploaders working together, it would take 3 h to remove the earth.   **7.** With both computers working, it would take 30 h to solve the problem.   **9.** It would take 30 min to cool the room with both air conditioners working.   **11.** It would take the second pipeline 90 min to fill the tank.   **13.** It would take the apprentice 15 h to construct the wall.   **15.** It will take the second technician 3 h to complete the wiring.   **17.** It would have taken one of the welders 40 h to complete the welds.   **19.** It would have taken one machine 28 h to fill the boxes.   **21.** The jogger ran 16 mi in 2 h.   **23.** The rate of travel in the congested area was 20 mph.   **25.** The rate of the jogger was 8 mph. The rate of the cyclist was 20 mph.   **27.** The rate of the jet is 360 mph.   **29.** Camille's walking rate is 4 mph.   **31.** The rate of the car is 48 mph.   **33.** The rate of the wind is 20 mph.   **35.** The rate of the gulf current is 6 mph.   **37.** The rate of the trucker for the first 330 mi was 55 mph. **39.** The bus usually travels 60 mph.

CHAPTER 6 REVIEW EXERCISES

1. $\dfrac{b^3y}{10ax}$ [6.1C]   **2.** $\dfrac{7x+22}{60x}$ [6.3B]   **3.** $\dfrac{2xy}{5}$ [6.1B]   **4.** $\dfrac{2xy}{3(x+y)}$ [6.1C]   **5.** $\dfrac{x-2}{3x-10}$ [6.4A]

6. $-\dfrac{x+6}{x+3}$ [6.1A]   **7.** $\dfrac{2x^4}{3y^7}$ [6.1A]   **8.** 62 [6.6A]   **9.** $\dfrac{(3y-2)^2}{(y-1)(y-2)}$ [6.1C]   **10.** $x=\dfrac{5}{3a-1}$ [6.7A]

11. 8 [6.5A]   **12.** $\dfrac{x^2+3y}{xy}$ [6.3B]   **13.** $y=-\dfrac{5}{4}x+5$ [6.7A]   **14.** $\dfrac{by^3}{6ax^2}$ [6.1B]   **15.** $\dfrac{x}{x-7}$ [6.4A]

16. $\dfrac{3x^2-x}{(6x-1)(2x+3)(3x-1)},\ \dfrac{24x^3-4x^2}{(6x-1)(2x+3)(3x-1)}$ [6.2B]   **17.** $a=\dfrac{T-2bc}{2b+2c}$ [6.7A]   **18.** 2 [6.5A]

19. $\dfrac{2x+1}{3x-2}$ [6.4A]   **20.** $\dfrac{x^2+5}{(x-5)(x-2)}$ [6.3B]   **21.** $c=\dfrac{100m}{i}$ [6.7A]   **22.** No solution [6.5A]   **23.** $\dfrac{1}{x^2}$ [6.1C]

24. $\dfrac{2y-3}{5y-7}$ [6.3B]   **25.** $\dfrac{1}{x+3}$ [6.3A]   **26.** $(5x-3)(2x-1)(4x-1)$ [6.2A]   **27.** $y=-\dfrac{4}{9}x+2$ [6.7A]

28. $\dfrac{2x+1}{x+2}$ [6.1B]   **29.** 5 [6.5A]   **30.** $\dfrac{3x-1}{x-5}$ [6.3B]   **31.** 10 [6.6A]   **32.** 12 [6.6A]   **33.** The length of QO

is 15 cm. [6.6C]   **34.** The area is $\dfrac{256}{3}$ in². [6.6C]   **35.** It would take 6 h to fill the pool. [6.8A]   **36.** The rate of

the car is 45 mph. [6.8B]   **37.** The rate of the wind is 20 mph. [6.8B]   **38.** The ERA is 1.35. [6.6B]

CHAPTER 6 TEST

1. $\dfrac{x^2-4x+5}{(x+3)(x-2)}$ [6.3B]   **2.** -1 [6.6A]   **3.** $\dfrac{(2x-1)(x-5)}{(x+3)(2x+5)}$ [6.1B]   **4.** $\dfrac{2x^3}{3y^3}$ [6.1A]   **5.** $t=\dfrac{d-s}{r}$ [6.7A]

6. 2 [6.5A]   **7.** $-\dfrac{x+5}{x+1}$ [6.1A]   **8.** $3(2x-1)(x+1)$ [6.2A]   **9.** $\dfrac{5}{(2x-1)(3x+1)}$ [6.3B]   **10.** $\dfrac{x+5}{x+4}$ [6.1C]

11. $\dfrac{x-3}{x-2}$ [6.4A]   **12.** $\dfrac{3x+6}{x(x-2)(x+2)},\ \dfrac{x^2}{x(x-2)(x+2)}$ [6.2B]   **13.** $\dfrac{2}{x+5}$ [6.3A]   **14.** $y=\dfrac{3}{8}x-2$ [6.7A]

15. No solution [6.5A]   **16.** $\dfrac{x+1}{x^3(x-2)}$ [6.1B]   **17.** The length of CE is 12.8 ft. [6.6C]   **18.** An additional 2 lb of

salt are needed. [6.6B]   **19.** It would take 4 h to fill the pool. [6.8A]   **20.** The rate of the wind is 20 mph. [6.8B]
21. 54 sprinklers are needed for a 3600 square-foot lawn. [6.6B]

CUMULATIVE REVIEW EXERCISES

1. $\dfrac{31}{30}$ [1.7A]   **2.** 21 [2.1A]   **3.** $5x-2y$ [2.2A]   **4.** $-8x+26$ [2.2D]   **5.** $-\dfrac{9}{2}$ [3.2A]   **6.** -12 [3.3B]

7. 10 [3.1D]   **8.** a^3b^7 [4.2A]   **9.** $a^2+ab-12b^2$ [4.3C]   **10.** $3b^3-b+2$ [4.5A]   **11.** x^2+2x+4 [4.5B]
12. $(3x-1)(4x+1)$ [5.3A]   **13.** $(y-6)(y-1)$ [5.2A]   **14.** $a(a+5)(2a-3)$ [5.3A]

15. $4(b+5)(b-5)$ [5.4B]   **16.** $-3,\dfrac{5}{2}$ [5.5A]   **17.** $\dfrac{2x^3}{3y^5}$ [6.1A]   **18.** $-\dfrac{x-2}{x+5}$ [6.1A]   **19.** 1 [6.1C]

20. $\dfrac{3}{(2x-1)(x+1)}$ [6.3B]   **21.** $\dfrac{x+3}{x+5}$ [6.4A]   **22.** 4 [6.5A]   **23.** 3 [6.6A]   **24.** $t=\dfrac{f-v}{a}$ [6.7A]
25. $5x-13=-8; x=1$ [3.4A]   **26.** The silver alloy is 70% silver. [3.6B]   **27.** The base is 10 in. The height is
6 in. [5.5B]   **28.** The cost of a $5000 policy is $80. [6.6B]   **29.** It would take both pipes 6 min to fill the
tank. [6.8A]   **30.** The rate of the current is 2 mph. [6.8B]

Answers to Chapter 7 Selected Exercises

PREP TEST

1. 3 [1.4B]   **2.** -1 [2.1A]   **3.** $-3x+12$ [2.2C]   **4.** -2 [3.2A]   **5.** $x=5$ [3.2A]   **6.** $y=-2$ [3.2A]

7. $-4x+5$ [4.5A]   **8.** 4 [6.6A]   **9.** $y=\dfrac{3}{5}x-3$ [6.7A]   **10.** $y=-\dfrac{1}{2}x-5$ [6.7A]

SECTION 7.1

1.

3.

5.

7. $A(2, 3)$, $B(4, 0)$, $C(-4, 1)$, $D(-2, -2)$

9. $A(-2, 5)$, $B(3, 4)$, $C(0, 0)$, $D(-3, -2)$ **11. a.** $2, -4$ **b.** $1, -3$ **15.** Yes **17.** No **19.** No **21.** No

23.

25.

27.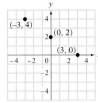

29. $\{(24, 600), (32, 750), (22, 430), (15, 300), (4.4, 68), (17, 370), (15, 310), (4.4, 55)\}$; No

31. $\{(390, 0.115), (591, 0.073), (517, 0.077), (576, 0.068), (605, 0.064)\}$; Yes **33.** Yes **35.** No **37.** Yes **39.** 8

41. 9 **43.** 2 **45.** -1 **47.** 22 **49.** $-\dfrac{3}{2}$ **51.** -7

SECTION 7.2

1.

3.

5.

7.

9.

11.

13.

15.

17.

19.

21.

23.

25.

27.

29.

31.

33.

35.

37. After flying for 3 min, the helicopter is 3.5 mi away from the victims.

39. A dog 6 years is equivalent in age to a human 40 years old.

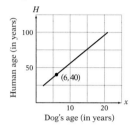

43. Increases; 3; 3   **45. a.** The cost is $.99.   **b.** The cost is $1.74.

SECTION 7.3

1. $(3, 0), (0, -3)$ **3.** $(2, 0), (0, -6)$ **5.** $(10, 0), (0, -2)$ **7.** $(-4, 0), (0, 12)$ **9.** $(0, 0), (0, 0)$ **11.** $(6, 0), (0, 3)$

13.

15.

17.

21. -2 **23.** $\dfrac{1}{3}$ **25.** $-\dfrac{5}{2}$

27. Undefined   **29.** 0   **31.** $-\dfrac{1}{3}$ **33.** Neither   **35.** Neither   **37.** Parallel   **39.** Neither   **41.** $m = 33$. The worldwide sales of camera-phones are increasing by 33 million units per year.   **43.** $m = -180$. The value of the car is decreasing $180 for each additional 1000 miles the car is driven.   **45.** $m = \dfrac{2}{3}$; $(0, -2)$ **47.** $m = -\dfrac{2}{5}$; $(0, 2)$

49. $m = \dfrac{1}{4}$; $(0, 0)$ **51.**

53.

55.

57.

59.

61.

63.

65.

67. Yes

SECTION 7.4

3. $y = 2x + 2$ **5.** $y = -3x - 1$ **7.** $y = \dfrac{1}{3}x$ **9.** $y = \dfrac{3}{4}x - 5$ **11.** $y = -\dfrac{3}{5}x$ **13.** $y = \dfrac{1}{4}x + \dfrac{5}{2}$

15. $y = 2x - 3$ **17.** $y = -2x - 3$ **19.** $y = \dfrac{2}{3}x$ **21.** $y = \dfrac{1}{2}x + 2$ **23.** $y = -\dfrac{3}{4}x - 2$ **25.** $y = \dfrac{3}{4}x + \dfrac{5}{2}$

27. The tennis player is using 1.55 g of carbohydrates per minute.

29. The percent of music purchased in stores is decreasing 3% per year.

Year (x = 0 corresponds to 1997)

31. No   **33.** Yes   **35.** $-\dfrac{3}{2}$ **37.** -5 **39.** $y = -\dfrac{2}{3}x + \dfrac{5}{3}$

CHAPTER 7 REVIEW EXERCISES

1. a.

2.

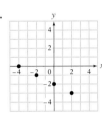

[7.1B]

3. $y = -\dfrac{8}{3}x + \dfrac{1}{3}$ [7.4B]   **4.** $y = -\dfrac{5}{2}x + 16$ [7.4A]

b. -2

c. -4 [7.1A]

5.

[7.2A]

6.

[7.2B]

7. Neither   [7.3B]   **8.** -1 [7.1D]

9. $y = -\dfrac{2}{3}x + \dfrac{11}{3}$ [7.4B]   **10.** Yes   [7.1C]   **11.** $\dfrac{7}{11}$ [7.3B]   **12.** $(8, 0), (0, -12)$ [7.3A]   **13.** 0 [7.3B]

14.

[7.3C]

15.

[7.2B]

16.

[7.3C]

17.

[7.2A]

18.

[7.3C]

19.

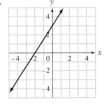

[7.2B]

20. {(55, 95), (57, 101), (53, 94), (57, 98), (60, 100), (61, 105), (58, 97), (54, 95)}; No   [7.1C]

21. The cost of 50 min of access time for one month is $97.50.

[7.2C]

22. The average annual telephone bill for a family is increasing by $34 per year.

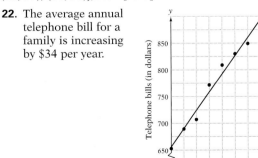

Year (x = 0 corresponds to 1999)   [7.4C]

CHAPTER 7 TEST

1. $(3, -3)$ [7.1B]   **2.**   **3.** Yes   [7.1C]   **4.** 6   [7.1D]   **5.** 3   [7.1D]

[7.1B]

6. $\{(3.5, 25), (4.0, 30), (5.2, 45), (5.0, 38), (4.0, 42), (6.3, 12), (5.4, 34)\}$; No   [7.1C]   **7.**

[7.2A]

8.   **9.**   **10.**

[7.2A]   [7.2B]   [7.2B]

11.   **12.** 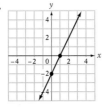   **13.** After 1 s, the speed of the ball is 96 ft/s.   [7.2C]

[7.3C]   [7.3C]

14. $m = 0.46$. The average hourly wage is increasing by \$.46 per year.   [7.3B]

15.   The average annual tuition for a private 4-year college is increasing \$809 per year.   [7.4C]

16. $(2, 0), (0, -3)$ [7.3A]   **17.** $(-2, 0), (0, 1)$ [7.3A]   **18.** 2   [7.3B]   **19.** Parallel   [7.3B]

20. Undefined   [7.3B]   **21.** $-\dfrac{2}{3}$ [7.3B]   **22.** $y = 3x - 1$ [7.4A]   **23.** $y = \dfrac{2}{3}x + 3$ [7.4A]

24. $y = -\dfrac{5}{8}x - \dfrac{7}{8}$ [7.4B]   **25.** $y = -\dfrac{2}{7}x - \dfrac{4}{7}$ [7.4B]

CUMULATIVE REVIEW EXERCISES

1. -12 [1.4B]   **2.** $-\dfrac{5}{8}$ [2.1A]   **3.** $f(-2) = -\dfrac{2}{3}$ [7.1D]   **4.** $\dfrac{3}{2}$ [3.2A]   **5.** $\dfrac{19}{18}$ [3.3B]   **6.** $\dfrac{1}{15}$ [1.7C]

7. $-32x^8y^7$ [4.2B]   **8.** $-3x^2$ [4.4A]   **9.** $x + 3$ [4.5B]   **10.** $5(x + 2)(x + 1)$ [5.2B]   **11.** $(a + 2)(x + y)$ [5.1A]

12. 4 and -2 [5.5A]   **13.** $\dfrac{x^3(x + 3)}{y(x + 2)}$ [6.1B]   **14.** $\dfrac{3}{x + 8}$ [6.3A]   **15.** 2   [6.5A]   **16.** $y = \dfrac{4}{5}x - 3$ [6.7A]

17. $(-2, -5)$ [7.1B]   **18.** Zero   [7.3B]   **19.** $y = \dfrac{1}{2}x - 2$ [7.4A]   **20.** $y = -3x + 2$ [7.4A]

21. $y = 2x + 2$ [7.4A]   **22.** $y = \frac{2}{3}x - 3$ [7.4A]   **23.** $62.30   [3.2B]   **24.** 46°, 43°, 91°   [3.5B]   **25.** The value

of the home is $110,000.   [6.6B]   **26.** It would take $3\frac{3}{4}$ h for both, working together, to wire the garage.   [6.8A]

27.   **28.**

[7.2A]   [7.3C]

Answers to Chapter 8 Selected Exercises

PREP TEST

1. $y = \frac{3}{4}x - 6$ [6.7A]   **2.** $1000   [3.3B]   **3.** $33y$ [2.2D]   **4.** $10x - 10$ [2.2D]   **5.** Yes   [7.1B]

6. $(4, 0), (0, -3)$ [7.3A]   **7.** Yes   [7.3B]   **8.** 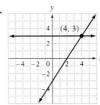   [7.3C]

9. 5 ml of 55% acetic acid are needed.   [3.6B]   **10.** It will take 1.5 h after the second hiker starts for the hikers to be side-by-side.   [3.6C]

SECTION 8.1

1. I: c; II: a; III: b   **3.** $(2, -1)$ **5.** The ordered-pair solutions of $y = -\frac{3}{2}x + 1$ **7.** No solution   **9.** $(-2, 4)$

11. Yes   **13.** No   **15.** Yes   **17.** Yes

19.   **21.**   **23.**   **25.**

27.   **29.**   **31.**   The ordered-pair solutions of $y = 2x - 2$

33.   **35.**   **37.**   **39. a.** Sometimes true   **b.** Always true
                                                                                                            c. Never true   **d.** Always true
                                                                                                    43. Answers will vary.

SECTION 8.2

3. (2, 1)   **5.** (4, 1)   **7.** (−1, 1)   **9.** No solution   **11.** No solution   **13.** $\left(-\dfrac{3}{4}, -\dfrac{3}{4}\right)$ **15.** (1, 1)   **17.** (2, 0)

19. (1, −2)   **21.** (0, 0)   **23.** Dependent. The solutions satisfy the equation $2x - y = 2$.   **25.** (−4, −2)   **27.** (10, 31)

29. (3, −10)   **31.** (−22, −5)   **33.** The amounts invested should be $1900 at 5% and $1600 at 7.5%.

35. The amounts invested were $2400 at 9% and $3600 at 6%.   **37.** The amounts invested should be $4400 at 8% and

$1600 at 11%.   **39.** The amount invested at 6.5% was $21,000.   **41.** The amounts invested were $12,000 at 8% and

$8000 at 7%.   **43.** The amount invested in the trust deed was $3750.   **45.** The gas dryer becomes more economical

after 185 loads of clothes.   **47.** 1   **49.** The assertion is not correct. The system of equations is independent. The

solution is (0, 2).   **51.** The research consultant's investment is $45,000.   **53.** Simple interest: $400; Compounded

monthly: $415.00; Compounded daily: $416.39

SECTION 8.3

1. (5, −1)   **3.** (1, 3)   **5.** (1, 1)   **7.** (3, −2)   **9.** Dependent. The solutions satisfy the equation $2x - y = 1$.

11. (3, 1)   **13.** Dependent. The solutions satisfy the equation $2x - 3y = 1$.   **15.** $\left(\dfrac{2}{3}, \dfrac{1}{2}\right)$ **17.** (2, 0)   **19.** (0, 0)

21. (5, −2)   **23.** $\left(\dfrac{32}{19}, -\dfrac{9}{19}\right)$ **25.** $\left(\dfrac{7}{4}, -\dfrac{5}{16}\right)$ **27.** (1, −1)   **29.** No solution   **31.** (3, 1)   **33.** (−1, 2)

35. (1, 1)   **39.** $A = 3; B = -1$ **41. a.** 1   **b.** $\dfrac{3}{2}$ **c.** 4

SECTION 8.4

1. The rate of the whale in calm water was 35 mph. The rate of the current was 5 mph.   **3.** The rate rowing in calm

water was 14 km/h. The rate of the current was 6 km/h.   **5.** The rate of the Learjet was 525 mph. The rate of the wind

was 35 mph.   **7.** The rate of the helicopter in calm air was 225 mph. The rate of the wind was 45 mph.   **9.** The rate

of the canoeist in calm water was 6 mph. The rate of the current was 1 mph.   **11.** The cost per pound of the wheat flour

was $.65. The cost per pound of the rye flour was $.70.   **13.** Reagent I is 25% hydrochloric acid. Reagent II is 35%

hydrochloric acid.   **15.** 12 gal of 87-octane gasoline and 6 gal of 93-octane gasoline must be used.   **17.** 1 nickel and

2 dimes or 3 nickels and 1 dime are in the bank.   **19.** 12.5 acres of good land and 87.5 acres of bad land were bought.

CHAPTER 8 REVIEW EXERCISES

1. Yes   [8.1A]   **2.** No   [8.1A]   **3.**

4.

The solutions are the ordered-pair solutions of $y = 2x - 4$.

[8.1A]   [8.1A]

5.

[8.1A]   **6.** (−1, 1)   [8.2A]   **7.** (1, 6)   [8.2A]   **8.** (−3, 1)   [8.3A]   **9.** $\left(-\dfrac{5}{6}, \dfrac{1}{2}\right)$ [8.3A]

10. No solution   [8.2A]   **11.** (1, 6)   [8.2A]   **12.** (1, −5)   [8.3A]   **13.** No solution   [8.3A]   **14.** Dependent.

The solutions satisfy the equation $y = -\dfrac{4}{3}x + 4$.   [8.2A]   **15.** (−1, −3)   [8.2A]   **16.** Dependent. The solutions satisfy

the equation $3x + y = -2$.   [8.3A]   **17.** $\left(\dfrac{2}{3}, -\dfrac{1}{6}\right)$ [8.3A]   **18.** The rate of the sculling team in calm water was

9 mph. The rate of the current was 3 mph.   [8.4A]   **19.** 1300 $6 shares were purchased, and 200 $25 shares were

purchased. [8.4B]   **20.** The rate of the flight crew in calm air was 125 km/h. The rate of the wind was 15 km/h. [8.4A]
21. The rate of the plane in calm air was 105 mph. The rate of the wind was 15 mph. [8.4A]   **22.** The number of ads
requiring $.25 postage was 130. The number of ads requiring $.45 postage was 60. [8.4B]   **23.** The amounts invested
are $7000 at 7% and $5000 at 8.5%. [8.2B]   **24.** There were originally 350 bushels of lentils and 200 bushels of corn in
the silo. [8.4B]   **25.** The amounts invested were $165,000 at 5.4% and $135,000 at 6.6%. [8.2B]

CHAPTER 8 TEST

1. Yes [8.1A]   **2.** Yes [8.1A]   **3.**   **4.** $(3, 1)$ [8.2A]   **5.** $(1, -1)$ [8.2A]

[8.1A]

6. $(2, -1)$ [8.2A]   **7.** $\left(\dfrac{22}{7}, -\dfrac{5}{7}\right)$ [8.2A]   **8.** No solution [8.2A]   **9.** $(2, 1)$ [8.3A]   **10.** $\left(\dfrac{1}{2}, -1\right)$ [8.3A]
11. Dependent. The solutions satisfy the equation $x + 2y = 8$. [8.3A]   **12.** $(2, -1)$ [8.3A]   **13.** $(1, -2)$ [8.3A]
14. The rate of the plane in calm air is 100 mph. The rate of the wind is 20 mph. [8.4A]   **15.** The price of a reserved-
seat ticket was $10. The price of a general-admission ticket was $6. [8.4B]   **16.** The amounts invested were $15,200 at
6.4% and $12,800 at 7.6%. [8.2B]

CUMULATIVE REVIEW EXERCISES

1. $\dfrac{3}{2}$ [2.1A]   **2.** $-\dfrac{3}{2}$ [3.1C]   **3.** 7 [7.1D]   **4.** $-6a^3 + 13a^2 - 9a + 2$ [4.3B]   **5.** $-2x^5y^2$ [4.4A]

6. $2b - 1 + \dfrac{1}{2b - 3}$ [4.5B]   **7.** $-\dfrac{4y}{x^3}$ [4.4A]   **8.** $4y^2(xy - 4)(xy + 4)$ [5.4B]   **9.** $4, -1$ [5.5A]

10. $x - 2$ [6.1C]   **11.** $\dfrac{x^2 + 2}{(x + 2)(x - 1)}$ [6.3B]   **12.** $\dfrac{x - 3}{x + 1}$ [6.4A]   **13.** $-\dfrac{1}{5}$ [6.5A]   **14.** $r = \dfrac{A - P}{Pt}$ [6.7A]

15. x-intercept: $(6, 0)$; y-intercept: $(0, -4)$ [7.3A]   **16.** $-\dfrac{7}{5}$ [7.3B]   **17.** $y = -\dfrac{3}{2}x$ [7.4A]

18. Yes [8.1A]   **19.** $(-6, 1)$ [8.2A]   **20.** $(4, -3)$ [8.3A]   **21.** The amounts invested should be $3750 at 9.6% and
$5000 at 7.2%. [8.2B]   **22.** The rate of the passenger train is 56 mph. The rate of the freight train is 48 mph. [3.6C]
23. The side of the original square is 8 in. [5.5B]   **24.** The rate of the wind is 30 mph. [8.4A]

25.   **26.**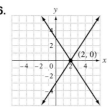

[7.2B]   [8.1A]

27. The rate of the motorboat in calm water is 14 mph. [8.4A]   **28.** The percent concentration of sugar in the mixture
is 35.3%. [3.6A]

Answers to Chapter 9 Selected Exercises

PREP TEST

1. $<$ [1.1A]   **2.** $-7x + 15$ [2.2D]   **3.** The same number can be added to each side of an equation without changing
the solution of the equation. [3.1B]   **4.** Each side of an equation can be multiplied by the same nonzero number without changing the solution of the equation. [3.1C]   **5.** There are 0.45 lb of fat in 3 lb of this grade of hamburger. [3.1D]

6. $-\dfrac{1}{2}$ [3.2A]   **7.** $-\dfrac{8}{3}$ [3.2A]   **8.** 2 [3.3B]   **9.**

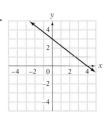

[7.2A]   **10.** [7.2B]

SECTION 9.1

3. $A = \{16, 17, 18, 19, 20, 21\}$ **5.** $A = \{9, 11, 13, 15, 17\}$ **7.** $A = \{b, c\}$ **9.** $A \cup B = \{3, 4, 5, 6\}$
11. $A \cup B = \{-10, -9, -8, 8, 9, 10\}$ **13.** $A \cup B = \{a, b, c, d, e, f\}$ **15.** $A \cup B = \{1, 3, 7, 9, 11, 13\}$
17. $A \cap B = \{4, 5\}$ **19.** $A \cap B = \varnothing$ **21.** $A \cap B = \{c, d, e\}$ **23.** $\{x \mid x > -5, x \in \text{negative integers}\}$
25. $\{x \mid x > 30, x \in \text{integers}\}$ **27.** $\{x \mid x > 5, x \in \text{even integers}\}$ **29.** $\{x \mid x > 8, x \in \text{real numbers}\}$
31. (number line from -5 to 5)   **33.** (number line from -5 to 5)   **35.** (number line from -5 to 5)
37. (number line from -5 to 5)   **39.** (number line from -5 to 5)   **41. a.** Never true   **b.** Always true
c. Always true   **43. a.** Yes   **b.** Yes

SECTION 9.2

1. $x < 2$ (number line from -5 to 5)   **3.** $x > 3$ (number line from -5 to 5)
5. $n \geq 3$ (number line from -5 to 5)   **7.** $x \leq -4$ (number line from -5 to 5)   **9.** $x < 1$ **11.** $x \leq -3$
13. $y \geq -9$ **15.** $x < 12$ **17.** $x \geq 5$ **19.** $x < -11$ **21.** $x \leq 10$ **23.** $x \geq -6$ **25.** $x > 2$ **27.** $d < -\dfrac{1}{6}$
29. $x \geq -\dfrac{31}{24}$ **31.** $x < \dfrac{5}{8}$ **33.** $x < \dfrac{5}{4}$ **35.** $x > \dfrac{5}{24}$ **37.** $x < -3.8$ **39.** $x \leq -1.2$ **41.** $x < 5.6$
43. (number line from -5 to 5) $x < 4$ **45.** (number line from -5 to 5) $y \geq 3$
47. (number line from -5 to 5) $x \leq 1$ **49.** (number line from -5 to 5) $x < -1$
51. (number line from -5 to 5) $b < -4$ **53.** $y \leq 0$ **55.** $x > \dfrac{2}{7}$ **57.** $x \leq -\dfrac{5}{2}$ **59.** $x < 16$ **61.** $x \geq 16$
63. $x \geq -14$ **65.** $x \leq 21$ **67.** $x > 0$ **69.** $x \leq -\dfrac{12}{7}$ **71.** $x > \dfrac{2}{3}$ **73.** $x \leq \dfrac{2}{3}$ **75.** $x \leq 2.3$ **77.** $x < -3.2$
79. $x \leq 5$ **81.** The team must win 11 or more games to be eligible for the tournament.   **83.** The service organization
must collect more than 440 lb on the fourth drive to collect the bonus.   **85.** The person needs 50 or more milligrams of
additional vitamin C to satisfy the recommended daily allowance.   **87.** No, the student cannot earn an A grade.
89. $\{c \mid c > 0\}$ **91.** $\{c \mid c > \in \text{real numbers}\}$ **93.** $\{c \mid c > 0\}$

SECTION 9.3

1. $x < 4$ **3.** $x < -4$ **5.** $x \geq 1$ **7.** $x < 5$ **9.** $x < 0$ **11.** $x < 20$ **13.** $y \leq \dfrac{5}{2}$ **15.** $x < \dfrac{25}{11}$ **17.** $n \leq \dfrac{11}{18}$
19. $x \geq 6$ **21.** In one month the agent expects to make sales totaling $20,000 or less.   **23.** A person must use more
than 60 min to exceed $10.   **25.** The amount of artificial flavors that can be added is less than or equal to 8 oz.
27. The distance to the ski resort must be greater than 38 mi.   **29.** $\{1, 2\}$ **31.** $\{3, 4, 5\}$ **33.** $\{x \mid x \in \text{real numbers}\}$

SECTION 9.4

1.   **3.**   **5.**   **7.**

9.   **11.**   **13.**   **15.**

17.   **19.**   **21.** 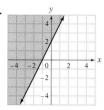 **23.** $x \le 3$

CHAPTER 9 REVIEW

1. $x > 18$ [9.2A]   **2.** $A \cap B = \varnothing$ [9.1A]   **3.** $\{x \mid x > -8, x \in \text{odd integers}\}$ [9.1B]

4. $A \cup B = \{2, 4, 6, 8, 10\}$ [9.1A]   **5.** $A = \{1, 3, 5, 7\}$ [9.1A]   **6.** $x \ge 4$ [9.3A]

7.   [9.1C]   **8.** $x \ge -4$ [9.3A]   **9.**

[9.4A]

10.   **11.** $\{x \mid x > 3, x \in \text{real numbers}\}$ [9.1B]   **12.** $x > 2$ [9.2A]

[9.4A]

13. $A \cap B = \{1, 5, 9\}$ [9.1A]   **14.**   [9.1C]   **15.**   [9.1C]

16. $x \ge -3$ [9.2B]   **17.** $x > -18$ [9.3A]   **18.** $x < \dfrac{1}{2}$ [9.3A]   **19.** $x < -\dfrac{8}{9}$ [9.2B]

20.   **21.** $x \ge 4$ [9.3A]   **22.** For florist B to be more economical, there must be 5 or more residents in the nursing home.   [9.3B]

[9.4A]

23. The minimum length is 24 ft.   [9.3B]   **24.** 32 is the smallest integer that satisfies the inequality.   [9.2C]

25. 72 is the lowest score that the student can receive and still attain a minimum of 480 points.   [9.2C]

CHAPTER 9 TEST

1.   [9.1C]   **2.** $\{x \mid x < 50, x \in \text{positive integers}\}$ [9.1B]   **3.** $A = \{4, 6, 8\}$ [9.1A]

4. $x \le -3$ [9.3A]   **5.** $x > \dfrac{1}{8}$ [9.2A]   **6.**   [9.1C]   **7.** $x < -1$ [9.3A]

8. $\{x \mid x > -23, x \in \text{real numbers}\}$ [9.1B]   **9.**   **10.**

[9.4A]   [9.4A]

11. $A \cap B = \{12\}$ [9.1A]   **12.** $x < -3$ 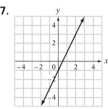 [9.2A]   **13.** $x \geq -\dfrac{40}{3}$ [9.2B]

14. $x < -\dfrac{22}{7}$ [9.3A]   **15.** $x \geq 3$ [9.2B]   **16.** $x \geq -4$ [9.3A]   **17.** The child must grow 5 in. or more. [9.2C]   **18.** The width must be less than or equal to 11 ft. [9.3B]   **19.** The diameter must be between 0.0389 in. and 0.0395 in. [9.2C]   **20.** The total value of the stock processed by the broker was less than or equal to $75,000. [9.3B]

CUMULATIVE REVIEW EXERCISES

1. $40a - 28$ [2.2D]   **2.** $\dfrac{1}{8}$ [3.2A]   **3.** 4 [3.3B]   **4.** $-12a^7b^4$ [4.2B]   **5.** $-\dfrac{1}{b^4}$ [4.4A]

6. $4x - 2 - \dfrac{4}{4x - 1}$ [4.5B]   **7.** 0 [7.1D]   **8.** $3a^2(3x + 1)(3x - 1)$ [5.4B]   **9.** $\dfrac{1}{x + 2}$ [6.1C]

10. $\dfrac{18a}{(2a - 3)(a + 3)}$ [6.3B]   **11.** $-\dfrac{5}{9}$ [6.5A]   **12.** $C = S + Rt$ [6.7A]   **13.** $-\dfrac{7}{3}$ [7.3B]

14. $y = -\dfrac{3}{2}x - \dfrac{3}{2}$ [7.4A]   **15.** $(4, 1)$ [8.2A]   **16.** $(1, -4)$ [8.3A]   **17.** $A \cup B = \{-10, -2, 0, 1, 2\}$ [9.1A]

18. $\{x \mid x < 48, x \in \text{real numbers}\}$ [9.1B]   **19.** ＋＋＋＋＋)＋(＋＋＋＋＋
−5−4−3−2−1 0 1 2 3 4 5 [9.1C]

20. ＋＋＋＋＋(＋＋＋＋＋＋
−5−4−3−2−1 0 1 2 3 4 5 [9.2B]   **21.** $x < -15$ [9.2B]   **22.** $x > 2$ [9.3A]

23. $\{x \mid x \leq -26, x \in \text{integers}\}$ [9.2C]   **24.** The maximum number of miles is 359 mi. [9.3B]   **25.** There are an estimated 5000 fish in the lake. [6.6B]   **26.** The angle measures are 65°, 35°, and 80°. [3.5B]

27.

[7.2A]

28.

[9.4A]

Answers to Chapter 10 Selected Exercises

PREP TEST

1. -14 [1.1B]   **2.** $-2x^2y - 4xy^2$ [2.2A]   **3.** 14 [3.1C]   **4.** $\dfrac{7}{5}$ [3.3A]   **5.** x^6 [4.2A]

6. $x^2 + 2xy + y^2$ [4.3D]   **7.** $4x^2 - 12x + 9$ [4.3D]   **8.** $4 - 9v^2$ [4.3D]   **9.** $a^2 - 25$ [4.3D]   **10.** $\dfrac{x^2y^2}{9}$ [4.4A]

SECTION 10.1

3. 4   **5.** 7   **7.** $4\sqrt{2}$ **9.** $2\sqrt{2}$ **11.** $18\sqrt{2}$ **13.** $10\sqrt{10}$ **15.** $\sqrt{15}$ **17.** $\sqrt{29}$ **19.** $-54\sqrt{2}$ **21.** $3\sqrt{5}$
23. 0   **25.** $48\sqrt{2}$ **27.** 15.492   **29.** 16.971   **31.** 16   **33.** x^3 **35.** $y^7\sqrt{y}$ **37.** a^{10} **39.** x^2y^2 **41.** $2x^2$
43. $2x\sqrt{6}$ **45.** $2x^2\sqrt{15x}$ **47.** $7a^2b^4$ **49.** $3x^2y^3\sqrt{2xy}$ **51.** $2x^5y^3\sqrt{10xy}$ **53.** $4a^4b^5\sqrt{5a}$ **55.** $8ab\sqrt{b}$
57. x^3y **59.** $8a^2b^3\sqrt{5b}$ **61.** $6x^2y^3\sqrt{3y}$ **63.** $4x^3y\sqrt{2y}$ **65.** $5a + 20$ **67.** $2x^2 + 8x + 8$ **69.** $x + 2$
71. $y + 1$ **73. a.** The speed of the car was $12\sqrt{15}$ mph.   **b.** 46 mph   **75.** No. For example, let $a = 16$ and $b = 9$.
$\sqrt{a + b} = \sqrt{16 + 9} = \sqrt{25} = 5$. $\sqrt{a} + \sqrt{b} = \sqrt{16} + \sqrt{9} = 4 + 3 = 7$.   **77.** No. 18 contains a perfect-square factor. $6\sqrt{2}$
79. a. 1   **b.** 3   **c.** $3\sqrt{3}$

SECTION 10.2

1. 2, 20, and 50   **5.** $3\sqrt{2}$ **7.** $-\sqrt{7}$ **9.** $-11\sqrt{11}$ **11.** $10\sqrt{x}$ **13.** $-2\sqrt{y}$ **15.** $-11\sqrt{3b}$ **17.** $2x\sqrt{2}$
19. $-3a\sqrt{3a}$ **21.** $-5\sqrt{xy}$ **23.** $8\sqrt{5}$ **25.** $8\sqrt{2}$ **27.** $15\sqrt{2} - 10\sqrt{3}$ **29.** $\sqrt{x}$ **31.** $-12x\sqrt{3}$

33. $2xy\sqrt{x} - 3xy\sqrt{y}$ **35.** $-9x\sqrt{3x}$ **37.** $-13y^2\sqrt{2y}$ **39.** $4a^2b^2\sqrt{ab}$ **41.** $7\sqrt{2}$ **43.** $6\sqrt{x}$ **45.** $-3\sqrt{y}$
47. $-45\sqrt{2}$ **49.** $13\sqrt{3} - 12\sqrt{5}$ **51.** $32\sqrt{3} - 3\sqrt{11}$ **53.** $6\sqrt{x}$ **55.** $-34\sqrt{3x}$ **57.** $10a\sqrt{3b} + 10a\sqrt{5b}$
59. $-2xy\sqrt{3}$ **61.** $5\sqrt{2}$

SECTION 10.3

3. 5   **5.** 6   **7.** x **9.** x^3y^2 **11.** $3ab^6\sqrt{2a}$ **13.** $12a^4b\sqrt{b}$ **15.** $2 - \sqrt{6}$ **17.** $x - \sqrt{xy}$ **19.** $5\sqrt{2} - \sqrt{5x}$
21. $4 - 2\sqrt{10}$ **23.** $x - 6\sqrt{x} + 9$ **25.** $3a - 3\sqrt{ab}$ **27.** $10abc$ **29.** $-2 + 2\sqrt{5}$ **31.** $16 + 10\sqrt{2}$
33. $6x + 10\sqrt{x} - 4$ **35.** $15x - 22y\sqrt{x} + 8y^2$ **37.** $x - y$ **41.** 3   **43.** 4   **45.** $6x^2$ **47.** $2x^2\sqrt{2y}$
49. $4x\sqrt{y}$ **51.** $\dfrac{2\sqrt{3x}}{3y}$ **53.** $\dfrac{-1 + \sqrt{2}}{2}$ **55.** $-\dfrac{5\sqrt{7} + 15}{2}$ **57.** $-\dfrac{5\sqrt{3} + 9}{2}$ **59.** $3 + \sqrt{6}$ **61.** $\dfrac{-12 + 3\sqrt{2}}{7}$
63. $\dfrac{14 - 9\sqrt{2}}{17}$ **65.** $\sqrt{15} + 2\sqrt{5}$ **67.** $\dfrac{x\sqrt{y} + y\sqrt{x}}{x - y}$ **69.** $4\sqrt{6} + 12$

SECTION 10.4

1. 25   **3.** 144   **5.** 5   **7.** No solution   **9.** -1 **11.** 1   **13.** 15   **15.** $\dfrac{7}{3}$ **17.** 2   **19.** 5   **21.** 4
23. The pitcher's mound is less than halfway between home plate and second base.   **25.** The periscope must be 16.67 ft
above the water.   **27.** The height of the screen is 21.6 in.   **29.** The distance of the child from the center is 18.75 ft.
31. The perimeter of the triangle is 30 units.   **35.** The area of the fountain is approximately 244.78 ft^2.

CHAPTER 10 REVIEW EXERCISES

1. 3 [10.3A]   **2.** $9a^2\sqrt{2ab}$ [10.1B]   **3.** 12 [10.1A]   **4.** $3a\sqrt{2} + 2a\sqrt{3}$ [10.3A]   **5.** $2\sqrt{6}$ [10.3B]
6. $-8\sqrt{2}$ [10.2A]   **7.** 2 [10.3A]   **8.** 1 [10.4A]   **9.** $-x\sqrt{3} - x\sqrt{5}$ [10.3B]   **10.** $-6\sqrt{30}$ [10.1A]
11. 20 [10.4A]   **12.** $20\sqrt{3}$ [10.1A]   **13.** $7x^2y^4$ [10.3B]   **14.** No solution [10.4A]
15. $18a\sqrt{5b} + 5a\sqrt{b}$ [10.2A]   **16.** $20\sqrt{10}$ [10.1A]   **17.** $7x^2y\sqrt{15xy}$ [10.2A]   **18.** $8y + 10\sqrt{5y} - 15$ [10.3A]
19. $26\sqrt{3x}$ [10.2A]   **20.** No solution [10.4A]   **21.** $\dfrac{8\sqrt{x} + 24}{x - 9}$ [10.3B]   **22.** $36x^8y^5\sqrt{3xy}$ [10.1B]
23. $2y^4\sqrt{6}$ [10.1B]   **24.** -1 [10.4A]   **25.** $-6x^3y^2\sqrt{2y}$ [10.2A]   **26.** $\dfrac{16\sqrt{a}}{a}$ [10.3B]   **27.** The distance across
the pond is approximately 43 ft. [10.4B]   **28.** The explorer weighs 144 lb on the surface of Earth. [10.4B]
29. The depth of the water is 100 ft. [10.4B]   **30.** The radius of the corner is 25 ft. [10.4B]

CHAPTER 10 TEST

1. $11x^4y$ [10.1B]   **2.** $6x^2y\sqrt{y}$ [10.3A]   **3.** $-5\sqrt{2}$ [10.2A]   **4.** $3\sqrt{5}$ [10.1A]   **5.** 9 [10.3B]
6. 25 [10.4A]   **7.** $4a^2b^5\sqrt{2ab}$ [10.1B]   **8.** $7ab\sqrt{a}$ [10.3B]   **9.** $\sqrt{3} + 1$ [10.3B]   **10.** $4x^2y^2\sqrt{5y}$ [10.3A]
11. 9 [10.4A]   **12.** $21\sqrt{2y} - 12\sqrt{2x}$ [10.2A]   **13.** $6x^3y\sqrt{2x}$ [10.1B]   **14.** $y + 2\sqrt{y} - 15$ [10.3A]
15. $-2xy\sqrt{3xy} - 3xy\sqrt{xy}$ [10.2A]   **16.** $\dfrac{17 - 8\sqrt{5}}{31}$ [10.3B]   **17.** $a - \sqrt{ab}$ [10.3A]   **18.** $5\sqrt{3}$ [10.1A]
19. The length of the pendulum is 7.30 ft. [10.4B]   **20.** The rope should be secured about 7 ft from the base of the
pole. [10.4B]

CUMULATIVE REVIEW EXERCISES

1. $-\dfrac{1}{12}$ [1.7A]   **2.** $2x + 18$ [2.2D]   **3.** $\dfrac{1}{13}$ [3.3B]   **4.** $6x^5y^5$ [4.2A]   **5.** $-2b^2 + 1 - \dfrac{1}{3b^2}$ [4.5A]
6. 1 [7.1D]   **7.** $2a(a - 5)(a - 3)$ [5.2B]   **8.** $\dfrac{1}{4(x + 1)}$ [6.1B]   **9.** $\dfrac{x + 3}{x - 3}$ [6.3B]   **10.** $\dfrac{5}{3}$ [6.5A]
11. $y = \dfrac{1}{2}x - 2$ [7.4A]   **12.** $(1, 1)$ [8.2A]   **13.** $(3, -2)$ [8.3A]   **14.** $x \le -\dfrac{9}{2}$ [9.3A]   **15.** $6\sqrt{3}$ [10.1A]
16. $-4\sqrt{2}$ [10.2A]   **17.** $4ab\sqrt{2ab} - 5ab\sqrt{ab}$ [10.2A]   **18.** $14a^5b^2\sqrt{2a}$ [10.3A]   **19.** $3\sqrt{2} - x\sqrt{3}$ [10.3A]
20. 8 [10.3B]   **21.** $-6 - 3\sqrt{5}$ [10.3B]   **22.** 6 [10.4A]   **23.** The cost of the book is \$24.50. [3.2B]
24. 56 oz of water must be added. [3.6B]   **25.** The numbers are 8 and 13. [5.5B]   **26.** It would take the small pipe,

working alone, 48 h.   [6.8A]   **27.**

[8.1A]

28.

[9.4A]

29. The smaller integer is 40.   [10.4B]   **30.** The height of the building is 400 ft.   [10.4B]

Answers to Chapter 11 Selected Exercises

PREP TEST

1. 41   [2.1A]   **2.** $-\dfrac{1}{5}$ [3.2A]   **3.** $(x + 4)(x - 3)$ [5.2A]   **4.** $(2x - 3)^2$ [5.4A]   **5.** Yes   [5.4A]   **6.** 3   [6.5A]

7.

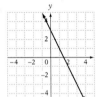

[7.2A]

8. $2\sqrt{7}$ [10.1A]   **9.** $|a|$ [10.1B]   **10.** 3.6 mi   [3.6C]

SECTION 11.1

1. $-3, 5$ **3.** $-\dfrac{5}{2}, \dfrac{1}{3}$ **5.** $-5, 3$ **7.** $1, 3$ **9.** $-2, -1$ **11.** 3 **13.** $-\dfrac{2}{3}, 0$ **15.** $-2, 5$ **17.** $\dfrac{2}{3}, 1$

19. $-3, \dfrac{1}{3}$ **21.** $\dfrac{2}{3}$ **23.** $-\dfrac{1}{2}, \dfrac{3}{2}$ **25.** $\dfrac{1}{2}$ **27.** $-3, 3$ **29.** $-\dfrac{1}{2}, \dfrac{1}{2}$ **31.** $-3, 5$ **33.** $1, 5$ **35.** $-6, 6$

37. $-1, 1$ **39.** $-\dfrac{7}{2}, \dfrac{7}{2}$ **41.** $-\dfrac{2}{3}, \dfrac{2}{3}$ **43.** $-\dfrac{3}{4}, \dfrac{3}{4}$ **45.** No real number solution   **47.** $-2\sqrt{6}, 2\sqrt{6}$ **49.** $-5, 7$

51. $-7, -3$ **53.** $-\dfrac{1}{3}, \dfrac{7}{3}$ **55.** $-\dfrac{12}{7}, -\dfrac{2}{7}$ **57.** $4 - 2\sqrt{5}, 4 + 2\sqrt{5}$ **59.** No real number solution

61. $-\dfrac{3}{4} - 2\sqrt{3}, -\dfrac{3}{4} + 2\sqrt{3}$ **63.** $-1, -\dfrac{\sqrt{6}}{3}, \dfrac{\sqrt{6}}{3}, 1$ **65.** The annual percentage rate is 8%.   **67.** Yes, because

$v \approx 73.5$ mph.

SECTION 11.2

1. $x^2 - 8x + 16, (x - 4)^2$ **3.** $x^2 + 5x + \dfrac{25}{4}, \left(x + \dfrac{5}{2}\right)^2$ **5.** $-3, 1$ **7.** $-2, 8$ **9.** 2 **11.** No real number

solution   **13.** $-4, -1$ **15.** $-8, 1$ **17.** $-2 - \sqrt{3}, -2 + \sqrt{3}$ **19.** $-3 - \sqrt{14}, -3 + \sqrt{14}$ **21.** $1 - \sqrt{2}, 1 + \sqrt{2}$

23. $\dfrac{-3 - \sqrt{13}}{2}, \dfrac{-3 + \sqrt{13}}{2}$ **25.** $1, 2$ **27.** $\dfrac{-1 - \sqrt{13}}{2}, \dfrac{-1 + \sqrt{13}}{2}$ **29.** $-5 - 4\sqrt{2}, -5 + 4\sqrt{2}$

31. $\dfrac{3 - \sqrt{29}}{2}, \dfrac{3 + \sqrt{29}}{2}$ **33.** $\dfrac{1 - \sqrt{17}}{2}, \dfrac{1 + \sqrt{17}}{2}$ **35.** No real number solution   **37.** $\dfrac{1}{2}, 1$ **39.** $-3, \dfrac{1}{2}$

41. $\dfrac{3}{2}, 2$ **43.** $-\dfrac{1}{2}, 1$ **45.** $-2, \dfrac{1}{3}$ **47.** $-2, -\dfrac{2}{3}$ **49.** $-\dfrac{3}{2}, \dfrac{1}{2}$ **51.** $-\dfrac{3}{2}, \dfrac{1}{3}$ **53.** $-\dfrac{1}{2}, \dfrac{4}{3}$ **55.** $-5.372, 0.372$

57. $-3.212, 1.712$ **59.** $-1.151, 0.651$ **61.** $1 - \sqrt{7}, 1 + \sqrt{7}$ **63.** $-1, 0$ **65.** $4 - \sqrt{3}, 4 + \sqrt{3}$ **67.** $0, 8$

69. $\dfrac{15 + 3\sqrt{17}}{2}$ **71.** The ball hits the ground approximately 4.81 s after it is hit.

SECTION 11.3

1. $-1, 5$ **3.** $-3, 5$ **5.** $-1, 3$ **7.** $-5, 1$ **9.** $-\dfrac{1}{2}, 1$ **11.** No real number solution   **13.** $0, 1$ **15.** $-\dfrac{3}{2}, \dfrac{3}{2}$

17. $-\dfrac{5}{2}, \dfrac{3}{2}$ **19.** No real number solution   **21.** $1 - \sqrt{6}, 1 + \sqrt{6}$ **23.** $-3 - \sqrt{10}, -3 + \sqrt{10}$

25. $2 - \sqrt{13}, 2 + \sqrt{13}$ **27.** $\dfrac{-1 - \sqrt{2}}{3}, \dfrac{-1 + \sqrt{2}}{3}$ **29.** $-\dfrac{1}{2}$ **31.** No real number solution   **33.** $\dfrac{-4 - \sqrt{5}}{2}, \dfrac{-4 + \sqrt{5}}{2}$

35. $\dfrac{1 - 2\sqrt{3}}{2}, \dfrac{1 + 2\sqrt{3}}{2}$ **37.** $\dfrac{-5 - \sqrt{2}}{3}, \dfrac{-5 + \sqrt{2}}{3}$ **39.** $-3.690, 5.690$ **41.** $-1.690, 7.690$ **43.** $-1.089, 4.589$

45. $-2.118, 0.118$ **47.** $-0.905, 1.105$ **51. a.** False   **b.** False   **c.** False   **d.** True   **53.** $-4, -3$

55. No solution   **57.** $-1, 11$ **59.** The planes are 1000 mi apart after 2 h.

SECTION 11.4

1. Down   **3.** Up   **5.** 4   **7.** -5 **9.** -42 **11.**   **13.**

15.   **17.**   **19.**   **21.**

23.   **25.** 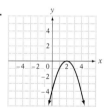   **27.** $(-6, 0), (1, 0); (0, -6)$ **29.** $(-6, 0); (0, 36)$

31. $(-2 - \sqrt{6}, 0), (-2 + \sqrt{6}, 0); (0, -2)$ **33.** No x-intercepts; $(0, 1)$ **35.** $\left(\dfrac{3}{2}, 0\right), (5, 0); (0, 15)$

37. $\left(\dfrac{-3 - \sqrt{33}}{6}, 0\right), \left(\dfrac{-3 + \sqrt{33}}{6}, 0\right); (0, 2)$ **39.** $y = 3x^2 + 6x + 5$ **41.** $y = 3x^2 - 6x$ **43.** $(-1, 0), (0, 0), (5, 0)$

45. $(-3, 0), (-1, 0), (1, 0)$

SECTION 11.5

1. The length is 4 m. The height is 10 m.   **3.** The width is 4 ft. The length is 6 ft.   **5.** The width is 50 ft. The length is 100 ft.   **7.** The hang time of the football is approximately 5.5 s.   **9.** The maximum velocity is 78 ft/s.   **11.** The large pizza has a radius of 7 in.   **13.** The first computer can solve the equation in 35 min. The second computer can solve the equation in 14 min.   **15.** Using the first engine it would take 12 h. Using the second engine it would take 6 h.   **17.** The rate of the plane in calm air is 100 mph.   **19.** The 16-inch pizza should cost $24.

CHAPTER 11 REVIEW EXERCISES

1. $-\dfrac{7}{2}, \dfrac{4}{3}$ [11.1A]   **2.** $-\dfrac{5}{7}, \dfrac{5}{7}$ [11.1B]   **3.** $-6, 4$ [11.2A]   **4.** $-6, 1$ [11.3A]   **5.** $-4, \dfrac{3}{2}$ [11.2A]

6. $\dfrac{5}{12}, 2$ [11.1A]   **7.** $-2 - 2\sqrt{6}, -2 + 2\sqrt{6}$ [11.1B]   **8.** $1, \dfrac{3}{2}$ [11.3A]   **9.** $-\dfrac{1}{2}, -\dfrac{1}{3}$ [11.1A]

10. No real number solution   [11.1B]   **11.** $2 - \sqrt{3}, 2 + \sqrt{3}$ [11.2A]   **12.** $\dfrac{3 - \sqrt{29}}{2}, \dfrac{3 + \sqrt{29}}{2}$ [11.3A]

13. No real number solution   [11.2A]   **14.** $-10, -7$ [11.1A]   **15.** $1, 2$ [11.1B]

16. $\dfrac{-4 - \sqrt{23}}{2}, \dfrac{-4 + \sqrt{23}}{2}$ [11.2A]   **17.** No real number solution   [11.3A]   **18.** $-2, -\dfrac{1}{2}$ [11.3A]

19.   **20.**   **21.**

[11.4A]   [11.4A]   [11.4A]

22. 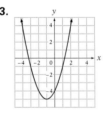   **23.**

[11.4A]   [11.4A]

24. $(-3, 0), (5, 0); (0, -15)$ [11.4A]   **25.** The rate of the hawk in calm air is 75 mph.   [11.5A]

CHAPTER 11 TEST

1. $-1, 6$ [11.1A]   **2.** $-4, \dfrac{5}{3}$ [11.1A]   **3.** $0, 10$ [11.1B]   **4.** $-4 - 2\sqrt{5}, -4 + 2\sqrt{5}$ [11.1B]

5. $-2 - 2\sqrt{5}, -2 + 2\sqrt{5}$ [11.2A]   **6.** $\dfrac{-3 - \sqrt{41}}{2}, \dfrac{-3 + \sqrt{41}}{2}$ [11.2A]   **7.** $\dfrac{3 - \sqrt{7}}{2}, \dfrac{3 + \sqrt{7}}{2}$ [11.2A]

8. $\dfrac{-4 - \sqrt{22}}{2}, \dfrac{-4 + \sqrt{22}}{2}$ [11.2A]   **9.** $-2 - \sqrt{2}, -2 + \sqrt{2}$ [11.3A]   **10.** $\dfrac{3 - \sqrt{33}}{2}, \dfrac{3 + \sqrt{33}}{2}$ [11.3A]

11. $-\dfrac{1}{2}, 3$ [11.3A]   **12.** $\dfrac{1 - \sqrt{13}}{6}, \dfrac{1 + \sqrt{13}}{6}$ [11.3A]   **13.**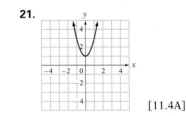

[11.4A]

14. $(-4, 0), (3, 0); (0, -12)$ [11.4A]   **15.** The width is 5 ft. The length is 8 ft.   [11.5A]
16. The rate of the boat in calm water is 11 mph.   [11.5A]

CUMULATIVE REVIEW EXERCISES

1. $-28x + 27$ [2.2D]   **2.** $\dfrac{3}{2}$ [3.1C]   **3.** 3 [3.3B]   **4.** $-12a^8b^4$ [4.2B]   **5.** $x + 2 - \dfrac{4}{x - 2}$ [4.5B]

6. $x(3x - 4)(x + 2)$ [5.3A/5.3B]   **7.** $\dfrac{9x^2(x - 2)^2}{(2x - 3)^2}$ [6.1C]   **8.** $\dfrac{x + 2}{2(x + 1)}$ [6.3B]   **9.** $\dfrac{x - 4}{2x + 5}$ [6.4A]

10. $(3, 0); (0, -4)$ [7.3A]   **11.** $y = -\dfrac{4}{3}x - 2$ [7.4A]   **12.** $(2, 1)$ [8.2A]   **13.** $(2, -2)$ [8.3A]   **14.** $x > \dfrac{1}{9}$ [9.3A]

15. $a - 2$ [10.3A]   **16.** $6ab\sqrt{a}$ [10.3B]   **17.** $\dfrac{-6 + 5\sqrt{3}}{13}$ [10.3B]   **18.** 5 [10.4A]   **19.** $\dfrac{1}{3}, \dfrac{5}{2}$ [11.1A]

20. $5 - 3\sqrt{2}, 5 + 3\sqrt{2}$ [11.1B]   **21.** $\dfrac{-7 - \sqrt{13}}{6}, \dfrac{-7 + \sqrt{13}}{6}$ [11.2A]   **22.** $-\dfrac{1}{2}, 2$ [11.3A]   **23.** The cost of the
mixture is $2.25 per pound.   [3.6A]   **24.** 250 additional shares are required.   [6.6B]   **25.** The rate of the plane in still
air is 200 mph. The rate of the wind is 40 mph.   [8.4A]   **26.** The score on the last test must be 77 or better.   [9.2C]

27. The middle integer can be −5 or 5. [11.5A]   **28.** The rate for the last 8 mi is 4 mph. [11.5A]

29.

[9.4A]

30.

[11.4A]

FINAL EXAM

1. −3 [1.1B]   **2.** −6 [1.2B]   **3.** −256 [1.4A]   **4.** −11 [1.4B]   **5.** $-\dfrac{15}{2}$ [2.1A]   **6.** $9x + 6y$ [2.2A]

7. $6z$ [2.2B]   **8.** $16x - 52$ [2.2D]   **9.** −50 [3.1C]   **10.** −3 [3.3B]   **11.** 12.5% [1.4B]   **12.** 15.2 [3.1D]

13. $-3x^2 - 3x + 8$ [4.1B]   **14.** $81x^4y^{12}$ [4.2B]   **15.** $6x^3 + 7x^2 - 7x - 6$ [4.3B]   **16.** $-\dfrac{x^4y}{2}$ [4.4A]

17. $\dfrac{3x}{y} - 4x^2 - \dfrac{5}{x}$ [4.5A]   **18.** $5x - 12 + \dfrac{23}{x + 2}$ [4.5B]   **19.** $\dfrac{4y^6}{x^6}$ [4.4A]   **20.** $\dfrac{3}{4}$ [7.1D]

21. $(x - 6)(x + 1)$ [5.2A]   **22.** $(3x + 2)(2x - 3)$ [5.3A/5.3B]   **23.** $4x(2x - 1)(x - 3)$ [5.4B]

24. $(5x + 4)(5x - 4)$ [5.4A]   **25.** $2(a + 3)(4 - x)$ [5.1B]   **26.** $3y(5 + 2x)(5 - 2x)$ [5.4B]   **27.** $\dfrac{1}{2}$, 3 [5.5A]

28. $\dfrac{2(x + 1)}{x - 1}$ [6.1B]   **29.** $\dfrac{-3x^2 + x - 25}{(x + 3)(2x - 5)}$ [6.3B]   **30.** $\dfrac{x^2 - 2x}{x - 1}$ [6.4A]   **31.** 2 [6.5A]   **32.** $a = b$ [6.7A]

33. $\dfrac{2}{3}$ [7.3B]   **34.** $y = -\dfrac{2}{3}x - 2$ [7.4A]   **35.** $(6, 17)$ [8.2A]   **36.** $(2, -1)$ [8.3A]   **37.** $x \le -3$ [9.2B]

38. $y \ge \dfrac{5}{2}$ [9.3A]   **39.** $7x^3$ [10.1B]   **40.** $38\sqrt{3a}$ [10.2A]   **41.** $\sqrt{15} + 2\sqrt{3}$ [10.3B]   **42.** 2 [10.4A]

43. $-1, \dfrac{4}{3}$ [11.1A]   **44.** $\dfrac{1 - \sqrt{5}}{4}, \dfrac{1 + \sqrt{5}}{4}$ [11.3A]   **45.** $2x + 3(x - 2), 5x - 6$ [2.3B]   **46.** The original value is $3000. [3.1D]   **47.** The markup rate is 65%. [3.2B]   **48.** $6000 must be invested at 11%. [8.2B]   **49.** The cost for the mixture is $3 per pound. [3.6A]   **50.** The percent concentration of acid in the mixture is 36%. [3.6B]   **51.** The distance traveled in the first hour was 215 km. [3.6C]   **52.** The angles measure 50°, 60°, and 70°. [3.5B]   **53.** The middle integer can be −4 or 4. [11.5A]   **54.** The width is 5 m. The length is 10 m. [5.5B]   **55.** 16 oz of dye are required. [6.6B]   **56.** Working together, it would take them 36 min or 0.6 h. [6.8A]   **57.** The rate of the boat in calm water is 15 mph. The rate of the current is 5 mph. [8.4A]   **58.** The rate of the wind is 25 mph. [11.5A]

59.

[7.3C]

60.

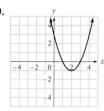

[11.4A]

Glossary

abscissa The first number in an ordered pair. It measures a horizontal distance and is also called the first coordinate. [7.1]

absolute value of a number The distance of the number from zero on the number line. [1.1]

acute angle An angle whose measure is between 0° and 90°. [3.5]

addend In addition, a number being added. [1.2]

addition The process of finding the total of two numbers. [1.2]

addition method An algebraic method of finding an exact solution of a system of linear equations. [8.3]

additive inverses Numbers that are the same distance from zero on the number line, but on opposite sides; also called opposites. [1.1/2.2]

adjacent angles Two angles that share a common side. [3.5]

alternate exterior angles Two angles that are on opposite sides of the transversal and outside the parallel lines. [3.5]

alternate interior angles Two angles that are on opposite sides of the transversal and between the parallel lines. [3.5]

analytic geometry Geometry in which a coordinate system is used to study the relationships between variables. [7.1]

angle Figure formed when two rays start from the same point. [1.8]

arithmetic mean of values Average determined by calculating the sum of the values and then dividing that result by the number of values. [1.3]

axes The two number lines that form a rectangular coordinate system; also called coordinate axes. [7.1]

base In exponential notation, the factor that is multiplied the number of times shown by the exponent. [1.4]

basic percent equation Percent times base equals amount. [3.1]

binomial A polynomial of two terms. [4.1]

binomial factor A factor that has two terms. [5.1]

center of a circle The point from which all points on the circle are equidistant. [1.8]

circle Plane figure in which all points are the same distance from its center. [1.8]

circumference The perimeter of a circle. [1.8]

clearing denominators Removing denominators from an equation that contains fractions by multiplying each side of the equation by the LCM of the denominators. [3.2/6.5]

coefficient The number part of a variable term. [2.1]

combining like terms Using the Distributive Property to add the coefficients of like variable terms; adding like terms of a variable expression. [2.2]

complementary angles Two angles whose sum is 90°. [1.8]

completing the square Adding to a binomial the constant term that makes it a perfect-square trinomial. [11.2]

complex fraction A fraction whose numerator or denominator contains one or more fractions. [6.4]

composite number A natural number greater than 1 that is not a prime number. [1.5]

conjugates Binomial expressions that differ only in the sign of a term. The expressions $a + b$ and $a - b$ are conjugates. [10.3]

consecutive even integers Even integers that follow one another in order. [3.4]

consecutive integers Integers that follow one another in order. [3.4]

consecutive odd integers Odd integers that follow one another in order. [3.4]

constant term A term that includes no variable part; also called a constant. [2.1]

coordinate axes The two number lines that form a rectangular coordinate system; also simply called axes. [7.1]

coordinates of a point The numbers in an ordered pair that is associated with a point. [7.1]

corresponding angles Two angles that are on the same side of the transversal and are both acute angles or are both obtuse angles. [3.5]

cost The price that a business pays for a product. [3.2]

decimal notation Notation in which a number consists of a whole-number part, a decimal point, and a decimal part. [1.6]

degree A unit used to measure angles. [1.8]

degree of a polynomial in one variable The largest exponent that appears on the variable. [4.1]

dependent system A system of equations that has an infinite number of solutions. [8.1]

dependent variable In a function, the variable whose value depends on the value of another variable known as the independent variable. [7.1]

descending order The terms of a polynomial in one variable arranged so that the exponents on the variable decrease from left to right. The polynomial $9x^5 - 2x^4 + 7x^3 + x^2 - 8x + 1$ is in descending order. [4.1]

diameter Line segment across a circle that passes through the circle's center. [1.8]

difference of two squares A polynomial of the form $a^2 - b^2$. [5.4]

discount The amount by which a retailer reduces the regular price of a product for a promotional sale. [3.2]

discount rate The percent of the regular price that the discount represents. [3.2]

domain The set of first coordinates of the ordered pairs in a relation. [7.1]

double root The two equal roots of a quadratic equation, which occurs when the discriminant $b^2 - 4ac$ equals zero. [11.1]

element of a set One of the objects in a set. [1.1/9.1]

empty set The set that contains no elements; also called the null set. [9.1]

equation A statement of the equality of two mathematical expressions. [3.1]

equilateral triangle A triangle in which all three sides are of equal length. [1.8]

equivalent equations Equations that have the same solution. [3.1]

evaluating a function Replacing x in $f(x)$ with some value and then simplifying the numerical expression that results. [7.1]

evaluating a variable expression Replacing each variable by its value and then simplifying the resulting numerical expression. [2.1]

even integer An integer that is divisible by 2. [3.4]

exponent In exponential notation, the elevated number that indicates how many times the base occurs in the multiplication. [1.4]

exponential form The expression 2^5 is in exponential form. Compare *factored form*. [1.4]

exterior angle of a triangle Angle adjacent to an interior angle of a triangle. [3.5]

factor by grouping Process of grouping and factoring terms in a polynomial in such a way that a common binomial factor is found. [5.1]

factor completely Refers to writing a polynomial as a product of factors that are nonfactorable over the integers. [5.2]

factor of a number In multiplication, a number being multiplied. [1.3]

factor a polynomial To write the polynomial as a product of other polynomials. [5.1]

factor a trinomial of the form $x^2 + bx + c$ To express the trinomial as the product of two binomials. [5.2]

factored form The expression $2 \cdot 2 \cdot 2 \cdot 2 \cdot 2$ is in factored form. Compare *exponential form*. [1.4]

first coordinate The first number in an ordered pair. It measures a horizontal distance and is also called the abscissa. [7.1]

first-degree equation in two variables An equation of the form $y = mx + b$, where m is the coefficient and b is a constant; also called a linear equation in two variables or a linear function. [7.2]

FOIL A method of finding the product of two binomials; the letters stand for First, Outer, Inner, and Last. [4.3]

formula A literal equation that states rules about measurements. [6.7]

function A relation in which no two ordered pairs that have the same first coordinate have different second coordinates. [7.1]

functional notation A function designated by $f(x)$, which is the value of the function at x. [7.1]

graph a point in the plane To place a dot at the location given by the ordered pair; also called plotting a point in the plane. [7.1]

graph of a relation The graph of the ordered pairs that belong to the relation. [7.1]

graph of an equation in two variables A graph of the ordered-pair solutions of the equation. [7.2]

graph of an integer A heavy dot directly above that number on the number line. [1.1]

graph of an ordered pair The dot drawn at the coordinates of the point in the plane. [7.1]

greater than A number a is greater than another number b, written $a > b$, if a is to the right of b on the number line. [1.1]

greater than or equal to The symbol $\geq$ means "is greater than or equal to." [1.1]

greatest common factor (GCF) The greatest common factor of two or more integers is the greatest integer that is a factor of all the integers. [1.5]

greatest common factor (GCF) of two or more monomials The product of the GCF of the coefficients and the common variable factors. [5.1]

half-plane The solution set of an inequality in two variables. [9.4]

hypotenuse In a right triangle, the side opposite the 90° angle. [10.4]

inconsistent system A system of equations that has no solution. [8.1]

independent system A system of equations that has one solution. [8.1]

independent variable In a function, the variable that varies independently and whose value determines the value of the dependent variable. [7.1]

inequality An expression that contains the symbol $>$, $<$, $\geq$ (is greater than or equal to), or $\leq$ (is less than or equal to). [9.1]

integers The numbers ..., -3, -2, -1, 0, 1, 2, 3, ... [1.1]

interior angle of a triangle Angle within the region enclosed by a triangle. [3.5]

intersecting lines Lines that cross at a point in a plane. [1.8]

intersection of sets A and B The set that contains the elements that are common to both A and B. [9.1]

irrational number The decimal representation of an irrational number never repeats or terminates and can only be approximated. [1.6/10.1]

isosceles triangle A triangle that has two equal angles and two equal sides. [1.8]

least common denominator The smallest number that is a multiple of each denominator in question. [1.6]

least common multiple (LCM) The LCM of two or more numbers is the smallest number that contains the prime factorization of each number. [1.5]

least common multiple (LCM) of two or more polynomials The polynomial of least degree that contains all the factors of each polynomial. [6.2]

legs The sides opposite the hypotenuse in a right triangle. [10.4]

less than A number a is less than another number b, written $a < b$, if a is to the left of b on the number line. [1.1]

less than or equal to The symbol $\leq$ means "is less than or equal to". [1.1]

like terms Terms of a variable expression that have the same variable part. [2.2]

line Having no width, it extends indefinitely in two directions in a plane. [1.8]

line of best fit A line drawn to approximate data that are graphed as points in a coordinate system. [7.4]

line segment Part of a line that has two endpoints. [1.8]

linear equation in two variables An equation of the form $y = mx + b$, where m and b are constants; also called a linear function or a first-degree equation in two variables. [7.2]

linear function An equation of the form $y = mx + b$, where m and b are constants; also called a linear equation in two variables or a first-degree equation in two variables. [7.2]

linear model A first-degree equation that is used to describe a relationship between quantities. [7.4]

literal equation An equation that contains more than one variable. [6.7]

markdown The amount by which a retailer reduces the regular price of a product for a promotional sale. [3.2]

markup The difference between selling price and cost. [3.2]

markup rate The percent of retailer's cost that the markup represents. [3.2]

monomial A number, a variable, or a product of numbers and variables; a polynomial of one term. [4.1]

moving average The arithmetic mean of the changes in the value of a stock for a given number of days. [1.3]

multiplicative inverse The reciprocal of a number. [2.2]

natural numbers The numbers 1, 2, 3, …. [1.1]

negative integers The numbers …, -4, -3, -2, -1. [1.1]

negative slope A property of a line that slants downward to the right. [7.3]

nonfactorable over the integers A polynomial that does not factor using only integers. [5.2]

null set The set that contains no elements; also called the empty set. [9.1]

numerical coefficient The number part of a variable term. When the numerical coefficient is 1 or -1, the 1 is usually not written. [2.1]

obtuse angle An angle whose measure is between 90° and 180°. [3.5]

odd integer An integer that is not divisible by 2. [3.4]

opposite of a polynomial The polynomial created when the sign of each term of the original polynomial is changed. [4.1]

opposites Numbers that are the same distance from zero on the number line, but on opposite sides; also called additive inverses. [1.1]

ordered pair Pair of numbers of the form that can be used to identify a point in the plane determined by the axes of a rectangular coordinate system. [7.1]

Order of Operations Agreement A set of rules that tell us in what order to perform the operations that occur in a numerical expression. [1.4]

ordinate The second number in an ordered pair. It measures a vertical distance and is also called the second coordinate. [7.1]

origin The point of intersection of the two coordinate axes that form a rectangular coordinate system. [7.1]

parabola The graph of a quadratic equation in two variables. [11.4]

parallel lines Lines that never meet; the distance between them is always the same. Parallel lines have the same slope. [1.8/7.3]

parallelogram Four-sided plane figure with opposite sides parallel. [1.8]

percent Parts of 100. [1.7]

perfect square The square of an integer. [10.1]

perfect-square trinomial A trinomial that is a product of a binomial and itself. [5.4]

perimeter The distance around a plane geometric figure. [1.8]

perpendicular lines Intersecting lines that form right angles. [1.8]

plane Flat surface that extends indefinitely. [1.8/7.1]

plane figure Figure that lies entirely in a plane. [1.8]

plot a point in the plane To place a dot at the location given by the ordered pair; to graph a point in the plane. [7.1]

point-slope formula If (x_1, y_1) is a point on a line with slope m, then $y - y_1 = m(x - x_1)$. [7.4]

polynomial A variable expression in which the terms are monomials. [4.1]

positive integers The integers, 1, 2, 3, 4, …. [1.1]

positive slope A property of a line that slants upward to the right. [7.3]

prime factorization Expressing a number as a product of its prime factors. [1.5]

prime number Number whose only factors are 1 and the number. [1.5]

prime polynomial A polynomial that is nonfactorable over the integers. [5.2]

principal square root The positive square root of a number. [10.1]

product In multiplication, the result of multiplying two numbers. [1.3]

proportion An equation that states the equality of two ratios or rates. [6.6]

Pythagorean Theorem The square of the hypotenuse of a right triangle is equal to the sum of the squares of the two legs. [10.4]

quadrant One of the four regions into which the two axes of a rectangular coordinate system divide the plane. [7.1]

quadratic equation An equation of the form $ax^2 + bx + c = 0$, where a, b, and c are constants and a is not equal to zero; also called a second-degree equation. [5.5/11.1]

quadratic equation in two variables An equation of the form $y = ax^2 + bx + c$, where a is not equal to zero. [11.4]

quadratic function A quadratic function is given by $f(x) = ax^2 + bx + c$, where a is not equal to zero. [11.4]

radical equation An equation that contains a variable expression in a radicand. [10.4]

radical sign The symbol $\sqrt{}$, which is used to indicate the positive, or principal, square root of a number. [10.1]

radicand In a radical expression, the expression under the radical sign. [10.1]

radius Line segment from the center of a circle to a point on the circle. [1.8]

range The set of second coordinates of the ordered pairs in a relation. [7.1]

rate The quotient of two quantities that have different units. [6.6]

rate of work That part of a task that is completed in one unit of time. [6.8]

ratio The quotient of two quantities that have the same unit. [6.6]

rational expression A fraction in which the numerator or denominator is a polynomial. [6.1]

rational number A number that can be written in the form a/b, where a and b are integers and b is not equal to zero. [1.6]

rationalizing the denominator The procedure used to remove a radical from the denominator of a fraction. [10.3]

ray Line that starts at a point and extends indefinitely in one direction. [1.8]

real numbers The rational numbers and the irrational numbers. [1.6]

reciprocal of a fraction Fraction that results when the numerator and denominator of a fraction are interchanged. [1.7]

reciprocal of a rational expression A rational expression in which the numerator and denominator have been interchanged. [6.1]

rectangle Parallelogram that has four right angles. [1.8]

rectangular coordinate system System formed by two number lines, one horizontal and one vertical, that intersect at the zero point of each line. [7.1]

relation Any set of ordered pairs. [7.1]

repeating decimal Decimal that is formed when dividing the numerator of its fractional counterpart by the denominator results in a decimal part wherein a block of digits repeats infinitely. [1.6]

right angle An angle whose measure is 90°. [1.8]

right triangle A triangle that contains a 90° angle. [10.4]

roster method Method of writing a set by enclosing a list of the elements in braces. [1.1/9.1]

scatter diagram A graph of collected data as points in a coordinate system. [7.4]

scientific notation Notation in which each number is expressed as the product of two factors, one a number between 1 and 10 and the other a power of ten. [4.4]

second coordinate The second number in an ordered pair. It measures a vertical distance and is also called the ordinate. [7.1]

second-degree equation An equation of the form $ax^2 + bx + c = 0$, where a, b, and c are constants and a is not equal to zero; also called a quadratic equation. [11.1]

selling price The price for which a business sells a product to a customer. [3.2]

set A collection of objects. [1.1/9.1]

set-builder notation A method of designating a set that makes use of a variable and a certain property that only elements of that set possess. [9.1]

similar objects Similar objects have the same shape but not necessarily the same size. [6.6]

simplest form of a fraction A fraction in which the numerator and denominator have no common factors other than 1. [1.6]

simplest form of a rational expression A rational expression is in simplest form when the numerator and denominator have no common factors. [6.1]

slope The measure of the slant of a line. The symbol for slope is m. [7.3]

slope-intercept form The slope-intercept form of an equation of a straight line is $y = mx + b$. [7.3]

solid An object that exists in space. [1.8]

solution of a system of equations in two variables An ordered pair that is a solution of each equation of the system. [8.1]

solution of an equation A number that, when substituted for the variable, results in a true equation. [3.1]

solution of an equation in two variables An ordered pair whose coordinates make the equation a true statement. [7.1]

solution set of an inequality A set of numbers, each element of which, when substituted for the variable, results in a true inequality. [9.1]

solving an equation Finding a solution of the equation. [3.1]

square Rectangle with four equal sides. [1.8]

square root A square root of a positive number x is a number a for which $a^2 = x$. [10.1]

standard form A quadratic equation is in standard form when the polynomial is in descending order and equal to zero. $ax^2 + bx + c = 0$ is in standard form. [5.5/11.1]

standard form of a linear equation in two variables An equation of the form $Ax + By = C$, where A and B are coefficients and C is a constant. [7.2]

straight angle An angle whose measure is 180°. [1.8]

substitution method An algebraic method of finding an exact solution of a system of equations. [8.2]

sum In addition, the total of two or more numbers. [1.2]

supplementary angles Two angles whose sum is 180°. [1.8]

system of equations Equations that are considered together. [8.1]

terminating decimal Decimal that is formed when dividing the numerator of its fractional counterpart by the denominator results in a remainder of zero. [1.6]

terms of a variable expression The addends of the expression. [2.1]

transversal A line intersecting two other lines at two different points. [3.5]

triangle A three-sided closed figure. [1.8]

trinomial A polynomial of three terms. [4.1]

undefined slope A property of a vertical line. [7.3]

uniform motion The motion of a moving object whose speed and direction do not change. [3.1/6.8]

union of sets A and B The set that contains all the elements of A and all the elements of B. [9.1]

value of a function at x The result of evaluating a variable expression, represented by the symbol $f(x)$. [7.1]

value of a variable The number assigned to the variable. [2.1]

variable A letter of the alphabet used to stand for a number that is unknown or that can change. [1.1]

variable expression An expression that contains one or more variables. [2.1]

variable part In a variable term, the variable or variables and their exponents. [2.1]

variable term A term composed of a numerical coefficient and a variable part. [2.1]

vertex Point at which the rays that form an angle meet. [1.8]

vertical angles Two angles that are on opposite sides of the intersection of two lines. [3.5]

x-coordinate The abscissa in an xy-coordinate system. [7.1]

x-intercept The point at which a graph crosses the x-axis. [7.3]

xy-coordinate system A rectangular coordinate system in which the horizontal axis is labeled x and the vertical axis is labeled y. [7.1]

y-coordinate The ordinate in an xy-coordinate system. [7.1]

y-intercept The point at which a graph crosses the y-axis. [7.3]

zero slope A property of a horizontal line. [7.3]

Index

A

Abscissa, 353
Absolute value, 5–6
Abundant number, 435
Acute angle, 161
Addends, 9
Addition, 9
  Associative Property of, 83
  Commutative Property of, 84
  of decimals, 38
  Distributive Property and, 83
  of fractions, 37–38
  of integers, 9–10
  Inverse Property of, 84
  negative numbers in, 9
  Order of Operations Agreement and, 25
  of polynomials, 193–194
  properties of, 83–84
  of radical expressions, 487–488
  of rational expressions, 301–305
  of rational numbers, 37–38
  verbal phrases for, 10, 93
  zero in, 84
Addition method, for solving systems of equations, 423–426
Addition Property of Equations, 116, 134
Addition Property of Inequalities, 453–454, 461
Addition Property of Zero, 84
Additive inverse, 5, 84
Adjacent angles, 161
Algebraic fraction(s), *see* Rational expression(s)
Al-Khwarismi, 543
Alternate exterior angles, 162
Alternate interior angles, 162
Amount
  in basic percent equation, 119
  in percent mixture equation, 120, 122
Analytic geometry, 353
Angle(s), 57–58
  acute, 161
  adjacent, 161
  complementary, 58
  corresponding, 162
  exterior, 162, 163
  interior, 162, 163
  measure of, 57–58
  obtuse, 161
  right, 58, 378, 499
  straight, 58
  supplementary, 58, 161
  of triangles, 163–164, 321, 499
Application problems
  area, 60–62
  average, 21, 469–470
  current or wind, 124–125, 429–430, 536
  decimals in, 40
  discount, 138

  factoring in, 269–270
  formulas in, 137–139, 149, 340–341
  inequalities in, 456, 462
  integers in, 12, 21, 68, 154–156, 269, 277
  investments, 120, 122, 417–418
  levers, 149
  linear functions in, 370
  line of best fit, 388
  markup, 137
  mixtures, 120, 122, 168–171, 431–432
  percent, 119–122
  percent mixtures, 120, 122, 170–171
  perimeter, 59–60
  profit, 540
  proportions in, 320
  Pythagorean Theorem in, 500, 504–505
  quadratic equations in, 269–270, 536–537, 540
  radical equations in, 499–500, 504–505
  rational numbers in, 40, 51
  simple interest, 120, 122, 417–418
  slope in, 378, 379
  systems of equations in, 429–432
  translating verbal problems, 96, 154–157
  triangles in, 59, 61, 320–322
  in two variables, 429–432
  uniform motion, 123–125, 172–173, 333–334, 393–394
  units in, 223–224
  value mixtures, 168–169
  work, 331–332, 537
Approximately equal to ($\approx$), 49
Area, 24, 60–62
  units of, 223
Arithmetic mean, 21, 469–470
Associative Property of Addition, 83
Associative Property of Multiplication, 85
Average, 21, 469–470
Average speed, 224
Axes, 353

B

Base
  in exponential expression, 24
  of parallelogram, 61
  in percent equation, 119
  of triangle, 61
Basic markup equations, 137
Basic percent equation, 119, 120, 121
Binomial(s), 193
  expanding powers of, 203, 225
  multiplication of, 202–204
  square of, 203, 225, 259–260, 518, 521
Binomial factor(s), 239–240
  of trinomials, 243–246, 251–254
Braces

  Order of Operations Agreement and, 25
  in roster method, 447
Bracket(s)
  Distributive Property and, 88
  on number line, 449
  Order of Operations Agreement and, 25

C

Calculator
  evaluating polynomials, 276
  graphing linear equations, 392–393
  negative key, 68
  plus/minus key, 68
  solving quadratic equations, 541–542
  solving system of equations, 436–437
  viewing window, 393
Candela (cd), 340, 341
Center of circle, 59
Chapter Review Exercises, 73, 106, 185, 229, 281, 345, 397, 439, 473, 507, 545
Chapter Summary, 69, 104, 182, 226, 278, 342, 394, 437, 472, 505, 543
Chapter Test, 75, 109, 187, 231, 283, 347, 399, 441, 475, 509, 547
Circle, 59, 61, 62
Circumference of circle, 59
Clearing denominators, 134–135
Coefficient, numerical, 79
Combining like terms, 83
Common denominator, 37, 298, 301, 302–303
Common factor(s)
  greatest common factor, 32, 46, 237–238, 245–246
  in numerator and denominator, 35, 46
  of polynomials, 237–240, 245–246
Common multiple, *see* Least common multiple (LCM)
Commutative Property of Addition, 84
Commutative Property of Multiplication, 85
Complementary angles, 58
Completing the square, 521
  in deriving quadratic formula, 527
  geometric construction of, 543
  in solving quadratic equations, 521–524
Complex fractions, 310–312
Composite number, 31, 103
Concentration, percent, 120, 122
Conjugate(s), 491
  of denominator, 494
Consecutive integers, 154
Constant, 79
Constant term, 79, 83
  of perfect-square trinomial, 521, 527
Conversion factors, 223

Coordinate axes, 353
Coordinates, 353–354
Coordinate system, rectangular, 353
Corresponding angles, 162
Cost, 137
  profit and, 540
  unit cost, 168, 431–432
Counterexample, 392
Cube
  of a number, 24
  of a variable, 93
Cumulative Review Exercises, 111,
  189, 233, 285, 349, 401, 443, 477,
  511, 549
Current or wind problems, 124–125,
  429–430, 536

D
Decimal(s)
  addition of, 38
  applications of, 40
  conversion to/from fractions or
    percents, 50–51
  conversion to/from scientific
    notation, 214
  division of, 48–49
  multiplication of, 47
  repeating, 36
  representing irrational numbers, 36,
    481
  representing rational numbers, 36
  subtraction of, 39
  terminating, 36
Decimal notation, 35–36
Deductive reasoning, 503–504
Deficient number, 435
Degree(s)
  of angle, 57–58
  of polynomial, 193
Denominator(s)
  clearing, 134–135
  common, 37, 298, 301, 302–303
  least common, 37
  of rational expressions, 301,
    302–303, 315
  rationalizing, 493–494
Dependent system of equations, 406,
  407, 415, 424
Dependent variable, 360
Descartes, René, 353
Descending order, 193
Diameter of circle, 59
Difference, 11, 93, 94
  of two squares, 259
Dimensional analysis, 223–224
Discount, 138
Discount rate, 138
Distance
  to horizon, 504–505
  units of, 223
Distance-rate problems, 123–125,
  172–173, 333–334
  wind or current in, 124, 429–430,
    536

Distance-time graphs, 393
Distributive Property, 83
  in factoring polynomials, 239
  in multiplying polynomials, 201–202
  in simplifying radical expressions,
    487
  in simplifying variable expressions,
    83, 86–88
  in solving equations, 147–148, 327
  in solving inequalities, 461
Dividend, 19
Division
  of decimals, 48–49
  of exponential expressions, 209–213
  of fractions, 48, 49
  of integers, 19–20
  of monomials, 209–213
  negative numbers in, 19
  one in, 20
  Order of Operations Agreement and,
    25
  of polynomials, 219–220
  of radical expressions, 492–494
  of rational expressions, 292
  of rational numbers, 48–49
  relation to multiplication, 19
  sign rules for, 19
  symbols for, 19
  verbal phrases for, 19, 93
  zero in, 20
Divisor, 19
Domain, 357, 358
Double root, 516

E
Element of a set, 3, 447
  symbol for, 4
Empty set, 447
Equals sign, phrases for, 154
Equation(s), 115
  Addition Property of, 116, 134
  basic markup equations, 137
  basic percent equation, 119, 120, 121
  checking the solution, 515
  equivalent, 116
  formulas, see Formula(s)
  fractions in, 134–135, 136, 315–316
  function given by, 358
  graphs of, see Graph(s)
  linear, see Linear equation(s)
  literal, 327–328
  Multiplication Property of, 117–118,
    134
  parentheses in, 147–148
  percent mixture equation, 120, 122,
    170
  quadratic, see Quadratic equation(s)
  radical, 497–500, 504–505
  relation given by, 358
  second-degree, 515
  simple interest equation, 120, 122,
    417
  solution(s) of, 115, 355–356, 541
  solving, see Solving equations

  squaring both sides of, 497
  systems of, see Systems of equations
  translating sentences into, 154–157
  in two variables
    quadratic, 531–533
    solution of, 355–356
  uniform motion equation, 123, 172,
    333
  value mixture equation, 168
Equilateral triangle, 59
Equivalent equations, 116
Equivalent fractions, 35
Eratosthenes, Sieve of, 102–103
Evaluating expressions
  absolute value, 5–6
  exponential, 24–27, 210, 211
  variable, 80
Evaluating functions, 360
Evaluating polynomials, 276
Even integer, 103, 154
Expanding a power of a binomial, 203,
  225
Exponent(s), 24
  negative, 210–213, 214
  rules of, 212
  in scientific notation, 214
  zero as, 209–210
  see also Power(s)
Exponential expression(s), 24
  base of, 24
  division of, 209–213
  evaluating, 24–27, 210, 211
  factored form of, 24
  multiplication of, 197–198
  Order of Operations Agreement and,
    25
  powers of, 198, 211
  as repeated multiplication, 24
  simplest form of, 210
  simplifying, 197–198, 209–213
Exponential form, 24
Expressions, see Exponential
  expression(s); Radical expression(s);
  Rational expression(s); Variable
  expression(s)
Exterior angle, 163

F
Factor(s), 30
  greatest common (GCF), 32, 46,
    237–238, 245–246
  in multiplication, 17, 18
  in numerator and denominator, 35,
    46
  prime, 31, 435–436
  proper, 435–436
  rules for finding, 30
Factored form, of exponential
  expression, 24
Factoring polynomials, 237
  applications of, 269–270
  checking, 245
  common factors, 237–238, 245–246
  completely, 245–246, 261–262

difference of two squares, 259
general strategy for, 261
by grouping, 239–240, 253–254
perfect-square trinomials, 259–260
solving equations by, 267–270,
 515–516
trial factors in, 251–252
trinomials, 243–246, 251–254,
 259–260
Factorization, prime, 31
Federal budget, 68
Figures, plane, 57, 59–62
Final Exam, 551
First coordinate, 353
First-degree equations, *see* Linear
 equation(s)
Focus on Problem Solving, 67, 101, 180,
 223, 275, 339, 392, 435, 469, 503, 540
FOIL method, 202–203
Formula(s), 327
 applications of, 137–139, 149,
  340–341
 point-slope, 385–386
 quadratic, 527–528
 slope, 376
Fraction(s), 35
 addition of, 37–38
 algebraic, *see* Rational expression(s)
 applications of, 40, 51
 common denominator of, 37, 298,
  301, 302–303
 complex, 310–312
 conversion to/from decimals or
  percents, 50–51
 division of, 48, 49
 equations containing, 134–135, 136,
  315–316
 equivalent, 35
 improper, 37
 multiplication of, 46, 47
 negative exponents on, 211
 negative sign in, 35, 38
 reciprocal of, 48
 simplest form of, 35
 subtraction of, 39
Fraction bar
 as division, 19, 35, 220
 Order of Operations Agreement and,
  25
Function(s), 357–358
 evaluating, 360
 linear, 365, 370
 quadratic, 531, 532, 541
 zero of, 541
Functional notation, 360

G

GCF, *see* Greatest common factor (GCF)
Geometry
 analytic, 353
 angles, 57–58, 161–164, 321
  right, 58, 378, 499
 areas, 24, 60–62, 223

lines, *see* Line(s)
 perimeters, 59–60
 planes, 57
 solids, 57
 volume, 24, 223
Go Figure, 2, 78, 114, 192, 236, 288,
 352, 404, 446, 480, 514
Graph(s)
 of data, 388, 469
 of equation in two variables,
  355–356
 of horizontal line, 368
 of inequality
  in one variable, 449–450
  in two variables, 465–466
 of integers, 3
 of linear equations, 365–369,
  375–380
  with graphing utility, 392–393
 line of best fit, 388
 of point in the plane, 353–354
 of quadratic equation in two
  variables, 531–533
 of a relation, 357
 scatter diagram, 388
 of system of equations, 405–408
 of vertical line, 368
Graphing calculator, *see* Calculator
Greater than, 4
Greater than or equal to, 4, 449
Greatest common factor (GCF), 32
 factoring out of polynomial,
  237–238, 245–246
 of monomials, 237–238
 in simplifying fractions, 46
Grouping, factoring by, 239–240,
 253–254
Grouping symbols, 25

H

Half-plane, 465
Height
 of parallelogram, 61
 of triangle, 61
Horizon, distance to, 504–505
Horizontal axis, 353
Horizontal line(s)
 graph of, 368
 perpendicular to vertical line, 378
 zero slope of, 377
Hypotenuse, 499

I

If…then sentences, 340
Illumination intensity, 340–341
Improper fractions, 37
Inconsistent system of equations, 406,
 415, 425
Independent system of equations, 406
Independent variable, 360
Inductive reasoning, 67
Inequality(ies), 449
 Addition Property of, 453–454, 461

applications of, 456, 462
 division by a constant, 455
 graphs of, 449–450, 465–466
 of integers, 4–5
 Multiplication Property of, 454–456,
  461
 parentheses in, 461–462
 solution set of, 449–450, 465–466
 symbols for, 4, 449
 in two variables, 465–466
Integers, 3, 36
 addition of, 9–10
 applications of, 12, 21, 68, 154–156,
  269, 277
 consecutive, 154
 division of, 19–20
 even, 103, 154
 as exponents, 209–213
 graph of, 3
 inequalities of, 4–5
 multiplication of, 17–18
 negative, 3
 odd, 103, 154
 positive, 3
 as rational numbers, 35
 subtraction of, 10–11
 word problems with, 154–156, 269
Intensity of illumination, 340–341
Intercept(s)
 of quadratic function, 532, 541
 of straight line, 375, 379–380, 385
 as zeros of a function, 541
Interest, simple, 120, 122, 417–418
Interior angles, 163
Internet, math information on,
 277–278
Intersecting lines, 57, 161–164
 system of equations and, 405, 406,
  407
Intersection of sets, 447
 graph of, 450
Inverse
 additive, 5, 84
 multiplicative, 85
Inverse Property of Addition, 84
Inverse Property of Multiplication, 85
Investment problems, 120, 122,
 417–418
Irrational numbers, 36, 481
Isosceles triangle, 59

L

Least common denominator, 37
Least common multiple (LCM), 32
 in adding fractions, 37–38, 298
 in adding rational expressions, 298,
  302–303
 of polynomials, 297–298
 in simplifying complex fractions,
  310–312
 in solving equations, 134–135,
  315–316
 in solving proportions, 319

Legs, of right triangle, 499
Length, 24
    units of, 223
Less than
    as inequality, 4
    as subtraction, 11, 93
Less than or equal to, 4, 449
Lever systems, 149
Light intensity, 340–341
Like terms, 83
Line(s), 57
    equations of, *see* Linear equation(s)
    horizontal, 368, 377, 378
    intercepts of, 375, 379–380, 385
    intersecting, 57, 161–164, 405, 406,
        407
    parallel, 57, 162, 377, 406, 415, 425
    perpendicular, 58, 378
    slope of, 376–380, 385–386
    vertical, 368, 377, 378
Linear equation(s), 365–369
    applications of, 370, 379, 388
    finding equation from two points,
        386–387
    graphing, 365–369, 392–393
    point-slope formula for, 385–386
    slope-intercept form of, 379–380, 385
    slope of, 376–379
    standard form of, 367
    systems of, *see* Systems of equations
Linear function, 365, 370
Linear inequality, 465–466
Linear model, 388
Line of best fit, 388
Line segment, 57
Literal equations, 327–328
Lumens (lm), 340, 341

M

Markdown, 138
Markup, 137
Markup rate, 137
Maximum profit, 540
Mean, 21, 469–470
Minus, 11, 93
Mixtures
    percent mixtures, 120, 122, 170–171
    value mixtures, 168–169, 431–432
Monomial(s), 193
    division of, 209–213
    division of polynomial by, 219
    factoring from polynomial, 237–238
    multiplication of, 197
    multiplication of polynomial by, 201
    simplifying powers of, 198
Motion, 123–125, 172–173
    average speed, 224
    graphs of, 393–394
    rational expressions and, 333–334
    with wind or current, 124–125,
        429–430, 536
Moving average, 21
Multiple, *see* Least common multiple
    (LCM)

Multiplication
    Associative Property of, 85
    of binomials, 202–204
    Commutative Property of, 85
    of decimals, 47
    Distributive Property and, 83
    of exponential expressions, 197–198
    of fractions, 46, 47
    of integers, 17–18
    Inverse Property of, 85
    of monomials, 197
    negative numbers in, 17–18
    one in, 85
    Order of Operations Agreement and,
        25
    of polynomials, 201–204
    properties of, 85–86
    of radical expressions, 491–492
    of rational expressions, 290–291
    of rational numbers, 46–47
    relation to division, 19
    sign rules for, 17, 18
    symbols for, 17
    verbal phrases for, 18, 93
    by zero, 267
Multiplication Property of Equations,
    117–118, 134
Multiplication Property of Inequalities,
    454–456, 461
Multiplication Property of One, 85
Multiplication Property of Zero, 267
Multiplicative inverse, 85

N

Natural numbers, 3, 36
Negation, 339
Negative exponent, 210–213, 214
Negative factor, 18
Negative integer, 3
Negative number(s)
    absolute value of, 6
    in addition, 9
    in division, 19
    in multiplication, 17–18
    square root of, 482
    in subtraction, 10–11
Negative sign, 5
    in fractions, 35, 38
    with parentheses, 87
Negative slope, 377
Negative square root, 481
Nielsen ratings, 181–182
Null set, 447
Number(s)
    absolute value of, 5–6
    abundant, 435
    composite, 31, 103
    deficient, 435
    integers, 3, 36
    irrational, 36, 481
    natural, 3, 36
    opposites, 5, 84
    perfect, 242, 435
    prime, 31, 102–103

    rational, 35
    real, 36, 83–85
    see also Integers; Rational numbers
Number line, 3
    inequalities on, 4, 449–450
Numerical coefficient, 79

O

Obtuse angle, 161
Odd integer, 103, 154
One
    in division, 20
    Multiplication Property of, 85
Opposite
    of a number, 5, 84
    of a polynomial, 194
Ordered pair(s)
    as coordinates, 353–354
    of a function, 357–358
    of a relation, 357
    as solution of an equation, 355–356
Order of Operations Agreement, 25–27
Ordinate, 353
Origin, 353

P

Parabola, 531–533
Parallel lines, 57
    slope of, 377
    systems of equations represented by,
        406
    transversal of, 162
    vertical, 377
Parallelogram, 59
    area of, 61, 62
Parentheses
    Associative Property and, 83, 85
    Distributive Property and, 83
    in equations, 147–148, 327
    in inequalities, 461–462
    as multiplication symbols, 17
    on number line, 449
    Order of Operations Agreement and,
        25
Pascal's Triangle, 225
Percent, 50
    basic percent equation, 119, 120, 121
    conversion to/from decimals or
        fractions, 50–51
    discount rate as, 138
    markup rate as, 137
Percent mixture problems, 120, 122,
    170–171
Perfect number, 242, 435
Perfect square, 481, 483
Perfect-square trinomial, 259–260,
    521, 527
Perimeter, 59–60
Periscope, 504–505
Perpendicular lines, 58, 378
Pixels, 392
Plane, 57
    coordinates in, 353–354
Plane figures, 57

areas of, 60–62
perimeters of, 59–60
Plot a point in the plane, 353–354
Point, coordinates of, 353–354
Point-slope formula, 385–386
Polynomial(s), 193
addition of, 193–194
applications of, 204
degree of, 193
division of, 219–220
evaluating with calculator, 276
factoring, *see* Factoring polynomials
fractions with, *see* Rational
expression(s)
least common multiple of, 297–298
multiplication of, 201–204
nonfactorable over the integers, 244,
245, 259, 261
prime, 244
subtraction of, 194
Positive integer, 3
Positive number, absolute value of, 6
Positive slope, 376
Power(s), 24
of a binomial, 203, 225
of exponential expressions, 198, 211
of a product, 198
of a quotient, 211
of ten, 214
verbal phrases for, 93
see also Exponent(s)
Premise, 340
Prep Test, 2, 78, 114, 192, 236, 288,
352, 404, 446, 480, 514
Price, selling, 137
Prime factorization, 31, 435–436
Prime numbers, 31, 102–103
Prime polynomial, 244
Principal, 120
Principal square root, 481
Principle of Zero Products, 267, 515
Problem Solving, Focus on, 67, 101,
180, 223, 275, 339, 392, 435, 469,
503, 540
Product, 17, 93, 94
power of, 198
of sum and difference of two terms,
203
Product Property of Square Roots,
481–482
Profit, 540
Projects and Group Activities, 68, 102,
181, 225, 276, 340, 392, 436, 469,
504, 541
Proper factors, 435–436
Property(ies)
of addition, 83–84
Associative
of Addition, 83
of Multiplication, 85
Commutative
of Addition, 84
of Multiplication, 85
Distributive, 83

in factoring polynomials, 239
in multiplying polynomials,
201–202
in simplifying radical expressions,
487
in simplifying variable expressions,
83, 86–88
in solving equations, 147–148, 327
in solving inequalities, 461
of Equations
Addition, 116, 134
Multiplication, 117–118, 134
Squaring Both Sides, 497
of Inequalities
Addition, 453–454, 461
Multiplication, 454–456, 461
Inverse
of Addition, 84
of Multiplication, 85
of multiplication, 85–86
of One
Division, 20
Multiplication, 85
of Square Roots
Product, 481–482
Quotient, 492–493
of Zero
Addition, 84
Division, 20
Proportion(s), 319
applications of, 320
similar objects and, 320–322
Pythagorean Theorem, 499
applications of, 500, 504–505

Q

Quadrants, 353
Quadratic equation(s), 267, 515
applications of, 269–270, 536–537,
540
graphing, 531–533, 541–542
solving
by completing the square,
521–524, 527
by factoring, 267–270, 515–516
with graphing calculator, 541–542
with quadratic formula, 527–528
by taking square roots, 517–518
standard form of, 267, 515
in two variables, 531–533
Quadratic formula, 527–528
Quadratic function, 531, 532, 541
Quantifiers, 339, 340
Quantity, in percent mixture, 120
Quotient, 19, 93, 94
Quotient Property of Square Roots,
492–493

R

Radical, 481
Radical equation(s), 497–498
applications of, 499–500, 504–505
Radical expression(s), 481
addition of, 487–488

applications of, 499–500, 504–505
division of, 492–494
multiplication of, 491–492
numerical, 481–482
rationalizing denominator of,
493–494
simplest form of, 481, 493
subtraction of, 487–488
variable, 483–484
Radical sign, 481
Radicand, 481
Radius of circle, 59
Range, 357
Rate, 319
of discount, 138
of interest, 120, 417
of markup, 137
of uniform motion, 123–125,
172–173, 333–334, 393–394
of wind or current, 124–125,
429–430, 536
of work, 331–332
Rating point, 181
Ratio, 319
as division, 19
similar objects and, 320
Rational expression(s), 289
addition of, 301–305
common denominator for, 301,
302–303
division of, 292
equations containing, 315–316
multiplication of, 290–291
simplifying, 289–290
subtraction of, 301–305
in uniform motion problems,
333–334
zero in denominator of, 315
Rationalizing the denominator,
493–494
Rational numbers, 35
addition of, 37–38
applications of, 40, 51
in decimal notation, 35–36
division of, 48–49
as fractions, 35
multiplication of, 46–47
as real numbers, 36
subtraction of, 39
see also Fraction(s); Decimal(s)
Ray, 57
Real numbers, 36
properties of, 83–85
Reciprocal
of a fraction, 48
of a rational expression, 292
of a real number, 85
Rectangle, 59
area of, 61, 62
perimeter of, 59
Rectangular coordinate system, 353
Relation, 357–358
Remainder, in polynomial division, 219
Repeating decimal, 36

Revenue, 540
Right angle, 58, 378, 499
Right triangle(s), 499
  applications of, 500, 504–505
Rise, 376
Root, double, 516
Roster method, 3, 4, 447–448
Rule for Dividing Exponential
  Expressions, 211
Rule for Multiplying Exponential
  Expressions, 197
Rule for Negative Exponents on
  Fractional Expressions, 211
Rule for Simplifying the Power of an
  Exponential Expression, 198
Rule for Simplifying the Power of a
  Product, 198
Rule for Simplifying the Power of a
  Quotient, 211
Run, 376

S

Scatter diagram, 388
Scientific notation, 214
Second coordinate, 353
Second-degree equation, 515
Selling price, 137
Sentences, translating into equations,
  154–157
Sequence, 67
Set(s), 3
  elements of, 3, 4, 447
  empty, 447
  inequalities and, 4–5
  intersection of, 447, 450
  null, 447
  of ordered pairs, 357
  of real numbers, 36
  union of, 447, 449
  writing with roster method, 3, 4,
    447–448
  writing with set-builder notation,
    448–449
Set-builder notation, 448–449
Share, of television market, 181
Sieve of Eratosthenes, 102–103
Sign rules
  for addition, 9
  for division, 19
  for multiplication, 17–18
  for subtraction, 10–11
Similar objects, 320
Similar triangles, 320–322
Simple interest, 120, 122, 417–418
Simplest form
  of exponential expression, 210
  of fraction, 35
  of radical expression, 481, 493
  of rate, 319
  of ratio, 319
  of rational expression, 289
Simplifying
  complex fractions, 310–312

exponential expressions, 197–198,
  209–213
Order of Operations Agreement and,
  25–27
radical expressions, 481–484,
  487–488, 491–494
rational expressions, 289–290
variable expressions, 83–88
Slope(s), 376–379
  of distance-time graph, 393
  equation of line and, 385–386
  graph of equation and, 379–380
  of horizontal lines, 377
  of parallel lines, 377
  of perpendicular lines, 378
  undefined, 377
  vertical lines and, 377
Slope formula, 376
Slope-intercept form, 379–380, 385
Solids, 57
Solution(s)
  of equation
    in one variable, 115
    in two variables, 355–356
  of quadratic equations, *see* Quadratic
    equation(s)
  of system of equations, 405–408
  as zeros of a function, 541
Solution set of inequality
  in one variable, 449–450
  in two variables, 465–466
Solving equations
  checking the solution, 116, 118
  containing fractions, 134–135, 136,
    315–316
  containing parentheses, 147–148,
    327
  containing radical expressions,
    497–500
  by factoring, 267–270, 515–516
  of form $ax = b$, 117–118
  of form $ax + b = c$, 134–137
  of form $ax + b = cx + d$, 146–147
  of form $x + a = b$, 116–117
  formulas, 137–139, 149, 327
  literal, 327–328
  proportions, 319
  quadratic, *see* Quadratic equation(s)
  systems of, *see* Systems of equations
  using Addition Property, 116–117,
    134
  using Distributive Property, 147–148,
    327
  using Multiplication Property,
    117–118, 134
  in verbal problems, 154–157
Solving inequalities
  with Addition Property, 453–454,
    461
  containing parentheses, 461–462
  with Multiplication Property,
    454–456, 461
Solving proportions, 319

Space, 57
Speed
  average, 224
  in uniform motion, 123–125,
    172–173, 333–334, 393–394
  of wind or current, 124–125,
    429–430, 536
Square(s), 59
  area of, 60–61
  of a binomial, 203, 225, 259–260,
    518, 521
  completing, 521–524, 527, 543
  difference or sum of, 259
  of a number, 24, 93, 481
  perfect, 481, 483
  perimeter of, 59
  of a variable, 483
Square root(s), 481
  of a negative number, 482
  of perfect squares, 481, 483
  principal, 481
  Product Property of, 481–482
  Quotient Property of, 492–493
  in solving quadratic equations,
    517–518
Squaring Both Sides of an Equation,
  Property of, 497
Standard deviation, 470–471
Standard form
  of linear equation in two variables,
    367
  of quadratic equation, 267, 515
Straight angle, 58
Straight line(s), *see* Line(s)
Substitution method, for solving
  systems of equations, 413–416
Subtraction
  of decimals, 39
  of fractions, 39
  of integers, 10–11
  negative numbers in, 10–11
  Order of Operations Agreement and,
    25
  of polynomials, 194
  of radical expressions, 487–488
  of rational expressions, 301–305
  of rational numbers, 39
  verbal phrases for, 11, 93
Sum, 9, 93, 94
Sum and difference of two terms,
  product of, 203, 259
Sum of two squares, 259
Supplementary angles, 58, 161
Symbols
  absolute value, 5
  angle, 57
  approximately equal to, 49
  degree, 57
  division, 19
  element of a set, 4
  empty set, 447
  equals sign, 154
  fraction bar, 19, 25, 35, 220

greater than, 4
grouping, 25, 26
inequalities, 4, 449
less than, 4
multiplication, 17
negative sign, 5, 35, 38, 87
not equal to, 20
perpendicular lines, 58
plus or minus, 517
radical sign, 481
right angle, 58
Systems of equations, 405
    applications of, 429–432
    approximating solution with
        calculator, 436–437
    dependent, 406, 407, 415, 424
    inconsistent, 406, 415, 425
    independent, 406
    solutions of, 405–408
    solving
        by addition, 423–426
        by graphing, 407, 408
        by substitution, 413–416

T

Table, in problem solving, 275, 435
Term(s)
    like terms, 83
    of sequence, 67
    of variable expression, 79, 83
Terminating decimal, 36
Time
    simple interest and, 120, 122
    of uniform motion, 123–125,
        172–173, 333–334, 393–394, 536
    in work problems, 331–332, 537
Times sign, 46
Total, 93
Translating
    sentences into equations, 154–157
    verbal expressions into variable
        expressions, 93–96
Transversal, 162, 163
Trial-and-error method, 180–181
Trial factors, 251–252
Triangle(s), 59
    angles of, 163–164, 321
    area of, 61
    equilateral, 59
    isosceles, 59
    perimeter of, 59

    right, 499–500, 504–505
    similar, 320–322
Trinomial(s), 193
    factoring, 243–246, 251–254, 260
    perfect-square, 259–260, 521, 527

U

Uniform motion, 123–125, 172–173
    graphs of, 393–394
    rational expressions and, 333–334
    with wind or current, 124–125,
        429–430, 536
Union of sets, 447
    graph of, 449
Unit cost, 168, 431–432
Units, 223–224
    of light intensity, 340–341
    rates and, 319
    ratios and, 319

V

Value
    of a function, 360
    of a variable, 79
Value mixture problems, 168–169,
    431–432
Variable(s), 4, 79
    assigning, 94–96, 154
    dependent, 360
    importance in algebra, 101–102
    independent, 360
    value of, 79
Variable expression(s), 79
    evaluating, 80
    like terms of, 83
    radicals in, 483–484
    simplifying, 83–88
    translating verbal expressions into,
        93–96
Variable part, 79
Variable terms, 79
Verbal expressions
    for addition, 10, 93
    for division, 19, 93
    for equals sign, 154
    for multiplication, 18, 93
    for powers, 93
    for subtraction, 11, 93
    translating into equations, 154–157
    translating into variable expressions,
        93–96

Vertex, of angle, 57
Vertical angles, 161
Vertical axis, 353
Vertical line(s)
    graph of, 368
    parallel, 377
    perpendicular to horizontal line, 378
    undefined slope of, 377
Viewing window, 393
Volume, 24
    units of, 223

W

Wind or current problems, 124–125,
    429–430, 536
Work problems, 331–332, 537
World Wide Web, math information
    on, 277–278

X

x-axis, 353
x-coordinate, 353
x-intercept(s)
    of lines, 375
    of quadratic functions, 532, 541
    as zeros of a function, 541
xy-coordinate system, 353

Y

y-axis, 353
y-coordinate, 353
y-intercept(s)
    of lines, 375, 379–380, 385
    of quadratic functions, 532

Z

Zero
    absolute value of, 6
    Addition Property of, 84
    in denominator of rational
        expression, 315
    in division, 20
    as exponent, 209–210
    as integer, 3
    Multiplication Property of, 267
    as slope, 377
    Web research on, 277–278
Zero of a function, 541
Zero Products, Principle of, 267, 515

TI-30X IIS

6 **Ab/c** 2 **Ab/c** 3 **+** 3 **Ab/c** 4 **ENTER**

Operations on fractions
$6\frac{2}{3} + \frac{3}{4} = 7\frac{5}{12}$

```
6⌐2⌐3+3⌐4
            7⌐5/12
```

Access operations in blue

.4 **2nd** **F◊D** **ENTER**

```
.4▸F◊D
            2/5
```
Change decimal to fraction or fraction to decimal

The value of π

π
```
π
3.141592654
```

3 **+** 2 **(** 10 **−** 6 **)** **ENTER**

```
3+2(10−6)
            11
```
Operations with parentheses

Power of a number (See Note 1 below.)

13 **∧** 4 **ENTER**

```
13⁴
            28561
```

11 **×** 25 **2nd** **%** **ENTER**

```
11*25%
            2.75
```
Operations with percent

2nd √ 36 **)** **ENTER**

```
√(36)
            6
```
Square root of a number

Used to complete an operation

(−) 12 **÷** 6 **ENTER**

Enter a negative number (See Note 2 below.)

7 **x²** **ENTER**

```
7²
            49
```
Square a number

```
−12/6
            −2
```

Photo courtesy of Texas Instruments Incorporated

fx-300MS

√ 36 **=**

```
√36
            6
```
Square root of a number

Access operations in gold

.4 **=** **SHIFT** **d/c**

```
.4
            2⌐5
```
Change decimal to fraction

6 **ab/c** 2 **ab/c** 3 **+** 3 **ab/c** 4 **=**

Operations on fractions
$6\frac{2}{3} + \frac{3}{4} = 7\frac{5}{12}$

```
6⌐2⌐3+3⌐4
            7⌐5⌐12
```

13 **∧** 4 **=**

```
13⁴
            28561
```
Power of a number (See Note 1 below.)

7 **x²** **=**

```
7²
            49
```
Square a number

3 **+** 2 **(** 10 **−** 6 **)** **=**

```
3+2(10−6)
            11
```
Operations with parentheses

(−) 12 **÷** 6 **=**

Enter a negative number (See Note 2 below.)

```
−12÷6
            −2
```

11 **×** 25 **SHIFT** **%** **=**

```
11×25%
            2.75
```
Operations with percent

Used to complete an operation

Photo courtesy of Casio, Inc.

SHIFT **π** **=**

```
π
3.141592654
```
The value of π

NOTE 1: Some calculators use the **yˣ** key to calculate a power.
For those calculators, enter 13 **yˣ** 4 **=** to evaluate 13^4.

NOTE 2: Some calculators use the **+/−** key to enter a negative number.
For those calculators, enter 12 **+/−** **÷** 6 **=** to calculate $-12 \div 6$.